BIOCHEMISTRY AND PHARMACOLOGY OF ETHANOL

VOLUME 2

BIOCHEMISTRY AND PHARMACOLOGY OF ETHANOL

VOLUME 2

Edited by

Edward Majchrowicz

National Institute on Alcohol Abuse and Alcoholism
Rockville, Maryland

and

Ernest P. Noble

University of California at Irvine
Irvine, California

PLENUM PRESS • NEW YORK AND LONDON

Library of Congress Cataloging in Publication Data

Main entry under title:

Biochemistry and pharmacology of ethanol.

Includes bibliographies and index.
1. Alcohol–Physiological effect. 2. Alcohol metabolism. 3. Alcohol–Toxicology. 4. Alcoholism–Chemotherapy. I. Majchrowicz, Edward. II. Noble, Ernest P. [DNLM: 1. Alcohol, Ethyl. QV84 B616]

| QP801.A3B56 | 612′.015 | 79-292 |

ISBN-13: 978-1-4684-3452-1 e-ISBN-13: 978-1-4684-3450-7
DOI: 10.1007/978-1-4684-3450-7

© 1979 Plenum Press, New York
Softcover reprint of the hardcover 1st edition 1979
A Division of Plenum Publishing Corporation
227 West 17th Street, New York, N.Y. 10011

On November 6, 1939, the Jagiellonian University (Copernicus' Alma Mater, founded 1364) in Cracow, Poland, planned to inaugurate the new academic year in Copernicus Hall. Before the inaugural ceremonies could begin however, the Gestapo arrested all persons present. A total of 185 members of the faculty were subsequently sent to concentration camps in Sachsenhausen and Dachau, where many of them perished.[†] Similar atrocities were committed against scholars in other universities in Poland and other countries.

These volumes are respectfully dedicated to their memory and to the memory of other scientists persecuted throughout the world.

E. M. and E. P. N.

*"Nevertheless, it does move!" — According to legend, these words were spoken quietly by Galileo after he was forced under the threat of torture by the Inquisition Court to renounce the Copernican system.

[†] *The Nazi Kultur in Poland* by several authors of necessity temporarily anonymous. Written in Warsaw under the German Occupation and published for the Polish Government in Exile by His Majesty's Stationery Office, London, 1945.

Contributors

Ronald L. Alkana • School of Pharmacy, University of Southern California, Los Angeles, California.

Samuel H. Barondes • Department of Psychiatry, University of California at San Diego, and Veterans Administration Medical Center, San Diego, California

Herbert Barry, III • Department of Pharmacology, University of Pittsburgh School of Pharmacy, Pittsburgh, Pennsylvania

James D. Beard • Departments of Physiology and Biophysics and of Psychiatry, Alcohol Research Center, University of Tennessee Center for the Health Sciences, and Memphis Mental Health Institute, Memphis, Tennessee

John C. Butte • Department of Biological Sciences, San Francisco State University, San Francisco, California

Jane H. Chin • Department of Pharmacology, Stanford University School of Medicine, Stanford, California

Theodore J. Cicero • Department of Psychiatry, Washington University School of Medicine, St. Louis, Missouri

Allan C. Collins • School of Pharmacy and Institute for Behavioral Genetics, University of Colorado, Boulder, Colorado

Frank H. Deis • Department of Biochemistry, Rutgers University, Piscataway, New Jersey

Richard A. Deitrich • School of Medicine, University of Colorado, Denver, Colorado

Robert L. Eskay • Laboratory of Preclinical Studies, National Institute on Alcohol Abuse and Alcoholism, Rockville, Maryland. *Present address:* Laboratory of Clinical Science, National Institute of Mental Health, Bethesda, Maryland

Morris D. Faiman • Department of Pharmacology and Toxicology, University of Kansas, Lawrence, Kansas

Gerhard Freund • Medical Service, Veterans Administration Medical Center, and Department of Medicine, College of Medicine, University of Florida, Gainesville, Florida

Irving Geller • Southwest Foundation for Research and Education, San Antonio, Texas

Peter K. Gessner • Department of Pharmacology and Therapeutics, School of Medicine, State University of New York at Buffalo, Buffalo, New York

R. Adron Harris • Department of Pharmacology, School of Medicine, University of Missouri, Columbia, Missouri

Walter A. Hunt • Behavioral Sciences Department, Armed Forces Radiobiology Research Institute, Bethesda, Maryland

Ryoko Kakihana • Department of Biological Sciences, San Francisco State University, San Francisco, California

W. R. Klemm • Department of Biology, Texas A & M University, College Station, Texas

David H. Knott • Department of Psychiatry, Alcohol and Drug Unit, University of Tennessee Center for the Health Sciences, and Memphis Mental Health Institute, Memphis, Tennessee

Yuh-Jyh Kuo • Department of Physiology, The University of Texas Medical School at Houston, Texas Medical Center, Houston, Texas

David Lester • Center of Alcohol Studies, Rutgers University, Busch Campus, New Brunswick, New Jersey

Edward Majchrowicz • Laboratory of Preclinical Studies, National Institute on Alcohol Abuse and Alcoholism, Rockville, Maryland

R. Massarelli • Centre de Neurochimie du CNRS and Université Louis Pasteur, Strasbourg, France

Wallace B. Mendelson • Laboratory of Clinical Psychopharmacology, Division of Special Mental Health Research, Intramural Research Program, National Institute of Mental Health, Saint Elizabeth's Hospital, Washington, D.C., and Unit on Sleep Studies, Biological Psychiatry Branch, Division of Clinical and Behavioral Research, Intramural Research Program, National Institute of Mental Health, Bethesda, Maryland

Ernest P. Noble • University of California, College of Medicine, Irvine, California

Dennis R. Petersen • School of Pharmacy and Institute for Behavioral Genetics, University of Colorado, Boulder, Colorado

Robert H. Purdy • Southwest Foundation for Research and Education, San Antonio, Texas

Mary K. Roach • Department of Biochemistry, University of Nairobi, Nairobi, Kenya

Henry L. Rosett • Boston University School of Medicine, and Maternal Health and Child Development Program, Boston City Hospital, Boston, Massachusetts

William Q. Sargent • Departments of Physiology and Biophysics and of Psychiatry, Alcohol Research Center, University of Tennessee Center for the Health Sciences, and Memphis Mental Health Institute, Memphis, Tennessee

Werner T. Schlapfer • Department of Psychiatry, University of California at San Diego, and Veterans Administration Medical Center, San Diego, California

Linda L. Shanbour • Department of Physiology, The University of Texas Medical School at Houston, Texas Medical Center, Houston, Texas

Albert Y. Sun • Sinclair Comparative Medicine Research Farm and Department of Biochemistry, University of Missouri, Columbia, Missouri

M. Elaine Traynor • Department of Psychiatry, University of California at San Diego, and Veterans Administration Medical Center, San Diego, California

Isabel J. Wajda • Research Institute for Neurochemistry, Rockland Research Institute, Ward's Island, New York

Paul B. J. Woodson • Department of Psychiatry, University of California at San Diego, and Veterans Administration Medical Center, San Diego, California

Preface

Alcohol abuse and alcoholism represents one of the major health, social, and economic issues facing not only America but much of the world. Problems with alcohol appear to be a common curse, afflicting almost all industrialized nations as well as the developing countries.

In the United States today at least 10 million people suffer the misuse. It is estimated that 205,000 individuals die prematurely each year from a variety of ethanol-induced factors, including cirrhosis, cancer, heart disease, suicide, homicide, and highway fatalities and other accidents. In purely economic terms, the alcohol-related cost to our society in 1975 is estimated at nearly $43 billion in lost production, medical expenses, motor vehicle accidents, fire losses, and the maintenance of social mechanisms to deal with these problems.

But the economic impact pales in comparision with the psychological pain and anguish brought to alcoholics and their family and friends. The disease of alcoholism bears a close relation to the topic of our dedication, for it is equally insidious and sadistic, it destroys humanity and rational thought, and it brings terror, pain, and death.

A nation increasingly concerned with improving the quality of life cannot tolerate a health and social problem of such magnitude, particularly when there is every prospect that it can be controlled and treated with increased understanding. Polio has succumbed to control; heart disease and certain cancers are yielding to concerted efforts. Alcohol abuse and alcoholism, our largest untreated disease, can also be controlled if we can generate the necessary national resources and commitment. The economic impact alone would warrant a far greater investment of national resources for the cure and treatment of this disease than now exists.

Fundamental to any measure of preventing or controlling alcohol-related problems is a broad base of rational knowledge upon which effective programs can be built. In the past, alcohol research has been something of a scientific stepchild, relegated to breweries and a few pharmacologists interested in its anesthetic properties; and for the most part the research was of ephemeral and casual fashion. But in recent years a growing number of basic scientists have been attracted to the subject. Some have found the effects of ethanol a useful tool in the study of fundamental biochemical processes, while others have become interested in such research for medical purposes. Consequently, the subject has expanded both in breadth and depth. For the simple chemical that it is, the effects of ethanol are remarkably far-reaching, and its study covers a multitude of scientific disciplines, ranging from the basic biochemistry of the liver, through the biophysics of membranes, to the electrophysiology of the central nervous system, not to mention its behavioral, psychological, and sociological consequences. From a practical standpoint, no single individual or even small group can possibly provide a meaningful analysis of the field, and consequently a large number of experienced investigators have been invited to write on selected topics in these volumes.

Editorially, our objective in the preparation of these volumes was to nurture the growing quality of alcohol research and to attract new scientists. Contributors were requested to provide in-depth, detailed, and critical reviews of their topics. While the authors were encouraged to draw their own conclusions and expand on their own theories and hypotheses, they were restricted to citation of the published data; references to unpublished data and abstracts as well as speculation were exceptional. In general, the contributions were subjected to critical review by experts in the field and revised accordingly by the authors. But as might be expected with so many contributors and topics, the critical and analytical quality of the articles does vary, in many cases reflecting the literature in a particular area. Often it was necessary to compromise quality for the relevance and interest of a topic; but such compromises were kept to a minimum.

The treatise is divided into thirteen sections, each containing several chapters dealing with various aspects of the biochemistry and pharmacology of ethanol. The sections are arranged in logical sequence, beginning with the absorption and distribution of ethanol within the body. The enzymatic mechanisms of ethanol elimination are then discussed and followed by sections dealing with the effects of ethanol on the intermediary metabolism of various tissues and body organs and the pathological changes accompanying acute and chronic administration of ethanol. The studies continue through the interaction of ethanol with drugs, neurotransmitters, and hormones, and conclude with the effects of ethanol on animal behavior. Thus, the topics progress logically from the actions of ethanol at the molecular level to its effects at higher levels of biological organization. Presumably, the behavioral effects of ethanol reflect the consequences of its molecular interactions.

We wish to thank the contributors for their efforts in writing these chapters and for their cooperation in responding to the critical comments made by the referees. Many collegues throughout the country were enlisted as referees and we wish to thank them for their invaluable suggestions and critiques, which contributed to the quality of the papers.

We owe special thanks to Drs. Joseph J. Higgins, Walter A. Hunt, Maria F. Majchrowicz, Albert A. Pawlowski, Mary K. Roach, Violet C. Sutherland, and Kenneth Warren, whose many helpful suggestions, continued support, and interest have effectively aided the preparation of these volumes.

Finally, we wish to express our sincere gratitude to Danuta and Wanda Majchrowicz for their cheerful secretarial assistance with the heavy burden of correspondence associated with this treatise. Their tireless efforts, performed during after-school hours, went beyond the call of filial duty.

E. M. and E. P. N.

Rockville, Maryland
July 1978

Contents

VII. Ethanol and Electrolyte Metabolism

Chapter 1

Water and Electrolyte Metabolism following Ethanol Intake and during Acute Withdrawal from Ethanol

> *James D. Beard and William Q. Sargent*

Chapter 2

Acid–Base Balance following Ethanol Intake and during Acute Withdrawal from Ethanol

> *William Q. Sargent, James D. Beard, and David H. Knott*

Chapter 3

Metabolism of Calcium and Magnesium during Ethanol Intoxication and Withdrawal

R. Adron Harris

VIII. Influence of Ethanol on Biological Membranes

Chapter 4

Effects of Ethanol on Membrane Transport

Linda L. Shanbour and Yuh-Jyh Kuo

Chapter 5

Changes in the Activity of Na$^+$,K$^+$-ATPase during Acute and Chronic Administration of Ethanol

Mary K. Roach

Chapter 6

Biochemical and Biophysical Approaches in the Study of Ethanol–Membrane Interaction

Albert Y. Sun

Chapter 7

Ethanol and the Blood–Brain Barrier

Jane H. Chin

IX. Interaction of Ethanol with Hormonal Functions

Chapter 8

Effects of Acute and Chronic Administration of Ethanol on Cyclic Nucleotides and Related Systems

Walter A. Hunt

Chapter 9

Interactions of Thyrotropin-Releasing Hormone and Somatostatin with Ethanol and Other Sedative/Hypnotics

   Robert L. Eskay

Chapter 10

Ethanol and Endocrine Function

   Ryoko Kakihana and John C. Butte

X. Interaction of Ethanol with Neurotransmitters

Chapter 11

Alterations in Neurotransmitter Function after Acute and Chronic Treatment with Ethanol

   Walter A. Hunt and Edward Majchrowicz

Chapter 12

Comparison between the Effects of Ethanol and Those of Opioid Drugs on the Metabolism of Biogenic Amines

Isabel J. Wajda

Chapter 13

Genetics of Ethanol Preference: Role of Neurotransmitters

Allan C. Collins

Chapter 14

Effects of Ethanol on the Cholinergic System

R. Massarelli

XI. Ethanol and Neuronal Electrophysiology

Chapter 15
Effects of Ethanol on Nerve Impulse Activity
   W. R. Klemm

Chapter 16
Tolerance to a Specific Synaptic Effect of Ethanol in *Aplysia*
   *M. Elaine Traynor, Werner T. Schlapfer, Paul B. J. Woodson, and
   Samuel H. Barondes*

XII. Interaction of Ethanol with Drugs

Chapter 17
Interaction of Ethanol with Other Drugs
   Richard A. Deitrich and Dennis R. Petersen

Chapter 18

Biochemical Pharmacology of Pyrazoles

Frank H. Deis and David Lester

Chapter 19

Biochemical Pharmacology of Disulfiram

Morris D. Faiman

Chapter 20

Amethystic Agents: Reversal of Acute Ethanol Intoxication in Humans

Ronald L. Alkana and Ernest P. Noble

Chapter 21
Drug Therapy of the Alcohol Withdrawal Syndrome
Peter K. Gessner

XIII. Physiology, Behavior, and Animal Models of Alcohol Dependence

Chapter 22
Ethanol-Induced Changes in Body Temperature and Their Neurochemical Consequences
Gerhard Freund

Chapter 23
Interrelationship between Ethanol Consumption and Circadian Rhythm
Irving Geller and Robert H. Purdy

Chapter 24

Pharmacologic and Electrophysiologic Effects of Ethanol in Relation to Sleep

Wallace B. Mendelson

Chapter 25

Clinical Pharmacology of the Fetal Alcohol Syndrome

Henry L. Rosett

Chapter 26

Behavioral Manifestations of Ethanol Intoxication and Physical Dependence

Herbert Barry, III

Chapter 27

A Critique of Animal Analogues of Alcoholism

Theodore J. Cicero

Contents of Volume 1

IV. Biochemistry of Ethanol-Induced Liver Damage and Disease

V. Effects of Ethanol on the Metabolism of Brain

VI. Metabolic Effects of Ethanol on Various Organs of the Body

Ethanol and Electrolyte Metabolism

VII

Water and Electrolyte Metabolism following Ethanol Intake and during Acute Withdrawal from Ethanol

1

James D. Beard and William Q. Sargent

A. INTRODUCTION—HISTORICAL PERSPECTIVE

The earliest observations on the effects of ethanol ingestion on fluid loss or gain were seemingly contradictory, and in retrospect were a clear delineation of the present-day controversy. Shakespeare commented in Act II, Scene iii, of *Macbeth* that consumption of alcoholic beverages prompted an increased urine flow.[1] On the other hand, Sutton,[2] Snowden,[3] and Hayward[4] reported evidence of cerebral edema, engorged cerebral blood vessels, and dural sinuses in patients dying of delirium tremens. They also reported abnormally large quantities of cerebrospinal fluid in these patients. Indeed, according to Benjamin Rush,[5] generalized edema of the "belly and limbs and finally of every cavity of the body" was a well-known characteristic malady of the alcoholic patient. Consequent to these observations, the earliest reported treatment programs for alcoholics were primarily aimed at reducing fluid volumes.[2-4,6-8]

The modern tenet that all alcoholics are dehydrated can be traced back to two simultaneously developed but independent sets of data. Bowman[9] reported that the clinical observations of dry mouth, increased thirst, and generalized dehydration were characteristics of all alcoholics. The short-term nutritional status of his study population, however, was not defined. In conjunction with this clinical trial, many basic science reports investigating the diuretic action of ethanol began to appear.[10-14] In these studies great emphasis was placed on the initial stages of intoxication without adequate attention to the long-term effects of

James D. Beard and William Q. Sargent • Departments of Physiology and Biophysics and of Psychiatry, Alcohol Research Center, University of Tennessee Center for the Health Sciences, and Memphis Mental Health Institute, Memphis, Tennessee 38104.

a single exposure. Therefore, the clinical observation of dehydration, together with the abundant data depicting the ethanol diuresis, fostered treatment modalities which focused on intravenous fluid and electrolyte replacement in all alcoholic patients. [15-19]

The primary objective of this chapter will be carefully analyze the pharmacological effects of ethanol on fluid balance. The secondary objective will be to evaluate the possible interaction of other variables on the pharmacological action of ethanol.

B. ACUTE ETHANOL INTAKE

1. Water Metabolism

a. Ascending Plasma Ethanol Concentration

Ethanol diuresis is probably the most thoroughly studied aspect of the action of ethanol on water balance. It has been demonstrated in both man[11,20] and the dog[13] that the diuresis is associated with an ascending blood ethanol concentration. Eggleton[11] further reported that there was a linear dose–response relationship between the amount of ethanol ingested and the magnitude of the diuresis. In addition, she suggested that the duration of the ascending blood ethanol concentration was the primary stimulus. It has also been strongly suggested that ethanol diuresis results from an inhibition of antidiuretic hormone release from the supraopticohypophyseal tract. Although actual vasopressin activity in the plasma has not been shown to be reduced after ethanol ingestion (see below), it has been shown repeatedly that injections of posterior pituitary extract to man will inhibit ethanol diuresis. [10-12] Rubini et al. [14] reported that ethanol diuresis produced when human subjects ingested 48 g ethanol (0.7 g/kg for a 70-kg subject) was characterized by an increase in renal free water clearance. These data also suggest that ethanol inhibits vasopressin release because it has been shown that as plasma vasopressin decreases free water clearance increases. [21] Van Dyke and Ames, [13] using a canine model, further substantiated the hypothesis that ethanol inhibits the release of vasopressin. They demonstrated that an intracarotid dose of ethanol (0.05 g/kg) produced a diuresis of magnitude similar to that of an oral dose of 0.6 g/kg. They also reported that ethanol ingestion or infusion did not produce greater diuresis in diabetes insipidus dogs.

It has been shown in a recent study that 32 g ethanol (0.5 g/kg for a 70-kg subject) ingested 30–45 min before smoking two cigarettes will inhibit the nicotine-produced increase in both vasopressin and the hormone-associated protein neurophysin. [22] Both neurophysin and vasopressin were measured by radioimmunoassay, thus supplying the first direct proof that ethanol can inhibit the release of vasopressin. Neither substance showed a decrease in concentration after ethanol, but such changes might be outside the sensitivity of the assay.

Another mechanism by which ethanol could alter urine flow would be by altering glomerular filtration rate. This mechanism, however, has been shown not to be operative during the ascending portion of the plasma ethanol curve. Strauss et al.[12] and Rubini et al.[14] both demonstrated that the administration of ethanol to human volunteers did not alter endogenous creatinine clearance within 4 hr. Kalbfleisch et al.[23] demonstrated that the inulin clearance was not changed following a small oral dose and constant infusion of ethanol by human subjects. These results were again confirmed by Ogata et al.[24] in human subjects and by Sargent et al.[25] in a canine model.

The actual mechanism by which rising blood ethanol concentration produces inhibition of vasopressin release has not been established. Either ascending plasma ethanol concentration (APEC) might depress vasopressin release via a normal physiological input, or the decreased vasopressin release may be a consequence of generalized central nervous system depression caused by APEC. The interaction of APEC with other stimuli known to alter vasopressin release was investigated by Strauss et al.,[12] Kleeman et al.,[26] and Husain et al.[22] (see above). Strauss et al.[12] reported that ingestion of 200 meq sodium chloride, a stimulus known to stimulate vasopressin release, just prior to or 20 min after the ingestion of 50 g ethanol (0.7 g/kg for a 70-kg subject) inhibited the diuresis. Further, the ingestion of 853 ml water 20 min after the same dose increased the magnitude and prolonged the duration of the diuresis. Kleeman et al.[26] reported that administration of approximately the same dose of ethanol to human volunteers did not appreciably increase urine flow rate already elevated by large oral loads of water. They also reported, contrary to Strauss et al.,[12] that the oral administration to previously hydrated subjects increased urine flow rate, although an intravenous infusion of sodium chloride (500 ml of 5–6%) was begun when the ethanol was ingested. The increased urine flow was primarily the result of an increased free water clearance, suggesting further inhibition of vasopressin release. In nonhydrated subjects, the ingestion of ethanol during a hypertonic saline infusion was not effective in producing diuresis. Regarding the latter two studies, it must be pointed out that the baseline plasma osmolality was very different between the hydrated and the nonhydrated subjects. Therefore, the baseline inputs to the receptors influencing vasopressin release differed greatly, so comparison is very difficult. Kleeman et al.[26] also reported that in hydrated male volunteers, the application of leg cuffs produced a prompt decrease in urine flow but that the prior administration of ethanol prevented the decrease. Further, the administration of ethanol during the application of cuffs was partially successful in preventing the decrease in urine flow. Both of these studies are subject to similar criticisms. First, the number of subjects in any one study was never greater than three. Therefore, valid statistical comparisons were not reported. Second, the stimulus—APEC—was not determined during these studies to ascertain if hypertonic fluid ingestion, leg cuffs, or hypertonic fluid infusion altered the rate of ethanol absorption from the gastrointestinal tract. Because of these problems and the recently demonstrated sophistication of the vasopressin releasing mechanism,[21] further studies delineating the interaction of physiological stimuli with APEC would certainly be fruitful.

The second alternative, that APEC inhibits vasopressin secretion because of a generalized central nervous system depression, is suggested primarily by way of peripheral evidence. For instance, the inhibition of oxytocin release also occurs during APEC,[27] but inhibition of anterior pituitary factors does not occur.[28] Therefore, it could be suggested that because both oxytocin and vasopressin are directly dependent upon nerve conduction for release and anterior pituitary factors are not, nerve depression produces the inhibition in release. It has also been shown that tachyphylaxis occurs with respect to the diuretic action of APEC.[13] This result might suggest acute tolerance, as has been shown to occur in reponse to other central nervous system depressant actions of APEC.[29]

b. Descending Plasma Ethanol Concentration

Nicholson and Taylor[30] were the first to report fluid balance studies on human volunteers after a single drinking episode. Each of three subjects ingested 2 g ethanol/kg over 7 hr. Fluid, food, and electrolyte intake were held constant and urine volumes and electrolyte excretory rates were measured over 8-hr intervals during the day of ethanol intake and the succeeding day. Independent of the type of ethanol administered, water excretion increased during the first 4 hr of ethanol intake. During the remaining 3 hr of continued ethanol ingestion, however, urinary flow rates decreased. Indeed, during the entire 24 hr, including the ethanol ingestion period, fluid loss was either the same as on the control day or less. During the next 24 hr, fluid loss decreased. Therefore, these investigators suggested that ethanol consumption produced a transient diuresis but that the long-term effect was fluid retention and expansion of body fluid volumes. In a later study, Nicholson and Taylor[31] demonstrated that either a single ethanol administration to the dog (3 g/kg) or a period of 4-hr ethanol intake in man produced significant elevations in plasma volume and blood volume. The largest increase in man occurred between 0 and 8 hr after the drinking period, but in the dog the largest increase was 24–26 hr after administration. This report has been confirmed by Sargent et al.,[32] who reported plasma volume expansion in the dog 18 hr after a single ethanol dose (3 g/kg). Most recently, fluid balance studies have been done on dogs after a 3-g/kg oral dose of ethanol.[33] Fluid intakes and outputs were measured from 0 to 6 hr, 6 to 18 hr, and 18 to 24 hr postadministration. During the initial period, which included APEC, the ethanol dose produced a greater urinary flow than did an isovolumetric water dose. The converse was found, however, during descending plasma ethanol concentration (DPEC) (i.e., the ethanol-treated animals had reduced urinary outputs). Indeed, for the 24-hr period, the mean urinary volume after the ethanol dose did not differ from that of after the water dose. Voluntary water intake was significantly elevated, however, in the ethanol-treated animals, so the animals had a significant positive water balance. The control treatment produced a mean negative balance. These studies all suggest that even though a diuresis occurs during APEC, a period of decreased urinary flow and a significant increase of water intake occurs during

DPEC. In the study of Nicholson and Taylor,[30] fluid intake was maintained constant during the study.

The overall effect of APEC, then DPEC, as attested to by all these studies, is fluid volume expansion. It is not known by what mechanism the fluid retention occurs, but it appears to be a coordinated effort between fluid intake and fluid loss regulatory systems. The importance of altered vasopressin release in the response has not been directly investigated, but it can be strongly suggested that increased release plays an important role. A possible mechanism for increased vasopressin release might be increased plasma osmolality that results from free water loss during APEC.

It has been reported by both Ogata et al.[24] and Beard et al.[34] that a single ethanol administration produced increases in the corrected osmolality near the peak plasma ethanol concentration. The corrected serum osmolality was determined by subtracting the mosmole/kg water contributed by the presence of ethanol. This was done because ethanol distributes across all biological barriers and therefore does not exert an osmotic force. In neither case, however, did the concentration of plasma electrolytes increase significantly. Therefore, although the results support the hypothesis that increases in plasma osmolality could inhibit the diuresis, the results do not support the theory that increased free water clearance during APEC produced the rise in osmolality. Millerschoen and Riggs[35] have investigated the responsiveness of the vasopressin releasing system to fluid administration and increased plasma osmolality during a constant plasma ethanol concentration. Their data suggest that the vasopressin releasing system (i.e., the osmolality receptors and the releasing center) had a decreased sensitivity to elevations in plasma osmolality after ethanol administration and during the maintenance of a constant plasma concentration. In essence, they suggest that ethanol administration shifted the set point for total body osmolality regulation upward. Initially, the resetting would produce a decrease in vasopressin release. Then, once the osmolality increased to the new set point, the release of vasopressin would be reestablished. The initial portion of this hypothesis would suggest that during APEC, either the osmolality receptors or vasopressin regulatory center was depressed and that it may indeed be a reflection of generalized central nervous system depression during APEC. A similar hypothesis has been proposed to explain respiratory depression during APEC (see Chapter 2). This hypothesis would also support the contention that increases in osmolality during DPEC inhibit the diuresis.

2. Electrolyte Metabolism

a. Ascending Plasma Ethanol Concentration

The effect of ascending plasma ethanol concentration on sodium and potassium excretion has been shown to be negligible. Strauss et al.[12] found no change, and Rubini et al.[14] found a slight decrease in monovalent ion excretion after an oral dose of 50 g ethanol to human subjects (0.7 g/kg for a 70-kg subject). Divalent ion excretion, however, appears to be altered by APEC. Heaton *et al.*[36] reported that the ingestion of 90 g ethanol (1.3 g/kg for a 70-kg subject) in divided doses over 3 hr by human volunteers produced an increase in magnesium excretion (260% in 3 hr) and calcium excretion (68% in 3 hr), but sodium and potassium excretions did not differ from those of controls. Kalbfleisch *et al.*[23] demonstrated that oral administration of ethanol followed by constant infusion of ethanol in glucose solution (total dose 0.6 g/kg for a 70-kg subject) produced increases in magnesium and calcium excretions in both normal and alcoholic subjects. In both groups, the changes began 20 min postadministration, were maximal between 60 and 80 min, and declined by 100–140 min. In 1968, Plaza *et al.*[37] again demonstrated an increase in magnesium excretion in male volunteers given either 0.5 or 0.75 g ethanol/kg. They reported further that the prior administration of 10 units of vasopressin blocked the increase in magnesium excretion. This result was of great interest, but further investigation has not been reported.

Divalent cation metabolism has also been investigated in animal models after a single ethanol dose. Peng *et al.*[38] reported that in both the dog and the rat, acute oral ethanol ingestion produced a statistically significant lowering of plasma calcium concentration. Furthermore, in the rat the response was shown to be dose-dependent up to a dosage of 4 g/kg ethanol, to be initiated within 30 min after ethanol ingestion, and to persist for at least 5 hr. Again in the rat, pyrazole pretreatment did not affect the response, whereas acetaldehyde administration (10, 20, or 40 mg/rat) alone was not effective in lowering plasma calcium concentrations. In addition, it was shown that neither acute nephrectomy, thyroidectomy, nor parathyroidectomy altered the hypocalcemic response. It was shown, however, that during the first 3 hr after ethanol administration, the urinary excretion of calcium (mg/3 hr) was statistically decreased. The authors concluded that the ethanol molecule *per se* was the causative agent for the hypocalcemia. Furthermore, because the hypocalcemia was concurrent with decreased urinary calcium excretion, they hypothesized that the primary effect of ethanol was to cause a shift of extracellular calcium into intracellular stores.

Sargent *et al.*[39] investigated divalent cation excretion in the dog after a single ethanol dose (3 g/kg). During APEC, they found a significant increase in the total plasma concentration of magnesium; a decrease in calcium, as found by Peng *et al.*[38]; and no change in zinc. Again, at 2 hr postdosage, they reported no change in the excretory rates of magnesium or calcium but did find a significant increase in the rate of zinc excretion. It appeared that the latter elevation was due to an intrarenal effect of the ethanol administration, because glomerular filtration rate and total plasma zinc concentration were unaltered. A more definitive statement, however, must await further evidence concerning the plasma protein binding of zinc after the ethanol. For example, zinc filtration could have increased if the free concentration of zinc in plasma had been elevated even though total plasma zinc and glomerular filtration rate remained constant.

b. Descending Plasma Ethanol Concentration

The balance study by Nicholson and Taylor[30] also included electrolyte balances. They found that sodium, potassium, and chloride retention occurred for 18–24 hr after the drinking session. Further, in the study of Sargent et al.,[33] urinary electrolyte excretions were measured. In this study food was withheld from the animals after ethanol or water was administered. The excretion of sodium and potassium was elevated, but not significantly, from 0 to 6 hr postadministration. Magnesium excretion, however, was significantly increased in the ethanol group. The latter data confirm earlier work (see above) demonstrating the action of APEC in increasing magnesium excretory rates. During the next two periods, 6–18 and 18–24 hr, the excretory rates of all three electrolytes tended to be less than the water controls. The total 24-hr excretory rate of sodium, as previously reported by Nicholson and Taylor,[30] was found to be significantly decreased in the ethanol-treated groups, and surprisingly, magnesium excretion was significantly decreased in the ethanol group as a result of the decreased excretory rate during DPEC. These studies both demonstrate that a third component of the volume regulatory systems, renal sodium excretion, is implicated in this fluid retention response.

The time course of sodium retention during DPEC has been specifically investigated. Sargent et al.[25] reported that in dogs, a 3 g ethanol/kg dose did not alter sodium excretion or reabsorption 2 hr after the dose was given. At 10 and 18 hr, however, glomerular filtration rate and sodium filtration rates were elevated 30–40%, but excretory rate was not different from control animals. Therefore, absolute sodium reabsorption (meq/min) was significantly increased and percent reabsorption was elevated. At 24 hr, although mean plasma ethanol concentration was essentially zero, the excretory rate of sodium was significantly less than in the controls and percent reabsorption was significantly elevated. This observation (i.e., that the effects on fluid and electrolyte balance are present and may be maximal when plasma ethanol concentration is near zero) has been reported previously. Nicholson and Taylor[31] and Sargent et al.[33] observed that plasma volume expansion and positive water balance were maximal 24–26 hr postadministration. It therefore seems reasonable to assume that the existence of a plasma ethanol concentration is not necessary for the response. Only with further investigation will the importance of these observations be appreciated.

In recent years one laboratory has investigated possible mechanisms by which sodium reabsorption might be increased after acute ethanol ingestion. They have demonstrated that the increases in glomerular filtration rate, effective renal plasma flow, and sodium reabsorption were not dependent upon ethanol's ability to produce depletion of peripheral catecholamine stores.[40] In a further study,[32] it was demonstrated that ethanol not only produced plasma volume expansion 18 hr after an acute ethanol administration (3 g/kg) in the dog, but also blunted the response to an acute extracellular volume expansion with Ringer's lactate solution. In the previous literature it has been reported that extracellular expansion prior to infusions of saline or hypooncotic albumin produced an augmented natriuretic response.[41–44] In the case of prior ethanol-induced volume expansion, however, the animals retained more of the infused Ringer's lactate solution than did the control animals. The results of this study strikingly demonstrated that ethanol administration did produce increased sodium reabsorption in the kidney and suggested a resetting of volume regulation.

The results of the Ringer's lactate infusion study[32] also implied that the proximal tubule, the loop of Henle, or both, were the sites responsible for increased sodium reabsorption after ethanol administration. To investigate the latter possibility, dogs were given furosemide (5 mg/kg initially and then 5 mg/kg per hr) 18 hr after acute ethanol administration (3 g/kg).[45] Furosemide has been shown to be a specific inhibitor of active chloride reabsorption, and thereby passive sodium reabsorption, in the ascending limb of the loop of Henle.[46,47] If furosemide increased sodium excretion in the ethanol-treated animals to a greater extent than in the control animals, it would be an indication that ethanol administration stimulated chloride and sodium reabsorption in the ascending loop of Henle. During the initial 30 min of furosemide infusion, there was an increase in sodium excretion above that produced in the control animals, but total kidney sodium reabsorption remained elevated in the ethanol-treated animals. Stimulation of sodium reabsorption in the ascending loop of Henle only partially accounted for the total increase in sodium reabsorption. In the second 30 min of furosemide infusion, ethanol-treated animals did not have elevated sodium excretory rates in comparison to control animals, and total kidney sodium reabsorption was increased, as was percent sodium reabsorption. These results suggest that possibly both the proximal tubule and the loop of Henle participated in the increased sodium reabsorption after ethanol administration. Therefore, proximal tubule sodium reabsorption was directly investigated by administering acetazolamide (5 mg/kg) 18 hr after acute ethanol administration (3 g/kg).[48] In recent years it has been demonstrated that bicarbonate reabsorption and hydrogen-ion secretion are effectors of proximal tubule sodium reabsorption.[49–51] The enzyme that catalyzes hydrogen ion secretion and bicarbonate reabsorption is carbonic anhydrase. The action of this enzyme can be blocked by the administration of acetazolamide. In this study the effect of acute ethanol administration on carbonic anhydrase-mediated sodium reabsorption was investigated. The results clearly indicated that acetazolamide blocked a significant amount of the increased total kidney sodium reabsorption caused by the acute administration of ethanol.

In summary, these data demonstrated that a single administration of ethanol or a single exposure produces an initial diuresis during the ascending plasma ethanol phase. Although the mechanism of action is unknown, it appears rather certain that the release of antidiuretic hormone is inhibited during APEC. Electrolyte excretions appear to be normal during this phase except for magnesium. The majority of human and animal work suggests that magnesium excretion is increased during APEC. During DPEC, there appears to be a coordinated triad of physiological actions producing fluid volume expansion. The urinary flow

Table I. Effect of Ethanol on Fluid and Electrolyte Metabolism

Dose (g/kg)	Species	Response	References
		Ascending Plasma Ethanol Concentration	
0.1–0.8	Man	Water diuresis	10–12, 14
0.6	Dog	Water diuresis	13
0.6–0.9	Man	No change in glomerular filtration rate	12, 14, 23, 24
3.0	Dog	No change in glomerular filtration rate	25
0.6, 0.7	Man	No change in sodium and potassium excretion	12, 14, 23
0.6, 1.3	Man	Increased magnesium excretion	23, 36, 37
		Descending Plasma Ethanol Concentration	
2.0	Man	Decreased urine volume; positive water balance	30
3.0	Dog	Decreased urine volume; positive water balance	33
3.0	Dog	Increased plasma volume and blood volume	31, 32
2.0	Man	Sodium, potassium, and chloride retention	30
3.0	Dog	Sodium retention; increased glomerular filtration rate and increased sodium reabsorption at both proximal tubule and loop of Henle	25, 32, 33, 45, 48

during this time is decreased and sodium reabsorption is increased. Further, voluntary intake of fluid is elevated. Therefore, the overall effect of a single ethanol exposure is to produce demonstrable increases in fluid volumes and a positive water balance (Table I).

C. CHRONIC ETHANOL INTAKE

1. Water Metabolism—Partition of Body Fluid Compartments

Few studies over any considerable length of time have been reported in this area (Table II). Beard et al.[60] investigated the dose-dependent effects of daily administrations on fluid volumes in dogs. In this study the animals were divided into four groups: a water group, a group receiving 2 g ethanol/kg per day, a group receiving 3 g ethanol/kg per day, and a group receiving 4 g ethanol/kg per day. Weekly fluid volume determinations (total body water, extracellular volume, and plasma volume were measured while interstitial and intracellular volumes were calculated) were made in each group for 8 weeks. The water group was not different from preadministration control values during the 8 weeks. The 2 g/kg per day group showed a tendency toward elevated volumes while the 3 g/kg per day group had significant increases in all the measured and calculated volumes after the first week of administration. It was also reported that the elevations remained essentially constant for the remaining 7 weeks of daily ethanol administrations. Of particular interest was the finding that the fluid volume expansions were isosmotic (i.e., increases in fluid retention were paralleled by increases in sodium retention). Also, the results seem to indicate that chronic ethanol ingestion pro-

Table II.   Effect of Chronic Ethanol Intake and Abstinence of Ethanol on Fluid and
Electrolyte Metabolism

Dose (g/kg)	Species	Response	References
		Chronic Ethanol Intake	
2.0, 3.0, 4.0 (8 wks)	Dog	Increase in all body fluid compartments 24 hr after last dose at 7-day intervals	52
Unknown	Man	Increase in all body fluid compartments upon admission to hospital	54, 55
		Acute Withdrawal from Ethanol	
None	Man	All body fluid compartments reduced to normal within 4 days due to increased urine excretion of water, sodium, and chloride	54, 55

duced a resetting of fluid volume regulation and that the resetting occurred within the first 7 days of treatment.

The animals receiving 4 g ethanol/kg per day were also found to have increased fluid volumes after 1 week of ethanol administration. Furthermore, fluid balances were determined daily on two animals in the group receiving 4 g/kg per day for the 8 weeks of the study. The results demonstrated that although urine volumes were greater than during the control period, fluid intake was increased to greater percentages, so that positive water balance ensued. Therefore, these results are in accord with the fluid balances reported after a single dose[48] (i.e., ethanol administration produces fluid volume expansion, and a primary factor in this response was increased voluntary intake). Also in 1968, Ogata et al.[24] investigated the urinary response of alcoholic subjects to long-term programmed (8–17 days) or free-selection (30 days) ethanol ingestion. The results demonstrated that mean 24-hr urine volumes in these subjects were not greater than the preethanol or the postethanol phase. Unfortunately, fluid intakes were not reported. This result does, however, agree with those of Beard et al.[52] and Sargent et al.[33] It was also observed in the Ogata et al.[24] study that corrected serum osmolality was increased in each subject during the ethanol-ingestion phase of the study. This increase in corrected serum osmolality was positively correlated with increases in serum sodium concentration. The increase in serum sodium, however, could not account for the entire increase in corrected serum osmolality. Therefore, this study supplies evidence that vasopressin release may be increased during chronic ethanol ingestion, and the increase in serum osmolality could be the stimulus.

It should be further stated that the Ogata et al.[24] did not conclude that the volunteers became dehydrated during the course of the study. Indeed, they showed that 24-hr urine volumes were not significantly different in the ethanol phase of the study in comparison to the preethanol phase. Therefore, increases in uncorrected and corrected serum osmolalities were not indicative of dehydration. Further, the most frequent cause of an elevated plasma osmolality has been reported to be the presence of ethanol in the serum.[53] Indeed, the elevation in plasma osmolality due to ethanol can be clinically used in the original diagnosis of alcoholic patients. Beard *et al.*[34] showed that additions of ethanol to plasma produced a linear increase in plasma osmolality (each 10 mg ethanol/100 ml added produced an increase in osmolality of 2.17 mosmole/kg water). Therefore, by applying this ratio, one can monitor the rate of ascent or decline of plasma ethanol concentration.

2. Electrolyte Metabolism

Beard *et al.*,[52] in their study with dogs, reported that two animals receiving the highest dose of ethanol (4 g/kg per day) showed a 16% decrease in mean 24-hr sodium excretion, an 18% decrease in mean 24-hr chloride excretion, and a decrease in mean 24-hr potassium excretion of 25%. This group also reported that plasma electrolyte concentrations of sodium, potassium, and chloride were unaltered even though the extracellular volume was increased in the 3 g/kg per day and 4 g/kg per day groups. These results suggest that chronic ethanol dosage produced an isosmotic expansion of the extracellular volume and that there was an increase in the extracellular amounts of sodium, potassium, and chloride. In human subjects, Ogata *et al.*[24] reported that urine 24-hr electrolyte excretions in the programmed drinking group were not different from those in the pre- or postethanol periods. To the contrary, however, sodium, potassium, and calcium excretory rates were decreased during the ethanol and postethanol periods in the free-drinking study. There were significant differences between these studies that might explain the dissimilar results. First, the length of ethanol consumption in the free-choice study was 20 days, while it was only 17 days in the programmed study. Second, the mean serum ethanol concentration in the subjects of the programmed study varied from 104 to 112 mg/dl, while in the free-choice study it varied from 174 to 212 mg/dl. It therefore appears that the altered electrolyte response pattern might be dependent upon both quantity of ethanol consumed and length of time of chronic consumption.

D. ACUTE WITHDRAWAL FROM ETHANOL

The free-choice model used by Ogata *et al.*[24] appears to reflect changes in long-term chronic alcoholics. The basis for this statement is the work of Beard and Knott.[54] In this study fluid volume and plasma electrolyte measurements were made on 30 male alcoholics when they were admitted to the treatment program. The basis for selection of these subjects was a 10-year alcoholism history; the diagnosis of acute withdrawal; a 1-week history of ethanol ingestion prior to admission; no signs or history of malnutrition, vomiting, or diarrhea; and no medical history of cardiovascular, renal, liver, or metabolic disease. The measured fluid volumes (i.e., total body water, extracellular fluid volume, and plasma volume) were found to be significantly increased upon admission, in comparison to the fourth day of hospitalization. During the same time, although *ad libitum* fluid intake was not increased, urinary output and the urinary excretion of sodium and chloride were significantly increased. Therefore, as in the animal studies,[52] chronic ethanol ingestion produced volume expansion and when intake ceased, excess water and solute were excreted via the kidneys (Table II).

In the Beard and Knott study, the patients were in stage I of acute withdrawal [i.e., psychomotor agitation, autonomic hyperactivity (tachycardia, systolic hypertension, hyperhidrosis), horizontal nystagmus, anorexia, and insomnia]. In a subsequent study, patients in stage II–III of acute withdrawal from ethanol [i.e., in addition to psychomotor agitation and autonomic hyperactivity, etc., there were hallucinations (one or a combination of visual, auditory, or tactile) and intermittent disorientation] were studied by Knott and Beard.[55] The entrance criteria for the second study were the same as in the previous study, and the same fluid volume measurements were determined.[54] Again, it was found that these patients had elevated fluid volumes upon admission and that fluid and solute loss ensued during 4 days of hospitalization. Most important, the degree of fluid volume expansion appeared to be related to the severity of the withdrawal symptomatology (i.e., the volume expansion in the patients in stage II or III of withdrawal was significantly greater than the group in stage I). This apparent correlation may, indeed, be an example of a more fundamental relationship. It is known that the content of fluid and electrolytes and their distribution across cell membranes are of paramount importance to the proper functioning of all excitable tissues.

E. SUMMARY AND CONCLUSIONS

The majority of evidence now clearly shows that both acute and chronic ethanol ingestion produce volume expansion with concommitant electrolyte retention. As to the possible mechanisms by which these phenomena occur, exciting new evidence is now beginning to appear in the literature so that the following hypotheses can be constructed. It should be quickly and clearly pointed out that these hypotheses are by no means final proof of any mechanism, but they should serve as an embarkation point for many exciting and fruitful scientific journeys.

The administration of ethanol initially produces a diuresis characterized by increases in free water clearance without major changes in solute excretion. It is relatively well established that during APEC, vasopressin release is inhibited, as indicated by water diuresis. At present, it does not appear that APEC is acting to inhibit vasopressin release through any proven physiological pathway. The inhibition may be due to an altered sensitivity or to an increase in the set point of the vasopressin releasing mechanism. If this were true, an increased afferent discharge would be required to reestablish vasopressin release. Therefore, the diuresis would not only be a result but, in turn, could act as a stimulus to increase afferent activity and thereby inhibit the diuresis.

This rationale would blend with three observations known to occur during DPEC: (1) urine flow is reduced, (2) solute reabsorption is increased, and (3) voluntary fluid intake is increased. Therefore, a triad of physiological pathways for isosmotic volume expansion have been shown to be operative during DPEC. In further support of this hypothesis it has been shown that the chronic administration of ethanol produces volume expansion followed by maintenance of the expansion as long as ethanol intake continues. It would then be expected that discontinuance of ethanol intake would result in increased water and solute loss. Indeed, this has been shown to occur.

Therefore, at the present time, it appears reasonable that ethanol does indeed alter total body fluid regulation by either decreasing the sensitivity of the sensory or integrative components of volume and osmolality regulation or by resetting the set point of the respective integrative areas.

ACKNOWLEDGMENTS. The authors thank Jackie M. Cash for typing the manuscript.

The preparation of this review was supported in part by a continuing grant from the State of Tennessee Department of Mental Health and Mental Retardation, Alcohol and Drug Abuse Section, and United States Public Health Service Grant AA02670 from the National Institute on Alcohol Abuse and Alcoholism.

F. REFERENCES

1. W. Shakespeare, *Macbeth*, II, iii.
2. T. Sutton, *Facts on Delirium Tremens, on Peritonitis, and on Some Other Inflammatory Affections, and on the Gout*, Thomas Underwood, London (1813).
3. I. C. Snowden, *Philadelphia J. Med. Phys. Sci.* **1**:191 (1820).
4. G. Hayward, *N. Engl. J. Med. Surg.* **11**:235 (1822).
5. B. Rush, *Q. J. Stud. Alcohol* **4**:321 (1943).
6. J. Klapp, *Eclectic Repert.* **7**:251 (1817).
7. T. H. Wright, *Am. J. Med. Sci.* **6**:17 (1830).
8. R. Steinbach, *Dtsch. Med. Wochenschr.* **41**:369 (1915).
9. K. M. Bowman, H. Wortis, and S. Keiser, *J. Am. Med. Assoc.* **112**:1217 (1939).
10. M. M. Murray, *J. Physiol. (Lond.)* **76**:379 (1932).
11. M. G. Eggleton, *J. Physiol. (Lond.)* **101**:172 (1942).
12. M. B. Strauss, J. D. Rosenbaum, and W. P. Nelson, III, *J. Clin. Invest.* **29**:1053 (1950).
13. H. B. van Dyke and R. G. Ames, *Acta Endocrinol.* **7**:110 (1951).
14. M. E. Rubini, C. R. Kleeman, and E. Lamdin, *J. Clin. Invest.* **34**:439 (1955).
15. J. Romano, *Ann. Med. Hist.* **3**:128 (1941).
16. J. A. Smith, *J. Am. Med. Assoc.* **152**:384 (1953).
17. R. W. McNichol, W. J. Cirksena, J. T. Payne, and M. C. Glasgow, *South. Med. J.* **60**:7 (1967).
18. T. M. Golbert, C. J. Sanz, H. D. Rose, and T. H. Leitschuh, *J. Am. Med. Assoc.* **201**:99 (1967).
19. S. M. Wolfe and M. Victor, *Ann. N.Y. Acad. Sci.* **162**:973 (1969).
20. H. W. Haggard, L. A. Greenberg, and R. P. Carrol, *J. Pharmacol. Exp. Ther.* **71**:349 (1941).
21. G. L. Robertson, E. A. Mahr, S. Athar, and T. Sinha, *J. Clin. Invest.* **52**:2340 (1973).

22. M. K. Husain, A. G. Frantz, F. Ciarochi, and A. G. Robinson, *J. Clin. Endocrinol. Metab.* **41**:1113 (1975).
23. J. M. Kalbfleisch, R. D. Lindeman, H. E. Ginn, and W. O. Smith, *J. Clin. Invest.* **42**:1471 (1963).
24. M. Ogata, J. H. Mendelson, and N. K. Mello, *Psychosom. Med.* **30**:463 (1968).
25. W. Q. Sargent, J. R. Simpson, and J. D. Beard, *J. Pharmacol. Exp. Ther.* **188**:461 (1974).
26. C. R. Kleeman, M. E. Rubini, E. Lamdin, and F. H. Epstein, *J. Clin. Invest.* **34**:448 (1955).
27. F. Fuchs, A. R. Fuchs, V. F. Poblete, Jr., and A. Risk, *Am. J. Obstet. Gynecol.* **99**:672 (1967).
28. J. Leppaluoto, M. Rapeli, R. Varis, and T. Ranta, *Acta Physiol. Scand.* **95**:400 (1975).
29. K. V. Tullis, W. Q. Sargent, J. R. Simpson, and J. D. Beard, *Life Sci.* **20**:875 (1977).
30. W. M. Nicholson and H. M. Taylor, *J. Clin. Invest.* **17**:279 (1938).
31. W. M. Nicholson and H. M. Taylor, *Q. J. Stud. Alcohol* **1**:472 (1940).
32. W. Q. Sargent, J. R. Simpson, and J. D. Beard, *J. Stud. Alcohol* **36**:1438 (1975).
33. W. Q. Sargent, J. R. Simpson, and J. D. Beard, in: *Currents in Alcoholism* (F. A. Seixas, ed.), Vol. III, pp. 419–430, Grune & Stratton, Inc., New York (1978); *Alcohol Clin. Exp. Res.*, Abstr., **1**:161 (1977).
34. J. D. Beard, D. H. Knott, and R. D. Fink, *South. Med. J.* **67**:271 (1974).
35. N. R. Millerschoen and D. S. Riggs, *Am. J. Physiol.* **217**:431 (1969).
36. F. W. Heaton, L. N. Pyrah, C. C. Beresford, R. W. Bryson, and D. F. Martin, *Lancet* **2**:802 (1962).
37. M. Plaza, R. Orozco, M. Rosemblitt, Y. Rendic, and M. Espinace, *Rev. Med. Chile* **96**:138 (1968).
38. T.-C. Peng, C. W. Cooper, and P. L. Munson, *Endocrinology* **91**:586 (1972).
39. W. Q. Sargent, J. R. Simpson, and J. D. Beard, *J. Pharmacol. Exp. Ther.* **190**:507 (1974).
40. W. Q. Sargent, J. R. Simpson, and J. D. Beard, *J. Pharmacol. Exp. Ther.* **193**:356 (1975).
41. M. Ladd, *J. Appl. Physiol.* **3**:379 (1951).
42. M. Ladd, *J. Appl. Physiol.* **3**:603 (1951).
43. F. G. Knox, E. G. Schneider, T. P. Dresser, and R. E. Lynch, *Am. J. Physiol.* **219**:904 (1970).
44. J. T. Higgins, Jr., *Am. J. Physiol.* **220**:1367 (1971).
45. J. D. Beard, W. Q. Sargent, and J. R. Simpson, *J. Stud. Alcohol* **38**:922 (1977).
46. J. H. Dirks and J. F. Seely, *Am. J. Physiol.* **219**:114 (1970).
47. M. Burg, L. Stoner, J. Cardinal, and N. Green, *Am. J. Physiol.* **225**:119 (1973).
48. W. Q. Sargent, J. R. Simpson, and J. D. Beard, *J. Stud. Alcohol* **38**:933 (1977).
49. K. J. Ullrich, H. W. Radtke, and G. Rumrich, *Pfluegers Arch. Eur. J. Physiol.* **330**:149 (1971).
50. J. P. Kokko, *J. Clin. Invest.* **52**:1362 (1973).
51. D. L. Maude, *Kidney Int.* **5**:253 (1974).
52. J. D. Beard, G. Barlow, and R. R. Overman, *J. Pharmacol. Exp. Ther.* **148**:348 (1965).
53. A. G. Robinson and J. N. Loeb, *N. Engl. J. Med.* **284**:1253 (1971).
54. J. D. Beard and D. H. Knott, *J. Am. Med. Assoc.* **204**:135 (1968).
55. D. H. Knott and J. D. Beard, *South. Med. J.* **62**:485 (1969).

Acid–Base Balance following Ethanol Intake and during Acute Withdrawal from Ethanol

2

William Q. Sargent, James D. Beard, and David H. Knott

A. INTRODUCTION

The regulation of acid–base balance in mammalian organisms is a highly complex and integrated homeostatic function. The hydrogen-ion sensitivity of all enzymes and other protein moieties is well known. It is primarily due to this sensitivity that the hydrogen ion concentration must be regulated at a sensitivity of 1 million times that of sodium regulation. To accomplish this feat requires the integration of intracellular and extracellular buffers, together with respiratory gas exchange and renal electrolyte loss or conservation.

The effects of ethanol on acid–base balance may be due to many different factors. First, ethanol can be a central nervous system depressant, thereby altering normal ventilatory control of blood gases. Second, cellular ethanol metabolism can produce increased quantities of hydrogen ion that can alter normal balance. Third, during certain events, such as binge drinking or withdrawal, other factors related to continuous heavy ethanol intake can severely compromise acid–base homeostasis.

In this chapter we will review the literature concerning the respiratory and metabolic consequences of ethanol intake that may alter acid–base balance. Further, literature concerning both binge drinking and the withdrawal syndrome will be reviewed.

William Q. Sargent and James D. Beard • Departments of Physiology and Biophysics and of Psychiatry, Alcohol Research Center, University of Tennessee Center for the Health Sciences, and Memphis Mental Health Institute, Memphis, Tennessee 38104. **David H. Knott** • Department of Psychiatry, Alcohol and Drug Unit, University of Tennessee Center for the Health Sciences, and Memphis Mental Health Institute, Memphis, Tennessee 38104.

B. EFFECTS OF A SINGLE ETHANOL DOSE

1. Respiratory Effects (Table I)

Raiha and Maenpaa[1] reported in 1964 that in both male and female rats intraperitoneal injections of ethanol in dosages of 1.2, 2.0, and 4.0 g/kg did not produce significant elevations in tail blood P_{CO_2}. Although there was a tendency for elevated P_{CO_2} at 30, 40, and 240 min after the lower doses, the 4.0-g/kg dose produced a tendency toward decreased levels at these times. Plasma bicarbonate was unaltered by the lower doses, but the 4.0-g/kg dose produced significant reductions at 30, 90, and 240 min. Therefore, only the highest dose caused a significant reduction in blood pH. They also found a similar constellation of results in humans given 150–200 ml absolute ethanol (approximately 1.7–2.3 g/kg to a 70-kg man) over 4 hr. It was found that at 2 hr, when the blood ethanol was 92 mg/dl, plasma bicarbonate was significantly reduced, but $P_{v_{CO_2}}$ and pH were unchanged. At 4 hr, when the blood ethanol was 140 mg/dl, bicarbonate was further reduced and pH was decreased. These results suggested that a metabolic acidosis occurred while the blood ethanol concentration was ascending. At these concentrations and dosages, a respiratory depression was not observed.

In 1968, however, Rosenstein *et al.*[2] did supply evidence that ethanol exerts a depressant action on the central respiratory center in cats. These authors first demonstrated that injections of ethanol into the lateral or third ventricle of the brain produced respiratory failure. The failure was not instantaneous but resulted from progressive decreases in respiratory rate and tidal volume. They further demonstrated that the response was not altered by vagotomy or section of buffer nerves from the carotid bifurcation. Therefore, it was suggested that ethanol had a direct effect on the central sensors and/or the central respiratory center. In an attempt to distinguish between these possible loci of action, the minute volume response to elevations in $P_{a_{CO_2}}$ was examined after ethanol injection into the subarachnoid space (0.05 or 0.08 ml of 95% ethanol). The results indicated that the minimum level of $P_{a_{CO_2}}$ required to sustain spontaneous breathing was unaltered by the ethanol (i.e., the intercept of the $P_{a_{CO_2}}$ versus

Table I. Effects of an Acute Ethanol Dose on Respiratory Function

Dose (g/kg)	Species	Response	Reference
1.2, 2.0, 4.0	Rat	No change in tail blood P_{CO_2} from 0–240 min postadministration	1
1.3–2.7	Man	$P_{v_{CO_2}}$ unchanged at 2 and 4 hr	1
0.78	Man	Decreased respiratory response to hypoxia and hypercapnia 20–60 min after dosage	4
3.0	Dog	$P_{v_{CO_2}}$ increased 1–6 hr postadministration	10

minute volume plot was unaffected). It was found, however, that there was a decreased responsiveness of the respiratory system to stepwise increases in $P_{a_{CO_2}}$ (i.e., the slope of the $P_{a_{CO_2}}$ versus minute volume plot was decreased after ethanol). From these results the authors suggested that ethanol did not affect the relay of sensory information to the respiratory center (unaltered intercept), but did affect the handling of the information by the center (decreased slope). These observations, although interesting, did not have a great deal of clinical significance until Johnstone and Reier[3] repeated some of their studies using human subjects. The subjects were infused with 0.35, 0.70, and 1.05 ml/kg ethanol. The infusions lasted for 1 hr and produced peak blood ethanol concentrations of 40, 99, and 121 mg/dl, respectively. The resting minute volume was decreased immediately after the infusion of 0.70 g/kg, but other changes in resting ventilatory function (end tidal P_{CO_2}, frequency, etc.) were unaffected 0, 60, and 120 min after any of the infusions. These authors also reported, however, a decreased responsiveness to stepwise elevations of $P_{a_{CO_2}}$. Because the latter results were obtained while oxygen tension was maintained at an elevated level, the authors suggested that ethanol also produced depression of the central respiratory center in man. Therefore, it appears that respiratory depression is not always indexed by altered $P_{a_{CO_2}}$ levels.

These results have been confirmed and expanded by the studies of Sahn *et al.*[4] Their eight volunteers ingested 0.78 g ethanol/kg within 30 min. The effect of hypoxia on minute volume was determined 0, 20, 40, and 60 min after ethanol consumption and the effects of hypercapnia were ascertained at 30 and 70 min. At the peak blood ethanol concentrations attained in these studies (112 mg/dl), alterations in resting ventilatory function were not found. The average minute volume response to hypoxia was decreased at 20 min but returned to normal by 60 min. It was also found that hypercapnic ventilatory drive was significantly diminished 70 min after the ethanol dose. These results are in agreement with the studies of Rosenstein *et al.*[2] (see above) because both found no change in the intercept of the $P_{a_{CO_2}}$ versus minute volume plot but did find significant reductions in slope, again indicating that ethanol affects the central response to sensory stimuli. The work of Johnstone and Reier[3] (see above) also appears to be in agreement, although a slope and intercept were not calculated.

One further study has singularly investigated respiratory depression after a single ethanol dose. Smith *et al.*[5] demonstrated that there was a dose-dependent increase in $P_{v_{CO_2}}$ in mice 0.5 hr after intraperitoneal injections of 2, 3, 4, and 5 g/kg. Furthermore, prior injections of disulfiram or α-methyl-p-tyrosine did not affect the response, but pretreatment with p-chlorophenylalanine (p-CPA) (300 mg/kg), 4 hr before ethanol, inhibited the increment in $P_{v_{CO_2}}$ at each ethanol dosage level. The rise in $P_{v_{CO_2}}$ owing to the administration of other alcohols was also inhibited by p-CPA; methylsergide, a compound that will competitively block serotonin at the receptor site, will also inhibit the response. Therefore, it was concluded that, at least in mice, serotonin modulated the respiratory depressant action of ethanol. The conclusion of this study must be questioned from two standpoints. First, it was reported in the latter study that although methylsergide blocked the respiratory depressant actions of ethanol, "sleeping times" in the animals, taken as an index of central nervous system depression, were unaffected. This result would suggest that the respiratory depressing effects of ethanol were separate from the generalized central nervous system depressant actions. These apparent contradictory results need further elucidation. Second, Tabakoff[6] has shown in the mouse that central nervous system steady-state levels and turnover of serotonin were unaltered 30 min after an intraperitoneal dose of ethanol (3 g/kg). If, indeed, serotonin is implicated in the respiratory depressant actions of ethanol, one must speculate that ethanol altered either the serotonin–receptor binding or the final intracellular action of serotonin. Indeed, further investigation in this area is warranted.

2. Effects on Bicarbonate Metabolism (Table II)

Blood bicarbonate concentrations have been measured in both man and animals after a single ethanol dose. Ylikahri *et al.*[7] examined acid–base balance after human volunteers ingested 1.5 g ethanol/kg. The ethanol solution was consumed over 3 hr and produced maximal blood ethanol concentrations of 99–230 mg/dl 4 hr after drinking began. At 16 hr the blood ethanol concentrations were zero. Changes were not found in P_{CO_2}, pH, or bicarbonate of fingertip blood samples at peak blood ethanol concentration. However, at 16 hr they did find significant decreases in bicarbonate and pH without a change in P_{CO_2}. The same group[8] has recently reported that a 1.75-g/kg dose in the paradigm mentioned previously also produced a decrease in pH and bicarbonate in human volunteers. They also demonstrated a correlation between metabolic acidosis and increase in plasma lactate and ketone bodies. Further, the administration of glucose or fructose (1.0 g/kg) during the 3 hr of drinking, or 12–13 hr later, produced significant amelioration of the increases in lactate and ketones and the decreases in pH and bicarbonate. The results confirm earlier work (see above) that demonstrates the effect of a single ethanol exposure producing a metabolic acidosis that is due primarily to alterations in glucose metabolism.

Table II. Effects of an Acute Ethanol Dose on Blood pH and Bicarbonate Concentration

Dose (g/kg)	Species	Response	References
1.5, 1.75	Man	Decrease in [HCO_3^-] and pH in fingertip blood 16 hr after dosage	7, 8
3.0	Dog	Venous [HCO_3^-] decreased 21–24 hr post-administration. Venous pH decreased 6–18 hr postadministration	10
3.0	Dog	Increased renal capacity for bicarbonate reabsorption 18 hr after dosage	11, 12

Renal bicarbonate metabolism has also been investigated after a single ethanol dose, by Sargent et al.[9] In these studies, male mongrel dogs were given orally either 3 g ethanol/kg or an isovolumetric quantity of water. Eighteen hours after dosage, when the mean plasma ethanol concentration was 114 mg/dl, the animals were given acetazolamide, an inhibitor of the enzyme carbonic anhydrase. This enzyme is found in many tissues, including the kidney, and reversibly catalyzes the hydration of carbon dioxide. In the kidney this process supplies hydrogen ions for secretion into the tubules, the first step required for bicarbonate reabsorption. It is also known that in the kidney, sodium and bicarbonate reabsorption are linked in the proximal tubule. Therefore, alterations in sodium excretion produced by acetazolamide may also be taken as indices of alterations in bicarbonate reabsorption. The results of this study clearly demonstrated that acetazolamide-sensitive renal sodium reabsorption was significantly increased 18 hr after a single ethanol dose (3 g/kg).

To further investigate acid–base balance after a single ethanol dose, Sargent et al.[10] gave male dogs 3 g ethanol/kg and obtained blood samples for 30 hr postdosage. Further, fluid intakes and outputs were measured for 24 hr. Plasma ethanol concentrations peaked at approximately 2 hr postdosage at 340 ± 16 mg/ dl. The $P_{v_{CO_2}}$ began to increase during the ascending portion of the plasma ethanol curve and was maximally increased at 6 hr. Plasma bicarbonate was unaltered at this time, so pH was significantly decreased. Beginning at approximately 8–10 hr, however, plasma bicarbonate did decrease, and it became significant at 21 hr postadministration. Bicarbonate remained decreased from 21 to 30 hr, but the expression of a metabolic acidosis was not observed, because of decreases in $P_{v_{CO_2}}$. It was also found that fluid retention and subsequent volume expansion began to appear 10 hr postadministration and reached a maximum at 30 hr postadministration. Therefore, a similar time course was found between volume expansion and decreases in plasma bicarbonate concentration.

The latter results, together with prior data (see above) suggesting increased renal sodium and bicarbonate reabsorption, prompted further study of renal bicarbonate reabsorbing capacity by Sargent et al.[11] Again, dogs were given 3 g ethanol/kg and 18 hr later infused with a hypertonic sodium bicarbonate solution. The hypertonic sodium bicarbonate infusion produced significantly greater increases in percent water, sodium, and bicarbonate excretions in the control animals. Therefore, it appeared that renal sodium bicarbonate reabsorbing capacity was increased after the ethanol dose because glomerular filtration rate and the individual filtration rates of sodium and bicarbonate were not different between groups. At this point, however, it could not be distinguished if the response was primarily dependent upon an increased capacity for sodium or upon bicarbonate reabsorption. To answer this question, in an identical paradigm, bicarbonate was totally replaced by chloride in the infusion solution. These results showed no differences between the water and ethanol groups in their renal water or electrolyte excretions. Therefore, it appears that bicarbonate ion was necessary for the response, but the possible involvement of carbonic anhydrase in the response was uncertain.

In an attempt to clarify the role of carbonic anhydrase in the response, Sargent *et al.*[12] infused dogs with a hypertonic sodium bicarbonate solution 18 hr after a single ethanol dose (3 g/kg) and then administered acetazolamide to block carbonic anhydrase activity. The results of the hypertonic infusion were the same as reported earlier[11]: the hypertonic infusion produced a greater increase in sodium, bicarbonate, and water excretions in the water-treated dogs in comparison to the ethanol-treated animals. After the administration of acetazolamide, bicarbonate and sodium excretory rates were not significantly different between groups. However, percent sodium reabsorption remained significantly elevated in the ethanol group. Although these results can be interpreted as supplying evidence that a single ethanol dose stimulated renal carbonic anhydrase activity, a great deal of further investigation will be necessary to substantiate this hypothesis.

C. EFFECTS OF "BINGE" DRINKING ON KETOACIDOSIS

In recent years, the observation of ketoacidosis in chronic alcoholic patients has been made repeatedly.[13–16] The effects of ethanol consumption on ketone metabolism, however, were first investigated by Lefevre *et al.*[17] It is known that excessive oxidation of fatty acids, which occurs during carbohydrate starvation, will cause an increased production of ketone bodies (acetoacetate and β-hydroxybutyrate). Lefevre *et al.*[17] examined the interrelations between ketogenesis and the presence of varying amounts of fat and/or ethanol in the diet. When the total calories from carbohydrate remained constant, it was found that a diet containing 48% ethanol and 36% fat produced greater ketogenesis than did a diet containing 82% fat. It was also found that the addition of ethanol *in vitro* to liver slices did not alter acetoacetate production. Liver slices from animals pretreated with ethanol, however, did show significant increases in acetoacetate production. The results of this study demonstrated that concurrent ethanol consumption potentiated the ketogenic effects of a low-carbohydrate diet. The results also suggested that the amount of fat in the diet affected the magnitude of the ketogenesis and that the ketogenic effect of ethanol required time for expression.

In recent years the clinical entity of alcoholic ketoacidosis has been defined. In these patients, hospitalization was preceded by increases in ethanol intake and decreases in food consumption. In most instances the patients developed hyperemesis and discontinued all fluid and food intake 24 hr prior to hospitalization. Some patients, however, drank ethanol until the time of admission. Upon examination it was reported that the majority of patients had a marked acidosis due to a disproportionate increase in β-hydroxybutyric acid (a ketone body) and a decrement in plasma bicarbonate. It was also found that generally $P_{v_{CO_2}}$ was also depressed, supposedly a result of either a respiratory compensation for the acidosis or the hyperexcitability of the withdrawal syndrome. Indeed, Fulop *et al.*[16] reported that in a large number of their patients, pH was near normal, although ketones were elevated and bicarbonate decreased. In these instances the metabolic acidosis had been almost totally compensated by respiratory adjustments.

In all the patients with alcoholic ketoacidosis, it was also found that plasma glucose was not elevated and that glycosuria was negligible. Insulin levels, however, were found to be low in this population; but insulin treatment was not necessary for the reversal of the signs and symptoms. Indeed, intravenous glucose solution has been shown to abate the acidosis within 12–14 hr.[14,15] Therefore, the syndrome has been clearly distinguished from the more common diabetes mellitus. Other consistent findings have included highly elevated plasma free fatty acids, a significant increase in the β-hydroxybutyrate/acetoacetate ratio, and Kussmaul breathing.[10–12]

At the present time it could be hypothesized that the following sequence of events could produce the syndrome[13–16]: Ethanol ingestion would first begin to deplete liver glycogen stores,[17] impede gluconeogenesis by decreasing the conversion of lactate to glucose,[18] and increase the release of free fatty acids from adipose tissue.[19] Carbohydrate starvation would be imposed upon this process due to decreased food intake. Under these conditions, not only would fatty acids serve as an increased source of fuel, but glycerol might become a more important precursor for gluconeogenesis; both events would lead to increased ketone production. In this general framework, it has also been proposed that elevations in plasma cortisol and growth hormone may cause further elevations in free fatty acids[16] and that ethanol-induced mitochondrial damage might enhance the ketogenesis.[13]

D. CHRONIC INTAKE AND WITHDRAWAL (TABLE III)

In recent years, alterations in acid–base balance during withdrawal have been investigated in human patients by two groups. Wolfe and Victor,[20] in a controlled study, administered 15 oz of 80-proof ethanol (approximately 2 g/kg for a 70-kg man) per day to four alcoholic patients for 60 days. Diet remained adequate during this time and supplemental vitamins were given. Within 8 hr after the last ethanol consumption, arterial pH was elevated in all the patients and remained increased for approximately 50 hr. The $P_{a_{CO_2}}$ was concomitantly

Table III. Effects of Acute Withdrawal on Blood Bicarbonate, pH, and P_{CO_2}

Hours after withdrawal	Effect	References
8–50	Decreased arterial pH and $P_{a_{CO_2}}$	20, 21
60–80	Normal arterial pH and reduced $P_{a_{CO_2}}$	21

decreased, thereby suggesting a predominantly respiratory alkalosis. In this study it was also observed that the period of alkalosis was associated with hypomagnesemia and hypersensitivity of the central nervous system.

In a second report, Wolfe and Victor[21] studied 31 alcoholic patients during withdrawal. Exclusion criteria for the study included pneumonia or other severe infections, advanced hepatic or renal disease, diabetes mellitus, chronic lung disease, trauma, acute pancreatitis, or gastrointestinal bleeding. After these patients were admitted to the hospital, arterial blood samples were obtained every 8–12 hr. The results of these studies suggested that there were two distinguishable periods of withdrawal: an early phase and a late phase. The early phase, 10–50 hr after admission, was characterized by hallucinations, convulsions, anorexia, insomnia, tremor, and respiratory alkalosis. Indeed, the magnitude of the alkalosis was positively correlated with the severity of the other withdrawal symptomatology. During the late withdrawal, 60–80 hr, which they classified as true delerium tremens, there was major disorientation and a decrease in $P_{a_{CO_2}}$, but arterial pH remained within normal limits. Therefore, although not reported, plasma bicarbonate must have also been reduced.

From these observations they hypothesized that the respiratory alkalosis might have been the initiating event in many of the signs and symptoms of withdrawal. They reasoned that respiratory alkalosis would produce cerebral hypoxia by (1) decreasing cerebral blood flow, which is known to be sensitive to $P_{a_{CO_2}}$, and (2) increasing the affinity of hemoglobin for oxygen. Therefore, cerebral hypoxia would be a major pathophysiological event in withdrawal. Although this is an attractive hypothesis, evidence has not been forthcoming to substantiate it.

Brooks and Adams[22] have also studied the acid–base balance of withdrawing subjects. They measured cerebrospinal fluid (CSF), plasma acid–base parameters, and lactate concentrations in normal patients, hyperventilation controls and idiopathic seizure patients, and ethanol-withdrawal seizure patients. Again, the alcoholic patients were found to be alkalotic and hypocapnic in comparison to nonseizure controls, and arterial lactate was elevated. Significant acid–base differences were not found in the CSF of the alcoholic patients, except that lactate was again significantly increased in comparison to nonseizure controls, hyperventilating patients, and idiopathic seizure patients. They also examined patients in delerium tremens 2–7 days after admission. The patients were again found to be alkalotic, but plasma bicarbonate was elevated in contrast to the findings of Wolfe and Victor (see above). It was also demonstrated that those patients had elevated CSF lactate concentrations. Interestingly, the magnitude of the increase in CSF lactate appeared to correlate positively with the severity of withdrawal.

The latter studies supply preliminary evidence that significant acid–base changes occur in withdrawal and that the resultant pathophysiology may be partially causative for the behavioral signs and symptoms of withdrawal. This area appears to be fertile ground for future research.

E. SUMMARY AND CONCLUSIONS

Changes in acid–base balance are pervasive of all aspects of ethanol abuse and withdrawal. The information presented in this chapter appears to suggest that during ethanol intake there is a tendency toward both metabolic and respiratory acidosis. During withdrawal, however, alkalosis, primarily of respiratory origin, appears to dominate. To date, the majority of studies have been concerned with establishing the existence of acid–base changes. Future studies, hopefully, will establish the relative importance of those changes in the entire gambit of phenomena associated with alcoholism (i.e., intoxication, adaptation, tolerance, abstinence).

ACKNOWLEDGMENTS. The authors wish to thank Jackie Cash for the typing of this manuscript.

The preparation of this review was supported in part by United States Public Health Service Grant AA02670 from the National Institute on Alcohol Abuse and Alcoholism, and a continuing grant from the State of Tennessee Department of Mental Health and Mental Retardation, Alcohol and Drug Abuse Section.

F. REFERENCES

1. N. Raiha and P. Maenpaa, *Scand. J. Clin. Lab. Invest.* **16**:267 (1964).
2. R. Rosenstein, L. E. McCarthy, and L. L. Borison, *J. Pharmacol. Exp. Ther.* **162**:174 (1968).
3. R. E. Johnstone and C. E. Reier, *Clin. Pharmacol. Ther.* **14**:501 (1973).
4. S. A. Sahn, S. Lukshminaragan, D. J. Pierson, and J. V. Weil, *Clin. Sci. Mol. Med.* **49**:33 (1975).
5. A. F. Smith, C. Engelsher, and M. Crofford, *Ann. N.Y. Acad. Sci.* **273**:256 (1976).
6. B. Tabakoff and W. O. Boggan, *J. Neurochem.* **22**:759 (1974).
7. R. H. Ylikahri, A. R. Poso, M. O. Huttanen, and M. E. Hillbom, *Scand. J. Clin. Lab. Invest.* **34**:327 (1974).
8. R. H. Ylikahri, T. Leino, M. O. Huttunen, A. R. Poso, C. J. P. Eriksson, and E. A. Nikkila, *Eur. J. Clin. Invest.* **6**:93 (1976).
9. W. Q. Sargent, J. R. Simpson, and J. D. Beard, *J. Stud. Alcohol* **38**:933 (1977).
10. W. Q. Sargent, J. R. Simpson, and J. D. Beard, in: *Currents in Alcoholism* (F. A. Seixas, ed.), Vol. III, pp. 419–430, Grune & Stratton, Inc., York (1978).
11. W. Q. Sargent, J. R. Simpson, and J. D. Beard, *Fed. Proc., Abstr.,* **36**:332 (1977).
12. W. Q. Sargent, J. R. Simpson, and J. D. Beard, *Physiologist, Abstr.,* **20**:83 (1977).
13. D. W. Jenkins, R. E. Eckel, and J. W. Craig, *J. Am. Med. Assoc.* **217**:177 (1971).
14. L. J. Levy, J. Duga, M. Girgis, and E. E. Gordon, *Ann. Intern. Med.* **78**:213 (1973).
15. M. T. Cooperman, F. Davidoff, R. Spark, and J. Pallotta, *Diabetes* **23**:433 (1974).
16. M. Fulop and H. D. Hoberman, *Diabetes* **24**:785 (1975).
17. A. Lefevre, H. Adler, and C. S. Lieber, *J. Clin. Invest.* **49**:1775 (1970).
18. R. A. Kreisberg, A. M. Siegal, and W. C. Owen, *J. Clin. Invest.* **50**:175 (1971).
19. J. B. Field, H. E. Williams, and G. E. Mortimore, *J. Clin. Invest.* **42**:497 (1963).
20. S. M. Wolfe and M. Victor, *Ann. N.Y. Acad. Sci.* **162**:973 (1969).
21. S. M. Wolfe and M. Victor, in: *Causation and Treatment of the Alcohol Withdrawal Syndrome* (P. G. Bourne and R. Fox, eds.), pp. 137–169, Academic Press, Inc., New York (1973).
22. B. R. Brooks and R. D. Adams, *Neurology* **25**:943 (1975).

Metabolism of Calcium and Magnesium during Ethanol Intoxication and Withdrawal

3

R. Adron Harris

A. INTRODUCTION

Acute and chronic ingestion of alcohol as well as withdrawal of alcohol can result in pronounced endocrine and renal effects leading to altered excretion of magnesium, calcium, and other ions. In chronic alcoholism, mineral deficiencies may also be aggravated by malnutrition, vomiting, and diarrhea. More than 20 years ago, depletion of serum magnesium was noted in chronic alcoholics undergoing withdrawal.[1] Since that time a number of investigators have confirmed this finding and have also noted a depletion of serum calcium in chronic alcoholics. Deficiencies of calcium and magnesium are known to produce tetany, tremor, irritability, delerium, and convulsions in humans.[2,3] These symptoms occur during ethanol withdrawal, suggesting that the hypomagnesemia and hypocalcemia noted during the withdrawal syndrome might be responsible for some of the behavioral alterations produced by alcohol withdrawal. This chapter will examine the evidence linking calcium and magnesium to the pharmacology of ethanol. The relationship of alcoholism to mineral deficiencies was last reviewed by Flink.[2,3] The involvement of calcium and magnesium in biological function has been extensively reviewed[3-7] and only a few of the most relevant observations will be treated briefly here.

1. Importance of Calcium and Magnesium in Body Function

Calcium and magnesium are essential in three types of biological phenomena: (1) regulation of enzyme activity, (2) induction of fluidity or phase changes

R. Adron Harris • Department of Pharmacology, School of Medicine, University of Missouri, Columbia, Missouri 65201.

in biological membranes by serving as counterions of charged groups on the membrane surfaces, and (3) alteration of electrical properties of excitable tissues by acting as charge carriers.

Calcium and magnesium affect the activities of several groups of enzymes critical to cellular and tissue functions. The first class includes the ATPases, which have important roles in cellular transport phenomenon. These enzymes are responsible for the liberation of the chemical energy contained in ATP. Several groups of ATPases which may be selectively stimulated by magnesium and calcium are known. The calcium- and magnesium-stimulated ATPases participate in muscle contraction whereas the magnesium-stimulated ATPases may also be involved in the release of neurotransmitters. The enzymes responsible for both the synthesis and degradation of cyclic nucleotides, adenylate cyclase, and phosphodiesterase are regulated by availability of calcium.[8] In addition, the accumulation and release of calcium in some tissues, such as sarcoplasmic reticulum, appear to be regulated by cyclic nucleotides.[8] Another class of enzymes regulated by calcium and magnesium are those involved in the synthesis of protein, RNA, and DNA. An additional role of magnesium in protein synthesis involves the aggregation of ribosomes and binding of messenger RNA to ribosomes.[7]

A number of cell surface phenomena are also regulated by these ions. Evidence that binding of calcium to membranes may be important for exocytotic processes, such as stimulus–secretion coupling, comes from experiments showing that this ion can promote the fusion of membranes.[9] An inward movement of calcium ions is required for stimulus–secretion coupling in a number of cells, including those of the nervous system. Smooth muscle contraction is also initiated by an influx of calcium ions. These actions of calcium may involve activation of intracellular enzymes as well as alteration of membrane properties by changes in ion binding.

Binding of calcium and magnesium alters the physical properties of membranes, such as phase-transition temperature and fluidity,[10–12] Calcium also induces phase separations in membranes.[13] Alterations in these properties of nerve membranes may affect the activity of membrane-bound enzymes and alter membrane conduction. A number of reports indicate that alteration of these same membrane properties by alcohols and related drugs may be relevant to their pharmacological actions.[14] For example, recent studies have shown that acute exposure to ethanol increased the fluidity of biological membranes, whereas chronic exposure to ethanol attenuated the fluidizing effect.[15,16] It has been suggested that alterations of membrane fluidity can alter neural transmission[17–19] and that the normalization of membrane fluidity after chronic ethanol treatment may represent a mechanism for tolerance.[15–18] The fluidizing effect of ethanol is generally attributed to a direct perturbation of the membrane bilayer,[14–16] whereas adaptation to this effect may be due to changes in membrane composition.[20] However, alterations in the availability of calcium or magnesium can greatly change the fluidity of membranes and could be responsible for some of the membrane effects of alcohols. The potential influence of calcium availability on the physical properties of brain membranes may be appreciated by noting that 0.1 mM calcium increased the order parameter (determined by electron paramagnetic resonance) of synaptosomal membranes by 5%,[21] whereas 80–160 mM ethanol decreased the order parameter of synaptosomal membranes by only 0.5%.[16] Interactions between the effects of divalent cations and ethanol at the membrane level promises to be an exciting research area.

In addition to altering the activities of enzymes and the properties of membranes, calcium may act as a charge carrier in a manner similar to sodium and potassium. This activity has been shown in cardiac tissue, smooth muscle, and squid and lobster axons but is probably of limited importance in mammalian neurons. It is not clear if magnesium acts as a charge carrier in biological systems. Much of our knowledge of calcium binding and movement is due to the availability of radioactive ^{45}Ca. Unfortunately, the only available isotope of magnesium, ^{28}Mg, has a half-life of only 21 hr and must be prepared by linear acceleration, which severely limits its availability.

2. Cellular Distribution of Calcium and Magnesium

The subcellular locations of calcium and magnesium within the body are quite different. Although both ions are extensively bound to proteins and acidic lipids, magnesium tends to be concentrated intracellularly, whereas calcium is often excluded from the intracellular fluids (Table I). For example, the concentration of calcium in plasma is about 2.5 times that of magnesium, whereas in erythrocytes the calcium concentration is only one thirtieth that of magnesium. In brain tissue, the calcium concentration is 1 mM and the magnesium concentration is sixfold higher, whereas in the cerebrospinal fluid the calcium and magnesium levels are similar. Thus, it can be seen that both erythrocytes and brain tissue accumulate magnesium, but not calcium, from the surrounding fluid. Although difficult to determine, it has been estimated that the intracellular concentration of free calcium ions is quite low (i.e., 10^{-8}–10^{-6} M) and that most of the intracellular calcium is sequestered in mitochondria and intracellular mem-

Table I. Concentrations of Calcium and Magnesium in
Various Human Tissues

Tissue	μmole/g or ml	
	Calcium	Magnesium
Plasma	2.5	1
Erythrocytes	0.1	3
Cerebrospinal fluid	1.25	1
Brain tissue	1.0	6
Skeletal muscle	1.2	11

branes. This localization allows the influx or release of a small amount of calcium to dramatically, but transiently, alter the intracellular concentration of ionized calcium. Such an arrangement is well suited for transmission and amplification of biological signals and is the basis for the regulation of events such as neurotransmitter release and muscular contraction.

B. CALCIUM ABSORPTION AND EXCRETION DURING ETHANOL INTOXICATION AND WITHDRAWAL

Acute administration of ethanol (0.34 g/kg orally plus 0.1 g/kg intravenously) has been found to increase the urinary excretion of calcium twofold in both alcoholic and nonalcoholic subjects.[22] However, in this study and in others using higher dosages (up to 1.5 g/kg orally) ethanol ingestion did not alter plasma calcium concentrations.[23–25] Acute administration of alcohol also failed to alter intestinal uptake of calcium in humans.[26] Using anesthetized dogs and rats, larger doses of ethanol (3 and 6 g/kg, orally, respectively) were found to significantly decrease plasma calcium concentrations; however, these decreases were small, being only about 10%.[27–29] These workers hypothesized that ethanol promotes a movement of calcium from extracellular spaces to intracellular sites. Acute administration of ethanol was reported to decrease rat brain calcium levels[64] and those effects will be discussed later in this chapter.

The effects of chronic alcohol consumption on plasma calcium levels were investigated by Ogata et al.[23] by giving three alcoholics a mixture of bourbon and ethanol (up to 4.4 g ethanol/kg per day) six times daily. After 6 days of this regimen, the dose of alcohol administered each 4 hr was chosen by each subject. This experiment was continued for a total of 17 days. Another group of three alcoholics was given the opportunity to obtain alcohol at any time by completing a simple operant task, and this experiment was continued for 30 days. This experimental design naturally resulted in considerable daily and individual variations in alcohol consumption which complicated interpretation of the results. However, examination of the data for each subject suggested that serum calcium decreased after several days of high alcohol intake (i.e., blood ethanol levels of about 200 mg/dl). When the data were averaged for the duration of the study, two of the six subjects showed significantly decreased plasma calcium as compared to the values obtained from 7 days before they were given access to alcohol. In addition, three of the six subjects showed a marked decrease in urinary excretion of calcium during chronic alcohol treatment.

In animal studies, daily administration of 3 g ethanol/kg to dogs for several days did not alter plasma calcium levels, although reabsorption of the ion in the kidney was increased.[27] This increased reabsorption may have been required to compensate for decreased absorption of calcium from the intestine, as duodenal calcium transport was decreased by 58% in rats chronically ingesting ethanol.[30] The release of corticosteroids by ethanol[31] may be responsible for the decreased calcium absorption.[32]

A number of investigators noted serum calcium levels decreased by 10–40% in hospitalized alcoholics.[33–35] Sullivan et al.[36] also found normal to slightly increased urinary excretion of calcium (5–7 mg/hr) in alcoholics despite a 10% reduction in serum calcium. Magnesium deficiency is known to produce hypocalcemia,[37] and in several studies[33,35] hypocalcemia was accompanied by hypomagnesemia. Interestingly, this hypocalcemia was corrected by administration of magnesium salts. Additional studies[38] suggested that depletion of magnesium inhibited synthesis or release of parathyroid hormone (PTH), whereas studies by Estep et al.[35] and Medalle et al.[39] suggested that alcoholics with hypomagnesemia may have a decreased response to PTH. Relevant to this point is the observation that, in rats, alcohol reduced the ability of PTH to raise serum calcium and may also have inhibited the secretion of PTH.[29]

The observation of hypocalcemia in hospitalized alcoholics suggested that depletion of calcium might be associated with withdrawal of alcohol. However, this suggestion was not supported by the experimental data of Ogata et al.[23] In the latter study, the cessation of chronic alcohol consumption did not alter plasma calcium concentrations. Although evaluation of the results from this study was complicated by the variability in alcohol ingestion discussed earlier, the data suggested that even those individuals steadily consuming large quantities of alcohol did not display a clear hypocalcemia during withdrawal, although they did display hypomagnesemia. Similarly, Sullivan et al.[36] noted only a slight hypocalcemia in subjects with marked hypomagnesemia during withdrawal. In considering the etiology of the hypocalcemia noted in the clinical studies, it should be realized that increased corticosteroid secretion, acute and chronic pancreatitis, as well as cirrhosis of the liver, may deplete calcium.[32,40–42] These conditions may, of course, be associated with chronic alcoholism.

C. MAGNESIUM ABSORPTION AND EXCRETION DURING ETHANOL INTOXICATION AND WITHDRAWAL

Ingestion of a single dose of ethanol (0.8–1.4 g/kg) increased the urinary excretion of magnesium.[22,43–45] Elevated magnesium excretion was more pronounced when blood alcohol levels were increasing than when they were decreasing.[46] Despite this increased excretion of magnesium, other investigators[23,24] noted that plasma magnesium concentrations were not affected by similar doses (1.0–1.5 g ethanol/kg). Thus, in humans, acute ethanol intoxication affects the metabolism of magnesium and calcium similarly; excretion of both ions is increased but plasma levels are not greatly altered. In dogs, acute ethanol administration (3 g/kg, orally) increased magnesium excretion and elevated serum magnesium slightly.[27]

Few studies have investigated the effects of chronic alcohol consumption on magnesium metabolism. In two studies, administration of alcohol in amounts up to 4.4 g/kg per day to volunteers for 8–30 days did not alter plasma magnesium concentrations.[23,47] However, in another study, Hines[48] reported a progressive fall in serum and erythrocyte magnesium concentrations in well-nourished alco-holic subjects given ethanol (about 5 g/kg per day) for 60 days. As would be expected, the subjects displayed a negative magnesium balance (i.e., excretion exceeded absorption).

In contrast to the paucity of information regarding magnesium metabolism during chronic alcohol ingestion, magnesium deficiency during alcohol with-drawal has been the subject of a number of reports. These findings, summarized in Table II, show a consistent decrease in plasma magnesium during ethanol withdrawal. Of particular importance are the studies from Mendelson's labora-tory[23,46,47] in which cessation of chronic alcohol consumption among well-nourished volunteers produced a decrease in serum magnesium concentrations. The decrease was maximal at about 24 hr after withdrawal was initiated and returned to prewithdrawal levels after 4–8 days. Similarly, Hines[48] found that cessation of chronic alcohol consumption in experimental subjects decreased both serum and erythrocyte magnesium concentrations. Flink et al.[49] have suggested that magnesium deficiency during alcohol withdrawal is due to increased plasma free fatty acids which chelate magnesium. Several reports have noted that the severity of the withdrawal syndrome may be related to the degree of hypomagnesemia during withdrawal.[36,46,50] In particular, Wolfe and Victor[51] found that the photomyoclonus during withdrawal was correlated with decreased serum magnesium levels.

However, the correlation of degree of hypomagnesemia with severity of withdrawal has been questioned by several investigators who have not found a clear relationship.[34,45,52] For example, Fankushen et al.[34] noted that six of seven patients undergoing alcohol withdrawal eventually improved without sup-plementation with magnesium. However, it should be noted that, with a normal diet, most alcoholics undergoing withdrawal will show a positive magnesium balance and may thus replenish their magnesium stores.[53,54] In addition, hypo-calcemia was present in several of the patients studied by Fankushen et al.,[34] and may have contributed to the severity of the withdrawal syndrome, since some signs and symptoms (impaired "mentation," positive Trousseau sign, and muscular weakness) were ameliorated after administration of calcium gluconate. In the study by Heaton et al.[45] clinical differences between alcoholics with hypomagnesemia and those with normal serum magnesium levels were not detected. However, these comparisons were based on serum magnesium con-centrations determined immediately upon hospitalization, and may be mislead-ing. The data of Wolfe and Victor[51] and Mendelson et al.[46] indicated that the degree of hypomagnesemia observed depended critically on how long after withdrawal determinations were made. These investigators found that maximal hypomagnesemia occurred 18–24 hr after withdrawal. Failure to determine accu-rately the maximal magnesium depletion is a shortcoming of most clinical studies.

Magnesium depletion has also been noted in skeletal muscle and erythro-cytes during withdrawal (Table II). In several studies, the decreases in muscle and erythrocyte magnesium were somewhat greater than the decrease in plasma

Table II. Plasma and Tissue Magnesium Concentrations during Ethanol Withdrawal Syndrome

Subjects	Number of subjects	Plasma Mg (mM, change[a])	Erythrocyte Mg (mM, change[a])	Muscle Mg (mM/kg, change[a])	Comments and references
Hospitalized alcoholics	7	0.54, −44%	2.6, −10%	4.8, −50%	Plasma and erythrocyte magnesium determined during first few days of hospitalization. Muscle magnesium value obtained from only one patient; determined on thirty-second day of hospitalization. [34]
Hospitalized alcoholics	20	0.76, −18%	—	9.0, −16%	Both determinations 1–4 days after treatment. [54]
Healthy volunteers	10	0.81, −16%	—	—	Subjects given up to 4.4 g/kg for 21 days. Plasma measured on first day of withdrawal. Subjects as own control. [47]
Healthy volunteers	6	0.78, −12%	—	—	Similar to Mendelson et al.,[47] except given alcohol for 8–30 days. [23]
Hospitalized alcoholics with delerium tremens	12	0.75, −17%	1.6, −40%	—	Thirteen healthy adults used as controls. [87]
Hospitalized alcoholics manifesting withdrawal syndrome	30	0.78, −20%	1.84, −32%	—	Determined upon hospitalization; no control determinations. Control values taken from other studies. [88]
Hospitalized alcoholics	10	0.83, −11%	2.2, −14%	6.4, −36%	All samples taken within 24 hr of hospitalization. [57]
Hospitalized alcoholics	14	0.81, −11%	—	—	Chronic alcoholics without delerium tremens. Magnesium determined within 3 days after hospitalization. [52]
	18	0.81, −11%	—	—	Chronic alcoholics with delerium tremens. Magnesium determined within 3 days after hospitalization. [52]
	13	0.69, −24%	—	—	Chronic alcoholics with Korsakoff's psychosis, Wernicke's syndrome, or alcoholic dementia. Magnesium determined within 3 days after hospitalization. [52]
Hospitalized alcoholics	80	0.79, −12%	—	—	Magnesium determined upon admission, control values taken from other studies. [60]
Various	213	0.57, −29%	—	—	Average from 11 studies compiled by Flink. [2]

[a] All values represent percent difference from normal control groups.

magnesium, although Sullivan *et al.*[36] did not find any depletion of erythrocyte magnesium during withdrawal despite the presence of hypomagnesemia. The mean magnesium concentrations of cerebrospinal fluid of 25 alcoholic patients was not significantly decreased during withdrawal, although some of the patients apparently had lower-than-normal cerebrospinal fluid magnesium.[55]

D. MAGNESIUM AND CALCIUM ADMINISTRATION DURING ETHANOL WITHDRAWAL

Despite numerous reports of hypomagnesemia during alcohol withdrawal, only a few investigations have evaluated the effects of magnesium administration on the severity of the withdrawal syndrome. Magnesium supplementation relieved tetany[33,56,57] and cardiac arrythmias[57–59] associated with hypomagnesemia during alcohol withdrawal. In addition, the photomyoclonus produced by the withdrawal of alcohol was suppressed promptly by administration of magnesium.[51] A recent study found magnesium to be as effective as diazepam in the treatment of alcohol withdrawal.[60] These effects were not likely due to neuromuscular blockade, since much higher doses of magnesium were required to produce this effect.[61] However, Vallee *et al.*[56] noted that although magnesium replacement relieved tetany, it did not prevent the appearance of delirium tremens, which occurred in the presence of normal serum concentrations of magnesium and calcium. Of course, altered metabolism or storage of magnesium may occur despite normal serum levels of the ion. In fact, McCollister *et al.*[53] found that alcoholics with normal serum magnesium concentrations had a positive magnesium balance during withdrawal, suggesting replenishment of depleted intracellular stores in tissues such as skeletal muscle and erythrocytes (Table II).

A 5-year study of 32 patients carried out by Stendig-Lindberg[52] suggested that long-term supplementation with magnesium reduced the development of alcoholic encephalopathy (Korsakoff's psychosis, Wernike's syndrome, and alcoholic dementia). Although this study cannot be considered conclusive, owing to the small number of subjects, it should be noted that availability of magnesium is required for utilization of thiamine since thiamine pyrophosphate acts as a coenzyme for α-ketoglutarate dehydrogenase, pyruvate decarboxylase, or pyruvate dehydrogenase only if magnesium is present. In addition, in calcium deficiency, brain thiamine stores are depleted in rats.[62] Thus, it is conceivable that chronic deficiencies of magnesium and calcium could inhibit the uptake and utilization of thiamine in brain, thus contributing to encephalopathies.

The relationship between tissue magnesium depletion and behavioral manifestations during magnesium deficiency may best be illustrated by several animal experiments. Experiments with rats have shown that consumption of a magnesium-deficient diet produced a hyperexcitability leading to convulsions.[63,64] This treatment also decreased serum, muscle, cerebrospinal fluid, and brain concentrations of magnesium.[65] After administration of magnesium, serum, muscle, and cerebrospinal fluid concentrations returned rapidly to normal levels while brain concentrations of magnesium increased much more slowly. Interestingly, the susceptibility to seizures persisted after the cerebrospinal fluid and serum deficits were corrected.[66] Thus, the relationship of clinical manifestations to magnesium depletion must be evaluated cautiously when serum magnesium is used as the sole measure of magnesium depletion.

Effects of administration of calcium during withdrawal of ethanol were detailed by Fankushen et al.[34] These workers reported effects of administration of calcium gluconate to two patients with hypocalcemia and neuromuscular irritability during withdrawal from alcohol. In both instances, the irritability disappeared with restoration of serum calcium.

E. INTERACTION OF CALCIUM WITH MEMBRANES: ALTERATION BY ETHANOL

Evidence for both altered binding and transport of calcium after acute and chronic treatment with alcohol was presented in several recent publications. Seeman et al.[67] first noted that addition of ethanol increased the binding of calcium to human erythrocyte membranes in vitro. This increased calcium binding was produced by low concentrations of ethanol (20 mM) and other short-chain aliphatic alcohols. More recently, Ross[68] has shown that acute administration of ethanol (2 g/kg, intraperitoneally) to rats increased the calcium binding capacity of brain synaptic membranes isolated from these animals. Since acute treatment with ethanol also decreased the calcium content of these membranes (discussed below), the increased in vitro calcium binding may reflect an increased availability of binding sites due to an ethanol-induced loss of endogenous calcium from these sites in vivo rather than an increased total number of sites.

In contrast to the increased binding of calcium found after acute exposure to alcohol, decreased calcium binding has been found after chronic administration of ethanol. Pachinger et al.[69] reported decreased binding of calcium to cardiac mitochondrial and sarcoplasmic reticulum membranes from dogs given ethanol for 3–6 months. The ability of these organelles to accumulate calcium was also reduced after chronic administration of alcohol; these alterations were suggested as a basis for the impaired cardiac function found in chronic alcoholism. In addition, alcoholic myopathy may be related to calcium changes, since Rubin et al.[70] have shown that chronic consumption of ethanol decreased calcium uptake into skeletal muscle sarcoplasmic reticulum and inhibited the calcium- and magnesium-stimulated ATPases of muscle actinomyosin. These studies were carried out with baboons which had consumed 4.5–8 g ethanol/kg per day for 6–9 months and with humans who had ingested 3–4 g/kg per day for 4 weeks but had not received alcohol for 12–24 hr before biopsy. Chronic consumption of ethanol was also found to reduce the binding of calcium to synaptic membranes isolated from rat brain.[71] In these experiments rats were given 15% (v/v) ethanol as their only fluid for 14 days.

Changes in calcium binding may be involved in the effects of alcohol on smooth muscle. Concentrations of ethanol as low as 17 mM have been shown to inhibit contractions of vascular and intestinal smooth muscle. These effects appear to be due to alterations in calcium influx.[72-74]

In summary, acute exposure to ethanol generally increased the binding of calcium to membranes while chronic ingestion of alcohol decreased binding and uptake of calcium in several types of tissues. It has been suggested that alterations in binding and transport of calcium may be related to the effects of alcohol on smooth, cardiac, and striated muscle, as well as on brain tissue.

F. EFFECT OF ALCOHOLS ON BRAIN CALCIUM

Acute administration of ethanol (1.5–2.5 g/kg, intraperitoneally, as a 50% solution) decreased brain calcium content in a dose-dependent manner.[71,75] Calcium depletion reached about 50% with 2.5 g/kg and was similar for all brain areas studied. Subcellular fractionation studies indicated that this decrease was restricted to the nerve-ending (synaptosomal) fraction.[68] Depletion of brain calcium was also found after injection of 2 g/kg of methanol, n-propanol, isopropanol, or t-butanol. Surprisingly, the depletion of brain calcium by ethanol was prevented completely by injection of the narcotic antagonist, naloxone (1 mg/kg). Naloxone also reduced the calcium-depleting effects of methanol and n-propanol, but did not alter the effects of isopropanol and t-butanol. Since isopropanol and t-butanol are not metabolized to aldehydes, it was suggested that the calcium-depleting effect of ethanol was due to the production of compounds such as tetrahydroisoquinolines (TIQ), which are condensation products of aldehydes and biogenic amines. This hypothesis was supported by the observation that salsolinol, a dopamine-derived TIQ, reduced brain calcium levels, although large doses (12–50 mg/kg, intraperitoneally) were required.[71] However, if we attribute the calcium-depleting effects of ethanol to the TIQs, it is difficult to explain why isopropanol and t-butanol also depleted brain calcium. In contrast to the calcium-depleting effects of acute alcohol treatment, chronic ethanol exposure [15% (v/v) in drinking water for 2 weeks] increased the calcium content of synaptic membranes from rat brain by 67%.[68] These clear changes in synaptic localization of calcium represent a mechanism by which alcohols can alter neurotransmission and constitute an exciting area of alcohol research. For example, depletion of brain calcium after acute ethanol administration might be expected to inhibit neurotransmitter release, and ethanol has indeed been found to inhibit the release of acetylcholine[76] and norepinephrine.[77] In addition, the decreased brain cyclic GMP found after ethanol treatment may be related to depletion of brain calcium.[78,79] However, the effects of ethanol on neurotransmitters and cyclic nucleotides remain controversial (see Secton 7). In addition, it should be noted that some of these results are disturbingly inconsistent with the pharmacology of alcohol. For example, naloxone effectively blocked the calcium-depleting action of ethanol but did not reduce the behavioral and physiological effects of ethanol such as loss of righting reflex, hypothermia, and disruption of operant behavior.[80] However, naloxone does block the facilitory effect of ethanol on self-stimulation behavior[80a] and administration of naloxone during chronic ethanol exposure may prevent the induction of ethanol dependence.[81] The potencies of several alcohols in lowering brain calcium did not appear to be related to their pharmacological potencies. It should be noted that another group of drugs, the narcotic analgesics, also depleted synaptic calcium stores from rat and mouse brain.[75,86] As would be expected from the observed depletion of calcium, elevation of brain calcium content antagonized the effects of opiates.[82] By analogy, increased brain calcium levels should also antagonize the effects of ethanol. However, administration of calcium was shown to increase, rather than decrease, the effects of ethanol.[83] Another surprising observation was that rats tolerant to the calcium-depleting effects of morphine were also tolerant to the calcium-lowering effects of ethanol,[71] although cross-tolerance between opiates and alcohols has not been found with behavioral and physiological measures. Other workers have failed to observe any alteration in the subcellular localization of calcium in brain after either acute or chronic ethanol treatments.[84] These results are difficult to reconcile with earlier reports.[68,71,75] Rather than altering calcium content, ethanol may instead interfere with synaptic function by altering the movement of calcium. For example, in vitro addition of ethanol to isolated synaptosomes inhibited the influx of calcium following membrane depolarization.[84] Such an action could be responsible for the inhibitory action of ethanol on the release of neurotransmitters.[76,77]

The effects of alcohol treatment on magnesium metabolism in the brain have received little attention and represent an important research area. Belknap et al.[85] have recently reported that chronic exposure of mice to ethanol by either the vapor inhalation method (4 days duration) or the liquid diet method (5 days duration) decreased the magnesium content of brain by 5–9%. These workers also noted that a similar reduction of brain magnesium, produced by feeding mice a magnesium deficient diet, resulted in behavioral signs strikingly similar to the signs which are characteristic of alcohol withdrawal.[85] The reduction of brain magnesium by chronic alcohol ingestion appears to be restricted to the subcellular fraction containing myelin.[84] These results, together with the clinical findings discussed earlier, suggest that magnesium depletion may be involved in the etiology of the alcohol withdrawal syndrome.

G. SUMMARY AND CONCLUSIONS

As can be appreciated from the observations presented above, the effects of ethanol on metabolism of calcium and magnesium are diverse and complex. Some important observations are summarized in this section. Acute ethanol intoxication in humans was shown to increase the urinary excretion of both calcium and magnesium without altering the plasma levels of these ions. In laboratory experiments, exposure to ethanol in vitro increased the affinity of erythrocyte membranes for calcium and altered movement of calcium in smooth muscle. Acute treatment with ethanol *in vivo* also increased the ability of rat brain synaptic membranes to bind calcium apparently by decreasing the endogenous calcium content of these membranes. These alterations in calcium binding and content represent a mechanism by which alcohols might alter synaptic transmission in the central nervous system.

Chronic ingestion of alcohol depleted serum calcium in some individuals. Chronic administration also reduced binding and transport of calcium in cardiac and skeletal muscles of humans and experimental animals. These latter changes may be involved in the myopathies and cardiopathies of chronic alcoholism. In rats, chronic ingestion of ethanol increased the calcium content of brain synaptosomes and decreased synaptic binding of calcium. These changes in brain content and binding of calcium after chronic administration of ethanol were the opposite of those produced by acute administration of alcohol and thus represent a possible adaptive mechanism for the development of tolerance and dependence. However, this remains speculative as it is not clear if the changes in brain calcium can be reproduced in all laboratories.

Of the many alterations of calcium and magnesium metabolism produced by alcohol, the most frequently studied has been the depletion of magnesium found during alcohol withdrawal. This depletion was noted in serum, erythrocytes, and skeletal muscle from patients and experimental subjects in withdrawal. Hypocalcemia was also noted during withdrawal from alcohol but did not appear to be as prevalent as hypomagnesemia. The importance of these deficiencies in the etiology of the alcohol withdrawal syndrome remains controversial. Magnesium and calcium deficiencies can themselves produce some of the symptoms of alcohol withdrawal, such as neuromuscular irritability, tetany, seizure susceptibility, and hallucinations. It is clear that these deficiencies should be corrected in alcoholic patients and that some signs of alcohol withdrawal may be ameliorated by administration of magnesium or calcium. However, the effects of supplementation with magnesium and calcium on the withdrawal syndrome have not been evaluated in any controlled clinical tests; thus, the possible value of these treatments remains uncertain. There are now several animal models of the alcohol withdrawal syndrome (see Chapter 27) which should prove most useful in evaluating the effects of supplementation with calcium and magnesium on withdrawal signs and the effects of alcohol withdrawal on brain calcium and magnesium levels. A careful comparison of the effects of acute and chronic administration of alcohol on localization of calcium and magnesium in the brain with the behavioral effects of these treatments is clearly needed. It is hoped that such investigations will lead to new treatments for the management of the alcohol withdrawal syndrome as well as an increased understanding of the pharmacology of ethanol.

ACKNOWLEDGMENTS. I wish to thank the Pharmaceutical Manufacturers Association Foundation for their generous support and Dr. V. C. Sutherland for her advice and encouragement.

H. REFERENCES

1. E. B. Flink, F. L. Stutzman, A. R. Anderson, T. Konig, and R. Fraser, *J. Lab. Clin. Med.* **43**:169 (1954).
2. E. B. Flink, in: *The Biology of Alcoholism* (B. Kissin and H. Begleiter, eds.), Vol. 1: *Biochemistry*, pp. 377–395, Plenum Press, New York (1971).
3. E. B. Flink, in: *Trace Elements in Human Health and Disease* (A. S. Prasad and D. Oberleas, eds.), Vol. II, pp. 1–21, Academic Press, Inc., New York (1976).
4. J. K. Aikawa, in: *Trace Elements in Human Health and Disease* (A. S. Prasad and D. Oberleas, eds.), Vol. II, pp. 47–78, Academic Press, Inc., New York (1976).
5. C. J. Duncan (ed.), *Calcium in Biological Systems* (Symposia of the Society for Experimental Biology, No. 30), pp. 1–218, Cambridge University Press, Cambridge, England (1976).
6. R. P. Rubin, *Calcium and the Secretory Process*, pp. 1–180, Plenum Press, New York (1974).
7. W. E. C. Wacker and A. F. Parisi, *N. Engl. J. Med.* **278**:658, 712, 772 (1968).
8. H. Rasmussen and D. P. B. Goodman, *Physiol. Rev.* **57**:421 (1977).
9. R. Schober, C. Nitsch, U. Rinne, and S. J. Morris, *Science* **195**:495 (1977).
10. D. Chapman, J. Urbina, and K. M. Keough, *J. Biol. Chem.* **249**:2512 (1974).
11. H. Trauble and H. Eibl, *Proc. Natl. Acad. Sci. USA* **71**:214 (1974).
12. S. A. Simon, L. J. Lis, J. W. Kauffman, and R. C. MacDonald, *Biochim. Biophys. Acta* **375**:317 (1975).
13. S. Ohnishi and T. Ito, *Biochemistry* **13**:881 (1974).
14. J. R. Trudell, *Anesthesiology* **46**:5 (1977).
15. J. H. Chin and D. B. Goldstein, *Science* **196**:684 (1977).
16. J. H. Chin and D. B. Goldstein, *Mol. Pharmacol.* **13**:435 (1977).
17. M. E. Traynor, P. B. J. Woodson, W. T. Schlapfer, and S. H. Barondes, *Science* **193**:510 (1976).
18. M. Curran and P. Seeman, *Science* **197**:910 (1977).
19. C. L. Stephens and M. Shinitzky, *Nature (Lond.)* **270**:267 (1977).
20. I. O. Ingram, *J. Bacteriol.* **125**:670 (1976).
21. J. Viret and F. Leterrier, *Biochim. Biophys. Acta* **436**:811 (1976).
22. J. W. Kalbfleisch, R. D. Lindeman, H. E. Ginn, and W. O. Smith, *J. Clin. Invest.* **42**:1471 (1963).
23. M. Ogata, J. H. Mendelson, and N. K. Mello, *Psychosom. Med.* **30**:463 (1968).
24. R. H. Ylikahri, A. R. Poso, M. D. Huttunen, and M. E. Hillbom, *Scand. J. Clin. Lab. Invest.* **34**:327 (1974).
25. J. M. Earll, K. Gaunt, L. A. Earll, and Y. Y. Djuh, *Aviat. Space Environ. Med.* **47**:808 (1976).
26. M. Verdy and D. Caron, *Biol. Gastro-enterol.* **6**:157 (1973).
27. W. Q. Sargent, J. R. Simpson, and J. D. Beard, *J. Pharmacol. Exp. Ther.* **190**:507 (1974).
28. T. C. Peng, C. W. Cooper, and P. L. Munson, *Endocrinology* **91**:586 (1972).
29. T. C. Peng and H. J. Gitelman, *Endocrinology* **94**:608 (1974).
30. E. L. Krawitt, *J. Lab. Clin. Med.* **85**:665 (1975).
31. F. W. Ellis, *J. Pharmacol. Exp. Ther.* **153**:121 (1966).
32. H. Rasmussen and P. Bordier, in: *The Physiological Basis of Metabolic Bone Disease* (H. Rasmussen and P. Bordier, eds.), pp. 201–205, The Williams & Wilkins Company, Baltimore (1974).
33. R. Medalle and C. Waterhouse, *Ann. Intern. Med.* **79**:76 (1973).
34. D. Fankushen, D. Raskin, A. Dimich, and S. Wallach, *Am. J. Med.* **37**:802 (1964).
35. H. Estep, W. A. Shaw, C. Watlington, R. Hobe, W. Holland, and S. G. Tucker, *J. Clin. Endocrinol. Metab.* **29**:842 (1969).
36. J. F. Sullivan, P. N. Wolpert, R. Williams, and J. D. Egan, *Ann. N.Y. Acad. Sci.* **162**:947 (1969).
37. M. E. Shils, in: *Trace Elements in Human Health and Disease* (A. S. Prasad and D. Oberleas, eds.), Vol II, pp. 23–46, Academic Press, Inc., New York (1976).

38. C. S. Anast, J. M. Mohs, S. L. Kaplan, and T. W. Burns, *Science* **177**:606 (1972).
39. R. Medalle, C. Waterhouse, and T. J. Hahn, *Am. J. Clin. Nutr.* **29**:854 (1976).
40. A. B. Gutman and E. B. Gutman, *J. Clin. Invest.* **16**:903 (1937).
41. M. D. Moore, *Gastroenterology* **60**:43 (1971).
42. A. D'Souza and M. H. Floch, *Am. J. Clin. Nutr.* **26**:352 (1973).
43. R. J. McCollister, A. S. Prasad, R. P. Doe, and E. B. Flink, *J. Lab. Clin. Med.* **52**:928 (1958).
44. R. J. McCollister, E. B. Flink, and M. D. Lewis, *Am. J. Clin. Nutr.* **12**:415 (1963).
45. F. N. Heaton, L. N. Pyrah, C. C. Beresford, R. W. Bryson, and D. F. Martin, *Lancet* **2**(7260):802 (1962).
46. J. H. Mendelson, M. Ogata, and N. K. Mello, *Ann. N.Y. Acad. Sci.* **162**:918 (1969).
47. J. H. Mendelson, L. LaDou, and C. Corbett, *Q. J. Stud. Alcohol,* Suppl. **4**:108 (1964).
48. J. D. Hines, in: *Erythrocyte Structure and Function* (G. J. Brewer, ed.), pp. 621–640, Alan R. Liss, Inc., New York (1975).
49. E. B. Flink, G. D. Marano, R. Morabito, S. R. Shane, and R. R. Scobbo, in: *Currents in Alcoholism* (F. A. Seixas, ed.), Vol. I, pp. 329–340, Grune & Stratton, Inc., New York (1977).
50. J. Nielsen, *Dan. Med. Bull.* **10**:225 (1963).
51. S. M. Wolfe and M. Victor, *Ann. N.Y. Acad. Sci.* **162**:973 (1969).
52. G. Stendig-Lindberg, *Acta Psychiatr. Scand.* **50**:465 (1974).
53. R. J. McCollister, E. B. Flink, and R. P. Doe, *J. Lab. Clin. Med.* **55**:98 (1960).
54. J. E. Jones, S. R. Shane, W. H. Jacobs, and E. B. Flink, *Ann. N.Y. Acad. Sci.* **162**:934 (1969).
55. L. S. Glickman, V. Schenker, S. Grolnick, A. Green, and A. Schenker, *J. Nerv. Ment. Dis.* **134**:410 (1962).
56. B. L. Vallee, W. E. C. Wacker, and D. D. Ulmer, *N. Engl. J. Med.* **262**:155 (1960).
57. P. Lim and E. Jacob, *Metab. Clin. Exp.* **21**:1045 (1972).
58. L. T. Iseri, J. Freed, and A. R. Bures, *Am. J. Med.* **58**:837 (1975).
59. K. Luomanmaki, J. Heikkila, and M. Harikainen, *Eur. J. Cardiol.* **3**:167 (1975).
60. O. Z. Shulsinger, P. J. Forni, and B. B. Clyman, in: *Currents in Alcoholism* (F. A. Seixas, ed.), Vol. I, pp. 319–327, Grune & Stratton, Inc., New York (1977).
61. G. Somjen, M. Hilmy, and C. R. Stephen, *J. Pharmacol. Exp. Ther.* **154**:652 (1966).
62. M. Kimura and Y. Itokawa, *J. Neurochem.* **28**:389 (1977).
63. H. D. Kruse, E. R. Orent, and E. V. McCollum, *J. Biol. Chem.* **96**:519 (1932).
64. D. R. Buck, A. W. Mahoney, and D. G. Hendricks, *Pharmacol. Biochem. Behav.* **5**:529 (1976).
65. J. G. Chutkow, *Neurology* **24**:780 (1974).
66. J. G. Chutkow, *Mayo Clin. Proc.* **49**:244 (1974).
67. P. Seeman, M. Chau, M. Goldberg, T. Sanks, and L. Sax, *Biochim. Biophys. Acta* **225**:185 (1971).
68. D. H. Ross, in: *Alcohol Intoxication and Withdrawal* (M. M. Gross, ed.), Vol. III, pp. 459–471, Plenum Press, New York (1977).
69. O. Pachinger, J. Mao, J. M. Fauvel, and R. J. Bing, *Recent Adv. Stud. Card. Struct. Metab.* **5**:423 (1975).
70. E. Rubin, A. M. Katz, C. S. Lieber, E. P. Stein, and S. Puszkin, *Am. J. Pathol.* **83**:499 (1976).
71. D. H. Ross, *Ann. N.Y. Acad. Sci.* **273**:280 (1976).
72. L. Hurwitz, S. Von Hagen, and P. D. Joiner, *J. Physiol.* **50**:1157 (1967).
73. B. M. Altura, H. Edgartian, and B. T. Altura, *J. Pharmacol. Exp. Ther.* **197**:352 (1976).
74. H. Edgarian and B. M. Altura, *Anesthesiology* **44**:311 (1976).
75. D. H. Ross, M. A. Medina, and H. L. Cardinas, *Science* **186**:63 (1974).
76. H. Kalant and W. Grose, *J. Pharmacol. Exp. Ther.* **158**:386 (1967).
77. A. Y. Sun, *Res. Commun. Chem. Pathol. Pharmacol.* **15**:705 (1976).
78. L. Volicer and B. P. Hurter, *J. Pharmacol. Exp. Ther.* **200**:298 (1977).
79. W. A. Hunt, J. D. Redos, T. K. Dalton, and G. N. Catravas, *J. Pharmacol. Exp. Ther.* **201**:103 (1977).
80. R. A. Harris, *Fed. Proc. Abstr.,* **36**:81 (1977).
80a. S. A. Lorens and S. M. Sainati, *Life Sci.* **23**:1359 (1978).
81. K. Blum, S. Futterman, J. E. Wallace, and H. A. Schwertner, *Nature (Lond.),* **265**:49 (1977).

82. R. A. Harris, H. H. Loh, and E. L. Way, *J. Pharmacol. Exp. Ther.* **195**:488 (1975).
83. C. K. Erickson, T. D. Tyler, and R. A. Harris, *Science* **199**:1219 (1978).
84. W. F. Hood and R. A. Harris, *Soc. Neurosci. Abstr.* **4**:425 (1978).
85. J. K. Belknap, J. H. Berg, and R. R. Coleman, *Pharmacol. Biochem. Behav.* **9**:1 (1978).
86. R. A. Harris, H. Yamamoto, H. H. Loh, and E. L. Way, *Life Sci.* **20**:501 (1977).
87. W. D. Smith and J. F. Hammarstein, *Am. J. Med. Sci.* **237**:413 (1959).
88. D. H. Knott and J. D. Beard, *South. Med. J.* **62**:485 (1969).

Influence of Ethanol on Biological Membranes

VIII

Effects of Ethanol on Membrane Transport

4

Linda L. Shanbour and Yuh-Jyh Kuo

A. INTRODUCTION

Transport is a fundamental process in all living cells and is particularly significant with respect to neural transmission, muscular contraction, respiration, excretion, secretion, hormonal control, and vision. The direct effect of ethanol on membranes and associated phenomena concerned with electrical activity was first demonstrated in the nervous system because of the general ethanol depression of this system. A number of review articles concerning ethanol and neural effects have been published.[1-6] Israel and Kalant[7,8] found that ethanol affected ion movement in a variety of organs and proposed that the inhibition of active transport of ions by ethanol could be a universal membrane phenomenon. If the primary action of ethanol is to alter membrane transport, this could lead to metabolic derangements, altered tissue function, and release of substances which could affect other tissues in the body.

The gastrointestinal tract is the first tissue that ingested ethanol contacts. The concentrations of ethanol to which this tissue is exposed are thus several times greater than levels attained in other tissues. This chapter will summarize our present knowledge regarding the influence of ethanol on membrane transport with the gastrointestinal system used as a primary model. In addition, the effect of ethanol on frog skin, which has transport characteristics very similar to the epithelium of the gastrointestinal mucosa, will be discussed. The relationship between function of the gastrointestinal system and alcoholism has been reviewed elsewhere[9-16] and in Volume 1, Chapter 22. Although information is scarce concerning the effects of ethanol on membrane transport in the liver,

Linda L. Shanbour and Yuh-Jyh Kuo • Department of Physiology, The University of Texas Medical School at Houston, Texas Medical Center, Houston, Texas 77025.

heart, muscle, kidney, and urinary bladder, these areas are summarized. Specific effects of ethanol on biological membranes in the nervous system are covered in Chapters 5, 6, and 7 of this publication.

B. EFFECTS OF ETHANOL ON MEMBRANE TRANSPORT IN GASTROINTESTINAL TISSUES

1. Salivary Glands and Oral Mucosa

There are early studies of the effects of ethanol on salivary glands and oral mucosa but there is a scarcity of recent articles. Chittenden *et al.*[17] in 1898 found that the local application of 50% ethanol to the oral mucosa and tongue stimulated salivary secretion. The introduction of ethanol into the stomach did not have this effect. These findings indicated that the stimulatory effect was not due to either the presence of ethanol in the bloodstream to stimulate secretory nerve endings in the salivary glands or to irritation of the stomach associated with salivation. Decrease in secretion was observed following the initial 10 min of stimulation.[18] In alcoholic patients, ulcerations of the oral mucosa and tongue have been reported.[19]

Information concerning the secretory function of the acini and the ducts of the salivary glands comes primarily from recent work with microelectrode, micropuncture, microcatheterization, and microperfusion techniques.[20] The technical difficulties have been considerable, owing to the small luminal diameter of these glands and to their heterogeneity. One electrophysiological sign of the link between stimulus and secretion is a change in transmembrane potential. The resting transmembrane potential is distinctly low compared to the levels in most excitable cells. Lundberg,[21-23] who first measured acinar transmembrane potentials, recorded values in cat submaxillary and sublingual glands of 20–35 mV (inside negative) across the luminal membrane. Similar values were also found in the submaxillary and parotid glands of the rat.[24] After the initiation of parasympathetic stimulation of a gland or administration of cholinergic drugs, the membrane potential of acinar cells usually changes. The changes involve a pronounced hyperpolarization (i.e., acinar cytoplasmic negativity increases).[21-23] Lundberg called this change in transmembrane potential the "secretory potential," which is mainly generated by the increase in membrane permeability of the acinar cell to K^+. In contrast, the relationship between the secretory potential and frank secretion is not clear. The final composition of the saliva (hypotonic, lower $[Na^+]$ and higher $[K^+]$ and $[HCO_3^-]$ than plasma) is attained in the ducts by reabsorptive and secretory mechanisms. Active transport is involved in reabsorption of Na^+ but not Cl^-, and in secretion of K^+ and probably HCO_3^-. Despite recent advances in our understanding of secretory processes in the salivary gland, information gaps remain. The exact mechanisms involved in the formation of the precursor fluid by the acini, secretion of K^+ and HCO_3^-, and transfer of water by the ducts remain to be clarified. After these problems are resolved, the mechanism of action of ethanol on salivary secretion may be determined.

2. Esophagus

Webb *et al.*[25] found that the absorption of several types of drugs through the esophagus increased after the esophagus was irritated with 50% ethanol. The only study to characterize the esophageal mucosa and ethanol effects was recently completed by Fromm and Robertson in rabbits.[26,27] They found that the mean electrical resistance of esophageal mucosa (1786 $\pm$ 168 $\Omega \cdot$ cm^2), was greater than that observed for fundic or antral mucosa (124 $\pm$ 8 $\Omega \cdot$ cm^2). The mean rate of luminal acid loss for the esophagus (0.28 $\pm$ 0.09 μeq/cm^2 per hr) (i.e., the mucosal permeability for H^+) was at least 50% less than that observed for the antrum (0.7 $\pm$ 0.1 μeq/cm^2 per hr) and a longer time (over 80 min) was required to achieve steady rates. No difference in the rate of luminal acid loss or electrical resistance was observed between tissue pairs of proximal and distal esophagus. Addition of 20% ethanol to the lumen did not significantly alter the mucosal permeability for the back diffusion of H^+ or affect electrical potential difference or electrical resistance across the mucosa. Serosal addition of ethanol also did not significantly alter the rate of luminal acid loss or the electrical resistance. Although negative results were obtained from the observations described above, the effects of higher concentrations of ethanol on the electrophysiological parameters and ion fluxes should be investigated since the esophagus is exposed to concentrations of ethanol greater than those attained in the stomach.

3. Stomach

a. Fundic Mucosa

One of the most important functions of the stomach is to secrete hydrochloric acid. Although it has been assumed that ethanol stimulates acid secretion,[28–33] the effects of ethanol on gastric acid secretion depend on several factors:

1. Species of animal
2. Techniques, experimental preparations, diet, anesthesia, and so on, employed
3. Route of ethanol administration (i.e., orally, intravenously, or intraperitoneally)
4. Dosages of ethanol
5. Duration of ethanol administration (i.e., acute or chronic)
6. Presence or absence of gastrointestinal disorders

7. Psychological factors
8. Use of adequate controls

Various effects of luminal application or intravenous administration of ethanol on the rate of acid secretion *in vivo* have been reported, and in many instances it is not clear whether or not these effects are the result of a direct action of ethanol on the parietal cells or indirect effects related to humoral or neurogenic factors. Rehm and Hokin,[34] using topical administration of ethanol in the exposed dog chambered gastric mucosa, were the first to demonstrate that ethanol inhibits active transport of H^+ in the gastric mucosa. The inhibitory effect of ethanol on acid secretion was further demonstrated by Sernka and co-workers[35,36] using the pH-stat method in both *in vivo* and *in vitro* dog stomach preparations. The advantages of this approach over the conventional 15-min collection method are the continuous measurement of acid secretion and the reduction of the gradient for back diffusion of acid if the end point for titration is set at pH 7.0. Subsequently, the inhibition of acid secretion by ethanol was also reported in rats[37] and rabbits.[27]

For more than two decades, attempts to elucidate the mechanism of hydrochloric acid formation by the oxyntic cell have focused on the electrophysiology of the gastric mucosa. In most vertebrates, there is a significant spontaneous electrical potential difference of about -40 mV across the fundic mucosa[38] (mucosal surface negative with respect to the reference serosal surface). Transmural potential difference is located in the mucosae[39,40] and has been considered essential to or influenced by acid secretion.[41,42] Geall *et al.*[43] have proposed that the maintenance of the gastric potential difference may be a sensitive indicator of mucosal integrity, although the precise interrelationships between potential difference and physiological or pathological factors affecting function or structure in the stomach are unknown.

The inhibitory effect of ethanol on potential difference was first observed by Rehm and Hokin[34] in the anesthetized dog stomach. These potentials recovered within minutes after the ethanol was removed from the mucosa, and no secretion of HCl was observed. The altered potential was not the result of an increase in liquid junction potential between gastric solution and chamber fluid. This phenomenon was later confirmed by Davenport *et al.*,[44,45] Geall *et al.*,[43] and Sernka *et al.*[35] in dogs, Shanbour *et al.* in the rat,[46] and Fromm and Robertson in the rabbit.[27] However, the decrease in potential difference may be either the result of nonselective leakage of ions owing to the increase in permeability of the membrane or the result of inhibition of ion transport. A preparation that incorporated tissue electrical resistance measurements was developed by Shanbour *et al.*[46] to distinguish between these two possibilities. They found an increase in electrical resistance after ethanol in the rat,[46] and this finding was later confirmed in the dog stomach[35] (Table I). The electrical resistance data suggest that the most likely explanation is that ethanol acts directly on the gastric mucosa to inhibit active transport of ions. If ethanol had primarily increased permeability, the electrical resistance would have decreased. These authors, using this differ-

Table I.   Effects of Ethanol on the Stomach (Percent Change from Control)

Potential difference	Electrical resistance	Acid secretion	ATP	Cyclic AMP
−75	+20	−68	−47	−5

ent approach, have seriously questioned the "barrier" concept,[44] which has been widely accepted and applied in gastrointestinal physiology. This concept proposed that the gastric mucosa acts as a "barrier" to prevent the movement of H^+ into the tissue and movement of Na^+ from the tissue into the lumen. Using the classical Heidenhain pouch preparation of chronic dogs, Davenport[45,47−49] has reported that ethanol and other gastric damaging agents (aspirin, bile salts, and urea) break this barrier, thus decreasing H^+ concentration in the lumen and increasing luminal Na^+ concentration. In other words, damaging agents increase the mucosal permeability for both H^+ and Na^+. However, there is no histological evidence for such a barrier. What is unique about the gastric mucosa is the capacity to actively secrete H^+ ion, as well as actively transport Na^+ and Cl^- ions.

Electrophysiological studies *in vivo* can reflect tissue function and acid secretion but have not yet been developed to the extent that specific ion transport, especially for Na^+ and Cl^-, can be determined. Present knowledge concerning ion transport in the gastric mucosa has principally been obtained in experiments on the isolated frog gastric mucosa with the short-circuit technique of Ussing and Zerahn.[50] This technique is used to identify the ionic currents responsible for the generation of the transmural potential difference. With this preparation, the electrochemical potential difference between the two bulk solutions bathing the stomach is maintained at zero and the concentrations of most ions are the same in both solutions. Net ionic movement under these circumstances is thus considered to result from active transport. It is well established that active electrogenic transport exists for only chloride ions[51,52] and hydrogen ions[53] in the frog gastric mucosa. Transport of ions other than chloride and hydrogen in the mucosa probably depends on passive diffusion. However, in the isolated mammalian gastric mucosa and intact stomach, the situation is different. In these preparations, an active electrogenic transport of Na^+ from the secretory (luminal or mucosal) to the nutrient (serosal) side, in addition to Cl^- secretion, has been demonstrated in several species including cat, dog, rhesus monkey, human,[54,55] rat and guinea pig,[56] and piglet.[57] The only exception suggesting that Na^+ may be actively secreted in the opposite direction from serosa to mucosa was reported by Fromm and his co-workers to occur in the rabbit gastric mucosa.[58]

In vitro studies by Kuo *et al.*[36] in the dog gastric mucosa confirmed the *in vivo* inhibition of active transport and demonstrated that ethanol inhibits Na^+,Cl^-, and H^+ transport. Later effects observed were increases in permeability as reflected by increases in unidirectional Na^+, Cl^-, and urea fluxes. The inhibitory effect of ethanol on Na^+, Cl^-, and H^+ transport has also been reported for the rabbit gastric mucosa.[27] This recently accumulated data clearly demonstrate that ethanol does inhibit active transport of ions in the gastric mucosa.

To further support or refute a "transport" concept, it is necessary to compare the effects of ethanol with other damaging agents to determine whether or not the effects of ethanol are unique. The initial effect of aspirin is inhibition of active transport (within minutes) followed by increase in tissue permeability in both intact and *in vitro* canine preparations.[59] Likewise, bile salts also primarily inhibit active transport and acid secretion.[60] These findings demonstrate that there is no reason to evoke a gastric "barrier" concept and that the mammalian gastric mucosa can be more accurately envisioned in terms of its unique active transport properties (i.e., H^+ and Cl^- secretion as well as Na^+ absorption). An active transport concept can explain other findings in the literature, including protein loss[44] and histological damage[45] produced by long-term administration of ethanol. Inhibition of active transport may disrupt the *milieu interieur,* resulting in histamine release from mast cells, increased capillary permeability, loss of protein, inhibition of enzymes, and progressive damage to the gastric mucosa. There apparently are no reports in the literature which cannot be explained on a transport basis. This concept is also consistent with the hyposecretion frequently observed in chronic alcoholics.[61]

Enzymes that have been implicated in active transport in the gastric mucosa have been studied in a further attempt to explain the mechanism of action of ethanol. Cyclic AMP has been suggested as a second messenger in acid secretion and bicarbonate ATPase in Cl^- transport.[62] Ethanol inhibits adenyl cyclase, phosphodiesterase, Mg^{2+}, and Mg^{2+},HCO_3-ATPase.[63,64] It is interesting that the threshold concentration of ethanol for enzyme inhibition (i.e., 10%, which is comparable to drinking one martini on an empty stomach) is the same as for active transport inhibition. Inhibition of enzymes could be responsible for or the result of the inhibition of ion transport. Ethanol does not alter tissue cyclic AMP content but does decrease ATP levels both *in vitro* and *in vivo*.[64,65] Other studies[65,66] have indicated that cyclic AMP is not a mediator in gastric acid secretion, although dibutyryl cyclic AMP potentiates histamine-stimulated acid secretion, probably through an increase in tissue permeability.[67] Of particular interest is the decrease in ATP content with ethanol, whereby ATP could be the energy source for gastric H^+ secretion.[68]

b. Antral Mucosa

Relatively little is known of the effects of ethanol on the distal stomach, compared to the studies on the fundus, although antral mucosal properties are of considerable clinical and physiological significance. It is this region of the stomach which undergoes peptic ulcer damage as well as providing an epithelium which separates the secretory epithelium of the fundus from the absorptive epithelium of the small intestine. In the isolated rabbit antral mucosa, Fromm and Robertson[27] found that the effects of 20% ethanol on the electrophysiological parameters and ion fluxes were similar to the observations in the fundic mucosa (i.e., decrease in potential difference, inhibition of active transport of Na^+ and Cl^-, and increase in mucosal permeability for Na^+, Cl^-, and erythritol). They also found that ethanol has a distinct polarity with respect to its application. Administration of ethanol on both luminal and serosal sides of mucosae produced the same results as when ethanol was present on only the luminal side. However, when ethanol was present on the serosal side only, potential difference, resistance, and ion fluxes were not siginificantly different from controls. These phenomena were consistent with studies on the fundus of canine stomach *in vivo*,[35] and further suggest a primary action of ethanol on active transport.

4. Intestine

The major function of the small intestine is to absorb water, electrolytes, and organic substances. During ethanol ingestion, some absorption of ethanol occurs in the stomach, but most of the ethanol is absorbed in the upper small intestine. Halsted *et al.*[69] found that ingested ethanol was completely absorbed in the jejunum before reaching the ileum. Israel *et al.*[70] have found that the ethanol concentration is about 1.0–3.0% in the human upper jejunum during moderate drinking. However, the concentration of ethanol in the jejunal lumen could increase to over 6.0% with prolonged ethanol consumption.[69]

Chronic alcoholism is frequently associated with malnutrition, vitamin deficiencies, and diarrhea.[14] Malabsorption of several substances, xylose,[71] D-glucose,[72-74] amino acids,[70,72,75,76] thiamine,[77,78] folic acid,[79-81] vitamin A,[82] vitamin B,[83] Ca^{2+},[84-86] Fe^{2+},[87] and Mg^{2+}[88] by ethanol has been reported during acute and chronic conditions in human subjects and laboratory animals. The absorption of most of these substances is by active transport processes. Investigators using an *in vitro* preparation, either Ussing-type flux chamber or inverted sac technique,[89] have demonstrated that ethanol has a direct effect on these transport processes.[70,72,73,75,78,84]

Considerable attention has been given to investigation on the mechanism of action of ethanol on the small intestine. In the absence of actively transported sugars or amino acids, *in vitro* preparations of rabbit jejunal mucosa are characterized by a small transepithelial potential difference, 2–3 mV, with the serosal solution electrically positive with respect to the mucosal solution.[90] Sodium flux studies indicate that the net flux of Na^+ is from the serosal to mucosal side, which is opposite in direction to the net flux of Na^+ across the ileum.[91] However, the addition of D-glucose to both mucosal and serosal bathing solutions results in the net absorption of Na^+ and an increase in the potential difference. Net absorption of Na^+ can also be stimulated by the addition of 3-O-methylglucose and L-alanine.[90,92] The active absorption of organic substances (i.e., glucose, amino acids, etc.) is also dependent to a considerable extent on the active transport of

Table II.   Effects of Ethanol on the Jejunum (Percent Change from Control)

Potential difference	Na^+ transport	3-O-Methylglucose transport	L-Alanine transport
−40	−75	−80	−60

Na^+ across the intestine.[93,94] Dinda *et al.*[73] were the first to determine that inhibition of glucose transport was linked with inhibition of Na^+ transport. They found that 2.6% (v/v) ethanol inhibited glucose transport and mucosal (M) to serosal (S) Na^+ flux but did not affect net Na^+ flux in the hamster jejunum. To eliminate the electrochemical gradients across the mucosa to further determine whether ethanol inhibits active absorption of Na^+, Kuo and Shanbour[92] used the *in vitro* Ussing chamber preparation to assess this problem. They found that ethanol at 1.8% produced no changes but that 3.0% (v/v, present on both sides of the mucosa to prevent an osmotic gradient) decreased the electrical potential difference and inhibited the active transport of Na^+, 3-O-methylglucose, and L-alanine (Table II). This concentration also increased the permeability of the mucosa for Cl^-, 3-O-methylglucose, and L-alanine. Ethanol, at 5.4%, potentiated the effects on potential difference, short-circuit current, and the permeability for electrolytes and organic substances. These results suggest that diarrhea associated with ethanol intake may be induced by the inhibition of active absorption of Na^+ and organic substances as well as possible potentiation through increase in mucosal permeability.

Since ethanol inhibits active transport of Na^+ and organic substances across the small intestine, the energy source, ATP,[95] and the enzyme, Na^+,K^+-ATPase,[96] for active transport become important. Carter and Isselbacher[97] found that ethanol given *in vivo* markedly decreased the ATP content of the small intestine of rats. These effects could not be prevented when ethanol metabolism was inhibited with pyrazole. Thus, it appears that ethanol exerts a direct effect on the ATP content of the small intestine. The inhibitory effect of ethanol on intestinal ATP was confirmed in the guinea pig.[98] In addition, Krasner *et al.*[98] reported that ethanol had no effect on the activity of Na^+,K^+-ATPase of whole tissue homogenate in guinea pig jejunum. However, Hoyumpa *et al.*[99] determined the activities of Na^+,K^+-ATPase in the brush border and basolateral membrane fractions of rat jejunal mucosa cells and found that the Na^+,K^+-ATPase activity of the basolateral membrane was 13-fold greater than in the corresponding brush border. Ethanol inhibited jejunal basolateral membrane Na^+,K^+-ATPase activity *in vitro* and *in vivo*.[99] This inhibition could be linked with the inhibitory action of ethanol on intestinal absorption.

The decrease in intestinal ATP by ethanol could be a secondary effect of ethanol on altered mucosal metabolism. Relatively few studies on the effects of ethanol on intestinal mucosal metabolism have been conducted. Greene *et al.*[100] studied the effects of ethanol on the activities of glycolytic enzymes in the small intestinal mucosa of volunteer subjects. During a constant diet regime, ethanol decreased the activities of a number of glycolytic enzymes: hexokinase, fructose-1-phosphate aldolase, and fructose-1,6-diphosphate aldolase. Ethanol also decreased the activity of the gluconeogenic enzyme, fructose-1,6-diphosphatase. Conversely, ethanol increased the activity of pyruvate kinase. Other studies to determine the effects of ethanol on fatty acid synthesis, cholesterogenesis, triglyceride synthesis, O_2 consumption, and CO_2 production have produced conflicting results,[101-105] depending on the duration of ethanol administration and the type of diet. Additional information on the effects of ethanol on cellular metabolism is necessary in order to fully understand the action of ethanol on the intestinal mucosa.

5. Pancreas—Exocrine Secretion

The pancreas secretes digestive enzymes, water, and electrolytes into the duodenum. Until recently, ethanol had been assumed to stimulate pancreatic secretion.[106] Intragastric, intraduodenal, or intravenous ethanol has been shown to produce either no change or a slight and highly variable increase in basal pancreatic secretion in dog and man. This is possibly due to the release of hormones or to a neurogenic effect. Mott et al.[107] using human subjects, and Bayer et al.[108] and Tiscornia et al.[109-111] using conscious dogs, have shown that ethanol markedly inhibits secretin and cholecystokinin-stimulated pancreatic secretion of water, bicarbonate, and protein. Since these preparations prevented acid from entering the duodenum, the inhibition was probably due to a direct effect of ethanol on pancreatic secretory cells. To test this hypothesis, Solomon et al.[112] used the isolated perfused rabbit pancreas and found that ethanol inhibited water, bicarbonate, and protein secretion independent of sphincteric constriction, indicating that ethanol has a direct effect on pancreatic secretion (Table III).

Inhibition of pancreatic secretion by ethanol could also be due to impairment of transport processes. However, information concerning pancreatic transport processes is limited. Few attempts have been made to determine the potential difference across pancreatic duct or acinar cells. Reber et al.[113] recorded a potential difference of −2 to −6 mV (lumen negative) across the rabbit pancreatic duct both in vivo and in vitro. In anesthetized cats, Way and Diamond[114] recorded a mean resting potential difference of 2.0 ± 0.3 mV (lumen positive) after correction for junction potentials. Application of the Nernst

Table III. **Effects of Ethanol on the Pancreas (Percent Change from Control)**

Volume	Bicarbonate output	Protein output	ATP	Cyclic AMP
−42	−40	−80	−34	−3

equation for concentrations of Na^+, K^+, Cl^-, and HCO_3^- in pancreatic juice and plasma (or bathing solution) indicated that, at least in the cat and rabbit, the observed concentrations of Na^+, K^+, and Cl^- in pancreatic juice were below electrochemical equilibrium, and HCO_3^- was above. This led to speculation that the secretion of HCO_3^- into the pancreatic juice must be associated with an active process. The presence of the other ions in pancreatic juice can be attributed to their electrochemical gradients.[115] However, sodium ion cannot be excluded as far as playing a participatory role in active transport. Using the *in vitro* preparation of rabbit pancreas, Rothman and Brooks[116] demonstrated that Na^+ in the bathing solution was absolutely necessary for pancreatic secretion. The mechanism by which ethanol impairs pancreatic transport has not been studied, since knowledge concerning membrane-involved pancreatic secretion is still rudimentary.

Studies have shown that ATP is necessary for pancreatic enzyme secretion[117] and that secretion of water and HCO_3^- depend on oxidative phosphorylation.[118] Thus, it seems reasonable that a decrease in ATP content would produce inhibition of secretion. Solomon *et al.*[112] supported this conclusion with studies in the *in vitro* perfused rabbit pancreas which demonstrated a decrease in ATP in response to ethanol. They also found that ethanol had no effect on cyclic AMP levels, although cyclic AMP may mediate secretin-stimulated pancreatic water and bicarbonate secretion.[119-121] Whether ethanol affects the activity of pancreatic Na^+,K^+-ATPase[122] has not been determined; neither has the effect of ethanol on cellular metabolism been carefully studied.

6. Gastrointestinal Motility

In contrast to secretion and absorption, little is known about the effects of ethanol on gastrointestinal motility. In man, ethanol ingestion results in a decrease in primary peristalsis in the esophagus and diminished contraction of the lower esophageal sphincter, while nonpropulsive contractions are increased in the distal esophagus.[123] The effect of ethanol on gastric emptying has been controversial, with reports showing delay,[124] augmentation,[125] or no effect.[126] In the human small intestine, Robles *et al.*[127] found that following the administration of ethanol, in a dose of 0.8 g/kg, either by the oral or intravenous route, there was a consistent decrease in type I wave motility, but no change in type III in the jejunum. In contrast, the ileum exhibited an increase in type III wave motility but no change in type I. This effect of ethanol on small intestinal motility appeared to be specific and was not simply the result of fluid load or of hypertonicity, since the administration of an equivolumetric amount of orange juice did not alter motility and the administration of an equivalent hypertonic solution of urea resulted in a totally different pattern of response. They suggested that the suppression following ethanol administration of type I (impeding) waves in the jejunum and the stimulation of type III (propulsive) waves in the ileum may contribute to the diarrhea frequently seen in binge-drinking alcoholic patients.

Considerable progress has been made in gastrointestinal motility research within recent years. A close interrelationship exists between gastrointestinal contraction and the electrical activity of smooth muscle cells. Smooth muscle cells of the caudal portion of the stomach and the intestine have a resting membrane potential resembling that of other smooth muscle cells (-60 mV).[128] The potential fluctuates rhythmically with a cyclic depolarization and repolarization of 5–15 mV. This phenomenon, reported by Alvarez and Mahoney[128] and by Bozler,[129] was called slow wave or basic electrical rhythm. Generation of the slow wave in smooth muscle cells depends upon the cyclic bidirectional flux of sodium ions. During the depolarization phase of the slow wave, a passive flux of sodium into the muscle cell occurs, followed by an active efflux of the ion during repolarization.[130] Job[131] has postulated that accumulation of ATP at the cell membrane leads to increased permeability to Na^+, producing depolarization. This, in turn, activates a sodium pump which transports the ion out of the cell. The pump depletes the level of ATP at the membrane, thereby reducing membrane permeability to Na^+ and producing repolarization of the membrane. After cessation of active transport, ATP is regenerated and the cycle repeats itself. In addition to the slow wave recorded from these smooth muscle cells, spike or action potentials (rapid depolarization of the membrane) are also recorded. Spike potentials which occur periodically initiate contraction.[132] Contractions do not normally occur without being preceded by spike potenials, and spike potentials occur during specific periods of slow wave. Thus, the slow wave determines the timing of the contractions. It is possible that if ethanol interferes with the sodium pump of the smooth muscle cell to change the amplitude and/or the frequency of the slow wave, eventually the contraction would be different than normal. This is reasonable since it has been reported that ethanol inhibits the action as well as the ^{22}Na efflux in the squid giant axon.[133]

C. EFFECTS OF ETHANOL ON ACTIVE TRANSPORT IN FROG SKIN

Frog skin is a type of epithelium which is very similar to the epithelium of the gastrointestinal tract. This system was first studied thoroughly by Ussing and Zerahn,[50] who showed that in the presence of equal concentrations of sodium on both sides, the short-circuit current across the skin is a function of the net transport of sodium from the outside to inside solutions. The system also serves as a valuable model in the study of active transport and of the effects of a number of substances thereon.[50,134–136]

Fuhrman[137] exposed the serosal surface of frog skin to 0.4 M ethanol. He found that there was a large decrease in the electrical resistance whereas the short-circuit current did not vary by more than 10%. He reported that in 1 hr there was a tenfold increase in chloride outflux. Israel and Kalant[138] reported that ethanol (0.04–0.35 M) added to the solution bathing the outside of frog skin produced an immediate decrease in short-circuit current proportional to the concentration of ethanol. Boyett and Vanbruggen,[139] using radioisotopic ^{22}Na, found that treatment of the outside of frog skin with ethanol decreased the net Na^+ transport as result of the inhibition of the unidirectional flux of Na^+ from outside to inside. Ethanol did not alter Na^+ outflux. The Cl^- permeability of the skin was increased severalfold when the skin was exposed to ethanol on either side. With 0.4 M ethanol in the inner bathing solution, all the unidirectional fluxes of Na^+, Cl^-, and urea increased. These studies suggest that ethanol may affect the Na^+,K^+-ATPase involved in Na^+ transport in frog skin. However, it is not clear whether the action of ethanol is due to decrease in the availability of Na^+ to the enzyme or to direct inhibition of the enzyme.

D. ETHANOL AND MEMBRANE TRANSPORT IN OTHER TISSUES

1. Liver

The liver is the primary site for the metabolism of ethanol. The source of energy, ATP, is involved in numerous hepatic reactions (e.g., carbohydrate, lipid and protein metabolism, and membrane transport). Therefore, a change in ATP could have adverse effects on liver metabolism. Reports on the acute effect of ethanol on hepatic ATP content are not consistent (Table IV),[97,140−144] probably because of variations in the dose used, the time observed, and the route of administration. In contrast, chronic treatment with ethanol in rats consistently decreased hepatic ATP and increased inorganic phosphate from 10 days to 16 months, with ethanol concentrations ranging from 5 to 30%.[140−142,145]

Table IV.   Changes in Hepatic ATP Levels after Acute Administration of Ethanol in Rats

Dose (g/kg)	Route	Time (hr)	Result[a]	Reference
7.5	Intragastric Intubation	8	↓	97
6.0	Intragastric Intubation	2	↔	140
6.0	Intragastric Intubation	8	↑	140
6.0	Intragastric Intubation	24	↔	140
2.4–3.0	Intraperitoneal	0.75	↑	141
1.5	Intravenous	0.5	↔	142
0.28–2.8	Intraperitoneal	0.5	↔	143
1.2–2.4	Intragastric Intubation	0.5	↔	144
1.2–2.4	Intravenous	0.5	↔	144

[a]Change in comparison with the controls: ↓, decrease; ↑, increase; ↔, no change.

A reduction in ATP may occur if the overall rate of processes consuming ATP exceeds the rate of those involved in its formation. Bernstein *et al.*[146] have demonstrated that Na^+,K^+-activated ATPase of liver homogenates and the active transport of ^{86}Rb (a functional analogue of K^+ transport) are markedly increased by 190% and 70%, respectively, in rats chronically treated with ethanol. In addition, increases in oxygen consumption and ethanol metabolism elicited by this chronic treatment[147-149] were completely abolished by ouabain or by removal of sodium from the incubation medium, two conditions known to block the sodium pump. This suggests that following chronic administration of ethanol, the sodium pump becomes the pacemaker for the decrease in hepatic ATP and increase in the rate of oxygen consumption, which in turn leads to an increase in the metabolism of ethanol and of other substrates.

The mechanisms by which chronic administration of ethanol leads to an increase in the activity of the sodium pump in the liver are not known. In contrast to the liver, ethanol inhibits Na^+,K^+-activated ATPase activity and ion transport in several other tissues.[150] Therefore, the increase in pump activity could be an adaptive and a compensatory mechanism to overcome the inhibition of the active transport of Na^+ and K^+ elicited by ethanol.

2. Heart

Impairment of cardiac function occurs in some chronic alcoholics.[151,152] This disorder is independent of nutritional factors, since myocardial damage is produced even when ethanol is given with adequate diets.[153] When administered acutely, ethanol produces impaired myocardial contractility in intact dogs.[154-156] The reported concentration range at which this negative inotropic effect is seen is between 15 and 65 mM. Similar findings have been reported for the isolated rat heart.[157] Gimeno *et al.*,[157] using fully oxygenated isolated rat atria, found rapid and reversible negative inotropic effects at ethanol concentrations ranging from 110 to 880 mg/dl. The relationship between contractile depression and concentration was essentially linear. Simultaneously with this concentration, there was a decrease in the action potential. Since the Na^+ transport system is the key factor for the production of the action potential and for the control of myocardial contractility, the negative inotropic effect of ethanol could be the result of impaired Na^+ transport. Although the effects of ethanol on ion distribution in cardiac muscle have not been investigated, Williams *et al.*[158] found that ethanol inhibited the Na^+,K^+-activated ATPase activity of plasma-membrane prepared from guinea pig heart. They also found that the degree of inhibition of Na^+,K^+-ATPase was dose-dependent (0.2–1.4 M) and was antagonized by the K^+ concentration in the reaction mixture.

The sarcoplasmic reticulum of the heart is another system that could be deleteriously affected by ethanol. This membrane system participates in the mediation of excitation–contraction coupling within the myocardium, presumably by releasing Ca^{2+} to the cardiac contractile proteins in response to a signal originating from the action potential. Calcium binding and calcium uptake, which represent two kinetically dissimilar mechanisms by which Ca^{2+} may be removed from the solution by the membranes of the sarcoplasmic reticulum, can be examined in preparations of cardiac microsomes. Ca^{2+} binding is a rapid but limited saturable mechanism by which Ca^{2+} becomes associated with cardiac microsomes. Ca^{2+} uptake is a slower but more extensive transport of Ca^{2+} into the interior of the microsomal vesicles where the cation is trapped by a precipitating agent such as oxalate.[159] Swartz et al.[160] found that ethanol inhibited calcium uptake and Ca^{2+} binding by cardiac microsomes enriched in fragmented sarcoplasmic reticulum from the heart of dogs. After 5 min of incubation, half-maximal inhibition of Ca^{2+} uptake occurred at approximately 1.3 M ethanol and that of Ca^{2+} binding at approximately 1.9 M ethanol. Prolonged exposure of the microsomes to ethanol increased the extent of inhibition of both Ca^{2+} uptake and Ca^{2+} binding. These inhibitory effects were almost completely reversed when microsomes were washed after exposure to ethanol. Calcium uptake and calcium binding by sarcoplasmic reticulum was also decreased in dogs chronically treated with ethanol for 6 months.[161] It is premature to attribute the negative inotropic actions of ethanol to its inhibitory effects on the transport of Ca^{2+} by the sarcoplasmic reticulum, since the concentrations of ethanol needed to inhibit Ca^{2+} uptake and Ca^{2+} binding exceeded those found in the blood of chronic alcoholic patients. However, the time dependence of these effects raises the possibility that prolonged exposure to ethanol might contribute to the myocardial weakness observed in chronic alcoholics by interfering with the retention of Ca^{2+} within the cardiac sarcoplasmic reticulum.

3. Muscle

Acute myopathy has been observed in some patients with chronic, severe alchoholism.[162] One possible mechanism for the injury to muscle cells in alcoholism is toxic inhibition of ion transport, which in turn affects muscle contraction. Knutsson,[163,164] using an intracellular microelectrode technique, found that ethanol at a concentration equal to or greater than 0.2 M significantly decreased the resting membrane potential and resistance in frog muscle fibers. The decrease in resistance was due to an increase in permeability of frog muscle fiber membrane to Na^+ and Cl^- ions rather than to an effect on the membrane permeability to K^+. In rats, following prolonged administration of ethanol, Mayer[165] also found that the resting muscle membrane potential becomes abnormally low. The resting membrane potential in alcoholic patients has been reported to be abnormally low.[166,167] This low potential was associated with an abnormally high muscle sodium content, indicating an increase in membrane permeability to sodium or an impairment of the electrogenic sodium pump. The resting membrane potential could also be influenced by the concentration ratio of K^+ in the cell to that prevailing in the extracellular water. Martin et al.[168] noted that hypokalemia occurs commonly in alcoholics. They also found that the K^+ concentration of muscle in alcoholic patients was subnormal, and proposed that K^+ deficiency may have been responsible for the myopathy.

4. Kidney and Urinary Bladder

Ethanol has been known to produce a water diuresis in man for many years; however, the exact site of the diuretic action is a matter of controversy.[169] Kleeman et al.[170] postulated that ethanol inhibits the release of antidiuretic hormone (ADH) from the neurohypophysis. The mechanism of this action is not clear, since the sensitive ADH assay has not been developed. It is possible that ethanol has a direct renal effect as well.[171]

Most studies suggest that ethanol ingestion does not alter renal hemodynamics during 6 hr postadministration in man.[172–173] However, with extended postingestion time, Sargent et al.[171] found that ethanol treatment produced elevations in glomerular filtration rate at 10 to 18 hr after the acute dosage and after chronic treatment in dog. Similarly, in man, acute ethanol ingestions have been reported to decrease sodium and potassium excretion, with no change in chloride excretion within the first 2 hr.[174] In contrast, Sargent et al.[171] reported that potassium excretion was increased while sodium excretion was not significantly altered at 10 and 18 hr following acute treatment with ethanol or after chronic ethanol treatment. This suggests that the increase in glomerular filtration rate and K^+ excretion after long-term treatment with ethanol were secondary effects of ethanol in response to a metabolite generated during the biotransformation of the ethanol molecule or in response to the declining plasma ethanol. The mechanism by which ethanol alters renal hemodynamics and electrolyte metabolism has not been investigated.

The toad urinary bladder is a good model epithelial membrane for transport studies. This tissue has similar electrical properties and active transmural transport of Na^+, as does the frog skin. Yorio and Bentley[175] found that ethanol (9%) decreased the potential difference and electrical resistance across the toad bladder when present on the mucosal surface while the short-circuit current remained unchanged. Unidirectional ^{22}Na and ^{36}Cl flux measurements showed an increase in the movement of Cl^- but no change in Na^+. The ADH-induced increase in Na^+ transport was also unaffected by the presence of ethanol. Meier and Mendoza[176] found that ethanol alone did not alter the flow of water along an osmotic gradient in the isolated toad urinary bladder. However, the increase in osmotic water flow produced by ADH, theophylline, or cyclic adenosine-3′,5′-monophosphate was inhibited by 0.5–5% ethanol in the mucosal or serosal bathing medium. This indicates that ethanol decreases the ability of the urinary bladder to respond to ADH. A similar effect on the collecting duct is proposed which might partially explain the water diuresis observed after ethanol ingestion in mammals.

E. CONCLUDING REMARKS

Active transport of ions and other substances is necessary for maintaining homeostases in the body, and these energy-dependent processes are altered by ethanol in nerve and muscle as well as in epithelial membranes such as the frog skin and gastrointestinal mucosa. The mechanisms of inhibition of ion transport may be (1) an effect on intracellular metabolism to decrease the energy supply for active transport, (2) altering membrane structure, and/or (3) blocking the transporting enzyme system, including enzyme and cofactor. Ethanol also affects passive permeability and receptor function in nerve and muscle, which may be the consequence of or precede alterations in active transport. Alteration of ion movement could induce functional changes of the target organs or tissues, and may interfere with function in other organs.

Compared to the studies on the liver, relatively few studies on the effect of ethanol on the metabolism of epithelial cells have been conducted. Some studies have demonstrated that ethanol inhibits the Na^+,K^+-ATPase involved in the Na^+ pump; however, the exact mechanism is still not clear. Several recent studies on the effects of ethanol on membrane structure (i.e., lipid bilayers, bacterial membranes, and red cell membranes), have not been able to distinguish between mechanisms of action involving the hydrocarbon chains of the membrane phospholipids, or protein components, or both.[177-179]

Although a considerable number of problems have not yet been solved, the techniques in electrophysiology, biochemistry, and biophysics are improving, which should facilitate future research in this area. If the basic mechanisms altered by ethanol can be accurately explained, this will further our understanding of the very clinically relevant problem of alcohol abuse and alcoholism.

F. SUMMARY

The effects of ethanol on transport in various tissues are summarized in Table V. The importance of fully understanding the mechanism of action of ethanol on membrane transport cannot be overemphasized. Maintenance of membrane transport of ions is necessary for most cells in the body to maintain the *milieu interieur* for normal physiological function. Sufficient data have accumulated to demonstrate that ethanol does not simply disrupt membranes but may reversibly stimulate or inhibit membrane transport processes. The gastrointestinal system affords an excellent tissue for examining the sequence of events which occur during acute and chronic administration of ethanol under both *in vivo* and *in vitro* conditions. Information obtained on the basic processes affected may then be extrapolated to other tissues where function may be different but basic processes similar. In addition, the gastrointestinal tract is the first tissue exposed to ethanol and alteration of its function, such as decreased acid secretion, may influence the possible release of substances, such as hormones, which may trigger off a sequence of events affecting the rest of the body.

Table V. Effects of Ethanol on Membrane Function in Various Organs

Species	Dosage (%)	Parameter	Results[a]	References
		Oral Mucosa		
Human	50	Salivary secretion	↑ then ↓	17
		Esophagus		
Rabbit	20	Permeability	↔	26, 27
		Fundic mucosa		
Dog	20	H^+ transport	↓	34–36
Rat	20	H^+ transport	↓	37
Rabbit	20	H^+ transport	↓	27
Dog	20	Na^+, Cl^- transport	↓	36
Rabbit	20	Na^+, Cl^- transport	↓	27
Dog	20	Potential difference	↓	34, 35, 43–45
Rat	20	Potential difference	↓	46
Rabbit	20	Potential difference	↓	27
Dog	20	Resistance	↑	35
Rat	20	Resistance	↑	46
		Antral mucosa		
Rabbit	20	Potential difference	↓	27
Rabbit	20	Na^+, Cl^- transport	↓	27
Rabbit	20	Permeability	↑	27
		Jejunum		
Hamster	2.6	Glucose transport	↓	73, 92
Rabbit	3.0	Alanine transport	↓	92
Rabbit	3.0	Potential difference	↓	92
Rabbit	3.0	Permeability	↑	92
		Pancreas		
Human	40.0	Stimulated secretion	↓	107
Dog	10–80	Stimulated secretion	↓	108–111
Rabbit	3.0	Basal secretion	↓	112
		Skin		
Frog	2.3	Na^+ transport	↓	139
Frog	2.3	Resistance	↓	137
Frog	2.0	Short-circuit current	↓	138
Frog	2.3	Permeability	↑	139
		Liver		
Rat	5–30	ATP	↓	140–142, 145
Rat	6	Na^+,K^+-ATPase	↑	146
Rat	6	Oxygen consumption	↑	146
		Heart		
Rat	0.14–1.1	Action potential	↓	157
Guinea pig	1.15–8.05	Na^+,K^+-ATPase	↓	158
Dog	7.5–11.0	Ca^{2+} transport	↓	160, 161
		Muscle		
Frog	1.2	Na^+, Cl^- permeability	↑	163, 164
		Urinary bladder		
Toad	9	Potential difference	↓	175
Toad	9	Resistance	↓	175
Toad	9	Cl^- permeability	↑	175

[a]Changes in comparison with the controls: ↓, decrease; ↑, increase; ↔, no change.

G. REFERENCES

1. R. F. Mayer and R. Gorcia-Mullin, in: *The Biology of Alcoholism* (B. Kissin and H. Begleiter, eds.), Vol. 2: *Physiology and Behavior,* pp. 21–65, Plenum Press, New York (1972).
2. H. Begleiter and A. Platz, in: *The Biology of Alcoholism* (B. Kissin and H. Begleiter, eds.), Vol. 2: *Physiology and Behavior,* pp. 293–343, Pelnum Press, New York (1972).
3. W. A. Hunt, in: *Biochemical Pharmacology of Ethanol* (E. Majchrowicz, ed.), pp. 195–210, Plenum Press, New York (1975).
4. H. Wallgren, P. Nikander, and P. Virtanen, in: *Alcohol Intoxication and Withdrawal: Experimental Studies* (M. M. Gross, ed.), Vol. II, pp. 23–36, Plenum Press, New York (1975).
5. Y. Israel, F. J. Carmichael, and J. A. MacDonald, in: *Alcohol Intoxication and Withdrawal: Experimental Studies* (M. M. Gross, ed.), Vol. II, pp. 55–64, Plenum Press, New York (1975).
6. A. K. Rawat, *Int. Rev. Neurobiol.* **19:**123 (1976).
7. Y. Israel-Jacard and H. Kalant, *J. Cell. Comp. Physiol.* **65:**127 (1965).
8. H. Kalant and Y. Israel, in: *Biochemical Factors in Alcoholism* (R. P. Maickel, ed.), pp. 25–38, Pergamon Press, Inc., Elmsford, N.Y. (1967).
9. J. M. Beazell and A. C. Ivy, *Q. J. Stud. Alcohol* **1:**45 (1940).
10. A. R. Cooke, *Australas. Ann. Med.* **19:**269 (1970).
11. F. L. Iber, *Gastroenterology* **61:**120 (1971).
12. W. Y. Chey, *Digestion* **7:**239 (1972).
13. S. H. Lorber, V. P. Dinoso, Jr., and W. Y. Chey, in: *The Biology of Alcoholism* (B. Kissin and H. Begleiter, eds.), Vol. 3: *Clinical Pathology,* pp. 339–357, Plenum Press, New York (1974).
14. E. Mezey, *Ann. N.Y. Acad. Sci.* **252:**215 (1975).
15. J. Lindenbaum and C. S. Lieber, *Ann. N.Y. Acad. Sci.* **252:**228 (1975).
16. H. Sarles, *Ann. N.Y. Acad. Sci.* **252:**171 (1975).
17. R. H. Chittenden, L. B. Mendel, and H. C. Jackson, *Am. J. Physiol.* **1:**164 (1898).
18. A. L. Winsor and E. I. Strongin, *J. Exp. Psychol.* **16:**589 (1933).
19. M. A. Blankenhorn and T. D. Spies, *J. Am. Med. Assoc.* **107:**641 (1936).
20. L. H. Schneyer and N. Emmelin, in: *Gastrointestinal Physiology* (A. C. Guyton, E. D. Jacobson, and L. L. Shanbour, eds.), pp. 183–226, University Park Press, Baltimore (1974).
21. A. Lundberg, *Acta Physiol. Scand.* **35:**1 (1955).
22. A. Lundberg, *Acta Physiol. Scand.* **40:**21 (1957).
23. A. Lundberg, *Physiol. Rev.* **38:**21 (1958).
24. L. H. Schneyer and C. A. Schneyer, *Am. J. Physiol.* **209:**1304 (1965).
25. W. W. Webb, R. B. Mullenix, and C. A. Dragstedt, *Proc. Soc. Exp. Biol. Med.* **29:**895 (1932).
26. D. Fromm and R. Robertson, *Surg. Forum* **26:**373 (1975).
27. D. Fromm and R. Robertson, *Gastroenterology* **70:**220 (1976).
28. A. C. Ivy and G. B. McIlvain, *Am. J. Physiol.* **67:**124 (1923).
29. B. I. Hirschowitz, H. M. Pollard, and S. W. Hartwell, Jr., and J. London, *Gastroenterology* **30:**244 (1956).
30. E. R. Woodward, C. Robertson, H. D. Ruttenberg, and H. Schapiro, *Gastroenterology* **32:**727 (1957).
31. W. T. Irvine, D. B. Watkin, and E. J. Williams, *Gastroenterology* **39:**41 (1960).
32. C.-E. Elwin, *Acta Physiol. Scand.* **75:**12 (1969).
33. W. Y. Chey, S. Kosay, and S. H. Lorber, *Am. J. Dig. Dis.* **17:**153 (1972).
34. W. S. Rehm and L. E. Hokin, *Am. J. Physiol.* **149:**162 (1947).
35. T. J. Sernka, C. W. Gilleland, and L. L. Shanbour, *Am. J. Physiol.* **226:**397 (1974).
36. Y.-J. Kuo, L. L. Shanbour, and T. J. Sernka, *Am. J. Dig. Dis.* **19:**818 (1974).
37. J. Puurunen and H. Karppanen, *Life Sci.* **16:**1513 (1975).
38. R. P. Durbin, in: *Handbook of Physiology* (C. F. Code, ed.), Vol. 2, pp. 879–888, American Physiological Society, Washington, D.C. (1967).
39. W. S. Rehm, *Am. J. Physiol.* **147:**69 (1946).
40. L. Villegas, *Biochim. Biophys. Acta* **64:**359 (1962).

41. W. S. Rehm, L. E. Hokin, T. P. DeGraffenried, II, F. J. Bajandas, and F. E. Coy, Jr., *Am. J. Physiol.* **164**:187 (1951).
42. S. Andersson and M. I. Grossman, *Gastroenterology* **49**:364 (1965).
43. M. G. Geall, S. F. Phillips, and W. H. J. Summerskill, *Gastroenterology*, **58**:437 (1970).
44. H. W. Davenport, H. A. Warner, and C. F. Code, *Gastroenterology* **47**:142 (1964).
45. H. W. Davenport, *Proc. Soc. Exp. Biol. Med.* **126**:657 (1967).
46. L. L. Shanbour, J. Miller, and T. K. Chowdhury, *Am. J. Dig. Dis.* **18**:311 (1973).
47. H. W. Davenport, *Gastroenterology* **46**:245 (1964).
48. H. W. Davenport, *N. Engl. J. Med.* **276**:1307 (1967).
49. H. W. Davenport, *Gastroenterology* **54**:175 (1968).
50. H. H. Ussing and K. Zerahn, *Acta Physiol. Scand.* **23**:110 (1951).
51. C. A. M. Hogben, *Proc. Natl. Acad. Sci. U.S.A.* **37**:393 (1951).
52. C. A. M. Hogben, *Am. J. Physiol.* **180**:641 (1955).
53. E. Heinz and R. Durbin, *Biochim. Biophys. Acta* **31**:246 (1959).
54. S. Kitahara, *Am. J. Physiol.* **213**:819 (1967).
55. S. Kitahara, K. R. Fox, and C. A. M. Hogben, *Am. J. Dig. Dis.* **14**:221 (1969).
56. T. J. Sernka and C. A. M. Hogben, *Am. J. Physiol.* **217**:1419 (1969).
57. J. G. Forte and T. E. Machen, *J. Physiol.* **244**:33 (1975).
58. D. Fromm, J. H. Schwartz, and F. Quijano, *Am. J. Physiol.* **228**:166 (1975).
59. Y.-J. Kuo and L. L. Shanbour, *Am. J. Physiol.* **230**:762 (1976).
60. Y.-J. Kuo and L. L. Shanbour, *Am. J. Physiol.* **231**:1433 (1976).
61. W. Y. Chey, O. Kusakcioglu, V. Dinoso, and S. H. Lorber, *Arch. Intern. Med.* **122**:399 (1968).
62. D. K. Kasbekar and R. P. Durbin, *Biochim. Biophys. Acta* **105**:472 (1965).
63. L. L. Tague and L. L. Shanbour, *Life Sci.* **14**:1065 (1974).
64. L. L. Tague and L. L. Shanbour, *Proc. Soc. Exp. Biol. Med.* **154**:37 (1977).
65. C. C. Mao, L. L. Shanbour, D. S. Hodgin, and E. D. Jacobson, *Gastroenterology* **63**:427 (1972).
66. C. C. Mao, E. D. Jacobson, and L. L. Shanbour, *Am. J. Physiol.* **225**:893 (1973).
67. J. C. Bowen, W. Pawlik, Y. M. Kuo, D. Williams, L. L. Shanbour, and E. D. Jacobson, *Gastroenterology* **69**:285 (1975).
68. G. Sachs, R. H. Collier, R. L. Shoemaker, and B. I. Hirschowitz, *Biochim. Biophys. Acta* **162**:210 (1968).
69. C. H. Halsted, E. A. Robles, and E. Mezey, *Am. J. Clin. Nutr.* **26**:831 (1973).
70. Y. Israel, J. E. Valenzuela, I. Salazar, and G. Ugarte, *J. Nutr.* **98**:222 (1969).
71. N. Krasner, K. M. Cochran, R. I. Russell, H. A. Carmichael, and G. G. Thompson, *Gut* **17**:245 (1976).
72. T. Chang, J. Lewis, and A. J. Glazko, *Biochim. Biophys. Acta* **135**:1000 (1967).
73. P. K. Dinda, I. T. Beck, M. Beck, and T. F. McElligott, *Gastroenterology* **68**:1517 (1975).
74. J. E. Fox, T. F. McElligott, and I. T. Beck, *Gastroenterology* **70**:884 (1976).
75. Y. Israel, I. Salazar, and E. Rosenmann, *J. Nutr.* **96**:499 (1968).
76. F. A. Jacobs and C. Overvold, *Fed. Proc.,* Abstr. **35**:463 (1976).
77. A. D. Thomson, H. Baker, and C. M. Leevy, *J. Lab. Clin. Med.* **76**:34 (1970).
78. A. M. Hoyumpa, Jr., K. J. Breen, S. Schenker, and F. A. Wilson, *J. Lab. Clin. Med.* **86**:803 (1975).
79. C. H. Halsted, R.C.. Griggs, and J. W. Harris, *J. Lab. Clin. Med.* **69**:116 (1967).
80. C. H. Halsted, E. A. Robles, and E. Mezey, *N. Engl. J. Med.* **285**:701 (1971).
81. C. A. J. Wardrop, R. V. Heatley, G. B. Tennant, and L. E. Hughes, *Lancet* **2**:640 (1975).
82. M. Small, A. Longarini, and N. Zamcheck, *Am. J. Med.* **27**:575 (1959).
83. J. Lindenbaum and C. S. Lieber, *Nature (Lond.)* **224**:806 (1969).
84. E. L. Krawitt, *J. Lab. Clin. Med.* **85**:665 (1975).
85. E. L. Krawitt, *Proc. Soc. Exp. Biol. Med.* **146**:406 (1974).
86. E. L. Krawitt, *Nature (Lond.)* **243**:88 (1973).
87. J. Wojcicki, *Q. J. Stud. Alcohol* **33**:958 (1972).
88. D. F. Schafer, D. V. Stephenson, A. J. Barak, and M. F. Sorrell, *J. Nutr.* **104**:101 (1974).
89. T. H. Wilson and G. Wiseman, *J. Physiol.* **123**:116 (1954).

90. D. Fromm, *Am. J. Physiol.* **224:**110 (1973).

91. M. Field, D. Fromm, and I. McColl, *Am. J. Physiol.* **220:**1388 (1971).

92. Y.-J. Kuo and L. L. Shanbour, *Am. J. Dig. Dis.* **23:**51–56 (1978).

93. S. G. Schultz and P. F. Curran, *Physiol. Rev.* **50:**637 (1970).

94. S. G. Schultz, R. E. Ruisz, and P. F. Curran, *J. Gen. Physiol.* **49:**849 (1966).

95. J. C. Skou, *Prog. Biophys. Mol. Biol.* **14:**133 (1964).

96. J. P. Quigley and G. S. Gotterer, *Biochim. Biophys. Acta* **173:**456 (1969).

97. E. A. Carter and K. J. Isselbacher, *Proc. Soc. Exp. Biol. Med.* **142:**1171 (1973).

98. N. Krasner, H. A. Carmichael, R. I. Russell, G. G. Thompson, and K. M. Cochran, *Gut* **17:**249 (1976).

99. A. Hoyumpa, Jr., S. Nichols, F. Wilson, and S. Schenker, *Clin. Res.* **24:**565A (1976).

100. H. L. Greene, F. B. Stifel, R. H. Herman, Y. F. Herman, and N. S. Rosensweig, *Gastroenterology* **67:**434 (1974).

101. E. Baraona, R. C. Pirola, and C. S. Lieber, *Biochim. Biophys. Acta* **388:**19 (1975).

102. W. R. J. Middleton, E. A. Carter, G. D. Drummey, and K. J. Isselbacher, *Gastroenterology* **60:**880 (1971).

103. E. A. Carter, G. D. Drummey, and K. J. Isselbacher, *Science* **174:**1245 (1971).

104. E. Baraona, R. C. Pirola, and C. S. Lieber, *Gastroenterology* **66:**226 (1974).

105. E. Baraona and C. S. Lieber, *Gastroenterology* **68:**495 (1975).

106. H. Kalant, *Gastroenterology* **56:**380 (1969).

107. C. Mott, H. Sarles, O. Tiscornia, and L. Gullo, *Am. J. Dig. Dis.* **17:**902 (1972).

108. M. Bayer, J. Rudick, C. S. Lieber, and H. D. Janowitz, *Gastroenterology* **63:**619 (1972).

109. O. Tiscornia, L. Gullo, and H. Sarles, *Digestion* **9:**231 (1973).

110. O. M. Tiscornia, L. Gullo, C. Barros Mott, M. A. Devaux, G. Palasciano, G. Hage, and H. Sarles, *Digestion* **9:**490 (1973).

111. O. M. Tiscornia, L. Gullo, G. H. Sarles, C. Barros Mott, A. Brasca, M. A. Devaux, G. Palasciano, and G. Hage, *Digestion* **10:**52 (1974).

112. N. Solomon, T. E. Solomon, E. D. Jacobson, and L. L. Shanbour, *Am. J. Dig. Dis.* **19:**253 (1974).

113. H. A. Reber, C. J. Wolf, and S. P. Lee, *Surg. Forum* **20:**382 (1969).

114. L. W. Way and J. M. Diamond, *Biochim. Biophys. Acta* **203:**298 (1970).

115. R. M. Preshaw, in: *Gastrointestinal Physiology* (A. C. Guyton, E. D. Jacobson, and L. L. Shanbour, eds.), pp. 265–291, University Park Press, Baltimore (1974).

116. S. S. Rothman and F. P. Brooks, *Am. J. Physiol.* **209:**790 (1965).

117. J. D. Jamieson and G. E. Palade, *J. Cell Biol.* **48:**503 (1971).

118. S. S. Rothman and F. P. Brooks, *Am. J. Physiol.* **208:**1171 (1965).

119. R. M. Case, M. Johnson, T. Scratcherd, and H. S. A. Sherratt, *J. Physiol.* **223:**669 (1972).

120. R. M. Case and T. Scratcherd, *J. Physiol.* **223:**649 (1972).

121. J. D. Gardner, T. P. Conlon, and T. D. Adams, *Gastroenterology* **70:**29 (1976).

122. A. S. Ridderstap and S. L. Bonting, *Am. J. Physiol.* **217:**1721 (1969).

123. S. R. Viegas de Andrade, W. J. Hogan, and D. H. Winship, *Gastroenterology* **56:**1204 (1969).

124. J. J. Barboriak and R. C. Meade, *Am. J. Clin. Nutr.* **23:**1151 (1970).

125. P. Harichaux and J. Moline, *C. R. Seances Soc. Biol. Fil.* **158:**1389 (1964).

126. A. R. Cooke, *Am. J. Dig. Dis.* **15:**449 (1970).

127. E. Robles, E. Mezey, and C. Halsted, *Johns Hopkins Med. J.* **135:**17 (1974).

128. W. C. Alvarez and L. J. Mahoney, *Am. J. Physiol.* **58:**476 (1922).

129. E. Bozler, *Am. J. Physiol.* **144:**693 (1945).

130. J. Liu, C. L. Prosser, and D. D. Job, *Am. J. Physiol.* **217:**1542 (1969).

131. D. D. Job, *Am. J. Physiol.* **220:**299 (1971).

132. N. W. Weisbrodt, in: *Gastrointestinal Physiology* (A. C. Guyton, E. D. Jacobson, and L. L. Shanbour, eds.), pp. 139–181, University Park Press, Baltimore (1974).

133. Y. Israel, E. Rosenmann, S. Hein, G. Colombo, and M. Canessa-Fischer, in: *Biological Basis of Alcoholism* (Y. Israel and J. Mardones, eds.), pp. 53–72, Wiley–Interscience, New York (1971).

134. V. Koefold-Johnsen, *Acta Physiol. Scand.* **42** (Suppl. 145):87 (1957).

135. J. C. Skou and K. Zerahn, *Biochim. Biophys. Acta* **35**:324 (1959).

136. P. F. Curran and J. R. Gill, Jr., *J. Gen. Physiol.* **45**:625 (1962).

137. F. A. Fuhrman, *Am. J. Physiol.* **171**:266 (1952).

138. Y. Israel and H. Kalant, *Nature (Lond.)* **200**:476 (1963).

139. J. D. Boyett and J. T. Vanbruggen, *Biochim. Biophys. Acta* **436**:686 (1976).

140. S. W. French, *Proc. Soc. Exp. Biol. Med.* **121**:681 (1966).

141. E. S. Higgins and W. L. Banks, *Biochem. Pharmacol.* **20**:1513 (1971).

142. H. P. T. Ammon and C. J. Estler, *Nature (Lond.)* **216**:158 (1967).

143. R. L. Veech, R. Guynn, and D. Veloso, *Biochem. J.* **127**:387 (1972).

144. L. L. Shanbour and C. P. Huang, *Alcohol. Clin. Exp. Res.* **1**:154 (1977).

145. J. E. C. Walker and E. R. Gordon, *Biochem. J.* **119**:511 (1970).

146. J. Bernstein, L. Videla, and Y. Israel, *Biochem. J.* **134**:515 (1973).

147. C. S. Lieber and L. M. DeCarli, *J. Biol. Chem.* **245**:2505 (1970).

148. E. Mezey, *Biochem. Pharmacol.* **21**:137 (1972).

149. E. Mezey and F. Tobon, *Gastroenterology* **61**:707 (1971).

150. Y. Israel, *Q. J. Stud. Alcohol* **31**:293 (1970).

151. G. E. Burch and T. D. Giles, *Am. J. Med.* **50**:141 (1971).

152. T. J. Regan, G. E. Levinson, and H. A. Oldewurtel, *J. Clin. Invest.* **48**:397 (1969).

153. D. S. Mierzwich, K. Wildenthal, and J. H. Mitchell, *Clin. Res.* **15**:215 (1967).

154. T. J. Regan, G. T. Koroxenisis, C. B. Moschos, H. A. Oldewurtel, P. H. Lehan, and H. K. Hellems, *J. Clin. Invest.* **45**:270 (1966).

155. L. C. Mendoza, K. Helberg, A. Rickart, G. Tillich, and R. J. Bing, *J. Clin. Pharmacol.* **11**:165 (1971).

156. J. Nakano and J. M. Kessinger, *Eur. J. Pharmacol.* **17**:195 (1972).

157. A. L. Gimeno, M. F. Gimeno, and J. L. Webb, *Am. J. Physiol.* **203**:194 (1962).

158. J. W. Williams, M. Tada, A. M. Katz, and E. Rubin, *Biochem. Pharmacol.* **24**:27 (1975).

159. D. I. Repke and A. M. Katz, *J. Mol. Cell. Cardiol.* **4**:401 (1972).

160. M. H. Swartz, D. I. Tepke, A. M. Katz, and E. Rubin, *Biochem. Pharmacol.* **23**:2369 (1974).

161. R. J. Bing, H. Tillmanns, J. M. Fauvel, K. Seeler, and J. C. Mao, *Clin. Res.* **35**:33 (1955).

162. R. Hed, H. Larsson, and F. Wahlgren, *Acta Med. Scand.* **152**:459 (1955).

163. E. Knutsson, *Acta Physiol. Scand.* **52**:242 (1961).

164. E. Knutsson and S. Katz, *Acta Pharmacol. Toxicol.* **25**:54 (1967).

165. R. F. Mayer, *Ann. N.Y. Acad. Sci.* **215**:370 (1973).

166. G. L. Bilbrey, L. Herbin, N. W. Carter, and J. P. Knochel, *J. Clin. Invest.* **52**:3011 (1973).

167. J. N. Cunningham, N. W. Carter, F. C. Rector, and D. W. Seldin, *J. Clin. Invest.* **50**:49 (1971).

168. J. B. Martin, J. W. Craig, R. E. Eckel, and J. Munger, *Neurology* **21**:1160 (1971).

169. H. B. Van Dyke and R. G. Ames, *Acta Endocrinol.* **7**:110 (1951).

170. C. R. Kleeman, M. E. Rubini, E. Lamdin, and F. H. Esptein, *J. Clin. Invest.* **34**:448 (1955).

171. W. Q. Sargent, J. R. Simpson, and J. D. Beard, *J. Pharmacol. Exp. Ther.* **188**:461 (1974).

172. M. B. Strauss, J. D. Rosenbaum, and W. P. Nelsen, III, *J. Clin. Invest.* **29**:1053 (1950).

173. J. M. Kalbfleisch, R. D. Lindeman, H. E. Ginn, and W. O. Smith, *J. Clin. Invest.* **42**:1471 (1963).

174. M. E. Rubini, C. R. Kleeman, and E. Lamdin, *J. Clin. Invest.* **34**:439 (1955).

175. T. Yorio and P. J. Bentley, *J. Pharmacol. Exp. Ther.* **197**:1 (1976).

176. K. E. Meier and S. A. Mendoza, *J. Pharmacol. Exp. Ther.* **196**:231 (1976).

177. C. M. Colley, S. M. Metcalfe, B. Turner, A. S. V. Burgen, and J. C. Metcalfe, *Biochim. Biophys. Acta* **233**:720 (1971).

178. V. A. Fried and A. Novick, *J. Bacteriol.* **114**:239 (1973).

179. S. J. Patterson, K. W. Butler, P. Huang, J. Labelle, I. C. P. Smith, and H. Schneider, *Biochim. Biophys. Acta* **266**:597 (1972).

Changes in the Activity of Na^+,K^+-ATPase during Acute and Chronic Administration of Ethanol

5

Mary K. Roach

A. INTRODUCTION

The association of Na^+,K^+-dependent–Mg^{2+}-activated adenosine triphosphatase (Na^+,K^+-ATPase) with the active transport of sodium and potassium ions is well documented.[1–5] The enzyme is present in every tissue that maintains transmembrane gradients of sodium and potassium and is especially active in excitable cells. A number of reports have appeared describing the influence of ethanol on this enzyme. This review will summarize these effects and attempt to evaluate their significance with regard to ethanol's pharmacological actions.

B. Na^+,K^+-ATPase: PROPERTIES

Before considering the effects of ethanol, a brief summary of the properties of Na^+,K^+-ATPase will be presented to provide a background for discussion. The reader is referred to recent reviews and monographs for detailed information about the enzyme.[4–6]

Na^+,K^+-ATPase is intimately associated with the plasma membrane, and evidence that it participates in the energy-requiring translocation of sodium and potassium is compelling. The enzyme consists of two polypeptide units closely associated with membrane lipids. The larger peptide (100,000 daltons) can be phosphorylated and probably contains the catalytic center; the smaller unit (50,000 daltons) is a sialoglycoprotein. Lipids are essential for the enzyme's activity, but there is controversy over which lipids are required.[7–9] Moreover,

Mary K. Roach • Department of Biochemistry, University of Nairobi, Nairobi, Kenya.

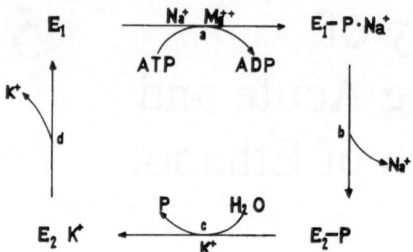

Fig. 1. Reaction sequence for Na$^+$,K$^+$-ATPase. E$_1$ and E$_2$ represent different conformational states of the enzyme. (a) Na$^+$-stimulated, Mg^{2+}-dependent phosphorylation; active center oriented to inside of cell. (b) Conformation change; Na$^+$ translocated to exterior. (c) K$^+$-stimulated dephosphorylation. (d) Conformation change returns enzyme to original orientation; K$^+$ translocated to interior.

Na$^+$,K$^+$-ATPase activity depends not only on the content but also on the physical state of the lipids, the activity seeming to increase with greater lipid fluidity.[10]

The reaction mechanism is under active investigation[6] and is seen most simply as a four-step sequence (Fig. 1). Kinetic studies of the reaction have shown that Na$^+$,K$^+$-ATPase is an allosteric enzyme with multiple binding sites for sodium and potassium, subject to regulatory control by both ions. The enzyme is specific for sodium, but potassium can be replaced by other monovalent cations. In addition to its sodium and potassium requirement, the enzyme can be distinguished from other ATPases by its sensitivity to inhibition by the cardiac glycoside, ouabain.

C. ACTIVITY *IN VITRO*

Evidence that sublethal concentrations of ethanol suppressed the active transport of sodium and potassium ions *in vitro*[11-13] prompted investigations into ethanol's effects on Na$^+$,K$^+$-ATPase activity. Most of these studies have dealt with brain or other nerve tissue.

1. Inhibition by Ethanol

Israel *et al.*[14] reported that ethanol inhibited Na$^+$,K$^+$-ATPase activity in microsomes from rat and guinea pig cerebral cortex and from eel electroplacque tissue. Inhibition was observed at ethanol concentrations that produce mild to severe intoxication *in vivo,* according to these authors, with 50% inhibition of activity at 0.22 M ethanol. The inhibition was competitive with respect to potassium; in contrast, raising the sodium concentration enhanced the ethanol inhibition. In a later study[15] inhibition of microsomal Na$^+$,K$^+$-ATPase was correlated with the suppression of potassium-ion transport by rat cerebral cortex slices. In these studies it was noted that ethanol inhibition in cortex slices incubated in 6 mM potassium was comparable to that in microsomes at 1 mM potassium. The authors concluded from this evidence that cortex slices presented a "diffusion barrier," which limited the access of potassium ions to the active transport site, and that this barrier was abolished in disrupted, microsomal preparations.

Using beef brain microsomes, Israel and Salazar[16] again demonstrated that the degree of inhibition varied with sodium- and potassium-ion concentrations. Erwin and co-workers,[17] however, failed to find substantial inhibition of Na⁺,K⁺-ATPase with a microsomal preparation of adult mouse brain. Ethanol (0.2 M) caused only 4% inhibition with 1 mM potassium. When potassium was raised to 20 mM, ethanol inhibition increased to 10% rather than decreasing as reported by Israel and co-workers.[14-16] The reasons for this difference are not apparent.

Other laboratories[18-23] have confirmed that ethanol inhibits Na⁺,K⁺-ATPase *in vitro* but have generally required higher ethanol concentrations to achieve inhibition comparable to that reported by Israel *et al.*[14] These results are summarized in Table I.

Sun and Samorajski[19] reported that ethanol inhibition of Na⁺,K⁺-ATPase increased significantly with age. In 3-month-old mice, inhibition was 27% with 0.34 M ethanol, increasing to 52% when the mice were over 2 years old. Inhibition by ethanol was also found to be age-dependent in postmortem samples of human brain, although this study was limited to five subjects. One subject had a history of chronic alcoholism, and it is intriguing that Na⁺,K⁺-ATPase from this person was relatively unresponsive to the inhibitory effects of ethanol.

In an investigation of the individual steps of the reaction sequence (Fig. 1), the potassium-dependent dephosphorylation was almost abolished with 1.6 M ethanol, a concentration that had little effect on the sodium-dependent step.[22] Whereas this ethanol concentration is at least four times greater than those used

Table I. Effect of Ethanol on Na⁺,K⁺-ATPase in Brain Tissue

Species	Cell fraction	Ethanol concentration (M)	Inhibition (%)	[K⁺] (mM)	[Na⁺] (mM)	Reference
Rat	Microsomes	0.22	50	2	100	14
			32	1	150	15
			12	10	150	15
	Synaptosomes	0.50	42	20	100	21
Mouse	Microsomes	0.20	4	1	100	17
			10	20	100	17
		0.22	50	0.5	100	20
			0	5	100	20
	Synaptosomes	0.34	27[a]	20	100	19
			44[b]	20	100	19
Guinea pig	Microsomes	0.22	50	1	100	14
	Synaptosomes	0.43	50	20	100	18
Beef	Microsomes	0.47	50	4	150	16

[a]Three months old.
[b]Eight months old (see the text).

in the studies cited above, the marked inhibition of dephosphorylation suggests that it would be suppressed at more moderate ethanol concentrations. These observations were extended by Hegyvary,[23] who used guinea pig kidney cortex to show that ethanol increased the affinity of ATPase for sodium and ATP, decreased the affinity for potassium, and suppressed the hydrolysis of the phosphorylated enzyme in the presence or absence of potassium. From this it is evident that ethanol inhibition is an allosteric event which alters the conformation of binding sites.

2. Specificity of Inhibition

Inhibition of Na^+,K^+-ATPase is not unique to ethanol but is a property shared by a number of compounds, including other short-chain alcohols (Table II). The inhibitory potency of the lower alcohols correlates directly with their lipid solubility.[15,18,23] A parallel between the degree of inhibition of membrane-bound Na^+,K^+-ATPase and the lipophilicity of the inhibitor has been reported for a number of other central nervous system (CNS) depressants[16,23-25] and is generally taken as evidence of a hydrophobic interaction between inhibitor and membrane complex that alters enzyme conformation. Interaction may occur with the lipid matrix of the membrane or with hydrophobic regions of the enzyme itself.

Table II. Inhibition of Brain Na^+, K^+-ATPase by CNS Depressants and Related Compounds

Inhibitor	Concentration (M)	Inhibition (%)	Species	Cell fraction	[K+] (mM)	Reference
Methanol	1.5	50	Guinea pig	Synaptosomes	20	18
n-Propanol	0.17	50	Guinea pig	Synaptosomes	20	18
	0.06	21	Rat	Microsomes	1	15
Isopropanol	0.18	50	Guinea pig	Synpatosomes	20	18
	0.10	26	Rat	Microsomes	1	15
n-Butanol	0.06	50	Guinea pig	Synaptosomes	20	18
	0.02	20	Rat	Microsomes	1	15
Isobutanol	0.09	50	Guinea pig	Synaptosomes	20	18
t-Butanol	0.07	42	Rat	Microsomes	1	15
n-Pentanol	6×10^{-3}	50	Beef	Microsomes	4	16
n-Octanol	8×10^{-5}	50	Beef	Microsomes	4	16
Acetone	0.17	50	Beef	Microsomes	4	16
Ethyl ether	0.16	50	Beef	Microsomes	4	16
Chloroform	3×10^{-3}	50	Beef	Microsomes	4	16
Octanoate	0.015	50	Rat	Synaptosomes	20	24
Acetaldehyde	0.02	42	Mouse	Microsomes	20	17
	0.02	74	Rat	Synaptosomes	20	21
	0.04	99	Rat	Synaptosomes	20	21

It seems improbable that nonspecific hydrophobic interactions would affect Na$^+$,K$^+$-ATPase to the exclusion of other membrane-bound enzymes. However, ethanol inhibition of ATPase activity is relatively specific for the ouabain-sensitive, Na$^+$,K$^+$-dependent enzyme; inhibition of Mg^{2+}-activated ATPase is negligible.[14,16,18] On the other hand, adenylate cyclase, another membrane-associated, lipid-dependent enzyme,[26] is reported to be stimulated by ethanol *in vitro*,[27] although other results dispute this.[28]

3. Relevance to Pharmacologic Effects of Ethanol

Many Na$^+$,K$^+$-ATPase inhibitors, including those in Tables I and II, can be categorized as anesthetics or CNS depressants, and a correspondence between pharmacologic efficacy and inhibitory potency has been noted.[16,24,29] This has prompted speculation that some of the CNS effects of these drugs may be mediated through inhibition of Na$^+$,K$^+$-ATPase.

This hypothesis is difficult to assess because, in the first place, it is not clear that inhibition occurs *in vivo*. With most drugs, including ethanol, the concentrations required to achieve marked inhibition *in vitro* are from 10 to 1000 times greater than pharmacologically effective blood levels. This objection is most often countered with the argument that the tissue level or membrane concentration may be appreciably greater than the plasma level of the drug. Although there is evidence that this may be valid for some drugs,[30,31] it is unlikely that ethanol would be concentrated in this manner.[31] Kalant and Israel[32] argue that there is indirect evidence that ethanol inhibits Na$^+$,K$^+$-ATPase, and, hence, active ion transport *in vivo*. They reiterate the fact that similar concentrations of ethanol gave similar inhibition of active transport by cortex slices in 6 mM potassium and of Na$^+$,K$^+$-ATPase preparations in 1–2 mM potassium.[14,15] They suggest that these potassium concentrations are effectively the same at the enzyme site and, presumably, approximate potassium ion availability *in vivo*. Nevertheless, the fact remains that there is no direct evidence to indicate that ethanol inhibits brain Na$^+$,K$^+$-ATPase *in vivo*.

A second cause for questioning the relationship between CNS depression and ATPase inhibition is that these inhibitors are not always CNS depressants. Such potent inhibitors as ouabain and ethacrynic acid produce convulsions when applied directly to the brain.[33] The intracerebral application of these inhibitors may not be directly comparable to the systemic administration of the depressants; nevertheless, these opposing effects must be explained in order for ethanol and other depressants to be considered to act through inhibition of Na$^+$,K$^+$-ATPase in the CNS.

To rationalize this conflict, one might speculate that a potent inhibitor such as ouabain suppresses the ''steady-state'' activity of Na$^+$,K$^+$-ATPase and decreases the ion gradients, lowering the resting potential and promoting depolarization. On the other hand, a weak inhibitor such as ethanol might be effective only after depolarization, when sodium transport is activated maximally to restore the resting potential.[34,35] Inhibition during this period could prolong the time before the cell is again able to be depolarized, thus depressing neuronal activity. This suggestion is contradicted, however, by evidence from Wallgren *et al.*[36] who used rat cerebral cortex slices to show that, during recovery after electrical stimulation, there was no inhibition of sodium extrusion and only a slight decrease in potassium accumulation in the presence of ethanol (0.11 M). These authors also reported a strong inhibition of sodium conductance during electrical stimulation and concluded that ethanol's effect on nerve function was primarily on the sodium conductance channels rather than on the transport mechanism.

D. ACTIVITY AFTER ACUTE ETHANOL ADMINISTRATION

Information relating to Na^+,K^+-ATPase activity after the acute administration of ethanol is very limited. Mice, injected intraperitoneally with ethanol (4 g/kg), were sacrificed after 1, 3, and 6 hr after which the total ATPase activity was assayed in various cell fractions of brain and liver.[37] Mitochondrial ATPase, which is the Mg^{2+} activated enzyme, increased significantly in brain at 1 and 3 hr and in liver at 3 hr, whereas microsomal and synaptosomal ATPase activities did not change. Synaptosomal ATPase is predominantly Na^+,K^+-dependent, and microsomes contain significant activities of both enzymes. Blood ethanol levels were 381, 230, and 51 mg/dl at 1, 3, and 6 hr, respectively.

Other workers reported a 37% increase in total ATPase activity of mouse brain microsomes 15 min after intraperitoneal injection of ethanol (4 g/kg), but did not differentiate between the Mg^{2+}-dependent and Na^+,K^+-dependent activities.[38]

These studies show an enhancement of total ATPase activity by ethanol *in vivo,* but it appears that the Mg^{2+}-activated enzyme may be responsible for the increase and that Na^+,K^+-dependent activity does not change. Although these studies give no evidence of the Na^+,K^+-ATPase inhibition seen *in vitro,* they may not accurately reflect *in vivo* conditions. Ethanol's action on Na^+,K^+-ATPase is reversible,[23] and preparation of subcellular fractions for assay would remove most ethanol from the membrane, erasing any inhibition. Therefore, these studies neither confirm nor deny an effect of ethanol on Na^+,K^+-ATPase *in vivo.*

E. ACTIVITY AFTER CHRONIC ETHANOL ADMINISTRATION

1. Brain

The molecular mechanisms of ethanol tolerance and physical dependence are not known, but these phenomena are generally thought to result from adaptive changes in the CNS that compensate for the inhibition of function by ethanol. After adaptation, ethanol withdrawal presumably results in hyperactivity of the altered systems which gives rise to abstinence signs. Na$^+$,K$^+$-ATPase activity has been reported to increase after chronic ethanol intake, and it has been suggested that this increase represents an adaptive change, related to tolerance or dependence.

Israel et al.[39] found a 65% elevation of Na$^+$,K$^+$-ATPase activity in homogenates of cerebral cortex from rats made alcohol tolerant by gastric intubation of increasing doses of ethanol, daily for up to 26 days. The rats were decapitated 24–32 hr after the last ethanol dose. Transport of ^{86}Rb, a radioactive analogue of potassium, increased by 25% in cerebral cortex slices of these rats. Transport activity returned to control values when measured 2 weeks after alcohol termination. The authors did not mention withdrawal signs, but said that the ethanol dosage produced ''a maximum degree of tolerance'' and speculated that the increased enzyme and transport activities reflected molecular processes of tolerance.

In rats inhaling ethanol vapor for 7 days, brain microsomal and synaptosomal Na$^+$,K$^+$-ATPase increased by as much as 40%.[40] The animals were sacrificed 12 hr after ethanol withdrawal, when abstinence signs, including convulsions, were evident. Elevated enzyme activity was correlated with mean daily blood ethanol levels and with the severity of alcohol withdrawal signs. Within 60 hr after withdrawal, Na$^+$,K$^+$-ATPase activity was normal.

Wallgren et al.[41] quantitated withdrawal hyperexcitability by measuring tail-twitch activity and found a direct relationship between excitability and cerebral microsomal Na$^+$,K$^+$-ATPase activity in rats receiving ethanol by stomach tube for 21 days. These authors also reported that Mg^{2+}-dependent ATPase activity decreased during withdrawal, whereas in other studies the Mg^{2+}-dependent enzyme did not change significantly.[39,40]

Cats, receiving oral doses of ethanol (1.5, 3.0, or 4.5 g/kg per day) for 5 weeks, had significantly increased Na$^+$,K$^+$-ATPase activity in the hippocampus and regions of the cortex but no elevation of activity in the caudate nucleus, reticular formation, and amygdala.[42] Mg^{2+}-dependent ATPase was elevated in this study. Although the cats were described as having increased tolerance to ethanol and withdrawal signs, they were not sacrificed until 4 days after the last ethanol dose.

In contrast, others failed to observe marked alteration in brain Na$^+$,K$^+$-ATPase activity after prolonged ethanol intake. Mice, consuming a 6% (w/v) ethanol liquid diet for 2 weeks, had elevated mitochondrial (i.e., Mg^{2+}-activated) ATPase, but there was no change in the microsomal or synaptosomal enzymes.[37] However, ethanol (148 mg/dl serum) was present in these animals when sacrificed and may have contributed to this finding. These authors reported a similar elevation of mitochondrial ATPase following an acute dose of ethanol, as discusssed in Section D.

Akera et al.[43] fed females rats a liquid diet containing ethanol (10%, w/v) for 24 days, which produced behavioral tolerance, measured by conditioned avoidance in a shuttle box and ''Rotarod'' performance. Although the authors commented that there was ''no remarkable change in Na$^+$,K$^+$-ATPase,'' there was, in fact, a slight (16–18%) but statistically significant increase in activity in homogenates of the medulla oblongata and cerebrum at 24 hr after withdrawal. Apparently, this increase was not the result of a change in enzyme content because there was no corresponding increase in ouabain-binding capacity, measured in the same preparations. In an additional experiment by Akera et al.[43] male rats received gastric intubation of ethanol in the manner of Israel et al.[39] and were described as "more irritable on handling" but showed no change in enzyme activity, either in whole brain homogenates or microsomes.

In a recent study,[44] two strains of rats, one bred for ethanol preference, the other for ethanol aversion, were given ethanol by stomach tube for 19 days, which produced behavioral tolerance, measured on the tilting plane. The ethanol-preferring strain had a significantly greater innate tolerance than the ethanol-avoiding strain. However, there was no difference in Na^+,K^+-ATPase activity between strains, before or after chronic ethanol, and ethanol did not elevate the enyzme's activity in either strain.

Both Wallgren et al.[41] and Roach et al.[40] also reported that there was no change in Na^+,K^+-ATPase when only behavioral tolerance[41] or minor withdrawal signs[40] were observed, in contrast to their findings in severely dependent, hyperexcitable rats. This seems to indicate that the enhancement of enzyme activity is associated more closely with ethanol dependence and withdrawal than with tolerance. This conclusion is contradicted, however, by the results of Goldstein and Israel,[20] who found Na^+,K^+-ATPase unaltered in brain homogenates of mice exhibiting withdrawal hyperexcitability and convulsions after inhaling ethanol vapor for 3–8 days. Moreover, Israel et al.[39] found a substantial increase in Na^+,K^+-ATPase activity in behaviorally tolerant rats, although Akera et al.,[43] using the same ethanol administration schedule, reported no increase in enzyme activity.

Reasons for these conflicting results are not obvious, despite the diverse experimental conditions. The mice used by Goldstein and Israel[20] may not be comparable to animals in other studies because they received pyrazole, which has been shown to have a depressant effect on the CNS, independent of its inhibition of ethanol oxidation.[45] Comparison of other experimental conditions does not reveal consistent differences that would explain all the discrepancies.

A similar controversy occurs in investigations of the effect of induced seizures on Na^+,K^+-ATPase. A number of studies report increased Na^+,K^+-ATPase activity following convulsions induced by such treatments as electrical stimulation,[46,47] ethyl chloride,[46] and pentylenetetrazole.[48,49] In contrast, others find no increase in this enzyme after induced convulsions.[50,51] Furthermore, Castillon and Rodnight[47] found differences in enzyme activity only when comparing particulate fractions, not whole homogenates, and only after the tissue had been frozen.

To explain this anomaly, Castillon and Rodnight[47] proposed that Na^+,K^+-ATPase activity may be controlled by membrane-bound factors that are released in varying amounts depending on animal treatment and tissue preparation. Whether or not such labile modulators exist, it is well known that the measurement of Na^+,K^+-ATPase activity is susceptible to artifactual alteration by such procedures as homogenization and freezing.[52] The discrepant results with ethanol, therefore, could be the result of unrecognized technical differences. For example, Goldstein and Israel[20] saw no change in the enzyme activity of whole homogenates, whereas had they examined particulate fractions as others did,[40,41] an increase might have been apparent.

The association of Na⁺,K⁺-ATPase activity with experimentally induced convulsions may also provide a plausible explanation for the increased enzyme activity in ethanol-dependent animals. It has been suggested that, during seizures, the ion transport mechanism is activated to compensate for changes in the distribution of electrolytes caused by repeated neuronal discharges.[46] Thus, the enhanced Na⁺,K⁺-ATPase activity during convulsions may be an immediate result of increased neuronal activity. This suggests the possibility that the close correspondence between Na⁺,K⁺-ATPase activity and withdrawal hyperexcitability[40,41] is also a consequence of greater neuronal activity during withdrawal, rather than that the enzyme is increased in an adaptive response to counteract prolonged ethanol inhibition. Considering the inconsistencies of the studies on which this proposal is based, its validity may be challenged. It is useful, however, as a reminder that changes occurring in parameters measured after prolonged ethanol intake are not necessarily "adaptive" changes, especially if measurements are made only when withdrawal signs are present. Before an effect is postulated as an adaptive mechanism, its time course should be followed to determine the extent to which the change coincides with the acquisition and subsidence of tolerance or dependence. That has not been done adequately in the studies reported here.

In addition, it would be useful to know whether elevated Na⁺,K⁺-ATPase activity results from net synthesis of new enzyme or from activation of existing Na⁺,K⁺-ATPase. This question is not clearly resolved by a recent report that the density of specific [³H]ouabain binding sites is markedly increased in certain brain regions of cats receiving ethanol, 1.5 g/kg, twice a day, for 5 weeks.[53] Specific ouabain binding is considered a measure of the availability of Na⁺,K⁺-ATPase molecules and would, of course, be elevated by greater net synthesis of the enzyme. However, as the authors point out, increased binding sites might also result from exposure of latent enzyme molecules by conformational changes in the plasma membrane. Conceivably, conformational changes could come about through alterations in membrane lipids. Other studies have shown that Na⁺,K⁺-ATPase activity changes when the membrane lipid composition is altered by dietary manipulation of lipid intake,[54−56] and limited evidence indicates that prolonged alcoholization may influence brain membrane lipids, as well.[41,57−60]

2. Liver

There are relatively few studies of the effect of chronic ethanol intake on the activity of liver Na⁺,K⁺-ATPase. In mice consuming a 6% (w/v) ethanol liquid diet for 2 weeks, Na⁺,K⁺-ATPase activity did not increase, although Mg²⁺-

ATPase was elevated by 30%.[37] A similar elevation of Mg^{2+}-ATPase was reported in rats consuming ethanol for 30 days, but Na^+,K^+-dependent activity was not mentioned.[61] However, beginning in 1973 a comprehensive series of papers by Bernstein et al.[62-67] has reported that consumption of a 5% (w/v) ethanol liquid diet for 3–4 weeks increased rat liver Na^+,K^+-ATPase by 190%.[63] The active transport of ^{86}Rb by liver slices also increased but by only 70%.[63] Mg^{2+}-ATPase activity decreased slightly. Various other metabolic changes were noted in these studies, including 50% increases in the rates of ethanol metabolism[62] and oxygen consumption,[63] both of which could be abolished by ouabain.[63,65] The authors suggested that the increase in the transport mechanism was responsible for the greater oxidative activity of the liver. This, in turn, increased the turnover of NADH, resulting in an enhanced rate of ethanol metabolism.[65] This association between the liver's oxidative capacity, NADH turnover, and ethanol oxidation is further supported by evidence that the mitochondrial uncoupling agent, 2,4-dinitrophenol, also increases both oxygen consumption and ethanol metabolism in the liver.[68]

Bernstein et al. speculated that the increase in liver Na^+,K^+-ATPase might be a compensatory mechanism to overcome the presumed inhibition of the enzyme by ethanol in vivo,[65] but they offered the alternative explanation that the increased activity could be a secondary effect, mediated by ethanol-induced hormonal changes.[64-67] The latter hypothesis is favored by evidence that the elevated metabolic activity in the liver after chronic ethanol resembles that produced by thyroid hormones[64,65] and by adrenergic stimulation.[65,66] Furthermore, the hypermetabolic state produced by chronic ethanol treatment was markedly diminished by thyroidectomy and abolished by adrenalectomy[67] and by pretreatment with the antithyroid drug, propylthiouracil.[69]

It should be emphasized that the increase in Na^+,K^+-ATPase activity in the liver, as well as in the brain and erythrocyte,[39] is markedly greater than the rise in active ion transport. From this evidence Bernstein et al.[65] proposed that the efficiency of the transport mechanism decreased after chronic ethanol. Furthermore, because there was no change in liver potassium content despite the 70% increase in transport, the authors suggested that the changes in transport activity were accompanied by increased membrane permeability to potassium ions.

F. OTHER TISSUES

The actions of ethanol on Na^+,K^+-ATPase seem to be general phenomena, since tissues other than brain and liver are affected, both in vitro[23,70-72] and after chronic ethanol ingestion.[39]

1. Heart

Williams et al.[70] showed that ethanol inhibited Na^+,K^+-ATPase of guinea pig cardiac membrane in a mechanism that again depended on potassium-ion concentration. There was 50% inhibition at 0.15 M ethanol and 5 mM potassium. The authors pointed out that ethanol has a negative inotropic effect on the heart, as opposed to the positive inotropic action of ouabain. Because of this, they said, it is difficult to attribute ethanol's inotropic effect to Na$^+$,K$^+$-ATPase inhibition, a conflict that seems analogous to that discussed earlier in which ethanol and ouabain have opposing effects in the CNS.

The authors also reported that the heart enzyme was inhibited by acetaldehyde.[70] However, the inhibition differs from that reported for brain.[17,21] Cardiac Na$^+$,K$^+$-ATPase was inhibited 30% with 8×10^{-4} M acetaldehyde, whereas the brain enzyme was not inhibited by 4×10^{-3} M acetaldehyde. Paradoxically, increasing the acetaldehyde concentration above 8×10^{-4} M did not greatly enhance inhibition of the heart enzyme, but the brain enzyme could be totally suppppressed with 0.04 M acetaldehyde.[21] Moreover, inhibition of the heart enzyme was reversible while brain Na$^+$,K$^+$-ATPase was said to be inhibited irreversibly.[21] The concentrations of acetaldehyde used in these studies are several times greater than those found *in vivo* after ethanol ingestion,[73] and thus these actions of acetaldehyde would not normally contribute to the pharmacological consequences of ethanol intoxication.

2. Intestine

Recently, ethanol was shown to inhibit Na$^+$,K$^+$-ATPase from the brush border of rat small intestine.[71] All straight-chain aliphatic alcohols from methanol to octanol were examined, and the inhibitory potency again correlated with chain length and lipophilicity. There was 50% inhibition with 0.44 M ethanol, a concentration within the range that can be found in the digestive tract after ethanol ingestion.[74] Consequently, the intestinal enzyme seems to be a likely candidate for ethanol inhibition *in vivo*. These authors suggest that suppression of Na$^+$,K$^+$-ATPase activity may contribute to the decreased transport of nutrients from the intestinal lumen that occurs after ethanol ingestion.

Chronic ethanol intake does not appear to alter intestinal Na$^+$,K$^+$-ATPase, however.[75] Guinea pigs were intubated for 14 days with 50% ethanol at a dose of 2.5 g/kg per day. Jejunal Na$^+$,K$^+$-ATPase was measured 24 hr after the last dose, but there was no significant difference in activity when compared to isocaloric glucose controls.

3. Erythrocytes

The Na$^+$,K$^+$-ATPase activity of erythrocytes was inhibited by ethanol *in vitro*[72] and increased after chronic ethanol intake.[39] Human alcoholic subjects, presenting signs of withdrawal hyperexcitability, had a 76% greater activity of red cell Na$^+$,K$^+$-ATPase and a 30% higher ^{86}Rb transport than did controls.[39]

G. SUMMARY AND CONCLUSIONS

There is no doubt that Na^+,K^+-ATPase is inhibited by ethanol *in vitro* if the potassium-ion concentration is sufficiently low, but the pharmacological significance of the inhibition remains uncertain. Inhibition of the enzyme apparently results from a lipophilic interaction of ethanol with the plasma membrane and requires at least 0.22 M ethanol for 50% inhibition. Therefore, it seems unlikely that ethanol concentrations occurring *in vivo* during moderate to moderately severe inebriation (e.g., 0.02–0.04 M) would be adequate to suppress the enzyme significantly. In view of this, the correspondence between lipophilicity and pharmacological potency of the lower aliphatic alcohols and other depressants suggests that other membrane activities, more sensitive to hydrophobic modification than is Na^+,K^+-ATPase, may be responsible for the depressant action of these agents. As evidence that membrane lipid properties are altered by moderate ethanol concentrations, Chin and Goldstein[76] showed recently that as little as 0.02 M ethanol increased the fluidity of synaptosomal and erythrocyte membranes *in vitro*. The functional correlates of this fluidizing effect are not known, however.

If Na^+,K^+-ATPase *per se* is not inhibited by ethanol *in vivo*, there is little justification for the hypothesis that the enzyme activity increases during chronic ethanol intake in an adaptive mechanism to counteract the inhibition. Rather, the reasons for the enhanced activity remain obscure. In the CNS, the increase seems related to the hyperexcitability and elevated neuronal activity that accompany withdrawal, with little correlation to the development of behavioral tolerance. On the other hand, the increased activity of the liver enzyme is interpreted as a hormonal stress response, with ethanol acting as the primary stressor. Although the elevation in erythrocyte Na^+,K^+-ATPase after chronic ethanol suggests that enhanced enzyme activity might be a generalized response, it is not clear that the increased activity has the same basis in all tissues. Data on the time course and dose response of changes in Na^+,K^+-ATPase in various tissues would help elucidate the pharmacological factors underlying the increased enzyme activity.

The nature of the biochemical changes in the Na^+,K^+-ATPase–membrane complex that result in greater enzyme activity are also unknown. There are at least three obvious alternatives:

1. There is increased net synthesis of the protein components of the complex.
2. Existing enzyme is activated or uncovered by alterations in membrane conformation, possibly brought about by changes in lipid configuration or composition.
3. There is a change in the availability of activators or inhibitors of the enzyme.

Recent evidence from [^{3}H]ouabain binding studies favor the first or second alternatives.

H. REFERENCES

1. J. C. Skou, *Physiol. Rev.* **45**:596 (1965).
2. R. Whittam and K. P. Wheeler, *Annu. Rev. Physiol.* **32**:21 (1970).
3. P. B. Dunham and R. B. Gunn, *Arch. Intern. Med.* **129**:241 (1972).
4. J. L. Dahl and L. E. Hokin, *Annu. Rev. Biochem.* **43**:327 (1974).
5. A. Schwartz, G. E. Lindenmayer, and J. C. Allen, *Pharmacol. Rev.* **27**:1 (1975).
6. "Properties and Functions of (Na⁺ + K⁺)-activated Adenosinetriphosphatase" (A. Askari, ed.), *Ann. N.Y. Acad. Sci.* **242** (1974).
7. B. Roelofsen and L. L. M. Van Deenen, *Eur. J. Biochem.* **40**:245 (1973).
8. E. Racker and L. W. Fisher, *Biochem. Biophys. Res. Commun.* **67**:1145 (1975).
9. S. Hilden and L. E. Hokin, *Biochem. Biophys. Res. Commun.* **69**:521 (1976).
10. R. E. Barnett and J. Palazzotto, *Ann. N.Y. Acad. Sci.* **242**:69 (1974).
11. Y. Israel and H. Kalant, *Nature (Lond.)* **200**:476 (1963).
12. D. H. Streeten and A. K. Solomon, *J. Gen. Physiol.* **37**:643 (1954).
13. Y. Israel-Jacard and H. Kalant, *J. Cell. Comp. Physiol.* **65**:127 (1965).
14. Y. Israel, H. Kalant, and I. Laufer, *Biochem. Pharmacol.* **14**:1803 (1965).
15. Y. Israel, H. Kalant, and A. E. LeBlanc, *Biochem. J.* **100**:27 (1966).
16. Y. Israel and I. Salazar, *Arch. Biochem. Biophys.* **122**:310 (1967).
17. V. G. Erwin, J. Kim, and A. D. Anderson, *Biochem. Pharmacol.* **24**:2089 (1975).
18. A. Y. Sun and T. Samorajski, *J. Neurochem.* **17**:1365 (1970).
19. A. Y. Sun and T. Samorajski, *J. Neurochem.* **24**:161 (1975).
20. D. B. Goldstein and Y. Israel, *Life Sci.* **11**:957 (1972).
21. B. Tabakof, *Res. Commun. Chem. Pathol. Pharmacol.* **7**:621 (1974).
22. R. Rodnight, *Biochem. J.* **120**:1 (1970).
23. C. Hegyvary, *Biochim. Biophys. Acta* **311**:272 (1973).
24. D. R. Dahl, *J. Neurochem.* **15**:815 (1968).
25. B. D. Roufogalis, *J. Neurochem.* **24**:51 (1975).
26. G. S. Levey, *Recent Prog. Horm. Res.* **29**:361 (1973).
27. H. L. Greene, R. H. Herman, and S. Kraemer, *J. Lab. Clin. Med.* **78**:336 (1971).
28. K. Kuriyama and M. A. Israel, *Biochem. Pharmacol.* **22**:2919 (1973).
29. P. W. Davis and T. M. Brody, *Biochem. Pharmacol.* **15**:703 (1966).
30. G. A. V. De Jaramillo and P. S. Guth, *Biochem. Pharmacol.* **12**:525 (1963).
31. S. Roth and P. Seeman, *Biochim. Biophys. Acta* **255**:207 (1972).
32. H. Kalant and Y. Israel, in: *Biochemical Factors in Alcoholism* (R. P. Maickel, ed.), pp. 25–27, Pergamon Press, Inc., Elmsford, N.Y. (1967).
33. F. Bergman, A. Costin, M. Chaimovitz, and A. Zerochia, *Neuropharmacology* **9**:441 (1970).
34. J. C. Keesey, H. Wallgren, and H. McIlwain, *Biochem. J.* **95**:289 (1965).
35. J. C. Keesey and H. Wallgren, *Biochem. J.* **95**:301 (1965).
36. H. Wallgren, P. Nikander, P. von Boguslawsky, and J. Linkola, *Acta Physiol. Scand.* **91**:83 (1974).
37. M. A. Israel and K. Kuriyama, *Life Sci.* **10**:591 (1971).
38. M. L. Jain, B. M. Curtis, and E. V. Bakutis, *Res. Commun. Chem. Pathol. Pharmacol.* **7**:229 (1974).
39. Y. Israel, H. Kalant, A. E. LeBlanc, J. C. Bernstein, and I. Salazar, *J. Pharmacol. Exp. Ther.* **174**:330 (1970).
40. M. K. Roach, M. M. Khan, R. Coffman, W. Pennington, and D. L. Davis, *Brain Res.* **63**:323 (1973).
41. H. Wallgren, P. Nikander, and P. Virtanen, in: *Alcohol Intoxication and Withdrawal: Experimental Studies* (M. M. Gross, ed.), Vol. II, pp. 23–36, Plenum Press, New York (1975).
42. W. H. Knox, R. G. Perrin, and A. K. Sen, *J. Neurochem.* **19**:2881 (1972).
43. T. Akera, R. H. Rech, W. J. Marquis, T. Tobin, and T. M. Brody, *J. Pharmacol. Exp. Ther.* **185**:594 (1973).
44. P. Nikander and L. Pekkanen, *Psychopharmacology* **51**:219 (1977).

45. A. E. LeBlanc and H. Kalant, *Can J. Physiol. Pharmacol.* **51:**613 (1973).
46. T. Harmony, R. Urba-Holmgren, C. M. Urbay, and S. Szava, *Brain Res.* **11:**672 (1968).
47. R. Castillon and R. Rodnight, *Biochem. J.* **127:**83P (1972).
48. A. Bignami, G. Palladini, and G. Venturini, *Brain Res.* **1:**413 (1966).
49. E. de Robertis, M. Alberici, and G. Rodriguez de Lores Arnaiz, *Brain Res.* **12:**461 (1969).
50. D. J. Brown and W. E. Stone, *J. Neurochem.* **20:**1461 (1973).
51. E. S. Marichich and A. G. Nasello, *Brain Res.* **57:**409 (1973).
52. F. E. Samson, Jr., and D. J. Quinn, *J. Neurochem.* **14:**421 (1967).
53. V. K. Sharma, M. G. Nagaraj, and S. P. Banerjee, *Brain Res.* **129:**183 (1977).
54. G. Y. Sun and A. Y. Sun, *J. Neurochem.* **22:**15 (1974).
55. R. P. Brivio-Haugland, S. L. Louis, K. Musch, N. Waldeck, and M. A. Williams, *Biochim. Biophys. Acta* **433:**150 (1976).
56. L. P. Solomonson, V. P. Liepkalns, and A. A. Spector, *Biochemistry* **15:**892 (1976).
57. H. M. Hakkinen and E. Kulonen, *Nature (Lond.)* **198:**995 (1963).
58. P. Lesch, E. Schmidt, and F. W. Schmidt, *Z. Klin. Chem. Klin. Biochem.* **11:**159 (1973).
59. A. K. Rawat, *Res. Commun. Chem. Pathol. Pharmacol.* **8:**461 (1974).
60. J. M. Littleton and G. John, *J. Pharm. Pharmacol.* **29:**579 (1977).
61. J. E. C. Walker and E. R. Gordon, *Biochem. J.* **119:**511 (1970).
62. L. Videla, J. Bernstein, and Y. Israel, *Biochem. J.* **134:**507 (1973).
63. J. Bernstein, L. Videla, and Y. Israel, *Biochem. J.* **134:**515 (1973).
64. Y. Israel, L. Videla, A. MacDonald, and J. Bernstein, *Biochem. J.* **134:**523 (1973).
65. J. Bernstein, L. Videla, and Y. Israel, *Ann. N.Y. Acad. Sci.* **242:**560 (1974).
66. Y. Israel, L. Videla, and J. Bernstein, *Fed. Proc. Symp.* **34:**2052 (1975).
67. J. Bernstein, L. Videla, and Y. Israel, *J. Pharmacol. Exp. Ther.* **192:**583 (1975).
68. H. Seiden, Y. Israel, and H. Kalant, *Biochem. Pharmacol.* **23:**2334 (1974).
69. Y. Israel, H. Kalant, H. Orrego, J. M. Khanna, L. Videla, and J. M. Phillips, *Proc. Natl. Acad. Sci. USA* **72:**1137 (1975).
70. J. W. Williams, M. Tada, A. M. Katz, and E. Rubin, *Biochem. Pharmacol.* **24:**27 (1975).
71. S. Mitjavila, C. Lacombe, and G. Carrera, *Biochem. Pharmacol.* **25:**625 (1976).
72. Y. Israel, *Q. J. Stud. Alcohol* **31:**293 (1970).
73. E. Majchrowicz and J. H. Mendelson, *Science* **168:**1100 (1970).
74. Y. Israel, J. E. Valenzuela, I. Salazar, and G. Ugarte, *J. Nutr.* **98:**222 (1969).
75. N. Krasner, H. A. Carmichael, R. I. Russell, G. G. Thompson, and K. M. Cochran, *Gut* **17:**249 (1976).
76. J. H. Chin and D. B. Goldstein, *Science* **196:**684 (1977).

Biochemical and Biophysical Approaches in the Study of Ethanol–Membrane Interaction

6

Albert Y. Sun

A. INTRODUCTION

Biological membranes not only form a closed compartment separating the cyto-plasmic materials from their surroundings, but also are functionally active through proper arrangement and interactions of their component molecules, mainly proteins, lipids, and polysaccharides. For example, cooperativity of mitochondrial oxidative phosphorylation depends on proper spatial organization and interaction of enzyme molecules on and in the mitochondrial membranes. The distribution of transport protein on membranes enables the transport of various chemicals into different compartments. Furthermore, the transport of information, such as neurotransmission at the synapses, contact inhibition and cell recognition at the cell surface, and so on, is also involved in the special type of membrane-mediated interaction with membrane components, especially gly-coprotein. It is no surprise that the problem of biological membranes has dominated our attention in almost every area of biological and biomedical research.

   Both the biochemical and biophysical properties of membranes are impor-tant for their function. Phospholipids are important biochemical components of plasma membrane. They provide the membranes with a proper molecular struc-ture for physiochemical activities. Phospholipids provide a relatively nonpolar environment in which they may assist in binding and orientation of ligands, such as substrates, inhibitors, and regulators, or may confer on the enzyme proteins an amphipathic character required for the active site. Strong circumstantial

Albert Y. Sun • Sinclair Comparative Medicine Research Farm and Department of Biochemistry, University of Missouri, Columbia, Missouri 65201.

evidence for participation of lipids in transport and catalytic processes at the molecular level has been tabulated.[1] Most functional proteins, such as mitochondrial ATPase involved in oxidative phosphorylation,[2,3] and Na^+,K^+-ATPase[4-7] and Ca^{2+}-ATPase[8,9] involved in transport of cations, are membrane-bound and can be activated by phospholipids.

The function of membrane-bound protein has also been linked with the physical state of membrane. Lipids of biomembranes may undergo a physical change with increase in the system temperature or added exogenous agents. The fatty acyl chains become more liquidlike and increase the lipid mobility. The change in fluidity of the lipid bilayer is known to be important for many properties and functions of cell membranes.[1,10] The phase transition of membrane lipids can have a major influence on the membrane functions; for example, oxygen uptake paralleled changes in fatty acid chain motion observed by electron paramagnetic resonance,[11] and the temperature dependence of sugar uptake was correlated to the fatty acid compositions of the membrane.[12,13]

It is well known that some exogenous agents, such as detergents and general anesthetics, including alcohols, can have profound effects on the biophysical and biochemical properties of membrane.[14,15] Ethanol is one of the most extensively investigated of these molecules because its effects on man at the organismal level are well known but not well understood. A number of review articles have already covered much of the information on the action of ethanol on biological membranes; for example, the action of ethanol on the nervous system,[16] the nonspecific action of ethanol on biomembranes,[14] and the partitioning of alcohols into membranes.[17] A critical review on the direct action of alcohols on membranes and associated phenomena concerned with electrophysiological activity was presented recently by Hunt.[18] The purpose of this review is to summarize some pertinent biophysical and biochemical approaches to our understanding of ethanol–membrane interactions that are important for the interpretation of acute and chronic effects of ethanol on membrane function, particularly in the central nervous system. Most of the biophysical techniques cited are either applicable to the study of ethanol–membrane interactions or used by researchers as a tool to probe the action of ethanol and other anesthetics on biological membrane systems. Some changes in membrane structure and properties after acute and chronic perturbations of membrane lipids by ethanol are also reviewed.

B. PHYSICAL APPROACHES

The fluidity of the membrane matrix is very important to membrane function in a living cell. The membrane fluidity is regulated by altered cholesterol content or fatty acid composition of phospholipids (unsaturation, chain length, and branching). Fluidity of membrane lipids may be determined by the mobility of chains, the lateral mobility of phospholipid molecules, and the extent of intermolecular interaction with other lipids and nonlipid molecules.

Any perturbation of membrane by temperature or incorporation of various agents, including ethanol, may modify the intermolecular interaction, alter dielectric permittivity when applying an electric field, or change the diamagnetic susceptibility for a magnetic field. Several physicochemical methods have been employed to study membrane structure and function related to the perturbation of membrane by ethanol or other anesthetics.

1. Spin-Labeling Method

Spin-labeled fatty acid or phospholipids having paramagnetic nitroxide groups at different positions in the fatty hydrocarbon chain are used as probes of the motion of lipid in the surrounding environment.[19] The shape of the paramagnetic spectrum is very sensitive to probe motion, and rotational correlation times can quantitate the rotational mobility of lipids. The lipid mobility decreases with decreasing temperature. A spin label such as TEMPO (2,2,6,6-tetramethylpiperidine-1-oxyl), which has a much more favorable partition coefficient for fluid than for rigid lipids, has been used as a probe for the occurrence of lateral phase separation in membrane. The high-field hyperfine signals in the paramagnetic spectrum have a different position in polar and nonpolar environments,[20] so that a clear splitting is evident when the probe is present in both environments at the same time. By plotting the spectral parameter f, given by the ratio of the two peaks, against temperature, an abrupt increase in the fluidity of the membrane occurs at a temperature characteristic of each phospholipid mixture. The temperature at which this sudden change from a gel phase to a lamellar smectic liquid crystalline phase occurs is defined as transition temperature. Trudell et al.[21] have found that inhalation anesthetics produce significant changes in the phase transition temperature of membranes. It is clear from their study that clinical concentrations of an inhalation anesthetic are capable of causing a large perturbation in membrane properties which have been shown to be important to the function of some membrane-bound proteins.

Spin labeling is also a useful probe technique for examining lipid–protein interactions. By calculating the order parameter, a commonly used index of membrane fluidity, from electron paramagnetic resonance spectra, several laboratories[20,22–25] were able to show that ethanol in vitro increased fluidity in several biological membranes. More recently, Chin and Goldstein,[26] using the spin-labeling technique, showed that membranes from mice that had been subjected to long-term ethanol treatment were relatively resistant to the fluidizing effect of ethanol.

2. Differential-Scanning Calorimetry

Several membrane functions depend on the transition temperature of membrane lipid.[1] A break in Arrhenius plots accompanies transition from high to low energy of activation upon temperature increase. The fluidization of the membrane lipid above the transition temperature may be responsible for the reduction of activation energy. Differential-scanning calorimetry detects the endothermic transition which accompanies "melting" of the lipid hydrocarbon chains.[27] It shows a sharp transition with pure lipid of only one fatty acid species but reveals wide endothermic transitions for biological membranes which are composed of heterogenous lipid molecules. Such an endothermic peak is probably associated with an increase in mobility of the polar head prior to the main transition.

A transition temperature of 20°C was determined with Na^+,K^+-ATPase preparation from rabbit kidney.[28] A change in activation energy near 20°C was also demonstrated.[29] Recently, a sharp transition break with Na^+,K^+-ATPase in a synaptic plasma membrane preparation from rat was demonstrated.[30] The Arrhenius plots indicated that Na^+,K^+-ATPase is more sensitive than Ca^{2+}-ATPase to change in fluidity of the synaptosomal membrane. Since low concentrations of ethanol increased fluidity of synaptosomal membranes,[24,25] the different responses of Na^+,K^+-ATPase and Ca^{2+}-ATPase to ethanol may be the result of increased membrane fluidity.

3. Nuclear Magnetic Resonance

Nuclear magnetic resonance spectroscopy is a valuable technique for studying various aspects of cell membrane structure and function. One can use selected nuclei for examination of the motion or conformational changes of a group containing those nuclei. It is possible to study ^1H nuclei while all other nuclei in the system remain transparent, or it is possible to look at ^{13}C, ^2H, ^{15}N, or ^{31}P nuclei when the motion of a group containing these nuclei is of interest. Two recent review articles have detailed the general principle and application.[31,32] Proton magnetic resonance is widely used in the study of membrane fluidity and the conformational change of membrane-bound polypeptides. When a phospholipid–water system such as egg yolk lecithin is examined by proton magnetic resonance, a fairly broad spectrum is obtained. Signals associated with the deuterated water group, $N^+(CH_3)_3$ group, and $[CH_2]$ chain are observed. Using sonicated 1,2-dihexadecyl-sn-glycero-3-phosphorylcholine liposomes, Shieh et al.[33] were able to show that the choline proton magnetic resonance peak appears above the phase transition temperature. However, the fatty acid methyl and methylene proton magnetic resonance peaks appear after general anesthetics are added. The linewidths of these proton magnetic resonance spectra decrease and the peak intensity increases when the amounts of anesthetics are increased. It appears that the general anesthetics act on the hydrophilic site (choline group) at low concentrations and then diffuse into the hydrophobic region with the addition of large amounts of anesthetics.

4. Fluorescent Probes

Fluorescent probes may be used in membrane studies. The hydrophobic chromophores, such as perylene or O-methylanthracene, are located in the nonpolar core of a lipid bilayer, and their spectral properties are sensitive to microviscosity of the microenvironment. The polarization of the fluorescence of these probes increases as a function of increasing viscosity and has been used as a sensitive probe for phase changes in biomembranes.[34,35] The ampholyte probes, such as 1-anilinonaphthalene-8-sulfonate (ANS), are located in the polar region of the lipid bilayer near the water interface. ANS is strongly fluorescent in hydrophobic media, but its fluorescence is quenched in an aqueous environment with a shift of emission maximum to a higher wavelength. ANS fluorescence is also sensitive to the viscosity of the microenvironment and has been used to detect phase changes in membranes.[36] It has been shown that higher alcohols are more effective in lowering ANS fluorescence.[37] This effect indicates that the action of alcohols is exerted at the lipid bilayer in biomembranes.

5. Circular Dichroism and Optical Rotatory Dispersion

Optical rotation measurements are unique techniques to investigate the secondary structure of polypeptides and membrane proteins.[38,39] In dealing with membrane protein, artifacts may arise from light scattering, absorption dampening, and differential light scattering. With the application of corrections to the spectra, the investigator must be careful not to introduce additional artifacts into the spectra, mostly below 233 nm.[40] Studies on the conformational changes of brain microsomes during active cation transport function were reported.[41] The corrected molar ellipticity of the membrane suspension is in line with axonal membranes and sarcoplasmic reticulum. However, the circular dichroism spectra of brain microsomes in the activated state are indistinguishable from those of the inactivated state. Ca^{2+} and Mg^{2+} cause a significant change in the state of aggregation of the membrane. However, no extensive changes were observed in major membrane polypeptides; perhaps less than 10% of the protein was actually involved in the transport process. Efforts have been made to correlate the changes in membrane fluidity by perturbation agents with alterations in the conformation and activity of membrane protein. Treatment with detergents results in a partial transition from disordered to α-helical conformation of the proteins.[42] Schoenborn and Featherstone[43] have indicated that anesthetics may act on protein conformation by perturbing lipid–protein interaction in membranes. However, very little is known about their effects on the conformational change of membrane protein, owing to the technical difficulties encountered since the functional protein is often embedded in a lipid environment.

C. MEMBRANE PERTURBATION AND PROTEIN CONFORMATION

There is much evidence that ethanol affects the nervous system by a primary direct action on the excitable membranes.[44] Because of its simplicity, ethanol, like most of the other anesthetics, appears to act on the membrane nonspecifically by physical means. Ethanol molecules may insert into the lipid membrane layers, thereby altering the lateral mobility of membrane lipids or interacting with the hydrophobic portion of the membrane proteins resulting in steric changes.[45–51] The effect of ethanol has been explained by the Overton–Meyer hypothesis that there is a correlation between the lipid solubility of anesthetic compounds, including aliphatic alcohols, and the potency of these compounds to suppress various biological functions.[52] We have shown that ethanol inhibited Na^+,K^+-ATPase activity, while other types of membrane-bound enzymes such as acetylcholinesterase remained essentially unaffected.[50] Furthermore, aliphatic alcohols with increasing molecular size and carbon chain length gave increasing inhibitory effect on the enzyme. Using the molar concentration which would produce 50% inhibition of the enzymic activity, a linear relationship between the degree of inhibition and the lipid solubility of the alcohols was obtained. The same type of ethanol–membrane interaction was also demonstrated with norepinephrine uptake, another membrane-dependent process.[53–56] Na^+,K^+-ATPase is responsible for the sodium-pump mechanism and thus regulates the repolarization of excitable neuronal membranes.[57] Norepinephrine uptake, on the other hand, is an important process for inactivation of the physiologic function of catecholamines at the synaptic level. The inhibitory effect of alcohols of varying chain lengths on the preceding two biochemical processes in brain correlated well with their anesthetic potency and lipid solubility.[58] These results have provided additional support for the Overton–Meyer hypothesis. Recent progress with modern biophysical techniques has helped us to understand some of the interactions of ethanol with biological membranes. It appears that membrane expansion, membrane fluidization, lipid disorganization, and conformational change of membrane protein are the main consequences of ethanol–membrane interaction.

1. Membrane Expansion

The expansion of membrane induced by ethanol can be determined with erythrocyte ghosts by measuring the density of membrane using a high-precision densitometer. At concentrations of ethanol known to block nerve fibers, the membrane expands on the order of 2–3%.[14,48] The increase in membrane volume is about 10 times greater than the volume of the ethanol incorporated into the membrane. However, this tenfold amplification was not observed with liposomes in which protein is not present. The effect of ethanol might be due to an ethanol-induced perturbation of the lipid matrix, which in turn could cause partial "unfolding" of the protein structure. The observed perturbation of protein structure may greatly affect lipid–protein interaction within the membrane. Since the lipid–protein interactions are very important in the modulation of catalytic activity of membrane-dependent processes, the action of ethanol or other general anesthetics, especially on neuronal membranes, may result in anesthesia and other behavioral changes. Seeman's[14] membrane expansion theory appears to explain satisfactorily the molecular action of anesthetics. Trudell et al.[21] have further demonstrated that a clinical concentration of an inhalant anesthetic produces changes in both the phase transition temperature of pure lipid bilayers and the lateral phase separation temperature of mixed dipalmitoyl and dimyristoyl phosphatidylcholine bilayers of a magnitude sufficient to influence protein function. They also showed that externally applied pressure is able to antagonize the effect of the anesthetic on these transition temperatures. The antagonism was shown to result from a reordering of fatty acid chains around the anesthetic molecule rather than an exclusion of the anesthetic from the bilayer.[59] Recently, Trudell[60] has questioned the tenfold volume increase as calculated by Seeman.[14] Based on theoretical and experimental research, he indicated that when a phospholipid bilayer membrane expands in surface area, it contracts in thickness. Thus, he maintained that the increase in membrane volume is approximately equal to the van der Waals volume occupied by the anesthetic molecules. However, with biological membranes, ethanol and other general anesthetics also interrupt the lipid–protein interaction, and this may contribute greatly to the increased volume observed. This point cannot be resolved until appropriate X-ray data of area versus thickness for intact biological membrane are available for an exact calculation of the increase in volume associated with the measured expansion of surface area. Nevertheless, the volume expansion theory has been used to interpret the molecular action of anesthetics in interfering with the gating mechanism and in causing deleterious change in the conformation of membrane protein.

2. Membrane Fluidization

Recent work has been reviewed extensively by Seeman.[14] It is recognized that anesthetics, including aliphatic alcohols, increase the fluidity of model lipid membranes.[21,24,25,61,62] Since the activity of membrane-bound enzymes appears to be related to the physical state of membrane lipid, the fluidization of membrane by temperature changes or exogenous agents such as ethanol may also be accompanied by more localized conformational changes near or at the active site of membrane-dependent protein. Several groups have used the modern physical techniques, especially electron paramagnetic resonance, to study the effect of alcohols on membrane fluidity.[24,25,63–68] Goldstein's laboratory has also studied the various membranes isolated from brain tissue, and the changes were related to changes in the physical properties of membranes after the animal was exposed to acute and chronic ethanol treatment.[24–26] Essentially, addition of low concentrations of ethanol (0.02–0.35 M) in vitro increased fluidity in erythrocyte, mitochondrial, and synaptosomal membranes.[24,25] The fluidizing effect of ethanol was dose-related in all membranes except myelin. They suggested that nonlethal concentrations of ethanol may increase membrane fluidity in vivo. They have also demonstrated that membranes from mice that had been subjected to long-term ethanol treatment were relatively resistant to the fluidizing effect of ethanol.[26] The continuous presence of a fluidizing agent such as ethanol may have provided the stimulus for an adaptive response by changing the membrane structure and function. Decreased resistance of synaptosomal Na^+,K^+-ATPase to ethanol after exposure of the animal to chronic ethanol administration has been demonstrated.[63] Whether results with ethanol may be related to the change in fluidity property of neuronal membranes as observed by Chin and Goldstein[26] remains to be seen.

3. Lipid Disorganization

Using electron spin resonance measurements with higher concentrations of ethanol[24,25] or higher alcohols,[64,65] the fluidity of the membrane increased, and the spin probe deviated away from its normal position perpendicular to the bilayers. The biphasic phenomenon observed recently in our laboratory and in others[1,30,63] may be related to lipid disorganization. At low concentrations (0.1–0.5%), ethanol enhanced mitochondrial Ca^{2+} transport and Na^+,K^+-ATPase, probably due to increasing membrane fluidity. However, at higher concentrations (>1.0%), ethanol may have caused lipid disorganization and disruption of the protein–lipid interactions in the membranes. The biphasic response to ethanol will be discussed in greater detail later in this chapter.

4. Protein Conformational Change

The conformation of protein molecules is characteristic of each protein species and is closely related to its biochemical activity. The conformation of membrane protein is maintained by hydrophobic and hydrogen bonding and requires a hydrophobic lipid environment for the formation and stability of hydrogen-bonded secondary structure. Conformational changes in membrane protein may be induced by detergent and anesthetics, including ethanol.[41,43] Schoenborn and Featherstone[43] have indicated that anesthetics may act on protein conformation by perturbing lipid–protein interactions in membranes. However, because of the limitation of experimental tools (circular dichroism and optical rotatory dispersion, or fluorescence probe), other approaches, such as kinetic analysis of enzymic reactions and other indirect methods, may have to be used in order to substantiate the hypothesis. Kinetic parameters of membrane enzymes are modified by ethanol and other solvents.[64,65] Ethanol hindered the conformational change of the phosphorylated enzyme of Na^+,K^+-ATPase in the presence or absence of K^+.[51,66] More recently, ethanol-induced conformational changes occurring in membrane protein were demonstrated after animals were exposed to chronic ethanol treatment.[67] Brain microsomal membrane isolated from animals chronically treated with ethanol increased significantly in the number of SH groups reactive to dithiobis-(2-nitrobenzoic acid) and N-ethylmaleimide over that from alcohol-naive animals. It is possible that ethanol embedded in the membrane environment may have caused a permanent alteration of the conformation of certain membrane proteins by adaptation.

D. BIOCHEMICAL APPROACHES

We have previously demonstrated that many membrane transport processes in the brain were altered due to the hydrophobic interaction of alcohol with membranes. The transport processes included in our *in vitro* studies were Na^+,K^+-ATPase,[50] norepinephrine (NE) uptake,[56] and Ca^{2+}-transport activity.[63] The effects of ethanol on these membrane-dependent processes may cause the modification of membrane structure and functions after prolonged ethanol intake. An increase in Na^+,K^+-ATPase activity has been demonstrated in rats, cats, and dogs after chronic ingestion of ethanol.[62,68–70] We have attributed the increase in Na^+,K^+-ATPase activity to an adaptive mechanism compensating for the stress imposed on the membrane by chronic exposure to ethanol. Other membrane-dependent processes, such as hormonal binding capacity, adenylate cyclase, and NE uptake activity, may also be altered as the result of a similar mechanism. As indicated earlier, ethanol may depress the nervous system by a primary direct action on the excitable membrane. Therefore, our attention is focused mainly on the brain membrane systems. Recent progress in alcohol research seems to be consistent with the membrane hypothesis that the cellular mechanism of action of ethanol is related to biophysical alterations in membrane properties.[46,71,72] We believe that all the membrane-dependent biochemical processes will be affected to some degree in the presence of alcohol. Furthermore, after chronic ethanol treatment, an adaptational mechanism may exist in order to offset the challenge resulting from continuous stimulation or depression.[7,49,71]

1. Membrane Functions

While there is abundant information regarding the effects of ethanol on membrane excitability, membrane potentials, and other physiological parameters,[46,74] advances in biochemical research on the effects of ethanol on membrane function have been modest, mainly because the effective concentration has been at least 0.5%. Table I summarizes some of the *in vitro* effects of ethanol on the membrane-dependent processes in brain. The effective concentrations vary according to a number of parameters. Some enzyme systems (e.g., Na^+,K^+-ATPase) appear to be more sensitive to ethanol than others.[75,76] Israel *et al.*[77,78] have also demonstrated that changes in K^+ concentration in the assay medium may affect the response of Na^+,K^+-ATPase to ethanol. Furthermore, enzyme preparations from younger animals were more resistant to ethanol.[79] Other conditions may also affect the sensitivity of membrane-dependent functions to ethanol.

Although some of the concentrations of ethanol used to effect changes in

Table I.   Effect of Ethanol on Brain Membrane-Dependent Processes *in Vitro*

Enzyme system	Enzyme source	Effective ethanol concentration (%, w/v)	Change in activity (%)	References
K$^+$ transport	Brain slices	0.4–0.5	−35	80
Na$^+$,K$^+$ transport	Brain slices	0.4–0.5	−45	81
Na$^+$,K$^+$-ATPase	Microsomes	2	−50	80, 83, 84
	Synaptosomes	2.2	−50	50
	Synaptic membrane	0.3–0.5	+15	30
	Synaptic membrane	2–3	−50	30
NE uptake	Brain slices	1	−25	54
	Synaptosomes	2.0–2.5	−50	53–56
Ca^{2+}-ATPase	Synaptic membrane	3	−50	30
[^{14}C]oleic acid incorporation into phospholipid	Synaptosomes	3	−50	7

activity as shown in Table I may be unrealistically high, they may have potential relevance based on the following: Other membrane-bound, but not membrane-dependent, enzymes are not as sensitive to low concentrations of ethanol (1–4%) as membrane-dependent enzymes are.[50] In addition, chronic administration of ethanol alters the activity of Na$^+$,K$^+$-ATPase,[73] and the effects of acute ethanol administration are dose dependent.[72] However, the question of the physiological relevance of these studies cannot be resolved unless sufficiently sensitive measurements can be developed.

Recent studies indicate a biphasic response of membrane function to various concentrations of ethanol which may be related to the biphasic phenomenon of spontaneous electrical activity of the brain under the influence of ethanol.[80,81] There appears to be an increase in the electrical activity at low doses, while at high doses, it is depressed. We have demonstrated that several membrane-bound systems exhibit the biphasic response to ethanol. At low concentrations (0.1–0.5%), ethanol enhanced mitochondrial Ca^{2+} accumulation[63] and Na$^+$,K$^+$-ATPase.[30] An *in vivo* NE uptake study also showed that at low blood ethanol level, NE uptake into nerve terminals was enhanced, although the biphasic effect was not apparent in the *in vitro* system.[56] Ethanol also elicited a biphasic effect on synaptosomal NE release.[66] However, this change may be related to the ethanol effect on Ca^{2+} transport, since neurotransmitter release depends upon the presence of available Ca^{2+} at the synaptic region. In the event of ethanol–membrane interaction, small amounts of ethanol entering the membrane may result in expansion of the molecular structure with a corresponding increase in the membrane fluidity.[24,25,48] Therefore, the enhancement of transport activities at low ethanol concentrations may be the result of an increase in membrane fluidity by the entering ethanol molecules. At higher concentrations (1–5%), however, ethanol may interfere with the transmembrane processes by interacting with the hydrophobic area of the membranes and altering the fine arrangement of the membrane structure. High concentrations of ethanol embedded in the membrane matrix may interrupt lipid–lipid or lipid–protein interaction, resulting in general inactivation of membrane enzymes. Recently, we have shown that different membrane-bound systems may respond differently to modification of the physical state of membrane lipids.[30] A biphasic mode of action with ethanol was observed for Na^+,K^+-ATPase but not for Ca^{2+}-ATPase or acetylcholinesterase. We have tried to correlate the sensitivity of these enzymes to temperature and detergents with their sensitivity to changes in membrane fluidity brought about by ethanol. Na^+,K^+-ATPase is more sensitive than Ca^{2+}-ATPase to fluidity changes. The biphasic effect of ethanol on the Na^+,K^+-ATPase activity of neuronal membranes may be related to membrane excitability and behavioral manifestation observed after ethanol administration.

More recently, we have demonstrated that the activity of synaptosomal Na^+,K^+-ATPase increased after a single large acute dose of ethanol (6–8 g/kg) by intragastric intubation.[72] The increase in Na^+,K^+-ATPase activity was dose dependent and did not appear until 1 hr after intubation and persisted for at least 16 hr after ethanol administration. Low doses of ethanol (1–4 g/kg) by intragastric intubation did not appear to induce a change in enzyme activity.[71,72] However, by using intraperitioneal injection of ethanol in mice (4 g/kg), Israel and Kuriyama[82] were able to demonstrate a 36% increase in brain mitochondrial ATPase activity 3 hr after ethanol administration. It appears that a minimum blood ethanol concentration may be required to induce a change in enzymic activity, and ethanol may reach the bloodstream faster and more efficiently by this technique. It is not yet clear whether this adaptational increase in enzymic activity is due to an increase in enzyme synthesis or to changes in the microenvironment of the synaptic membrane after acute ethanol administration.

We have also observed an increase in Na^+,K^+-ATPase activity in brain homogenates as well as the synaptic plasma membrane fraction with regard to chronic ethanol administration.[73] The increase of Na^+,K^+-ATPase was more obvious in the purified synaptic plasma membrane than in the brain homogenates. Although this effect was not evident in some other studies using brain microsomes[82,83] or homogenates,[69,84] a similar increase in enzymic activity has been confirmed.[68,69,85] There is also a relative increase in Na^+,K^+-ATPase with increasing withdrawal excitability.[86] As shown in Table II, the failure to demonstrate an increase in enzymic activity after prolonged alcohol drinking may be due to differences in animal species, in the regional and subcellular fractions used for the assay, and in the methods used to induce alcohol addiction. Furthermore, special precautions must be taken when comparing membrane properties and functions. Akera et al.[83] treated the membrane preparation with deoxycholate before assaying the enzymic activity. Since deoxycholate has a detergent effect, such treatment may have altered the membrane structure, thus affecting the Na^+,K^+-ATPase activity. This may account for the discrepancy between the reports of Israel et al.[68] and Akera et al.[83] The induction of Na^+,K^+-ATPase depends on a high dose level of ethanol. In rat,[72] >6 g/kg is required for an acute dose and 6–8 g/kg daily for 21 days is required for chronic treatment.[62,71] Therefore, ethanol may have caused lipid disorganization, altered the lipid–protein

Table II. Effect of Chronic Ethanol Administration on Na^+,K^+-ATPase from Brain

Species	Method of induction[a]	Duration (days)	Enzyme preparation[b]	Change in activity (%)	Reference
Rat	3–8 g/kg per day, ig	21–26	Whole brain	+ 65	68
Cat	1.5–4.5 g/kg per day, ig	35	Frontal cortex	+ 21	69
			Associate cortex	+ 23	69
			Caudate nucleus	+ 19	69
			Hippocampus	+ 21	69
			Amygdala	—	69
Dog	3–6 g/kg per day, ig	21	Synaptosomes	+ 15	73
			Synaptic membrane	+ 40	73
Rat	EVI	7	Synaptosomes	+ 30	85
Mouse	EVI (pyrazole added)	3–8	Whole brain	—	84
Mouse	LD (6% ethanol)	14	Synaptosomes	—	82
Rat	LD (10% ethanol)	24	Microsomes	—	83
Rat	3–10 g/kg per day, ig	24	Microsomes	—	83

[a]Abbreviations: ig, intragastric intubation; EVI, ethanol vapor inhalation; LD, liquid diet.
[b]Preparations are either homogenates or subcellular fractions from brain.

interaction in the membrane, and modified the protein conformation. Changes in the kinetic properties of the enzyme, such as an altered K_m and V_{max} with respect to chronic ethanol administration, were observed. In addition, Na^+,K^+-ATPase was more sensitive to the inhibitory action of ethanol upon prolonged ethanol administration.[73] Chin and Goldstein[26] have also indicated an increase in resistance to the fluidizing effect of ethanol after mice were exposed to long-term ethanol treatment.

Other membrane-dependent processes may be affected similarly to Na^+,K^+-ATPase after acute and chronic ethanol administration. Those which are involved in neurotransmission and hormonal action are particularly worthy of investigation. Several groups have investigated the effect of ethanol on receptor sites with particular reference to adenylate cyclase[87–90] and protein kinase.[89–90] Adenylate cyclase may play an important role in the function of the central nervous system or other hormone-responsive target organs. The enzyme is embedded in the lipid matrix of a membrane, and the role of phospholipid has been shown to be critical to the functioning of the hormone-responsive adenylate cyclase. For example, solubilization of membrane-bound adenylate cyclase activated by norepinephrine abolished the responsiveness to norepinephrine. However, addition of highly purified monophosphatidyl-inositol restored NE responsiveness to the solubilized enzyme.[91] Adenylate cyclase also has many other characteristics similar to Na^+,K^+-ATPase, such as temperature dependence and sensitivity to detergent and phospholipase treatment with respect to hormone responsiveness.[92,93] Gorman and Bitensky[94] also showed a biphasic response of glucagon-sensitive adenylate cyclase activity with C_1–C_4 alcohols, but not the NE-sensitive enzyme. Apparently, the interchain interaction and the fluidity of the membranes are also important for the activity of this enzyme. Acute ethanol treatment showed no change in adenylate cyclase activity. However, the cyclic AMP level was lower in brain tissue after acute ethanol administration.[87,95,96] Chronic administration of ethanol increased the adenylate cyclase activity in the brain with corresponding increase in cyclic AMP level.[89] In another study,[97] no change in cyclic AMP levels was found with either acute or chronic ethanol treatment. The discrepancy may be due to the technique by which the animals were killed. Cyclic AMP-dependent protein kinase activity in brain also increased after chronic, but not acute, ethanol administration.[89] Apparently, the adenylate cyclase and protein kinase, like Na^+,K^+-ATPase, were suppressed with high doses of ethanol *in situ*. The increase in activity after chronic ethanol administration is a reflection of biological adaptation, probably the result of the change in the microenvironment or protein conformation as suggested earlier. After chronic treatment with ethanol, an altered sensitivity of dopamine receptor together with a decreased response of dopamine-sensitive adenylate cyclase was also demonstrated.[98]

Another important group of compounds involved in cellular function are glycoprotein- and ganglioside-containing sialic acid. These carbohydrate-containing moieties are localized at the cell surface[99–101] and are responsible for immunogenicity,[102,103] Ca^{2+} binding,[104,105] interaction of hormone with target cells,[106] blood clot formation,[107] reproduction processes,[108] and certain surface enzyme activities.[109] Neuraminidase, which hydrolyzes sialic acid from glycoprotein and ganglioside, was localized in the synaptosomal membrane.[110] Acute ethanol treatment leads to an increase in neuraminidase activity with a corresponding decrease in sialic acid content.[111] Chronic ethanol treatment, on the other hand, decreased neuraminidase activity and increased sialic acid content. Although its membrane dependency has not yet been established, the location of neuraminidase and its response to ethanol treatment may have great impact on the neurotransmission process. The ethanol treatment also affected the Ca^{2+}-binding property of synaptic plasma membrane.[112] Ca^{2+} binding decreased after prolonged ethanol administration, probably as a result of the structural change of membrane glycoprotein as demonstrated by Noble *et al.*[113] It is clear from these studies that the intrinsic properties of the neuronal membrane associated with acute and chronic ethanol administration need to be investigated.

2. Binding Properties

Using microwave absorption techniques, Grenell[17] demonstrated that ethanol could bind to both refined cortical membranes and synaptic membranes. Ethanol binding appears to be associated with a change in the physical state of membrane water. We have investigated the uptake and release of [^{14}C]ethanol using purified synaptic plasma membrane from squirrel monkey.[114] Ethanol uptake increased with time of interaction and was affected by temperature changes. The bound [^{14}C]ethanol could be readily replaced by other aliphatic alcohols, such as n-propanol and n-butanol. Also, the amount of ethanol bound to the membrane was inversely correlated to the membrane-bound Na^+,K^+-ATPase activity. The embedding of ethanol molecules may have altered the fine arrangement of lipid molecules of the synaptic plasma membrane, consequently hindering the transmembrane intermediate step of the Na^+,K^+-ATPase, which activity involves conformational change of the carrier protein.[7,114]

By studying the uptake and release of [^{14}C]ethanol in synaptic plasma membrane, we have also demonstrated that the ethanol–membrane interaction is readily reversible. These results further indicate that the reversible inhibition of Na^+,K^+-ATPase is caused by the simple dissolution of ethanol into the lipid phase, which results in a disturbance of the enzyme's microenvironment.[114] Therefore, the reversible depressant effect of alcohol shown in man may be related to its inhibitory action on Na^+,K^+-ATPase at the synaptic level.

3. Membrane Metabolism

As mentioned earlier, changes in physical and biochemical properties of membranes have been observed after animals were exposed to chronic ethanol treatment. The decrease in membrane fluidity[26] and sensitivity of Na^+,K^+-ATPase to ethanol[73] represent an adaptive response to the fluidizing effect of alcohols. Since membrane fluidity depends on the biochemical arrangement of the membrane, including lipid composition (chain length and degree of unsaturation), cholesterol/phospholipid ratios, and lipid–protein interaction, investigations on membrane composition and metabolism after acute and chronic administration of ethanol should provide useful information.

a. Membrane Lipid Composition

Although most evidence indicates that enthanol effects are related to changes in the integrity of the membranes, studies on ethanol-induced changes in membrane phospholipids and acyl group composition have not been extensive. Studies by Lesch[115,116] on brain lipid composition from autopsies of patients with a history of cerebrohepatic degeneration indicated a reduction in the lipid components. The results suggested general myelin loss, which was not confined to a particular brain region. Changes in acyl group composition of membrane lipids were found together with other degenerative changes. Since these studies were done using brain tissues, it was not possible to conclude anything about acyl group changes in specific membranes.

Littleton and John[117] demonstrated a general decrease in polyunsaturated fatty acids in synaptic plasma membranes of mice of the TO Swiss strain exposed to ethanol vapors. The results seemed to agree well with the decrease in membrane fluidity observed with chronic ethanol treatment.[26] However, no change in the fatty acid composition of synaptosomal membranes of DBA mice was indicated after induced ethanol dependence.[26] The inconsistency is due in part to the genetic differences between these two strains of mice.

b. Lipid Metabolism

It is generally recognized that the acyl groups of membrane phosphoglycerides in brain are engaged in active turnover.[26] This dynamic equilibrium between the esterified acyl groups and free fatty acids is thought to be maintained by an intricate cyclic mechanism which may be regulated by neurotransmitter substances.[118] Several hormones, including epinephrine in dog heart,[119] acetylcholine in pancreas,[120] and thyrotropin in thyroid,[121] have been found to increase the synthesis of monophosphatidylinositol in their target tissues. The influence of these hormones on phospholipid metabolism may have some bearing on the activation–inactivation process as modulated by adenylate cyclase following hormone binding. As reviewed by Lieber,[122] ethanol may produce a variety of effects on lipid metabolism in a number of body organs. However, the effects of ethanol administration on acyl group composition of membrane phospholipid are poorly defined, and the effects on lipid metabolism remain to be explored. Reactivation of the lipid-dependent enzyme is apparently more effective with lipid having a greater proportion of unsaturated fatty acids.[123,124] The transfer of arachidonate to synaptosomal diacylphosphatidylcholine in a group of rats ($n = 14$) administered 30% ethanol (8 g/kg) daily for 21 days by intragastric intubation was 24% higher than controls.[125] In the same study, another group of rats was on 15% ethanol as its sole drinking source *ad libitum,* and the transferase activity increased even more (58%) after 13 months on ethanol. The increase in acylation of membrane phosphoglyceride may be regarded as an adaptive change in order to facilitate the increased transport activities and neural stimulation under the influence of ethanol intoxication.

Another aspect of lipid metabolism which may be affected by ethanol is the mobility and availability of cholesterol for membrane structure. Cholesteryl ester hydrolase is an important enzyme which responds to stress and hormonal manipulation. It may be involved in cholesteryl ester pooling, adrenocortical functions, and some membrane transport processes. We have studied the effect of acute and chronic ethanol administration on this enzyme in adrenal tissue.[126] Results indicated that 1 hr after an acute dose of ethanol (2–8 g/kg) there was significant enhancement of the activity, which persisted for 24 hr after ethanol treatment. However, with prolonged administration of ethanol (6 g/kg daily for 14 days by intragastric intubation technique), a 50% decrease in cholesteryl ester hydrolase activity was observed. This change in enzymic activity may alter the hormonal balance and physical properties of plasma membranes.

E. SUMMARY AND CONCLUSIONS

Information available indicates that ethanol binds to the biomembrane and alters the physical state of membranes by changing the fluidity and lipid–protein interactions. The binding is nonspecific and reversible. Several physical methods were employed to study the changes caused by ethanol or other general anesthetics. Since biological membrane is a much more complicated system than the lipid bilayer membrane model or soluble system, loss of sensitivity and contributions by artifacts are often encountered in various physical techniques employed in membrane research. Often, one must employ two or three methods in order to obtain clearer results. Among those physical methods mentioned in this paper, the electron spin resonance technique appears to be very useful in probing the change in membrane fluidity. However, the spin probes, like fluorescence probes, which are foreign molecules, are likely to perturb the lipid bilayers. In studying the perturbing effect of ethanol, one must use extremely low concentrations of probes to avoid affecting the original packing of the membranes.

It appears that several membrane-dependent biochemical processes, such as Na^+,K^+-ATPase, NE uptake, and adenylate cyclase, are affected by the presence of ethanol. Objections were raised in the past that high concentrations of ethanol (1–4%, w/v) are needed to produce a significant change in biochemical activity. It is true that the concentrations required to inhibit Na^+,K^+-ATPase activity or NE uptake are unrealistically high as compared with the physiologically relevant levels of ethanol. However, a biochemical system can only serve as a model to study a certain type of reaction or interaction. Conditions for the assay are often optimized and may not be the same as for the intact membrane in the physiological situation. For example, if the K^+ concentration in the incubation medium for Na^+,K^+-ATPase assay is lowered, much higher inhibition occurs as compared to the regular control. Furthermore, the effect of ethanol on in vivo NE uptake after an acute ethanol dose by intragastric intubation was demonstrated. At low doses up to 5 g/kg when the blood ethanol was below 200 mg/dl, NE uptake was enhanced approximately 20%. But as the dose level increased to 8 g/kg, giving blood ethanol levels of about 500 mg/dl, a decline in NE uptake was observed. It is possible that under physiological conditions a much lower ethanol concentration will affect membrane processes than is required in vitro. Therefore, a high dose level of ethanol may actually inhibit those membrane-dependent processes in situ, causing the adaptive change in their activity after acute and chronic ethanol treatment. However, inhibition of NE uptake by high doses of ethanol was not clear cut. Perhaps, a more sensitive system needs to be developed in order to clearly demonstrate the interaction of ethanol and membrane-dependent processes in vivo. It is of interest to note that all three membrane-dependent enzymes we studied are sensitive to the inhibitory effect of ethanol at concentrations of 1–4% in vitro, and their activities were enhanced after chronic ethanol administration, perhaps by some adaptive mechanism.[71]

The biphasic response of some membrane-dependent biochemical processes to ethanol parallels some of the behavioral manifestations after alcohol drinking. It is possible that at low concentrations (<0.5%), ethanol increases the fluidity of membrane and thus facilitates the transport process, while at high concentrations (1–4%), ethanol causes lipid disorganization, disrupts lipid–protein interaction, and alters the membrane-bound protein conformation.

Since ethanol can be viewed as a simple molecule acting nonspecifically on biological membrane systems, all membrane-dependent processes may have responses similar to those of Na^+,K^+-ATPase, which has been studied extensively. Thus, the active transport systems, hormonal interaction, the immune response, and the cell–cell recognition process may be affected by ethanol–membrane interaction. However, not every membrane-dependent process responds to ethanol in the same manner. Apparently, there are differences in sensitivity to ethanol which are related to differences in sensitivity to membrane perturbations.

The increase in activities of membrane-dependent enzymes such as Na^+,K^+-ATPase and adenylate cyclase and the changes in the intrinsic properties of membranes such as fluidity brought about by ethanol may be related to the development of tolerance and physical dependence. Information on the composition of membrane phospholipids, proteins, cholesterol, and carbohydrates, as well as on the metabolism of membrane lipid, protein, and sialic acid, may contribute to our understanding of ethanol–membrane interaction and eventually the biochemical basis of alcoholism.

ACKNOWLEDGMENTS. The author thanks Drs. L. Foudin and G. Y. Sun for their help with the preparation of the manuscript and Mrs. D. Torres and Mrs. D. Tumulty for typing this paper. This work was supported in part by USPHS Grant AA-02054.

F. REFERENCES

1. B. Fourcans, and M. K. Jain, *Adv. Lipid Res.* **12**:147 (1974).
2. Y. Kagawa and E. Racker, *J. Biol. Chem.* **241**:2461 (1966).
3. A. F. Knowles, R. J. Guillory, and E. Racker, *J. Biol. Chem.* **246**:2672 (1971).
4. M. K. Jain, A. Strickholm, and E. H. Cordes, *Nature (Lond.)* **222**:871 (1969).
5. M. K. Jain, F. P. White, E. Williams, A. Strickholm, and E. H. Cordes, *J. Membr. Biol.* **8**:363 (1972).
6. R. Tanaka and K. P. Strickland, *Arch. Biochem. Biophys.* **111**:583 (1965).
7. A. Y. Sun and G. Y. Sun, in: *Function and Metabolism of Phospholipids in CNS and PNS* (G. Porcellati, L. Amaducci, and G. Galli, eds.), pp. 169–197, Plenum Press, New York (1976).
8. A. Martonosi, J. Donley, and R. A. Halpin, *J. Biol. Chem.* **243**:61 (1968).
9. D. H. MacLennan, P. Seeman, C. H. Iles, and C. Yip, *J. Biol. Chem.* **246**:2702 (1971).
10. R. Cherry, in: *Biological Membranes* (D. Chapman, ed.), Vol. 3, pp. 47–102, Academic Press, Inc., New York (1976).

11. S. Elter, M. A. Williams, T. Watkins, and A. D. Keith, *Biochim. Biophys. Acta* **339**:190 (1974).
12. C. F. Fox, in: *Membrane Molecular Biology* (C. F. Fox and A. D. Keith, eds.), pp. 145–161, Academic Press, Inc., New York (1972).
13. G. Wilson and C. F. Fox, *J. Mol. Biol.* **55**:49 (1971).
14. P. Seeman, *Pharmacol. Rev.* **24**:583 (1972).
15. G. Lenaz, E. Bertoli, G. Curatola, L. Mazzanti, and A. Bigi, *Arch. Biochem. Biophys.* **172**:278 (1976).
16. H. Kalant, *Fed. Proc. Symp.* **34**:1930 (1975).
17. R. G. Grenell, in: *The Biology of Alcoholism* (B. Kissin and H. Begleiter, eds.), Vol. 2: *Physiology and Behavior,* pp. 1–19, Plenum Press, New York (1972).
18. W. A. Hunt, in: *Biochemical Pharmacology of Ethanol* (E. Majchrowicz, ed.), pp. 195–210, Plenum Press, New York (1975).
19. I. C. P. Smith, in: *Biological Applications of Electron Spin Resonance* (A. M. Swartz, T. R. Bolton, and D. C. Berg, eds.), pp. 483–539, Wiley–Interscience, New York (1972).
20. E. J. Shimshick and H. M. McConnell, *Biochemistry* **12**:2351 (1973).
21. J. R. Trudell, D. G. Payan, J. H. Chin, and E. N. Cohen, *Proc. Natl. Acad. Sci. USA* **72**:210 (1975).
22. J. R. Trudell, W. L. Hubbell, E. N. Cohen, and J. J. Kendig, *Anesthesiology* **38**:207 (1973).
23. S. J. Paterson, K. W. Bulter, P. Huang, J. Labelle, I. C. P. Smith, and H. Schneider, *Biochim. Biophys. Acta* **266**:597 (1972).
24. J. H. Chin, and D. B. Goldstein, in: *Alcohol Intoxication and Withdrawal: Experimental Studies* (M. M. Gross, ed.), Vol. IIIa, pp. 111–122, Plenum Press, New York (1977).
25. J. H. Chin and D. B. Goldstein, *Mol. Pharmacol.* **13**:435 (1977).
26. J. H. Chin and D. B. Golstein, *Science* **196**:684 (1977).
27. D. Chapman, *Lipids* **4**:251 (1969).
28. J. S. Charnock, D. M. Dotty, and J. C. Russell, *Arch. Biochem. Biophys.* **142**:633 (1971).
29. N. Gruener and Y. Avi-Dor, *Biochem. J.* **100**:762 (1966).
30. A. Y. Sun, Eighth Annu. Meet., Soc. Neurosci., Abstr. (1978).
31. D. Chapman, *Ann. N.Y. Acad. Sci.* **195**:179 (1972).
32. D. W. Urry, in: *The Enzymes of Biological Membranes* (A. Martonosi, ed.), Vol. 1, pp. 31–69, Plenum Press, New York (1976).
33. D. D. Shieh, I. Ueda, H. Lin, and H. Eyring, *Proc. Natl. Acad. Sci. USA* **73**:3999 (1976).
34. V. Cogan, M. Shimitzky, G. Weber, and T. Nishita, *Biochemistry* **12**:521 (1973).
35. J. F. Faucon and C. Lussan, *Biochim. Biophys. Acta* **307**:459 (1973).
36. H. Traüble and P. Overath, *Biochim. Biophys Acta* **307**:491 (1973).
37. G. Lenaz, G. P. Castelli, and A. M. Sechi, *Arch. Biochem. Biophys.* **167**:72 (1975).
38. D. W. Urry, *Proc. Natl. Acad. Sci. USA* **265**:15 (1972).
39. D. W. Urry, *Methods Enzymol.* **32**:220 (1974).
40. D. F. H. Wallach, D. A. Low, and A. V. Bertland, *Proc. Natl. Acad. Sci. USA* **70**:3235 (1973).
41. M. M. Long, L. Masotti, G. Sachs, and D. W. Urry, *J. Supramol. Struct.* **1**:259 (1973).
42. B. Jirgensons, *J. Biol. Chem.* **242**:912 (1967).
43. B. P. Schoenborn and R. M. Featherstone, *Adv. Pharmacol.* **5**:1 (1967).
44. H. Wallgren, in: *International Encyclopedia of Pharmacology and Therapeutics* (J. Tremolieres, ed.), Section 20: *Alcohols and Derivatives,* Vol. 1, pp. 161–188, Pergamon Press, Inc., Elmsford, N.Y. (1970).
45. R. G. Grenell, in: *Alcohol Intoxication and Withdrawal: Experimental Studies* (M. M. Gross, ed.), Vol. II, pp. 11–23, Plenum Press, New York (1975).
46. H. Kalant, in: *The Biology of Alcoholism* (B. Kissin and H. Begleiter, eds.), Vol. 1: *Biochemistry,* pp. 1–62, Plenum Press, New York (1971).
47. L. J. Mullins, in: *Handbook of Neurochemistry* (A. Lajtha, ed.), Vol. 6, pp. 395–421, Plenum Press, New York (1971).
48. P. Seeman, *Experientia* **30**:759 (1974).
49. M. W. Hill and A. D. Bangham, in: *Alcohol Intoxication and Withdrawal: Experimental Studies* (M. M. Gross, ed.), Vol. II, pp. 1–9, Plenum Press, New York (1975).

50. A. Y. Sun and T. Samorajski, *J. Neurochem.* **17**:1365 (1970).
51. G. Y. Sun and A. Y. Sun, *Biochim. Biophys. Acta* **280**:306 (1972).
52. K. H. Meyer and H. Hemmi, *Biochem. Z.* **277**:39 (1935).
53. M. K. Roach, D. L. Davis, W. Pennington, and E. Nordyke, *Life Sci.* **12**:433 (1973).
54. Y. Israel, F. J. Carmichael, and J. A. McDonald, *Ann. N.Y. Acad. Sci.* **15**:38 (1973).
55. A. Y. Sun, *Fourth Int. Meet. Int. Soc. Neurochem.*, Abstr. **476** (1973).
56. A. Y. Sun, S. Shrader, and R. N. Seaman, in: *Currents in Alcoholism* (F. A. Seixas, ed.), Vol. 3, p. 141, Grune & Stratton, Inc., New York (1978).
57. J. C. Skou, *Physiol. Rev.* **45**:596 (1965).
58. A. Goldstein, L. Aronow, and M. Kalman, *Principles of Drug Action*, pp. 153–156, Harper & Row Publishers, New York (1969).
59. J. R. Trudell, W. L. Hubbell, and E. N. Cohen, *Biochim. Biophys. Acta* **291**:328 (1973).
60. J. R. Trudell, *Biochim. Biophys. Acta* **470**:509 (1977).
61. J. R. Trudell, W. L. Hubbell, and E. N. Cohen, *Biochim. Biophys. Acta* **291**:321 (1973).
62. M. W. Hill, *Biochim. Biophys. Acta* **356**:117 (1974).
63. J. P. Burke, M. E. Tumbleson, R. N. Seaman, and A. Y. Sun, *Res. Commun. Chem. Pathol. Pharmacol.* **18**:569 (1977).
64. C. M. Grisham and R. E. Barnett, *Biochim. Biophys. Acta* **311**:417 (1973).
65. C. Hegyvary, *Biochim. Biophys. Acta* **311**:272 (1973).
66. A. Y. Sun, *Res. Commun. Chem. Pathol. Pharmacol.* **15**:705 (1976).
67. B. Gruber, E. C. Dinovo, E. P. Noble, and S. Tewari, *Biochem. Pharmacol.* **26**:2181 (1977).
68. Y. Israel, H. Kalant, A. E. LeBlanc, J. C. Bernstein, and I. Salazar, *J. Pharmacol. Exp. Ther.* **174**:330 (1970).
69. W. H. Knox, R. G. Perrin, and A. K. Sen, *J. Neurochem.* **19**:2881 (1972).
70. M. E. Post and A. Y. Sun, *Res. Commun. Chem. Pathol. Pharmacol.* **6**:887 (1973).
71. A. Y. Sun, R. N. Seaman, and C. C. Middleton, in: *Alcohol Intoxication and Withdrawal: Experimental Studies* (M. M. Gross, ed.), Vol. IIIa, pp. 123–138, Plenum Press, New York (1977).
72. A. Y. Sun, S. L. Seaman, and R. N. Seaman, in: *Currents in Alcoholism* (M. Gallanger, ed.), Vol. 5, p. 63, Grune & Stratton, Inc., New York, in press (1979).
73. A. Y. Sun, G. Y. Sun, and C. C. Middleton, in: *Currents in Alcoholism* (F. A. Seixas, ed.), Vol. 1, p. 81, Grune & Stratton, Inc., New York (1977).
74. H. Kalant, in: *International Encyclopia of Pharmacology and Therapeutics* (J. Tremolieres, ed.), *Alcohols and Derivatives*, Section 20: Vol. 1, pp. 189–236, Pergamon Press, Inc., Elmsford, N.Y. (1970).
75. Y. Israel, H. Kalant, and A. E. LeBlanc, *Biochem. J.* **100**:27 (1966).
76. Y. Israel-Jacard and H. Kalant, *J. Cell. Comp. Physiol.* **65**:127 (1966).
77. Y. Israel, H. Kalant, and I. Laufer, *Biochem. Pharmacol.* **14**:1803 (1965).
78. Y. Israel and I. Salazar, *Arch. Biochem. Biophys.* **122**:310 (1967).
79. A. Y. Sun and T. Samorajski, *J. Neurochem.* **17**:1365 (1970).
80. W. J. Horsey and K. Aker, *Q. J. Stud. Alcohol* **14**:363 (1953).
81. A. A. Hadji-Dimo, R. Edberg, and D. H. Ingvar, *Q. J. Stud. Alcohol* **29**:828 (1968).
82. Y. Israel and K. Kuriyama, *Life Sci.* **10**:591 (1971).
83. T. Akera, R. H. Rech, W. J. Marquis, T. Tobin, and T. M. Brody, *J. Pharmacol. Exp. Ther.* **185**:594 (1973).
84. D. B. Goldstein and Y. Israel, *Life Sci.* **11**:957 (1972).
85. M. K. Roach, M. M. Khan, R. Coffman, W. Pennington, and D. L. Davis, *Brain Res.* **63**:323 (1973).
86. H. Wallgren, P. Nikander, and P. Virtanen, in: *Alcohol Intoxication and Withdrawal: Experimental Studies* (M. M. Gross, ed.) Vol. II, pp. 23–36, Plenum Press, New York (1975).
87. L. Volicer and B. P. Hurter, *J. Pharmacol. Exp. Ther.* **200**:298 (1977).
88. S. W. French, D. S. Palmer, and K. D. Wiggers, in: *Alcohol Intoxication and Withdrawal: Experimental Studies* (M. M. Gross, ed.), Vol. IIIa, pp. 515–538, Plenum Press, New York (1977).

89. K. Kuriyama, K. Nakagawa, M. Muramatsu, and K. Kakita, *Biochem. Pharmacol.* **25**:2541 (1976).
90. K. Kuriyama, K. Nakagawa, and N. Miki, in: *Alcohol Intoxication and Withdrawal: Experimental Studies* (M. M. Gross, ed.), Vol. IIIa, pp. 173–191, Plenum Press, New York (1977).
91. G. S. Levey, *J. Biol. Chem.* **246**:7405 (1971).
92. G. S. Levey and D. C. Lehotay, in: *The Enzymes of Biological Membranes* (A. Martonosi, ed.), Vol. 4, pp. 259–282, Plenum Press, New York (1976).
93. R. Coleman, *Biochim. Biophys. Acta* **300**:1 (1973).
94. R. E. Gorman and M. W. Bitensky, *Endocrinology* **87**:1075 (1970).
95. L. Volicer and B. I. Gold, in: *Biochemcial Pharmacology of Ethanol* (E. Majchrowicz, ed.), pp. 211–237, Plenum Press, New York (1975).
96. E. K. Orenberg, J. Renson, and J. D. Barchas, *Neurochem. Res.* **1**:659 (1976).
97. J. D. Redos, W. A. Hunt, and G. N. Catravas, *Life Sci.* **18**:989 (1976).
98. P. L. Hoffman and B. Tabakoff, *Nature (Lond.)* **268**:551 (1977).
99. M. C. Glick, C. A. Comstock, M. A. Cohen, and L. Warren, *Biochim. Biophys. Acta* **233**:247 (1971).
100. D. F. H. Wallach and E. H. Eylar, *Biochim. Biophys. Acta* **52**:594 (1961).
101. E. G. Lapetina, E. F. Soto, and E. DeRobertis, *Biochim. Biophys. Acta* **135**:33 (1967).
102. K. D. Bagshawe and G. A. Currie, *Nature (Lond.)* **218**:1254 (1968).
103. S. Oxley and W. O. Griffin, Jr., *Surg. Forum* **22**:113 (1971).
104. D. J. Triggle, *Prog. Cell Surf. Sci.* **5**:267 (1972).
105. A. Vaccari, R. Vertno, and A. Furlani, *Biochem. Pharamcol.* **20**:2603 (1971).
106. A. Haksar, S. Bamirkiewicz, and F. G. Peron, *Biochem. Biophys. Res. Commun.* **52**:956 (1973).
107. J. Vermyler, G. de Gaetano, M. B. Donati, and M. Verstraeti, *Br. J. Haematol.* **26**:645 (1974).
108. K. G. Gould, P. N. Srivastava, E. M. Cline, and W. L. Williams, *Contraception* **3**:261 (1971).
109. V. Stefanovic, P. Mandel, and A. Rosenberg, *J. Biol. Chem.* **251**:493 (1976).
110. C. L. Schengrund and A. Rosenberg, *J. Biol. Chem.* **245**:6196 (1970).
111. T. L. Mutchler, H. L. Cardenas, and D. H. Ross, *Fed. Proc.*, Abstr. **37**:318 (1978).
112. D. H. Ross, in: *Alcohol Intoxication and Withdrawal: Experimental Studies* (M. M. Gross, ed.), Vol. IIIa, pp. 459–471, Plenum Press, New York (1977).
113. E. P. Noble, P. J. Syapin, R. Vigran, and A. Rosenberg, *J. Neurochem.* **27**:217 (1976).
114. A. Y. Sun, *Ann. N.Y. Acad. Sci.* **273**:295 (1976).
115. P. Lesch, *Klin. Wochenschr.* **50**:663 (1972).
116. P. Lesch, E. Schmidt, and F. W. Schmidt, *Z. Klin. Chem. Klin. Biochem.* **11**:159 (1973).
117. J. M. Littleton and G. John, *J. Pharm. Pharmacol.* **29**:579 (1977).
118. R. H. Michell, D. Allan, and J. B. Finean, in: *Function and Metabolism of Phospholipids in CNS and PNS* (G. Porcellati, L. Amaducci, and C. Galli, eds.), pp. 3–13, Plenum Press, New York (1976).
119. Z. N. Gaut and C. G. Huggins, *Nature (Lond.)* **212**:612 (1966).
120. L. E. Hokin, *Ann. N.Y. Acad. Sci.* **165**:695 (1969).
121. T. W. Scott, S. C. Mills, and N. Frienkel, *Biochem. J.* **109**:325 (1968).
122. C. S. Lieber, *Lipids* **9**:103 (1974).
123. P. D. Jones and S. J. Wakil, *J. Biol. Chem.* **242**:5267 (1967).
124. L. Sartorelli, L. Galzigna, C. R. Rossi, and D. M. Gibson, *Biochem. Biophys. Res. Commun.* **26**:90 (1969).
125. G. Y. Sun, D. M. Creech, D. R. Corbin, and A. Y. Sun, *Res. Commun. Chem. Pathol. Pharmacol.* **16**:753 (1977).
126. G. Y. Sun, D. M. Creech, and A. Y. Sun, in: *Currents in Alcoholism* (M. Gallanger, ed.), Grune & Stratton, New York, in press (1979).

Ethanol and the Blood–Brain Barrier 7

Jane H. Chin

A. INTRODUCTION

The primary effect of ethanol is on the central nervous system (CNS). Like other centrally acting agents, ethanol eventually passes into the blood and must be effectively delivered to its site(s) of action in the brain. The exchange of material from the blood to the CNS is more complex than that taking place in other tissues of the body. At least two characteristic features of the brain contribute to these differences. First, the brain contains a special barrier system in its capillaries that selectively controls the type and amount of substances that can enter its environment from the bloodstream. In addition, the brain is surrounded both internally and externally by cerebrospinal fluid (CSF). The composition of this CSF is regulated by specific barriers located in the choroid plexus of the ventricles and in the arachnoid membranes lying over the fluid surrounding the outer surface of the CNS. A drug thus has the possibility of gaining direct access to the CNS through the capillary circulation or through the CSF.

In the last decade, with the use of electron microscopy and electron-dense tracers, much has been learned about the structural nature of this barrier. Recent studies clearly demonstrate that the blood–brain barrier (BBB) system can no longer be considered as a single entity that physically restricts substances from the brain only on the basis of their molecular size or lipid solubility. This barrier differs depending upon the brain site studied. It encompasses multiple transport systems, which in turn regulate which materials enter or leave the brain or CSF, and it is dynamically changing in order to maintain an ultrastable environment in the CNS. This barrier can thus function as a protective mechanism to exclude foreign material from the brain, can maintain homeostasis by preventing large changes in brain composition, and can even influence cerebral metabolism

---

Jane H. Chin • Department of Pharmacology, Stanford University School of Medicine, Stanford, California 94305.

through the supply and control of substrates and their metabolic products. It should be emphasized that the barrier from blood to brain is not absolute but only relatively restrictive. In addition, the barriers between blood and brain may be multiple. The term "brain barrier" used in this chapter refers to any factor or factors limiting the exchange of an agent along its transport route from blood to its site of action in the CNS.

This review will briefly summarize some of the recent structural features and functions of the BBB system and the effects of ethanol on this system. For greater details of the BBB system, the interested reader is referred to several symposia,[1-4] review articles,[5-8] and a book.[9]

B. STRUCTURAL BARRIERS

1. Blood to Brain

The cerebral capillaries serve as the major site of exchange of materials between blood and brain (see Fig. 1, I). Brain capillaries are surrounded by basement membranes (BM). 20–30 nm thick, and by a dense network of astrocytic processes known as glial feet (A). These glial sheaths around the capillaries were once thought to be the structural basis for the BBB.[11] However, this view has now been changed as a result of more recent electron microscope studies of ultrastructure that have used protein tracers such as horseradish peroxidase.[12] These studies clearly show that the barriers are located between the cells in the capillary lining (Fig. 1, E). The capillary endothelial cells overlapped and are joined to form tight junctions (TJ), which excluded protein markers with molecular weight as low as 1800 from both directions, either when they were in the capillary lumen (Fig. 1, open arrow on left) or when they reached the extracellular space from the ventricular system[12,13] (Fig. 1, dashed line). Developmentally, these tight junctions, which are networks of fibrillar strands, are present in the newborn.[14] In contrast to brain, particles up to 4 nm in diameter can readily pass from blood to cardiac and skeletal muscles since the capillaries in these tissues contain spaces between their endothelial cells known as gap junctions.[15] Large molecules can also be transported as pinocytotic vesicles through the cells in most general capillaries throughout the body.[15] However, transport by pinocytosis through the brain endothelial cell seldom occurs, except in special arterioles in cerebral cortex and scattered in the diencephalon and midbrain, where the BBB is circumvented.[16] Tight junctions between endothelial cells in brain capillaries form a continuous layer that serves as a major barrier to restrict large blood-borne, water-soluble molecules from entering the brain. This barrier also extends to the blood vessels supplying the spinal cord and its roots, the arachnoid space around the CNS,[9] and the endoneurium of the peripheral nerve.[17]

Protein markers that gain access to the extracellular space can enter most of the other structures intervening between capillary endothelial cells and the

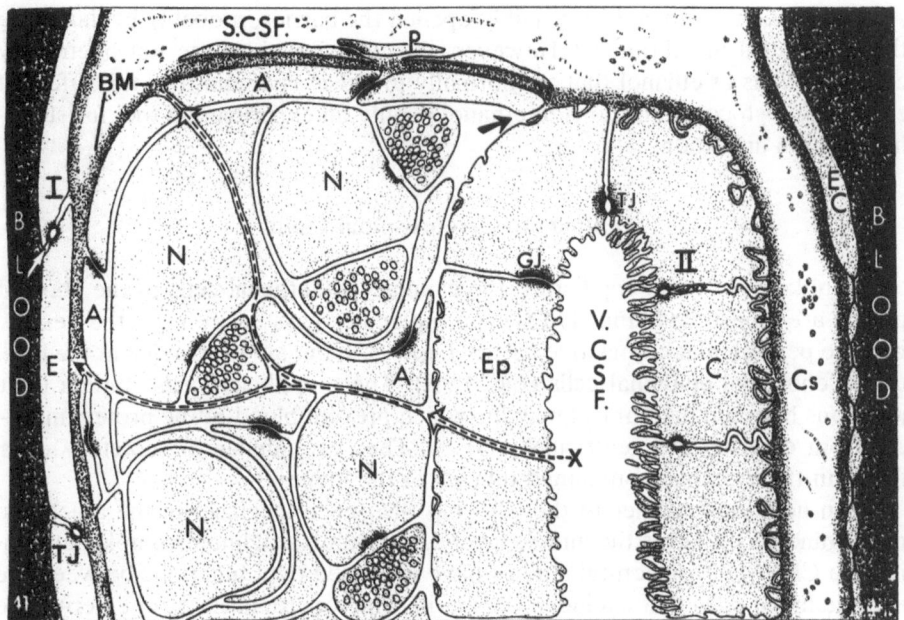

Fig. 1. Schema of blood, brain and cerebrospinal fluid relationships in the mouse. Tight junctions (TJ) between the endothelial cells (E) in the brain capillaries (I) and between the epithelial cells (C) in the choroid plexus (II) form barriers that prevent the protein marker peroxidase from entering brain from blood. In contrast, when peroxidase (X) is placed directly into the CSF of the ventricles (V.CSF.), it can freely move, as shown by the dashed lines, through gap junctions (GJ) between ependymal cells (Ep), traverse the extracellular space surrounding neurons (N) and glia or astrocytic processes (A), and reach the basement membrane (BM) of both the brain capillaries and the surface of the brain. There is also some free exchange across pial cells (P) between the CSF of the subarachnoid space (S.CSF.) and the surface of the brain. The thick closed arrow (top) indicates a "functional leak" whereby substances crossing the fenestrated endothelial cells (EC) of the choroid plexus blood vessels can pass along the choroidal stroma (Cs) to enter the parenchyma of the brain at the root of the choroid plexus. A "leak" in the opposite direction could also occur, as indicated by the arrow point. [Modified from Brightman and Reese,[10] Fig. 41; by permission of the authors and Rockefeller University Press.]

neuronal membranes (see Fig. 1, dashed lines), although some of the pathways may be circuitous. When placed into the ventricles, these tracers can move through the extracellular space,[10,18] through the basement membranes,[19] and through gap junctions that separate the adjacent astrocytic glial feet from the capillary walls.[10] Transport through plasma and subcellular membranes of glia or neurons is regulated as in capillary endothelial cells, to be discussed below.

In pathological conditions the BBB may be altered. Experimentally, hypertonic solutions, when applied directly to the brain surface for 10 min or when injected into the carotid arteries, have opened tight junction barriers.[20] These solutions are thought to widen the tight junctions after shrinking the endothelial cells so that blood-borne Evans blue albumin can move out of the pia vessels.[21]

A 40% solution of ethanol irreversibly opened the barrier, probably by damaging the endothelial membrane.[20] However, the tight junctions are not opened by concentrations of ethanol that are normally found *in vivo* and thus this effect of extremely high concentrations of ethanol is only an experimental tool for studying junction distortion.

2. Blood to Cerebrospinal Fluid

Unlike most brain capillaries, the capillary endothelial cells in the choroid plexus are fenestrated (Fig. 1, EC) and molecules pass easily from the vessels into the perivascular connective tissue (Fig. 1, Cs) but are obstructed at the level of the choroidal epithelial cells (Fig. 1, II). These epithelial cells contain tight junctions between their apical margins which prevent blood-borne protein markers from entering the ventricular CSF.[22] They are similar to secretory cells found in the proximal convoluted tubules of the kidney.

An additional barrier to the CSF has also been localized in the arachnoid membrane which forms the middle layer of the dural–arachnoid complex enclosing the CNS. Ultrastructural studies have identified tight junctions between the flat arachnoid cells that excluded the marker microperoxidase from entering the CSF when it is injected into the blood or from reaching the dural tissue when it is placed in the subarachnoid space.[23]

CSF is continuously formed as an active secretion, mainly from the choroid plexuses of the lateral, third, and fourth ventricles[5] (Fig. 2). About 30% can also be formed at extrachoroidal sites.[25,26] In mammals the CSF is replaced at a rate of about 0.4–0.6% of its total volume per minute.[5] CSF that is formed in the lateral ventricles passes caudally into the third ventricle and then into the fourth ventricle, which is connected to the cisterna magna and the central spinal canal, from which the fluid circulates into the cerebral and spinal subarachnoid spaces (Fig. 2). It is also constantly being removed at multiple sites. It returns to the venous blood by bulk flow from the subarachnoid space into the superior sagittal sinus in the dura (Fig. 2) or is absorbed in the cranial and spinal nerve root sheaths. In humans CSF formed at the choroid plexus requires about 4 to 6 hr before it returns to the venous blood by bulk flow through the arachnoid granulations or villi (Fig. 2). Thus, the concentration of drugs, metabolites, or endogenous substances in CSF may depend upon both the time and the site of sampling. For example, ethanol concentration in human lumbar CSF was lower during the first 80 min after ethanol consumption than in the cisterna magna or in the blood,[27] owing to the lag in the CSF flow to the lumbar area, rather than to an effect of ethanol on the barrier to the CSF.

3. Cerebrospinal Fluid to Brain

The basis for analyzing CSF fluid as an index of cerebral metabolism in humans depends upon free exchange of material between the brain and the CSF.

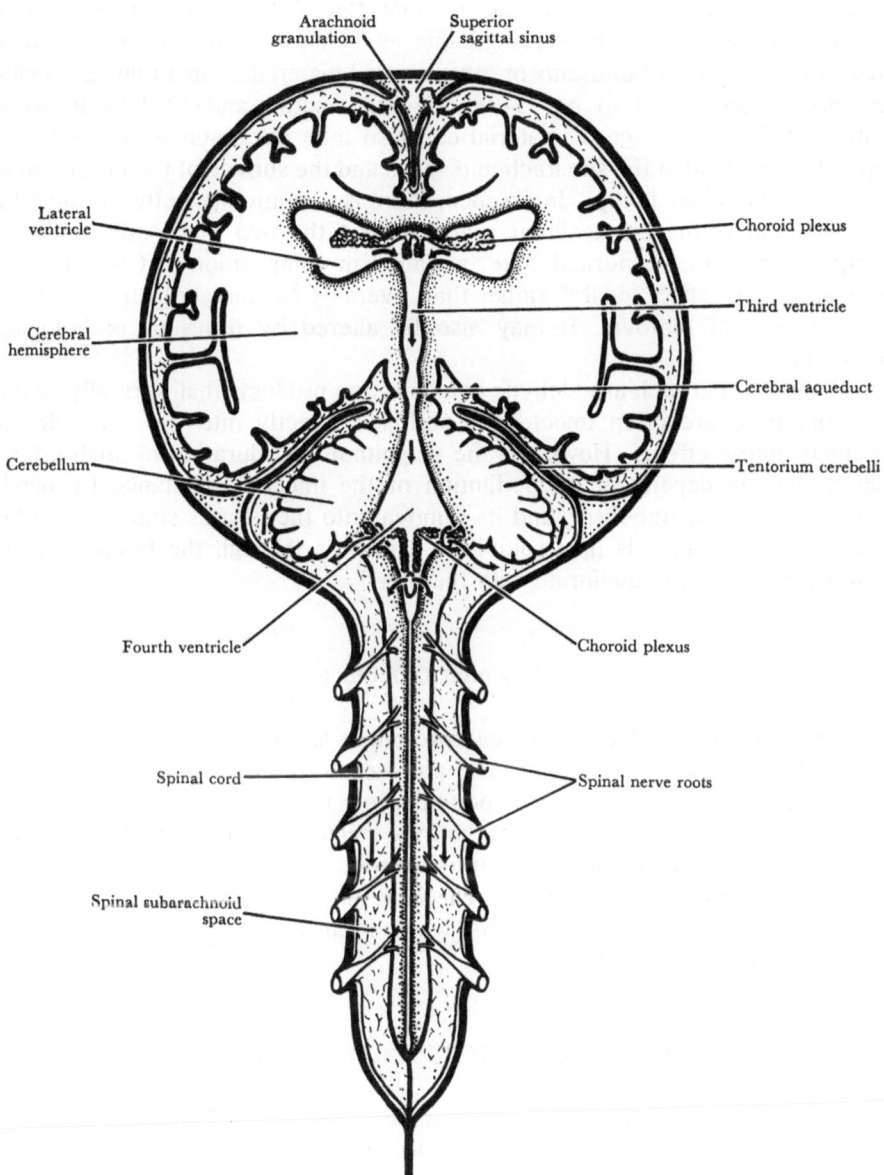

Fig. 2. Pathways of the cerebrospinal fluid circulation. [From Millen and Wollam,[24] Fig. 12; by permission of Oxford University Press.]

The ependymal cells lining the ventricles (Fig. 1, Ep) are freely permeable to large molecules that can pass from the extracellular space of the brain into the ventricles or vice versa.[28] In this manner the CSF can maintain a stable environment in the brain by serving either as a sink to remove excess metabolites[5] or as a source of nutrients or other needed materials. The brain extracellular space is estimated to be 15–18% in gray matter and 12–13% in white matter.[29] Free exchange of material can also take place across the pial cells separating the fluid in the subarachnoid space and the surface of the brain, but to a lesser extent (Fig. 1, top). In essence, there is a continuity between the CSF and the fluid in the extracellular space so that the two fluids are similar in composition.[30] In the normal state, variation in composition of CSF along its circulation can reflect local[31] rather than overall CNS metabolism, depending upon its site of removal. It may also be altered by drugs or pathological conditions.

Those agents such as aldehyde condensation products that normally cannot pass the BBB are often injected or perfused directly into the ventricles to produce central effects. However, the magnitude and duration of such effects will be largely dependent upon dilution of the injected substance by newly formed CSF, its metabolism, and its removal into the venous sinuses. In addition, if the substance is lipid-soluble, it will pass through the brain capillary endothelial cells and equilibrate with the blood.

4. Regional Differences

Tight junctions are not uniformly present in all brain capillaries. In certain regions the capillary endothelial cells are fenestrated and allow proteins and organic molecules to move from blood to nervous tissue. These areas include the area postrema, median eminence of the hypothalamus, line of attachment of the choroid plexus, pineal gland, preoptic region, supraoptic crest, subfornical organ, and intercolumnar tubercle. The lack of a BBB may be related to the specialized functions of these regions, which in turn may be influenced directly by the plasma composition.

C. FUNCTIONAL BARRIERS AND ETHANOL EFFECTS

The structural barriers that have been described above are mainly intercellular while the functional barriers to be discussed in this section are intra- or transcellular. Tight junctions join adjacent cells at each of the three anatomical barrier sites, in the brain capillary walls, in the choroid plexus, and in the arachnoid membrane, so that the ultimate exchange between plasma and the brain is highly selective and must take place through the cells themselves. Thus, under normal circumstances, most molecules entering the brain must first pass through the barrier cells, rather than between them, either by the process of passive diffusion or by some transport mechanism. After passing these external barriers of the nervous system into the extracellular space, the substance may again encounter other barriers that are present in the plasma membranes of neurons, glia cells, or nerve endings. Inside the cell the subcellular membranes may further limit free access. These multiple barriers must thus be traversed before a drug such as ethanol or a substrate arrives at its final destination or site of action. Similar barriers may also be present during the reverse process when a substrate returns to the blood from the cell. These functional barriers may thus operate in either direction, presumably as a homeostatic mechanism to maintain an ultrastable environment in the brain. They supply substrates or remove metabolites, depending upon the needs of the organism. They may allow simple exchange across the barrier cells with the same compound or structurally related agents (heteroexchange). The mechanisms controlling each of these functions need not be identical and may depend upon the experimental conditions and the population of cells being examined. For example, transport of amino acids across the capillary walls *in vivo* is sodium independent, but the *in vitro* uptake of amino acids by brain slices is sodium dependent.[32]

The following sections will examine the effects of ethanol on these transcellular transport mechanisms between blood and brain.

1. Diffusion

Diffusion is the passive movement of solutes from an area of higher to lower concentrations. As in other cells, the membranes of the barrier endothelial and epithelial cells contain mainly lipids and protein. The individual components that contribute to the lipids and proteins, however, may vary. For example, the lipid composition of brain capillary endothelial cells[33] has been shown to have higher sphingomyelin levels than neurons. Perhaps, variation in chemical composition of the membranes according to their type and regional distribution may relate to their function[34] and differential interaction with drugs.[35] To pass across the cerebral capillary endothelial cell, a solute from the plasma must first penetrate the membrane facing the capillary lumen, traverse the cell cytoplasm without being degraded enzymatically, and then pass through the outer membrane into the brain extracellular space. The probability of a substance entering these capillary cells or the corresponding epithelial cells at the other blood–CSF barrier sites is influenced by many factors. The degree of ionization of the substance at blood pH as well as the plasma concentration is important. Drugs in the ionic form will tend to remain in the plasma aqueous phase. Lipid-soluble substances such as bilirubin that bind strongly to plasma protein become polar and are restricted from crossing the barrier by both their size and polarity. In general, the un-ionized form of drugs normally penetrates cells in proportion to its lipid/aqueous partition coefficient. Lipid-soluble drugs and water[36] can diffuse passively from plasma across the capillary endothelial cells and equilibrate with the brain extracellular space.

It is generally agreed that ethanol freely diffuses across the BBB system and that its concentration in the brain depends upon the plasma concentration, the vascularization, the local blood flow, and the water content of the area. The high *in vivo* permeability of the cerebral capillary walls to ethanol has been differentiated from the slower rate of cell uptake by two independent tracer methods in anesthetized animals.[37,38] Ethanol was shown to be readily removed from the blood within the first 15 sec after simultaneous injection of a rapid intracarotid bolus of the test substance and a tracer. Crone[37] used an indicator dilution technique in which rapid serial samples of arteriovenous differences in dogs showed that ethanol, propanol, and butanol were at least 90% extracted from blood within the first 5 sec after injection, when compared to a nondiffusible dye. Oldendorf[38] compared the actual brain uptake of [^{14}C]ethanol to a highly permeable tracer, [^{3}H]water, that freely diffuses into the brain. Ethanol, which was found to have an olive oil/water partition coefficient of only 0.04,[38] was completely cleared from the blood of the rat during a single brain passage. The influence of the anesthetic agents used in these preparations have not been assessed.

Kalant[39] has reviewed the earlier data showing that ethanol at concentrations up to 468 mg/dl did not increase the cerebral capillary permeability to normally impermeant dyes in cats, rats, and guinea pigs. During intravenous infusions, ethanol equilibration time between blood and brain was measured to be 30–120 sec by the arteriovenous method[40] and 30–60 sec by the brain uptake measurements.[41] High concentrations of ethanol (1 M), *n*-butanol (125 mM), and *t*-butanol (500 mM) also crossed the barriers in the nerve cords of insects and reduced the amplitude of electrophysiological action potentials.[42]

Since ethanol is highly soluble in water, it is distributed more to gray matter, which has a higher water content than white matter of brain.[43–45] Differential rates of distribution may also depend upon the blood supply to the area.[44] After injection of [1-^{14}C]ethanol (0.5 g/kg, iv) into monkeys during the last 30 days of pregnancy, radioactivity in the brain of the fetus was greater in cortex than in white matter, with high concentrations localized in visual cortex, cerebellum, hippocampus, putamen, caudate, pons nuclei, and trigeminal and facial nerves.[45] This study showed that ethanol readily diffused from maternal blood through the placenta and entered the brain as well as other organs of the fetus. The relationship of prolonged ethanol exposure of different parts of the fetal brain during chronic maternal alcoholism to the development of the fetal alcohol syndrome,[46] however, is not yet known. In freely moving rats implanted with an intracerebral cannula that perfused local brain areas, a low dose of ethanol (1 g/kg) was reported to produce higher brain concentrations of ethanol in the reticular formation, cerebral cortex, and amygdala than in the caudate and lateral hypothalamus within the first 15 min after intraperitoneal injection.[47] Since these findings are somewhat unexpected, further studies are required to substantiate these results.

The rapid equilibration of ethanol between blood and brain indicates that ethanol passes the external barrier cells surrounding the brain and readily penetrates the plasma and subcellular membranes of neurons. These neuronal membranes, which are also included as part of the brain barrier, do not limit, but freely allow ethanol to diffuse into their bilayers. This nonspecific ethanol action of dissolving in the membrane may be sufficient to change the physical properties of the membrane in such a way as to disrupt normal functioning of some cells. In support of such a hypothesis are our ethanol results from biophysical experiments.[34] Since an optimal range of membrane fluidity (i.e., molecular motion within the bilayer) appears to be important for many biological functions,[48,49] we have measured the fluidity of mouse erythrocyte and brain membranes by a sensitive electron paramagnetic resonance (EPR) technique.[34] These membranes were spin-labeled with a fatty acid spin-label probe, 5-doxylstearic acid. The EPR spectrum is affected by the motion of the probe, which in turn reflects the fluidity of its environment. From the EPR spectrum we have calculated an order parameter, S,[50] a commonly used index of membrane fluidity. In model membranes, the order parameter can vary between the limits of 1 (completely ordered as in a solid) and 0 (completely fluid). In biological membranes, the order parameter indicate intermediate fluidity. The *in vitro* addition of low concentrations of ethanol (100–200 mg/dl) increased the fluidity (decreased the order parameter) of erythrocyte and synaptosomal plasma membranes and the subcellular mitochondrial membranes but had no effect on myelin. These differential ethanol effects may reflect differences in the chemical composition of the various brain membranes. The fluidizing effect of ethanol was dose-related up to 1600 mg/dl in all the membranes except myelin. The small decrease in order parameter caused by ethanol at 400 mg/dl in these experiments agree with the changes in fluidity predicted for the same concentration of anesthetic agents in phospholipid model membranes.[51] These results suggest that nonlethal concentrations of ethanol may increase the fluidity of specific membranes *in vivo* to disturb normal lipid protein interactions by the same mechanisms proposed for the anesthetic agents.[52,53]

Tolerance also develops to this *in vitro* membrane fluidizing effect of ethanol.[54] DBA/2J male mice were maintained for 8 days on a liquid diet (chocolate-flavored Slender, Carnation) to which ethanol was added to provide 33% of the calories. Membranes were prepared from animals whose effective alcohol intake pattern met previously defined criteria for the development of physical dependence[55] as scored by convulsions on handling. Ethanol added *in vitro* was consistently less effective in increasing the fluidity of synaptosomal and erythrocyte membranes from the chronic ethanol-treated mice than their pair-fed sucrose controls. By definition, this is tolerance. The results from these experiments suggest that the animals had adapted to the chronic presence of ethanol *in vivo* by changing their membrane composition to offset the fluidizing effect of ethanol so that the *in vitro* addition of the drug produced less of an effect. Such regulation of membrane fluidity in response to continuous exposure to ethanol may be a novel form of drug tolerance in mammalian cells.

There is still considerable controversy as to whether metabolites of ethanol, in particular, acetaldehyde, reach the brain during normal ethanol oxidation. Experimental variables such as species, strain, sex, ethanol dose, time of measurements, diet, or method of analysis may account for some of the reported differences. Acetaldehyde has been reported in the brain following acute doses of ethanol in rats,[56] guinea pigs,[57] and mice.[58] In contrast, when corrections are made for blood contamination in brain tissue and when nonenzymatic formation of acetaldehyde was inhibited, acetaldehyde was not detected in the brain following a single dose of ethanol (3 g/kg, ip) until the cerebral blood acetaldehyde level exceeded 250 nmole/ml.[59,60] An aldehyde dehydrogenase in the cerebral capillary wall has been postulated to degrade acetaldehyde before it enters the brain.[60] However, the presence of acetaldehyde in the CSF[61] argues against this proposal.

2. Carrier-Mediated Transport

Carrier-mediated transport is a downhill process from high to low concentrations and does not require energy. A carrier molecule is postulated to have specific sites for many of the polar metabolic substrates which are then moved in the complexed form back and forth across the plasma membrane. The kinetics of the carrier complex transport are the same as that of the enzyme–substrate complex. Most carrier sites are saturable, stereospecific, competitively inhibited, and may be bidirectional. Eight independent carrier-mediated transport systems have been identified across the cerebral capillaries, mainly by the Oldendorf brain uptake method. Separate carrier systems transport monosaccharides,[62,63] neutral amino acids,[64] basic amino acids,[64] acidic amino acids,[65] monocarboxylic acids,[66] nucleosides,[67] purines,[67] and choline.[68] When the carrier sites are saturated, transport into the brain becomes independent of plasma concentration. The sites have steric specificity for the D form of glucose and normally the L form of amino acids. Competitive inhibition among substrates with close chemical and steric similarities that share the same transport system may also limit the availability of essential precursors needed by the brain. This feature is important in regulating amino acid transport but not glucose transport. For example, tryptophan, the serotonin precursor, shares the same transport system as other neutral amino acids. Its effective transport into the brain for synthesis of serotonin may thus be dependent upon not only its own plasma concentration but also the dietary intake of other amino acids that compete for the carrier molecule.[69]

The neutral essential amino acids and precursors for amine neurotransmitters are rapidly transported across the capillary walls in vivo and in vitro in both neonates[32] and adults,[63] and this transport is sodium independent.[32] In contrast, the uptake of both the nonessential amino acids that can be readily formed from glucose (glutamate, glycine, GABA, taurine, and proline)[63] and the amine transmitters[63] is extremely slow or negligible across the capillary walls but is rapid in brain slices.[70] The carrier-mediated transport systems for neuronal tissue may have quite different kinetics and energy requirements from those in the capillary wall.[71] The lack of carrier systems in the BBB for the polar neurotransmitters themselves probably serves as a means of conserving and localizing these agents in the brain.[63]

α-Aminoisobutyric acid, an unnatural amino acid, has been useful in studying BBB functions, since it is considered to be transported from blood to brain like other amino acids but is not metabolized.[72] The transport of this amino acid into the brain was decreased in mice by blood ethanol concentrations of 100 mg/dl,[73] but the inhibitory effect was found to be secondary to the hypothermia produced by acute administration of ethanol.[74] No significant change was found in the brain/plasma ratio of α-aminoisobutyric acid 2–3 weeks following 4 months of chronic treatment on a liquid diet containing ethanol as 35% of the total caloric intake.[73] Butanol (100 mg/dl) or very high concentrations of ethanol (1000 mg or 4000 mg/dl) *in vitro* had no effect on the exit of this same amino acid when it was preloaded in cerebral cortex slices.[75] Ethanol (500–4000 mg/dl) was reported to decrease the efflux of the dopamine precursor, dihydroxyphenylalanine, so that there was net accumulation of the amino acid in cerebral cortical slices.[76] The hypothermic effect of acute ethanol administration (5 g/kg) may also be involved in the decreased brain uptake in the rat of δ-aminolaevulinic acid, a porphyrin precursor.[77]

Transport of the short-chain monocarboxylic acids which are more than 99% ionized at body pH has been shown to be carrier-mediated, saturable, and stereospecific, at least for L-lactate.[66] Lactate, pyruvate, acetate, propionate, butyrate,[66] and the ketone body, 3-hydroxybutyrate,[78] share the same carrier system. It has been postulated that this carrier system functions as a countertransport system to regulate the influx and efflux of ionized monocarboxylic acids across the blood–brain barrier.[78] Fasting[79] and a high-fat diet[80] increase the capacity of this system to transport ketone bodies.

When [1-^{14}C]acetate was injected into the carotid artery of anesthetized rats and the brain was removed 15 sec later, radioactivity was already incorporated into cholesterol, palmitic acid, and the polar lipids.[81] These results indicate that *de novo* synthesis of complex lipids from acetate that may be derived from oxidation of ethanol[82] can be extremely rapid in the brain. At low concentrations, acetate is transported by the monocarboxylic acid carrier system, but at higher concentrations self-inhibition occurs and the nonionized form of acetate probably diffuses across the blood–brain barrier like the more lipid-soluble longer-chain fatty acids. [^{14}C]Palmitate is also rapidly incorporated into rat brain total lipids by intracarotid injection, and its distribution into the individual lipid components was similar to that of the slower labeling by intraperitoneal injection.[83] In mice, chronic ethanol feeding on a liquid diet for 4 weeks did not alter [1-^{14}C]palmitate incorporation into brain total lipids or cholesterol.[84]

3. Active Transport

Active transport implies movement uphill from a low to a high concentration and requires the expenditure of energy. It may be stereospecific, saturable, and competitively inhibited like mediated transport. The active transport of cations, inorganic anions, and organic acids from CSF has been demonstrated. Evidence is now accumulating that substances once thought to be removed from the CSF only by the choroid plexus can also be cleared by brain tissue and cerebral capillaries.[85,86] The relative importance of these two mechanisms in CSF homeostasis is yet to be resolved. The maintenance of ionic gradients between plasma and brain extracellular space or CSF implies an active process, at least in part. The concentration of K^+ in brain extracellular space, thought to be controlled by glial cells,[87] has been determined by selective ion-exchange electrodes to be approximately 3 mM, a value similar to CSF but lower than that for plasma.[30] The choroid plexus secretion of approximately 3 mM K^+ appears to involve a saturable mediated carrier.[88] Potassium is actively removed from the CSF, and this clearance was blocked by the ATPase inhibitor, ouabain.[89] In contrast, the reduction of CSF secretion by ouabain is variable,[30] although very high concentrations of acetazolamide that block 99.5% of the carbonic anhydrase activity in the choroidal epithelial cells[30] reduced both net sodium transport and CSF secretion.[90] The effects of ethanol on Na^+,K^+-ATPase and active transport are discussed in detail in Chapters 4 and 5 of this volume.

Clinically, the concentrations of metabolites of dopamine and serotonin in the spinal fluid have been of great interest to monitor the level of brain amines. The same transport system that actively removes the organic acid penicillin from the CSF also clears the dopamine metabolite homovanillic acid (HVA) and the serotonin metabolite 5-hydroxyindoleacetic acid (5-HIAA). In cats the clearance of 5-HIAA, which was probenecid-sensitive, was much greater from the subarachnoid space than from the ventricles.[86]

The dopamine metabolites, HVA and 3,4-dihydroxyphenylacetic acid (DOPAC), were increased in brains of rats intoxicated with ethanol either by acute (5 g/kg, ig) or chronic treatment (9–15 g/kg per day for 4 days).[91] Ethanol at blood concentrations of 150–250 mg/dl did not block the transport of these metabolites across the BBB, since their elimination rate from the brain was the same in both control and acute ethanol-treated animals that were injected with pargyline (50 mg/kg, ip) to inhibit monoamine oxidase.

The effects of a single dose of ethanol on the brain concentration of the serotonin metabolite, 5-HIAA, are conflicting, possibly due to time of sampling, strain, or species differences. A decrease in rats[92] to a transient late increase in C57B1/6J male mice[93] has been reported. In the latter study, elevated brain levels of 5-HIAA[93] were not associated with peak ethanol brain concentrations or with hypothermia,[74] which itself has been shown to increase 5-HIAA during pentobarbital anesthesia.[94] In contrast to the lack of effect of ethanol on the transport of dopamine metabolites across the BBB in rats,[91] it has been proposed that ethanol inhibits the transport of this serotonin metabolite out of brain in mice[95] and out of CSF in cats.[96] Animals were first given ethanol *in vivo* and then anesthetized so that [^{14}C]-5-HIAA could be injected into the lateral ventricle[95] or in the artificial CSF perfusing the spinal subarachnoid space.[96] Maximal decreases of 5-HIAA transport from brain or CSF was observed at ethanol concentrations above 460 mg/dl (100 mM).[95,96] The active uptake of labeled 5-HIAA by mouse isolated choroid plexus was also decreased by the *in vitro* addition of 50–400 mg/dl of ethanol to the suspending medium.[95]

Chronic ethanol treatment of C57 mice by inhalation also elevated brain 5-HIAA at 10 min up to 20 hr during the withdrawal period,[93] but the increase was not correlated to the severity of the withdrawal reactions as measured by pentylenetratrazol-induced seizures. In rats maintained on alcohol for 18 days the brain level of 5-HIAA was also increased during ethanol or t-butanol intoxication and withdrawal.[97]

The level of 5-HIAA in lumbar CSF has been determined in human subjects who had a history of 3–32 years of alcoholism but had been off of alcohol for at least 2 weeks prior to the study.[98,99] After chronic alcohol treatment for 7 days (1 ml/kg four times a day) the level of 5-HIAA was increased on the twelfth day after withdrawal, with the largest effects in 5 of 8 subjects with low basal 5-HIAA values.[98] After a single dose of ethanol (3 g/kg divided into five doses), the 5-HIAA concentration was decreased 29–58% of baseline values in 7 of 11 subjects.[99] Cyclic AMP was concomittantly decreased 27 to 92% in 10 of 11 patients.

The significance of these changes in 5-HIAA in intoxication, physical dependence, hangover, withdrawal, or other effects of ethanol is not known.

4. Regulatory Enzymes

Some amino acids and amines may be excluded from the brain, since they are enzymatically degraded upon passage through the capillary endothelial cells. The enzymes monoamine oxidase (MAO), DOPA decarboxylase, GABA-α-ketoglutarate transaminase, and succinic semialdehyde dehydrogenase have been identified in the endothelial cells. In general, the amine precursors readily enter the capillary endothelial cells through the appropriate carrier-mediated transport system. For example, when the precursors L-3,4-dihydroxyphenylala-nine (L-dopa) or L-5-hydroxytryptophan are injected into mice or rats, they are decarboxylated by DOPA decarboxylase (aromatic-L-amino-acid decarboxylase) in the cerebral capillary endothelia to dopamine or serotonin. The amine is subsequently degraded by MAO, which is also present in these cells.[100,101] Thus, the amines do not normally enter the brain. In the presence of MAO blocking agent or high concentrations of precursor, the amine or amino acid can saturate the degradation or trapping system in the capillary cell and enter the brain. Ethanol (2 g/kg, iv) elevated brain catecholamine uptake in the neonate chick when it was injected simultaneously with large doses of exogenous norepinephrine (5 g/kg, iv) or epinephrine (5 g/kg, iv).[102] Ethanol may have facilitated brain entry of amines by acting as a solvent. Since ethanol has no effect on monoamine oxidase activity,[103] the mechanism for amine accumulation by ethanol remains to be clarified.

The low rate of transport of GABA by the neutral amino acid carrier system into the brain is also attributed to enzyme degradation in the capillary walls. GABA-α-ketoglutarate transaminase and succinic semialdehyde dehydrogenase in the endothelial cells convert GABA to succinic acid and thus limit the entry of this amino acid into the brain.[104]

Membrane bound γ-glutamyl transpeptidase has been found in choroid plexus, cerebral capillaries, and neuron and glia in the brain.[105,106] This enzyme is thought to couple amino acid transport to glutathione metabolism,[107] but its role in transporting amino acid at the BBB is yet to be established. Ethanol (100–300 mg/dl) *in vitro* increased the activity of a high-K_m isoenzyme of γ-glutamyl transpeptidase prepared from whole rat brain.[108] Similar increased enzymatic activity was observed in rats treated with pyrazole and ethanol[109] by the inhalation method. Basal enzyme levels in all brain areas were higher in long-sleep mice than the short-sleep mice,[110] two strains of mice with different sensitivity to the intoxicating effects of ethanol.

5. Interaction with Other Drugs

It has been proposed that ethanol, when combined with certain substances, may act as a solvent.[111] It may facilitate penetration through the BBB of aromatic monoamines[112] or acetylcholine derivatives[113] that do not normally cross the BBB of adult animals or it may enhance their brain uptake in neonates[102] that have incomplete BBB. Competitive inhibition of monoamine metabolism at the aldehyde step[114] or the enzymatic barriers already discussed may also be involved. In contrast, the ethanol solvent effects may be more peripheral. The higher concentration of [¹⁴C]phenobarbital in rat blood and brain when combined with 15% ethanol (3 g/kg) was attributed to the faster absorption across the peritoneum rather than an effect on the BBB.[115]

Pretreatment of rats with ethanol (0.6–7.2 g/kg orally in drinking water) shortened the duration of tremors induced by harmine and also lowered the concentrations of the alkaloid in the brain.[116] These effects have been attributed to an ethanol-induced decreased permeability of the BBB to harmine, but a hypothermic effect of ethanol[74] has not been ruled out.

D. SUMMARY AND CONCLUSIONS

The blood–brain barrier is considered not as a single entity, but as an elaborate, dynamic system that maintains a constant environment in the central nervous system. This system consists of both structural barriers located between the cells and functional barriers within the cells. Morphologically, the barriers consist of tight junctions between cells at three specific sites: (1) between endothelia of capillaries to the brain, spinal cord, and arachnoid space; (2) between epithelia of the choroid plexus in the ventricles; and (3) between cells in the arachnoid membranes covering the subarachnoid fluid bathing the CNS. These barriers also extend to the peripheral nerve, where tight junctions are found in the blood vessels to the endoneurium and between the cells of the perineurium surrounding nerve bundles. Specialized areas of the CNS lack this BBB system.

Since most substances cannot penetrate the structural barriers between the cells, they must bypass this route and traverse the barrier cells, either by the process of passive diffusion or by some transport mechanism, without being degraded by regulatory enzymes located in these cells. Brain substrates such as glucose and amino acids are transported from blood to brain by carrier-mediated systems that are stereospecific, saturable, and competitively inhibited. Similar transport systems for influx and efflux, but with different kinetic properties and energy requirements, have also been shown at additional barrier sites in neurons and glia. Diet may influence brain uptake of those substrates that share a common carrier transport. CSF is actively secreted, mainly by the choroid plexus. CSF may act as a "sink" along its circulation to remove brain metabolites by bulk flow. In addition, inorganic halides and organic acids are actively cleared from CSF across the choroid plexus and from brain across the capillary walls. The ionic environment in all cells, extracellular space, and the CSF is maintained, at least in part, by active transport mechanisms.

Ethanol, like other lipid-soluble agents, rapidly diffuses and equilibrates across the BBB cells. Its brain concentration depends upon the plasma concentration, vascularization, local blood flow, and water content of the area. The tight junctions between cells at the structural barriers are not disrupted at concentrations of ethanol found *in vivo*.

Very few experiments deal specifically with the effects of ethanol on the transport systems of the BBB, with the exception of ATPase, which is reviewed in another chapter. The decrease in brain amino acid transport observed after ethanol has been attributed to its hypothermic effect after acute administration. Ethanol *in vitro* had no effect or decreased the amino acid efflux from brain slices at high concentrations. Brain 5-HIAA, the serotonin metabolite, was reported to be increased by acute or chronic administration of ethanol, which may inhibit the active clearance of this organic acid from the CNS. The passage of drugs or their metabolites across the BBB may also be altered by ethanol. Ethanol may disrupt function by affecting membrane fluidity in some of the brain barriers.

A knowledge of the actions of ethanol on the membranes of the BBB system may lead to a better understanding of the mechanism of intoxication, tolerance, and physical dependence on this agent.

ACKNOWLEDGMENTS. Supported by USPHS Grant AA01066 to D. B. Goldstein, whom I thank for helpful suggestions during her critical reading of this review.

E. REFERENCES

1. A. Lajtha and D. Ford, *Brain Barrier Systems,* Elsevier Publishing Company, New York (1968).
2. H. F. Cserr, J. D. Fenstermacher, and V. Fencl, *Fluid Environment of the Brain,* Academic Press, Inc., New York (1975).
3. G. Levi, L. Battistin, and A. Lajtha, *Transport Phenomena in the Nervous System—Physiological and Pathological Aspects,* Plenum Press, New York (1976).

4. L. Z. Bito, H. Davson, and J. D. Fenstermacher, *The Ocular and Cerebrospinal Fluids*, Academic Press, Inc., New York (1977).
5. H. Davson, in: *The Structure and Function of Nervous Tissue* (G. H. Bourne, ed.), pp. 321–445, Academic Press, Inc., New York (1972).
6. J. S. Dunn and G. M. Wyburn, *Scott. Med. J.* **17**:21 (1972).
7. D. H. Ford, in: *Reviews of Neuroscience* (S. Ehrenpreis and I. J. Kopin, eds.), Vol. 2, pp. 1–42, Raven Press, New York (1976).
8. W. M. Pardridge and W. H. Oldendorf, *J. Neurochem.* **28**:5 (1977).
9. S. I. Rapoport, *Blood–Brain Barrier in Physiology and Medicine*, Raven Press, New York (1976).
10. M. W. Brightman and T. S. Reese, *J. Cell Biol.* **40**:648 (1969).
11. E. A. Maynard, R. L. Schultz, and D. C. Pease, *Am. J. Anat.* **100**:409 (1957).
12. T. S. Reese and M. J. Karnovsky, *J. Cell Biol.* **34**:207 (1967).
13. T. S. Reese, N. Feder, and M. W. Brightman, *J. Neuropathol. Exp. Neurol.* **30**:137 (1971).
14. G. D. Pappas and V. M. Tennyson, *J. Cell Biol.* **15**:227 (1962).
15. M. J. Karnovsky, *J. Cell Biol.* **35**:213 (1967).
16. E. Westergaard and M. W. Brightman, *J. Comp. Neurol.* **152**:17 (1973).
17. Y. Olsson and T. S. Reese, *J. Neuropathol. Exp. Neurol.* **30**:105 (1971).
18. N. H. Becker, A. Hirano, and H. M. Zimmerman, *J. Neuropathol. Exp. Neurol.* **27**:439 (1968).
19. M. W. Brightman, *Am. J. Anat.* **117**:193 (1965).
20. S. I. Rapoport, *Am. J. Physiol.* **219**:270 (1970).
21. S. I. Rapoport, M. Hori, and I. Klatzo, *Am. J. Physiol.* **223**:323 (1972).
22. M. W. Brightman, in: *Brain Barrier Systems* (A. Lajtha and D. Ford, eds.), pp. 19–40, Elsevier Publishing Company, New York (1968).
23. S. Nabeshima, *Anat. Rec.* **169**:384 (1971).
24. J. W. Millen and D. H. M. Woollam, *The Anatomy of the Cerebrospinal Fluid*, Oxford University Press, Oxford (1962).
25. M. Pollay and F. Curl, *Am. J. Physiol.* **213**:1031 (1967).
26. T. H. Milhorat, M. K. Hammock, J. D. Fenstermacher, D. P. Rall, and V. A. Levin, *Science* **173**:330 (1971).
27. H. G. Mehrtens and H. W. Newman, *AMA Arch. Neurol. Psychiatry* **30**:1092 (1933).
28. D. P. Rall, W. W. Oppelt, and C. S. Patlak, *Life Sci.* **1**:43 (1962).
29. J. D. Fenstermacher and C. S. Patlak, in: *Fluid Environment of the Brain* (H. F. Cserr, J. D. Fenstermacher, and V. Fencl, eds.), pp. 201–214, Academic Press, Inc., New York (1975).
30. R. Katzman, in: *The Nervous System* (D. B. Tower, ed.), Vol. 1, pp. 291–297, Ravens, New York (1975).
31. J. A. Kessler, C. S. Patlak, and J. D. Fenstermacher, *Brain Res.* **116**:471 (1976).
32. H. Sershen and A. Lajtha, *Exp. Neurol.* **53**:465 (1976).
33. A. N. Siakotos, G. Rouser, and S. Fleischer, *Lipids* **4**:234 (1969).
34. J. H. Chin and D. B. Goldstein, *Mol. Pharmacol.* **13**:435 (1977).
35. K. W. Miller and K. Y. Pang, *Nature (Lond.)* **263**:253 (1976).
36. E. A. Bering, Jr., *J. Neurosurg.* **9**:275 (1952).
37. C. Crone, *Acta Physiol. Scand.* **64**:407 (1965).
38. W. H. Oldendorf, *Proc. Soc. Exp. Biol. Med.* **147**:813 (1974).
39. H. Kalant, in: *The Biology of Alcoholism* (B. Kissin and H. Begleiter, eds.), Vol. 1: *Biochemistry*, pp. 1–62, Plenum Press, New York (1971).
40. E. Fischer and H. Wallgren, *Physiologist* **1**:27 (1957).
41. H. R. Hulpieu and V. V. Cole, *Q. J. Stud. Alcohol* **7**:89 (1946).
42. M. V. Thomas, *Experientia* **31**:1194 (1975).
43. E. Kulonen and O. Forsander, *Arch. Int. Pharmacodyn. Ther.* **123**:1 (1959).
44. M. Scherrer-Etienne and J. M. Posternak, *Schweiz. Med. Wochenschr.* **93**:33 1016 (1963).
45. B. T. Ho, G. E. Fritchie, J. E. Idanpaan-Heikkila, and W. M. McIsaac, *Q. J. Stud. Alcohol* **33**:485 (1972).
46. C. N. Ulleland, *Ann. N.Y. Acad. Sci.* **197**:167 (1972).

47. C. K. Erickson, *Life Sci.* **19**:1439 (1976).
48. D. Chapman, *Q. Rev. Biophys.* **8**:185 (1975).
49. D. L. Melchior and J. M. Stein, in: *Annual Review of Biophysics and Bioengineering* (L. J. Mullins, W. A. Hagins, L. Stryer, and C. Newton, eds.), pp. 205–238, Annual Reviews Inc., Palo Alto, Calif. (1976).
50. W. L. Hubbell and H. M. McConnell, *J. Am. Chem. Soc.* **93**:314 (1971).
51. J. R. Trudell, W. L. Hubbell, and E. N. Cohen, *Biochim. Biophys. Acta* **291**:321(1973).
52. J. R. Trudell, D. G. Payan, J. H. Chin, and E. N. Cohen, *Proc. Natl. Acad. Sci. USA* **72**:210 (1975).
53. J. R. Trudell, *Anesthesiology* **46**:5 (1977).
54. J. H. Chin and D. B. Goldstein, *Science* **196**:684 (1977).
55. D. B. Goldstein and V. W. Arnold, *J. Pharmacol. Exp. Ther.* **199**:408 (1976).
56. K. H. Kiessling, *Exp. Cell Res.* **27**:367 (1962).
57. S. Tsukamoto, *Jpn. J. Stud. Alcohol* **7**:168 (1972).
58. J. M. Littleton, P. J. Griffiths, and A. Ortiz, *J. Pharm. Pharmacol.* **26**:81 (1974).
59. H. W. Sippel, *J. Neurochem.* **23**:451 (1974).
60. C. J. P. Eriksson and H. W. Sippel, *Biochem. Pharmacol.* **26**:241 (1977).
61. H. Pettersson and K. H. Kiessling, *Biochem. Pharmacol.* **26**:237 (1977).
62. C. Crone, *J. Physiol. (Lond.)* **181**:103 (1965).
63. W. H. Oldendorf, *Am. J. Physiol.* **221**:1629 (1971).
64. J. J. Richter and A. Wainer, *J. Neurochem.* **18**:613 (1971).
65. W. H. Oldendorf and J. Szabo, *Am. J. Physiol.* **230**:94 (1976).
66. W. H. Oldendorf, *Am. J. Physiol.* **224**:1450 (1973).
67. E. M. Cornford and W. H. Oldendorf, *Biochim. Biophys. Acta* **394**:211 (1975).
68. W. H. Oldendorf and L. D. Braun, *Brain Res.* **113**:219 (1976).
69. R. J. Wurtman and J. D. Fernstrom, *Biochem. Pharmacol.* **25**:1691 (1976).
70. A. Lajtha and J. Toth, *J. Neurochem.* **10**:909 (1963).
71. A. Lajtha and M. Banay-Schwartz, in: *Transport Phenomena in the Nervous System— Physiological and Pathological Aspects* (G. Levi, L. Battistin, and A. Lajtha, eds.), pp. 415–434, Plenum Press, New York (1976).
72. R. Kuttner, J. A. Sims, and M. W. Gordon, *J. Neurochem.* **6**:311 (1961).
73. G. Freund, *Brain Res.* **46**:363 (1972).
74. G. Freund, *Life Sci.* **13**:345 (1973).
75. A. Cherayil, J. Kandera, and A. Lajtha, *J. Neurochem.* **14**:105 (1967).
76. K. Kaniike and H. Yoshida, *Jpn. J. Pharmacol.* **13**:292 (1963).
77. F. B. McGillion, G. G. Thompson, M. R. Moore, and A. Goldberg, *Biochem. Pharmacol.* **23**:472 (1974).
78. J. E. Cremer, L. D. Braun, and W. H. Oldendorf, *Biochim. Biophys. Acta* **448**:633 (1976).
79. A. Gjedde and C. Crone, *Am. J. Physiol.* **229**:1165 (1975).
80. T. J. Moore, A. P. Lione, M. C. Sugden, and D. M. Regen, *Am. J. Physiol.* **230**:619 (1976).
81. G. A. Dhopeshwarkar, C. Subramanian, and J. F. Mead, *Biochim. Biophys. Acta* **248**:41 (1971).
82. H. Casier, *Q. J. Stud. Alcohol* **23**:529 (1962).
83. G. A. Dhopeshwarkar and J. F. Mead, *Adv. Lipid Res.* **11**:109 (1973).
84. A. K. Rawat, *Res. Comm. Chem. Pathol. Pharmacol.* **8**:461 (1974).
85. H. Davson and J. R. Hollingsworth, *J. Physiol. (Lond.)* **233**:327 (1973).
86. L. I. Wolfson, R. Katzman, and A. Escriva, *Neurology* **24**:772 (1974).
87. R. K. Orkand, J. G. Nicholls, and S. W. Kuffler, *J. Neurophysiol.* **29**:788 (1966).
88. A. Ames, III, K. Higashi, and F. B. Nesbett, *J. Physiol. (Lond.)* **181**:506 (1965).
89. M. W. B. Bradbury and B. Stulcova, *J. Physiol. (Lond.)* **208**:415 (1970).
90. A. Ames, III, K. Higashi, and F. B. Nesbett, *J. Physiol (Lond.)* **181**:516 (1965).
91. F. Karoum, R. J. Wyatt, and E. Majchrowicz, *Br. J. Pharmacol.* **56**:403 (1976).
92. D. J. Palaic, J. Desaty, J. M. Albert, and J. C. Panisset, *Brain Res.* **25**:381 (1971).
93. B. Tabakoff and W. O. Boggan, *J. Neurochem.* **22**:759 (1974).
94. L. Isaac, *Nature (London) New Biol.* **243**:269 (1973).

95. B. Tabakoff, R. F. Ritzmann, and W. O. Boggans, *J. Neurochem.* **24**:1043 (1975).
96. B. Tabakoff, M. Bulat, and R. A. Anderson, *Nature (Lond.)* **254**:708 (1975).
97. H. Wallgren, A. L. Kosunen, and L. Ahtee, *Isr. J. Med. Sci.* Suppl. **9**:63 (1973).
98. V. Zarcone, J. Barchas, E. Hoddes, J. Montplaisir, R. Sack, and R. Wilson, in: *Alcohol Intoxication and Withdrawal: Experimental Studies* (M. M. Gross, ed.), Vol. II, pp. 431–451, Plenum Press, New York (1975).
99. E. K. Orenberg, V. P. Zarcone, J. F. Renson, and J. D. Barchas, *Life Sci.* **19**:1669 (1976).
100. A. Bertler, B. Falck, and E. Rosengrenn, *Acta Pharmacol. Toxicol.* **20**:317 (1963).
101. A. Bertler, B. Falck, C. Owman, and E. Rosengrenn, *Pharmacol. Rev.* **18**:369 (1966).
102. J. P. Hanig, J. M. Morrison, Jr., and S. Krop, *Eur. J. Pharmacol.* **18**:79 (1972).
103. R. A. Lahti and E. Majchrowicz, *Q. J. Stud. Alcohol* **35**:1 (1974).
104. N. M. VanGelder, in: *Brain Barrier Systems* (A. Lajtha and D. Ford, eds.), pp. 259–271, Elsevier Publishing Company, New York (1968).
105. Z. Albert, M. Orlowski, Z. Rzucidlo, and J. Orlowska, *Acta Histochem.* **25**:312 (1966).
106. E. Reyes and G. C. Palmer, *Res. Comm. Chem. Pathol. Pharmacol.* **14**:759 (1976).
107. A. Meister, *Science* **180**:33 (1973).
108. E. Reyes, *West. Pharmacol. Soc.* **20**:345 (1977).
109. E. Reyes, *Fed. Proc. Abstr.* **36**:332 (1977).
110. E. Reyes, *Res. Comm. Chem. Pathol. Pharmacol.* **17**:353 (1977).
111. H. Wallgren and H. Barry, III, *Actions of Alcohol*, Vols. I and II, Elsevier Publishing Company, Amsterdam (1970).
112. G. Rosenfeld, *Q. J. Stud. Alcohol* **21**:584 (1960).
113. H. Wallgren, *Biochem. Pharmacol.* **6**.195 (1961).
114. R. A. Lahti and E. Majchrowicz, *Biochem. Pharmacol.* **18**:535 (1969).
115. B. B. Coldwell, H. L. Trenholm, B. H. Thomas, and S. Charbonneau, *J. Pharm. Pharmacol.* **23**:947 (1971).
116. G. Back, G. Seidel, and W. Endell, *Pharmacology* **14**:67 (1976).

Interaction of Ethanol with Hormonal Functions

IX

Effects of Acute and Chronic Administration of Ethanol on Cyclic Nucleotides and Related Systems

8

Walter A. Hunt

A. INTRODUCTION

Over the past 10–20 years the possible role of cyclic nucleotides in biological function has received considerable attention. It is now believed that cyclic nucleotides act as a "second messenger" in the actions of a variety of hormones. Since a physiological response could result from the disruption of any one of a chain of events, the possible role of cyclic nucleotides in the actions of ethanol both after acute and chronic administration have been explored.[1] It will be the purpose of this review to collate and analyze the existing data on ethanol–cyclic-nucleotide interactions and to determine if these effects could be responsible for some of the actions of ethanol.

In order to gain perspective it will be necessary to summarize the present theories on the role of cyclic nucleotides in biological function and how they accomplish their actions. Many excellent reviews[2–4] have been written exhaustively detailing what has been reported on cyclic nucleotides. Hence, our purpose here will be only to present basic concepts in order to help understand the potential relevance of the actions of ethanol on this system.

B. ETHANOL AND CYCLIC NUCLEOTIDES IN THE NERVOUS SYSTEM

1. General Aspects of Cyclic Nucleotide Function

Evidence to date supports a role of cyclic nucleotides in synaptic transmission. Cyclic nucleotides have been implicated in the actions of a number of neurotransmitters, including norepinephrine, dopamine, serotonin, acetylcho-

---

Walter A. Hunt • Behavioral Sciences Department, Armed Forces Radiobiology Research Institute, Bethesda, Maryland 20014.

line, GABA, and glutamate. Adenosine-3',5'-cyclic monophosphate (cyclic AMP) appears to be related to the effects of norepinephrine,[5-7] dopamine,[8,9] and serotonin,[10] while guanosine-3',5'-cyclic monophosphate (cyclic GMP) is related to the effects of acetylcholine,[11-14] GABA,[15] and glutamate.[16]

The current belief concerning the possible role of cyclic nucleotides in synaptic transmission involves the initial stimulation of a receptor by the neurotransmitter substance. In the case of biogenic amines, interaction with their receptors involves adenylate cyclase which converts ATP to cyclic AMP,[17] while acetylcholine, GABA, and glutamate act through guanylate cyclase, which converts GTP to cyclic GMP.[18] When these enzymes are stimulated, an accumulation of the corresponding cyclic nucleotide occurs. This elevated cyclic nucleotide level increases the activity of cyclic-nucleotide-dependent protein kinases, which appear to be separate entities, one specific for cyclic AMP and one for cyclic GMP.[19] These protein kinases catalyze the phosphorylation of proteins located in synaptosomal membranes,[20,21] and this process is believed to alter the permeability of the membrane to ions, which in turn changes the resting membrane potential.

The most detailed account of the interaction of neurotransmitters and cyclic nucleotides and the concomitant changes in the membrane potential has been with the superior cervical ganglion. Evidence accumulated here has been easier to interpret because of its relatively simple structural organization when compared to the brain. It now appears that dopamine is released from interneurons in the ganglion and stimulates the formation of cyclic AMP postsynaptically.[8,22,23] These events coincide with the development of the slow inhibitory postsynaptic potential.[24] On the other hand, acetylcholine released from preganglionic fibers stimulates the formation of cyclic GMP postsynaptically[23] and may be related to the slow excitatory postsynaptic potential.[25] This effect seems to be mediated through muscarinic receptors, since it can be blocked by muscarinic antagonists but not by nicotinic antagonists.[26]

Most of the research on the effect of ethanol on cyclic nucleotides has been on cyclic AMP and related systems in brain tissue. Measurements of brain cyclic AMP are very tricky especially in the cerebellum because of a rapid postmortem accumulation of this nucleotide.[27] Consequently, rapid inactivation of brain enzymes is very important. To accomplish this, high-intensity microwave irradiation has been used. Initially, a conventional microwave oven was tried,[27] but it required 20 sec to inactivate adenylate cyclase and phosphodiesterase. More recently a microwave oven was developed which focuses the beam on the head of an animal.[28] With this modification the enzymes are inactivated in 2 sec.

2. Effects of Ethanol on Cyclic AMP

a. Acute Treatment

In an early study Volicer and Gold[29] reported, using conventional microwave irradiation for 45 sec, that a single 1- to 6-g-ethanol/kg dose, given orally, depressed cyclic AMP levels up to 60% in whole brain of rats in a dose-dependent manner. This response was predominantly in the cerebellum. Orenberg *et al.*[30] examined cyclic AMP levels in several areas of the mouse brain after 9 sec of conventional microwave irradiation. In the cerebral cortex, cyclic AMP levels were depressed up to 65% 1 hr after treatment by 0.4- to 3.2-g-ethanol/kg doses, given intraperitoneally. However, under similar conditions, subcortical and cerebellar levels were significantly elevated by 60 and 170%, respectively, 10 min after treatment. When mice were killed by immersion in liquid nitrogen, ethanol had no effect on cyclic AMP levels in the cerebral hemispheres.[31] Recently, using a 9-sec exposure of focused microwave irradiation, Volicer and Hurter[32] found a dose-dependent reduction in cyclic AMP levels after 1–6 g ethanol/kg, given orally, in the cerebral cortex, cerebellum, and brain stem of the rat 1 hr after treatment. On the other hand, Redos *et al.*[33] were unable to demonstrate any alteration in cyclic AMP levels in any area of the brain 2 hr after treatment with a 6-g-ethanol/kg oral dose and using 3.5 sec of focused microwave irradiation.

The preceding discussion illustrates the uncertainty of the effect of single doses of ethanol on cyclic AMP levels in the brain. The source of these ambiguities is not clear but may be related to methodological considerations. A summary of some of the data can be found in Table I. As pointed out earlier, rapid inactivation of the synthetic and degradative enzymes for cyclic AMP is very important to minimize its postmortem accumulation in the brain and to reflect more closely the actual levels *in vivo*. In the studies to date, exposure durations vary from 3.5 to 9.0 sec for focused microwave irradiation and 9 to 45 sec using conventional microwave ovens. Although the enzymes can be inactivated fairly quickly, it is unclear whether the existing technology does it fast enough, especially in the cerebellum, where postmortem accumulation is most pronounced.[27] Also, ethanol inhibits this accumulation in rats[29] and could induce artifacts by making it appear that ethanol-depleted cyclic AMP, when in

Table I. Effect of a Single Dose of Ethanol on Cyclic AMP Levels in Brain

	Dose	Time after treatment	Mode of killing[a]	Cyclic AMP		Reference
				Control	Treated	
Cerebellum	4 g/kg, po	3 hr	CMI	2.2[b]	1.5	29
	4 g/kg, po	1 hr	FMI	1.4[b]	0.7	32
	6 g/kg, po	2 hr	FMI	3.4[c]	3.2	33
	3.2 g/kg, ip	10 min	CMI	0.6[b]	1.5	30
Cerebral cortex	4 g/kg, po	3 hr	CMI	8.0[b]	7.0	29
	4 g/kg, po	1 hr	FMI	1.4[b]	0.9	32
	6 g/kg, po	2 hr	FMI	7.2[c]	7.2	33
	3.2 g/kg, ip	10 min	CMI	1.4[b]	1.0	30
	4 g/kg, ip	1.5 hr	ILN	0.7[b]	0.7	31

[a]CMI, conventional microwave irradiation; FMI, focused microwave irradiation; ILN, immersion in liquid nitrogen.
[b]Values expressed as pmole/mg tissue.
[c]Values expressed as pmole/mg protein.

fact it might only have blocked the postmortem accumulation that could take place within the first few seconds after death. In further support of this possibility is the report of Jones and Stavinoha[34] that a significant elevation of cyclic AMP can occur in the brain under these circumstances. They compared the cyclic AMP levels in various parts of the brain of mice exposed either to 4 sec of 1.5 kW or 0.3 sec of 6 kW of focused microwave radiation. In all cases the cyclic AMP levels were higher with the lower intensity, longer duration irradiation. In the cerebellum, for example, the values increased by 100%.

Most of the studies using microwave irradiation expose the brain longer than is necessary to inactivate enzymes. The potential problems of overexposure have not been explored. For example, does prolonged high heat decompose cyclic AMP? Does the progressive cellular damage induced by continued exposure alter the distribution of cyclic AMP in a manner that might ultimately affect its quantitation? Or does interaction of ethanol and other drugs with the brain alter the dynamics of inactivation?

Different responses in cyclic AMP levels have been obtained in different areas of the brain after ethanol treatment and are a function of the dose and time after exposure. Most studies do not report blood ethanol concentrations, making any meaningful comparisons between studies difficult.

In summary, in order to determine what single doses of ethanol do to cyclic AMP levels in the brain, it is necessary to employ methods of killing animals that are well understood and standardized. Experiments need to be carried out carefully providing complete dose–response and time-course relationships accompanied by blood ethanol concentrations. Finally, the possibility of species differences should not be ignored.

b. Chronic Treatment

Chronic treatment with ethanol has produced results on cyclic AMP levels and related systems that are less controversial. If animals are treated for at least 8 days, cerebral cyclic AMP is elevated 70%.[31,32] However, after only 4 days of treatment, no increases in cyclic AMP are observed.[33] These increases in cyclic AMP appear to be related to an elevation in adenylate cyclase activity. Kuriyama and Israel[31] found that at a time when cyclic AMP levels were elevated, there was also an increase in adenylate cyclase activity. Neither response was observed after a single dose of ethanol. Also, phosphodiesterase activity was unaffected after either treatment.[31]

More interesting and possibly more functionally important findings are the studies determining the sensitivity of the cyclic AMP system to neurotransmitters. Present evidence is quite compelling that chronic administration of ethanol alters the sensitivity of cyclic-AMP-generating systems to norepinephrine. An early study demonstrated that treatment with ethanol for 14 days decreased the sensitivity of this system to norepinephrine.[35] French and co-workers have subsequently reported some very interesting results on changes of sensitivity of cyclic AMP accumulation to neurotransmitters in rats after 16 weeks of ethanol treatment. The results observed depended on the time after withdrawal the measurements were made. In cortical slices obtained from animals withdrawn for 2 hr, the sensitivity of the cyclic-AMP-generating system to norepinephrine was depressed, with the dose–response curve shifting 4.3-fold to the right.[36] This development of noradrenergic subsensitivity could be explained by hyperactivity of noradrenergic nerve terminals.[37,38] However, it was not established whether a single dose of ethanol might produce the same result on cyclic AMP accumulation. Three days after withdrawal, a time corresponding to the development of delirium tremens,[39] the opposite response was observed. The sensitivity of the cyclic-AMP-generating system to norepinephrine increased 2.4-fold.[40,41] In addition, the sensitivity to histamine and 5-HT was also enhanced.[42] The response to 5-HT might result from a chronic depression of serotonergic function.[43]

The significance of alterations in the sensitivity of the cyclic-AMP-generating system to specific aspects of alcoholism has not been demonstrated. At the least they reflect changes induced by long-term consumption of ethanol. It is not clear whether they are related to the development of physical dependence, which was minimal with the methods used,[44] or whether they are expressions of some nonspecific neural toxicity. The whole question of adaptive mechanisms has considerable significance for explaining tolerance and physical dependence. If ethanol treatment induces alterations in neurotransmitter function (see Chapter 11) chronic changes can lead to compensatory changes in other parameters in an attempt to maintain homeostasis. The role of cyclic nucleotides in adaptive processes has received considerable attention and has been concisely reviewed recently by Dismukes and Daly.[2] Because of the potential importance of these findings, it is imperative that the studies of French and co-workers be repeated using an animal model that takes only a few days for inducing physical dependence. In this way, nonspecific neural toxicity can be discounted, and the relationship between neurotransmitter sensitivity and overt, spontaneous withdrawal signs will be a more convincing possibility.

Adaptive changes can take place not only in the sensitivity of receptors to neurotransmitters, but possibly also with alterations in the activity of the receptor for cyclic AMP, the cyclic-AMP-dependent protein kinase. As mentioned earlier, this enzyme catalyzes the phosphorylation of protein in synaptosomal membranes. Recently Kuriyama et al.[45] reported a study in which they measured cyclic-AMP-dependent protein kinase activity in mouse cerebral cortex after acute and chronic administration of ethanol. They found that after treatment for 2 weeks, synaptosomal enzymatic activity was elevated fourfold but returned to control levels 7 days after withdrawal. No alteration in protein kinase activity was observed after a single dose of ethanol. One problem with this work is that the conditions under which the animals were killed after chronic ethanol treatment were not defined, and the appearance of possible withdrawal signs was not reported. Thus, it is difficult to determine whether this alteration is related to the development of physical dependence. It appears, however, that the elevation in cyclic-AMP-dependent protein kinase activity might be a consequence of the increased activity of adenylate cyclase and cyclic AMP levels[31] and not to an adaptive change in the sensitivity of protein kinase to cyclic AMP analogous to norepinephrine-induced cyclic AMP accumulation.

3. Effects of Ethanol on Cyclic GMP

The possible role of cyclic GMP in the actions of ethanol has been recently reported. Redos et al.[46] found that ethanol depletes cerebellar cyclic GMP in the rat in a dose-dependent manner. The maximum effect was observed 1 hr after oral administration of 6 g ethanol/kg, which resulted in a loss of 95% of the cyclic GMP. In addition, the degree of depletion was directly proportional to the blood ethanol concentration with control values obtained when ethanol had been eliminated. Similar results were found by Volicer and Hurter.[32]

Studies of the effects of ethanol on cyclic GMP have been expanded to include measurements in other areas of the brain and after both acute and chronic administration. Sample results can be found in Table II. Single doses of ethanol were found to deplete cyclic GMP in the cerebral cortex, caudate nucleus and thalamus as well as in the cerebellum.[32,47] In chronically treated rats rendered ethanol dependent, similar results were observed if the animals were still intoxicated. However, in the cerebellum and the brain stem, tolerance develops to the effect of ethanol on cyclic GMP levels[47] corresponding to the development of behavioral tolerance under similar experimental conditions.[48] When acutely and chronically treated animals are compared at similar blood ethanol concentrations, cyclic GMP levels were significantly higher in the chronically treated animals. If cyclic GMP levels are depressed for a long period, one might expect to see an overshoot after the elimination of ethanol. This was not observed after 4 days of treatment. Cyclic GMP levels in all areas of the brain studied were at control levels during the ethanol withdrawal syndrome.[47] When treatment was extended to 8 days, a significant elevation of cerebellar cyclic GMP was found.[32] However, since withdrawal signs develop with both treatment regimens, elevation of cerebellar cyclic GMP by itself does not appear to be a prerequisite for the expression of an ethanol withdrawal syndrome.

Table II.   Effect of Ethanol Treatment on Cyclic GMP Levels in Brain[a]

	Control	Acute[b]	Chronic intoxicated	Chronic withdrawal	References
Cerebellum	4.0[b]	0.2	1.2	3.3	46, 47
	0.6[c]	0.3	0.3	1.5	32
Cerebral cortex	0.5[b]	0.1	0.2	0.4	46, 47
	0.1[c]	0.05	0.05	0.1	32

[a]Cyclic GMP levels were determined 1 hr after a single 6-g-ethanol/kg dose given orally.
[b]Values expressed as pmole/mg protein.
[c]Values expressed as pmole/mg tissue.

The mechanism by which ethanol depletes cyclic GMP is unknown. An attempt to show a direct effect of ethanol in concentrations up to 200 mM on guanylate cyclase or cyclic GMP-phosphodiesterase activity *in vitro* was not successful.[47] However, the assays were performed under optimal enzymatic conditions which do not necessarily reflect what occurs *in vivo*. For example, calcium may regulate the activity of guanylate cyclase, and changes in the actions of calcium might depress the synthesis of cyclic GMP.[49–51] Ethanol has been shown to induce several alterations in calcium that might restrict the intracellular concentration of calcium and suppress guanylate cyclase activity. Ethanol can block the inward calcium currents in *Aplysia* neurons,[52] enhance calcium binding to erythrocyte ghost membranes,[53] and reduce brain calcium.[54] Although all these studies may not be relevant to an effect on guanylate cyclase in brain, they do indicate a need for more direct experiments into the possible role of calcium in the actions of ethanol.

Alterations in neurotransmitter activity might be involved in the ethanol-induced depletion of cyclic GMP. As indicated earlier, cyclic GMP appears to be related to the actions of acetylcholine, GABA, and glutamate. Ethanol has been demonstrated to inhibit cortical and reticular acetylcholine release *in vivo*[55] and depresses cerebellar glutamate levels.[56] Whether these changes could explain, at least in part, the depletion of cyclic GMP has yet to be determined.

Alterations in systems involving cyclic GMP could be involved in the expression of an ethanol withdrawal syndrome. Collier *et al.*[57] have studied the effect of agents which have actions related to cyclic nucleotides on ethanol withdrawal head twitches in mice. They found that dibutyryl cyclic GMP and GTP, compounds which elevate brain cyclic GMP, increase the incidence of head twitches over twofold, while dibutyryl cyclic AMP, ATP, and prostaglandins, compounds which elevate brain cyclic AMP, antagonize head twitches. These results support the view that an alteration in the balance of cyclic GMP and cyclic AMP in favor of the former might play a role in the expression of the withdrawal syndrome. Further, there is a parallel between harmaline- and withdrawal-induced tremors. Harmaline affects the cerebellum by stimulating the excitatory climbing fibers that synapse on the Purkinje cells[58] in a manner resembling activation of climbing fibers.[59] The tremor induced by harmaline is accompanied by an elevation in cerebellar cyclic GMP and can be blocked by prior administration of benzodiazepines.[60] These drugs have been used to treat the ethanol withdrawal syndrome in man.[61] Consequently, future research is warranted to try to find a possible role of cyclic GMP or related systems in ethanol dependence, possibly through alterations resulting from chronic cyclic GMP depletion.

C. ETHANOL AND CYCLIC NUCLEOTIDES IN THE STOMACH

Ethanol is known to stimulate gastric acid secretion in a number of species, including man.[62] The response obtained is similar to that elicited by histamine,[63] but subsequent work ruled out a role of histamine in the effect of ethanol.[64,65] Cyclic AMP has been implicated as a mediator in acid secretion,[66,67] and in addition the gastric mucosa contains a cyclic-AMP-dependent protein kinase which may be involved in translocation of ions.[68] Therefore, it is possible that changes in cyclic AMP metabolism may be related to the ability of ethanol to alter acid secretion.

The possible relationship between the effect of ethanol on gastric acid secretion and cyclic AMP metabolism has been explored in several species, including man. In the rat, local application of ethanol on the gastric mucosa in concentrations of 1–10% reduced acid secretion in a concentration-dependent manner from 30 to 100% and decreased cyclic AMP levels up to 50%.[69] In the dog, ethanol has a biphasic effect on acid secretion. In concentrations below 20% ethanol stimulates secretion, while at concentrations above 20% it depresses secretion.[70] Ethanol also exerts the same biphasic effect on mucosal cyclic AMP levels.[71] In man, ethanol concentrations that stimulate acid secretion are accompanied by increases in mucosal cyclic AMP levels.[72] It would appear, then, that alterations in gastric acid secretion as affected by ethanol are directly correlated with changes in cyclic AMP content.

In determining the mechanism by which ethanol alters cyclic AMP levels, a first step is to study the effect of ethanol on mucosal adenylate cyclase and phosphodiesterase activities. In the rat, ethanol inhibits both enzymes in a concentration-dependent fashion but with different sensitivities.[69,73] For example, at about 10% ethanol, phosphodiesterase activity is reduced by 50%, but adenylate cyclase is nearly completely blocked. In the dog and man, ethanol stimulates adenylate cyclase activity at low concentrations[71,72] but is ineffective at high concentrations when cyclic AMP levels and acid secretion are reduced.[71] As with the rat, phosphodiesterase in the dog is inhibited. These findings suggest that ethanol may elevate acid secretion by stimulating adenylate cyclase, but they do not explain how ethanol can depress secretion and cyclic AMP levels.

As pointed out earlier, assays in vitro are not always reliable in reflecting the true status in the living animal. Other factors often contribute to the rate of synthesis and degradation of a compound. Cyclic AMP falls into this category. It is well known that cyclic AMP levels in a variety of tissues are affected by a number of hormones, neurotransmitters, and ions. Ethanol may induce alterations in one of these other factors.

One factor that affects the content of cyclic AMP is the availability of its precursor ATP if adenylate cyclase is not saturated. Puurunen et al.[74] perfused the stomach with 10% ethanol and measured acid secretion and cyclic AMP and ATP levels in the whole gastric mucosa and in the superficial mucosa, where most of the acid-secreting parietal cells are located. After 40 min, all three parameters were significantly reduced in the superficial mucosa but not in the whole mucosa. When perfusion was discontinued, acid secretion and cyclic AMP and ATP returned to control levels in about 60 min. Since no additional cyclic AMP was detected in the gastric perfusate, the reduced cyclic AMP in the superficial mucosa cannot be explained by leakage of cyclic AMP from the mucosal cells. A reduction of mucosal ATP levels has also been observed in the dog after ethanol perfusion.[75]

Prostaglandins interact with cyclic AMP metabolism[76,77] and inhibit gastric acid secretion.[78,79] In an attempt to determine if prostaglandins might play a role in ethanol's depressant effect on acid secretion in the rat, Karppanen and Puurunen[80] pretreated their animals with indomethacin, an inhibitor of prostaglandin synthesis.[81] They found that the ability of ethanol to depress acid secretion was antagonized, suggesting that an increase in prostaglandin synthesis may mediate, in part, this effect of ethanol. However, they did not measure cyclic AMP levels, so conclusions on an interaction of prostaglandins and cyclic AMP in this system cannot be made, although cyclic AMP levels have been reported to be elevated in rat fundic muscle after intraperitoneal injection of prostaglandin E_1.[82] Also, it is unfortunate that these experiments were not done in dogs, where the mechanism of the inhibition of acid secretion at high ethanol concentrations is less clear.

In summary, the effects of ethanol on gastric acid secretion are accompanied by changes in superficial mucosal cyclic AMP levels. These changes reflect, in part, alterations in the activities of adenylate cyclase and phosphodiesterase. It is not clear how ethanol exerts its effect, but considering the high ethanol concentrations used, partial denaturation or structural alterations of the enzymes are possible. Also, part of the reduction in acid secretion that can be induced by ethanol in some species might be mediated through a reduction in ATP levels or an increased synthesis and release of prostaglandins. It would be interesting to determine whether ethanol actually stimulates prostaglandin release and whether it is more effective in species where ethanol has a pronounced inhibitory effect— for example, in the rat. At this stage, no cause-and-effect relationship has been established between the ethanol-induced reduction of acid secretion and prostaglandins. Further research will be required to clarify this point.

Under normal circumstances where systemic effects of ethanol are considered, the high concentrations of 1–50% used in the experiments with the stomach are far in excess of what would be compatible with life. However, we are dealing with localized actions of ethanol under conditions similar to normal drinking of alcoholic beverages. Taking into account the damage to the stomach known to result from high concentrations of ethanol,[62] the preceding discussion further emphasizes the need for caution in drinking beverages with a high alcoholic content.

D. ETHANOL AND CYCLIC NUCLEOTIDES IN OTHER TISSUES

1. Liver

Compared to the central nervous system and the stomach, very little research has been done on the possible involvement of cyclic nucleotides in the actions of ethanol on other organs. A few reports exist but generally are not enlightening as to their possible significance.

Ethanol has been reported to induce alterations in cyclic AMP metabolism in the liver. Short-chain aliphatic alcohols stimulate glucagon-responsive adenylate cyclase.[83] The potency increases with each additional carbon in the chain. However, no differences in potency are observed among primary, secondary, and tertiary butanols. The enhanced activity of adenylate cyclase by ethanol is not apparent until the concentrations are at least 2%, with the maximum effect reached at 5% ethanol. These alcohols, also, stimulate adenylate cyclase *in vitro* in other tissues, including kidney, intestine, fat, and brain.[84,85] In these studies only unphysiologically high concentrations could elicit the response.

Because of the peripheral adrenergic hyperactivity observed after chronic ethanol administration,[37,86] French *et al.*[87] explored the possibility that similar treatment might alter the sensitivity of the cyclic AMP system to norepinephrine in the liver. They found that after 18 weeks of ethanol ingestion, the ED_{50} for norepinephrine stimulation had increased sixfold in liver homogenate particulate fractions, indicating the development of adrenergic subsensitivity. This altered sensitivity returned to control values 3 days after withdrawal, unlike that observed in brain.[40] However, if the measurements were made in liver mitochondria, an increased sensitivity was seen.[88] Finally, the effects of chronic ethanol treatment appear to be mediated through β-receptors, since β-adrenergic antagonists block the ability of norepinephrine to stimulate cyclic AMP accumulation, with no effect by α-adrenergic antagonists.[87]

2. Skin

One study has appeared studying the effect of short-chain alcohols on cyclic AMP in the skin. Yoshikawa *et al.*[89] reported that cyclic AMP levels are increased up to 2.5-fold in a concentration-dependent manner in epidermal slices incubated in 1–5% ethanol. Similar results were found with 1-propanol and acetone, but not with methanol and 1-butanol. This effect appears to be related to an activation of epidermal adenylate cyclase. As yet, no physiological significance can be attributed to these findings.

E. SUMMARY AND CONCLUSIONS

It is now clear that acute and chronic treatment with ethanol exerts a number of effects on the cyclic nucleotides in several organ systems. In the brain, for instance, ethanol intoxication results in lower levels of cyclic nucleotides, especially cyclic GMP. However, after chronic administration of ethanol, indicators of cyclic nucleotide function are elevated. These changes appear to reflect changes in the activity of the nervous system. During intoxication, the brain is depressed, whereas during ethanol withdrawal syndrome, hyperactivity can be observed. It is yet to be determined whether changes in cyclic-nucleotide-mediated systems are responsible for intoxication, dependence, or other biological alterations for which they have been implicated after ethanol consumption. At the present state of research, changes in cyclic nucleotides can be shown to be concomitants of certain effects of ethanol. However, no one has been able to demonstrate that these changes in themselves can explain ethanol-induced behavioral abnormalities. In fact, because of the effect ethanol has on neurotransmitters and their relationship to cyclic nucleotides, it is possible that alterations in cyclic nucleotides are secondary to some other response of ethanol. The exact role of cyclic nucleotides in the actions of ethanol and its potential significance will require further research.

F. REFERENCES

1. L. Volicer and B. I. Gold, in: *Biochemical Pharmacology of Ethanol* (E. Majchrowicz, ed.), pp. 181–196, Plenum Press, New York (1975).
2. R. K. Dismukes and J. W. Daly, *J. Cyclic Nucleotide Res.* **2**:321 (1976).
3. J. A. Nathanson, *Physiol. Rev.* **57**:157 (1977).
4. M. Williams and R. Rodnight, *Prog. Neurobiol.* **8**:183 (1977).
5. G. R. Siggins, B. J. Hoffer, and F. E. Bloom, *Science* **165**:1018 (1969).
6. J. Schultz and J. W. Daly, *J. Neurochem.* **21**:573 (1973).
7. J. P. Perkins and M. M. Moore, *J. Pharmacol. Exp. Ther.* **185**:371 (1973).
8. J. W. Kebabian and P. Greengard, *Science* **194**:1346 (1971).
9. J. W. Kebabian, G. L. Petzold, and P. Greengard, *Proc. Natl. Acad. Sci. USA* **69**:2145 (1972).
10. J. A. Nathanson and P. Greengard, *Proc. Natl. Acad. Sci. USA* **71**:797 (1974).
11. J. A. Ferrendelli, A. L. Steiner, D. B. McDougal, Jr., and D. M. Kipnis, *Biochem. Biophys. Res. Commun.* **41**:1061 (1970).
12. J. F. Kuo, T. P. Lee, P. L. Reyes, K. C. Walton, T. E. Donnelly, Jr., and P. Greengard, *J. Biol. Chem.* **247**:10 (1972).
13. T. P. Lee, J. F. Kuo, and P. Greengard, *Proc. Natl. Acad. Sci. USA* **69**:3287 (1972).
14. T. W. Stone, D. A. Taylor, and F. E. Bloom, *Science* **187**:845 (1975).
15. C. C. Mao, A. Guidotti, and E. Costa, *Mol. Pharmacol.* **10**:736 (1974).
16. C. C. Mao, A. Guidotti, and E. Costa, *Brain Res.* **79**:510 (1974).
17. E. W. Sutherland, T. W. Rall, and T. Menon, *J. Biol. Chem.* **237**:1220 (1962).
18. J. G. Hardman and W. W. Sutherland, *J. Biol. Chem.* **244**:6363 (1969).
19. J. F. Kuo and P. Greengard, *J. Biol. Chem.* **245**:2493 (1970).
20. E. M. Johnson, H. Maeno, and P. Greengard, *J. Biol. Chem.* **246**:7731 (1971).
21. E. M. Johnson, T. Veda, H. Maeno, and P. Greengard, *J. Biol. Chem.* **247**:5650 (1972).
22. P. Kalix, D. A. McAfee, M. Schorderet, and P. Greengard, *J. Pharmacol. Exp. Ther.* **188**:676 (1974).
23. J. W. Kebabian, F. E. Bloom, A. L. Steiner, and P. Greengard, *Science* **190**:157 (1975).
24. D. A. McAfee and P. Greengard, *Science* **178**:310 (1972).
25. P. Greengard and J. W. Kebabian, *Fed. Proc. Symp.* **33**:1059 (1974).
26. J. W. Kebabian, A. L. Steiner, and P. Greengard, *J. Pharmacol. Exp. Ther.* **193**:474 (1975).
27. M. J. Schmidt, D. E. Schmidt, and G. A. Robinson, *Science* **173**:1142 (1971).
28. A. Guidotti, D. L. Cheney, M. Trabucchi, M. Doteuichi, C. Wang, and R. A. Hawkins, *Neuropharmacology* **13**:1115 (1974).
29. L. Volicer and B. I. Gold, *Life Sci.* **13**:269 (1973).
30. E. K. Orenberg, J. Renson, and J. D. Barchas, *Neurochem. Res.* **1**:659 (1976).
31. K. Kuriyama and M. A. Israel, *Biochem. Pharmacol.* **22**:2919 (1973).
32. L. Volicer and B. P. Hurter, *J. Pharmacol. Exp. Ther.* **200**:298 (1977).
33. J. D. Redos, W. A. Hunt, and G. N. Catravas, *Life Sci.* **18**:989 (1976).
34. D. J. Jones and W. B. Stavinoha, *J. Neurochem.* **28**:759 (1977).
35. M. A. Israel, H. Kimura, and K. Kuriyama, *Experientia* **28**:1322 (1972).

36. S. W. French, P. E. Reid, D. S. Palmer, M. E. Narod, and C. W. Ramey, *Res. Commun. Chem. Pathol. Pharmacol.* **9**:575 (1974).
37. W. A. Hunt and E. Majchrowicz, *J. Neurochem.* **23**:549 (1974).
38. L. A. Pohorecky, *J. Pharmacol. Exp. Ther.* **189**:380 (1974).
39. M. Victor and R. D. Adams, *Assoc. Res. Nerv. Dis. Proc.* **32**:526 (1953).
40. S. W. French and D. S. Palmer, *Res. Commun. Chem. Pathol. Pharmacol.* **6**:651 (1973).
41. S. W. French, D. S. Palmer, N. E. Narod, P. E. Reid, and C. W. Ramey, *J. Pharmacol. Exp. Ther.* **194**:319 (1975).
42. S. W. French, D. S. Palmer, and N. E. Narod, *Can. J. Physiol. Pharmacol.* **53**:248 (1975).
43. W. A. Hunt and E. Majchrowicz, *Brain Res.* **72**:181 (1974).
44. S. W. French and J. R. Morris, *Res. Commun. Chem. Pathol. Pharmacol.* **4**:221 (1972).
45. K. Kuriyama, K. Nakazawa, M. Muramatsu, and K. Kakita, *Biochem. Pharmacol.* **25**:2541 (1976).
46. J. D. Redos, G. N. Catravas, and W. A. Hunt, *Science* **193**:58 (1976).
47. W. A. Hunt, J. D. Redos, T. K. Dalton, and G. N. Catravas, *J. Pharmacol. Exp. Ther.* **201**:103 (1977).
48. E. Majchrowicz and W. A. Hunt, *Psychopharmacology* **50**:107 (1976).
49. J. A. Ferrendelli, D. A. Kinscherf, and M. M. Chang, *Mol. Pharmacol.* **9**:445 (1973).
50. K. Nakazawa and M. Sano, *J. Biol. Chem.* **249**:4207 (1974).
51. D. R. Olson, C. Kon, and B. M. Breckinridge, *Life Sci.* **18**:935 (1976).
52. M. C. Bergmann, M. R. Klee, and D. S. Faber, *Pfluegers Arch. Eur. J. Physiol.* **248**:139 (1974).
53. P. Seeman, M. Chan, M. Goldberg, T. Sanko, and L. Sax. *Biochim. Biophys. Acta* **225**:185 (1971).
54. D. H. Ross, M. A. Medina, and H. L. Cardenas, *Science* **186**:63 (1974).
55. C. K. Erickson and D. J. Graham, *J. Pharmacol. Exp. Ther.* **185**:583 (1973).
56. I. A. Syntinsky, B. M. Guzikov, M. U. Gomanko, V. R. Eremin, and N. N. Konovalova, *J. Neurochem.* **25**:43 (1975).
57. H. O. J. Collier, M. D. Hammond, and C. Schneider, *Br. J. Pharmacol.* **58**:9 (1976).
58. R. Llinás and R. A. Volkind, *Exp. Brain Res.* **18**:64 (1973).
59. J. C. Eccles, R. Llinás, and K. Sasaki, *J. Physiol. (Lond.)* **182**:268 (1966).
60. C. C. Mao, A. Guidotti, and E. Costa, *Brain Res.* **83**:516 (1975).
61. G. Sereny and H. Kalant, *Br. Med. J.* **1**:92 (1965).
62. S. H. Lorber, V. P. Vincente, Jr., and W. Y. Chey, in: *The Biology of Alcoholism* (B. Kissin and H. Begleiter, eds.), Vol. 3: *Clinical Pathology*, pp. 339–357, Plenum Press, New York (1974).
63. C. A. Dragstedt, J. S. Gray, A. H. Lawton, and M. Ramires de Abellano, *Proc. Soc. Exp. Biol. Med.* **43**:26 (1940).
64. W. T. Irvine, D. B. Watkin, and E. J. Williams, *Gastroenterology* **39**:1 (1960).
65. W. T. Irvine, H. D. Ritchie, and H. M. Adam, *Gastroenterology* **41**:258 (1961).
66. M. S. Amer, *Am. J. Dig. Dis.* **17**:945 (1972).
67. D. V. Kimberg, *Gastroenterology* **67**:1023 (1974).
68. J. F. Kuo, B. K. Krueger, J. R. Sanes, and P. Greengard, *Biochim. Biophys. Acta* **212**:79 (1970).
69. J. Puurunen and H. Karppanen, *Life Sci.* **16**:1513 (1975).
70. J. M. Beazell and A. C. Ivy, *Q. J. Stud. Alcohol* **1**:45 (1940).
71. J. Puurunen, H. Karppanen, M. Kairaluoma, and T. Larmi, *Eur. J. Pharmacol.* **38**:275 (1976).
72. H. Karppanen, J. Puurunen, M. Kairaluoma, and T. Larmi, *Scand. J. Gastroenterol.* **11**:603 (1976).
73. L. L. Tague and L. L. Shanbour, *Clin. Res.*, Abstr. **20**:736 (1972).
74. J. Puurunen, K. Hiltunen, and H. Karppanen, *Eur. J. Pharmacol.* **42**:85 (1977).
75. L. L. Tague and L. L. Shanbour, *Proc. Soc. Exp. Biol. Med.* **154**:37 (1977).
76. R. W. Butcher and C. E. Baird, *J. Biol. Chem.* **243**:1713 (1968).
77. G. A. Robinson, A. Arnold, and R. C. Hartmann, *Pharmacol. Res. Commun.* **1**:325 (1969).
78. A. Bennett and B. Fleshler, *Gastroenterology* **59**:790 (1970).
79. D. E. Wilson, *Prostaglandins* **1**:281 (1972).
80. H. Karppanen and J. Puurunen, *Eur. J. Pharmacol.* **35**:221 (1976).

81. J. R. Vane, *Nature (Lond.)* **231**:232 (1971).
82. N. L. Shearin and W. L. Pancoe, *Experientia* **32**:1553 (1976).
83. R. E. Gorman and M. W. Bitensky, *Endocrinology* **87**:1075 (1970).
84. H. L. Greene, R. H. Herman, and S. Kraemer, *J. Lab. Clin. Med.* **78**:336 (1971).
85. K. Mashiter, G. D. Mashiter, and J. B. Field, *Endocrinology* **94**:370 (1974).
86. M. Ogata, J. H. Mendelson, N. K. Mello, and E. Majchrowicz, *Psychosom. Med.* **33**:159 (1971).
87. S. W. French, D. S. Palmer, and M. E. Narod, *Res. Commun. Chem. Pathol. Pharmacol.* **9**:575 (1976).
88. S. W. French, D. S. Palmer, and M. Narod, *Am. J. Pathol.* **74**:76a (1974).
89. K. Yoshikawa, K. Adachi, K. M. Halprin, and V. Levin, *Br. J. Dermatol.* **94**:611 (1976).

Interactions of Thyrotropin-Releasing Hormone and Somatostatin with Ethanol and Other Sedative/Hypnotics

9

Robert L. Eskay

A. INTRODUCTION

Regulatory peptides in the brain exert either a stimulatory or inhibitory effect on the release of pituitary hormones *in vivo*. These regulatory peptides gain entry into the hypophysial portal circulation and activate specific receptors on appropriate pituitary gland cells. The isolation, characterization, and synthesis of peptides were dependent upon *in vitro* or *in vivo* bioassay systems. In such systems, the ability of diencephalic extracts and/or synthesized peptides to alter the basal release of pituitary hormones was resolved. The naming of these regulatory compounds was determined by the pituitary hormone secretion most affected or first examined following the administration of the presumed regulatory compounds. Thyrotropin-releasing hormone (TRH, TRF, Thyroliberine), luteinizing hormone-releasing hormone (LHRH, LRF, GnRH, Gonadoliberine), and somatostatin (GIF, SRIF, SS) were named after they were found to alter the release of thyroid-stimulating hormone, luteinizing hormone, and growth hormone, respectively. However, TRH, LHRH, and somatostatin have each been found, under appropriate conditions, to affect the release of at least one additional pituitary hormone. (For recent reviews, see Refs. 1 and 2.)

The traditional hypothesis that hypothalamic peptides are confined to the diencephalon and involved only in the regulation of pituitary function has given

Robert L. Eskay • Laboratory of Preclinical Studies, National Institute on Alcohol Abuse and Alcoholism, Rockville, Maryland 20852. *Present address:* Laboratory of Clinical Science, National Institute of Mental Health, Bethesda, Maryland 20014.

way to a broader concept that implicates the hypothalamic regulatory peptides as neurotropic substances—peptidergic neurotransmitters or general modulators of central nervous system (CNS) function. Brain peptides such as TRH,[3-6] LHRH,[7,8] and somatostatin[9,10] are found in extrahypothalamic regions of many species, and numerous investigators have demonstrated that these peptides can act directly on the CNS. For example, it has been demonstrated that the administration of TRH reduces depression in humans,[11-13] that the iontophoresis of TRH, LHRH, and somatostatin alters electrical activity of neurons,[14-17] and that membrane depolarization induces the release of various peptides *in vitro* from synaptosome-enriched homogenates of the brain.[18-20] In short, all of the characterized oligopeptides present in the CNS modify, by direct action, some aspect of brain function. For the interested reader, several recent monographs have appeared that point out the multiplicity of effects of all the known oligopeptides on brain function.[21-24]

The discussion that follows will examine first the findings that the administration of TRH and related substances modify general conditions, such as narcosis and hypothermia, induced by a class of compounds referred to as sedative/hypnotics. Second, available information on the possible mechanism(s) and site(s) of action of TRH in the CNS will be addressed. Third, circumstantial evidence linking TRH with some aspects of the ethanol withdrawal syndrome will be examined. Fourth, a review of the central action of somatostatin on locomotor, behavioral, and electrophysiological events in the conscious or sedated animal will be provided. Fifth, the paucity of information on potential molecular mechanisms mediating the action of somatostatin in the CNS will be evaluated. Finally, a summary section will be included with recommendations for the purpose of stimulating additional research aimed at understanding the role of TRH and somatostatin in CNS homeostasis.

B. INTERACTIONS OF THYROTROPIN-RELEASING HORMONE WITH ETHANOL AND OTHER SEDATIVE/HYPNOTICS

To further clarify the role of TRH as a neurotropic substance, animal studies have ensued which support the suggested antidepressant qualities of TRH, as noted in humans. For example, TRH administration antagonizes several testable central effects of a variety of CNS depressants. Breese *et al.*[25,26] demonstrated that intravenously, intraperitoneally, or intracisternally administered TRH antagonized ethanol narcosis and ethanol-induced hypothermia in rodents without affecting the metabolism rate of ethanol. Neither the deamidated metabolite of TRH nor the constituent amino acids of TRH had any effect on either action of ethanol.[25] Cott *et al.*,[26] however, found that several TRH analogues, as well as the deamidated derivative, did affect ethanol-induced sleep time and/or hypothermia. In contrast to the observed effect of TRH, the administration of amphetamine prolonged ethanol-induced sleep time and resulted in a marked increase in locomotor activity, therefore suggesting that TRH and amphetamine have different mechanisms of action.[25] Prasad *et al.*[27] found that acid TRH did alter ethanol-induced narcosis and reported that another metabolite of TRH (histidyl-proline diketopiperazine) was more potent than TRH in antagonizing ethanol-induced sleep time. Others[28] have shown that TRH and a related tripeptide lessened ethanol-induced hypothermia and modified in part the depressant action of ethanol on a variety of behavioral tasks. Numerous studies have demonstrated that the ability of TRH to antagonize the depressant effects of ethanol is not unique and that TRH lessens the CNS depression associated with barbiturates, ether, chloral hydrate, reserpine, chlorpromazine, and diazepam.[26,29–34] In addition, Horita *et al.*[35] found that TRH, in doses of 0.1–100 μg given intracerebroventricularly, decreased pentobarbital-induced narcosis in rabbits, and that TRH in doses greater than 10 μg produced behavioral excitation and hyperthermia. However, TRH antagonism of phenobarbital-induced narcosis was not as apparent.

C. POSSIBLE MECHANISM(S) OF ACTION(S) OF THYROTROPIN-RELEASING HORMONE IN THE CENTRAL NERVOUS SYSTEM

Although it is known that the administration of TRH results in profound behavioral changes and alterations of other testable parameters by direct action on the CNS, the biochemical mechanisms mediating these changes are not clear. The involvement of acetylcholine, biogenic amines, and cyclic nucleotides as possible mediators in the central action(s) of TRH has been examined. TRH administration did not alter brain levels of norepinephrine, dopamine, or serotonin[30,36,37] or the uptake or metabolism of intracisternally administered [^{3}H]norepinephrine.[36] Keller *et al.*[38] suggested that TRH administration activates noradrenergic neurons throughout the brain via enhanced turnover and release of noradrenaline. In another study,[39] TRH administration enhanced the release and turnover of norepinephrine in the rat brain but did not alter endogenous norepinephrine levels. TRH-induced release of [^{3}H]norepinephrine and [^{3}H]dopamine from hypothalamic and striatal synaptosomes suggests that TRH has a direct effect on catecholaminergic neurons. Furthermore, the decrease in [^{3}H]norepinephrine and the increase in [^{3}H]normetanephrine observed in rat brains treated with TRH indicates that TRH is causing an increased release of norepinephrine. These results are consistent with the findings of elevated concentrations of norepinephrine at postsynaptic receptor sites, and they provide a tentative explanation for the observed analeptic and antidepressant properties of TRH. Constantinidis *et al.*[40] suggest that the administration of TRH enhances norepinephrine release. These workers found that TRH (20–40 mg/kg), given ½, 2, or 3 hr before α-methyl-p-tyrosine (AMPT), enhanced the decrease of green fluorescence in norepinephrine terminals of the cerebral cortex; and a similar effect was noted in the hypothalamus only after TRH (40 mg/kg) was adminis-

tered ½ or 1 hr prior to AMPT. However, Reigle et al.[41] were unable to demonstrate a change in metabolism or disposition of [³H]norepinephrine following
acute or chronic TRH treatment.

Several investigators have used biogenic amine depletors and antagonists in
their studies of the CNS action(s) of TRH. Breese et al.[30] provided both
supportive and conflictive evidence for catecholaminergic involvement in the
central action of TRH. Kulig[42] demonstrated that the depressant effects of
AMPT on motor activity and conditioned avoidance behavior were antagonized
by TRH. In a study by Horita et al.,[43] AMPT, 6-hydroxydopamine, and
phenoxybenzamine were found to lessen the behavioral excitatory actions of
rabbits administered 25 μg of TRH, intracerebroventricularly, but the administration of 100 μg of TRH was not affected by catecholamine antagonists.
Furthermore, in pentobarbital-narcotized rabbits, treatment with AMPT, 6-
hydroxydopamine, or phenoxybenzamine lessened the arousal effect of intracerebroventricularly infused TRH (100 μg). It appears that the dose of TRH
administered, if large enough, overrides the effects of catecholamine antagonists
in conscious but not sedated animals. In support of this idea, Horita and
Carino[44] earlier reported that the effects of intracerebroventricularly infused
TRH (100 μg) were not antagonized by adrenergic or serotonergic blockers,
amine depletors, or most depressants. Although the authors made only slight
reference to this manuscript in their follow-up paper,[43] their initial findings
suggest that the dose of TRH administered affects the outcome of a particular
experiment and the conclusions derived thereof. In fact, in contrast to their later
findings,[43] they concluded that catecholaminergic pathways are not essential for
TRH-induced hyperthermia and modification of behavior.[44] Thus, both catecholamine depletors (AMPT, 6-hydroxydopamine) and blockers (phenoxybenzamine) may affect the actions of TRH. However, Cott et al.[26] suggest alternative
explanations, for they found that prior treatment of mice with antiadrenergic
drugs (AMPT, phentolamine, propanolol) did not interfere with TRH-induced
lessening of sleep in ethanol-narcotized mice.

The site(s) and mechanism(s) of action(s) of TRH on catecholamine pathways responsible for the effect of TRH-induced arousal are not clear; however,
Horita et al.[44] point out the analeptic action of TRH is similar to that produced
by d-amphetamine and suggest[43] that the reticular formation of the brain stem
may be involved. The authors further point out[43] that TRH probably acts at
some site(s) outside the reticular formation. They theorize that signals reach the
reticular formation causing norepinephrine release from nerve terminals, and
that this, in turn, leads to arousal. The findings of Cohn et al.[45] suggest also that
some of the central effects of TRH may be mediated via catecholaminergic pathways in the brain. Similar results produced by the infusion of TRH and d-
amphetamine (indirect-acting dopamine releaser) suggest that TRH acts also as
an indirect dopamine releaser in nigrostriatal dopamine pathways. In support of
this finding, they found that previous treatment with haloperidol (dopaminergic
inhibitor) blocked d-amphetamine or TRH-induced rotations. Additional work is
necessary to determine if TRH-induced changes are mediated through altered

dopamine metabolism or through altered receptor sensitivity. However, TRH does appear to influence locomotor activity, not unlike amphetamine, through well-defined dopaminergic pathways in the striatum.

In addition to the purported involvement of catecholaminergic pathways in the action of TRH in the CNS, several investigators[26,30,35,46] have suggested that some of the TRH-induced effects on the CNS may be acting through cholinergic systems. The intracisternal administration of anticholinergic agents (atropine sulfate and atropine methyl nitrate) in mice antagonized the action of TRH on pentobarbital- or ethanol-induced sleep time, but these drugs were additive with TRH in shortening ethanol-induced sleep time.[26] Furthermore, the administration of carbachol (a cholinergic agonist) shortened ethanol-induced sleep time; however, the combination of TRH and carbachol did not. Pharmacological studies indicate that cholinergic systems in the brain may be involved in the action of TRH, but insufficient evidence is available at this time to prove or disprove this theory. In addition to a role for TRH as a neurotransmitter linking various cholinergic neuronal systems, Yarbrough[46] demonstrated that TRH applied by iontophoresis to somatosensory cortical neurons did not alter cellular excitability. However, TRH enhanced the excitability of acetylcholine or carbachol, suggesting that TRH may function indirectly as a modulator of various transmitters.

Finally, since dibutyryl cyclic AMP, a phosphodiesterase-resistant analogue of cyclic AMP, is considered an arousal factor and the administration of TRH shortens the sleep time of a variety of CNS depressants, the possibility of TRH acting as an activator of cyclic AMP in the regulation of narcosis has been studied.[47–49] In short, the dissimilarities of effects on narcosis, body temperature, and locomotor activity following the intracerebroventricular administration of TRH or dibutyryl cyclic AMP do not support the notion that the action of TRH is via the cyclic AMP system.

D. EFFECT OF THYROTROPIN-RELEASING HORMONE ON ETHANOL WITHDRAWAL SYNDROME

Although the mechanisms of actions of TRH on the CNS are obscure, TRH administration appears to ameliorate some aspects of the ethanol withdrawal syndrome in humans. In clinical studies, investigators[50,51] have reported that TRH lessens depression associated with the ethanol withdrawal syndrome. Loosen et al.[52] have reported antidepressant actions of TRH only on the day of injection in humans experiencing the ethanol withdrawal syndrome. A complex array of signs and symptoms are characteristic of the ethanol withdrawal syndrome[53] and, most likely, involve multiple brain areas and biochemical events; therefore, any conclusions derived from these studies that might be useful adjunct therapy for the depressed alcoholic must await further investigations. Since similar behavioral signs, like shaking and tremors, are common in withdrawal from addictive compounds such as ethanol or morphine, the findings of

Wei *et al.*[54] are worth mentioning. Microinjections of TRH into areas of the CNS [where naloxone (opiate antagonist) precipitates withdrawal shaking in morphine-dependent rats] produce shaking, and these areas of the brain are known to contain endogenous TRH. Future studies utilizing the technique of microinjections of TRH into discrete areas of the brain in ethanol-dependent rats may further our understanding of the possible specific areas of the brain responsible for the increased locomotor activity typical of the ethanol withdrawal syndrome.

E. INTERACTIONS OF SOMATOSTATIN WITH ETHANOL AND OTHER SEDATIVE/HYPNOTICS

In contrast to the established effects of TRH as an antidepressant, studies with somatostatin indicate that the administration of somatostatin results in reduced motor activity,[48,55,56] enhanced pentobarbital-induced sleep time, and hypothermia,[29,31] as well as increased pentobarbital-induced mortality.[34] These findings suggest that the primary influence of somatostatin on the CNS is inhibitory; however, Plotnikoff *et al.*[57] have found that somatostatin produces excitatory effects as well.

Recently, Rezek *et al.*[58–60] in a series of methodologically sound reports, have indicated that the primary effect of somatostatin on the brain is one of excitation and only at larger, possibly pharmacological doses does somatostatin give the impression of inducing tranquility or sedation. The intraventricular application of somatostatin (1 μg/μl per min for 10 min) resulted in stereotyped circular running, catatonia, paraplegia-in extension, and/or tonic–clonic seizures; whereas, the infusion of somatostain at a lower rate (0.5 μg/μl per min for 20 min) did not precipitate the more profound behavioral changes such as seizures or paraplegia-in extension.[58] Furthermore, alterations in sleep–wake patterns with reductions in slow-wave and REM sleep were noted following infusion of somatostatin or TRH. As mentioned previously, the infusion of TRH (1 μg/μl per min for 10 min) increased motor activity and stereotyped behavior; however, somatostatin was a more potent excitant than TRH at the same dose. In fact, a dose of glutamic acid (a putative neurotransmitter with known excitatory action) 100 times greater than the amount of somatostatin required to produce seizures was needed to induce similar activation. These findings are at odds with several earlier reports which suggested that somatostatin was a CNS depressant. This discrepancy is probably due to differences in routes of administration (systemic versus intraventricular), dose, and/or duration of contact (of the active agent, somatostatin) with areas of the brain sensitive to the effects of this peptide.

Presumably, substances are distributed rapidly and evenly throughout the CNS following intraventricular injection; therefore, Rezek and co-workers have labored to determine if various brain loci have different thresholds for somatostatin. Cortical infusion of somatostatin (1 μg/μl per min for 10 min) in normal or hypophysectomized rats resulted in similar behavioral, motor, and electrophysiological effects, as noted following intracerebroventricular administration.[59]

The stereotyped behavior which was produced was not as long lasting or as intense as that seen with intracerebroventricular infusions; however, cortical infusions of somatostatin produced prolonged coordination difficulties, along with drowsiness. This contrasted to intracerebroventricular administration, which produced activation with only slight coordination problems. Furthermore, overall sleep reduction was less after cortical application of somatostatin. Next, since the hippocampus and striatal complex have large surface areas in contact with the lateral ventricles and are likely areas of action of somatostatin, both areas were examined by direct cerebral infusion of somatostatin.[60,61] Infusion into the hippocampus produced behavioral and electrophysiological changes that were dose-related to the amount of somatostatin given; whereas, several somatostatin analogues failed to induce similar results.[60] Injection of somatostatin into the basal ganglia produced effects similar to intracerebroventricular infusions of somatostatin. Slight changes in behavioral and electrophysiological indices were probably related to the specific area of infusion in the striatal complex.[61] Behavioral excitation was typical at low doses of somatostatin (0.01 or 0.1 μg); whereas, at higher, possibly pharmacological doses, the excitatory effect was masked temporarily by disturbances of motor control. The amygdala was also implicated as one of the functionally active sites of somatostatin action in the brain.[62] Infusion of lower doses of somatostatin (0.01 or 0.1 μg) into the amygdala stimulated behavioral activity, tremors, and stereotyped movements, whereas higher doses (1.0 or 10.0 μg) resulted in serious motor disturbances, which resulted in behavioral inhibition. Both high and low doses of somatostatin reduced slow-wave and rapid-eye-movement sleep, but somatostatin analogues had virtually no effect on behavior or sleep–wake cycles.

Clearly, somatostatin produces a plethora of behavioral and electrophysiological changes by direct action on the CNS; and upon application to specific loci in the brain, the presence, intensity, duration, and sequence of appearance of specific changes can be altered. Furthermore, the finding that strict, structural requirements associated with endogenous somatostatin are needed to elicit the various changes outlined above is further evidence that somatostatin is playing some physiologically important role in the CNS. Although the interaction of somatostatin and several CNS depressants has been evaluated, modifications in the activity of somatostatin-containing neurons in the presence of ethanol has not been examined. Knowing that ethanol alters sleep–wake cycles and that animals undergoing withdrawal from ethanol exhibit spontaneous seizures, tremors, body rigidity, and stereotyped behavior,[63] the possible involvement of somatostatin in the manifestation of these signs becomes more believeable.

F. POSSIBLE MECHANISM(S) OF ACTION(S) OF SOMATOSTATIN IN THE CENTRAL NERVOUS SYSTEM

As discussed in an earlier section, several possible mechanisms of action for TRH have been proposed, although available hypotheses require more rigorous scientific documentation. Unfortunately, even less is known with regard to the biochemical events that lead to observed behavioral and electrophysiological changes following somatostatin administration. Cohn and Cohn[56] suggest that the CNS actions of somatostatin may be mediated in part through cholinergic mechanisms. Cerebral infusions of somatostatin (25–50 μg) induced a behavioral response referred to as "barrel rotation," whereas intraperitoneal injection of atropine prior to or during rotation blocked these somatostatin-induced changes. Pretreatment of rats with haloperidol, reserpine, or apomorphine (drugs that alter dopamine-mediated activity) had no effect on somatostatin-induced barrel rotations. Therefore, a single somatostatin-induced behavioral change (barrel rotation) requires a functional cholinergic system, whereas a similar TRH-induced change in locomotor activity apparently operated through a dopaminergic mechanism.[45]

Since the postsynaptic actions of several putative neurotransmitters involve cyclic AMP as a second messenger, changes in cyclic AMP levels in the neostriatum, cerebral cortex, and hippocampus following intracerebroventricular infusion of somatostatin have been examined by Herchl et al.[64] Cyclic AMP levels increased in the cerebral cortex, neostriatum, and hippocampus at 5 and/or 15 min after the intracerebroventricular infusion of somatostatin (10 μg). Sotalol (β-adrenergic blocker) pretreatment eliminated or lessened somatostatin-induced increases in cyclic AMP in the hippocampus and cerebral cortex but not in the neostriatum. However, the behavioral effects of intracerebroventricular-injected somatostatin appear not to involve cyclic AMP as the second messenger, since the typical locomotor excitation induced by somatostatin was unaffected by pretreatment with sotalol, whereas cyclic AMP levels were reduced.

G. SUMMARY AND CONCLUSIONS

In addition to the unequivocal role of TRH and somatostatin as regulators of pituitary hormone secretion and to reports implicating somatostatin as a regulator of various peripheral hormone secretions, the direct action(s) of somatostatin and TRH on the CNS as neurotropic agents is apparent. Clearly, the central administration of somatostatin and TRH induces changes in temperature, narcosis, locomotion, and various electrophysiological events. Unfortunately, an understanding of the biochemical basis for these changes is not evident. It appears that some of the central actions of TRH involve biogenic amine pathways and/or enhanced release and turnover of norepinephrine. In addition, cholinergic mediation of the effects of TRH has been suggested; however, the evidence is inconclusive. At present, results do not support the hypothesis that cyclic AMP is the second messenger for TRH or somatostatin. Although even less is known about the molecular events responsible for somatostatin-induced effects in the CNS, cholinergic pathways are implicated in at least one behavioral event (barrel rotation).

The intent of this review was to focus on the ability of TRH and somatostatin to modify several aspects of intoxication. Insufficient information was available to make even a fledgling attempt at understanding the biochemical changes responsible for the observed analeptic actions of TRH in ethanol-sedated animals without reference to papers concerning the possible relationship between other sedative/hypnotics and this peptide. Justification for this general approach may be found in the knowledge that, in general, all the sedative/hypnotics are synergistic or additive with the effects of ethanol.

In reviewing the rapidly burgeoning area of the effects of brain peptides on CNS function, one empathizes with the worker using various behavioral or locomotor changes which depend on the continuity of multisynaptic connections in his or her attempt to unravel precise molecular mechanisms of actions of the various putative peptidergic neurotropic substances. For example, duration of sleep has been used extensively to evaluate the effect of TRH and somatostatin on narcosis, but agreement among workers is lacking on the location of sleep centers and the molecular events that lead to sleep or arousal. The importance of early studies utilizing gross behavioral and physiological changes to establish that the various brain peptides can act as neurotropic substances should not be overlooked. Putative neurotransmitters (biogenic amines and acetylcholine) have been implicated as mediators in the action of TRH and somatostatin in the CNS, although considerable disagreement prevails among workers. The need to examine in greater detail alterations in turnover rates and/or release of putative neurotransmitters (nonpeptidergic) in the CNS of conscious animals following intracerebroventricular infusion or precise microinjections of TRH and somatostatin into areas containing these endogenous peptides is obvious.

Further clarification of the phsyiological activities of TRH and somatostatin in the conscious animal is necessary. Lacking are studies designed to test the hypothesis that ethanol and other sedative/hypnotics alter the actions of brain cells containing TRH and somatostatin. Determination of turnover rates, synthesis, and/or release of TRH and somatostatin *in vivo* in the CNS in the conscious animal and intoxicated animal have not been accomplished because rapid, reliable methods to assess these end points are not known. However, radioimmunoassays for TRH and somatostatin are common, so determination of these peptides in various regions of the brain from acutely and chronically treated or ethanol-dependent animals may provide suggestive information as to the physiological role of TRH and somatostatin in ethanol abuse. High- and low-affinity binding receptors for TRH on plasma membranes have been demonstrated in the CNS[65]; therefore, the effects of long- and short-term ethanol consumption on TRH and somatostatin-binding membranes in the brain could be performed.

Without exception, whenever the interactions of TRH and somatostatin with ethanol or other sedative/hypnotics were examined, mention was not made of the biphasic action of the latter agents on the brain. It is probable that considerable confusion exists in the ethanol literature as it relates to brain function because of the lack of attention to this fact. In short, ethanol is a stimulant at low doses (less than 1 g/kg) and a depressant at moderate to high doses (greater than 1 g/kg). Furthermore, if large doses of ethanol are administered, CNS depression is followed by a period of excitation. This is mentioned with the hope that additional biochemical, physiological, and behavioral studies designed to elucidate the interaction of brain peptides with ethanol, in particular, and sedative/hypnotics, in general, will consider the dose of drug used, as well as the time frame of animal sacrifice after large doses of ethanol are administered. Finally, general problems beset peptide pharmacologists in the design and execution of studies aimed at understanding the neurotropic effects of brain peptides. The shortcomings of not considering in detail the metabolic, dose–response, specificity of, and pharmacological versus physiological effects of peptides are discussed.

H. REFERENCES

1. S. Reichlin, R. Saperstein, I. Jackson, A. Boyd, and Y. Patel, *Annu. Rev. Physiol.* **38**:389 (1976).
2. W. Vale, C. Rivier, and M. Brown, *Annu. Rev. Physiol.* **39**:473 (1977).
3. A. Winokur and R. Utiger, *Science* **185**:265 (1974).
4. C. Oliver, R. Eskay, N. Ben-Jonathan, and J. C. Porter, *Endocrinology* **95**:540 (1974).
5. M. Brownstein, M. Palkovits, J. M. Saavedra, R. Basseri, and R. D. Utiger, *Science* **185**:267 (1974).
6. I. Jackson and S. Reichlin, *Endocrinology* **95**:854 (1974).
7. T. Hokfelt, K. Fuxe, M. Goldstein, O. Johansson, H. Frazer, and S. Jeffcoate, in: *Anatomical Neuroendocrinology* (W. E. Stumpf and L. D. Grant, eds.), pp. 240–262, S. Karger, Basel (1976).
8. C. Beattie, A. Corbin, T. Foell, V. Garsky, W. McKinley, R. Rees, D. Sarantakis, and J. Yardley, *J. Med. Chem.* **17**:1016 (1974).
9. W. Vale, P. Brazeau, C. Rivier, M. Brown, B. Boss, J. Rivier, R. Burgus, N. Ling, and R. Guillemin, *Recent Prog. Horm. Res.* **31**:365 (1975).
10. M. Brownstein, A. Arimura, H. Sato, A. Schally, and J. Kizer. *Endocrinology* **96**:1456 (1975).
11. A. J. Kastin, R. H. Ehrensing, D. S. Schalch, and M. S. Anderson, *Lancet* **10**:740 (1972).
12. A. J. Kastin, N. P. Plotnikoff, R. Hall, and A. V. Schally, in: *Hypothalamic Hormones: Chemistry, Physiology, Pharmacology and Clinical Uses* (M. Motta, P. G. Crosignani, and L. Martini, eds.), pp. 261–268, Academic Press, Inc., New York (1975).
13. A. J. Prange, Jr., I. C. Wilson, P. Lara, L. B. Alltrop, and G. R. Breese, *Lancet* **11**:999 (1972).
14. R. G. Dyer and R. E. Dyball, *Nature (Lond.)* **252**:486 (1974).
15. L. Renaud, J. Martin, and P. Brazeau, *Nature (Lond.)* **255**:233 (1975).
16. R. L. Moss, in: *Hypothalamus and Endocrine Functions* (F. Labrie, J. Meities, and G. Pelletier, eds.), pp. 95–128, Raven Press, New York (1976).
17. R. A. Nicoll, *Nature (Lond.)* **265**:242 (1977).
18. G. Bennett, J. Edwardson, D. Holland, S. Jeffcoate, and N. White, *Nature (Lond.)* **257**:323 (1975).
19. J. Warberg. R. L. Eskay, A. Barnea, R. Reynolds, and J. C. Porter, *Endocrinology* **100**:814 (1977).
20. J. Warberg, C. Oliver, R. L. Eskay, C. R. Parker, A. Barnea, and J. C. Porter, in: *Frontiers of Hormone Research* (U. van Wimeroma Greidanus, ed.), pp. 162–167, S. Karger, Basel (1976).
21. L. Miller, C. Sandman, and A. Kastin, *Neuropeptide Influences on the Brain and Behavior*, pp. 1–298, Raven Press, New York (1977).
22. S. Reichlin, R. Baldessarini, and J. Martin, *The Hypothalamus*, pp. 1–490, Raven Press, New York (1978).
23. U. S. von Euler and B. Pernow, *Substance P*, pp. 1–360, Raven Press, New York (1977).
24. H. Gainer, *Peptides in Neurobiology*, pp. 1–310, Plenum Press, New York (1977).
25. G. Breese, J. Cott, B. Cooper, A. J. Prange, Jr., and M. Lipton, *Life Sci.* **14**:1053 (1974).
26. J. Cott, G. Breese, B. Cooper, T. Barlow, and A. Prange, Jr., *J. Pharmacol. Exp. Ther.* **196**:594 (1976).

27. C. Prasad, T. Matsui, and A. Peterkofsky, *Nature (Lond.)* **268**:142 (1977).
28. C. Porter, V. Lotti, and M. DeFelice, *Life Sci.* **21**:811 (1977).
29. A. J. Prange, Jr., G. Breese, G. Jahnke, B. Martin, B. Cooper, J. Cott, I. Wilson, L. Alltop, M. Lipton, G. Bissette, C. Nemeroff, and P. Loosen, *Life Sci.* **16**:1907 (1975).
30. G. Breese, J. Cott, B. Cooper, A. Prange, Jr., M. Lipton, and N. Plotnikoff, *J. Pharmacol. Exp. Ther.* **193**:11 (1975).
31. A. J. Prange, Jr., G. Breese, G. Jahnke, B. Cooper, J. Cott, I. Wilson, M. Lipton, and N. Plotnikoff, *Neuropsychobiol.* **1**:121 (1975).
32. A. Prange, Jr., G. Breese, J. Cott, B. Martin, B. Cooper, I. Wilson, and N. Plotnikoff, *Life Sci.* **14**:447 (1974).
33. J. M. Stolk and B. C. Nisula, in: *Hormones, Homeostasis and the Brain* (W. H. Grupen, T. B. van Wimersma Greidanus, B. Bohus, and D. de Wied, eds.), pp. 47–56, Elsevier Publishing Company, Amsterdam (1975).
34. M. Brown and W. Vale, *Endocrinology* **96**:1333 (1975).
35. A. Horita, M. Carino, and R. Chesnut, *Psychopharmacologia* **49**:57 (1976).
36. G. Breese, B. Cooper, A. Prange, Jr., J. Cott, and M. Lipton, in: *The Thyroid Axis, Drugs, and Behavior* (A. J. Prange, Jr., ed.), pp. 115–127, Raven Press, New York (1974).
37. N. Plotnikoff, *Prog. Brain Res.* **42**:11 (1975).
38. H. H. Keller, G. Bartholini, and A. Pletscher, *Nature (Lond.)* **248**:528 (1974).
39. W. Horst and N. Spirt, *Life Sci.* **15**:1073 (1974).
40. J. Constantinidis, F. Geissbuhler, J. Gaillard, T. Hovaguimian, and R. Tissot, *Experientia* **30**:1182 (1974).
41. T. Reigle, J. Avni, P. Platz, J. Schildkraut, and N. Plotnikoff, *Psychopharmacologia* **37**:1 (1974).
42. B. Kulig, *Neuropharmacology* **14**:489 (1975).
43. A. Horita, M. A. Carino, and H. Lai, *Prog. Neuropsychopharmacol.* **1**:107 (1977).
44. A. Horita and M. Carino, *Psychopharmacol. Commun.* **1**:403 (1975).
45. M. L. Cohn, M. Cohn, and F. H. Taylor, *Brain Res.* **96**:134 (1975).
46. G. G. Yarbrough, *Nature (Lond.)* **263**:523 (1976).
47. M. L. Cohn, H. Yamaoka, and B. Kraynack, *Neuropharmacology* **12**:401 (1973).
48. M. L. Cohn, in: *Molecular Mechanisms of Anesthesia* (R. B. Fink, ed.), pp. 485–500, Raven Press, New York (1975).
49. M. L. Cohn, M. Cohn, B. Krysik, and F. Taylor, *Pharmacol. Biochem. Behav.* **5**(1):129 (1976).
50. A. J. Prange, Jr., I. C. Wilson, G. Breese, and M. Lipton, *Prog. Brain Res.* **42**:1 (1975).
51. L. I. Huey, D. S. Janowsky, A. J. Mandell, L. C. Judd, and M. Pendery, *Psychopharmacol. Bull.* **11**:29 (1975).
52. Von P. T. Loosen, I. C. Wilson, P. Lara, A. J. Prange, Jr., and C. Pettus, *Arzneim. Forsch. Drug Res.* **26**:1164 (1976).
53. M. Victor and R. D. Adams, *Res. Publ. Assoc. Res. Nerv. Ment. Dis.* **32**:526 (1953).
54. E. Wei, S. Sigel, H. Loh, and E. Way, *Nature (Lond.)* **253**:739 (1975).
55. D. S. Segal and A. J. Mandell, in: *The Thyroid Axis, Drugs and Behavior* (A. J. Prange, Jr., ed.), pp. 129–133, Raven Press, New York (1974).
56. M. L. Cohn and M. Cohn, *Brain Res.* **96**:138 (1975).
57. N. Plotnikoff, A. Kastin, and A. Schally, *Pharmacol. Biochem. Behav.* **2**:693 (1974).
58. V. Havlicek, M. Rezek, and H. Friesen, *Pharmacol. Biochem. Behav.* **4**:455 (1976).
59. M. Rezek, V. Havlicek, K. Hughes, and H. Friesen, *Pharmacol. Biochem. Behav.* **5**:73 (1976).
60. M. Rezek, V. Havlicek, K. Hughes, and H. Friesen, *Neuropharmacology* **15**:499 (1976).
61. M. Resek, V. Havlicek, L. Leybin, C. Pinsky, E. Kroeger, K. Hughes, and H. Friesen, *Can. J. Physiol. Pharmacol.* **55**:234 (1977).
62. M. Rezek, V. Havlicek, K. Hughes, and H. Friesen, *Neuropharmacology* **16**:157 (1977).
63. E. Majchrowicz, *Psychopharmacologia* **43**:245 (1975).
64. R. Herchl, V. Havlicek, M. Rezek, and E. Kroeger, *Life Sci.* **20**:821 (1977).
65. D. Burt and S. H. Snyder, *Brain Res.* **93**:309 (1975).

Ethanol and Endocrine Function   10

Ryoko Kakihana and John C. Butte

A. INTRODUCTION

In the past two decades there has been a rapid increase in the knowledge of endocrine function in man and animals. It is clear that the central nervous system (CNS) plays a major role in regulating endocrine function. Therefore, drugs affecting the CNS have gained renewed interest in studies of neuroendocrine interactions. Studies of ethanol on endocrine systems have been variable in coverage. Although clinical observations associating such endocrinopathies as hypogonadism and gynecomastia with male alcoholics were made in the early 1900s, the hormonal alterations responsible for these conditions have been demonstrated only recently. The availability of sensitive and specific assays will undoubtedly play a major role in understanding the pathological consequences of ethanol on neuroendocrine function.

Although the prime objective of ethanol studies is to understand the human disease state alcoholism, animal studies are essential in elucidating the biochemical and physiological processes which lead to the observed pathologies of ethanol abuse. The variability due to genetic susceptibility, nutritional status, duration of alcohol use, type of alcoholic beverage consumed, and the advanced state of the disease itself makes the alcoholic an undesirable experimental subject. Both animal and human studies will be reviewed and, whenever possible, will proceed from the acute effects of ethanol on naive experimental subjects to the effect on chronic alcohol users.

A point crucial to understanding the deleterious effect of ethanol or other drugs on any hormonal system is to determine the site of action of the drug in question. For example, for each neuroendocrine system there exists a balance between the secretions of the brain, pituitary gland, and peripheral target organs;

Ryoko Kakihana and John C. Butte • Department of Biological Sciences, San Francisco State University, San Francisco, California 94132.

and a perturbation at any point in this axis will alter hormone release or synthesis that will, in turn, alter the entire system. In addition, secondary factors, including altered hormonal metabolism and excretion, could result in disturbances of normal endocrine function. Therefore, a complete understanding of the effects of ethanol on each neuroendocrine system will be realized only when the action of ethanol at each level is known. An attempt will be made to specify the site of action of ethanol and the consequences of a particular change on the remainder of each neuroendocrine loop.

B. ENDOCRINE HORMONES OF WATER AND ELECTROLYTE BALANCE AND METABOLISM

1. Vasopressin

The inhibitory effect of ethanol on vasopressin release was one of the earliest ethanol–endocrine effects investigated. Edkins and Murray,[1] while studying the effect of ethanol on glucose tolerance in humans, noted an ethanol-induced diuresis in most of their subjects. Since it had been demonstrated earlier[2] that ethanol decreased urine production in a dog heart–lung–kidney preparation, they concluded that the effect was not a direct effect on the kidneys. The similarity of ethanol diuresis to the diuresis associated with diabetes insipidus and water-induced diuresis led them to suggest that ethanol depressed the activity of the posterior pituitary gland.[1] These results were confirmed and extended by subsequent studies[3,4] demonstrating that ethanol-induced diuresis in man could be antagonized by extracts of the posterior pituitary. In addition, Strauss et al.[5] suggested that ethanol acts on the supraoptico nucleus to block the release of vasopressin. More direct evidence for a central effect of ethanol was provided by van Dyke and Ames,[6] who found that an intracarotid infusion of ethanol (50 mg/kg) into dogs produced a prompt diuresis at a time when the peripheral blood contained no detectable level of ethanol. It was concluded from this that the diuretic effect of ethanol was due to the central inhibition of antidiuretic hormone.

Eggleton[4] noted that urine flow decreased to nearly normal levels after a period of time, even in the presence of relatively high blood ethanol levels (BEL). A second ingestion of ethanol produced only slight diuresis but greatly elevated BEL. A subsequent investigation[6] demonstrated that repeated oral doses of ethanol caused a gradual increase in BEL to 260 mg/dl, whereas the diuretic response decreased to zero. Whether this diminished response to repeated doses of ethanol is due to an increase in extracellular osmolarity or to some other mechanism allowing vasopressin to be released is not apparent.

2. Aldosterone–Renin

The retention of sodium, potassium, and chloride during ethanol-induced diuresis was demonstrated by Nicholson and Taylor.[7] The initial interpretations of these findings suggested that the effect of ethanol was a direct one on renal tubular function. Rubini *et al.*[8] suggested that a possible stimulus might be the redistribution of blood volume caused by peripheral vasodilation. However, changes observed in alcoholic subjects, including sodium retention and low serum potassium levels 3–5 days following ethanol withdrawal, suggested that elevated aldosterone levels might account for these observations.[9]

Fabre *et al.*[10] examined the effect of ethanol administration on the hemorrhage-induced increase of aldosterone secretion in pentobarbital-anesthetized dogs. Under normal experimental conditions there was a biphasic increase in the aldosterone response to terminal hemorrhage. Alterations in the secretion rate of aldosterone were found when the animals were infused with varying amounts of ethanol in saline 1 hr prior to initiation of hemorrhaging. BEL of 100 mg/dl stimulated aldosterone secretion. However, inhibition of aldosterone secretion occurred when BELs were 300 mg/dl or higher. Studies on human alcoholics[10] suggested that urinary aldosterone increased the first 2 days after the initiation of ethanol consumption, but in two subjects urinary aldosterone returned to normal levels even during the period of highest BEL.

Farmer and Fabre[11] administered vodka (14 oz in 5 hr) to five normal male humans, aged 21–25 years, from 0800 to 1200 hr and determined plasma levels of aldosterone, cortisol, and renin at 2-hr intervals. All three hormones increased as BEL declined. Linkola *et al.*[12] demonstrated in normal male humans that aldosterone decreased during the intoxication phase and increased during the withdrawal phase, whereas plasma renin levels increased during both the intoxication and withdrawal phases. The changes in urinary Na^+/K^+ ratios agreed with the plasma aldosterone levels during ethanol intoxication and withdrawal. Oral ethanol administration to rats in a positive water balance produced an increase in urinary sodium and potassium concentration as well as an increase in urine.[13] Therefore, Na^+ retention or loss following ethanol consumption is dependent upon the hydration state of the animal. In addition, Linkola *et al.*[12] suggested that increased renin release following ethanol intoxication and hangover was due to ethanol-induced diuresis under conditions of normal water balance followed by dehydration.

C. ENDOCRINE HORMONES OF ENERGY BALANCE AND METABOLISM

1. Thyroid

Few studies exist concerning the acute and chronic effects of ethanol ingestion on the hypothalamic–pituitary–thyroid axis. Bleecker *et al.*[14] demonstrated a significantly greater ^{131}I uptake following $[^{131}I]$triiodothyronine injection in several tissues of rats acutely treated with ethanol. The thyroid of ethanol-treated rats showed a decreased level of ^{131}I. Chronic ethanol consumption in rats resulted in an increased ^{131}I uptake by the thyroid gland.[15] Whether these findings indicate altered plasma levels of thyroxine or triiodothyronine was not investigated.

Augustine[16] determined the plasma level of protein-bound iodine and triio-dothyronine in 103 acutely intoxicated male and female alcoholics and did not find any abnormality in thyroid function. Ingestion of a moderate amount of ethanol (1.5 g/kg) had no effect on thyroid-stimulating hormone in nine healthy male humans with no history of excessive drinking.[17] Thyroid-stimulating hormone levels after ethanol ingestion were found to be similar to control levels. The response of the pituitary to thyrotropin-releasing hormone injection was similar in the control and ethanol-consuming groups.

The number of studies concerning the acute or chronic effect of ethanol on thyroid function is limited. Although available studies fail to suggest any major changes in thyroid function due to ethanol consumption, further studies would be appropriate to confirm or alter this view.

2. Adrenal Medulla

a. Acute Studies

Studies[18] using dogs demonstrated that an increased urinary output of epinephrine and norepinephrine resulted following the administration of 3.2 g ethanol/kg via stomach tube. The adrenal medulla with intact innervation was required for the increased urinary output of these hormones to occur. Quite high levels of ethanol were used in this study, probably producing a general stress reaction as a result of the circulatory and respiratory effects produced. In humans, urinary excretion of catecholamine was shown to increase after the ingestion of whiskey (0.51–1.27 g ethanol/kg)[19] or whiskey or wine (0.27–0.54 g ethanol/kg).[20]

The direct measurement of catecholamines in adrenal venous blood following ethanol infusion has been done on cats and dogs. Perman,[21] using nembutal-anesthetized cats, found that infusion of ethanol equal to or less than 0.5 g/kg produced little or no changes in catecholamine output. At infusion levels equal to and greater than 0.6 g ethanol/kg there were increases in both epinephrine and norepinephrine output, although the relative amounts of each varied markedly. Similar results were found by Hirose et al.[22] Although ethanol can cause an increase in adrenal–medullary secretion, relatively high doses are necessary.

b. Chronic Studies

The effect of chronic ingestion of ethanol on the medullary output of catecholamines has received little attention, although it is known that epinephrine and norepinephrine secretion are elevated during ethanol withdrawal in alcoholic subjects.[23,24] Alcoholics not showing withdrawal signs and symptoms showed no increase in urinary epinephrine following a single dose of ethanol.[25] A similar effect has been shown in rats[26] chronically consuming 10% aqueous ethanol (v/v). Chronic ethanol consumption reduced urinary catecholamines, although an acute injection of ethanol (1.2 g/kg) produced an increase in urinary catecholamines. The increase was less than that found in control rats not previously exposed to ethanol, thus suggesting a possible adaptation of the adrenal medulla. Pohorecky,[27] however, has provided evidence that adrenal catecholamine turnover is increased in rats chronically treated with ethanol.

The recent development of sensitive radioenzymatic assays for epinephrine and norepinephrine in small amounts of plasma[28,29] will allow a more direct approach to the study of the effect of ethanol on the adrenal–medullary system.

3. Growth Hormone

Growth hormone is released from the pituitary gland in response to a variety of stimuli that can be divided into three general categories[30]: (1) energy substrate deficiency, (2) increase in plasma levels of certain amino acids, and (3) reaction to stressful stimuli,[31,32] which include epinephrine and histamine injection, hemorrhage, and emotional stress. Growth hormone also has a clearly observable circadian secretory rhythm.[33]

Available studies on the effect of ethanol on growth hormone levels are almost totally on human subjects. These studies began with a clinical observation that severe hypoglycemia is frequently associated with alcoholic individuals.[34] According to Freinkel et al.,[35] there have been over 100 cases of ethanol-induced hypoglycemia. This has been shown to be a direct inhibitory effect of ethanol on gluconeogenesis without either an increase in insulin level or a major hepatic malfunction.[36,37] The homeostatic regulation of glucose is very complex, and controlled by a network of many hormones (i.e., insulin, glucagon, growth hormone, and adrenal hormones). Therefore, it is beyond the scope and space allotted to the present chapter to discuss the effect of ethanol on growth hormone in the total context of glucose homeostasis. Consequently, this section will be limited to a brief review of available studies pertaining directly to the changes in growth hormone after ethanol administration. For a review of carbohydrate metabolism in alcoholism, see Ref. 37.

Bellet et al.[38] demonstrated in healthy human volunteers who fasted overnight a significant and sustained elevation of growth hormone after an oral administration of ethanol, 1.2 g/kg in a 20% solution. A concurrent, but slightly delayed decrease in free fatty acids and an increase in plasma 11-hydroxycorticosteroids were also observed in these subjects. Insulin-induced hypoglycemia is a potent stimulus for the release of growth hormone in humans. The effect of ethanol administration on this response was investigated by Priem et al.[39] In normal human volunteers, 50 ml of ethanol "suitably" diluted with water was administered 30 min prior to an insulin injection. In the presence of ethanol, the growth hormone response to the insulin-induced hypoglycemia was significantly reduced in terms of its peak height and duration of response. Since there was no difference in the hypoglycemic effect following insulin injection in the presence or absence of ethanol, it was concluded that ethanol exerted a direct attenuating effect on the brain centers controlling growth hormone secretion. Consistent with this interpretation is the finding[40] that growth hormone response to arginine infusion was reduced in three normal subjects if ethanol was administered prior to the arginine test.

Few studies are available concerning growth hormone levels in chronic ethanol-treated animals or alcoholics. The retarded growth observed in patients suffering from fetal alcohol syndrome will undoubtedly increase interest in this area. A recent paper by Tze *et al.*[41] investigated growth hormone levels in five patients, 0.4–7.6 years of age, who exhibited this syndrome. The blood levels of growth hormone were normal or slightly elevated following insulin-induced hypoglycemia and arginine infusion tests. In the two patients tested for serum somatomedin, the levels were found to be normal. This indicated that the growth hormone was biologically active and suggested the possibility of peripheral tissue insensitivity to the hormone. Obviously, more studies are needed to elucidate the effect of chronic or acute ethanol administration on the release of growth hormone in normal and alcoholic subjects.

4. Glucocorticoids

The major glucocorticoids released from the adrenal cortex of humans and rodents are cortisol and corticosterone. A number of physical, chemical, and psychological stresses are known to activate the hypothalamic–pituitary–adrenocortical system, resulting in the release of glucocorticoids.[42] In the past two decades evidence has been accumulating to indicate that ethanol also activates the release of glucocorticoids from the adrenal glands.

In early studies[43–49] thymic involution and the depletion of adrenal ascorbic acid or cholesterol were used as an index of adrenal–cortical activation. In recent studies[50–53] direct measurements of the elevation of circulating glucocorticoids after ethanol administration have been accomplished.

a. Site of Action

Ellis[51] has shown that the normal secretion of corticosterone following ethanol administration (2 or 4 g/kg) was inhibited in hypophysectomized rats or in pentobarbital/morphine-treated rats. Morphine is a potent inhibitor of adrenocorticotrophic hormone (ACTH) release from the pituitary gland and is often used in combination with pentobarbital to block hypothalamic–pituitary–adrenal activation in experimental preparations.[54] Injection of Dexamethasone, a synthetic glucocorticoid, inhibits the normal release of corticosterone after ethanol injection in mice,[55] which also supports the notion that ethanol exerts its effect directly on the CNS. Furthermore, ethanol challenge results in a significant reduction in pituitary ACTH content[55] as measured by bioassay, which indirectly implies the release of the hormone. In contrast, Leppäluoto *et al.*[17] were unable to demonstrate any significant change in plasma ACTH following the administration of 1.5 g ethanol/kg in normal male subjects. It is likely that the lack of ACTH elevation was caused by the low BEL attained.

It appears that ethanol does not directly stimulate the adrenal–cortical cells. However, whether ethanol stimulates the hypothalamus or pituitary gland, or whether it stimulates (or inhibits) higher loci in the CNS which in turn send the appropriate stimuli to the hypothalamic centers, is an issue still to be resolved.

b. Acute Studies

Ellis[50] has shown an elevation of plasma 17-hydroxycorticosteroids in dogs over a period of 6 hr following ethanol infusion (1, 2, and 3 g/kg). A similar dose–response effect was found in a study on rats.[51] The response has been shown to vary according to the strain of animals involved.[52] Corticosterone levels measured 60 min after intraperitoneal injection of 1.6 g/kg were higher in DBA/2J and Balb/c than C57BL/J mice. The former two strains are characterized as ethanol-avoiding and more sensitive to ethanol measured by ethanol-induced sleep time.[52,56] Suzuki et al.[53] demonstrated a dose-related increase in the rate of 17-hydroxycorticosteroid secretion produced by ethanol in unanesthetized dogs.

Early studies in humans[57,58] presented equivocal results with regard to the stimulatory effect of ethanol on adrenal glucocorticoids. In these studies the plasma corticosteroids actually fell slightly a few hours following the ingestion of ethanol. Later studies by Mendelson and Stein[59] and Fazekas[60] provided the first direct evidence for the activation of the adrenal gland by moderately intoxicating levels (70–100 mg/dl) of ethanol in human subjects. In the former study, serum cortisol levels were measured prior to, during, and after ethanol ingestion in four nonalcoholic psychiatric patients and in four alcoholics. A significant increase of serum cortisol was found in nonalcoholic controls during the drinking phase (the average intake was 2.4–3.7 g/kg per day). In nonalcoholics the cortisol rise was also associated with the development of gastrointestinal illness due to ethanol ingestion, whereas the alcoholics showed the largest increase in serum cortisol during the ethanol withdrawal phase. In Fazekas's study,[60] serial blood samples were obtained in nonalcoholic subjects every hour up to 9 hr after the ingestion of ethanol in the form of wine (1 and 2 g/kg). A biphasic response of blood cortisol to the ethanol ingestion was noted. At 1 hr postingestion, a significant increase occurred and a subsequent decrease of the hormone level below the initial level followed. Studies of healthy human volunteers which have been subsequently reported[61–63] suggest that ethanol at a BEL of about 100 mg/dl or higher stimulates the hypothalamic–pituitary–adrenal system.

c. Chronic Studies

The effect of chronic ethanol administration on the adrenocortical function has been the subject of several animal studies.[51,64–67] Kalant and Czaja[64] found no difference between the levels of ascorbic acid in the adrenals of the ethanol-treated Wistar rats (intubated with 2 g ethanol/kg daily for 1 month) and the adrenals of the controls. In addition, Ellis[51] treated male rats with ethanol once daily for 7 days (2 g/kg). Plasma corticosterone did not differ from controls 30 min after the injection of ethanol on the eighth day.

Pretreatment of C57BL/J male mice with ethanol [10% (v/v) in drinking water] for 16 weeks resulted in the alteration of the temporal pattern of the plasma corticosterone response to ethanol injection (1.6 g/kg).[65] Twenty minutes following ethanol injection, the corticosterone levels were significantly higher in the ethanol-consuming mice. Thereafter (30 and 60 min) corticosterone levels were significantly higher in the control mice. A significant alteration of the adrenocortical response, measured by plasma corticosterone levels 60 min post-ethanol injection, occurred after only 1 week of ethanol consumption. Deadaptation after 6 weeks of ethanol consumption required 2–4 weeks. Ethanol metabolism and adrenal sensitivity to ACTH were shown to be unaltered by chronic ethanol consumption in these animals. The results suggested a direct effect of ethanol on the CNS. Ratcliffe[66] also found diminished corticosterone levels in chronically ethanol consuming female Wistar rats 30–360 min following an acute intraperitoneal injection of ethanol (2 g/kg). A general decrease in hypothalamic sensitivity in ethanol-consuming animals was indicated by a diminished corticosterone response to sound stress and to the hypothermic effects of chlorpromazine. Unfortunately, no mention was made concerning the time of blood sampling. If the samples were obtained during the trough in the circadian rhythm as is normally done, the control levels of corticosterone would be abnormally high.[67]

A recent report by Kakihana and Moore[68] demonstrated that the chronic consumption of ethanol in DBA/2J mice resulted in the alteration of the normal diurnal rhythm of corticosterone. The corticosterone levels were higher during the normal trough period but lower during the daily peak period compared to those of the controls. The results from the above studies described indicate that an appreciable alteration of the CNS has occurred following chronic ethanol consumption in rats and mice.

d. Hypothalamic–Pituitary–Adrenal Responsiveness in Alcoholics

Previous studies on mice[65] and rats[66] suggested that chronic ethanol consumption attenuated the hypothalamic–pituitary–adrenal response to ethanol. It has been shown that serum cortisol is elevated in alcoholics following ethanol consumption.[69] Whether or not an attenuation occurs in alcoholics is an interesting question. Merry and Marks[62] reported that elevated basal levels of cortisol (using a nonspecific fluorometric procedure) decreased in "acutely withdrawn alcoholics" following ethanol ingestion, in contrast to the increase found in the control subjects. These investigators noted that the alleviation of the anxiety normally accompanying ethanol withdrawal in alcoholics may not be the reason for this decrease. Diazepam, a drug which relieved the anxiety symptoms in the alcoholics during ethanol withdrawal, had no effect on cortisol levels. A similar effect not related to ethanol has been observed by Stokes.[70] It was found that in psychiatric patients undergoing electroconvulsive treatment, the plasma cortisol levels fell significantly following treatment in those patients who had initially high values. There was an inverse relationship between the plasma cortisol levels before the electroconvulsive treatment and the changes found 15 and 60 min following treatment. Whether there is a common mechanism for these two somewhat unexpected results, in which high resting levels of glucocorticoid are found to decrease following what is normally a stimulus to the hypothalamic–pituitary–adrenal system, is an intriguing question.

Margraf et al.[71] noted the possible alteration of the diurnal rhythm of hypothalamic–pituitary–adrenal axis. Three alcoholics studied exhibited accentuated variations in corticosteroids with two daily peaks rather than the normal single peak. Whether the two-peaked diurnal pattern represents a desynchronization of the system or a shift toward the normal diurnal rhythm is not clear from the study. Other investigators have alluded to higher morning levels of cortisol in alcoholics as compared to control subjects in the context of possible circadian changes.[59,62]

e. Summary

Available evidence indicates that ethanol activates the hypothalamic–pituitary–adrenal system in various species of animals, with elevation of the circulating glucocorticoids, and that the hormonal changes observed are dose-related to ethanol. In addition, chronic ethanol consumption has been shown to alter the normal circadian rhythm and the responsiveness of the hypothalamic–pituitary–adrenal system to various stimuli.

D. ENDOCRINE HORMONES OF REPRODUCTIVE PROCESSES

1. Testosterone, Luteinizing Hormone, Prolactin, Follicle-Stimulating Hormone

a. Acute Studies

Chaudhury and Matthews[72] prevented pregnancy in estrus rabbits by ethanol intubation (4.0 g/kg) 30 min prior to mating, whereas copper acetate-induced ovulation was not prevented by ethanol, suggesting that ethanol did not act directly on the ovary to prevent ovulation. Kieffer and Ketchel[73] demonstrated that a subcutaneous injection of ethanol (7.9 g/kg), at 1:00 P.M. on the day of proestrus significantly inhibited ovulation in rats and that subcutaneous administration of luteinizing hormone at 2:30 P.M. following the ethanol injection reversed the inhibition of ovulation. From this evidence it was postulated that ethanol inhibited the ovulatory surge of luteinizing hormone.

In the male rat Symons and Marks[74] demonstrated that intubation with 1.87 g ethanol/kg lowered plasma luteinizing hormone levels at 60 min. Cicero and Badger[75] found that an intraperitoneal injection of ethanol (2.5 g/kg) lowered plasma luteinizing hormone and testosterone levels. Luteinizing hormone levels were significantly lower at 2 hr, while the testosterone levels were significantly lower at 8 hr. Although the time course of these two changes was nearly parallel during the first 3 hr, the investigators claimed that the fall in luteinizing hormone levels preceded the drop in testosterone levels by approximately 1 hr and used this to establish a cause-and-effect relationship between the luteinizing hormone and testosterone decreases. Cicero and Badger[75] also demonstrated the possibility of opposite hormonal effects, depending on the dose of ethanol used. Ethanol injections of less than 1 g/kg were stimulatory for luteinizing hormone and testosterone (measured at 3 and 4 hr, respectively), whereas injections more than 1 g/kg were inhibitory. Comparison of the two hormonal responses is confounded somewhat by hormonal measurement at a single time point differing by 1 hr. Although the times of sampling were based on the temporal response to 2.5 g/kg ethanol, it is possible that the temporal response is dose-dependent. Cicero et al.[76] were able to diminish the 24-hr postcastration rise in luteinizing hormone by injecting ethanol (2.5 g/kg, intraperitoneal) at 6-hr intervals starting 2 hr after surgery. Furthermore, the responsiveness of pituitary gonadotrophins in ethanol-treated rats to exogenously administered luteinizing hormone-releasing hormone was normal.

Ylikahri et al.[77] studied the acute effect of ethanol on both testosterone and luteinizing hormone in normal male humans. These investigators did not find any change in plasma testosterone or luteinizing hormone levels during the first few hours following ethanol ingestion (1.5 g/kg). However, approximately 15 hr after the cessation of drinking, there was a definite decrease in plasma testosterone and a concurrent increase in luteinizing hormone. Mendelson and Mello[78] demonstrated that testosterone levels decreased as BEL increased. At the highest BEL, testosterone levels were low and luteinizing hormone levels were high. These overall results are in support of a proposed direct testicular effect of ethanol on testosterone levels.[79] In contrast, Leppaluoto et al.[11] reported that administration of 1.5 g ethanol/kg to nine healthy male subjects between 1800 and 2100 hr produced significantly reduced luteinizing hormone levels at 2400 and 0700 hr. Luteinizing hormone levels after the injection of luteinizing hormone-releasing hormone were similar in ethanol-treated and untreated subjects. It was thus suggested that the inhibitory effect of ethanol on luteinizing hormone was at a suprapituitary site.

b. Chronic Studies

Badr and Bartke[80] reported an experiment in which DBA/2J male mice were intubated once daily for 5 days with ethanol doses ranging from 0.16 to 1.24 g/kg. Testosterone levels determined by radioimmunoassay were drastically reduced 1 hr after the last intubation. Whether the effect was directly on the gonad, or via the suppression of the gonadotrophic hormones from the pituitary gland, was not obvious. Symons and Marks[74] demonstrated that ethanol consumption (15% aqueous ethanol) during neonatal development in the male Wistar rat inhibited the increase in the basal level of luteinizing hormone normally observed. Furthermore, it was found that although the basal levels of luteinizing hormone in the ethanol-treated animals were lower than in the control animals, the responsiveness of the pituitary gland to an injection of exogenous luteinizing hormone-releasing hormone was shown to be normal in the ethanol-treated animals. Thus, it was concluded that chronic treatment of the rat with ethanol had impaired the release of luteinizing hormone-releasing hormone from the hypothalamus.

In a study designed to develop a model for ethanol-induced hypogonadism, Van Thiel *et al.*[81] investigated the effect of chronic ethanol consumption in male rats using a liquid ethanol diet with isocaloric sucrose control from day 20 to day 61. Testicular, prostatic, and seminal vesicle atrophy as well as lowered plasma testosterone levels occurred in the ethanol-consuming animals. Although both sucrose and ethanol liquid diet groups showed a significant decrease in weight gain when compared to *ad libitum* rat chow controls, the various parameters measured when normalized for body weight were not different between the *ad libitum* and the isocaloric sucrose controls. This demonstrated that the caloric deprivation found with ethanol consumption was not responsible for the hormonal and organ weight changes observed.

It has recently been shown that plasma estradiol as well as hepatic microsomal aromatase activity were elevated in rats following chronic ethanol consumption.[82] At the same time the *in vitro* rate of gonadal estradiol secretion was unchanged. This was interpreted as evidence for a direct effect of ethanol on estradiol synthesis in the liver and perhaps other peripheral tissues.

The effect of chronic consumption of ethanol on gonadal hormones in normal human subjects was recently reported by Gordon *et al.*[82] Subjects were given continued oral administration of ethanol with an adequate diet fortified with vitamins. An average of 3 g ethanol/kg was consumed as 15% solution (w/v) in flavored beverage every 3 hr. In one group consisting of five men, liver biopsies were performed for histological examination and measurement of testosterone A-ring reductase activity before and after ethanol administration lasting 24–26 days. In the second group of six subjects, some were tested for the metabolic clearance rate of testosterone, others for plasma levels of testosterone, androstenedione, estradiol-17β, luteinizing hormone, follicle-stimulating hormone, and plasma protein-binding capacity for testosterone. The results clearly indicated that plasma testosterone was significantly decreased by 29–55% after ethanol administration for a period up to 24 days. The decrease of plasma testosterone levels following ethanol intake was partially explained by the increased metabolism, which was attributed to the increased activity of hepatic testosterone A-ring reductase, the rate-limiting enzyme in the metabolism of testosterone. But these investigators noted that the reduced testosterone levels might be better accounted for by the reduced production rate of the hormone, which was calculated as the product of metabolic clearance and plasma concentration. The continued oral administration of ethanol additionally resulted in the reduced plasma-binding capacity for the androgen, which could have contributed to the increased rate of androgen clearance. The lack of a consistent pattern of luteinizing hormone suppression during periods of decreased episodic secretion of plasma testosterone in the early part of the study suggested that ethanol was acting directly on the gonads. Since recent studies by Rubin et al.[84,85] have implicated a role for prolactin in secretion of testosterone by the testis, it is premature to associate a decrease in plasma testosterone, independent of a similar change in luteinizing hormone, with a direct effect of ethanol on gonadal function.

c. Studies on Alcoholics

The hypogonadism and hyperestrogenization often observed in chronic alcoholic males were first documented more than 50 years ago.[86] Clarification of the causative factors has occurred mainly in the last few years, with the development of rather elegant analytical procedures. Initially, the problem was attacked by investigating the levels of steroids likely to be responsible for the effects. Such an approach fails to distinguish between an ethanol-caused change in hormone release pattern and metabolic changes in existing hormone levels.

Korenman et al.[87] demonstrated an elevation of estradiol in humans with gynecomastia and cirrhosis. Others, however, have not.[88,89] An increased urinary excretion of testosterone glucuronide was first noted by Fabre et al.[90] This was interpreted as indicative of a shift from oxidative pathways of androgen metabolism to reductive, since no evidence pointed to increased testosterone synthesis or increased renal clearance of testosterone glucuronide. Coppage and Cooner[91] reported a lowered level of testosterone in several male alcoholics, which was confirmed by Southren et al.,[92] who additionally provided evidence for a lowered rate of testosterone synthesis. On the basis of a prompt rise in testosterone following administration of human chorionic gonadotrophin to male alcoholics, Southren et al.[92] proposed that hypogonadism is probably secondary to hypothalamic–pituitary suppression rather than primary testicular dysfunction. Van Thiel et al.[89] confirmed the lowered testosterone levels and normal responsiveness of alcoholics to human chorionic gonadotrophin. They also demonstrated inadequate increases of follicle-stimulating hormone and luteinizing hormone following clomiphene stimulation. Van Thiel et al.[93] also studied the plasma level of estrone, prolactin, neurophysin, and sex steroid-binding globulin in chronic alcoholic men. Both estrone and prolactin levels were significantly increased in the chronic alcoholics. Estrogen-stimulated neurophysin and sex steroid-binding globulin, both indicators of estrogenization, were elevated. Significantly higher levels of estrone and prolactin were present in alcoholic subjects displaying gynecomastia than in those who did not. Subjects with spider angiomata had significantly higher plasma estrone levels than did alcoholic subjects without this clinical feature. Many of the changes that occur in the male alcoholic leading to the clinical appearance of hypogonadism and hyperestrogenization have been defined. Many of the studies have provided conflicting evidence, however. In contrast to Van Thiel *et al.*,[93] Turkington,[94] using a bioassay for prolactin, failed to find elevations in subjects with gynecomastia and hepatic cirrhosis.

Perhaps the most important point(s) to be established concerning the impact of ethanol on the reproductive system is the initial site(s) of action of ethanol. Whether it is a direct effect on the testis as suggested by Van Thiel and Lester[86] or an inhibition of gonadotrophin release leading to depressed testosterone levels is not clear at this time. It is possible that ethanol might be altering another hormone, such as prolactin, which in conjunction with a relatively constant level of luteinizing hormone might alter testosterone levels. It is likely that the hormonal changes observed in alcoholics are in part the result of direct effects of ethanol on the CNS as well as peripheral target organs.

2. Oxytocin

Fuchs and Wagner[95] determined the effect of ethanol on increased uterine motility caused by the suckling-induced release of oxytocin in lactating rabbits. Oxytocin release as determined by uterine motility was inhibited by ethanol. Cobo and Quintero[96] were able to demonstrate a differential inhibition by ethanol and by water overload on vasopressin activity and oxytocin response to suckling in humans. Vasopressin response measured by free water clearance was inhibited by water overload and by BEL between 50 mg and 100 mg/dl, while no modification of the mammary oxytocin response was observed under these conditions. A somewhat discordant observation of the effect of ethanol on milk ejection in the rat was recorded by Lincoln.[997] Using both intramammary pressure and nursing behavioral characteristics of the pups, it was demonstrated that although the latency between the onset of nursing and milk ejection was increased following a 5-g ethanol/kg injection (anesthetic dose), the average time between milk ejections and the milk yields were about the same in the ethanol-treated and control animals. However, as the author points out, if one considers the time period of the difference in latencies as the time period for measurement of oxytocin release, then inhibition was nearly complete. It might be considered that the eventual oxytocin release is an adaptation of the central effect to the inhibiting property of ethanol.

Fuchs later extended her studies to the effects of ethanol on oxytocin release during parturition in the rabbit.[98] Using a uterine sensitivity test to oxytocin as the indication of the closeness to the onset of spontaneous labor, she was able to demonstrate abnormal and prolonged delivery with BEL between 0.25 and 0.35 g/dl. With BEL about 0.2 g/dl, labor was less intense and delivery somewhat prolonged. Lower BEL were without effect. This inhibition of oxytocin release was tested clinically to prevent premature labor, with some success.[99]

E. DISCUSSION

The studies mentioned in the foregoing sections have demonstrated that a large number of endocrine systems are altered by acute and chronic doses of ethanol. Some of the endocrine changes are relatively specific and isolated in their actions. For example, ethanol-induced changes in vasopressin alter water metabolism and changes in the gonadal hormones alter sexual characteristics. In contrast, ethanol-induced changes in the hypothalamic–pituitary–adrenal system have potentially a multitude of effects on normal homeostasis. General effects attributed to glucocorticoids. such as hyperglycemia through increased protein catabolism and gluconeogenesis, permissive effects on the responsiveness of smooth muscle to catecholamines, increased muscle endurance, and alterations in the electroconvulsive threshold, are attributed to the presence of normal or elevated levels of glucocorticoids.[30,100] Glucocorticoids have been shown to have stabilizing effects on biomembranes[101] as well as modulating effects on the sensitivities of the sensory systems, such as taste and olfaction.[102] Enzyme systems concerned with ethanol metabolism[103] and neurotransmitter turnover[104–106] are reported to require the presence of glucocorticoids for the maintenance of their normal activity levels. Catecholamine turnover was increased in mice following corticosterone injection.[107] Whether ethanol-induced alteration in glucocorticoid levels can in turn alter neurotransmitter metabolism by changing enzyme levels remains to be seen. One interesting interaction of glucocorticoids, ethanol, and the CNS function has recently been demonstrated. Sze et al.[108] were able to prevent by adrenalectomy the development of audiogenic seizures observed in mice during ethanol withdrawal. Injection of corticosterone into the adrenalectomized ethanol-treated mice completely restored the increase of brain tryptophan hydroxylase activity and the occurrence of withdrawal audiogenic seizures. There is also evidence that glucocorticoids have direct effects on gonadal function. Engel and Frowein[109] have shown that the decrease in testicular Leydig cell alcohol dehydrogenase activity, which is normally observed during early postnatal development in the rat, can be prevented by corticosterone injections.

With such a wide range of effects produced by a single hormonal system, alterations in this system by ethanol might prove to have far-reaching consequences. Further investigations of ethanol and the hypothalamic–pituitary–adrenal system are especially needed.

In several of the systems studied, certain factors have become clear and need to be considered in future studies of ethanol–hormonal interrelationships. Cicero and Badger[76] have demonstrated that dose and time relationships are important in determining whether ethanol causes an increase or decrease in plasma luteinizing hormone and testosterone. Studies of the hypothalamic–pituitary–adrenal system in ethanol-adapted mice have indicated either an increased level in plasma corticosterone to ethanol at 20 min or a decreased level at 30–60 min.[65] Much of the conflicting data in the literature can be attributed to differences in the doses of ethanol used, the route of administration, and the time of sampling for hormonal determination. Subtle differences such as duration of chronic ethanol use and the possible effects of ethanol withdrawal in the various studies could alter circadian rhythms to varying degrees that could confuse comparisons between studies. With increased recognition of the critical nature of these variables, ambiguities which presently exist in studies of ethanol–endocrine interrelationships should diminish.

F. CONCLUSIONS

Acute ethanol administration has been shown to produce perturbations of many endocrine systems. Those endocrine systems affected by ethanol tend to adapt somewhat following chronic ethanol consumption. However, ethanol-induced alterations in endocrine function and their interaction with other biological systems cannot be related in any cogent manner at the present time to the development of alcoholism. It is possible that further studies might elucidate significant facts, allowing an endocrine-based theory of alcoholism to be developed. Ethanol–endocrine interrelationships are significantly interesting independent of this goal. Ethanol is a useful tool in the basic study of any endocrine system affected by it, although certain areas have received inadequate attention. Clinical uses can possibly be made of the stimulatory or inhibitory effects of ethanol on a given endocrine system if these effects are understood in detail. Finally, endocrine-related pathological conditions which develop in the alcoholic might be lessened or prevented as the exact mechanism of their development becomes known.

ACKNOWLEDGMENTS. We wish to record our gratitude to the editorial staff for their criticism of this chapter. We also wish to thank Charlene Butte for her excellent and dedicated assistance in the course of preparing this chapter.

This work was supported by a grant from the National Institute on Alcohol Abuse and Alcoholism (AA03074).

G. REFERENCES

1. N. Edkins and M. M. Murray, *J. Physiol.* **71**:403 (1931).
2. A. Bornstein and A. Loewy, *Biochem. Z.* **191**:271 (1927).
3. M. M. Murray, *J. Physiol.* **76**:379 (1932).
4. M. G. Eggleton, *J. Physiol.* **101**:172 (1942).
5. M. B. Strauss, J. D. Rosenbaum, and W. P. Nelson, *J. Clin. Invest.* **29**:1053 (1950).
6. H. B. van Dyke and R. G. Ames, *Acta Endocrinol.* **7**:110 (1951).
7. W. M. Nicholson and H. M. Taylor, *J. Clin. Invest.* **17**:279 (1938).
8. M. E. Rubini, C. R. Kleeman, and E. Lamdin, *J. Clin. Invest.* **34**:439 (1955).
9. G. Sereny, A. Rapoport, and H. Husdan, *Metabolism* **15**:896 (1966).
10. L. F. Fabre, R. W. Farmer, and H. W. Davis, in: *Biological Aspects of Alcohol* (M. K. Roach, W. M. McIsaac, and P. J. Creaven, eds.), pp. 418–440, University of Texas Press, Austin (1971).
11. R. W. Farmer and L. F. Fabre, Jr., in: *Biochemical Pharmacology of Ethanol* (E. Majchrowicz, ed.), pp. 277–289, Plenum Press, New York (1975).

12. J. Linkola, F. Fyhrquist, M. M. Nieminen, T. H. Weber, and K. Tontti, *Eur. J. Clin. Invest.* **6**:191 (1976).
13. J. Linkola, *Acta Physiol. Scand.* **92**:212 (1974).
14. M. Bleecker, D. H. Ford, and R. K. Rhines, *Life Sci.* **8**:267 (1969).
15. H. R. Murdock, *Q. J. Stud. Alcohol* **28**:419 (1967).
16. J. R. Augustine, *Can. Med. Assoc. J.* **96**:1367 (1967).
17. J. Leppäluoto, M. Rapeli, R. Varis, and T. Ranta, *Acta Physiol. Scand.* **95**:400 (1975).
18. G. I. Klingman and M. Goodall, *J. Pharmacol.* **121**:313 (1957).
19. I. Abelin, C. Herren, and W. Berli, *Helv. Med. Acta* **25**:591 (1958).
20. E. S. Perman, *Acta Physiol. Scand.* **44**:241 (1958).
21. E. S. Perman, *Acta Physiol. Scand.* **48**:323 (1960).
22. T. Hirose, R. Higashi, H. Ikeda, K. Tamura, and T. Suzuki, *Tohoku J. Exp. Med.* **109**:85 (1973).
23. E. Giacobini, S. Izikowitz, and A. Wegmann, *Arch. Gen. Psychiatry* **3**:289 (1960).
24. M. Ogata, J. H. Mendelson, N. K. Mello, and E. Majchrowicz, *Psychosom. Med.* **33**:159 (1971).
25. E. Giacobini, S. Izikowitz, and A. Wegmann, *Experientia* **15**:467 (1960).
26. J. P. von Wartburg, W. Berli, and H. Aebi, *Helv. Med. Acta* **28**:89 (1961).
27. L. A. Pohorecky, *J. Pharmacol. Exp. Ther.* **189**:380 (1974).
28. K. Engelman and B. Portnoy, *Cir. Res.* **26**:53 (1970).
29. J. T. Coyle and D. Henry, *J. Neurochem.* **21**:61 (1973).
30. W. F. Ganong, *Review of Medical Physiology*, 4th ed., Lange Medical Publications, Los Altos, Calif. (1969).
31. V. Meyer and E. Knobil, *Endocrinology* **80**:163 (1967).
32. I. C. Greenwood and J. Landon, *Nature (Lond.)* **210**:540 (1966).
33. E. D. Weitzman, *Annu. Rev. Med.* **27**:225 (1976).
34. T. M. Brown and A. M. Harvey, *J. Am. Med. Assoc.* **117**:12 (1941).
35. N. Freinkel, D. L. Singer, R. A. Arkey, S. J. Bleicher, J. B. Anderson, and C. K. Silbert, *J. Clin. Invest.* **42**:1112 (1963).
36. P. Steer, R. Marnell, and E. E. Werk, *Ann. Intern. Med.* **71**:343 (1969).
37. R. A. Arky, in: *The Biology of Alcoholism* (B. Kissin and H. Begleiter, eds.), Vol. 1: *Biochemistry*, pp. 197–227, Plenum Press, New York (1971).
38. S. Bellet, N. Yoshimine, O. A. P. DeCastro, L. Roman, S. Parmar and H. Sandberg, *Metabolism* **20**:762 (1971).
39. H. A. Priem, B. C. Shanley, and C. Malan, *Metabolism* **25**:397 (1976).
40. G. Tamburrano, S. Tamburrano, S. Gambardella, and D. Andreani. *J. Clin. Endocrinol. Metab.* **42**:193 (1976).
41. W. J. Tze, H. G. Friesen, and P. M. MacLeod, *Arch. Dis. Child.* **51**:703 (1976).
42. G. Mangili, M. Motta, and L. Martini, in: *Neuroendocrinology* (L. Martini and W. F. Ganong, eds.), Vol. I, p. 297, Academic Press, Inc., London (1966).
43. J. J. Smith, *J. Clin. Endocrinol.* **11**:792 (1951).
44. J. C. Forbes and G. M. Duncan, *Q. J. Stud. Alcohol* **12**:355 (1951).
45. J. C. Forbes and G. M. Duncan, *Q. J. Stud. Alcohol* **14**:19 (1953).
46. J. C. Forbes and G. M. Duncan, *Endocrinology* **55**:822 (1954).
47. G. A. Santisteban and C. A. Swinyard, *Endocrinology* **59**:391 (1956).
48. G. A. Santisteban, *Q. J. Stud. Alcohol* **22**:1 (1961).
49. C. Czaja and H. Kalant, *Can. J. Biochem. Physiol.* **39**:327 (1961).
50. F. W. Ellis, *Proc. Soc. Exp. Biol. Med.* **120**:740 (1965).
51. F. W. Ellis, *J. Pharmacol. Exp. Ther.* **153**:121 (1966).
52. R. Kakihana, E. P. Noble, and J. C. Butte, *Nature (Lond.)* **218**:360 (1968).
53. T. Suzuki, R. Higashi, T. Hirose, H. Ikeda, and K. Tamura, *Acta Endocrinol.* **70**:736 (1972).
54. F. N. Briggs and P. L. Munson, *Endocrinology* **57**:205 (1955).
55. E. P. Noble, in: *Recent Advances in Studies of Alcoholism* (N. K. Mello and J. H. Mendelson, eds.), pp. 72–106, U.S. Government Printing Office, Washington, D.C. (1971).
56. D. A. Rodgers and G. E. McClearn, in: *Roots of Behavior* (E. L. Bliss, ed.), p. 68, Paul B. Hoeber Inc., (Harper & Row, Publishers), New York (1962).

57. F. E. Krusius, K. O. Vartia, and O. Forsander, *Ann. Med. Exp. Biol. Fenn.* **36:**424 (1958).
58. B. Kissin, V. Schenker, and A. C. Schenker, *Am. J. Med. Sci.* **239:**690 (1960).
59. J. H. Mendelson and S. Stein, *Psychosom. Med.* **28:**616 (1966).
60. G. Fazekas, *Q. J. Stud. Alcohol* **27:**439 (1966).
61. J. S. Jenkins and J. Connolly, *Br. Med. J.* **2:**804 (1968).
62. J. Merry and V. Marks, *Lancet* **2:**990 (1972).
63. S. Bellet, L. Roman, O. DeCastro, and M. Herrera, *Metabolism* **19:**664 (1970).
64. H. Kalant and C. Czaja, *Can. J. Biochem. Physiol.* **40:**975 (1962).
65. R. Kakihana, J. C. Butte, A. Hathaway, and E. P. Noble, *Acta Endocrinol.* **67:**653 (1971).
66. F. Ratcliffe, *Arch. Int. Pharmacodyn. Ther.* **197:**305 (1972).
67. J. C. Butte, R. Kakihana, and E. P. Noble, *J. Endocrinol.* **68:**235 (1976).
68. R. Kakihana and J. A. Moore, *Psychopharmacologia* **46:**301 (1976).
69. J. H. Mendelson, M. Ogata, and N. K. Mello, *Psychosom. Med.* **33:**145 (1971).
70. P. E. Stokes, in: *Recent Advances in the Psychobiology of the Depressive Illness* (T. A. Williams, M. M. Katz, and J. A. Shield, eds.), pp. 199–220, National Institute of Mental Health DHEW Publ. No. (HSM) 70-9053 (1972).
71. H. W. Margraf, C. A. Moyer, L. E. Ashford, and L. W. Lavalle, *J. Surg. Res.* **7:**55 (1967).
72. R. R. Chaudhury and M. Matthews, *J. Endocrinol.* **34:**275 (1966).
73. J. D. Kieffer and M. M. Ketchel, *Acta Endocrinol.* **65:**117 (1970).
74. A. M. Symons and V. Marks, *Biochem. Pharmacol.* **24:**955 (1975).
75. T. J. Cicero and T. M. Badger, *J. Pharmacol. Exp. Ther.* **201:**427 (1977).
76. T. J. Cicero, D. Bernstein, and T. M. Badger, *Alcohol. Clin. Exp. Res.* **2:**249 (1978).
77. R. Ylikahri, M. Huttunen, M. Harkonen, U. Seuderling, S. Onikki, S.-L. Karonen, and H. Adlercreutz, *J. Steroid Biochem.* **5:**655 (1974).
78. J. H. Mendelson and N. K. Mello, *Annu. Rev. Med.* **27:**321 (1976).
79. D. H. Van Thiel, J. Gavaler, and R. Lester, *Science* **186:**941 (1974).
80. F. Badr and A. Bartke, *Steroids* **23:**921 (1974).
81. D. H. Van Thiel, J. S. Gavaler, R. Lester, and M. D. Goodman, *Gastroenterology* **69:**326 (1975).
82. G. G. Gordon, A. L. Southren, and C. S. Lieber, *Alcohol. Clin. Exp. Res.* **2:**259 (1978).
83. G. G. Gordon, K. Altman, A. L. Southren, E. Rubin, and C. S. Lieber, *N. Engl. J. Med.* **295:**793 (1976).
84. R. T. Rubin, P. R. Gouin, A. Lubin, R. E. Poland, and K. M. Pirke, *J. Clin. Endocrinol. Metab.* **40:**1027 (1975).
85. R. T. Rubin, R. E. Poland, and B. B. Tower, *J. Clin. Endocrinol. Metab.* **42:**112 (1976).
86. D. H. Van Thiel and R. Lester, *Gastroenterology* **71:**318 (1976).
87. S. G. Korenman, L. E. Perrin, and T. McCallum, *J. Clin. Invest.* **48:**45 (1960).
88. J. R. Kent, R. J. Scaramuzzi, W. Lauwers, A. F. Parlow, M. Hill, R. Penardi, and J. Hilliard, *Gastroenterology* **64:**111 (1973).
89. D. H. Van Thiel, R. Lester, and R. J. Sherins, *Gastroenterology* **67:**1188 (1974).
90. L. F. Fabre, Jr., P. J. Pasco, J. M. Liegel, and R. W. Farmer, *J. Stud. Alcohol* **34:**57 (1973).
91. W. S. Coppage and A. E. Cooner, *N. Engl. J. Med.* **273:**902 (1965).
92. A. L. Southren, G. G. Gordon, J. Olivo, F. Rafii, and W. S. Rosenthal, *Metabolism* **22:**695 (1973).
93. D. H. Van Thiel, J. S. Gavaler, R. Lester, D. L. Loriaux, and G. D. Braunstein, *Metabolism* **24:**1015 (1975).
94. R. W. Turkington, *J. Clin. Endocrinol.* **34:**62 (1972).
95. A.-R. Fuchs and G. Wagner, *Acta Endocrinol.* **44:**593 (1963).
96. E. Cobo and C. A. Quintero, *Am. J. Obstet. Gynecol.* **105:**877 (1969).
97. D. W. Lincoln, *Nature (Lond.)* **243:**227 (1973).
98. A.-R. Fuchs, *J. Endocrinol.* **35:**125 (1966).
99. F. Fuchs, A.-R. Fuchs, V. F. Poblete, Jr., and A. Risk, *Am. J. Obstet. Gynecol.* **99:**627 (1967).
100. A. B. Eisenstein (ed.), *The Adrenal Cortex,* Little, Brown and Company, Boston (1967).
101. G. Weissmann, *Circulation* **53:**171 (1976).

102. R. I. Henkin, in: *Progress in Brain Research* (D. DeWied and J. A. W. M. Weijnen, eds.), Vol. 32, pp. 270–294, Elsevier Publishing Company, Amsterdam (1970).
103. P. Y. Sze, *Biochem. Med.* **14:**156 (1975).
104. E. C. Azmitia, Jr., and B. S. McEwen, *Science* **166:**1274 (1969).
105. E. C. Azmitia, Jr., and B. S. McEwen, *Brain Res.* **78:**291 (1974).
106. P. Y. Sze and L. Neckers, *Brain Res.* **72:**375 (1974).
107. P. M. Iuvone, J. Morasco, and A. Dunn, *Brain Res.* **120:**571 (1977).
108. P. Y. Sze, J. Yanai, and B. E. Ginsburg, *Brain Res.* **80:**155 (1974).
109. W. Engel and J. Frowein, *Nature (Lond.)* **251:**146 (1974).

Interaction of Ethanol with Neurotransmitters

X

Alterations in Neurotransmitter Function after Acute and Chronic Treatment with Ethanol

11

Walter A. Hunt and Edward Majchrowicz

A. INTRODUCTION

Several hormones found in the brain and in the circulation influence behavior and a variety of physiological processes. A number of findings confirm this statement. The distributions of several biogenic amines in specific anatomical regions of the brain differ extensively, suggesting the relation of those amines to specific physiological function. The property of centrally active drugs that affect behavior in a specific manner is related to their ability to alter either the metabolism or storage of biogenic amines in particular brain areas. The increasing interest in these substances associated with acute and chronic administration of ethanol is based on abundantly available data suggesting that they have an active role in various emotional and behavioral states. For example, there is an increased urinary excretion of norepinephrine in persons engaged in competitive sports. Aggressive behavior is accompanied by an increase of norepinephrine excretion in urine, and in anxious subjects there is an increased excretion of epinephrine. Furthermore, numerous studies concerning the role of biogenic amines in mental disease culminated in the postulation of a catecholamine hypothesis of affective disorders and in the development of a variety of therapeutic agents that have specific effects on neurotransmitters. The apparent potentiation of schizophrenic symptoms by the use of monoamine oxidase inhibitors and by loading with tryptophan or methionine also suggests the involvement of biogenic amines in

Walter A. Hunt • Behavioral Sciences Department, Armed Forces Radiobiology Research Institute, Bethesda, Maryland 20014. **Edward Majchrowicz** • Laboratory of Preclinical Studies, National Institute on Alcohol Abuse and Alcoholism, Rockville, Maryland 20852.

mental disorders. The beneficial effects of monoamine oxidase inhibitors on depression further implicates a role of biogenic amines in thought formation and mental processes.

Interest in the role of biogenic amines in acute ethanol intoxication and physical dependence has increased over the years since the appearance of early reports of an ethanol-induced shift in peripheral biogenic amine metabolism from oxidative to reducive pathways and an increased urinary excretion of catecholamines. The overt behavioral changes observed during acute and chronic consumption of alcoholic beverages may be reflections of physiological changes which are mediated by a number of specific changes in the metabolism and distribution of neurotransmitters in the brain and in the peripheral nervous system. In addition, a distinctly characteristic spectrum of signs and symptoms observed during an ethanol withdrawal syndrome may be an expression of changes on the molecular level in the central neurotransmitters which are entirely different from the changes during an acute episode of ethanol intoxication. With the advent of animal analogues of ethanol dependence, these possibilities have been explored extensively. However, considerable controversy has developed on what specific effects ethanol has on the biogenic amine systems. In the course of this review, studies reported to date will be surveyed to (1) determine where consistencies of the data exist, and (2) to see if the changes observed can explain the acute and chronic effects of ethanol on the nervous system. Discussion will center on the neurotransmitters norepinephrine, dopamine, serotonin, and γ-aminobutyric acid. Acetylcholine will not be covered here but, rather, in Chapter 14.

B. IMPORTANCE OF EXPERIMENTAL DESIGN

In reviewing the acute and chronic effects of ethanol and the development of tolerance and physical dependence, it is necessary to first keep in mind the potential problems in interpreting reported results and how those problems might be related to experimental design. Traditionally, pharmacological studies generally deal with the mechanism of action of a single dose of a drug. Attention is given to some alteration in a parameter as it relates to dose and to time after treatment and to whether this effect can be correlated to the concentration of the drug at the effector site. Presumably, the action of the drug declines as it is removed from its receptor.

When studying the chronic effects of a drug and especially addicting ones, a wide variety of problems exists in determining the relevance of measured alterations. First, one must be concerned with the cumulative effects of successive acute doses and how the body reacts to these repeated insults. The question here involves general toxicity at the cellular level. In the case of ethanol and other addictive agents, the development of tolerance and physical dependence, presumably at the molecular level, confounds the problem of interpretation of data. The difficulty is determining whether the measured change is related to successive acute effects, tolerance, physical dependence, or the presence of intoxication during chronic treatment. At present there is no convincing evidence on ethanol that links these various phenomena, and therefore the phenomena should be considered mediated by different mechanisms until proven otherwise. Consequently, care must be taken in an experimental design in an attempt to distinguish between these effects of ethanol.

The problem has been addressed further by Victor,[1] who pointed out that the literature relating to the ethanol withdrawal syndrome in humans has been characterized by confusion, inexactness, and contradictions. The literature's major shortcomings are that no clear distinction has been made between the problems of acute ethanol intoxication, ethanol addiction, mild withdrawal signs and symptoms, and delerium tremens. Therefore, when trying to determine the potential biological changes in ethanol-dependent animals, measurements should be made at definitively characterized stages appearing during the course of the entire withdrawal period. Recent studies in animal analogues of ethanol dependence help to clarify these phenomena. Traditionally, both in humans and animals, the initiation of the ethanol withdrawal period has been defined as the administration of the last dose of ethanol. Subsequently, during the following several hours the humans and animals show sequential overt signs of intoxication that are essentially similar to those observed during acute intoxication with ethanol, with the exception that these signs appear at a rather higher blood ethanol concentration due to the acquired tolerance to ethanol.[2–4] As the blood ethanol concentrations descend to approximately 100–150 mg/dl, there is a gradual but relatively rapid onset of ethanol withdrawal signs and reactions that attain their maximum severity when the blood ethanol concentration descends to approximately the zero level. On the basis of this clearly defined transition from behavioral signs of intoxication to the overt signs and reactions of the ethanol withdrawal syndrome, it has been possible to distinguish two phases in the entire ethanol withdrawal period: (1) prodromal detoxication phase, and (2) ethanol dependence phase (or ethanol withdrawal syndrome).[3,5] These two successive clusters of distinctly different behavioral signs associated with the respective phases of the ethanol withdrawal period represent two different functional states of the central nervous system: depression (prodromal detoxication phase), when ethanol is still present in the body, and later hyperexcitability (ethanol dependence phase), when the blood ethanol concentration approaches the zero level or is absent. Consequently, it is of the utmost importance to define the time of measurements both in regard to the behavioral state and in relation to the concurrent blood ethanol concentration. The time elapsed after the last dose of ethanol has relatively little direct significance. The role of tolerance can be examined either by treating animals in a manner that would induce tolerance but not physical dependence[2] or by comparing the effect found in chronically intoxicated animals to that found after a single dose of ethanol at equivalent blood ethanol concentrations.[6]

C. CATECHOLAMINES

1. Content

The effect of ethanol on catecholamine levels in the brain has been controversial ever since Gursey and Olson[7] reported a reduction in norepinephrine content after a single dose of ethanol. For the most part, subsequent investigations were unable to detect any changes in either norepinephrine or dopamine content.[8-14] A few exceptions have been reported under certain conditions. A small reduction of whole brain and hypothalamic norepinephrine can be observed 2–6 hr after ethanol treatment.[10-12] Also a transient increase in striatal dopamine levels was apparent 1 hr after ethanol treatment.[15]

Varying results have been obtained on the effect of chronic administration of ethanol on catecholamine levels, and they were dependent on duration of treatment. When animals received ethanol for 3–5 days, no change in norepinephrine or dopamine content could be observed.[11,16] After 10 days, either a lack of effect or an elevation of norepinephrine and dopamine has been reported.[17-19] On the other hand, after 2–3 weeks of ethanol treatment, norepinephrine was found to be reduced[13] and striatal dopamine was elevated.[19] After 7 weeks of treatment neither whole brain norepinephrine nor dopamine levels were affected.[20] However, 1 year of treatment has resulted in an elevated norepinephrine content in the brain stem and dopamine in the caudate nucleus.[21]

Parenthetically, it is of interest to mention that significant elevations in catecholamines can be seen after 3–4 days of ethanol administration if the animals are treated concurrently with pyrazole to stabilize blood ethanol concentrations.[16,18] Also, pyrazole itself alters catecholamine levels.[22] This illustrates the care needed when using pyrazole treatments in studying catecholamines. It is possible to be misled by alterations that are induced by the combination of two drug treatments even when each one is ineffective on its own.

2. Turnover

Possibly a more meaningful approach to the question of whether ethanol has an effect on brain catecholamines is through measurements of transmitter turnover which are generally assumed to reflect the activity of neurons using a given transmitter. The earliest report studied the effect of a single dose of ethanol on catecholamine turnover using the rate of its depletion after treatment with α-methyl-p-tyrosine (AMPT), a tyrosine hydroxylase inhibitor.[23] Ethanol was found to accelerate the rate of norepinephrine depletion but not that of dopamine when AMPT was given 15 min after a 2-g/kg dose of ethanol, intraperitoneally. However, subsequent studies demonstrated that at later times after ethanol administration, norepinephrine turnover was depressed whether using as an index norepinephrine depletion after AMPT or the conversion of [³H]norepinephrine to [³H]metabolites.[11,12,14,24,25] In fact, the response to ethanol was shown to be biphasic, with accelerated norepinephrine turnover observed early after ethanol administration but reduced norepinephrine turnover at later times.[11,14] This depression is apparently a result of inhibition of noradrenergic neurons originating in the locus coeruleus.[26] In addition, ethanol administration appears to have a biphasic effect on dopamine turnover, with no effect early after treatment but depressed turnover later, when using the rate of AMPT as the index of turnover.[11] Dopamine turnover approached control values when ethanol had been eliminated. This depression appears to be localized in the caudate nucleus and substantia nigra.[27] Using another approach, a single dose of ethanol increased the rate of synthesis of [³H]catecholamines from [³H]tyrosine[10] and of tyrosine hydroxylase activity[28] in a dose-dependent manner.

A few investigations have addressed the question of turnover of catecholamines by measuring the endogenous levels of their metabolites, changes which appear to accompany alterations in turnover in a parallel fashion.[29,30] A single intragastric dose of ethanol (5 g/kg) elevated whole brain 3-methyl-4-hydroxyphenylglycol (MHPG), a metabolite of norepinephrine, homovanillic acid (HVA), and dihydroxyphenylacetic acid (DOPAC) levels 3 hr later.[31] Similar results were observed for striatal DOPAC, which was dose-dependent.[32] Only one period after treatment was examined.

There appears to be general agreement on the effect of chronic administration of ethanol on norepinephrine. Ethanol treatment for 4–21 days increased norepinephrine turnover whether inferred by the rate of depletion of norepinephrine after AMPT treatment,[11] reduction of injected [³H]norepinephrine,[12,24,33] or elevation of norepinephrine metabolites.[31] A biphasic response of dopamine turnover may occur after ethanol withdrawal. Dopamine turnover is accelerated during the prodromal detoxication phase, but during the withdrawal syndrome it is reduced.[11,31] However, after ethanol treatment for 270 days, striatal tyrosine hydroxylase activity was elevated.[34]

An explanation is not altogether apparent for the discrepancies reported on effects of ethanol on catecholamines. Some of the reasons may lie with the differences in dose used, the time after ethanol administration, the route of administration, and the blood ethanol concentration. The problem with measurements of catecholamine turnover after ethanol treatment may revolve around the basic assumption that a steady state is present throughout the period during which measurements are taken. This period can be up to 5 hr when using AMPT. Although measurements of metabolites would obviate some of this problem, their levels would at best reflect an event that occurred sometime prior to their determination. Blood ethanol concentrations do not remain stable for a prolonged period of time after a single dose, especially when given intraperitoneally. They rise rapidly, reach a relatively short-lived plateau, and then decline at a relatively constant rate. If any change in catecholamine function does occur with ethanol treatment and is dependent on the blood ethanol concentration, the degree to which the variable is altered can be affected over a short period of time. If there are biphasic effects, the situation is compounded. Ideally, the approach taken should reflect the actual change induced by a particular blood ethanol concentration at the exact time studied. With the multitude of apparent effects induced by ethanol, the turnover studies to date on this subject are probably partially invalid, since a steady state does not seem to be present under the experimental conditions explored.

3. Uptake and Release

Another approach in the study of the effect of ethanol on dopamine has been the measurement of K^+-stimulated dopamine release in vitro from slices derived from the caudate nucleus. Darden and Hunt[35] found that a single dose of ethanol accelerated dopamine release in a dose-dependent manner until blood ethanol concentrations reached about 300 mg/dl. Above that level, dopamine release declined in a dose-dependent manner until K^+ was virtually incapable of stimulating release. K^+-stimulated dopamine release returned to control values when ethanol was eliminated. In animals rendered ethanol-dependent over a 4-day period, dopamine release was accelerated twofold during the prodromal detoxication phase, but rapidly declined to 50% of controls during signs of a severe withdrawal syndrome.

The mechanism of the alterations in ethanol-induced dopamine release is unknown. It does not appear, however, to be a direct effect on dopaminergic terminals, since addition of ethanol to the medium containing the brain slices does not significantly alter dopamine release in sublethal concentrations.[36–38] Similar results were found for norepinephrine uptake and release in vitro[37,39,40] after a single dose of ethanol[25] and in ethanol-dependent rats.[41]

One interesting set of experiments has revolved around the ability of K^+ to stimulate striatal dopamine synthesis in vitro. Ethanol has been shown to inhibit this response when concentrations of 0.2–0.8% were added to the incubation medium.[38] On further examination, ethanol inhibited tyrosine hydroxylase in the presence but not the absence of K^+ and was related to alterations in the affinity of the pteridine cofactor.[42] Finally, evidence suggests that the K^+-induced increase in dopamine synthesis may be mediated by adenosine-3′,5′-cyclic monophosphate (cyclic AMP).[43,44] This is accomplished presumably by the accumulation of cyclic AMP during depolarization.[45] Ethanol was found to block the elevated rate of dopamine synthesis elicited by dibutyryl cyclic AMP.[46] The data tend to suggest that ethanol treatment might block striatal dopamine synthesis by inhibiting tyrosine hydroxylase through a reduction in cyclic AMP levels. However, most evidence in vivo would argue to the contrary. As mentioned before, a single dose of ethanol increases dopamine synthesis[10] and tyrosine hydroxylase activity.[28] Furthermore, striatal cyclic AMP is not affected by ethanol administration.[47] Consequently, the findings that ethanol reduces dopamine synthesis *in vitro* appear to be of only academic interest.

4. Pharmacological Studies

Pharmacological evidence has been accumulated in an attempt to implicate catecholamines in the effects of ethanol. However, the results have been contradictory. In studies with a single dose of ethanol, the behavioral signs have been divided based on apparent excitation with low doses and depression with high doses. Most reports have dealt with excitation. Early experiments have shown that AMPT can inhibit the increased motor activity induced by low doses of ethanol.[48] This suggests that an increased utilization of catecholamines is somehow involved in the enhanced locomotor activity. However, upon closer examination, variable results have been observed with other agents effective in blocking catecholamines. Both dopaminergic agonists and antagonists as well as α-adrenergic antagonists were capable of reversing ethanol's stimulation of locomotor activity.[49,50] β-Adrenergic blockers were ineffective.[50]

In the study of the depressive effects of ethanol, results are equally controversial. β-Adrenergic blockers and dopaminergic agonists have been reported to enhance, antagonize, or be ineffective in relieving ethanol-induced depression.[50-57] On the other hand, α-adrenergic blockers neither augmented or antagonized it.[50,55]

One problem with interpreting these experiments is that a multitude of different behavioral end points were used in evaluating the effects of ethanol. These have included ethanol sleep times, locomotor activity, startle response, motor coordination, memory, attention, mood, and EEG. Since these responses might be expected to be expressed by different mechanisms, drugs might also have variable actions. In addition, if a single dose of ethanol has a biphasic effect on catecholamine turnover and release, the results with drugs could depend on the dose of ethanol, the route of administration, and the time after injection. Therefore, such studies need to be done under a variety of conditions.

Pharmacological experiments in ethanol-dependent animals have been more consistent, mostly because they generally used the same behavioral endpoint. Withdrawal convulsions elicited by handling were exacerbated by reserpine and AMPT, which deplete catecholamines.[58,59] Both α- and β-adrenergic blockers, norepinephrine, and dopamine antagonists potentiated withdrawal convulsions[58-60] and L-dopa and dopamine inhibited them.[61] On the other hand, AMPT and FLA-63, a depletor of norepinephrine, blocked the increase in locomotor stimulation during withdrawal.[17]

It should be pointed out that care needs to be taken in interpreting these data even though they are basically consistent. The problem with using the drugs mentioned above is their known ability to affect the seizure threshold. Compounds such as reserpine increase seizure susceptibility,[62-65] but dopamine can act as an anticonvulsant.[66] In fact, reserpine alone can facilitate the production of convulsions on handling.[67] Furthermore, reserpine and similar drugs are well-known depressants. Therefore, most of the effects of the drugs which modify catecholamine function would be expected to have the observed effects on the withdrawal syndrome. It must be emphasized that data of this type should be interpreted with caution. The drugs may modify behavior through mechanisms unrelated to those that are responsible for expression of the withdrawal syndrome but that are still capable of facilitating or antagonizing it. This approach could have potential therapeutic value but can be misleading when it comes to elucidating mechanisms.

An interesting set of recent experiments has been able to dissociate ethanol tolerance and physical dependence. It has often been assumed that tolerance and dependence are different expressions of the same biological phenomenon. However, this view has been questioned in recent years.[68-71] It was shown that pretreatment of mice with 6-hydroxydopamine could prevent development of tolerance as measured by sleep times and hypothermia but not physical dependence.[72,73] This response was associated with a depletion of norepinephrine without affecting dopamine or serotonin. The authors concluded that the integrity of noradrenergic neurons is necessary for development of tolerance although not the expression of it. Treatment with 6-hydroxydopamine after the development of tolerance had no effect.

It is difficult to visualize how the activity of noradrenergic neurons could be involved in initiating tolerance without altering the ability of ethanol to depress the brain. This is particularly true in light of the finding that tolerance can be related to reversal of changes in the fluidity of membrane fragments *in vitro* where no noradrenergic connections are intact.[74] Nevertheless, the finding that ethanol tolerance and physical dependence can be dissociated may have considerable importance in elucidating the mechanisms by which each acts.

In summary, present results suggest that acute and chronic administration of ethanol alters catecholamine function. A single dose of ethanol exerts a biphasic effect on both norepinephrine and dopamine turnover and dopamine release with predominant stimulation, accompanied possibly by a transient change in catecholamine levels. Consistent and convincing data are lacking to support a role of catecholamines in ethanol-induced intoxication except possibly the increased locomotor activity observed with low doses of ethanol. In ethanol-dependent animals, divergent results are obtained with the catecholamines, with norepinephrine turnover increased and dopamine turnover and release reduced. Whether these changes have a role in expression of the withdrawal syndrome has not been conclusively demonstrated, but the extremely close parallel between increases in norepinehprine turnover and the startle response is a very promising development.[55] More experiments need to be conducted in which precise and continuously variable behavioral parameters related to ethanol intoxication and withdrawal can be compared to biochemical or biophysical measurements in specific areas of the brain through complete time courses. This might reveal more convincing and relevant correlations. Cause-and-effect relationships may prove to be difficult to demonstrate.

D. SEROTONIN

1. Content

Initial investigations on the effect of ethanol on serotonergic systems involved measurements of serotonin content in the brain. Early studies of Gursey and Olson[7] suggested that a single dose of ethanol reduces serotonin levels in the brain; however, numerous studies since then have not been able to find such a change.[8,9,13,75-81] Similar negative findings were found in most studies after chronic administration of ethanol for 4–21 days.[13,16,78-82] A few exceptions have been reported. Griffiths et al.[17] and Littleton et al.[18] found that serotonin levels were elevated after 10 days of ethanol treatment by inhalation but not after 4 and 7 days. The reason for this finding is unclear, but it may reflect the fact that increasing amounts of ethanol were administered during the 10-day period, while the other longer-term studies used the same concentration of ethanol. Another exception was that when pyrazole was given concurrently with ethanol, the effect could be seen after only 4 days of treatment even though pyrazole alone had no effect on serotonin levels.[16,18]

2. Turnover

Although serotonin levels in the brain are generally unaltered by ethanol treatment, the serotonergic system could be modified through effects on turnover, which is presumably a better index of neurotransmitter activity. Turnover is determined by measuring the rate of synthesis or degradation at steady state. The rate of serotonin synthesis has been determined after ethanol administration by either measuring tryptophan hydroxylase activity in vitro[83] or in vivo[28] or by measuring the synthesis of [^{14}C]serotonin from injected [^{14}C]-5-hydroxytryptophan, a serotonin precursor.[84] Ethanol has not been found to affect serotonin synthesis even at high, sublethal doses. Another approach is to block monoamine oxidase to prevent catabolism of serotonin and then measure the rate of accumulation of serotonin in the brain. Results have been variable, showing an increase,[85] decrease,[77,80] and no change[78,79] in serotonin accumulation. Alterations in turnover can also be inferred by quantifying 5-hydroxyindoleacetic acid levels (5-HIAA). 5-HIAA levels have been reported to be elevated after a single dose of ethanol,[81] but the changes appear to be due to an inhibition of the active transport of 5-HIAA from the brain.[86,87]

These differences are not easily reconciled, but they may be related to experimental design. In all the studies cited above, only one dose of ethanol was used, and the monoamine oxidase inhibitor was administered at one specific time. The doses varied from 1.6 to 5 g/kg, and were administered either orally or intraperitoneally. The time of inhibitor treatment ranged from 15 min before to 2 hr after ethanol administration. It is possible that ethanol has effects on serotonin metabolism which are dependent on dose, blood ethanol concentration, time after ethanol treatment, and route of administration. These questions have been recently addressed.[88] It has been discovered that a single dose of ethanol given either orally or intraperitoneally to rats has a biphasic effect on brain tryptophan, serotonin, and 5-HIAA. After 2–4 hr all three substances were elevated, and after 7 hr they were reduced. These responses appear to be related to concomitant and parallel changes in free serum tryptophan. Under normal circumstances tryptophan hydroxylase, the rate-limiting enzyme in the synthesis of serotonin, is unsaturated, and the free serum concentrations of tryptophan appear to play a role in the regulation of serotonin synthesis and turnover.[89,90] It seems, therefore, that the changes in the serotonergic substances observed here might be related to the effects of ethanol on free serum tryptophan levels.

Experiments on the effects of chronic administration of ethanol on serotonin turnover are contradictory. Hunt and Majchrowicz[80] induced physical dependence over 5 days of ethanol treatment and found that serotonin turnover was depressed in the animals during the prodromal detoxication phase but returned to control values during overt withdrawal signs. On the other hand, Kuriyama et al.[78] found an elevation in serotonin turnover in intoxicated mice after placing them on a liquid diet containing ethanol for 14 days. After only 8 days of treatment, no effect was observed. In a longer-term study for 32 days, Frankel et al.[79] were unable to detect any change in serotonin turnover. The reasons for these divergent results are not readily apparent but they may be related to a biphasic effect on serotonin metabolism when ethanol is present in the body.[88] Also, the length of exposure to ethanol may affect the dynamics of the response. More detailed and carefully designed studies are needed to clarify this point.

3. Uptake and Release

The foregoing discussion suggests that whatever effect ethanol might have on serotonergic neurons is indirect. This is further supported by the fact that ethanol added *in vitro* to brain slices or synaptosomes was unable to alter serotonin uptake,[37] release,[39] or metabolism[91] in physiologically compatible concentrations. Also, serotonin uptake is not changed in physically dependent rats.[41]

4. Pharmacological Studies

Attempts have been made to implicate serotonergic mechanisms in the development of ethanol tolerance and dependence with the use of pharmacological agents which modify serotonergic transmission. Frankel *et al.*[92] reported that chronic administration of *p*-chlorophenylalanine, a serotonin depletor, with ethanol for up to 26 days reduced the amount of tolerance developed when compared to that observed in animals treated with ethanol alone. They suggested that the integrity of serotonergic neurons must be intact for the full expression of tolerance. Biochemical support for this possibility is generally lacking. As discussed earlier, the effect of chronic administration of ethanol on serotonin levels has been inconsistent. The complexity of interpretation increases with the finding that pretreatment with *p*-chlorophenylalanine augments the depressed startle response induced by ethanol.[55] Also, methysergide, a serotonin antagonist, increases ethanol sleep times.[93] Opposite responses might have been expected if these results were to be consistent with those of Frankel *et al.*[92] In experiments with physically dependent animals, *p*-chlorophenylalanine, L-tryptophan, and 5-hydroxytryptophan were unable to block the signs and reactions of the ethanol withdrawal syndrome.[58,94] However, blockade of serotonin receptors with methysergide potentiated withdrawal convulsions.[95] The data cited above are not convincing for a role of serotonin in the chronic effects of ethanol.

In summary, it has been difficult to convincingly demonstrate a role of serotonin in the acute and chronic consumption of ethanol. This has been due to a possible transient effect on serotonergic neurons. A single dose of ethanol may induce an indirect biphasic alteration in serotonin metabolism through changes in free serum tryptophan levels of unknown origin. However, no relation to ethanol intoxication has been ascribed to it. A role of serotonin in ethanol dependence has yet to be revealed.

E. SHIFT IN THE METABOLISM OF BIOGENIC AMINES

It is well established that both acute and chronic administration of ethanol results in a shift in the peripheral metabolism of biogenic amines from an oxidative to a reductive pathway. During consumption of alcoholic beverages, the urinary concentrations of 5-hydroxyindoleacetic acid (5-HIAA) and vanilylmandelic acid (VMA) are decreased and those of 5-hydroxytrytophol (5-HTOH) and 3-methoxy-4-hydroxyphenylglycol (MHPG) are elevated.[96–99]

This shift in biogenic amine metabolism from the oxidative to the reductive pathway may be related to an alteration in the liver NADH/NAD ratio during the oxidation of ethanol. Or it could be the result of competitive inhibition of aldehyde dehydrogenase by acetaldehyde, the first metabolic product of ethanol oxidation.[101] The evidence in favor of competitive inhibition seems to be more plausible since it is supported by direct experimental results. In liver homogenates incubated with serotonin, acetaldehyde causes suppression of 5-HIAA formation with the concurrent increase of 5-HTOH.[100] This effect of acetaldehyde is not altered by a severalfold increase of NAD concentration in the incubation medium. Further support of these results was provided by an *in vivo* demonstration in rats that exogenous acetaldehyde had a pronounced effect on

[^{14}C]norepinephrine metabolism, causing a decrease in acidic metabolites and an increase in reduced metabolites.[102]

Whether a shift in biogenic amine metabolism has any relevance in brain depends on the presence of alcohol dehydrogenase and sufficient quantities of acetaldehyde. In general, the major site of ethanol metabolism is in the liver. However, small quantities of alcohol dehydrogenase have been detected in brain,[103] suggesting that an elevated NADH/NAD ratio might be possible from localized ethanol oxidation. Such results have been found after a single dose of ethanol.[104,105] The problem with these findings is that they were obtained from animals which were decapitated or immersed in liquid nitrogen—techniques that do not sufficiently protect against postmortem changes in metabolites.[106] In experiments in which animals were killed by the freeze-blowing technique, these differences disappear.[107] Coupled with the inability of excessive NAD to antagonize the acetaldehyde-induced blockade of 5-HIAA formation *in vitro*,[100] this effectively rules out the likelihood that ethanol metabolism in brain could lead to a shift in biogenic amine metabolism by this mechanism.

The presence of acetaldehyde in the brain has undergone considerable debate. Since little acetaldehyde would be formed within the brain and since the level of alcohol dehydrogenase is low, any that might accumulate would have to come from the blood. Recent evidence has suggested that there is little acetaldehyde in the brain after a single dose of ethanol.[108–110] Apparently, previous measurements of acetaldehyde have detected amounts derived from nonenzymatic conversion of ethanol to acetaldehyde.

There is direct experimental evidence to suggest that no shift in metabolism from oxidative to reductive pathways occurs in brain after ethanol treatment. Karoum *et al.*[31] measured several metabolites of norepinephrine and dopamine after acute and chronic administration of ethanol. The neutral metabolite of norepinephrine, MHPG, was significantly elevated in both cases. However, the acidic metabolites of both norepinephrine and dopamine were also increased. The authors suggested that these alterations were more likely related to an acceleration of catecholamine turnover rather than a shift in metabolism.

In summary, there is no evidence that ethanol treatment can shift biogenic amine metabolism from oxidative to reductive pathways in brain, nor are the prerequisites for such an effect—elevation of the NADH/NAD ratio and sufficient acetaldehyde—likely to exist.

F. γ-AMINOBUTYRIC ACID

1. Content

Substantial evidence indicates that γ-aminobutyric acid (GABA) and other amino acids may be neurotransmitters in the nervous system.[111,112] This has led investigators to explore the possibility that GABA may have some role in the actions of ethanol.

There has been considerable disagreement on the effect of ethanol on GABA levels in the brain. In part, this may be due to the manner in which the animals were killed. It is known that a significant postmortem alteration in GABA content in the brain can occur[113–115]; therefore, the rapidity of inactivation of brain enzymes for the synthesis and metabolism of GABA could be important in assessing the effect of ethanol on the GABA system.

The existing studies can be divided into two groups based on the method by which the animals are killed. In one group the animals are killed by decapitation and then processed; in the other, after decapitation the heads are submerged in liquid nitrogen to be frozen. Although freezing the brain might be expected to increase the validity of GABA measurements, the results between the two groups were similar and inconclusive.

In early experiments Hakkinen and Kulonen[116,117] found a 34% increase in brain GABA levels after administration of 4.5 g ethanol/kg. Similar results were reported in cats and mice.[13,118] However, most other work has contradicted these studies by not finding a significant effect of ethanol on GABA levels in the brain.[119–125]

One problem in finding a possible change has been the use of a single dose of ethanol at only one time period after administration. Most of the studies cited above used similar experimental designs and may have missed alterations induced under different conditions. The work of Syntinsky et al.[126] supports this possibility. They injected rats with 2–8 g ethanol/kg, intraperitoneally, killed them 30 min later by decapitation, and allowed their heads to fall into liquid nitrogen. Increasing doses altered cerebral and cerebellar GABA levels in a biphasic manner. At low doses GABA was elevated and at high doses it was reduced. The authors could not attribute these changes to actions on either glutamate decarboxylase (GAD) or GABA-γ-oxoglutamate transaminase (GABA-T) activities. Sutton and Simmonds,[124] on the other hand, found a 10% increase in GAD activity and no change in GABA levels. The relationship between GABA levels and GAD activity relative to the acute effects of ethanol clearly needs to be reexamined.

As discussed earlier, the relevance of chronic studies on the actions of ethanol depends on their experimental design and on whether physical dependence is induced. In one set of experiments, ethanol was administered from 3 to 26 weeks with no apparent attempt to induce physical dependence. In the other set, physical dependence was induced over a shorter period of treatment from 3 to 10 days. Interestingly, the two groups of experiments netted opposite results. In the longer-term studies, ethanol was consumed as a liquid diet[13] or in the drinking fluid.[124,126] Irrespective of the length of treatment, an elevation in brain GABA levels of 30–60% was observed and remained elevated for 2 days. On the other hand, when physical dependence was induced over a short period of treatment, a reduction in brain GABA was seen.[16,127,128] In these experiments, ethanol was administered by inhalation for 3–10 days with or without daily pyrazole injections to stabilize blood ethanol concentrations. Although pyrazole in itself did not affect GABA levels, pyrazole treatment increased the action of physical dependence, reducing GABA levels up to 60%.[16,127] When pyrazole was not used, depletion was only 20–25%.[16,128] This is yet another example of the risk in using pyrazole in the induction of ethanol dependence.

2. Relevance of Changes in γ-Aminobutyric Acid Levels

The possible significance of alterations in GABA levels has to be examined from two different perspectives: when GABA is viewed as a neurotransmitter or as a metabolic intermediate. A few studies exist which deal with the effect of ethanol on uptake and release of GABA, two parameters related to neurotransmission. In this regard, ethanol is not particularly potent. A concentration of 450 mM ethanol added to an incubation medium containing brain slices is needed to inhibit [^{3}H]-GABA uptake just 25%.[39] In addition, 660 mM is required to block electrically stimulated [^{3}H]-GABA release from brain slices.[37] These concentrations exceed lethal levels. After induction of physical dependence, GABA uptake is unaffected.[41] These data do not support a direct effect of ethanol on GABA as a neurotransmitter, although changes in GABA levels could have functional significance on neurotransmission *in vivo* possibly through an altered availability of GABA for its receptor.

GABA is synthesized from glutamate, an intermediate in the citric acid cycle, and could be influenced, therefore, by changes in glucose metabolism. Ethanol inhibits glucose metabolism, as demonstrated by an increase in glucose and a reduction in lactate and pyruvate in the brain[107,129,130] and the reduction in $^{14}CO_2$ formation, total CO_2, and oxygen uptake.[131] Also, the percentage incorporation of [^{14}C]glucose into these intermediates is reduced in a dose-dependent manner.[130]

Chronic depression of glucose utilization could result in the attempt by the brain to derive energy from another source. One way could be through the GABA shunt, which converts GABA first into succinate and then to glutamate with the reduction of NAD and the production of ATP.[132] During the induction of ethanol dependence, blood ethanol concentrations are high and glucose utilization is reduced during that period[13] just as it is after a single dose of ethanol. For the brain to maintain sufficient energy to function, GABA may in part be used as an energy source through the GABA shunt.

Indirect evidence exists that might lend support to the idea that chronic ethanol treatment increases the utilization of the GABA shunt. With the chronic reduction of glucose metabolism, ATP levels are reduced,[13] possibly forcing the brain to find an alternative energy source. After short-term ethanol administration with high doses, GABA levels are reduced which may reflect increased catabolism rather than inhibited synthesis. GABA-T activity is elevated after chronic ethanol treatment, with no changes in GAD activity.[124] Further glutamate levels are reduced after a single dose of ethanol[107,130] but not after chronic treatment.[13,128] This would be expected with an increase in GABA catabolism.

3. Pharmacological Studies

Some effort has been made to implicate GABA pharmacologically in the effects of ethanol with the use of drugs which manipulate the GABA system. Using the information that low doses of ethanol induce behavioral stimulation[48,133] and that ethanol dependence enhances the susceptibility to seizures in experimental animals,[134–137] certain drugs which increase GABA levels or mimic its action in the brain were assessed on their ability to antagonize those altered physiological responses. Cott *et al.*[138] pretreated mice with γ-hydroxybutyrate, baclophen, or aminooxyacetic acid, and then injected them intraperitoneally with 2.4 g ethanol/kg. They found that all three drugs were capable of antagonizing the ethanol-induced locomotor stimulation. Attempts have also been made to inhibit ethanol withdrawal signs with aminooxyacetic acid[58] and dipropylacetate,[139,140] both GABA-T inhibitors.[141,142] These two drugs were shown to antagonize audiogenic seizures and seizures on handling as well as sound-induced hyperactivity. These data suggest that alterations in GABA function may be involved in both the hyperactivity observed after low doses of ethanol and the seizures during the ethanol withdrawal syndrome.

As pointed out earlier, inferences of the mechanism of behavioral and neurological alterations using pharmacological agents must be done with care, especially if the drugs employed might affect the response under study. In the experiments described here, this possibility is particularly important. Considerable evidence has accumulated demonstrating a relationship between GABA function and the susceptibility to seizures.[143] Both aminooxyacetic acid and dipropylacetate are effective against convulsions produced by several means, presumably because of their ability to elevate GABA levels in the brain.[144,145] It is possible that these drugs, when used to antagonize withdrawal seizures, lower the general susceptibility to seizures regardless of whether ethanol treatment causes them and regardless of whether ethanol interacts with the GABA system. However, during conditions when withdrawal seizures are observed, there is a reduction in GABA levels in the brain.[16,127,128] This lends validity to the conclusion that the ability of some drugs to increase GABA levels and antagonize withdrawal seizures does suggest a role of the GABA system in this phenomenon.

In summary, the evidence presented suggests that GABA may be involved in some of the actions of acute and chronic ethanol treatment. A reduction in GABA appears to accompany the increased locomotor activity after a low dose of ethanol and the seizures observed during an ethanol withdrawal syndrome. The mechanisms of the ethanol-induced changes in GABA levels are unknown and do not appear to be a direct effect on GABA neurons *per se*. These responses may occur secondarily to other alterations after ethanol treatment. Because the available data are not altogether consistent, further research is needed to clarify a more exact role of GABA in the effects of ethanol and to determine whether localized changes in specific regions of the brain will correlate more closely with observed behavioral effects. Finally, more detailed experiments determining dose–response and time–response relationships need to be performed using a very rapid means of inactivating brain enzymes such as focused microwave irradiation.

G. SUMMARY AND CONCLUSIONS

The evidence so far indicates that both acute and chronic administration of ethanol have significant effects on various aspects of neurotransmitter function. Norepinephrine and dopamine are the transmitters most consistently altered. Biphasic changes are found on norepinephrine and dopamine turnover and release, and they are dose- and time-dependent. There is no general agreement on what effect ethanol has on serotonin and GABA. Although alterations in transmitter function can be ascribed to chronic ethanol treatment, it has not yet been possible to conclusively demonstrate their role in the expression of the various signs and responses of the ethanol withdrawal syndrome. This is mainly due to the inability to accurately quantitate the withdrawal syndrome with objective, continuously variable measures. Although it can be anticipated that the mechanisms of ethanol-induced alterations in the brain will be elucidated, it may prove to be considerably more difficult to show a cause-and-effect relationship between these changes and specific behavioral signs of intoxication and withdrawal.

H. REFERENCES

1. M. Victor, *Psychosom. Med.* 33:636 (1966).
2. A. E. LeBlanc, H. Kalant, P. J. Gibbins, and N. D. Berman, *J. Pharmacol. Exp. Ther.* 168:244 (1969).
3. E. Majchrowicz, *Psychopharmacologia* 43:245 (1975).
4. E. Majchrowicz and W. A. Hunt, *Psychopharmacology* 50:107 (1976).
5. F. W. Ellis and J. R. Pick, *J. Pharmacol. Exp. Ther.* 176:88 (1970).
6. W. A. Hunt, J. D. Redos, T. K. Dalton, and G. N. Catravas, *J. Pharmacol. Exp. Ther.* 201:103 (1977).
7. D. Gursey and R. E. Olson, *Proc. Soc. Exp. Biol. Med.* 104:280 (1960).
8. D. H. Efron and G. L. Gessa, *Arch. Int. Pharmacodyn. Ther.* 142:111 (1963).
9. G. Duritz and E. B. Truitt, Jr., *Biochem. Pharmacol.* 15:711 (1966).
10. A Carlsson, T. Magnusson, T. H. Svensson, and B. Waldeck, *Psychopharmacologia* 30:27 (1973).
11. W. A. Hunt and E. Majchrowicz, *J. Neurochem.* 23:549 (1974).
12. L. A. Pohorecky, *J. Pharmacol. Exp. Ther.* 189:380 (1974).
13. A. K. Rawat, *J. Neurochem.* 22:915 (1974).
14. L. A. Pohorecky and L. S. Jaffe, *Res. Commun. Chem. Pathol. Pharmacol.* 12:433 (1975).
15. I. J. Wajda, I. Manigault, and J. P. Hudick, *Biochem. Pharmacol.* 26:655 (1977).
16. C. T. Chopde, D. M. Brahmankar, and V. N. Shripad, *J. Pharmacol. Exp. Ther.* 200:314 (1977).
17. P. J. Griffiths, J. M. Littleton, and A. Ortiz, *Br. J. Pharmacol.* 50:489 (1974).
18. J. M. Littleton, P. J. Griffiths, and A. Ortiz, *J. Pharm. Pharmacol.* 26:81 (1974).

19. L. Ahtee and M. Svartström-Fraser, *Acta Pharmacol. Toxicol.* **36**:289 (1975).
20. M. Mazur and A. Szmigielski, *Acta Physiol. Pol.* **27**:281 (1976).
21. M. E. Post and A. Y. Sun, *Res. Commun. Chem. Pathol. Pharmacol.* **6**:887 (1973).
22. E. McDonald, M. Marselos, and U. Nousiainen, *Acta Pharmacol. Toxicol.* **37**:106 (1975).
23. H. Corrodi, K. Fuxe, and T. Hökfelt, *J. Pharm. Pharmacol.* **18**:821 (1966).
24. P. V. Thadani, B. M. Kulig, F. C. Brown, and J. D. Beard, *Biochem. Pharmacol.* **25**:93 (1976).
25. P. V. Thadani and E. B. Truitt, Jr., *Biochem. Pharmacol.* **26**:1147 (1977).
26. L. A. Pohorecky and J. Brick, *Brain Res.* **131**:174 (1977).
27. N. G. Bacopoulos, R. K. Bhatnager, and L. S. VanOrden, III, *J. Pharmacol. Exp. Ther.* **204**:1 (1978).
28. A. Carlsson and M. Lindqvist, *J. Pharm. Pharmacol.* **25**:437 (1973).
29. J. L. Meek and N. H. Neff, *J. Pharmacol. Exp. Ther.* **184**:570 (1973).
30. D. F. Sharman, *Br. Med. Bull.* **29**:110 (1973).
31. F. Karoum, R. J. Wyatt, and E. Majchrowicz, *Br. J. Pharmacol.* **56**:403 (1976).
32. G. Bustos and R. H. Roth, *J. Pharm. Pharmacol.* **28**:580 (1976).
33. L. A. Pohorecky, L. S. Jaffe, and H. A. Berkeley, *Life Sci.* **15**:427 (1974).
34. S. Liljequist, S. Ahlenius, and J. Engel, *Naunyn-Schmiedebergs Arch. Pharmakol.* **300**:205 (1977).
35. J. H. Darden and W. A. Hunt, *J. Neurochem.* **29**:1143 (1977).
36. P. Seeman and T. Lee, *J. Pharmacol. Exp. Ther.* **190**:131 (1974).
37. F. J. Carmichael and Y. Isreal, *J. Pharmacol. Exp. Ther.* **193**:824 (1975).
38. K. Gysling, G. Bustos, I. Concha, and G. Martinez, *Biochem. Pharmacol.* **25**:157 (1976).
39. M. K. Roach, D. L. Davis, W. Pennington, and E. Nordyke, *Life Sci.* **12**:433 (1973).
40. A. Y. Sun, *Res. Commun. Chem. Pathol. Pharmacol.* **15**:705 (1976).
41. M. K. Roach, M. M. Khan, R. Coffman, W. Pennington, and D. L. Davis, *Brain Res.* **63**:323 (1973).
42. G. Bustos, R. H. Roth, and V. H. Morgenroth, III, *Biochem. Pharmacol.* **25**:2493 (1976).
43. J. E. Harris, R. J. Baldessarini, V. H. Morgenroth, III, and R. H. Roth, *Proc. Natl. Acad. Sci. USA* **72**:789 (1975).
44. R. L. Patrick and J. D. Barchas, *J. Pharmacol. Exp. Ther.* **197**:97 (1976).
45. H. Shimizu, C. R. Creveling, and J. W. Daly, *Mol. Pharmacol.* **6**:184 (1970).
46. K. Gysling and G. Bustos, *Biochem. Pharmacol.* **26**:559 (1977).
47. J. D. Redos, W. A. Hunt, and G. N. Catravas, *Life Sci.* **18**:989 (1976).
48. A. Carlsson, J. Engel, and T. H. Svensson, *Psychopharmacologia* **26**:307 (1972).
49. A. Carlsson, J. Engel, V. Strombom, T. H. Svensson, and B. Waldeck, *Naunyn-Schmiedebergs Arch. Pharmakol.* **283**:117 (1974).
50. J. A. Matchett and C. K. Erickson, *Psychopharmacology* **52**:201 (1977).
51. A. Smith, K. Hayashida, and Y. Kim, *J. Pharm. Pharmacol.* **22**:644 (1970).
52. K. Blum, W. Calhoun, J. Merritt, and J. E. Wallace, *Nature (Lond.)* **242**:407 (1973).
53. J. H. Mendelson, A. M. Rossi, J. G. Bernstein, and J. Kuehnle, *Clin. Pharmacol. Ther.* **15**:571 (1974).
54. R. L. Alkana, E. S. Parker, H. B. Cohen, H. Birch, and E. P. Noble, *Psychopharmacology* **51**:29 (1976).
55. L. A. Pohorecky, M. Cagan, J. Brick, and L. S. Jaffe, *Pharmacol. Biochem. Behav.* **4**:311 (1976).
56. G. H. Wimbish, R. Martz, and R. B. Fortney, *Life Sci.* **20**:65 (1977).
57. R. L. Alkana, E. S. Parker, H. B. Cohen, H. Birch, and B. P. Noble, *Psychopharmacologia* **55**:203 (1977).
58. D. B. Goldstein, *J. Pharmacol. Exp. Ther.* **186**:1 (1973).
59. K. Blum and J. E. Wallace, *Br. J. Pharmacol.* **71**:109 (1974).
60. K. Blum, M. G. Hamilton, M. Hirst, and J. E. Wallace, *Alcohol. Clin. Exp. Res.* **2**:113 (1978).
61. K. Blum, J. D. Eubanks, J. E. Wallace, and H. A. Schwertner, *Experientia* **32**:493 (1976).
62. G. Chen, C. Ensor, and B. Bohner, *Proc. Soc. Exp. Biol. Med.* **86**:507 (1954).
63. A. W. Lessin and M. W. Parks, *Br. J. Pharmacol.* **14**:108 (1959).

64. A. J. Azzaro, G. R. Wenger, C. R. Craig, and R. E. Stitzel, *J. Pharmacol. Exp. Ther.* **180**:558 (1972).
65. P. C. Jobe, R. E. Stull, and P. F. Geiger, *Neuropharmacol.* **13**:961 (1974).
66. B. J. Jones and D. J. Roberts, *Br. J. Pharmacol.* **34**:27 (1968).
67. D. B. Goldstein, *Psychopharmacologia* **32**:27 (1973).
68. B. Kissin, in: *Biology of Alcoholism* (B. Kissin and H. Begleiter, eds.), Vol. 3: *Clinical Pathology*, pp. 1–36, Plenum Press, New York (1974).
69. R. F. Ritzman and B. Tabakoff, *J. Pharmacol. Exp. Ther.* **199**:158 (1976).
70. F. Huidobro, J. P. Huidobro-Toro, and E. L. Way, *J. Pharmacol. Exp. Ther.* **198**:318 (1976).
71. M. J. Turnbull and J. W. Watkins, *Eur. J. Pharmacol.* **36**:15 (1976).
72. R. F. Ritzman and B. Tabakoff, *Nature (Lond.)* **263**:418 (1976).
73. B. Tabakoff and R. F. Ritzman, *J. Pharmacol. Exp. Ther.* **203**:319 (1977).
74. J. Chin and D. B. Goldstein, *Science* **196**:684 (1976).
75. J. Häggendal and M. Lindqvist, *Acta Pharmacol. Toxicol.* **18**:278 (1961).
76. G. R. Pscheidt, B. Issekutz, and H. E. Himwich, *Q. J. Stud. Alcohol* **22**:550 (1961).
77. G. M. Tyce, E. V. Flock, W. F. Taylor, and C. A. Owens, Jr., *Proc. Soc. Exp. Biol. Med.* **134**:40 (1970).
78. K. Kuriyama, G. E. Rauscher, and P. Y. Sze, *Brain Res.* **26**:450 (1971).
79. D. Frankel, J. M. Khanna, H. Kalant, and A. E. LeBlanc, *Psychopharmacologia* **37**:91 (1974).
80. W. A. Hunt and E. Majchrowicz, *Brain Res.* **72**:181 (1974).
81. B. Tabakoff and W. O. Boggan, *J. Neurochem.* **22**:754 (1974).
82. E. A. Moscatelli, K. Fujimoto, and T. C. Gilfoil, *J. Neurochem.* **25**:273 (1975).
83. M. A. Rogawski, K. Suzanne, and A. J. Mandell, *Biochem. Pharmacol.* **23**:1955 (1974).
84. A Feldstein and C. M. Sidel, *Int. J. Neuropharmacol.* **8**:347 (1969).
85. D. J. Palaić, J. Desarty, J. M. Albert, and J. C. Panisset, *Brain Res.* **25**:381 (1971).
86. B. Tabakoff, M. Bulat, and R. A. Anderson, *Nature (Lond.)* **254**:708 (1975).
87. B. Tabakoff, R. F. Ritzman, and W. O. Boggan, *J. Neurochem.* **24**:1043 (1975).
88. A. A. Badawy and M. Evans, *Biochem J.* **160**:315 (1976).
89. J. D. Fernstrom and R. J. Wurtman, *Science* **173**:149 (1971).
90. A. Tagliamonte, G. Biggio, L. Vargin, and G. L. Gessa, *Life Sci.* **12**:277 (1973).
91. D. Eccleston, W. H. Reading, and I. M. Ritchie, *J. Neurochem.* **16**:274 (1969).
92. D. Frankel, J. M. Khanna, A. E. LeBlanc, and H. Kalant, *Psychopharmacologia* **44**:247 (1975).
93. K. Blum, J. E. Wallace, W. Calhoun, R. G. Tabor, and J. D. Eubanks, *Experientia* **30**:1053 (1974).
94. P. J. Griffiths, J. M. Littleton, and A. Ortiz, *Br. J. Pharmacol.* **51**:307 (1974).
95. K. Blum, J. E. Wallace, H. A. Schwertner, and J. D. Eubanks, *J. Pharm. Pharmacol.* **28**:837 (1976).
96. A. Feldstein, H. Hoagland, H. Freeman, and O. Williamson, *Life Sci.* **6**:53 (1967).
97. V. E. Davis, H. Brown, J. A. Huff, and J. L. Cashaw, *J. Lab. Clin. Med.* **69**:132 (1967).
98. V. E. Davis, H. Brown, J. A. Huff, and J. L. Cashaw, *J. Lab. Clin. Med.* **69**:787 (1967).
99. M. Ogata, J. H. Mendelson, N. K. Mello, and E. Majchrowicz, *Psychosom. Med.* **33**:159 (1971).
100. R. A. Lahti and E. Majchrowicz, *Life Sci.* **6**:1399 (1967).
101. R. A. Lahti and E. Majchrowicz, *Biochem. Pharmacol.* **18**:535 (1969).
102. M. J. Walsh, E. B. Truitt, Jr., and V. E. Davis, *Mol. Pharmacol.* **6**:416 (1970).
103. N. H. Raskin and L. Sokoloff, *J. Neurochem.* **19**:273 (1972).
104. A. K. Rawat and K. Kuriyama, *Science* **176**:1133 (1972).
105. A. K. Rawat, K. Kuriyama, and J. Mose, *J. Neurochem.* **20**:23 (1973).
106. R. L. Veech, R. L. Harris, D. Veloso, and E. H. Veech, *J. Neurochem.* **20**:183 (1973).
107. D. Veloso, J. V. Passaneau, and R. L. Veech, *J. Neurochem.* **19**:2679 (1972).
108. H. W. Sippel, *J. Neurochem.* **23**:451 (1974).
109. B. Tabakoff, R. A. Anderson, and R. F. Ritzman, *Biochem. Pharmacol.* **25**:1305 (1976).
110. C. J. P. Eriksson and H. W. Sippel, *Biochem. Pharmacol.* **26**:241 (1977).
111. S. H. Snyder, A. B. Young, J. P. Bennett, and A. H. Mulder, *Fed. Proc. Symp.* **32**:2039 (1973).
112. E. Roberts, *Biochem. Pharmacol.* **23**:2637 (1974).

113. R. A. Lovell, S. J. Elliott, and K. A. C. Elliott, *J. Neurochem.* **10**:479 (1963).
114. F. N. Minard and I. K. Mashawar, *Life Sci.* **5**:1409 (1966).
115. R. P. Shank and M. H. Aprison, *J. Neurobiol.* **2**:145 (1971).
116. H. M. Hákkinen and E. Kulonen, *Nature (Lond.)* **182**:726 (1959).
117. H. M. Hákkinen and E. Kulonen, *Biochem. J.* **78**:588 (1961).
118. M. Mouton, C. Lefornier-Contensou, and J. Chalopin, *C. R. Hebd. Seances Acad. Sci.* **264**:2649 (1967).
119. R. A. Ferrari and A. Arnold, *Biochim. Biophys. Acta.* **52**:361 (1961).
120. E. S. Higgins, *Biochem. Pharmacol.* **11**:394 (1962).
121. E. R. Gordon, *Can. J. Physiol. Pharmacol.* **45**:915 (1967).
122. E. V. Flock, G. M. Tyce, and C. A. Owen, Jr., *Proc. Soc. Exp. Biol. Med.* **131**:214 (1969).
123. C. A. Leslie, Z. Gottesfeld, and K. A. C. Elliott, *Can. J. Physiol. Pharmacol.* **49**:833 (1971).
124. I. Sutton and M. A. Simmonds, *Biochem. Pharmacol.* **22**:1685 (1973).
125. H. M. Hákkinen and E. Kulonen, *J. Neurochem.* **27**:631 (1976).
126. I. A. Syntinsky, B. M. Guzikov, M. U. Gomanko, V. R. Eremin, and N. N. Knonvalova, *J. Neurochem.* **25**:43 (1975).
127. G. J. Patel and J. Lal, *J. Pharmacol. Exp. Ther.* **186**:625 (1973).
128. P. J. Griffiths and J. M. Littleton, *Br. J. Exp. Pathol.* **58**:19 (1977).
129. E. V. Flock, G. M. Tyce, and C. A. Owen, Jr., *Proc. Soc. Exp. Biol. Med.* **135**:325 (1970).
130. M. K. Roach and W. N. Reese, Jr., *Biochem. Pharmacol.* **20**:2805 (1971).
131. E. Majchrowicz, *Can. J. Biochem.* **43**:1041 (1965).
132. C. F. Baxter, in: *Handbook of Neurochemistry* (A. Lajtha, ed.), Vol. 3, pp. 289–353, Plenum Press, New York (1970).
133. G. W. Read, W. Cutting, and A. Furst, *Psychopharmacologia* **1**:346 (1960).
134. D. G. McQuarrie and E. Fingl, *J. Pharmacol. Exp. Ther.* **124**:264 (1958).
135. D. B. Goldstein and N. Pal, *Science* **172**:288 (1971).
136. F. Ratcliff, *Arch. Int. Pharmacodyn. Ther.* **196**:146 (1972).
137. W. A. Hunt, *Neuropharmacology* **12**:1097 (1973).
138. J. Cott, A. Carlsson, J. Engel, and K. Lindqvist, *Naunyn-Schmiedebergs Arch. Pharmakol.* **295**:203 (1976).
139. M. E. Hillbom, *Neuropharmacology* **14**:755 (1975).
140. E. P. Noble, R. Gillies, R. Vigran, and P. Mandels, *Psychopharmacologia* **46**:127 (1976).
141. Y. Godin, L. Heiner, J. Mark, and P. Mandel, *J. Neurochem.* **16**:869 (1969).
142. S. Simler, L. Ciesielski, M. Maitre, H. Randrianarisoa, and P. Mandel, *Biochem. Pharmacol.* **22**:1701 (1973).
143. J. D. Wood, *Prog. Neurobiol.* **5**:77 (1975).
144. H. Meunier, G. Carraz, Y. Meunier, P. Eymard, and M. Airmard, *Therapie* **18**:435 (1963).
145. K. Kuriyama, E. Roberts, and M. K. Rubinstein, *Biochem. Pharmacol.* **15**:221 (1966).

Comparison between the Effects of Ethanol and Those of Opioid Drugs on the Metabolism of Biogenic Amines

12

Isabel J. Wajda

A. INTRODUCTION

Morphine and ethanol produce profound alterations in human behavior, but exact biochemical mechanisms underlying their central actions are still unknown. Narcotic drugs and ethanol alter neuronal activity of the central nervous system, and changes in neurotransmission continue to be investigated. The multiplicity of central effects of ethanol and especially of morphine suggests that different neurotransmitters are involved. Biogenic amines such as norepinephrine, dopamine, and serotonin are of particular interest, since changes in the levels of those amines may lead to alterations in behavior.

In spite of a considerable effort in research dealing with changes induced by opiates or ethanol on the steady-state levels, turnover, and metabolism of biogenic amines, the elucidation of the role they play in the action of those addictive drugs still eludes us. This chapter discusses the difficulties which underlie such research; in addition, an attempt is made to explain why contradictory results are often obtained.

This review is limited to only a few aspects of the investigations dealing with the effects of opiates and ethanol on central neurotransmitters: namely, to the changes in the steady-state levels and in the turnover and metabolism of three particular biogenic amines—norepinephrine, dopamine, and serotonin. Forma-

---

Isabel J. Wajda • Research Institute for Neurochemistry, Rockland Research Institute, Ward's Island, New York 10035.

tion of the condensation products of aldehydes and biogenic amines will also be discussed, as an illustration of one of the attempts to bridge the gap between the two groups of drugs.

B. COMPARATIVE STUDIES OF ACTIONS OF MORPHINE AND ETHANOL

Alcoholic beverages and opiate alkaloids belong to the oldest group of drugs. Their therapeutic (morphine) and social (alcohol) values depend entirely on their central actions. Both belong to the depressant and addictive type of drugs, both cause euphoria, and in long-term uninterrupted use they produce strong dependence. Tolerance develops differently with the two drugs; it is very high in the case of morphine and not very strong with ethanol. Also, in both short-term effects and withdrawal signs and symptoms, the drugs show more differences than similarities.

These differences become evident when one compares the toxic effects of morphine and ethanol in addicts and in nonusers. The minimum lethal dose of morphine (200 mg) is about 10 times higher in addicts.[1] Similarly in rats, a dose of 200 mg/kg produces 80% mortality in normal rats while for chronically treated rats a dose of 300 mg/kg produces only 10–12% mortality.[10] Moreover, a dose of 200 mg/kg could be maintained without any mortality in rats treated chronically with morphine. The lethal dose of ethanol, on the other hand, differs very little between addicts or chronically treated animals and the normal case.

There is little evidence for long-sought connections between the two drugs. Ethanol-dependent mice were challenged with the opiate antagonist naloxone, but the characteristic opiate-dependent jumping was not evoked.[2] Many human opiate addicts have previous records of heavy drinking, and when narcotics are not available they tend to replace them with ethanol.[3] Such a connection between the two drugs was tested in rats, and measurements of the voluntary consumption of ethanol and morphine were made. Morphine was found to suppress drinking of ethanol,[4] but ethanol was not able to influence the oral consumption of saccharin-adulterated morphine.[5] Another connection in the action of ethanol and morphine was found when acute administration of morphine to mice suppressed convulsions produced by withdrawal from ethanol[6]; this effect persisted longer than the analgesia. The fact that convulsions during ethanol withdrawal were also suppressed by dopamine suggested to the authors that such action of morphine or of ethanol may be due to a release or increased activity of dopamine in the central nervous system.

In some experiments ethanol was administered together with pyrazole according to the method developed by Goldstein.[7] However, each of these drugs has its own characteristic pharmacological effects. Pyrazole, a potent inhibitor of alcohol dehydrogenase, retards the disappearance of ethanol from the blood of experimental animals,[8] providing an additional powerful tool in the

studies of pharmacological properties of ethanol. However, the delayed toxico-
logical properties of pyrazole, such as hyperplasia of the liver and the thyroid,
atrophy of testes and seminal vesicles, and deposition of iron in the spleen and
bone marrow in mice and rats, limits its use in such studies.[9] Anemia and
moderate leukopenia were observed in rats after 150 mg/kg per day of chronic
administration of pyrazole.[10] Catecholamine levels were estimated after pyrazole treatment in rats (50–300 mg/kg, ip). A dose-dependent fall in norepinephrine content of the brain was observed, but no changes occurred in the levels of
dopamine or 5-hydroxytryptamine.[11]

C. EFFECTS ON THE STEADY-STATE LEVELS OF BIOGENIC AMINES

1. Norepinephrine

a. Effects of Morphine

The effects of morphine on the levels of biogenic amines are quite complex.
They depend on the dose of morphine employed, the species used, and the amine
in question; they also differ in different parts of the brain. In 1954, Vogt[12]
demonstrated that a single injection of morphine was capable of depleting the
catecholamines of the cat brain. Generally, it has been observed that in rats
morphine will reduce norepinephrine levels, but increases have also been
reported (Table I).

Gunne found[13] that acute administration of 20 and 30 mg morphine/kg to
rats was without effect on the levels of brain norepinephrine. Two consecutive
injections of 30 mg/kg, given at two intervals, reduced norepinephrine levels
from 0.56 μg/g to 0.40 μg/g. Chronic treatment for 3 weeks, with increasing doses
of morphine given four times a day (1800 mg/kg per day), increased the norepinephrine levels of the brain from 0.53 to 0.61 μg/g. Withdrawal from morphine
did not change the increased levels of norepinephrine. Similarly, Maynert and
Klingman[14] found a decrease from 736 ng/g to 614 ng/g in rat brain norepinephrine levels after a single dose of 200 mg/kg; the decrease was never larger than
20%. Chronic treatment with 100 mg/kg (twice a day for 5 days) did not change
the normal levels of norepinephrine, but the same treatment with double the dose
of morphine showed a considerable increase (834 ng/kg) in norepinephrine levels.
He also observed that although 200 mg/kg represented approximately the LD_{50}
for control rats, this dose could be maintained without any mortality in chronic
experiments.

Using mice as experimental animals, Rethy et al.,[15] measuring the effect of
morphine on the total catecholamine content of the brain, found that doses of 30
mg/kg lowered the content to 80% of controls, 100 mg/kg to 70%, and 300 mg/kg
to only 65%. Similar results were reported by Takagi and Nakama,[29] who found
a 23% decrease in mouse brain norepinephrine levels after 20 mg/kg of morphine.
Fennessy and Lee[27] obtained a dose–response curve (doses of morphine from

Table I. Levels of Catecholamines in the Brain of Animals Treated with Morphine

Biogenic amine	Animal	Region of the brain	Route of administration	Dose (mg/kg)	Changes[a]	Reference
NE	Rat	Brain	Acute, ip	30	No change	13
			Acute, ip	30 + 30	↓ 29%	
			Chronic, ip	1800/day	↑ 15%	
			Withdrawal	—	No change	
NE	Rat	Brain	Acute, ip	200	↓ 17%	14
			Chronic, ip	200/day	No change	
			Chronic, ip	400/day	↑ 13%	
Total catecholamines	Mouse	Brain	Acute, ip	30	↓ 80%	15
			Acute, ip	100	↓ 70%	
			Acute, ip	300	↓ 65%	
NE	Mouse	Brain	Acute, sc	20	↓ 23%	29
NE	Mouse	Brain	Acute, sc	2.5–10	↓ 30%	27
			Acute, sc	20	No change	
			Acute, sc	100	↑ 15%	
DA	Mouse	Brain	Acute, sc	0.1	↑ 20%	27
			Acute, sc	10	↓ 50%	
			Acute, sc	100	↑ 60%	
DA	Mouse	Brain	Acute, sc	20	↓ 23%	29
DA	Rat	Corpus striatum	Acute, sc	10	No change	30
DA	Rat	Corpus striatum	Acute, sc (1 hr)	30	No change	31
			Acute, sc (2 hr)	30	↑ 47%	
			Acute, sc (4 hr)	30	↑ 39%	
DA	Rat	Corpus striatum	Acute, ip	30	↑ 21%	32
DA	Rat	Corpus striatum	Acute, sc	30	↑ 37%	33

[a]Changes: ↓, decrease; ↑, increase.

0.1 to 100 mg/kg) and found a gradual decrease in norepinephrine levels calculated as per cent of control values. A 30% decrease was caused by 5 mg/kg, but 100 mg/kg increased the levels of norepinephrine to 5% above control. Withdrawal from long-term morphine treatment was either without effect[13] or reduced the elevated norepinephrine content to the normal levels.[16] The search for changes in norepinephrine levels in the brain produced by administration of morphine was initiated because norepinephrine per se exhibits analgesic properties and enhances the analgesic effects of morphine.[20] The evidence for the direct involvement of norepinephrine in the analgesic action of morphine is nevertheless circumstantial, and was usually obtained after using various manipulations to elevate or lower norepinephrine levels in the brain. One of the unresolved problems related to changes in the brain levels of biogenic amines was the use of reserpine in morphine analgesia. Potentiation of morphine effects[17] and opposite results as well[18] were reported. Reserpine is a long-acting substance; the depletion of biogenic amines lasts for days after a single injection. This prolonged depletion might influence the effects of morphine, according to the time of administration of both substances. Intraventricular administration of biogenic amines was found to attenuate the antinociceptive effects of morphine,[19] although neither intraventricular nor intravenous injections of norepinephrine produced significant effects on the nociceptive threshold of rats. Reserpine pretreatment for 15 hr before injection of morphine antagonized the analgesic effects of morphine, and in rats so treated, the intraventricular injection of serotonin, but not of norepinephrine, restored the antinociceptive effect of morphine. From those results the authors concluded that the analgesic effects of morphine depend on a dynamic balance between the concentrations of norepinephrine and serotonin in the brain, with serotonin promoting and norepinephrine antagonizing the antinociceptive effects of morphine.

Estimations of the steady-state levels of norepinephrine in the brains of animals treated with morphine suggested that norepinephrine plays an insignificant role in mediating morphine effects.[20] There are several findings that support this opinion. Changes in norepinephrine levels are inconsistent; they vary according to the species used. During an abrupt withdrawal from morphine, depletion in brain norepinephrine levels occurs in dogs but rats show little change in this respect. Gunne[13] ascribes these differences to different reactions to withdrawal from morphine according to the species used; dogs appear to be very excited, while the rats react with a mixed state of drowsiness and irritability. In rats and mice lower doses administered acutely produced mostly depression of norepinephrine brain levels (Table I); withdrawal left the steady-state levels unchanged, and only very high doses of morphine in either chronic or acute treatment resulted in a 13–15% increase in norepinephrine.[13,14]

b. Effects of Ethanol

Equally complex pictures emerge from the studies of the steady-state levels of norepinephrine recorded in ethanol-treated animals. Corrodi et al.[21] reported that ethanol alone (2 g/kg, ip) did not change the levels of norepinephrine in the brain of rats, but in animals pretreated with α-methyl-p-tyrosine (250 mg/kg, ip) ethanol enhances the release and the synthesis of norepinephrine, resulting after 4 hr in a 14% decrease in the levels of norepinephrine. Using the same inhibitor of catecholamine synthesis, Hunt and Majchrowicz[22] reported an unaltered steady-state level of norepinephrine in acutely treated rats (5 g/kg, po) after 15 min but a decrease of 18% in animals killed 2 hr after administration of ethanol. Blood levels of ethanol at the time of decapitation ranged from 100 to 300 mg/dl.

Post and Sun[23] have pointed out that chronic administration of ethanol leads to a slight increase in norepinephrine concentration in the hypothalamus. Rats were maintained on 15% ethanol (v/v) as their sole drink for 1 year. The blood level of ethanol at the time of sacrifice was 30 mg/dl. The levels of norepinephrine in the hypothalamus were 4.1 μg/g in the ethanol group and 3.6 μg/g in the controls. The difference was not significantly higher in the ethanol group when compared with the controls. In the same year Sun[92] also demonstrated that the reuptake of norepinephrine by isolated brain synaptosomes is inhibited by low concentrations of ethanol (0.086–1.225 M). The author suggests that the inhibition of norepinephrine uptake at the synaptosomal site may produce a temporary accumulation of norepinephrine at the synaptic junction and the lack of uptake may lead to depressive states such as those observed in prolonged periods of ethanol intake in man.

Gursey and Olson[25] reported a decrease in norepinephrine levels, of 50% in chronically (2 g/kg per day) and 55% in acutely (2 g/kg) treated rabbits with a blood ethanol concentration of 90 mg/dl. Ahtee and Svartstrom-Fraser[26] were not able to detect any changes in norepinephrine levels in chronically treated rats. The blood ethanol concentration in those animals was 91.5 mmole/liter, and the treatment consisted of 8–11 g ethanol/kg daily for 7–10 days by oral administration.

2. Dopamine

a. After Morphine

Estimates of the steady-state levels of dopamine after morphine are also far from being consistent. Using mice, Fennessy and Lee[27] recorded an initial increase of 20% after 0.1 mg morphine/kg. But for injection of 0.5, 1.0, 5.0, and 10.0 mg/kg they found a gradual decrease, reaching 50% of control values after 10.0 mg/kg. Still higher amounts (100 mg/kg) increased the levels of brain dopamine by 60% of control values. Serotonin, dopamine, and norepinephrine are of particular interest, since these amines play a vital role in brain function.[24] The changes in norepinephrine and dopamine levels reported by those authors do not parallel the development of analgesia; they are similar to but are not in complete agreement with the activity dose–response curve. Dopamine, which is considered mainly as the precursor of norepinephrine. plays an independent role in striatal neural transmission. Corpus striatum, which represents predominantly the cholinergic and dopaminergic center of integration in the central nervous system contains about 10 times as much dopamine as the other parts of the brain. The next highest concentration of dopamine is found in the olfactory tubercle which has only half of the striatal content of dopamine. The investigation of changes in dopamine levels, therefore, focused mainly on corpus striatum.

Changes in dopamine levels in the striatum after administration of morphine cannot be implicated in the analgesic action of opiates, but they can be of importance in motor responses that are affected by morphine. Striatum represents an old, important motor center with an inhibitory influence on motor activity[28]; inhibitory reactions such as akinesia, arrest response, and catalepsy can also be triggered by high doses of morphine. Cataleptic reaction can easily influence tests on analgesia in animals and in man they can have an influence on emotional behavior.

Studies on the steady-state levels of dopamine are, like those of norepinephrine, inconclusive (Table I). Takagi and Nakama[29] found a 23% decrease in the brain content of dopamine (20 mg morphine/kg), whereas Kuschinsky reported normal levels.[30] Increase in rat striatal dopamine levels after a single administration of morphine was reported by us previously. This was a delayed response; after 1 hr the levels of dopamine were close to normal (after saline: 7.74 μg/g; morphine: 8.26 μg/g); the increase occurred 2 and 4 hr after subcutaneous injection of morphine (10.43 and 10.22 μg dopamine/g, respectively).[31] Increased levels of dopamine in the striatum were also reported by other laboratories. Injection of 30 mg morphine/kg resulted in a 21% increase in rat striatal levels of dopamine,[32] while 60 mg morphine/kg increased the levels to 18.5 nmole/g relative to control levels of 13.5 nmole/g.[33] Those findings suggest that the dopaminergic system is involved in whatever action morphine exerts on the striatum. It is also known for rats that repeated morphine administration reverses the acute effects of morphine, such as catalepsy and akinesia, suggesting an activation of the dopaminergic mechanism.

The involvement of dopaminergic neurotransmission after acute administration of opiates is represented by a decrease in dopaminergic activity, resulting in akinesia, catalepsy, and muscular rigidity. The lack of normal release of dopamine results in accumulation of the neurotransmitter in the striatum. On the other hand, after repeated injections of morphine, locomotor stimulation and stereotypic behavior similar to that observed after central stimulants become evident, indicating increased dopaminergic activity. Finally, the withdrawal from chronically administered morphine was shown to produce hyperactivity of dopaminergic neurotransmission.[34] Compensatory supersensitivity of dopamine receptors after morphine administration was suggested.[35] Using striatal slices from morphine-withdrawn rats, Bosse and Kuschinsky[36] measured K^+-induced release of [14C]dopamine and were able to show that chronic administration of morphine to rats increases the dopaminergic transmission in the striatum. The activation of the dopamine mechanism was shown by an increase in K^+-induced release of [14C]dopamine from striatal slices of morphine-withdrawn animals as compared with that from striatal slices of saline-treated controls.

It seems, therefore, that some of the central actions of morphine can be based on morphine-induced local alterations in the dopaminergic neurotransmission which can be detected even in the steady-state levels of dopamine, provided that one deals with parts of the brain rich in the dopaminergic mechanism. Changes in the striatal neurotransmission could influence other parts of the brain and could also be reflected in abnormal motor activity (catalepsy, arrest reaction). They could, on the other hand, represent an aftereffect of increased neuronal activity.

Elevation in brain dopamine, with greatest increase in the corpus striatum, was observed during naloxone-precipitated withdrawal in the morphine-dependent jumping of mice and rats.[37] The increase was noted within 2 min in dependent mice after administration of naloxone, and ranged from 2.0 μg/g to 2.6 μg/g tissue. We repeated those experiments and included measurements of dopamine levels in mice in which jumping was induced by high ambient temperature. In both cases we found increases in dopamine levels in the brain; in total brain the increase was from 1.0 μg/g to 1.5 μg/g tissue, at 10 min after injection, but the increase did not coincide with jumping.[38] This, again, indicates that the increase of 50% in dopamine levels, although related to the injection of the drug, might be an aftereffect of the increased neuronal activity and not necessarily the cause of it. Such an explanation might be valid, especially in cases in which the levels of the transmitter change much later than the appearance of behavioral or physiological signs following administration of the drug.

b. Effects of Ethanol

Narcotic action of ethanol seems to be influenced by neuroamines. Effects such as prolongation of ethanol-induced sleeping time occur after injection of dopamine, serotonin, or neuroamine precursors or metabolites, including unusual metabolites (possibly isoquinolines derived from the condensation of acetaldehyde and phenylethylamines).[39] Several laboratories have been engaged in establishing the levels of dopamine neurotransmitters in the brains of animals intoxicated with ethanol. As with the studies on the levels of biogenic amines after morphine, the results from different laboratories do not agree. Post and Sun[23] reported in chronically treated rats [15% ethanol (v/v) as a sole drink for 1 year] a large increase in dopamine levels in the caudate nucleus and a less pronounced rise in norepinephrine levels in the hypothalamus. The values for dopamine content in the caudate nucleus were 7.4 $\mu g/g$ for the control group and 18.6 $\mu g/g$ for the ethanol group. Norepinephrine content of the hypothalamus was 3.6 $\mu g/g$ for the control and 4.1 $\mu g/g$ for the ethanol group. The authors came to the conclusion that the dopaminergic neurons are more sensitive to ethanol than are noradrenergic neurons.

Using mice, Griffith et al.[40] found a transient fall in indole and catecholamines after acute administration of ethanol. Chronic treatment, however, resulted in increased levels of all biogenic amines. An increase in norepinephrine levels was also reported by Pohorecky et al.[41] in brains and hearts of rats during the period of withdrawal from ethanol. Still different results were obtained in our laboratory when we examined the steady-state levels of dopamine in the striatum of rats subjected to ethanol treatment. We found an increase in the levels of dopamine (from 7.9 $\mu g/g$ to 10.7 $\mu g/g$) after intoxicating doses of ethanol (4 g/kg, ip).[42] The increase was short-lived: 4 hr after injection the striatal dopamine was at normal levels. We also noticed that lower doses of ethanol (1 g/kg) reduced the levels of striatal dopamine. Even higher levels of dopamine were found in the striatum of rats maintained for 2 months on ethanol (16.2%) in drinking water. In those animals we found an increase in striatal dopamine from 8.0 $\mu g/g$ to 13.1 $\mu g/$g wet wt. These results are summarized in Table II.

The differing results cannot be compared, however, since in some cases mice instead of rats were used,[40] and in others the material taken for analysis was not identical. For example, Pohorecky et al.[41] used telencephalon, midbrain, and brainstem, whereas in our experiments[42] only striatum was examined. The reports from different laboratories rarely agree on the steady-state

Table II. Levels of Biogenic Amines in the Brain of Animals Treated with Ethanol

Biogenic amine	Animal	Brain region	Route of administration	Dose (g/kg)	Changes	Reference
NE and DA	Rat	Brain	Acute, ip	2	↓14% and 0	21
NE and 5-HT	Rabbit	Brain	Acute, iv	2	↓55% and ↓40%	25
NE and 5-HT			Chronic, iv	2/day	↓50% and ↓50%	
NE and DA	Rat	Brain	Chronic, po	8–11	No change	26
5-HT	Rat	Brain	Acute, po	4.5	↑56%	54
NE	Rat	Brain	Acute, ip	2	No change	55
5-HT	Rat	Brain	Acute, ip	2	No change	
5-HT	Rat	Brain	Acute, ip	4	↑16.6%	56
			Chronic	Liquid diet	↑20%	
DA	Rat	Caudate nucleus	Chronic	15% ethanol for 1 year	↑150%	23
		Hypothalamus	Chronic		↑14%	
5-HT	Rat[b]	Brain	Acute, po	5	No change	57
			Chronic	15 3/day	No change	
			Withdrawal	—	No change	
NE	Rat[c]	Brain	Acute, po	5	↓18%	22
DA	Rat	Brain	Acute, po	5	No change	
NE, DA, 5-HT	Mice	Brain	Acute (3 hr)	Inhalation: 10–15 mg/liter	↓20–40%	40
NE, DA, 5-HT	Mice	Midbrain brainstem	Chronic (10 days)	2 liters/min	↑40–50%	
NE	Rat		Withdrawal	Liquid diet (35% ethanol)	↑28%	41
DA	Rat	Caudate nucleus	Acute, ip	4	↑35%	42
		Caudate nucleus	Chronic	16% ethanol for 2 months	↑63%	
		Caudate nucleus	Withdrawal	Drinking fluid	↑60%	

[a]Changes: ↓, decrease; ↑, increase.
[b]Rats treated with pargyline (75 mg/kg).
[c]Rats treated with α-methyl-p-tyrosine (250 mg/kg).

levels of biogenic amines after opiates or ethanol. There are many reasons for the discrepancies. Very often the results are gathered from different experimental animals; the methods of killing or of extraction, and even of estimations vary; diurnal changes in the control and experimental animals are not taken into account; and finally changes not detectable in some areas of the brain might occur in others. The situation is even more complicated because of the treatment with addictive drugs. The effects can be evaluated after a single exposure to the drug; after prolonged treatment, of which there are numerous variations, or during a withdrawal period from the drug, where again the timing, the nature of the treatment, and partial or complete clearing of the drug may play an important role. These seem to be some of the reasons why the turnover of biogenic amines, their metabolism, and the elimination of the metabolites are of greater interest than the steady-state levels in the studies of the effects of addictive drugs.

In spite of the different results regarding the steady-state levels of biogenic amines in general, there is some agreement about the levels of dopamine in the striatum. When the animals were exposed to ethanol only, without pyrazole or any other pharmacological agent, large increases in striatal dopamine were found.[23,42] Similar results, although less pronounced, were found for the effects of morphine on the levels of striatal dopamine.[27,31-33] A possible explanation for this similarity could relate to actions that might not be specific, especially in the case of ethanol, but might involve disturbances in the normal state of membranes. This, in turn, could lead to accumulation of a neurotransmitter and change its normal release. We see changes in the striatum because of the high concentration of dopamine in this part of the brain.

3. Serotonin

a. Morphine Action

The role of 5-hydroxytryptamine in central morphine actions has been studied extensively, but the results are contradictory. While there is agreement about the lack of changes in brain steady-state levels of serotonin, suggestions were made that increased turnover of brain serotonin was associated with tolerance and physical dependence in mice.[43] Subsequent studies were mostly contradictory, and while some did agree with increased turnover of serotonin in tolerant animals,[44] others failed to confirm even those findings.[45] Similar experiments performed in rats confirmed small but significant increase in the turnover of brain serotonin, and increases were also reported in the levels of 5-hydroxyindoleacetic acid (5-HIAA), with no alteration of brain serotonin.[46]

It seems, therefore, that morphine has a greater effect on the turnover of catecholamines, and some perhaps insignificant influence on the steady-state levels of dopamine, while the levels of serotonin are not changed at all. One of the reasons that the levels of biogenic amines remain unchanged or change very insignificantly in spite of increased biosynthesis is probably the increase in their acid metabolites. Higher-than-normal levels of homovanillic acid (HVA), dihydroxyphenylacetic acid (DOPAC), and 5-hydroxyindoleacetic acid have been found in the brain of morphine-treated animals.[46,60,61] The concentration of biogenic amines in different parts of the brain also has an influence on the effects of narcotic drugs. Corpus striatum, the dopaminergic center, with the highest concentrations of dopamine, also shows the major increase in dopamine biosynthesis after opiates.[33]

b. Effects of Ethanol

A decrease in norepinephrine and serotonin levels in the brains of rabbits treated with ethanol was reported by Gursey and Olson[25] (Table II). This finding was not confirmed by Pscheidt et al.,[49] who also used rabbits in their experiments. In the early 1960s several studies on serotonin brain levels after ethanol recorded mostly conflicting results.[50-53] Bonnycastle et al.[54] reported an increase in serotonin levels after ethanol in rats, while changes no greater than those observed after injection of saline were recorded by Duritz and Truit.[55] Pohorecky et al.,[56] on the other hand, found a 16.6% increase in brain serotonin levels in rats injected (ip) with 4 g ethanol/kg, and after chronic treatment, consisting of a liquid diet for 2 weeks, they recorded an increase of 20%. Hunt and Majchrowicz[57] were not able to find any changes in the steady-state levels of serotonin (Table II).

D. TURNOVER OF BIOGENIC AMINES

1. After Morphine

The steady-state level of biogenic amines did not provide a satisfactory answer to the question about the exact relationship between biogenic amines and narcotic drugs. Additional information, such as alteration in the rate of turnover of amines, began to accumulate, starting in 1970. Changes in the rate of catecholamine synthesis were reported by Clouet and Ratner,[58] who used intracisternal injection of [^{14}C]tyrosine. They found an increase in [^{14}C]dopamine in the hypothalamus (from 194 to 238 cpm/mg protein) within 1 hr after injection of 60 mg morphine/kg. In the striatum the increase was highest in rats treated for 5 days with morphine (from 70 to 97 cpm/mg protein). The conversion of [^{14}C]tyrosine to norepinephrine was highest in the hypothalamus in chronically treated rats (from 53 to 176 cpm/mg protein).

Similar studies conducted on mice[59] found morphine-induced increases in catecholamines formed from [^{14}C]tyrosine (from 1800 dpm, a rise to 2900 dpm/g brain tissue). The same authors reported later[60] that 100 mg morphine/kg and 10 mg levorphanol/kg nearly doubled the incorporation of [^{14}C]tyrosine into both [^{14}C]dopamine and [^{14}C]norepinephrine in the brains of mice. The

[^{14}C]catecholamine synthesis was measured in the cerebral cortex, diencephalon, striatum, brain stem, and cerebellum, and these effects could be blocked by a pretreatment with naloxone.

In subsequent years (1973 and 1974) several publications appeared confirming morphine-induced increases in the turnover rate of dopamine in the striatum.[32,33,47,48] Alterations in dopaminergic transmission in the striatum[36] and increases in dopamine utilization in the striatum after morphine[30] were reported.

It seems, therefore, that although the levels of catecholamines are relatively unaffected by morphine, the turnover of amines is increased affecting mainly the biosynthesis of dopamine in the striatum. The steady-state levels depend also on the rate of catabolism of amines, which in turn regulates the levels of metabolites.

2. After Ethanol

The subject of this section, like that of the previous one, is reviewed in other parts of this volume. Only the information needed for comparison of changes in the metabolism of biogenic amines after opiates and ethanol will be mentioned here.

Hunt and Majchrowicz found that the turnover rates of serotonin were unchanged in ethanol-treated rats.[57] This was true for rats treated acutely (5 g/kg), chronically, or during the withdrawal syndrome. Another series of experiments from the same laboratory, dealing with the effects of ethanol on the turnover of catecholamines, revealed a very important fact—that the effects on the turnover of catecholamines are biphasic.[22] In rats given a single dose of ethanol (5 g/kg) the turnover of norepinephrine, which is increased during the first few hours after ethanol, is somewhat reduced later (5 hr), whereas that of dopamine is at first unaffected but later it is decreased. From the point of view of physical dependence and the alterations in catecholamine function in the central nervous system after ethanol, these findings shed more light on the complicated neuronal processes and the varied results from different laboratories.

A transient rise in 5-HIAA levels in the brain of mice injected with 3 g/kg (ip) of ethanol was recorded by Tabakoff et al.,[61] with no change in serotonin levels, where serotonin and 5-HIAA were estimated by a modified method of Maickel et al.[62] using o-phthalaldehyde. Simultaneously, a decrease in serotonin turnover and an inhibition of 5-HIAA uptake by isolated choroid plexus were noted. An increase in catecholamine metabolites was recorded by Karoum et al.[65] after ethanol treatment in rats. They found no evidence of changes in the transport of phenolic acids across the blood–brain barrier in rats acutely treated with ethanol, and the increase in catecholamine metabolites was ascribed to the increase in their turnover. The estimation of acidic metabolites of catecholamines was done using combined gas chromatography/mass spectrometry.

It is conceivable that the increase in acidic metabolites of biogenic amines is a final result of many causes, which can be of a different nature for any particular amine. Comparing the effects of opiates and ethanol on the metabolism of biogenic amines, one finds that there is more information available about the acidic metabolites of biogenic amines in ethanol-treated animals. After opiates the turnover of catecholamines, especially that of dopamine, is increased, whereas that of serotonin is still controversial. Nevertheless, the levels of acidic metabolites of serotonin and catecholamines increase. In ethanol-treated animals the turnover of catecholamines is less affected, while that of serotonin is either decreased or unchanged. Again, the metabolite levels are higher.

How these changes, especially those in the levels of metabolites, are reflected in the normal functioning of the central nervous system is difficult to evaluate. It is also impossible to assess how much of the aftereffect of those drugs depends on the accumulation of the acidic metabolites of biogenic amines, which probably exert some regulatory influence on the levels of neuroamines. However, increases in the content of acidic metabolites of biogenic amines must be interpreted with caution because they very often occur after the use of substances that have varied effects on the central nervous system. Increases in the levels of HVA in the striatum of mice were found after treatment with neuroleptics such as chlorpromazine and haloperidol, analgesics such as methadone,[66] or anorectic drugs such as amphetamine, which also was found to elevate the HVA concentration in the striatum of rats.[67] The antiemetic drug, metoclopramide (20 mg/kg), induced a sixfold increase in striated HVA concentration in rats.[68] There are also differences according to the species, the region of the brain, and even to the strain of the animals.[46]

E. ACID METABOLITES OF BIOGENIC AMINES

Release of serotonin, dopamine, or norepinephrine is followed by either the subsequent uptake of amines or the destruction of their activity, which is catalyzed either by monoamine oxidase, followed by the formation of acidic metabolites of biogenic amines, or in the case of catecholamines, by catechol-O-methyltransferase. Dopamine is metabolized mainly into homovanillic acid and partially into dihydroxyphenylacetic acid, whereas serotonin gives rise to 5-hydroxyindoleacetic acid. The concentration of the acid metabolites in the cerebrospinal fluid can be influenced by drugs acting on the main routes of elimination, which include diffusion, the bulk flow of fluid, and the active transport mechanism.

Levels of the acid metabolites of biogenic amines higher than normal have been recorded in brains of morphine-treated animals.[46,63,64] A direct relationship between morphine-induced catalepsy and a dose-dependent increase in the concentration of HVA in the striatum was reported.[69]

Similar results were published by Kuschinsky and Hornykiewicz.[70] HVA was also measured in corpus striatum and in two limbic structures, nucleus accumbens and olfactory tubercle; again, significant increases after morphine were recorded in all three brain areas.[71] The levels of HVA can be considered as a qualitative reflection of dopamine metabolism[72]; this is of clinical importance because HVA is the only dopamine metabolite detectable in human cerebrospinal fluid. However, to obtain a more complete picture of changes in dopamine metabolism, dihydroxyphenylacetic acid, another acid metabolite of dopamine, is also considered. Changes in brain levels of DOPAC were found to provide a useful index of alterations of the functional activity of control dopaminergic neurons,[73] and the presence of free and conjugated forms of this metabolite in different areas of the brain was recently reported.[74]

Since choroid plexus is considered to be one of the active sites of the transport of acid metabolites of biogenic amines, we examined the effects of morphine on this system.[75] Earlier we had noticed changes in the appearance of the choroid plexus after morphine, finding on histological examination numerous vacuoles together with pronounced ischemia.[93] Accumulation of HVA and 5-HIAA in the choroid plexus was not changed, however, by either acute or chronic treatment with morphine, suggesting that morphine does not influence the elimination of those metabolites from the choroid plexus.[75] In another series of experiments we examined the effect of DOPAC on the uptake of 5-HIAA in the choroid plexus and found that DOPAC behaves differently from other acidic metabolites of biogenic amines. While HVA and other derivatives of phenylacetic acid inhibited the uptake of 5-HIAA in the choroid plexus, DOPAC had a stimulating effect on this process.[94] All those observations suggest that the changes in acidic metabolites produced by drugs might be a composite of many events, not only the direct effects of drugs on metabolism of biogenic amines, but also interactions of metabolites themselves on their uptake, accumulation, or elimination.

F. ACETALDEHYDE, BIOGENIC AMINES, AND THEIR CONDENSATION PRODUCTS

An interesting relationship of ethanol and opium alkaloids has been postulated on the basis that aldehydes derived from biogenic amines and acetaldehyde can react with biogenic amines to produce various alkaloids. Acetaldehyde and dopamine or norepinephrine can form tetraisoquinolines. Acetaldehyde and serotonin give rise to harmaline alkaloids, while 3,4-dihydrophenylacetaldehyde reacts with dopamine to give tetrahydropapaveroline (THP).

Formation of tetrahydroisoquinolines (TIQ) occurs easily *in vitro*, as described by Pictet and Spengler.[76] Unstable Schiff's base intermediates are formed by nonenzymatic condensation of aldehydes with amines, and after cyclization they form TIQs (Fig. 1A). A cyclic product of the condensation of dopamine, salsolinol, has been studied in rat brain and liver homogenates.[77] Even more intriguing is a condensation product of dopamine and 3,4-dihydroxyphenylacetaldehyde (dopaldehyde) tetrahydropapaveroline (Fig. 1B).

Tetrahydropapaveroline (THP), a demethylated and hydrogenated derivative of papaverine, was first described by Holtz *et. al.*[78] in 1964, who studied its

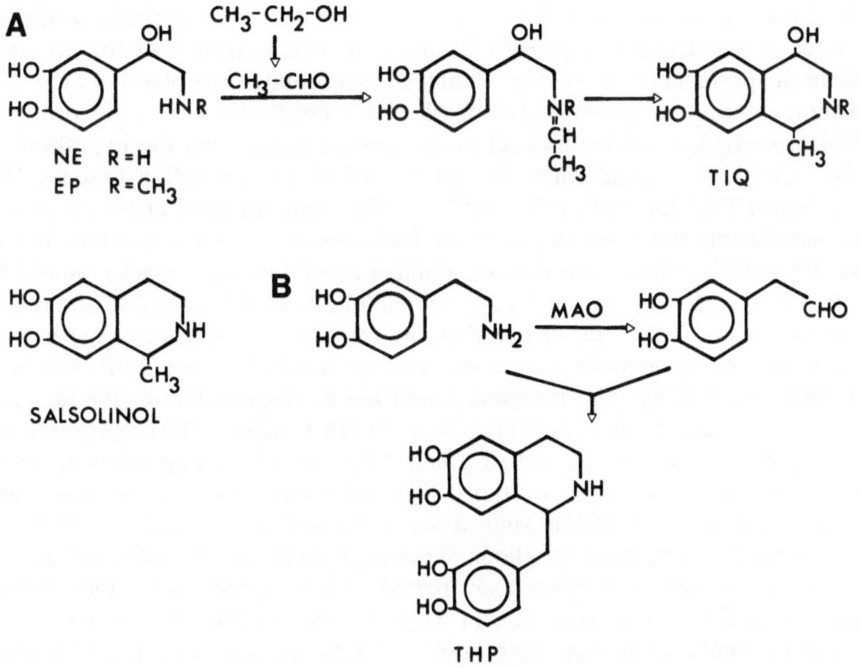

Fig. 1. (A) Formation of tetrahydroisoquinolines (TIQ) from acetaldehyde and biogenic amines. NE, norepinephrine; EP, epinephrine; salsolinol, the condensation product of dopamine and acetaldehyde. (B) Formation of tetrahydropapaveroline (THP) from dopamine and dopaldehyde.

pharmacological properties and compared them to those of papaverine. The condensation reaction was found to be augmented by ethanol (100 mM) or acetaldehyde (0.5–4.0 mM).[79,80] Davis and Walsh put forward a hypothesis that THP formed in mammalian cells could lead to formation of morphinelike alkaloids. THP is an intermediate in the synthesis of morphine in the opium poppy. During the experiments of Davis and Walsh[85] it was found that acetaldehyde (0.5–2.0 mM) condensed directly with dopamine (5 mg in a total volume of 4 ml), forming another derivative of tetraisoquinoline, salsolinol. This type of reaction has been known to occur for a number of years.[81]

Formation of tetrahydropapaverolines depends on the presence of biogenic aldehydes, which normally are rapidly oxidized. The oxidation step in their catabolism was found to be competitively inhibited by acetaldehyde at the level of aldehyde dehydrogenase.[82] The ethanol metabolite, acetaldehyde, is therefore involved in the formation of both TIQ and THP. In one case it takes part directly in the condensation reaction, and in the other it prevents catabolism of biogenic aldehydes.

Condensation products of epinephrine and acetaldehyde were demonstrated by Cohen and Collins[83] in the perfusates of isolated bovine adrenal glands. Perfusion was done with buffered acetaldehyde solution (1 μg/ml in isotonic saline buffered to pH 7.4 with 0.01 M sodium phosphate). Formation of such alkaloids was intriguing, especially because the concentration of the acetaldehyde in the perfusing solution was similar to that found in the blood of man after ingestion of intoxicating doses of alcoholic beverages.[84] However, the formation of THP *in vivo* has not been conclusively demonstrated, and the hypothesis of Davis and Walsh[85] concerning the etiology of alcohol addiction based on the formation of THP has been criticized.[86,87] The main points of criticism were[87] that experiments designed to prove the formation of such compounds *in vivo* were inconclusive; that even *in vitro*, milligrams of dopamine were required for significant THP formation, and that after inhibition of aldehyde dehydrogenase by disulfiram no THP could be found in the liver, heart, and urine of experimental animals. The same authors state that neither metabolites nor THP itself have high addictive liability, and therefore could not be responsible for the development of the syndrome of alcohol addiction. Another polemic developed between Seevers *et al.*[86] on the subject of ethanol and morphine dependence, where similarities and differences in those two addictive drugs were discussed. Recently, a minireview of this subject was published by Cohen,[88] in which he raised several unanswered questions about a possible role of TIQs not only in alcoholism, but also in Parkinsonism treated with large doses of L-dopa, a condition which favors the formation of TIQs.[89] The questions[86] related to the possible connection between the presence of TIQ compounds in the adrenergic system and behavioral changes in subjects under the influence of ethanol, and the possibility of understanding the addictive aspects of alcoholism after understanding the pharmacology of TIQ compounds.

The role of THP in ethanol dependence was recently investigated by O'Neil and Rahwan[90] in mice. Animals were subjected to ethanol vapors for 6 days (14 mg/liter of inspired air on day 1, increasing to 20–30 mg/liter on day 6), and evaluated for withdrawal convulsions on handling. The brains were analyzed for salsolinol, and no evidence was obtained for the *in vivo* formation of salsolinol, even in dopamine-rich areas of the brain. They were not able to find salsolinol even in mice in which blood ethanol levels reached 700 mg/dl.

Bigdelli and Collins,[91] on the other hand, using the same method of estimation, were able to demonstrate formation of salsolinol in dopamine-rich areas of the rat brain (17 ng/g brain tissue) after acute administration of ethanol (9 g/kg in three intraperitoneal doses over 7 hr), but the rats were pretreated with pyrogallol, which increases the levels of acetaldehyde in blood, and therefore the conditions were unphysiological.

Formation of the condensation products of aldehydes and biogenic amines led to many hypothetical explanations of alcohol addiction, based on the formation of morphinelike alkaloids in the brains of ethanol-treated animals. It seems, however, that as with the steady-state levels of biogenic amines, the investigations of the role of biogenic amines and their condensation products with aldehydes is still an open question and the results depend very much on the methods of treatment and the methods of estimation.

G. SUMMARY AND CONCLUSIONS

Comparison of the effects of opioid alkaloids and ethanol shows considerable similarity in their action on the steady-state levels of biogenic amines. The results from different laboratories vary greatly, ranging from depletion to no change or slight rise, except for dopamine levels, which often were found to increase. This is probably due to many factors that are difficult to control. In the regions of the brain where there are high concentrations of some biogenic amines, the changes are more pronounced.

There is a similarity in the effects of morphine and ethanol on the steady-state levels of dopamine in the corpus striatum. In both cases, when only one pharmacological agent is used (either morphine or ethanol), the levels of dopamine in the striatum increase. The increase produced by morphine is smaller than that after ethanol. Disturbances in the physicochemical state of the membranes could be responsible for the changes, and the extent of the increase might depend on the membrane solubility of the drug used.

Morphine has a stimulating effect on the turnover of biogenic amines. There is definite agreement as to the increase of biosynthesis of catecholamines, but some discrepancy in the case of the turnover of serotonin. Increased biosynthesis is reflected in an increase of the acidic metabolites. The role of the choroid plexus and the transport of metabolites seems to be unchanged in morphine-treated animals.

Ethanol has a different effect on the turnover of catecholamines, and shows a biphasic action, different for norepinephrine and dopamine; in the final analysis, the biosynthesis of all three biogenic amines seems to be little affected. There is, nevertheless, an increase in acidic metabolites.

The condensation products of biogenic amines and aldehydes presents a tempting explanation of the addictive properties of ethanol. However, although theoretically feasible, the formation of substances such as salsolinol remains to be supported by results *in vivo*.

ACKNOWLEDGMENT. The original research reported here and the preparation of the manuscript were supported by the Grant DA00130 from NIDA.

H. REFERENCES

1. H. E. Himwich, *J. Am. Med. Assoc.* **163**:545 (1975).
2. A. Goldstein and B. A. Judson, *Science* **172**:290 (1971).
3. H. Kalant and P. E. Dews, in: *Experimental Approaches to the Study of Drug Dependence* (H. Kalant and R. D. Hawkins, eds.), pp. 100–134, University of Toronto Press, Toronto (1969).
4. J. D. Sinclair, J. Adkins, and S. Walker, *Nature (Lond.)* **246**:425 (1973).
5. R. Galfant and Z. Amit, *Nature (Lond.)* **259**:415 (1976).
6. K. Blum, J. E. Wallace, H. A. Schwerter, and J. D. Eubanks, *Experientia* **32**:79 (1976).
7. D. B. Goldstein and N. Pal, *Science* **172**:288 (1971).

8. C. S. Lieber, E. Rubin, L. M. DeCarli, P. Misra, and H. Gang, *Lab. Invest.* **22**:615 (1970).
9. G. Magnusson, J. A. Nyberg, N. O. Bodin, and E. Hanson, *Experientia* **15**:1198 (1972).
10. W. L. Wilson and N. G. Bottiglieri, *Cancer Chemother. Rep.* **21**:137 (1972).
11. E. MacDonald, M. Marselos, and U. Nousiainen, *Acta Pharmacol. Toxicol.* **37**:106 (1975).
12. M. Vogt, *J. Physiol (Lond.)* **123**:451 (1954).
13. L. M. Gunne, *Acta Physiol. Scand.* **58** (Suppl. 204):1 (1963).
14. E. W. Maynert and G. J. Klingman, *J. Pharmacol. Exp. Ther.* **135**:285 (1962).
15. C. R. Rethy, C. B. Smith, and J. E. Villarreal, *J. Pharmacol. Exp. Ther.* **176**:472 (1971).
16. T. Akera and T. M. Brody, *Biochem. Pharmacol.* **17**:675 (1968).
17. L. J. Garcia and M. Rocha Silva, *J. Pharm. Pharmacol.* **13**:734 (1961).
18. R. A. Verri, F. G. Graeff, and A. P. Corrado, *Int. J. Neuropharmacol.* **7**:283 (1968).
19. C. G. Sparkes and P. S. J. Spencer, *Br. J. Pharmacol.* **42**:230 (1971).
20. E. L. Way and F. H. Shen, in: *Narcotic Drugs* (D. H. Clouet, ed.), pp. 229–253, Plenum Press, New York (1971).
21. H. Corrodi, K. Fuxe, and T. Hokfelt, *J. Pharm. Pharmacol.* **18**:821 (1966).
22. W. A. Hunt and E. Majchrowicz, *J. Neurochem.* **23**:549 (1974).
23. M. E. Post and A. Y. Sun, *Res. Commun. Chem. Pathol. Pharmacol.* **6**:887 (1973).
24. E. Costa and N. H. Neff, in: *Handbook of Neurochemistry* (A. Lajtha, ed.), Vol. 4, pp. 45–90, Plenum Press, New York (1970).
25. D. Gursey and R. E. Olson, *Proc. Soc. Exp. Biol. Med.* **104**:280 (1960).
26. L. Ahtee and M. Svartstrom-Fraser, *Acta Pharmacol. Toxicol.* **36**:289 (1974).
27. M. R. Fennessy and J. R. Lee, *Br. J. Pharmacol.* **45**:240 (1972).
28. G. Stille and A. Sayers, *Experientia* **23**:1028 (1967).
29. H. Takagi and M. Nakama, *Jpn. J. Pharmacol.* **16**:483 (1966).
30. K. Kuschinsky, *Experientia* **29**:1365 (1973).
31. I. J. Wajda and I. Manigault, in: *Tissue Responses to Addictive Drugs* (D. Ford and D. H. Clouet, eds.), pp. 171–185, Spectrum Publications, Inc., New York (1976).
32. C. Gauchy, Y. Agid, J. Glowinski, and A. Cheramy, *Eur. J. Pharmacol.* **22**:311 (1973).
33. J. C. Johnson, M. Ratner, G. Y. Gold, and D. H. Clouet, *Res. Commun. Chem. Pathol. Pharmacol.* **9**:41 (1974).
34. S. K. Puri and H. Lal, *Psychopharmacologia* **32**:113 (1973).
35. D. H. Clouet and K. Iwatsubo, *Life Sci.* **17**:35 (1975).
36. A. Bosse and K. Kuschinsky, *Naunyn-Schmiedebergs Arch. Pharmakol.* **294**:17 (1976).
37. E. T. Iwamoto, J. K. Ho, and E. L. Way, *J. Pharmacol. Exp. Ther.* **187**:558 (1973).
38. J.-T. Huang and I. J. Wajda, *J. Pharm. Pharmacol.* **27**:940 (1975).
39. A. Marchall and M. Hirst, *Experientia* **32**:201 (1976).
40. P. J. Griffiths, J. M. Littleton, and A. Ortiz, *Br. J. Pharmacol.* **50**:489 (1974).
41. J. A. Pohorecky, L. S. Jaffe, and H. A. Berkely, *Life Sci.* **15**:427 (1974).
42. I. J. Wajda, I. Manigault, and J. P. Hudick, *Biochem. Pharmacol.* **26**:653 (1977).
43. E. L. Way, H. H. Loh, and F. H. Shen, *Science* **162**:1290 (1968).
44. Y. Maruyama, G. Hayashi, S. E. Smits, and A. E. Takemori, *J. Pharmacol. Exp. Ther.* **178**:20 (1971).
45. D. L. Chaney, A. Goldstein, S. Algeri, and E. Costa, *Science* **171**:1169 (1971).
46. G. G. Yarbrough, D. M. Buxbaum, and E. Sanders-Bush, *Life Sci.* **10**:977 (1971).
47. E. Costa, A. Carenzi, A. Guidotti, and A. Revuelta, *Biochem. Pharmacol.*, Suppl. **2**:833 (1973).
48. M. F. Sugrue, *Br. J. Pharmacol.* **52**:159 (1974).
49. G. R. Pscheidt, B. Isekutz, and H. E. Himwich, *Q. J. Stud. Alcohol.* **22**:550 (1961).
50. D. H. Effron and G. L. Gessa, *Biochem. Pharmacol.* **8**:172 (1961).
51. J. Haggendal and M. Linquist, *Acta Pharmacol. Toxicol.* **18**:278 (1961).
52. D. H. Effron and G. L. Gessa, *Arch. Int. Pharmacodyn. Ther.* **142**:111 (1961).
53. N. Rudas and L. Vacca, *Acta Neurol.* **19**:848 (1964).
54. D. D. Bonnycastle, M. D. Bonnycastle, and E. G. Anderson, *J. Pharmacol. Exp. Ther.* **135**:17 (1962).

55. G. Duritz and E. B. Truitt, Jr., *Biochem. Pharmacol.* **15**:711 (1966).
56. L. A. Pohorecky, L. S. Jaffe, and H. A. Berkeley, *Res. Commun. Chem. Pathol. Pharmacol.* **8**:1 (1974).
57. W. A. Hunt and E. Majchrowicz, *Brain Res.* **72**:181 (1974).
58. D. H. Clouet and M. Ratner, *Science* **168**:854 (1970).
59. C. B. Smith, J. E. Villarreal, J. H. Bednarczyk, and M. I. Sheldon, *Science* **170**:1106 (1970).
60. C. B. Smith, M. J. Sheldon, J. H. Bednarczyk, and J. E. Villarreal, *J. Pharmacol. Exp. Ther.* **180**:547 (1971).
61. B. Tabakoff, R. F. Ritzman, and W. O. Bogan, *J. Neurochem.* **24**:1043 (1975).
62. R. P. Maickel, R. H. Cox, Jr., J. Saillant, and F. Miller, *Int. J. Neuropharmacol.* **7**:275 (1968).
63. K. Fukui and H. Takagi, *Br. J. Pharmacol.* **44**:45 (1972).
64. K. Kuschinsky and O. Hornykiewicz, *Eur. J. Pharmacol.* **19**:119 (1972).
65. F. Karoum, R. J. Wyatt, and E. Majchrowicz, *Br. J. Pharmacol.* **56**:403 (1976).
66. G. L. Gessa, L. Vargiu, G. Biggio, and A. Tagliamonte, *Biochem. Pharmacol.*, Suppl. **2**:841 (1973).
67. A. Jori and E. Monti, *J. Pharm. Pharmacol.* **26**:993 (1974).
68. L. Ahtee, *Br. J. Pharmacol.* **55**:381 (1975).
69. L. Ahtee and J. Kaarianen, *Eur. J. Pharmacol.* **22**:206 (1973).
70. K. Kuschinsky and O. Hornykiewicz, *Eur. J. Pharmacol.* **26**:41 (1974).
71. B. H. C. Westerink and J. Korf, *Eur. J. Pharmacol.* **33**:31 (1975).
72. F. A. Wiesel, C. G. Fri, and G. Sedvall, *Eur. J. Pharmacol.* **23**:104 (1973).
73. R. H. Roth, L. C. Murin, and J. R. Walters, *Eur. J. Pharmacol.* **36**:163 (1976).
74. M. A. Elchisak, J. W. Maas, and R. H. Roth, *Eur. J. Pharmacol.* **41**:369 (1977).
75. J.-T. Huang and I. J. Wajda, *Br. J. Pharmacol.* **60**:363 (1977).
76. A. Pictet and T. Spengler, *Chem. Ber.* **44**:2030 (1911).
77. Y. Yamanaka, M. J. Walsh, and V. E. Davis, *Nature (Lond.)* **227**:1143 (1970).
78. P. Holtz, K. Stock, and E. Westermann, *Arch. Exp. Pathol. Pharmakol.* **248**:387 (1964).
79. M. J. Walsh, V. E. Davis, and Y. Yamanaka, *J. Pharmacol. Exp. Therap.* **174**:388 (1970).
80. V. E. Davis, M. J. Walsh, and Y. Yamanaka, *J. Pharmacol. Exp. Therap.* **174**:401 (1970).
81. C. Shopf and F. Bayerle, *Am. Chem.* **513**:190 (1934).
82. R. A. Lahti and E. Majchrowicz, *Life Sci.* **6**:1399 (1967).
83. G. Cohen and M. Collins, *Science* **167**:1749 (1970).
84. G. Cohen, *Biochem. Pharmacol.* **20**:1757 (1971).
85. V. E. Davis and M. J. Walsh, *Science* **167**:1005 (1970).
86. M. H. Seevers, V. E. Davis, and M. J. Walsh, *Science* **170**:1113 (1970).
87. P. V. Halushka and P. C. Hoffmann, *Science* **169**:1104 (1970).
88. G. Cohen, *Biochem. Pharmacol,* **25**:1123 (1976).
89. M. Sandler, S. B. Carter, K. R. Hunter, and G. M. Stern, *Nature (Lond.)* **241**:439 (1973).
90. P. J. O'Neil and R. G. Rahwan, *J. Pharmacol. Exp. Therap.* **200**:306 (1977).
91. M. G. Bigdelli and M. A. Collins, *Biochem. Med.* **12**:55 (1975).
92. A. Y. Sun, Proc. 4th Int. Meeting Int. Soc. Neurochem., Abstr., p. 476 (1973).
93. I. J. Wajda, S. H. Wajda, I. Manigault, and L. Steiner, *Fed. Proc.*, Abstr. **32**:488 (1974).
94. I. J. Wajda, J.-T. Huang, and A. Lajtha, *Trans. Am. Soc. Neurochem.*, Abstr. **7**:156 (1976).

Genetics of Ethanol Preference 13
Role of Neurotransmitters

Allan C. Collins

A. INTRODUCTION

The development of a suitable animal model for alcoholism has been the focus of an intensive research effort for the last quarter century or longer. Although a number of techniques have proved to be of value in generating animals which are physically dependent upon ethanol, none of these models fully mimics the human condition of alcoholism. Lester and Freed[1] have argued that a satisfactory animal model of alcoholism should meet several criteria. Among these is that the model should be typified by the oral ingestion of pharmacologically effective doses of ethanol, without food deprivation, and that this ingestion be maintained in the presence of competing fluids. The ingestion of ethanol in the presence of competing fluids has come to be known as ethanol preference. Although no animal models have been fully developed in which such voluntary consumption is sufficient to produce physical dependence, there are a number of models available in which a marked consumption of ethanol can be easily observed. This review will discuss some of these models, as well as some of the information currently available which may aid in explaining those factors which influence voluntary ethanol consumption. In addition, the value of genetic analyses as a tool to test hypotheses concerning the underlying biochemical processes which influence this ethanol-related behavior will be discussed. Particular emphasis will be placed on those studies which have explored the role of neurotransmitters in ethanol preference.

Allan C. Collins • School of Pharmacy and Institute for Behavioral Genetics, University of Colorado, Boulder. Colorado 80309.

B. METHODS OF ASSESSING ETHANOL PREFERENCE

Most ethanol preference studies have utilized laboratory animals, princi-pally the mouse or rat, as the experimental subject. As a consequence, the results obtained must be viewed cautiously as to their meaning with respect to human ethanol consumption. It would seem unlikely that animals consume ethanol for the cultural and sociological reasons that humans do. Nevertheless, the data obtained may provide valid explanations of the pharmacological and physiologi-cal bases for ethanol consumption in man.

1. Two-Bottle Choice Method

Two general procedures have been developed and are in routine use for the assessment of ethanol preference. The simpler of the two involves providing the test animal with a choice of two fluid sources—one containing water and the other a solution of ethanol which does not vary in concentration from day to day. In general, bottle position is reversed on a prearranged schedule to avoid the development of a position preference. Perhaps the most critical step in develop-ing this procedure is the selection of ethanol concentration. Valid results cannot be easily obtained if a single ethanol concentration is chosen arbitrarily. Rather, a preliminary experiment is necessary to ascertain a concentration which best suits the purpose of the experimenter. Many studies which have utilized mice have been carried out with 10% solutions. This concentration has come in favor because of the early observations of McClearn and Rodgers,[2] who noted that this concentration allowed the differentiation of ethanol preference among a number of inbred mouse strains. Since a definite advantage of inbred strains is a relative constancy over time and space, it is highly likely that investigators who have used 10% ethanol solutions in preference tests (utilizing C57BL mice, in particular) have obtained valid results. Nevertheless, since environmental influ-ences such as temperature, photoperiod, and season of the year seem to influ-ence ethanol consumption in inbred mice, it is recommended that all experiments in which preference is to be measured be preceded by a pilot study to aid in selection of the ethanol solution to be used. When test animals are of an outbred strain (one maintained by systematically avoiding the mating of relatives), a preliminary experiment seems mandatory, perhaps for every animal. This is necessary because outbred strains have considerable genetic variability, which will serve to increase variability in ethanol choice. The increased variability may be so great as to decrease seriously the reliability of data obtained from the two-bottle single-concentration technique.

2. Variable-Concentration Method

The other technique commonly used in preference testing was developed principally in the laboratory of R. D. Myers. This technique involves increasing the concentration of ethanol on a periodic basis, usually daily, in a stepwise fashion from a low initial value to a high concentration. Generally, three tubes are used—one containing water, another the ethanol solution, and the third being empty. The latter, referred to as the "dummy" tube, aids in decreasing the development of a position preference. A definite advantage of this technique is that it provides a dose–response curve for each animal. Thus, it should provide reliable data with outbred as well as inbred strains. The major disadvantage of the technique is it is time-consuming and laborious. It is also difficult to carry out studies which would measure short-term influences on ethanol consumption. In spite of these difficulties, the variable-concentration technique has provided a considerable portion of the utilizable information concerning those factors which may influence ethanol preference.

C. GENETIC EFFECTS ON ETHANOL PREFERENCE

A variety of evidence indicates that ethanol preference is controlled by genetic factors. Many investigations have compared voluntary ethanol consumption with various behavioral and biochemical traits in inbred mouse strains. Inbred strains are derived by mating individuals more closely related than would occur by chance in the population.[3] This has been achieved classically in laboratory mice and rats by the mating of full siblings. Inbreeding results in a substantial increase in genetic uniformity within a strain. If one member of an inbred strain is homozygous for a specific allele at a given genetic locus, other members of the strain will also be homozygous at that locus. For all practical purposes, a strain which has been maintained by inbreeding for 20 or more generations may be regarded as being genetically uniform. Inbreeding is nondirectional, and the genetic configuration achieved in a given strain is fortuitous.

1. Differences among Inbred Strains

Strain differences in ethanol preference have been reported for both rats and mice, although by far the most information concerning genetic influences on preference obtained by this method comes from the mouse. In the first report of an influence of genotype on ethanol preference, published in 1949, Williams and co-workers[4] described differences in ethanol consumption by inbred strains of both rat and mouse. McClearn and Rodgers,[5] using the two-bottle choice method, found wide variances in ethanol consumption among inbred mouse strains, with the C57BL strain consuming about two-thirds of its daily fluid intake from a 10% ethanol solution. Other strains, most notably the DBA/2, avoid ethanol entirely. These observations have been reaffirmed in a number of laboratories throughout the world. The conclusion that this differential is due to genetic rather than environmental influences is strongly supported by the recent studies of Randall and Lester,[6] who noted that ova transplant did not influence ethanol preference in either the C57BL or DBA/2 strain. Environment is not without influence, however, since these investigators noted that cross-fostering DBA mice with C57BL mothers increased ethanol preference in the DBA strain, although it had no effect on ethanol preference in the C57BLs. Thus, a genotype–environment interaction appears to exist.

Studies of the genetics of ethanol preference using inbred rat strains are not nearly as numerous as are those using the mouse. Nonetheless, the literature clearly supports the notion that rat strains differ in ethanol preference. For example, Russell and Stern[7] noted that the Tryon maze-bright strain has a high ethanol preference, whereas the Tryon maze-dull, Wistar, and Hooded strains have low preference. More recently, Melchior and Myers[8] observed significant differences in four rat strains not only in control ethanol consumption but also in response to various drugs which alter neurotransmitter function.

2. Advantages and Disadvantages of Inbred Strains

The use of inbred strains has both advantages and disadvantages. The major advantage is that results are generally easier to replicate (e.g., the repeated observations that C57BL mice have a high alcohol preference and that DBA mice avoid alcohol). Clearly, this replicability facilitates the testing of hypotheses, which may explain differing preference. The major disadvantage related to working with inbred strains relates to the fact that homozygosity is being approached at all genetic loci. This results in a loss in generalizability. It is possible that a gene may be of major importance in determining a behavioral trait (such as alcohol preference) in an inbred population, while that same gene may be of minimal importance in a heterogeneous population such as man. Results obtained from homogeneous animal stocks should therefore be viewed cautiously, and other approaches should be used as further tests of each hypothesis. In particular, if experiments are being carried out to ascertain whether a specific biochemical trait is responsible for influencing a behavior such as ethanol preference, adequate proof has *not* been obtained even if the biochemical and behavioral traits occur together in one or more inbred strains in a fashion which appears rational. More rigorous methods of testing the hypotheses are necessary, for the genes which influence the behavioral and biochemical traits may simply be linked on the same chromosome rather than being related to one another in a cause-and-effect relationship.

3. Selective Breeding

Selective breeding is another genetic approach used in the study of ethanol preference. Selective breeding is a mating procedure which is used to produce populations that differ in a particular behavioral attribute. Selection is successful only if a genetically heterogeneous population is used as the foundation stock. All animals in the foundation population are measured for the trait in question (e.g., ethanol preference). Females showing high levels of the trait are mated with males showing high levels and, similarly, low females are mated with low males. This procedure is carried out in successive generations. In general, mating with siblings is avoided in order to maintain the advantage of outbreeding. As selection proceeds, the two selected lines approach homogeneity at those genetic loci which affect the trait being measured; at the same time, heterogeneity is maintained at all other genetic loci. This situation decreases the likelihood that spurious correlations would be observed in a study that attempted to provide a biochemical explanation for a behavioral trait. Since the possibility of spurious correlations is a major disadvantage of using inbred strains, it is obvious that the use of selected lines is more appropriate for such research. This advantage is frequently outweighed by the time and effort required to generate selected lines. However, once available, selected lines which have been produced by systematic outbreeding offer an extremely powerful research tool.

The first attempt at selective breeding for ethanol preference was carried out by Mardones,[9] who bred for differing ethanol preference by starting with a single mating pair of high-preferring and a single pair of low-preferring rats. Succeeding generations were obtained by brother–sister matings, so selection in this case was accompanied by inbreeding. Eriksson[10] has developed two rat lines which differ in ethanol preference by using the previously described selection-outbreeding technique. The high-preferring line is designated AA (alcohol-addicted); the low-preferring line is referred to as ANA (alcohol-nonaddicted). Although the two lines have not actually been demonstrated to differ in addiction liability, the designations AA and ANA continue to be used by the Finnish workers and will be used here to avoid unnecessary confusion. Eriksson has conducted a genetic analysis of the AA and ANA rats and has concluded that ethanol preference is a polygenic additive trait and that its heritability is low. (A polygenic additive trait is one which is influenced by a number of genes, the effects of which add together to determine the final dimensions of the trait; heritability refers to the percent of variance in the trait which is due to additive genetic variance.) The conclusion that ethanol preference is a polygenic additive trait which has a low heritability is supported by the observation that many generations of continued selection were required to completely separate the two lines with respect to ethanol preference.

4. Other Genetic Techniques

The observations that inbred strains differ in ethanol preference and that lines which differ in preference can be obtained by selective breeding provide unequivocal proof that ethanol preference is influenced by genetic factors. This finding, in addition to being of interest in itself, also indicates that further genetic analyses can be used to test hypotheses concerning the underlying mechanisms which influence ethanol preference. Although seldom used in pharmacological research, such genetic techniques are extremely powerful tools for testing hypotheses. If two strains or lines are found to differ in a logical fashion with respect to some biochemical or physiological parameter which other data have suggested may be involved in ethanol preference, presumptive evidence will have been obtained to support the notion that these parameters are involved in preference. The next step would be to cross these strains or lines to obtain derived F_1 and F_2 generations. Should preference and the biochemical or physiological measure segregate together into the F_2 generation, strong support will have been obtained for the hypothesis. This technique is used frequently in genetic research; it allows the determination of whether two traits are related, and also permits an estimate of the importance of variation in one to variation in the other. Such information is invaluable for ascertaining the mechanisms which underlie a polygenically determined trait such as ethanol preference.

D. METABOLISM AND ETHANOL PREFERENCE

1. Ethanol Metabolism

The genetic method has been of significant value in evaluating the effects of differential ethanol and acetaldehyde metabolism on preference. Much of the earlier work in this area concentrated on the potential role of hepatic alcohol dehydrogenase. The first study in this area was that of Rodgers et al.,[11] who measured ethanol preference and *in vitro* hepatic alcohol dehydrogenase activity in six inbred mouse strains (RIII/Crgl, DBA/2Crgl, BALB/cCrgl, A/Crgl/3, C3H/Crgl/2, and C57BL/Crgl) and found a positive correlation between the two measures. Subsequent studies utilizing inbred mouse strains have confirmed this finding.[12-14] Since, as noted earlier, such observations could be entirely fortuitous, McClearn[15] tested the correlation between ethanol preference and hepatic alcohol dehydrogenase both in the F_2 generation derived from a C57BL × DBA cross and in a heterogeneous stock of mice (HS) systematically derived from crossing eight inbred strains. No significant correlation was seen between preference and enzyme activity in the F_2 animals, while a low correlation was found in the HS mice. Although these data may have been affected by the relatively small number of animals used in the F_2 study, they indicate that only a small portion of the variance in preference can be explained by alcohol dehydrogenase activity. Thus, the results of this study suggest that the role of this enzyme in preference is considerably smaller than the inbred strain studies suggest. These data clearly illustrate the potential hazards of inferring relationships from inbred strain studies alone. An explanation for the relatively low correlation between hepatic alcohol dehydrogenase activity and preference comes from a study by Schlesinger,[16] who noted that the ethanol preferring C57BL mice have considerably greater hepatic alcohol dehydrogenase activity than do the ethanol avoiding DBAs, but that circulating blood ethanol levels are virtually identical in the two strains.

Data obtained by Koivula et al.,[17] using Eriksson's AA and ANA rats, also shed doubt on the hypothesis that high alcohol dehydrogenase activity promotes increased ethanol consumption. These investigators detected a greater hepatic alcohol dehydrogenase activity in the ANA line than in AA animals. Another investigation[18] revealed no correlation between *in vitro* enzyme activity in these lines and *in vivo* ethanol disappearance rates. Taken together, these data suggest that hepatic alcohol dehydrogenase activities and ethanol disappearance rates are of minimal importance in determining ethanol preference.

2. Acetaldehyde Metabolism

The data currently available clearly suggest that the second step in ethanol metabolism, acetaldehyde oxidation by hepatic aldehyde dehydrogenases, is critically involved in determining preference. This argument receives support both from studies utilizing the inbred mouse strains which differ in preference and from studies using AA and ANA rats. Schlesinger *et al.*[19] noted that C57BL mice have considerably less circulating acetaldehyde following ethanol than do DBA animals. This differential is explainable on the basis of greater hepatic aldehyde dehydrogenase activity in the C57BL strain.[20] Similarly, Koivula *et al.*[17] noted that AA rats have significantly greater mitochondrial and microsomal aldehyde dehydrogenase activities than do ANA animals. No difference in cytosolic activities of this enzyme was observed. Since several studies[21,22] suggest that the mitochondrial enzymes are most important in acetaldehyde metabolism in rats, these data further strengthen the notion that high acetaldehyde levels serve to inhibit ethanol consumption. Further support comes from the observation that female C57BL mice consume more ethanol than do males[13,14] and have lower acetaldehyde levels.[23] While a complete genetic analysis has not been carried out in order to make a more accurate assessment of the contribution of variation in acetaldehyde metabolism to variation in ethanol preference, data currently at hand indicate that higher blood acetaldehyde levels decrease the likelihood that ethanol consumption will occur.

E. SEROTONIN AND ETHANOL PREFERENCE

The other major type of explanation as to why some animals prefer ethanol and others do not is the theory that various neurotransmitters may play a role in mediating preference. This area of research is characterized by inconsistencies in findings from various laboratories.

The greater portion of this research has focused on the potential role of serotonin. This interest arose initially when Myers and Veale[24] reported that male Long-Evans rats exhibited a decreased ethanol preference, as measured using the variable-concentration technique, when treated chronically with the tryptophane hydroxylase inhibitor, *p*-chlorophenylalanine. Although some aversion to ethanol was seen during chronic drug treatment, a greater effect was seen after treatment was discontinued. The effect seemed to be relatively specific to the serotonergic system, since depletion of catecholamines, as induced by α-methyl-*p*-tyrosine treatment, had minimal effects on preference. A subsequent study,[25] also carried out with Long-Evans rats, revealed that p-chlorophenyl-alanine treatment reversed stress-induced increases in ethanol preference. A possible genetic influence was reported in that animals that drank more ethanol were markedly affected by p-chlorophenylalanine, whereas low drinkers were minimally affected. Interestingly, the treatment did not decrease ethanol consumption to less than that seen in control groups; it simply reversed the stress-induced increase in ethanol consumption. It should be noted that the p-chloro-phenylalanine dose used in these experiments, 300 mg/kg, is a high one. This may be of considerable significance in that a study by Frey et al.[26] noted a marked decrease in ethanol preference with this dose in the Leo rat strain, whereas only a small and variable decrease in preference was seen when d-p-chloroamphetamine, a weak inhibitor of tryptophane hydroxylase, was used. These authors suggest that a large depletion in serotonin may be necessary to decrease preference. This may explain the inability of Geller[27] to obtain a consistent decrease in ethanol preference when he studied the influence of 75- to 150-mg/kg doses of p-chlorophenylalanine on preference in Sprague-Dawley rats. Rather than finding a decrease in preference, Geller detected increased ethanol consumption during p-chlorophenylalanine administration. After drug treatment was discontinued, aversion to ethanol developed. Variation in p-chlo-rophenylalanine dosage is probably not the only reason for inconsistency in results obtained in various laboratories. The strain of rat may also be critical, since Myers and Tytell[28] failed to find an effect of 300-mg/kg doses of p-chlo-rophenylalanine on preference in Royal Victoria rats during the time of administration. A decrease in preference was seen, however, after drug treatment was discontinued.

Several recent studies shed doubt on the hypothesis that the effect of p-chlorophenylalanine on preference is related to alteration of brain serotonin levels. Holman et al.[29] noted that a 316-mg/kg dose of p-chlorophenylalanine given intraperitoneally every 3 days caused the same degree of depletion of brain serotonin in Wistar rats, as did a 316-mg/kg dose given orally each day. This study failed to detect an alteration in ethanol preference even though brain serotonin was decreased substantially. Parker and Radow[30] also failed to detect an effect of daily 300-mg/kg doses of p-chlorophenylalanine on preference in Wistar rats. It seems doubtful that this lack of effect is due to a strain difference, since Nachman et al.[31] used the same strain in a study in which they observed a decrease in ethanol preference following 200-mg/kg-per-day doses.

There is also some question as to the specificity of p-chlorophenylalanine's actions. Parker and Radow[30] noted that this drug also decreased saccharin consumption and concluded that it acted by causing the acquisition of a nonspe-cific conditioned aversion. This possibility had been suggested earlier by Nach-man et al.[31] Jofre de Breyer[32,33] provided a possible explanation as to how p-chlorophenylalanine could serve as an aversive stimulus for ethanol consumption independently of its actions on serotonin when he detected a considerable increase in circulating acetaldehyde levels following p-chlorophenylalanine and ethanol administration. However, the reason why some strains of rat would be affected and others not is not readily evident.

A number of other compounds which may alter peripheral or central stores of serotonin have been tested for their effects on ethanol preference. Several studies have investigated the effect of 5-hydroxytryptophan, the immediate metabolic precursur to serotonin, on preference. Myers et al.[34] injected 50-mg/kg doses of 5-hydroxytryptophan every 8 hr during one of their standard preference testing cycles. This treatment caused a long-term (more than 3-month) depression in ethanol consumption in Royal Victoria rats. A decreased consumption was also seen if 5-hydroxytryptophan was injected intraventricularly. Geller[27] reported that 5-hydroxytryptophan decreased ethanol consumption in Sprague-Dawley rats. In a subsequent study, Myers and Melchior[35] added tryptophane, the precursor to 5-hydroxytryptophan, in relatively high concentrations to the diet of several rat strains. In one, the Royal Victoria, tryptophane induced a significant elevation in ethanol consumption; only a small increase was seen in the Long-Evans strain; and Sprague-Dawley rats were unaffected. Once again, the reasons for these strain differences are not immediately obvious. It should also be noted that the Sprague-Dawley strain, which did not respond to tryptophane treatment, is the strain used by Geller[27] that failed to respond to p-chlorophenylalanine treatment with a decrease in ethanol consumption.

A number of other drugs have been used to alter serotonergic function in an attempt to assess further the role of serotonin in preference. Geller et al.[36] examined the effect of the serotonergic receptor blocker, cinanserin, on ethanol preference in Sprague-Dawley rats. Cinanserin increased ethanol intake. Myers and Melchior[37] observed a significant increase in ethanol consumption following the lesioning of central serotonergic neurons by intraventricular 5,6-dihydroxytryptamine injection. Conversely, a similar destruction of catecholamine-containing neurons by 6-hydroxydopamine treatment decreased ethanol consumption.

Genetic influences appear to control the response to 5,6-dihydroxytryptamine and 6-hydroxydopamine.[8] In Sprague-Dawley and Long-Evans rats, the former increases ethanol consumption and the latter decreases it. Holtzman rats increase their ethanol consumption after 5,6-dihydroxytryptamine but are unaffected by 6-hydroxydopamine, whereas 5,6-dihydroxytryptamine is without effect and 6-hydroxydopamine decreases ethanol preference in Wistar rats. Thus, once again, genotype appears to be of prime importance in determining the response to drug treatment.

A consistent difficulty with all these studies has been determining whether an observed result is due to an influence on serotonergic pathways or to some other undefined effect. This point was brought home to us in a series of experiments recently completed in our laboratory. An initial study by Sanders et al.,[38] utilizing C57BL/6J mice, had detected a significant and long-lasting decrease in ethanol preference accompanying pargyline treatment. No effect was seen when the animals were treated with p-chlorophenylalanine. Shortly after this study was completed, Cohen et al.[39] reported that pargyline and several other, but not all, monoamine oxidase inhibitors increase circulating acetaldehyde levels if administered prior to ethanol. As a result, Sanders et al.[40] investigated the effects on preference of pargyline; of another monoamine oxidase inhibitor, Lilly 51641 (N-[2-(o-chlorophenoxy)-ethyl]-cyclopropylamine), which also increases acetaldehyde levels; and of nialamide, which does not alter acetaldehyde metabolism. Both pargyline and Lilly 51641 decreased ethanol consumption in C57BL mice, whereas nialamide was without effect. The most straightforward explanation of these results is that ethanol preference is not altered as a result of monoamine oxidase inhibition and that the effects of pargyline and Lilly 51641 on preference are related to their effect on acetaldehyde metabolism.

While reviewing this literature, a recurring theme was seen in discussions of the results of the various studies. Continual concern was expressed as to whether the effect on preference of those drugs which modify brain serotonin was indeed mediated via an effect on serotonin. This concern is obviated in those studies which have used genetic analysis as a tool to assess the role of serotonin in preference. Several such studies, using the AA and ANA rats which have been selectively bred for differing preference, have been carried out by Finnish investigators. Ahtee and Eriksson[41] observed that both serotonin and 5-hydroxyindole acetic acid, the acid metabolite serotonin, were 15–20% higher in brains obtained from AA rats. Continued exposure of the AA rats to ethanol for 1 month resulted in an increase in brain serotonin content in the AA line. No effect was seen in ANA animals.

A subsequent investigation[42] revealed the differences in serotonin were restricted principally to the hypothalamus, with a smaller difference in the thalamus–midbrain and cortex. Although these data support the hypothesis that preference is influenced by the serotonergic system, the Finnish workers have yet to carry out a genetic analysis which would allow an estimation of the relative importance of differential serotonin levels in ethanol preference. Pickett and Collins[43] carried out such an analysis using C57BL/Ibg and DBA/Ibg mice, as well as the F_1 and F_2 generations derived from crossing these two strains. Ethanol preference and brain serotonin content segregated independently of one another into the F_2 generation, suggesting that absolute brain serotonin level is not important in influencing preference. These results do not preclude the possibility that differential serotonin turnover rate or differential influence of ethanol on serotonin level or turnover are involved in determining ethanol consumption. In addition, since this study utilized inbred strains which have limited genetic variability, it is possible that a small role of serotonin in ethanol preference was missed. The differences in serotonin seen in the AA and ANA rats argues that further studies are necessary to clarify this point.

F. OTHER NEUROTRANSMITTERS AND ETHANOL PREFERENCE

Realtive to the accumulation of information with respect to serotonin, very few data have been obtained on the role of other neurotransmitters in preference. However, some information is available concerning the catecholamines. As noted previously, inhibition of catecholamine synthesis with α-methyl-p-tyrosine is without effect on ethanol preference.[24] The conclusion that catecholamines are not important in preference is partially supported by our observation that preference is not altered by nialamide,[40] but it is contradicted by the observation the 6-hydroxydopamine decreases preference in some rat strains.[8]

A recent investigation by Kiianmaa,[44] using Erikkson's AA rats, provided evidence to support the notion that serotonin has a minimal role in determining ethanol consumption, whereas the role of norepinephrine may be more important. Kiianmaa measured the effect on preference of lowering brain serotonin levels by three different means: electrocoagulation of the dorsal and median raphe nuclei of the midbrain, injection of 5,6-dihydroxytryptophane into the lateral cerebral ventricle, and oral administration of p-chlorophenylalanine. The first two methods did not alter ethanol consumption despite the fact that they caused a considerable decrease in brain serotonin; p-chlorophenylalanine caused a slight reduction in ethanol intake coupled with a significant increase in water consumption. On the other hand, stereotaxic injections of 6-hydroxydopamine, which caused degeneration of the coerulocortical noradrenergic pathways, produced an increase in ethanol consumption 5–6 weeks after 6-hydroxydopamine administration. This increase was short-lived, however. Amit et al.[45] examined the effects of calcium carbimide, FLA-63, and disulfiram (inhibitors of dopamine β-hydroxylase and aldehyde dehydrogenase) on ethanol consumption in Wistar rats. The suppression of ethanol consumption by these drugs appeared to be at least partially correlated with their ability to disrupt catecholamine synthesis. Further support for the notion that ethanol consumption is influenced by catecholamines comes from the observation that lithium carbonate suppresses ethanol consumption in AA rats.[46] Since lithium is believed to enhance the reuptake of norepinephrine,[47] lithium could serve to decrease available norepinephrine via a mechanism different from that seen following inhibition of dopamine β-hydroxylase.

The only study using genetic methods in an attempt to test the hypothesis that catecholamines influence ethanol consumption was carried out in the AA and ANA lines.[48] This study detected a slightly greater concentration of dopamine but not norepinephrine in brains of AA rats. No attempts have been made to ascertain whether ethanol consumption and brain catecholamine content segregate together in the F_2 generation produced by an AA by ANA cross.

The potential role of cholinergic neurons in ethanol preference has received only minimal attention. Cicero and Myers[49] noted that the cholinergic agonist, carbachol, produced a polydipsia in Wistar rats when injected into the nucleus reuniens, preoptic region, septum, or anterior hypothalamus when water was offered in the test situation. If ethanol was offered, these same rats totally rejected even normally preferred ethanol solutions. Ho et al.[50] compared C57BL and DBA mice with respect to brain content of acetylcholine and uptake of [^{14}C]acetylcholine by whole brain homogenates. Both measures were considerably higher in C57BL mice, whereas brain acetylcholinesterase was higher in the DBAs. In addition, inhibition of acetylcholine synthesis decreased ethanol preference in the C57BL mice. These data, in contrast to those of Cicero and Myers, support the notion that cholinergic systems function to increase ethanol consumption. It is obvious that further studies are necessary to define the role of cholinergic systems in ethanol preference.

G. MELATONIN AND ETHANOL PREFERENCE

Ethanol consumption varies throughout the day, occurring mostly during the dark hours. Geller[51] was the first to conduct a careful study of the circadian rhythmicity of ethanol consumption. He noted that consumption by Sprague-Dawley rats increased if they were maintained in a dark room. A similar increase in ethanol consumption was achieved by injecting the animals over a 2- to 4-week period with the pineal hormone, melatonin. Burke and Kramer[52] obtained similar results in a study which also utilized Sprague-Dawley rats. However, Reiter *et al.*[53] were unable to elicit an increase in ethanol consumption in congenitally blind Wistar rats, although pinealectomy did cause a decrease in consumption by these animals. A subsequent investigation by Blum *et al.*,[54] using Sprague-Dawley rats, confirmed the observation that pinealectomy decreases ethanol preference, while the administration of melatonin was without effect. However, it should be noted the animals which were injected with melatonin were already consuming over 80% of their daily fluid intake from an ethanol solution. Thus, an increase in preference may not have been detectable. The effect of pinealectomy on ethanol preference is not restricted to rats, since Reiter *et al.*[55] observed that pinealectomized hamsters exhibited a reduced preference for ethanol. Daily subcutaneous implants of metalonin did not influence preference in these animals. Thus, it appears that some factor related to the pineal gland is related to increased ethanol preference. However, since the effects of melatonin are not consistent from one study to another, it remains ambiguous as to whether this substance is the responsible agent.

H. ROLE OF TETRAHYDROISOQUINOLINES AND β-CARBOLINES IN ETHANOL PREFERENCE

The formation of condensation products of the biogenic amines and aldehydes has been implicated as a potential mechanism underlying alcoholism. This hypothesis is reviewed elsewhere in this volume. Several studies have attempted to assess the role of amine–aldehyde condensation products in ethanol consumption. Using male Sprague-Dawley rats, Geller and Purdy[56] tested the effect on ethanol preference of the β-carboline condensation product derived from 5-methoxytryptamine and acetaldehyde, 1-methyl-6-methoxy-1,2,3,4-tetrahydro-2-carboline. Preliminary evidence indicated that this condensation product caused a rapid decrease in ethanol preference which was accompanied by an increase in brain serotonin levels. Since serotonin also induced a decrease in ethanol consumption, these investigators suggested that the effect might be mediated via an influence on serotonin content or turnover.

A very recent study by Myers and Melchior[57] examined the effect of intraventricular injections of tetrahydropapaveroline, the condensation product of dopamine and its deaminated aldehyde product, on ethanol consumption. Three to six days after initiation of tetrahydropapaveroline infusion, the Sprague-Dawley rats, which normally reject ethanol, began drinking large amounts. This was accompanied by signs and reactions of intoxication and by blood ethanol concentrations as high as 200 mg/dl. As noted by the authors, the amount of ethanol consumed exceeded that seen in any previous experiment using a free-choice experimental regime. When the animals were tested for susceptibility to audiogenic seizures during the daylight hours, while ethanol consumption was minimal, an enhanced sensitivity was seen. Thus, the animals may have been physically dependent upon either ethanol or tetrahydropapaveroline.

A subsequent study[58] found that the excessive intake of ethanol persisted as long as 9 months after discontinuing treatment with tetrahydropapaveroline, suggesting that the treatment caused an irreversible change in those neuronal systems which influence ethanol consumption. This study also noted that as little as 0.4 ng of S-(−)-tetrahydropapaveroline injected intraventricularly every 30 min increased ethanol consumption, although this dose was not as effective as either 40 ng or 4 μg. Results indicated the existence of a complicated dose-response relationship. Direct injection into the brain seems to be necessary to induce an increase in ethanol consumption, since chronic intraperitoneal injection was without effect on consumption. On the other hand, some degree of chronic exposure seems necessary, inasmuch as a single 40-μg injection of tetrahydropapaveroline also had no effect. Tetrahydropapaveroline does not seem to alter a taste factor, since chronically injected rats prefer alcohol–saccharin mixtures to a saccharin solution. This study also noted that a withdrawal syndrome was evident when infusion with tetrahydropapaveroline was discontinued, even at blood ethanol levels as high as 100 mg/dl.

More recently, Myers and Melchior[59] and Myers and Oblinger[60] examined the effects of a number of tetrahydroisoquinolines and a β-carboline (tryptoline) on ethanol consumption. Tryptoline was found to increase consumption. Some of the tetrahydroisoquinolines were virtually without effect, whereas others (such as salsolinol) were nearly as effective as tetrahydropapaveroline. All these condensation products caused a long-lasting increase in the consumption of ethanol.

There are several perplexing aspects of these exciting studies. The formation of these condensation products is clearly enhanced in vitro by elevated acetaldehyde levels, a condition which appears to be a strong in vivo deterrent to ethanol consumption. An additional concern relates to the problem of physical dependence. If the opiate model of dependence is satisfactory for alcohol, a peculiar phenomen occurred in these studies. The opiate literature suggests that an animal made dependent upon morphine will work to obtain the drug; then, once the drug is obtained, the animal will cease to work for it. In these studies, even though the condensation products were supplied continuously, the animals also ingested high doses of alcohol. An explanation for this phenomenon is not readily apparent and seems necessary before the meaning of these experiments will be clear.

I. SUMMARY AND CONCLUSIONS

Despite considerable effort, the data currently at hand do not allow any definitive statements to be made concerning the role of any neurotransmitter in ethanol preference. A number of factors may explain this predicament. First, variable results have generally been obtained when drugs have been used in attempts to effect changes in specific neurotransmitters. This may be a result of the fact that different doses and treatment regimens have frequently been used, or it may have resulted because no drug is likely to affect only one biochemical system. Second, most studies have used only one type of laboratory animal as experimental subjects. This approach is very likely to produce data that are not generalizable, especially when inbred strains are used. It seems to this reviewer that the role of neurotransmitters in ethanol consumption will be ascertained only by the application of a number of approaches using a variety of techniques.

The preceding discussions show that a considerable effort has been expended in an attempt to assess the underlying mechanisms which explain voluntary ethanol consumption. A continual frustration of the researchers who work in this area has been the lack of a clear-cut relationship between ethanol consumption by animal models and consumption by man. One problem may be that, with the exception of the recent studies carried out by Myers and Melchoir, a degree of consumption adequate to develop physical dependence has not been achieved in animals. The fact that a definite relationship has not been established has shed some doubt on the relevance of animal research. On the other hand, ethanol preference studies have clearly demonstrated the role of genetic factors in this ethanol-related behavior. It is to be hoped that this review will inspire investigators who conduct research related to alcoholism to use more than one strain of animal in their studies and to use further genetic analyses to test hypotheses concerning the underlying mechanisms which may explain various ethanol-related behaviors.

J. REFERENCES

1. D. Lester and E. X. Freed, *Pharmacol. Biochem. Behav.* **1**:103 (1973).
2. G. E. McClearn and D. A. Rodgers, *Q. J. Stud. Alcohol* **20**:691 (1959).
3. G. E. McClearn, *Ann. N.Y. Acad. Sci.* **197**:26 (1972).
4. R. J. Williams, L. J. Beery, and E. Beerstecher, Jr., *Arch. Biochem.* **23**:275 (1949).
5. G. E. McClearn and D. A. Rodgers, *J. Comp. Physiol. Psychol.* **54**:116 (1961).
6. C. L. Randall and D. Lester, *J. Stud. Alcohol* **36**:973 (1975).
7. K. E. Russell and M. H. Stern, *Physiol. Behav.* **10**:641 (1973).
8. C. L. Melchior and R. D. Myers, *Pharmacol. Biochem. Behav.* **5**:63 (1976).
9. J. Mardones, *Int. Rev. Neurobiol.* **2**:41 (1960).
10. K. Eriksson, *Science* **159**:739 (1968).
11. D. A. Rodgers, G. E. McClearn, E. L. Bennett, and M. Hebert, *J. Comp. Physiol. Psychol.* **56**:663 (1963).

12. G. E. McClearn and D. A. Rodgers, *J. Comp. Physiol.* **54**:116 (1961).
13. K. Eriksson and P. H. Pikkarainen, *Jpn. J. Stud. Alcohol* **5**:1 (1970).
14. K. Eriksson and P. H. Pikkarainen, *Metabolism* **17**:1037 (1968).
15. G. E. McClearn, in: *Nebraska Symposium on Motivation* (W. J. Arnold, ed.), pp. 47–83, University of Nebraska Press, Lincoln, Neb. (1968).
16. K. Schlesinger, *Am. J. Psychiatry* **122**:767 (1966).
17. T. Koivula, M. Koivusalo, and K. O. Lindros, *Biochem. Pharmacol.* **24**:1807 (1975).
18. C. J. P. Eriksson, *Biochem. Pharmacol.* **22**:2283 (1973).
19. K. Schlesinger, R. Kakihana, and E. L. Bennett, *Psychosomat. Med.* **28**:514 (1966).
20. J. R. Sheppard, P. Albersheim, and G. E. McClearn, *Biochem. Genet.* **2**:209 (1968).
21. N. Grunnet, *Eur. J. Biochem.* **35**:236 (1973).
22. S. O. C. Tottmar, H. Pettersson, and K. H. Kiessling, *Biochem. J.* **135**:577 (1973).
23. G. P. Redmond and G. Cohen, *Nature* **236**:117 (1972).
24. R. D. Myers and W. L. Veale, *Science* **160**:1469 (1968).
25. R. D. Myers and T. J. Cicero, *Psychopharmacology* **15**:373 (1969).
26. H.-H. Frey, M. P. Magnussen, and C. K, Nielsen, *Arch. Int. Pharmacodyn. Ther.* **183**:165 (1970).
27. I. Geller, *Pharmacol. Biochem. Behav.* **1**:361 (1973).
28. R. D. Myers and M. Tytell, *Physiol. Behav.* **8**:403 (1972).
29. R. B. Holman, V. Hoyland, and E. E. Shillito, *Br. J. Pharmacol.* **53**:299 (1975).
30. L. F. Parker and B. L. Radow, *Pharmacol. Biochem. Behav.* **4**:535 (1976).
31. M. Nachman, D. Lester, and J. Le Magnen, *Science* **168**:1244 (1970).
32. I. J. Jofre de Breyer, C. Acevedo, and M. Torrelio, *Arzneim.-Forsch.* **22**:2140 (1972).
33. I. J. Jofre de Breyer, *Arzneim.-Forsch.* **23**:954 (1973).
34. R. D. Myers, J. E. Evans, and T. L. Yaksh, *Neuropharmacology* **11**:539 (1972).
35. R. D. Myers and C. L. Melchior, *Psychopharmacology* **42**:109 (1975).
36. I. Geller, R. J. Hartmann, and F. S. Messiha, *Proc. West. Pharmacol. Soc.* **18**:141 (1975).
37. R. D. Myers and C. L. Melchior, *Res. Commun. Chem. Pathol. Pharmacol.* **10**:363 (1975).
38. B. Sanders, A. C. Collins, and V. H. Wesley, *Psychopharmacology* **46**:159 (1976).
39. G. Cohen, D. MacNamee, and D. Dembiec, *Biochem. Pharmacol.* **24**:313 (1975).
40. B. Sanders, A. C. Collins, D. R. Petersen, and B. S. Fish, *Pharmacol. Biochem. Behav.* **6**:319 (1977).
41. L. Ahtee and K. Eriksson, *Physiol. Behav.* **8**:123 (1972).
42. L. Ahtee and K. Eriksson, *Ann. N.Y. Acad. Sci.* **215**:126 (1973).
43. R. A. Pickett, II, and A. C. Collins, *Life Sci.* **17**:1291 (1976).
44. K. Kiianmaa, in: *The Effects of Centrally Active Drugs on Voluntary Alcohol Consumption* (J. D. Sinclair and K. Kiianmaa, eds.), pp. 73–84 (1975).
45. Z. Amit, D. E. Levitan, and K. O. Lindros, *Arch. Int. Pharmacodyn. Ther.* **223**:114 (1976).
46. J. D. Sinclair, *Med. Biol.* **52**:133 (1974).
47. R. W. Colburn, F. K. Goodwin, W. E. Bunney, Jr., and J. M. Davis, *Nature (Lond.)* **215**:1395 (1967).
48. L. Ahtee and K. Eriksson, *Acta Physiol. Scand.* **93**:563 (1975).
49. T. J. Cicero and R. D. Myers, *Physiol. Behav.* **4**:559 (1969).
50. A. K. S. Ho, C. S. Tsai, and B. Kissin, *Pharmacol. Biochem. Behav.* **3**:1073 (1975).
51. I. Geller, *Science* **173**:456 (1971).
52. L. P. Burke and S. Z. Kramer, *Pharmacol. Biochem. Behav.* **2**:459 (1974).
53. R. L. Reiter, K. Blum, J. E. Wallace, and J. H. Merritt, *J. Stud. Alcohol* **34**:937 (1973).
54. K. Blum, J. H. Merritt, R. L. Reiter, and J, E. Wallace, *Curr. Ther. Res.* **15**:25 (1973).
55. R. J. Reiter, K. Blum, J. E. Wallace, and J. H. Merritt, *Comp. Biochem. Physiol.* **47A**:11 (1974).
56. I. Geller and R. Purdy, *Ann. N.Y. Acad. Sci.* **215**:54 (1973).
57. R. D. Myers and C. L. Melchior, *Science* **196**:554 (1977).
58. C. L. Melchior and R. D. Myers, *Pharmacol. Biochem. Behav.* **7**:19 (1977).
59. R. D. Myers and C. L. Melchior, *Pharmacol. Biochem. Behav.* **7**:381 (1977).
60. R. D. Myers and M. M. Oblinger, *Drug Alcohol Depend.* **2**:469 (1977).

Effects of Ethanol on the Cholinergic System

14

R. Massarelli

A. INTRODUCTION

Acetylcholine (AcCh) was the first molecule implicated in nervous transmission. It may then seem surprising that we still do not know the regulatory steps of AcCh metabolism.

In comparison, a large body of evidence has been published on such other putative neurotransmitters as the biogenic amines, and their metabolic regulation and pharmacological manipulation are well known. Technical difficulties may be the cause of our ignorance of the regulation of cholinergic metabolism, since specific and sensitive methods for AcCh, choline (Ch), choline acetyltransferase (ChAcT, E.C. 2.3.1.6), and acetyl cholinesterase (AcChE, E.C. 3.1.1.7) have appeared in the literature only recently.

In this chapter we will examine the effects of ethanol on the cholinergic system, keeping in mind that the rate-limiting step of AcCh turnover is not known. Consequently, these effects of ethanol should be regarded with caution as well as interest. Unfortunately, the literature on the subject is meager and sometimes contradictory, making the task of this reviewer rather difficult.

B. CHOLINERGIC TRANSMISSION AND METABOLISM OF ACETYLCHOLINE

Historically, the chemical theory of nerve transmission originated with AcCh, and its role was first suggested in 1914 by Dale[1] and later demonstrated by Loewi.[2] AcCh attracted the interest of a number of scientists until it was found that the molecule occurred in many tissues, not always of nervous, nor

---

R. Massarelli • Centre de Neurochimie du CNRS and Université Louis Pasteur, Strasbourg, France.

even animal, origin.[3] These findings understandably cooled the enthusiasm for AcCh. There was, however, also a purely technical reason for the decline in interest; the bioassay exclusively employed to measure AcCh until the 1960s was not specific, mainly because of the nonselective extraction procedures.

Interest was stimulated again when Nachmansonn and Machado[4] discovered the enzyme that synthesizes AcCh, choline acetyltransferase, and further when AcCh was implicated in the "quantal" theory of chemical nerve transmission.[5] Only 50 years after Dale's and Loewi's discovery, the first metabolic study of AcCh was done in the cat superior cervical ganglion.[6] At present we still do not know precisely the rate-limiting step in AcCh synthesis, but a number of new ideas and hypotheses, as well as the development of greatly improved analytic techniques, suggest that this point will soon be clarified.

Let us now consider what is currently accepted or discussed about AcCh metabolism and then consider the effects of ethanol on the same events.

1. Analysis of Acetylcholine Concentrations

In studying the metabolism of cholinergic systems, several difficulties were encountered that only recently have been overcome by the development of specific and sensitive methods of detection and rapid methods of sacrifice.

Specificity and sensitivity have been achieved with several radioenzymatic and gas chromatographic methods that are presently utilized with success in most laboratories (see, for reviews, the book edited by I. Hanin, Ref. 7).

It has been more difficult to establish a method of sacrifice that eliminates the postmortem effects so important in the cholinergic system. Endogenous concentrations of Ch in rat brain range from 50 to 600 nmole/g after decapitation,[8] while AcCh concentrations range from 15 nmole/g with total freezing of the animal in liquid nitrogen to 30 nmole/g using decapitation or the "near freezing" technique (immersion of the animal in liquid nitrogen for 10–11 sec).[9] Recently, two techniques have been developed, freeze-blowing[10] and focused microwave irradiation.[11] The latter is especially useful because it rapidly inactivates hydrolyzing enzymes[13] while maintaining intact the basic structures of nerve tissue. (Table I summarizes some recent values of AcCh and Ch concentrations.)

2. Acetylcholine Synthesis

There are three main components of AcCh synthesis.

a. Choline Acetyltransferase

Present evidence suggests that AcCh synthesis takes place in the cytoplasm since most ChAcT is solubilized during nerve impulse transmission.[19] Figure 1 depicts cholinergic metabolism at the nerve ending. Early work with partially

Table I. Acetylcholine and Choline in Rat and Mouse Cerebra[a]

	Acetylcholine (nmole/g)		Choline		Reference
Rat					
Microwave	24.8 ± 1.3		26.3 ± 1.6		12
	26.1 ± 0.7		25.4 ± 4.2		12
Decapitation	17.2 ± 2.1		156.7 ± 14.2		12
Decapitation	14.9 ± 2.8		148.4 ± 13.4		12
Near freezing	14.4 ± 0.7		71.2 ± 4.8		14
Different experimental	17.4 ± 0.99[c]	23.2 ± 0.70[d]	331 ± 21[c]	328 ± 17[d]	
conditions[b]	15.8 ± 0.77[e]	28.9 ± 2.95[f]	313 ± 41[e]	274 ± 16[f]	15
Immersion in liquid nitrogen	19.8 ± 0.7		25.3 ± 1.8		16
Freeze blowing	27.8 ± 1.0		28.6 ± 5.4		16
Mouse					
Decapitation	16.1 ± 0.3		77 ± 22		17
	15.34		43.4		18
Near freezing	16.4 ± 0.7		62.5 ± 1.9		14

[a]For a more complete list of values, see Saelens and Simke.[134]

[b]Rats were maintained in a dark/light cycle for several weeks before sacrifice. The values obtained for AcCh concentrations reflect a diurnal oscillation of the transmitter.

[c]Animals sacrified by total freezing in liquid nitrogen during dark.

[d]Animals sacrified by decapitation in liquid nitrogen during dark.

[e]Animals sacrified by total freezing in liquid nitrogen during light.

[f]Animals sacrified by decapitation in liquid nitrogen during light.

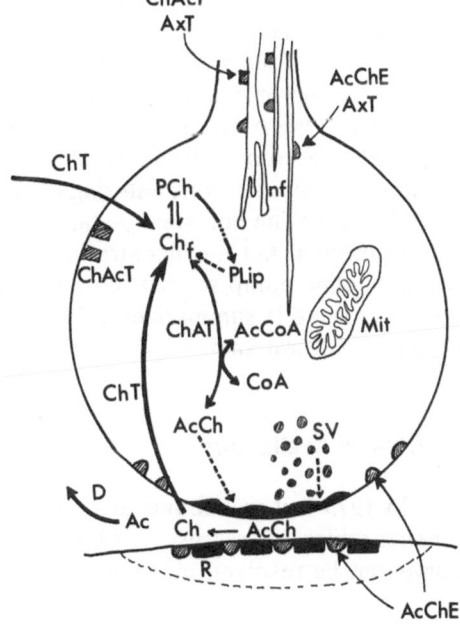

Fig. 1. Cholinergic metabolism at a nerve ending. Abbreviations: Ac, acetate; AcCh, acetylcholine; AcChE, acetylcholinesterase; AcCoA, acetyl coenzyme A; AxT, axonal transport; ChAcT, choline acetyltransferase; Ch_f, free choline compartment; ChT, choline transport mechanism; mit, mitochondrion; nf, neurofilaments; PCh, phosphorylcholine; PLip, phospholipids; R, receptor; SV, synaptic vesicles; D, diffusion. Ch entering the cell is transformed into PCh and through several steps (— · · · →) into PLip. It is under discussion whether part of the PLip pool may be a source for free choline. Ch is also acetylated by ChAcT and AcCh is then liberated, after nervous impulse, as free AcCh from the cytosol or as AcCh bound to SV. After interaction with R it is rapidly hydrolyzed into Ac and Ch, which is taken up for reutilization. The acetate precursor possibly belongs to an extramitochondrial AcCoA compartment.

Table II. Choline Acetyltransferase Activity in Several Organs of Different Animal Species

Species	Organ	Activity (nmole/min per mg protein)	Reference
Mammals			
Rat	Brain	1.0	25
	Brain	0.7	26
	Brain synaptosomes	37.0	21
Rabbit	Heart	24.0	27
	Brain [a]	20,000	27
	Caudate nucleus	3.8	28
Guniea pig	Brain synaptosomes	22.0 [b]	21
Ox	Caudate	1.0	20
Sheep	Caudate	3.2	28
Human	Brain	0.23	27
	Caudate	3.0	29
	Heart	5.6	27
	Placenta, immature (24 weeks)	10.0	30
	Placenta, at term	0.36	35
Fish			
Torpedo	Electric organ	15.0	31
Birds			
Chicken	Ciliary ganglia	30	32
	Cerebral hemispheres	0.2	33
Pigeon	Brain synaptosomes	15 [b]	20

[a] Partially purified.
[b] Activity expressed as nmole/min per g fresh tissue of a synaptosomal P2 fraction. [34]

purified preparations indicated that ChAcT is not rate-limiting, and AcCh synthesis may be regulated by mass action of the precursors, choline and AcCoA. [20]

Several findings [21] pointed out that the enzyme is compartmentalized and that under appropriate ionic strength conditions it is solubilized and activated. Moreover, it has been shown in synaptosomes that Cl^- ions act more specifically than Na^+ or K^+ ions to bring about solubilization and activation. After passage of the nervous impulse, ions entering the cell solubilize the enzyme for synthesis of new AcCh. [22] The non-steady-state levels of AcCh (after release of AcCh, there is a decrease in AcCh concentrations) may in turn change the kinetic parameters of the choline pump so that more choline becomes available for AcCh synthesis. [23,24] (Table II summarizes values of ChAcT activity obtained from different organs of several animal species.)

b. Choline Availability

In 1973 [36] it was found that Ch is taken up in synaptosomes with complex kinetics which are represented by a curve on a Lineweaver–Burk diagram. The curve can be resolved into two straight lines with different kinetic parameters representing two transport mechanisms. The component with the lower apparent K_m (commonly called the high-affinity uptake) seems to be involved in the regulation of AcCh.[36-39] It should be mentioned that the endogenous synthesis of choline in the central nervous system (CNS)[40-42] seemingly provides a small portion of the choline needed by nervous tissue. Choline thus has to be transported to the brain across the blood–brain barrier (possibly as phosphatidylcholine)[43-45] and then across cell membranes.

c. Acetyl-CoA Availability

The availability of the acetyl precursor, AcCoA, is less understood. It seems established that, *in vivo,* good labeling of AcCh is obtained from radioactive pyruvate,[46] but exactly how the cell accomplishes this metabolic transformation is not known.

A recent hypothesis, based on experiments with synaptosomes, suggests that a cytoplasmic pool of AcCoA, rapidly turning over, would label AcCh[47] preferentially. The studies on the acetyl moiety for AcCh synthesis are, however, complicated by the fact that it is difficult to measure AcCoA concentrations accurately. Furthermore, turnover studies with radiolabeled compounds that are not direct precursors of the compound to be studied are invalid.[48,49]

Moreover, it has been found that acetate is better than pyruvate for labeling AcCh in lobster nerve,[50] in the electric organ of the electric fish *Torpedo marmorata,*[51,52] and in rat diaphragm.[53] This may mean that neuromuscular junctions differ in their cholinergic metabolism from other types of synapses. This suggestion is intriguing considering the pharmacologic difference that exists between neuromuscular and other cholinergic effector sites (i.e., nicotinic and muscarinic).

3. Acetylcholine Degradation

The degradation of AcCh is due mainly to enzymatic hydrolysis of the ester. AcChE is a very stable enzyme, mainly membrane-bound and concentrated on the postsynaptic membrane, where it terminates the action of AcCh on the postsynaptic receptor.

Being ubiquitous, AcChE is not a good marker for cholinergic neurons.[54] Neurons, however, contain a higher proportion of AcChE than glial cells, which have more butyrylcholinesterase (E.C. 3.1.1.8 , BuChE), also called nonspecific cholinesterase or pseudocholinesterase.[55] The function of BuChE is rather mysterious, since no other ester of choline can be found in nervous tissue.[56] A recent hypothesis suggests that BuChE is a precursor of AcChE.[57]

Several forms of AcChE have been found that differ in their sedimentation coefficients and localization,[58] which may imply different metabolic roles, although it is premature to draw any definite conclusions.

4. Acetylcholine Compartmentation and Turnover

The first study of AcCh turnover is relatively recent,[59,60] as are the first studies in which the simultaneous detection of choline- and acetylcholine-specific activities[61] disclosed AcCh compartmentation.[62]

Homogenization of nervous tissue in mild acidic solution releases a certain amount of AcCh and stronger acidic conditions cause further release.[19] It has been concluded that AcCh is localized in at least two compartments: free in the cytosol and bound to synaptic vesicles. AcCh would then be synthesized by ChAcT in the cytosol and stored in the synaptic vesicles.

There is conflicting evidence regarding which compartment of AcCh is preferentially released upon stimulation. What would seem to be an easy subject of experimentation has instead proved to be highly difficult. Recently, Whittaker's group presented evidence that in the electric organ of Torpedo,[63] which is purely cholinergic, bound AcCh is released preferentially supporting the exocytotic hypothesis of nerve transmission. On the other hand, Dunant et al., have challenged this finding by showing, in the same system, that cytoplasmic AcCh is released preferentially upon nerve stimulation.[64]

Most evidence, however, points to the existence of a small and very rapidly turning over compartment of AcCh, which is released preferentially upon stimulation, that contains newly synthesized AcCh,[65−67] but whose morphological localization is difficult to define.

5. Regulation of Acetylcholine

It has been shown that AcCh may regulate its own synthesis and that the synthesis and release of AcCh are linked.[61,68,69] As mentioned earlier, ChAcT does not seem to be limiting in AcCh synthesis, and AcCh is a poor inhibitor of the enzyme under low ionic strength conditions[69,71] [see, however, Rossier[22]]. It has been suggested also that the CoA/AcCoA ratio is a rate-limiting factor.[70]

AcCh synthesis may also be controlled by the transport of choline, and experiments tend to confirm this possibility, even though the existence of a high-affinity transport site, specifically situated in cholinergic neurons, is disputed.[72] A single carrier mechanism which changes conformation with substrate concentration has also been proposed to explain the apparently contrasting experimental evidence.[24]

It seems established that after hydrolysis by released AcCh, choline is taken back into the nerve terminal and directed toward AcCh synthesis.[69,73]

C. EFFECTS OF ETHANOL ON CHOLINERGIC TRANSMISSION

The most consistent data in the literature have been obtained on the action of ethanol on cholinergic transmission. A remarkable difference has been obtained, however, between neuromuscular junction and nervous tissue.

The effects of ethanol on membrane permeability are well known. Na^+-K^+ transport[74,75] and, consequently, Na^+,K^+-ATPase activity[74,76,77] are strongly reduced by ethanol *in vitro*. This might call for a change as well in the release of neurotransmitter from the presynaptic terminal. In fact two opposite effects have been observed.

1. Neuromuscular Junction

In neuromuscular junction it was found that aliphatic alcohols have an effect that can be explained through either an inhibition of AcChE[78] or an increase of neurotransmitter release.[78-84] In the phrenic nerve–diaphragm preparation,[79] ethanol (8 mM) displayed a presynaptic effect, increasing the quantal content of end-plate potentials (EPPS), and the frequency of miniature end-plate potentials (MEPPS), and a postsynaptic effect, lengthening the time course of EPPS and MEPPS and increasing MEPP amplitude. The effect on MEPP was shown to be due to the prolongation of the decay phase of miniature end plate currents.[80]

Methyl and *n*-propyl alcohols had similar effects.[79] Similarly, there exists a relationship between chain length and concentration of alcohols, giving a decrease in the peak of sodium conductance in squid giant axons.[85] Gage[79] suggests that the increased ouput of transmitter produced by the alcohols is a function of their activity on the nerve-ending membrane and can be related to changes in membrane conductance.

The possibility exists, however, that ethanol's action lies in an increase of receptor sensitivity to AcCh or in the receptor–response coupling.[86] Moreover, as noted by Kalant,[87] any deductions on the release of AcCh obtained with changes in electric activity may be invalid, owing to the nonspecificity of the technique.

2. Nervous Tissue

Different results have been obtained in studying the effects of ethanol in brain tissue compared to the neuromuscular junction. When AcCh concentrations were measured in the bathing fluid of *in vitro* experiments[94-96] or *in vivo* with cups or push–pull cannulae,[97-99] a decrease of AcCh release was observed after ethanol treatment. This action seems related to the effects obtained with other CNS depressants, such as barbiturates or anesthetics, which increase the content of extractable AcCh in the brain of several animal species[100-104] while decreasing AcCh release.[105-106]

In the early 1950s Larrabee *et al.*[107] showed that ethanol decreased synaptic transmission in the cat superior cervical ganglion at concentrations that did not affect the respiration of ganglion cells, while several alcohols were shown to diminish sodium conductance in squid giant axons[85,108] and ethanol depressed spinal reflexes in the cat.[109]

All these data suggested a possible involvement of AcCh in the effects of

Table III. Effect of Ethanol on AcCh Release from Brain *in Vivo* and *in Vitro*

Species	AcCh release[a]	Dose ethanol	Route	Observations	Reference
In vivo					
Rabbit	↓	1.0–2.0 g/kg	iv	Collecting cups and push–pull cannulae[b]	97
Cat	↓	1.0–1.5 g/kg	ip	Push–pull cannulae, freely mobile	98
Cat	↓	1.0–2.0 g/kg	ip	Immobilized push–pull cannulae	99
In vitro					
Rat, brain cortex slices	↓	0.11 M			94
Rat, brain cortex slices	↓	(LD_{50}) 0.17 M			96
Rat, brain slices	↓	0.11 M		No effect when K^+ is raised from 5 mM to 15 mM	95

[a] ↓ , Decreased release of acetylcholine.
[b] Anaesthetized animals with pentobarbital (15 mg/kg, iv).

ethanol in the CNS, without indicating whether these effects were the result of a specific interference with AcCh metabolism or to a nonspecific action by ethanol on nerve cell membranes.

a. *In Vitro* Studies

In vitro studies[94] have shown that ethanol (0.11 M) inhibits the spontaneous release of AcCh from brain slices but not the potassium-stimulated release (even at 0.22 M ethanol). Moreover, inhibition of AcCh release by ethanol was preferentially higher when compared to the release of serotonin, dopamine, noradrenaline, glutamate, and GABA, suggesting a direct effect of ethanol on cholinergic neurons.[96]

However, slices from rats made tolerant to ethanol by intubation with 4 g ethanol/kg body wt for the first 3 days (and then 5 g/kg) show a decreased release of AcCh after ethanol treatment,[95] which suggests that cells may develop a resistance to the direct action of ethanol (Table III).

b. *In Vivo* Studies

AcCh release measured by cortical collecting cups or push–pull cannulae is depressed after ethanol treatment in anesthetized[97] or immobilized,[99] as well as in freely mobile animals.[98] The release of AcCh from the reticular formation is more depressed than cortical release,[97] a finding that is interesting to correlate with the suggestion that many of the actions of ethanol on the CNS result from decreased activity of the midbrain reticular formation[86] (Table III).

D. EFFECT OF ETHANOL ON ACETYLCHOLINE CONCENTRATIONS

Based on the effect of ethanol on the release of AcCh and the inverse relationship that exists between endogenous AcCh concentrations and release,[110] it would be reasonable to expect that levels of AcCh are increased upon ethanol treatment. In fact, this has been proved experimentally, but ethanol has a biphasic action on AcCh metabolism as it has on catecholamine turnover.[111] An initial increase in AcCh concentrations, localized mainly in brain stem and striatum, is followed by a significant decrease in the same brain areas of chronic intoxicated animals.[91] It is interesting to note that the decrease of AcCh in intoxicated animals may be correlated with the increased choline uptake in chronically cultured nerve cells.[112] The latter case seems to involve increased choline phosphorylation, which may suggest a decrease in free choline and, consequently, in AcCh synthesis. The biphasic action of ethanol on AcCh metabolism may explain the different results obtained by other authors (Table IV).

E. ACTION OF ETHANOL ON ACETYLCHOLINE SYNTHESIS

As already mentioned, it is possible to interfere with AcCh synthesis at the following steps: the availability of choline, the availability of AcCoA, and the activity of ChAcT.

Table IV. Effects of Ethanol on Acetylcholine Concentrations

Species	[AcCh][a]	[Ch][a]	Dose ethanol	Route	Observations	Reference
Mouse	↓	→	3 g/kg	ip	Single dose	88
Mouse	↓	→	6% (v/v) (liquid diet)	po	Repeated doses	88
Rat	↓		15% (v/v)	po	Repeated doses	89
Rat	↓		15% (v/v)	po	Repeated doses	90
Rat	↑		10–20 g/kg	iv	Single dose	97
Rat	↑		6 g/kg	Intubation	Single dose	91
Rat	↓		9–15 g/kg	Intubation	Repeated doses	91
Rat	↑		6 g/kg	Intubation	Single dose	92
Mouse	→	→	0.18 g/kg	ip	[b]	93

[a] ↓, Decrease in concentrations; ↑, increase; →, no change.
[b] Acetylcholine and choline were analyzed from brain slices. It was also shown that the uptake of [³H]-AcCh in slices was increased after ethanol treatment.

1. Choline Acetyltransferase Activity

A recent report[22] again gives to ChAcT the possibility of being rate-limiting in AcCh regulation. One would thus expect that ethanol, on the basis of its effects on AcCh release, might act on the enzyme, reducing its activity. There are three different lines of experimental evidence in which no change, a decrease, and an increase in ChAcT activity have been reported.

No change in ChAcT activity was found when the experiments were performed on a mitochondrial preparation from rat brain slices.[94] This is certainly not the ideal fraction in which to study ChAcT activity, especially because of the presence of carnitine acetyltransferase (E.C. 2.3.1.7), which may acetylate choline under certain conditions.[27]

Decreased ChAcT activity was observed in a crude mitochondrial fraction of mouse brain after 2 weeks of treatment with 1 M ethanol,[88] but this experiment used an old method which gave control values higher than usual (40.0 nmole AcCh synthetized/min per mg protein compared with 1.0 nmole/min per mg, Table II).

Increased activity was obtained both in homogenates and in a partially purified preparation of rat cerebra.[113] The amount of ethanol (0.17 M) necessary to affect the enzyme activity was in the range of lethal doses *in vivo*, however, in the author's words, "this gave only a slight stimulation of ChAcT activity." Since the activity of ChAcT is very dependent on the ionic conditions of the medium, we would suspect that the changes that ethanol causes in the Na^+-K^+ pump and ATPase activities might, at least at the regional level, influence the activity of ChAcT. For example, the number of ChAcT molecules released from the cellular membrane as a result of ATPase inhibition might be decreased because fewer ions would move across the membrane.

2. Choline Availability

Ethanol does not alter the steady-state concentrations of brain choline.[88,92] The effect then observed on AcCh release and brain AcCh concentrations may come from changes of the transport of choline across the cell membrane. Alternatively, ethanol might influence the turnover of the free choline compartment while maintaining its steady state. It has been shown recently that there is endogenous synthesis of choline in nervous tissue. Dross and Kewitz[40] have demonstrated, in rats, an arteroveinous difference of plasma choline and have shown that the higher venous concentration of choline originates from methylation of bound ethanolamine (glycerophosphorylethanolamine).

There are no reports of ethanol's effects on the endogenous synthesis of choline in the brain. It would be of interest to study the effect that ethanol may have on this synthesis; perhaps there might be a similarity with the effects of ethanol on liver choline metabolism.

There are several reports that stress the necessity of dietary choline in the synthesis of brain AcCh, and a diet rich in choline will increase AcCh levels in the brain.[114–116] These findings further confirm the importance of choline transport across the nerve cell membrane.

Choline, however, has first to cross the blood–brain barrier. We know that after iv injection radioactive choline is very rapidly phosphorylated and possibly crosses the blood–brain barrier as phosphatidyl choline.[117] Very shortly after injecting large doses of radioactive choline, however, it is possible to find free labeled choline in the brain.[59,60,118–120] Moreover, choline transport into the brain does not seem to be a saturable process but rather is mediated by diffusion.[40,118]

Once choline is in the brain it has to be taken into nerve cells, and it has been shown that neurons as well as glial cells in culture have rather similar transport processes.[72] If ethanol is added to the culture medium, there is no immediate effect on the uptake of choline in nerve cell cultures. However, growing the cells with ethanol will increase the uptake at low K_m, that is, at an external choline concentration below 2 μM.[112] A difference exists with respect to the cell origin, and neuronal cells increase choline uptake earlier than glia. The increased uptake seems to influence the metabolism of choline, since increased radioactivity was found in the phosphorylcholine compartment. This experiment was performed at only one incubation time, however, so that no inference can be made about endogenous choline metabolism. In contrast to nerve cell cultures, ethanol inhibits choline uptake in Novikoff hepatoma cell cultures in a competitive fashion.[121] This finding is interesting in relation to differences in membrane permeability among nerve and hepatic cells.

In vivo, ethanol decreases the incorporation of choline into AcCh in mice under both acute and chronic conditions,[88] while pretreatment with various concentrations of ethanol (0.8–22 M, 1–100% ip in mice) protects mice against hemicholinium-3 poisoning.[122] Hemicholinium-3 is a powerful inhibitor of choline uptake,[123] and as such it has been used frequently in studies in nerve tissue.

3. Acetyl-CoA Availability

AcCoA and CoA levels have been reported to decrease in mouse brain after acute and chronic treatment with ethanol.[88] However, the values reported for the coenzymes (40 and 412 nmole/g, respectively) disagree with values obtained with more sensitive methods, which are between 5 and 8 nmole/g for AcCoA and about 18 nmole/g for CoA.[124] Using a more sensitive method of detection, Guynn was unable to observe any change of CoA or AcCoA levels after treatment of rats with 3 g ethanol/kg 120 min before sacrifice.[124]

To explain these results we should reexamine the controversy about the origin of the acetyl group for AcCh synthesis. Two facts now seem established: (1) AcCh is labeled preferentially by pyruvate[46] *in vivo* as well as *in vitro* in synaptosomes,[47] and (2) radioactive acetate preferentially labels AcCh in the neuromuscular junction.[50-53] We can discuss only the first point, since there is no extensive experimental evidence regarding the neuromuscular synthesis of AcCh from the acetyl moiety.

It is not known where pyruvate is transformed to AcCoA, but there are some theoretical possibilities. Pyruvate may be transformed into AcCoA in mitochondria and then the AcCoA may be taken into the cytoplasm (the site of AcCh synthesis) via one or more of the following possible steps: (1) direct outward transport of AcCoA; (2) transformation of AcCoA into acetylcarnitine, which crosses the mitochondrial membrane and gives its acetate to cytoplasmic CoA; (3) formation of citrate, which is carried out of the mitochondrion and converted back to AcCoA; and (4) breakdown of AcCoA in the mitochondrion to acetate and CoA with acetate diffusing to the cytosol, where it is retransformed to acetyl-CoA by acetyl-CoA synthetase (E.C. 6.2.1.1). As one can easily judge, the difficulty stems from the tranport of the acetyl moiety outside the mitochondrion. At the moment, none of the preceding hypotheses has been convincingly shown to be suitable.[125]

Another possibility is that pyruvate does not enter the mitochondrion but is transformed to AcCoA in the cytoplasm by a pyruvate dehydrogenase (E.C. 1.2.4.1). This seems to be the case,[47] and it is possible to conceive of a very small pool of AcCoA turning over very rapidly, which participates in the acetylation of choline and which might be influenced by ethanol treatment.

F. ACTION OF ETHANOL ON ACETYLCHOLINE DEGRADATION

Once released in the synaptic cleft, a neurotransmitter must be rapidly inactivated. At the present time, two main systems are known which act with sufficient rapidity: (1) enzymatic inactivation, or (2) reuptake of the transmitter. Most of the attempts to show an active uptake of AcCh into the nerve ending have been inconclusive,[126,127] especially because of the presence of AcChE. To study the uptake of AcCh, anticholinesterase drugs must be used to prevent hydrolysis, and inhibitors of AcChE are known to influence directly the uptake of choline.[128] In any case AcChE action represents a very efficient way to stop the action of AcCh, hydrolyzing the mediator to acetate the choline very rapidly (in 0.004 sec). The choline is taken up for reutilization in AcCh synthesis, while acetate does not seem to be transported, at least in synaptosomes.[47]

As AcChE is membrane bound, one would suspect that the effects of ethanol on cell membranes could alter, even indirectly, the function of the enzyme. The few studies done show a certain discrepancy, however. It has been known since 1940[129] that low concentrations of ethanol activate AcChE, while inhibition is observed at higher concentrations. Later it was shown that the capacity of alcohols to activate the enzyme from rat brain homogenates decreases as the chain length of the alcohol increases.[130] In these studies, however, the alcohol concentrations used to activate AcChE were very high

(0.19 M). More recent studies[76] in isolated nerve-ending fractions have shown inhibition of AcChE with ethanol concentrations as low as 0.043 M. Parallel with this effect another membrane-bound enzyme, Na^+,K^+-ATPase, was also inhibited by the ethanol treatment.

Denaturation of AcChE by ethanol has been shown on bovine erythrocyte preparations with an LD_{50} of 4.3 M.[131] However, other authors[88,94,96] found no difference in AcChE activity in the supernatant from brain homogenates[94] incubated with 0.22 M ethanol or in brain homogenates from mice fed with a 6% (v/v) ethanol liquid diet over a 2-week period.[88]

This concentration is difficult to explain, considering that both results were obtained with reliable methods. Perhaps the differences stem from the extraction of the enzyme or, as usual, from differences in alcohol treatments.

G. SUMMARY AND CONCLUSIONS

A summary of the data in the literature on the effects of ethanol on the cholinergic system should stress the following points (see Fig. 2):

1. Ethanol affects cholinergic transmission but has opposite effects on neuromuscular junctions and nervous tissue. AcCh release increased in motor end plates after ethanol treatment and decreased in brain slices and *in vivo* experiments with perfusion cortical cups or push–pull cannulae. This difference might be correlated with different metabolic properties of AcCh in the two types of synaptic junctions.

2. Ethanol seems to affect AcCh concentrations *in vivo* in a biphasic fashion. After ethanol treatment an initial increase in AcCh concentrations, mainly localized in the brain stem and striatum, is followed by a significant decrease, in the same areas, when the animals become intoxicated.

3. Contradictory findings have been published on the effects of ethanol on the activities of ChAcT and of AcChE. It seems that there is an increase in ChAcT activity and a decrease in AcCh activity, although the concentrations of ethanol used are at or above the lethal dose level. More work has to be done in order to clarify this point *in vivo*.

4. The availability of choline is affected by ethanol in nerve tissue cultures when an increase in the uptake of choline was shown in cells cultivated for various times in the presence of ethanol.

5. No effect was observed on CoA or AcCoA concentrations in rat brain when sensitive techniques were used.

It is certainly difficult to draw any definite conclusions on the basis of the results reviewed here. Many of the contradictory findings on the effects of ethanol on AcCh metabolism may be related to differences in doses, rates, and routes of ethanol administration; others to methodological differences. We have already mentioned some of the technical difficulties inherent in the study of AcCh metabolism like the postmortem effects. To these we must add other

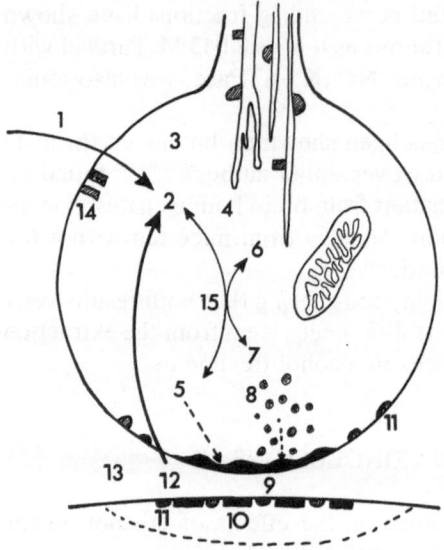

Fig. 2. Sites of possible action of ethanol on the cholinergic metabolism of a nerve ending. Chronic ethanol may act on the transport of Ch [either on its uptake (1) or on its reuptake (12)] without changing the steady-state levels of Ch (2). It increases the radioactive labeling in PCh (3), however. AcCh (5 and 8) levels are increased without changes in AcCoA (6) or CoA (7) levels. Released AcCh (9) is diminished in nerve tissue but increased in neuromuscular junctions. Ethanol may affect as well the receptor (10) response or the action of AcChE (11) as well as the solubility (14) and the activity (15) of ChAcT. Ethanol may further affect phospholipid (4) metabolism through its direct effect on membranes while no action would be expected on the resorption of acetate (13) coming from the hydrolysis of AcCh.

factors which aggravate a situation *per se* complicated by the use of a drug which has so many general effects:

1. The multicompartmentation of AcCh and of Ch.

2. The behavior of ChAcT, which, in the presence of ions at suitable ionic strength, is activated and solubilized and otherwise is partly attached to the plasma membranes.

3. A carrier system for choline, which is complex in nature and ion-dedependent as well.

4. Receptors with different affinities for AcCh, which are also influenced by ions.

With this in mind, we study the action of a drug that changes the membrane permeability to ions.

We may expect changes in the cholinergic metabolism, which, however, may be difficult to measure. It is known that mice that show an alcohol preference (C57 B1/6J strain) have higher levels of AcCh and a higher uptake of choline than other mouse strains and that inhibition of ChAcT shifts ethanol preference to water.[132,133] This relationship between ethanol and the cholinergic system further confirms a metabolic interaction.

Finally, the discrepancies noted in the effects of ethanol on AcCh metabolism might be explained if one considers the possibility of two effects of ethanol, one obtained after an acute dose and the other after chronic treatment. The acute effect of ethanol could be general and, in reducing the firing rate of presynaptic terminals, could increase the endogenous content of AcCh without really affecting the parameters of AcCh regulation. Under chronic treatment, a continuous action on the cell membrane will necessarily affect AcCh metabolism, possibly by shifting the metabolism of choline.

In the neuromuscular junction, the overall effect of ethanol might be different since, as already mentioned, acetate is the precursor for the compartment of AcCoA directed to AcCh synthesis and ethanol has an action opposite to that in ganglia or in the CNS. It is possible that acetate, as product of ethanol metabolism, has an influence yet to be determined in the metabolism of AcCh at the level of the acetyl moiety in neuromuscular junctions.

H. REFERENCES

1. H. Dale, *J. Pharmacol. Exp. Ther.* **6**:147 (1914).
2. O. Loewi, *Pfluegers Archiv. Gesamte Physiol. Menschen Tiere* **189**:239 (1921).
3. H. C. Chang and J. H. Gaddum, *J. Physiol. (Lond.)* **79**:155 (1933).
4. D. Nachmansohn and A. L. Machado, *J. Neurophysiol.* **6**:397 (1943).
5. B. Katz, *The Release of Neural Transmitter Substances,* Liverpool University Press, Liverpool (1969).
6. R. Birks and F. C. MacIntosh, *Can. J. Biochem.* **39**:787 (1961).
7. I. Hanin, *Choline and Acetylcholine: Handbook of Chemical Assay Methods* (I. Hanin, ed.), Raven Press, New York (1974).
8. I. Eade, E. Hebb, and S. P. Mann, *J. Neurochem.* **20**:1499 (1973).
9. M. Toru and M. H. Aprison, *J. Neurochem.* **13**:1533 (1966).
10. R. L. Veech and R. A. Hawkins, in: *Research Methods in Neurochemistry* (N. Marks and R. Rodnight, eds.), pp. 171–182, Plenum Press, New York (1974).
11. W. B. Stavinoha, S. T. Weintraub, and A. T. Modak, *J. Neurochem.* **20**:361 (1973).
12. W. B. Stavinoha and S. T. Weintraub, *Science* **183**:964 (1974).
13. A. Guidotti, D. L. Cheney, M. Trabucchi, M. Doteuchi, and C. Wang, *Neuropharmacology* **13**:1115 (1974).
14. H. Ladinsky and S. Consolo, in: *Choline and Acetylcholine: Handbook of Chemical Assay Methods* (I. Hanin, ed.), pp. 1–18, Raven Press, New York (1974).
15. I. Hanin, R. Massarelli, and E. Costa, in: *Drugs and Cholinergic Mechanisms in the CNS* (E. Heilbronn and A. Winter, eds.), pp. 33–54, Research Institute of National Defense, Stockholm (1970).
16. S. P. Mann and C. Hebb, *J. Neurochem.* **28**:241 (1977).
17. B. Karlen, E. Lundgren, B. Lundholm, I. Nordgren, and B. Holmstedt, in: *Cholinergic Mechanisms* (P. G. Waser, ed.), Raven Press, New York (1975).
18. D. J. Jenden, L. Choi, R. W. Silverman, J. A. Steinborn, M. Roch, and R. A. Booth, *Life Sci.* **14**:55 (1974).
19. F. Fonnum, in: *Cholinergic Mechanisms* (P. G. Waser, ed.), pp. 145–160, Raven Press, New York (1975).
20. V. A. S. Glover and L. T. Potter, *J. Neurochem.* **18**:571 (1971).
21. F. Fonnum, *Biochem. J.* **103**:262 (1967).
22. J. Rossier, in: *Cholinergic Mechanisms and Psychopharmacology* (D. J. Jenden, ed.), Plenum Press, New York (1977).
23. J. R. Simon, S. Atweh, and M. J. Kuhar, *J. Neurochem.* **26**:909 (1976).
24. R. Massarelli, in: *Cholinergic Mechanisms and Psychopharmacology* (D. J. Jenden, ed.), Plenum Press, New York (1977).
25. L. T. Potter, V. A. S. Glover, and J. K. Saelens, *J. Biol. Chem.* **243**:3864 (1968).
26. J. Rossier, A. Baumann, F. Rieger, and P. Benda, in: *Cholinergic Mechanisms* (P. G. Waser, ed.), pp. 283–292, Raven Press, New York (1975).
27. H. L. White and J. C. Wu, *Biochemistry* **12**:841 (1973).
28. S. Tucek, *J. Neurochem.* **14**:519 (1967).
29. G. Bull, C. Hebb, and D. Ratkovic, *J. Neurochem.* **17**:1505 (1970).
30. D. Morris, *Biochem. J.* **98**:754 (1966).

31. G. Bull, C. Hebb, and D. Morris, *Comp. Biochem. Physiol.* **28**:11 (1969).
32. M. Sorimachi and K. Kataoka, *Brain Res.* **70**:123 (1974).
33. A. Ebel, R. Massarelli, M. Sensenbrenner, and P. Mandel, *Brain Res.* **76**:461 (1974).
34. V. P. Whittaker, I. A. Michaelson, and R. J. A. Kirkland, *Biochem. J.* **90**:293 (1964).
35. R. Roskoski, C. T. Lim, and L. M. Roskoski, *Biochemistry* **14**:5105 (1975).
36. H. Yamamura and S. H. Snyder, *J. Neurochem.* **21**:1355 (1973).
37. P. Guyenet, P. Lefresne, J. C. Beaujouan, and J. Glowinski, *Brain Res.* **62**:523 (1973).
38. P. Guyenet, P. Lefresne, J. Rossier, J. C. Beaujouan, and J. Glowinski, *Mol. Pharmacol.* **9**:630 (1973).
39. T. Haga and N. Noda, *Biochim. Biophys. Acta* **291**:564 (1973).
40. K. Dross and H. Kewitz, *Naunyn-Schmiedeberg's Arch. Pharmacol.* **274**:91 (1972).
41. H. Kewitz, O. Pleul, K. Dross, and F. Schwartzkopff, in: *Cholinergic Mechanisms* (P. G. Waser, ed.), pp. 131–135, Raven Press, New York (1975).
42. H. Kewitz and O. Pleul, *Proc. Natl. Acad. Sci. USA* **73**:2181 (1976).
43. G. B. Ansell and S. Spanner, *J. Neurochem.* **14**:873 (1967).
44. G. B. Ansell and S. Spanner, *Biochem. J.* **110**:201 (1968).
45. G. B. Ansell and S. Spanner, *Biochem. J.* **122**:741 (1971).
46. D. L. Cheney, C. J. Gubler, and A. W. Jaussi, *J. Neurochem.* **16**:1283 (1969).
47. P. Lefresne, M. Hamon, J. C. Beaujouan, and J. Glowinski, *Biochimie* **59**:197 (1977).
48. J. M. Reiner, *Arch. Biochem. Biophys.* **46**:53 (1953).
49. D. B. Zilversmit, *Am. J. Med.* **29**:832 (1960).
50. S. C. Cheng and R. Nakamura, *Biochem. J.* **118**:451 (1970).
51. M. Israel and S. Tucek, *J. Neurochem.* **22**:487 (1974).
52. N. Morel, *C. R. Hebd. Seances Acad. Sci.* **280**:999 (1975).
53. P. Dreyfus, *C. R. Hebd. Seances Acad. Sci.* **280**:1893 (1975).
54. J. R. Cooper, E. E. Bloom, and R. H. Roth *The Biochemical Basis of Neuropharmacology*, pp. 70–74, Oxford University Press, New York (1974).
55. G. B. Koelle, in: *Handbuch der experimentellen Pharmakolgie* (G. B. Koelle, ed.), Vol. 15, pp. 187–298, Springer-Verlag, Berlin (1963).
56. C. G. Hammar, I. Hanin, B. Holmstedt, R. J. Kitz, D. J. Jenden, and B. Karlén, *Nature (Lond.)* **220**:915 (1968).
57. W. A. Koelle, E. G. Symrl, G. A. Ruch, V. E. Siddons, and G. G. Koelle, *J. Neurochem.* **28**:307 (1977).
58. B. Wermuth, P. Ott, R. Gentinetta, and V. Brodbeck, in: *Cholinergic Mechanisms* (P. G. Waser, ed.), pp. 299–308, Raven Press, New York (1975).
59. J. Schuberth, B. Sparf, and A. Sundwall, *J. Neurochem.* **16**:695 (1969).
60. J. Schuberth, B. Sparf, and S. Sundwall, *J. Neurochem.* **17**:461 (1970).
61. I. Hanin, R. Massarelli, and E. Costa, *J. Pharmacol. Exp. Ther.* **181**:10 (1972).
62. R. Massarelli, Thèse de Doctorat d'Etat ès Sciences pp. 38–42, Université Louis Pasteur, Strasbourg (1975).
63. M. J. Dowdall and V. P. Whittaker, in: *Cholinergic Mechanisms* (P. G. Waser, ed.), pp. 23–42, Raven Press, New York (1975).
64. Y. Dunant, J. Gautron, M. Israel, B. Lesbats, and R. Manaranche, *J. Neurochem.* **19**:1987 (1972).
65. B. Collier, *J. Physiol.* **205**:341 (1969).
66. P. C. Molenaar, V. J. Nickolson, and R. L. Polak, *Br. J. Pharmacol.* **47**:97 (1973).
67. L. W. Chakrin, R. M. Marchbanks, J. F. Mitchell, and V. P. Whittaker, *J. Neurochem.* **19**:2727 (1972).
68. L. T. Potter, *J. Physiol. (Lond.)* **206**:145 (1970).
69. B. Collier and F. C. MacIntosh, *Can. J. Physiol. Pharmacol.* **47**:127 (1969).
70. H. L. White and J. C. Wu, *J. Neurochem.* **20**:297 (1973).
71. A. A. Kaita and A. M. Goldberg, *J. Neurochem.* **16**:1185 (1969).
72. R. Massarelli and P. Mandel, in: *Transport Phenomena in the Nervous System* (G. Levi, L. Battistin, and A. Lajtha, eds.), pp. 199–209, Plenum Press, New York (1976).

73. B. Collier and B. Katz, *J. Physiol. (Lond.)* **238**:639 (1974).
74. Y. Israel, H. Kalant, and A. E. Le Blanc, *Biochem. J.* **100**:27 (1966).
75. J. L. Barker, *Brain Res.* **93**:77 (1975).
76. A. Y. Sun and T. Samorajski, *J. Neurochem.* **17**:1365 (1970).
77. Y. Israel and I. Salazar, *Arch. Biochem. Biophys.* **122**:310 (1967).
78. T. P. Feng and T. H. Li, *Clin. J. Physiol.* **16**:317 (1941).
79. P. W. Gage, *J. Pharmacol. Exp. Ther.* **150**:236 (1965).
80. P. W. Gage, R. N. McBurney, and G. T. Schneider, *J. Physiol.* **244**:409 (1975).
81. P. W. Gage and R. N. McBurney, *J. Physiol.* **226**:79 (1972).
82. F. Inoue and G. B. Frank, *Br. J. Pharmacol.* **30**:186 (1967).
83. K. Okada, *Jpn. J. Physiol.* **17**:245 (1967).
84. K. Okada, *Jpn. J. Physiol.* **20**:97 (1970).
85. C. M. Armstrong and L. Binstock, *J. Gen Physiol.* **48**:265 (1964).
86. H. Kalant, *Int. J. Neurol.* **9**:111 (1974).
87. H. Kalant, *Fed. Proc. Symp.* **34**:1930 (1975).
88. A. K. Rawat, *J. Neurochem.* **22**:915 (1974).
89. J. N. Moss, R. D. Smyth, H. Beckman, and G. J. Martin, *Arch. Int. Pharmacodyn.* **168**:235 (1967).
90. R. D. Smyth and H. Beck, *Arch. Int. Pharmacodyn. Ther.* **182**:295 (1969).
91. W. A. Hunt and T. K. Dalton, *Brain Res.* **109**:628 (1976).
92. O. M. Brown, M. E. Post, and S. Mallov, *J. Stud. Alcohol* **38**:603 (1977).
93. V. G. Carson, D. J. Jenden, and E. P. Noble, *Proc. West. Pharmacol. Soc.* **19**:341 (1976).
94. H. Kalant, Y. Israel, and M. A. Mahon, *Can. J. Physiol. Pharmacol.* **45**:172 (1967).
95. H. Kalant and W. Grose, *J. Pharmacol. Exp. Ther.* **158**:386 (1967).
96. F. J. Carmichael and Y. Israel, *J. Pharmacol. Exp. Ther.* **193**:824 (1975).
97. C. K. Erickson and D. T. Graham, *J. Pharmacol. Exp. Ther.* **185**:583 (1973).
98. E. P. Morgan and J. W. Phillis, *Comp. Gen. Pharmacol.* **6**:281 (1975).
99. J. W. Phillis and K. Jhamandas, *Comp. Gen. Pharmacol.* **2**:306 (1971).
100. J. Crossland and A. J. Merrick, *J. Physiol. (Lond.)* **188**:83 (1967).
101. K. A. C. Elliott, R. L. Swank, and N. Henderson, *Am. J. Physiol.* **162**:469 (1950).
102. W. B. Stavinoha and L. C. Ryan, *J. Pharmacol. Exp. Ther.* **150**:231 (1965).
103. D. Richter and J. Crossland, *Am. J. Physiol.* **159**:247 (1949).
104. C. L. Malhotra and P. G. Pundlik, *Br. J. Pharmacol.* **24**:348 (1965).
105. D. Beleslin and R. L. Polak, *J. Physiol. (Lond.)* **177**:411 (1965).
106. D. Beleslin, R. L. Polak, and D. H. Sproull, *J. Physiol. (Lond.)* **177**:420 (1965).
107. M. G. Larrabee, J. G. Ramos, and E. Bulbring, *J. Cell Comp. Physiol.* **40**:461 (1952).
108. J. W. Moore, W. Ulbricht, and M. Takata, *J. Gen. Physiol.* **48**:279 (1964).
109. G. M. Kolmodin, *Acta Physiol. Scand.* **26** (suppl. 106):530 (1953).
110. J. C. Szerb, H. Malik, and E. G. Hunter, *Can. J. Physiol. Pharmacol.* **48**:780 (1970).
111. W. R. Hunt and E. Majchrowicz, *J. Neurochem.* **23**:549 (1974).
112. R. Massarelli, P. J. Syapin, and E. P. Noble, *Life Sci.* **18**:397 (1976).
113. R. B. Reisberg, *Life Sci.* **14**:1965 (1974).
114. D. R. Haubrich, P. F. L. Wang, and P. Wedeking, *Fed. Proc.* **33**:1503 (1974).
115. G. Racagni, M. Trabucchi, and D. L. Cheney, *Naunyn-Schmiedeberg's Arch. Pharmacol.* **290**:99 (1975).
116. E. L. Cohen and R. L. Wurtman, *Life Sci.* **16**:1095 (1975).
117. G. B. Ansell and S. Spanner, in: *Cholinergic Mechanisms* (P. G. Waser, ed.), pp. 117–129, Raven Press, New York (1975).
118. J. J. Freeman, R. L. Choi, and D. J. Jenden, *J. Neurochem.* **24**:729 (1975).
119. S. M. Aquilonius, in: *Metabolic and Deficiency Diseases of the Nervous System* (P. J. Vinken, G. W. Brugh, and H. L. Klawans, eds.), pp. 435–458, North-Holland Publishing Company, Amsterdam, (1977).
120. B. Sparf, *Acta Physiol. Scand. Suppl.* **397**:19 (1973).
121. P. G. W. Plagemann and D. P. Richey, *Biochim. Biophys. Acta* **344**:263 (1974).

122. A. R. Boobis, A Gibson, and R. W. Stevenson, *Biochem. Pharmacol.* **24**:485 (1975).
123. F. C. MacIntosh, R. I. Birks, and P. B. Sastry, *Nature (Lond.)* **179**:1181 (1956).
124. R. W. Guynn, *J. Neurochem.* **27**:303 (1976).
125. S. Tucek, in: *Drugs and Cholinergic Mechanisms in the CNS* (E. Heilbronn and A. Winter, eds.), pp. 117–131, Research Institute of National Defense, Stockholm (1970).
126. H. S. Katz, S. Salehmoghaddam, and B. Collier, *J. Neurochem.* **20**:569 (1973).
127. R. L. Polak, *Biochem. Pharmacol.* **21**:144 (1971).
128. R. Massarelli, V. Stefanovic, and P. Mandel, *Brain Res.* **112**:103 (1976).
129. F. Hein and A. Fahr, *Arch. Exp. Pathol. Pharmakol.* **195**:59 (1940).
130. A. Todrich, K. P. Fellowes, and J. P. Rutland, *Biochem. J.* **48**:360 (1951).
131. R. M. Dawson and H. D. Crone, *J. Neurochem.* **24**:411 (1975).
132. A. K. S. Ho, C. S. Tsai, R. C. A. Chen, H. Begleiter, and B. Kissin, *Psychopharmacologia* **40**:101 (1974).
133. A. K. S. Ho, C. S. Tsai, and B. Kissin, *Pharmacol. Biochem. Behav.* **3**:1073 (1975).

Ethanol and Neuronal
Electrophysiology

XI

Effects of Ethanol on Nerve Impulse Activity

15

W. R. Klemm

A. INTRODUCTION

The study of ethanol effects on nerve impulses (action potentials) in experimental animals is a convenient way to evaluate mechanisms of intoxication and tolerance in humans. The behavioral effects of ethanol are caused by what it does to the activity of neurons, and impulse discharge is among the more obvious and readily quantifiable properties of neurons and one of the few properties that can be observed directly in the intact, living animal. Ethanol can dramatically alter impulse discharge, and evaluation of these changes in various brain loci and under various neurochemical environmental conditions can give some insight into mechanism of intoxication and tolerance development.

Certain advantages of studying impulse discharge are shared by the commonly used electrophysiological index of neuronal function, the electroencephalogram. However, the continuously fluctuating electroencephalographic potentials are compounded from many neuronal, and perhaps glial, sources and are subject to certain interpretive ambiguities that do not exist with impulse discharge. Additionally, the literature on ethanol action on impulses has not been surveyed extensively, as has been the case for the electroencephalogram.

This review will survey five decades of research on ethanol effects on nerve impulse activity. Fortunately for reading convenience, the quantity of such research is nowhere near as voluminous as might be expected. Indeed, I submit that this area of ethanol research has been neglected, especially at the *in vivo* mammalian level and that many opportunities for discovery are being lost by default.

W. R. Klemm • Department of Biology, Texas A & M University, College Station, Texas 77843.

1. Relevance to Basic Mechanisms of Ethanol Effect

There are two basic consequences of neuronal activity: the secretion of neurotransmitters and the electrogenesis of discontinuous voltage pulses (impulses). Neurotransmitters in the synapse either excite or inhibit the genesis of impulses in a postsynaptic neuron. Thus, impulse activity reflects underlying transmitter synthesis storage and release processes and their associated metabolic functions.

Many theoretical issues can be tested readily by use of an impulse assay. As one example, there is the controversy over the role of calcium in the mediation of ethanol effects. Calcium by itself influences membrane excitability along axons and, moreover, is necessary for the release of certain neurotransmitters. Thus, it seems natural to test hypotheses on calcium's role in intoxication with physiological studies that have a nerve impulse–transmitter interrelation emphasis, as opposed to more gross physiological, psychological, or behavioral analysis.

There are many biochemical ways to investigate ethanol actions on membranes and intra- and extracellular chemicals; but the key issue, which can only be addressed by looking at functional consequences (such as impulse discharge), is the effect that ethanol has on the brain's processing and output functions that underlie the ultimate behavioral changes.

2. Convenience for Assay of Acute and Chronic Effects

Impulses have several properties that make them ideal for studying ethanol effects: quantification is relatively unambiguous, and effects can be studied under diverse experimental conditions, such as in isolated axons, cells, or ganglia *in vitro,* or in a variety of *in vivo* preparations. This review illustrates the convenience of a nerve impulse assay approach for the study of ethanol and, moreover, reveals many unexploited opportunities for discovery.

B. EFFECTS ON PERIPHERAL NERVOUS SYSTEM

1. Impulse Propagation in Somatic Nerve

Ethanol was among the first drugs ever studied for the effect on impulses, once the necessary technical advances had been made, specifically the Einthoven string galvanometer (circa 1901), the biomedical amplifier and recording camera by Forbes and Thacher in 1920, and the oscilloscope by Erlanger and Gasser in 1922.

Working in Forbes' laboratory in the early 1920s, Davis and collaborators conducted a classic experiment[1]; they tested the hypothesis that the amplitude of an impulse would decrease as it traversed a narcotized region of nerve, and they chose ethanol as the narcotizing agent. They recorded impulses at various points along an isolated cat peroneal nerve, a section of which was exposed to an air stream that was bubbled through 5–6% ethanol. Ethanol depressed the amplitude of impulses uniformly along the exposed segment of nerve, and the size was normal beyond the exposed segment; recovery occurred in about 30 min. At about the same time, the same experiment had been performed independently in Japan.[2]

While these studies are of great historical interest, they and many subsequent studies of *in vitro* preparations involved unduly high concentrations of ethanol (see Section B.3), the relevance of which to human intoxication is unclear. Therefore, these reports will only be cited without discussion (Table I).

There have been some human clinical studies of ethanol effects on nerve impulse conduction velocity; this research has already been reviewed.[17] In general, impaired conduction is most likely to occur when ethanol ingestion has been extensive over a long period of time; the mechanisms are difficult to interpret, however, because the results were often confounded by coexisting nutritional deficiencies, especially vitamin B deficiencies.

Several other topics are related to ethanol effects on somatic nerves, but these have been reviewed elsewhere (see below) and are beyond the intended scope of this review. Some of this literature indicates that the primary events associated with ethanol action in the brain occur within neuronal membranes, wherein excitability is reduced by such changes as increased membrane resistance and reduced early inward sodium current during impulse generation.[18–20] Calcium levels appear to alter the effect of ethanol on membranes,[19] and ethanol decreases acetylcholine release in brain.[19,20] Other literature[20–21] deals with ethanol's metabolic effects, such as inhibition of Na^+,K^+-ATPase and sodium–potassium transport, decreased brain protein turnover, and reduced intermediary metabolism and high-energy phosphate turnover; many of these effects may be secondary reflections of ethanol depression of neurons. A few other reports deal with ethanol effects on sensory receptors,[21] including observations of altered processing reactions in the retina, depression of carotid baroceptors and muscle spindles, and enhanced activity in carotid chemoceptors and in muscle spindles (at concentrations less than 0.417 M). Finally, there are studies of the myopathies and neuropathies that develop in chronic human alcoholics; the neuropathies include demyelination of peripheral nerve, causing reduced conduction velocity.[17]

2. Ganglionic Effects

The research on ganglionic preparations has generally been confined to invertebrate, *in vitro* preparations, as a matter of convenience. But, as with the studies on somatic nerve, the concentrations of ethanol that have produced significant effects were higher than that commonly produced from oral ingestion by humans. Thus, these reports are also referenced, without discussion, in Table I.

Table I. Electrophysiological Studies of Ethanol in *in Vitro* Preparations

Species	Nature of preparation	General effect of ethanol	Ethanol concentration	Reference
Rabbit/cat	Peroneal and tibial nerve	Depolarization of resting potential; decreased size of action potential; depressed excitability	Unquantified	3
Frog	Sciatic nerve	As in Ref. 3	0.5–2 M	4
		Blockade of conduction (reversible by raising membrane potential with external current)	1–2 M	4
	Sciatic/peroneal nerve	Differential blockade of sensory versus motor fibers	0.65–1 M	5
	Sympathetic ganglia	Induction of post ganglionic repetitive discharge after single-pulse preganglionic stimulation	0.2–0.8 M	6
		Delayed depression of synaptic transmission	0.4–1 M	6
Toad	Sciatic (compound action potentials)	Decreased size of action potentials; increased rheobase and chronaxie; prolonged refractory periods	>0.513 M	7
		Increased size of action potentials; decreased refractory periods	0.085 M	7
Squid	Giant axon (voltage clamp)	Decreased size of action potential; depolarization of resting potential; depressed Na^+-K^+ conductances	0.51 M	8
	Giant axon (sucrose gap voltage clamp)	As in Ref. 8	3–6% (0.65–1.3 M)	9
Lobster	Giant axon	Depolarization of resting potential; conduction blockade; increased after potential; increased spontaneous repetitive discharge	1–2.5 M	10
Land snail	Ganglionic neurons	Depression of acetylcholine-induced depolarization	Not stated	11
	Ganglionic neuron	Selective depression of acetylcholine-induced excitation without altering membrane properties; depression caused by a series of depressants correlated with hydrophobicity	~0.05–0.15 M	12

Table I. (*Continued*)

Species	Nature of preparation	General effect of ethanol	Ethanol concentration	Reference
Tritonia	Ganglionic neurons	As in Ref. 3	2–6%	13
		Depression of evoked excitatory postsynaptic responses	(0.434–1.3 M)	13
Aplysia	Ganglionic neurons (voltage clamp)	As in Ref. 3	0.5–4%	14
		Some neurons hyperpolarized; reduction of early inward current; blockade of both de- and hyperpolarizing responses to acetylcholine	(0.109–0.868 M)	14
	Ganglionic neurons	Depression of evoked excitatory postsynaptic potentials, when no effect noted on initiation or propagation of impulses in presynaptic axon	2-6% (0.434–1.3 M)	13
	Ganglionic neurons	Some cells depolarized, others hyperpolarized and others not affected; decreased size of action potential	0.5–5% (0.109–1.085 M)	14
	Ganglionic neurons	Depression of posttetanic potentiation	0.8 M	15
	Ganglionic neurons	Tolerance development of ethanol-induced depression of posttetanic potentiaion	Repeated exposure	16

3. The *in Vitro* Dilemma

Many data have come from the *in vitro* studies just cited. However, certain key questions are not readily answered by such methods; the basic dilemma is in not knowing how accurately *in vitro* effects reflect the *in vivo* situation. For example, ethanol-induced depolarization and reduced impulse size are not commonly observed in an intact nervous system; even direct microiontophoretic injection of ethanol on single neurons in intact vertebrate brain does not reduce spike amplitude at concentrations that markedly alter discharge rate.[22] Some of the *in vitro* data are supported by one *in vivo* study in anesthetized cats; resting membrane potential of spinal motoneurons was increased by low doses (~0.5 g/kg), whereas larger doses (1 g/kg) tended to depolarize the membrane and to decrease spike height and slopes.[23] If such decreased spike height were typical *in vivo,* it could decrease transmitter release and thus depress transmission across synapses (see below).

Most *in vitro* studies have neglected the effects of ethanol metabolites and reaction products. No studies have compared ethanol effects with those of acetaldehyde or of various biogenic aldehydes that result from the spontaneous condensation reaction of acetaldehyde and catecholamines.

A most serious reservation about the reported *in vitro* data is that ethanol concentrations were usually greater than levels associated with behavioral intoxication. Many of the reported effects are toxic effects and not pharmacologically pertinent for humans. As an example, ethanol, given intravenously causes general anesthesia of humans at blood concentrations of about 50 mM (0.23%) (v/v),[24] but much higher concentrations (0.5–1.0 M) (2.3–4.6%) are needed for blockade along axons.[25] This difference is not explained by the differences in fiber diameters in peripheral nerve and in the brain, because even though ethanol is more potent for thin-diameter fibers,[25] that potency difference is only ~2-fold, not the 10- to 20-fold potency difference for general versus local anesthesia. Thus, the major effects of ethanol on impulse activity in the intact nervous system probably do not reflect a direct action on the impulse propagation mechanisms that are reflected in the usual *in vitro* methods.

This survey of the literature supports the conclusion drawn in another review[26]: namely, that ethanol is much less effective on the impulse discharge of any one isolated neuron than it is on the integrated functions of neuronal populations. We shall see that low levels of ethanol (<50 mM) alter the normal functions of neurons in complex circuits, thus underlying ethanol's effects on complex human behaviors.

C. EFFECTS ON CENTRAL NERVOUS SYSTEM

Most of what has been learned about actions of ethanol on the central nervous system has come from studies of experimental animals, especially the common laboratory animals. Some preliminary comment on ethanol dosage is necessary to help relate these studies to ethanol consumption by humans. Following the common convention in *in vivo* pharmacological studies with animals, doses of ethanol have usually been expressed in terms of g/kg, and blood alcohol levels have not usually been monitored. However, to help relate animal doses to human intoxication conditions, it is approximately correct to say that overt signs of intoxication in animals such as rats, rabbits, and cats do not become evident until the dose exceeds 1 g/kg. In cats, for example, doses below 1 g/kg have no behavioral effect, a dose of 2 g/kg produces clear ataxia followed by sleeping, and a dose of 3–5 g/kg causes a deep stupor that lasts 2–5 hr.[27] The peak blood level achieved by 2 g/kg, given intravenously or intraperitoneally, is approximately 250 mg/100 ml, which is sustained above the 200-mg/100-ml level for at least 2 hr.[27] By comparison, the legal definition of intoxication in humans is based on a blood level of 100 mg/100 ml, which is created by an approximate dose of 0.9 g/kg (based on rapid intake on an empty stomach in a 150-lb person of 4 oz of 90-proof whiskey).[28]

The common view is that intoxication results mostly from a preferential effect of ethanol on synaptic transmission. The view can be illustrated by some reported observations of compounded impulse activity associated with a spinal reflex.[29] Low, intravenously administered doses of ethanol (0.2–0.9 g/kg) consistently depressed the ventral root potential of motoneurons, during both mono- and polysynaptic sensory stimulation, and decreased the spontaneous discharge rate of interneurons of the spinal cord (23 of 24 cells at 0.1–0.6 g/kg). But even large doses (2 g/kg) had no effect on the dorsal root potentials of the sensory neurons in the reflex, in which activity depended only on axonal conduction. Since all these observations occurred in spinal transected cats, the effects were direct and did not involve secondary, supraspinal influences.

This special vulnerability of synapses, at mildly intoxicating concentrations, thus makes it especially pertinent for an understanding of human intoxication to discuss ethanol action on synapses.

1. Depression of Postsynaptic Potentials

Study of postsynaptic potentials in the intact mammalian central nervous system is most conveniently done by evaluating activity within simple, defined, reflex arcs. Observing first the behavioral effects of ethanol on specific reflexes helps to define the locus for more detailed electrophysiological analysis. One early mammalian study revealed that such reflexes as the monosynaptic masseteric reflex and the polysynaptic soleus reflex are readily affected by moderate doses of ethanol.[30]

Postsynaptic potential effects of ethanol are usually studied by intracellular recording techniques, which provide the least equivocal way to evaluate ethanol effects on the neurotransmitters, hormones, and metal ions that regulate postsynaptic potentials. (For any other purposes, however, extracellular recording seems just as useful and is certainly more convenient technically.)

Unfortunately, ethanol effects on mammalian reflexes have not been studied much at the level of postsynaptic potential changes. In anesthetized cats, ethanol in doses of about 1 g/kg, iv, depressed inhibitory as well as excitatory postsynaptic potentials of spinal motoneurons for both mono- and polysynaptic pathways; the polysynaptic pathways were more susceptible.[23,31]

The demonstration of differential susceptibity of polysynaptic paths is consistent with the studies, including some on humans, on evoked EEG responses. Ethanol in small or moderate doses has its greatest effect on the long-latency components of averaged evoked EEG responses[20]; these components are generally attributed to polysynaptic activity, although some portion of the total effect could derive from ethanol's preferential depression of slow-conducting, fine-diameter fibers.

In several invertebrate preparations (mollusks and crustaceans), ethanol shared with many central nervous system depressants the property of depressing excitatory postsynaptic potentials without altering the inhibitory events.[11]

However, some interesting contrasting observations have been made on the well-defined Mauthner cell (M cell) system that regulates startle reflexes of fish.[18] Gills were perfused with 1–2% (217–434 mM) ethanol, a level which caused behavioral excitability (brain concentration of ethanol was 0.003–0.005 mg/g). This level did not alter resting membrane potentials or action potentials that were evoked by axonal stimulation, but did interfere with recurrent collateral monosynaptic inhibition of the M cell; both the electrical and chemical components of the inhibition were depressed (Fig. 1). The depression of inhibitory postsynaptic potentials seemed to have been due to an impaired release of

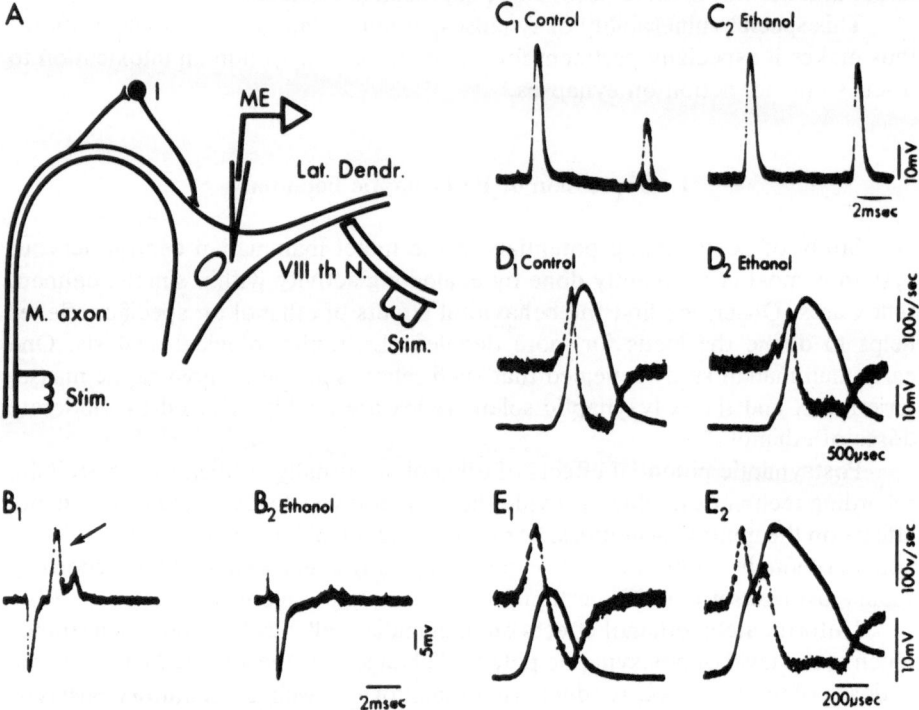

Fig. 1. Selective effect of ethanol on synaptic activity, as opposed to resting membrane potential or action potential. (A) Experimental preparation, showing intracellular recording from Mauthner cell in brainstem of the fish, which involves a recurrent inhibitory pathway (I). (B) Extracellular field near the soma induced by antidromic stimulation of the Mauthner cell axon; this potential is hyperpolarizing and electrically inhibits the cell. Ethanol interferes with this potential and its electrical inhibition. (C) Ethanol blockade of the recurrent postsynaptic inhibition. In this preparation the inhibition is best evaluated by paired antidromic stimulation in which the conductance increase associated with the inhibitory postsynaptic potential attenuates the second spike; note that ethanol blocks this effect. (D) Lack of ethanol effect of resting membrane potential and only minor effects on the intracellular antidromic spike; upper trace of each pair is the electronically differentiated version of the actual spike in the lower trace. (E) Both records obtained as in (D) but under conditions of high ethanol concentration; distinct alterations of antidromic spike are evident. [From Faber and Klee.[18]]

acetylcholine from axon collaterals of the M cell. Such results indicate that synaptic transmission mechanisms were more sensitive to ethanol than either spike electrogenesis or the membrane properties that govern the resting membrane potential.[31] The known depression by ethanol of spike electrogenesis would thus seem to be secondary to synaptic depression.

The demonstration of a selective depression of neuronal inhibitory processes should spur a search for similar phenomena in higher animals. Some investigators believe that the behavioral disinhibition which occurs in human intoxication could have its origins in cellular disinhibition.

2. Potentiation of Presynaptic Inhibition

Presynaptic inhibition is an important type of inhibition in which impulse generation in a postsynaptic neuron is impaired because impulses in one presynaptic terminal cause a relatively persistent depolarization of another presynaptic terminal, reducing the size of its impulses and consequently reducing excitatory transmitter release. Some early work suggested that ethanol might facilitate presynaptic inhibition in the spinal cord,[32,33] but one report in cats indicated no such effect in doses up to 0.9 g/kg.[29] In a follow-up of this issue,[34] ethanol (0.15–0.45%) (33–98 mM, w/v) was perfused in an isolated frog spinal cord preparation and found to increase the size of the dorsal root potentials, which are considered to arise from the electrical field generated by presynaptic terminal depolarization. Since presynaptic depolarization is a cardinal sign of presynaptic inhibition, ethanol presumably enhanced presynaptic inhibition in this preparation. An interrelation with GABA was indicated by the observation that perfusion of GABA depolarized dorsal root terminals and simultaneous alcohol perfusion increased the GABA effect.[34]

Another study on intact cats is also consistent.[35] Low concentrations of ethanol (0.05–0.16%) (11–35 mM) apparently selectively increased the depolarization of the terminals of the primary sensory neurons in the trigeminal nerve. Additionally, the ethanol effect was not present when the cats were transected rostral to the trigeminal nerve nucleus, suggesting that the presynaptic depolarization was actually mediated indirectly via ethanol's action on some rostral brain region. These results also suggest that the common clinical observation of facial "numbing" during acute intoxication in humans is actually due to a depressed sensory input that is accomplished at the terminals of these first-order sensory neurons.

3. Depression of Impulse Discharge

The depression of impulse discharge in mammalian brain by systemically administered ethanol is well documented. In addition to the illustration of this effect that was already given,[29] there are other important examples. One of the early studies focused on cerebellar unit activity, on the assumption that the ataxic behavior associated with intoxication was mediated via cerebellar dysfunction. In decerebrate, unanesthetized cats, intravenously administered ethanol (0.6 g/kg) depressed 11 of the 12 cells which had been electrophysiologically identified as Purkinje cells in the cerebellar cortex; similar results were noted in neurons of the lateral vestibular (Deiter's) nucleus.[36]

Responses of lateral hypothalamic cells to ethanol have also been of interest, because the lateral hypothalamus is known to be involved in regulation of taste, thirst, and ingestion behavior and therefore might be involved in regulating ethanol intake. In urethane-anesthetized rats, the intravenous injection of very low doses of ethanol (0.04 g/kg) depressed the spontaneous discharge of certain neurons.[37,38]

Several studies of "multiple-unit activity," which is the impulse discharge from a restricted population of neurons, have also confirmed that many neurons in a wide variety of brain areas are depressed by ethanol. For example, one study in acute, unanesthetized rat preparations revealed that marked depression of unit activity occurred in a variety of brain areas within 2–10 min after intraperitoneal injection of moderate doses (0.2–1 g/kg).[39] In another study of chronically prepared unanesthetized rabbits, a monitoring survey of 14 brain areas revealed statistically significant depression in at least seven areas after low to moderate doses (0.3–1.2 g/kg, intraperitoneally administered); median latencies until clear changes developed were short (5–10 min after injection).[40] In another study[41] ethanol effects were evaluated in several closely linked limbic system structures of chronically prepared unanesthetized rabbits; in most electrode loci, ethanol depressed spontaneous discharge within 5 min after low intraperitoneal doses (0.15–0.6 g/kg); the depression was particularly evident in the septum, entorhinal cortex, CA_1 zone of the hippocampus, and the fimbria-fornix. Particularly noteworthy was the observation that such depression could occur even with the lowest dose (0.15 g/kg).

The mechanism of the depression observed in these studies is not known, but it probably arises from the depression of postsynaptic potentials and enhanced presynaptic inhibition. As recorded from any given neuron, decreased discharge rate can arise directly from ethanol depression of the recorded neuron, from ethanol depression of activity in neurons that normally supply excitatory drive to the recorded neuron (disfacilitation), or even from a direct activiation of inhibitory neurons that normally inhibit the recorded neuron.

4. "Excitation" of Impulse Discharge

For certain neurons in a network, reduced impulse discharge is not the universal effect of ethanol. Even in peripheral nerve fibers, for example, recall the report that the compound action potential amplitude of toad sciatic nerve was actually increased by low concentrations of ethanol.[7] Systemically administered ethanol certainly can produce increased unit activity, and it may be useful to mention some of this literature in the hope that it will spur research to determine whether the effect is direct or indirect.

Certain aspects of biphasic effects have been reviewed[20] in the context of EEG and evoked response studies and of spinal reflex actions. Here, we focus on a probable underlying cause of such biphasic effects, namely the biphasic actions on neuronal impulse discharge rate.

Among the good examples of ethanol-induced increases in impulse activity is the demonstration that low doses of ethanol (0.1–0.6 g/kg, intravenously administered to cats) often caused an initial transient burst of accelerated firing of spontaneously discharging interneurons in the dorsal horn of the spinal cord; the increase lasted about 15 sec, after which neurons became depressed.[29] In another study of cats, moderate intravenous doses of ethanol (0.6 g/kg) tended to accelerate discharge rate of cerebellar cortex interneurons (6 of 10 cells).[36] In the lateral hypothalamus of rats, certain sodium- and/or glucose-sensitive neurons responded to microelectroosmotically applied ethanol with increased discharge.[37] Marked excitability to intravenously injected ethanol has been reported from urethane-anesthetized rats whereby doses as low as 0.04 g/kg increased the discharge rate in 53% of lateral hypothalamic neurons (and depressed in 32% and unaffected in 15%).[38]

Many examples of increased unit activity have also been observed in studies of intact, unanesthetized animals. For example, in acutely prepared rats, intraperitoneal doses of ethanol (0.2–1.0 g/kg) quite commonly produced statistically significant increases in multiple-unit activity, occurring in at least one rat for each of five brain areas (cerebral and cerebellar cortex, hippocampus, septum, and medial forebrain bundle).[39] The increased impulse discharge, when present, involved a substantial number of neurons, because in a multiple-unit recording, several to many neurons in a population of units around an electrode must increase their discharge before becoming distinguishable from the background firing of others in the population. In another multiple-unit study that involved screening of many brain areas in chronically prepared intact rabbits, statistically significant trends for increased unit activity were evident in the cerebral cortex, lateral geniculate body, and septum at higher doses (>0.9 g/kg) and in the medial forebrain bundle, substantia nigra, thalamus, cerebellar cortex, and hippocampus at lower doses (<0.9 g/kg).[40] In a similar study of chronically prepared rabbits, in which various limbic system sites were monitored, doses in the range 0.15–0.6 g/kg often caused clear increases in the entorhinal cortex, hippocampal commissure, and CA_1 region of the hippocampus.[41]

One question which often arises when ethanol is given intraperitoneally (in most studies in concentrations of 10–30%) is that unit activity could change because of nociceptive effects. However, if that were the case, one would expect responses immediately, and this has not been observed in those studies with experimental designs that could reveal that; responses were consistently delayed for several minutes.[39–41] Also, these studies disclosed a greater effectiveness of higher doses, yet the concentrations were the same. The basis for ethanol-induced increases in unit activity is not clear, but an indirect mechanism seems plausible, whereby ethanol may depress activity of inhibitory neurons, thus releasing other neurons from such inhibition (i.e., disinhibition). Since some inhibitory systems are well defined and accessible to study, such as the cerebellar pyramidal cell projection to the deep nuclei, this hypothesis appears to be testable.

In fact, this hypothesis was tested (but not explicitly) 7 years ago in a study of Renshaw cells in the spinal cord; these cells, when activated, inhibit spinal motoneurons. Moderate doses of ethanol (0.16–2.4 g/kg, given intravenously) enhanced activity of Renshaw cells and in turn increased the depression on spinal motoneurons.[42] Since the only known input to Renshaw cells is the recurrent excitatory cholinergic axon branch of motoneurons, it seems that ethanol's effect in this situation was excitatory to Renshaw cells.

There is also some other evidence for direct excitation of neurons by ethanol. For example, in one study where ethanol was applied directly to neurons (by microelectroosmosis), ethanol caused clear excitation in 28 of 29 cells that showed a response (58 others showed no response).[43] Paradoxically, all 12 of the cells in the lateral hypothalamic area were excited, yet previous results indicated that at least 32% of those neurons should have decreased activity if the results had paralleled those involving intravenous administration of ethanol.[38] Another microelectroosmotic study of cerebral cortex cells of anesthetized rats revealed that of the 30 cells given ethanol, 3 were excited, 7 were depressed, and 20 did not respond.[22]

The basis for any such direct excitation, assuming that it is real and not a methodological artifact, is open to speculation. One possibility is that sodium permeability across neuronal membranes is actually increased by low doses of ethanol. Ethanol antagonism of calcium (see below) could lead to increased permeability.

Excitatory effects could also be produced by any tendency of ethanol to enhance neurotransmitter systems. In the case of acetylcholine, for example, there are data to support this possibility. For example, in a study[44] of transmission in the cat stellate ganglion, which is cholinergic, synaptic transmission was facilitated by perfusion of ethanol at levels of 1 and 2% (217 and 434 mM); depression was evident at 3%. Also, as mentioned earlier, ethanol enhances activity of the cholinergic motoneuron–Renshaw cell system.[42] A final example of direct ethanol excitation is found in the neuromuscular junction, also cholinergic. At low concentrations (from 8 mM to about 217 mM) there is a transient enhancement of junction activity, but at larger concentrations, there is clear depression.[45,46]

Note, however, that there are clear demonstrations of cholinergic synaptic transmission in which ethanol suppressed the cholinergic transmission (the fish M-cell system and the cat spinal motoneuron system).[31] Thus, it would appear that ethanol has differing actions even when the same transmitter system is involved. Ethanol may have differential effects on either transmitter or receptor. These conclusions may be premature, because direct studies of ethanol effects on synaptic transmission in mammalian brain have not yet been performed.

One can only speculate on the relevance of human intoxication of the observations of increased impulse activity in certain neurons. If the cell-level response is reflecting disinhibiton, we have a basis for beginning to understand the behavioral indices of dishinibition which occur during intoxication such as ataxia and uninhibited social behavior. The absence of clear enhancement of any behaviors during intoxication would seem to suggest that direct excitation of unit activity, if present, might interfere with behavioral performance by the excitation being restricted to inhibitory neurons. For example, ethanol excitation of Renshaw cells has the behavioral effect of depressing skeletal muscle, a depression which is additive to the *direct* depression by ethanol of spinal motoneurons.[42,47]

Almost all the studies surveyed here have used a simple rate measure of ethanol's effect. However, discharge rate is not the only meaningful measure, and in fact, it can be misleading. Discharge rate is notoriously unstable over time, and many changes can occur that are unrelated to any ethanol effect. Some good examples of this problem have been reported.[48]

The temporal distribution of successive spikes in a train of impulses is an important index of ethanol effect. It is possible that ethanol can alter the pattern of intervals between and among impulses, thus drastically altering function, without having any major effect on the total number of intervals. In fact, an observation of this kind has been reported.[48]

5. Interactions with Hormones, Ions, and Transmitters

In the previously mentioned study in which microelectroosmotic application of ethanol increase discharge of certain cells in the hypothalamus,[43] it was also noted that many of these cells were also excited by angiotensin. When one of those cells was examined for the combined effects of ethanol, sodium ions, and angiotensin, ethanol mildly increased the cell's activity, which was enhanced by simultaneous ejection of sodium ions and enhanced drastically when angiotensin was also simultaneously delivered. Similar multiplicative effects were observed in studies from that same laboratory involving systemic injection of drugs.[37]

The theoretical basis for another recent electroosmotic study was the preexisting evidence that ethanol anatagonized calcium and that depressant actions of biogenic amines on cortical neurons seemed to be mediated via a calcium-dependent mechanism.[22] Thus, the investigators anticipated an interaction between ethanol and transmitters. Ethanol injected alone onto single cells of the somatosensory cortex of anesthetized rats produced mixed effects on the spontaneous or glutamate-induced impulse discharge: 23% of neurons tested decreased in firing rate, 10% increased, and 67% were unaffected. These neurons had been previously depressed by iontophoretic injection of amine transmitters (acetylcholine, norepinephrine, serotonin), but these amines had no effect when injected again after the ethanol treatment; recovery of sensitivity to the amines requried 4–15 min. The interaction with GABA was different; its depression was not blocked by ethanol.

Such data were reasonably interpreted to indicate that ethanol may act on neuronal calcium, because similar results had been reported in other studies that employed specific calcium antagonists. Other evidence reviewed in that paper includes a finding that calcium could reverse the inhibition by ethanol of K^+-induced contractures of guinea pig ileum and that ethanol increases binding of calcium to erythrocyte membranes.

In another study, iontophoretic application of several specfic calcium antagonists (manganese, nickel, cobalt, and lanthanum, and verapamil) prevented or inhibited the inhibitory effects of biogenic amine transmitters on cortical impulse activity, while not affecting GABA depression or acetylcholine excitation.[49] Ethanol had the same action, suggesting that it acts as a calcium antagonist. Ethanol (20–3000 mM) reportedly increases binding of calcium to erythrocyte membranes[50]; it is clearly important to know if that is also true of neuronal membranes; and, indeed, it has recently been shown that synaptosmal binding of calcium does increase, after a transient decrease, after acute or chronic ethanol (0.001–0.05 mM).[51] Data of this kind suggest that human intoxication could be influenced significantly by the tissue levels of calcium. Such possibilities seem readily testable.

6. Regional and Temporal Differences

Intuitively, we could expect regional differences in ethanol action among brain areas simply from observing the behavior of inebriated humans. The earliest signs of intoxication clearly involve functions that are subserved by the higher nervous system: the speech slurring and cognitive impairments suggest an action on the cerebral cortex; the ataxia which occurs at still higher doses suggests cerebellar involvement. Finally, it is quite clear that the vital brainstem functions of consciousness and cardiorespiratory regulation are not affected until very large amounts have been ingested.

In the literature reviewed thus far, there is abundant evidence that the impulse activity of all neurons is not affected by ethanol in the same way. Even at the simplest levels of organization, several *in vitro* studies suggested basic differences among neurons in their sensitivity to ethanol; of special note was the demonstration of greater effect on axonal conduction blockade in sensory than in motor neurons,[5] and the variable ability of ethanol to depolarize or hyperpolarize *Aplysia* ganglion cells.[14] Regional differences have also been observed in the differential sensitivity of cells in the retina of frogs[52]; recordings of ganglion cell spikes under different stimulus conditions revealed that direct application of ethanol (2 μl, 5–10%) selectively caused the amacrine cell-mediated inhibitory surround effect to disappear, suppressed the green rod-mediated responses, and concurrently decreased sensitivity in the dark-adapted eye while increasing sensitivity for the cone system in eyes receiving background lighting. This increased cone sensitivity may be a specific illustration of ethanol disinhibition, as mentioned earlier. It is known that the output of every cone seems to be inhibited by the outputs of large populations of cones in distant surrounds; thus, ethanol could increase cone sensitivity merely by depressing this lateral surround inhibition.

We might therefore expect that ethanol causes different actions in given areas of the brain. As one known example, ethanol depressed the inhibitory effects of interneurons that act on fish Mauthner cells via axon collaterals, while not affecting either excitatory or inhibitory effects of other inputs to the same target cell.[18] Regional differences in sensitivity to ethanol have been reported at the mammalian spinal level; at doses (0.1–0.6 g/kg, given intravenously to cats) that did not affect spontaneous impulse discharge in motoneurons, a clear depression was observed in spinal interneurons.[29] This differential effect on interneurons was also suggested by a previous study.[47] Intracellular recordings of spinal motoneurons revealed that at 0.5–0.9 g/kg, the resting membrane voltage was reduced slightly (~8 mV), and the size and waveform of the antidromic action potential was drastically altered.[29] (Since antidromic responses are nonsynaptic, *in vitro* data on axonal impulses may have some relevance to the intact mammalian situation.) The simplest explanation for the greater effect of ethanol on interneurons is that these are much smaller than motoneurons, and thus have greater surface/volume ratio. However, differences could exist in the synaptic transmitter systems affecting the two cell types.

Because ethanol depresses sodium conductance across the membrane, Wayner and colleagues postulated that ethanol would be especially potent for those sodium-sensitive cells in the hypothalamus that regulate thirst. These cells, in rats, responded to ethanol doses as low as 40 mg/kg,[38] a dose that generally does not affect impulse activity.

Regional and temporal differences have been demonstrated among cells in four brain areas in a recent microelectroosomotic study.[43] Lateral hypothalamic cells were the most responsive in that lower electrophoretic currents were needed to alter cell discharge. Thalamic cells, however, were more responsive in the sense that their onset of response was much shorter and peak response rates were reached sooner. Duration of effect after the ejection current stopped was longest in zona incerta cells.

It is quite logical to suspect the cerebellum as being one of the primary target areas for ethanol action. Motor signs of intoxication, such as ataxia and nystagmus, usually are attributed to cerebellar and/or vestibular function. The first specific test of this possibility demonstrated that the moderate dose of 0.6 g/kg, given intravenously to cats, tended to accelerate cerebellar interneuron discharge rate, while clearly depressing cerebellar Purkinje cell and vestibular nuclei discharge.[36] Such depression is apparently direct, because ethanol (1 g/kg) caused marked depression of cerebellar cortex multiple-unit activity in rats in which the cerebellar cortex was denervated by large electrolytic lesions of the peduncles.[53]

At least one other study[36] has tested the issue of direct versus indirect ethanol action; ethanol depression of vestibular nucleus neurons of cats was demonstrated also in decerebellate preparations. Thus, the action was not mediated via the cerebellum, although the vestibular nucleus still received some other inputs.

The key issue, when asking questions about regional differences, is that ethanol action on cells in a given area must be compared with action in other brain areas. In other words, when ethanol is observed to be very effective in altering impulse activity in a certain brain area, we should compare that effect with impulse activity that is simultaneously recorded from many other areas.

Few studies have addressed this issue. Given the extraordinary sampling bias involved in monitoring single-unit activity, it does not seem feasible, in general, to pursue any research that attempts to compare activity of an unidentified single neuron in one brain area with that of many other such neurons elsewhere. Therfore, a reasonable alternative is to monitor impulse activity from small populations of neurons simultaneously from several different brain areas of unanesthetized animals.

In the first such study,[36] on paralyzed and artificially respired rats, there were clear signs of regional and temporal differences in unit-activity responses to ethanol, despite the fact that such differences were not evident from the electroencephalogram (EEG) obtained from the same electrode tips. Intraperitoneal doses of 0.2–1.0 g/kg (20%) caused unit-activity changes in certain locations within 2–10 minutes. In 53% of the rats, ethanol caused a clear change in the unit activity in one electrode location, which occurred before any changes were noted at any of the other six electrode locations. Temporal order was difficult to evaluate, because identical locations were not monitored in all the rats; nonetheless, some brain areas were conspicuous for developing short-latency responses. The hippocampus, for example, was the first site to show a unit-activity change 65% of the time (11 times out of 17 opportunities), and the cerebellar cortex was first 50% of the time (6 of 12). By comparing whether a given brain area responded or not, statistically significant sensitivity to these doses of ethanol was evident for the cerebral cortex, cerebellar cortex, and several limbic system sites.

In the next study of multiple locations in chronically prepared, unanesthetized rabbits,[40] such regional and temporal unit-activity differences were also noted (Fig. 2); as before these were not evident in the EEG. At the largest dose (1.2 g/kg) all brain areas were affected, but at low doses, several areas were conspicuously insensitive to ethanol: caudate nucleus, septum, fornix, and medial forebrain bundle. Several areas were very sensitive, responding in less than 5 min to doses as low as 0.3 g/kg: cerebellar cortex, cerebral cortex, hippocampus, lateral and medial geniculate nuclei, midbrain reticular formation, and pyriform cortex.

Even within a given brain area, regional and temporal unit-activity differences can occur. In rabbits with nine electrode placements in and near the hippocampus, low doses (0.15–0.6 g/kg) had relatively greater effect on spontaneous activity in the septum, CA_1 of the hippocampus, and the entorhinal cortex, and significantly less effect in the CA_3 and dentate zones of the hippocampus.[41]

Regional differences in the hippocampus have also been examined more specifically by evaluating unit activity in specific neuronal pathways[54] (Fig. 3). Evoked unit activity was recorded in five specific layers of the hippocampus in response to electrical stimulation of three major input structures. For all stimu-

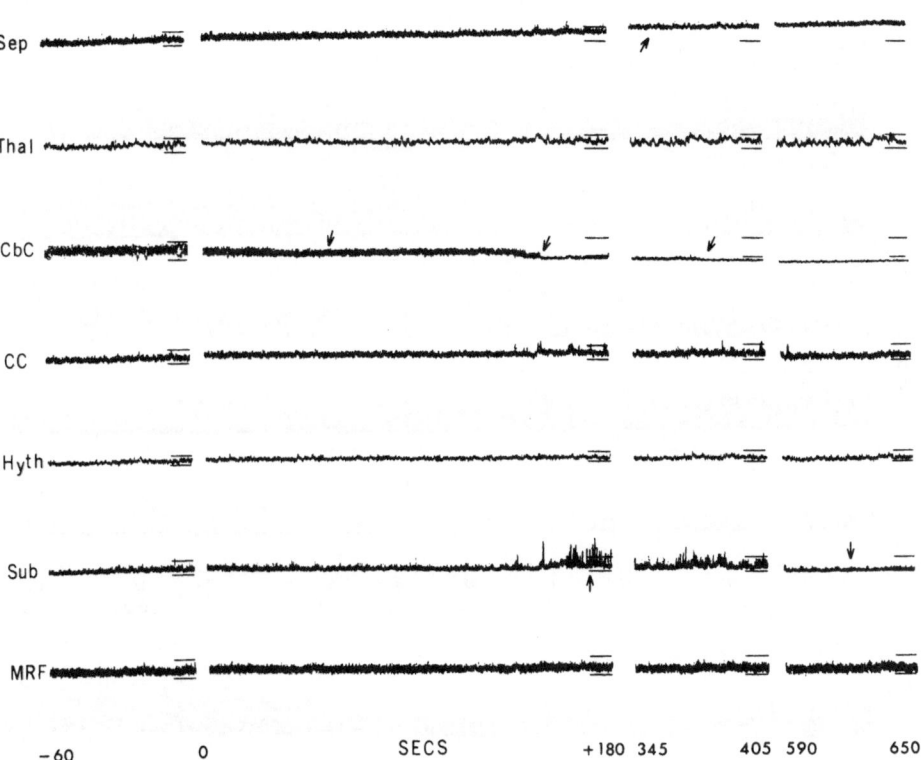

Fig. 2. Mixed excitatory–depressive actions and regional–temporal differences in ethanol effect on spontaneous impulse activity in unanesthetized rabbit. Tracings are continuous analogue integration displays of the multiple-unit activity derived from chronically implanted semimicroelectrodes. On the left is a 1-min preinjection record, followed by 3 consecutive minutes immediately after injection of 0.3 g/kg of 20% ethanol given intraperitoneally. The two sections on the right are 1-min samples at later times. The two horizontal lines (10 sec long) at the end of traces indicate the statistically determined range which would contain all parts of the traces under spontaneous and saline-injected conditions. Activity from two brain areas, the posterior hypothalamus (Hyth) and the midbrain reticular formation (MRF), in this rabbit never exceeded the change range. A very dramatic and short latency response appeared in the cerebellar cortex (CbC) which had progressive decreases in unit activity (arrows). The septum (Sep) had a delayed, sustained increase (arrow). The thalamus (Thal) (anterior central nucleus) had no quantitative change, but there was a qualitative change toward long-duration integrator waveforms in the last two panels. The cerebral cortex (CC) and the subiculum (Sub) developed transient unit increases about 2½–10 min after injection, with Sub activity becoming subsequently depressed (arrow). [From Klemm et al.[40]]

lus–response site pairing, ethanol (0.8 g/kg, given slowly intravenously) caused two- to threefold increases in stimulus threshold needed to evoke statistically significant changes in impulse discharge. Ethanol elevated brainstem reticular formation stimulation threshold more in the hippocampal molecular, and to a lesser extent in the dentate granule cell layer. Since investigators believe that the hippocampus is critical in memory formation, ethanol depression of its respon-

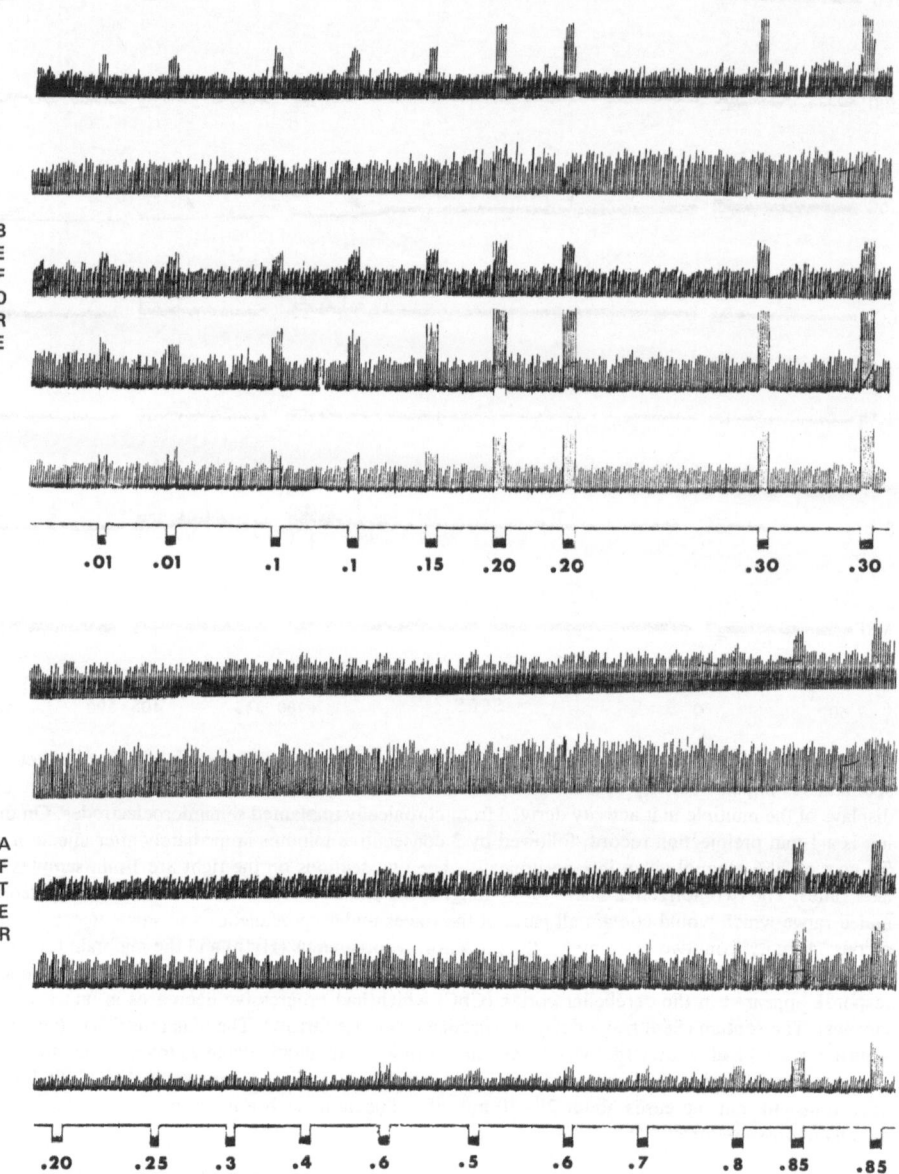

Fig. 3. Ethanol depression of evoked unit activity. Strip-chart traces for five areas of the hippocampus reflect the total unit activity, expressed in terms of the voltage that has accumulated in each integrator channel during the preceding second. The evoked activity results from electrical stimulation of the entorhinal cortex, before and after ethanol injection (0.8 g/kg, iv). Before ethanol, the stimulus intensities illustrated (voltage and delivery time shown in lower trace of each section) evoked statistically significant increases in each area, except for the second trace. After ethanol, these voltages were ineffective, and evoked increases did not occur until much larger intensities were employed. [From Folger and Klemm.[54]]

sivity could well explain the commonly observed memory-disrupting effects of ethanol.

The few studies which have evaluated unit activity simultaneously in more than one brain area[39-41] have made it abundantly clear that ethanol does not act in just one place nor at the same time in all places. Thus, there are many "target areas" of the ethanol's action, and these have many dynamic interactions with each other and with other brain areas. Thus complete understanding of ethanol action will probably require not only simultaneous monitoring of activity in many points along complex, but defined circuits, but also some sort of systems analysis or other mathematical modeling approach.

The demonstration of regional differences must be interpreted carefully. If ethanol acts principally on a given anatomical type of neuron or on a given neurochemical transmitter system which is not uniformly distributed in the brain, the observed regional differences merely reflect the distribution of the neurons or transmitter system. Such observations are nonetheless helpful because they help identify brain areas where subsequent research should be concentrated.

Brain areas in which clear differential changes in unit activity occur are not necessarily the proximate sites of that action; several indirect influences can change unit activity at a given recording site. Ethanol can act on neurons that supply input to the recording site. The indirect effect may be a result of action on general level of arousal, emotional state, or motor activity.

Another problem is that there may be differences in "receptors" for ethanol, which in turn may be differentially distributed, as seems to be the case with morphine receptors. Even more perplexing is that we do not even know if there are receptors for ethanol; in all likelihood, there are not, at least not in the conventional sense of stereospecific interaction. As mentioned earlier,[12,25] electrophysiolgical responses seem to be affected more by ethanol's hydrophobic properties than by its steric properties. But even given the argument that ethanol's action is on or in the membrane, it is still clear that membrane properties are not the same in all neurons.[31] Moreover, those properties may not even be the same at different loci in the same neuron; ethanol depression of electrogenesis has been localized to specific areas of neuronal membrane in two different preparations (the axon hillock region of the fish M cell and the soma–dendritic membranes of cat spinal motoneurons).[31]

D. UNDEREXPLOITED AREAS OF RESEARCH

Use of the nerve impulse assay approach is underexploited in ethanol research. In addition to elucidating the role of specific neurotransmitter systems, an impulse assay system would seem to be convenient for testing hypotheses, involving calcium or second messengers for example, about the membrane-level mechanisms of ethanol action. Processes involved in the genesis of tolerance could be profitably investigated, including determination of the role of acetaldehyde and biogenic aldehydes. Some specific strategems and methods are discussed below.

1. Intracerebral Microinjection

There has been no interest in intraventricular injections of ethanol, because it readily crosses the blood–brain barrier without alteration. However, some value might be found for intracerebral injection, via implanted cannulae in specific brain areas.

There is an amazing lack of literature involving intracerebral injection of ethanol,[55] and none of it has involved recording of unit activity. Perhaps a major reason is that investigators fear that effects could not be achieved without using such large doses that nonspecific toxic effects would dominate the response. In pilot experiments, we demonstrated that intracannular injections of doses up to 50 μg (in 1 μl) can cause dose-related, reversible effects; however, the effects thus far have not been very reproducible, and we suspect that there are enormous unsolved problems with the geometrical relationship to the neurons being recorded, especially in terms of mechanical trauma of the cannula and attached electrodes and in terms of such variables as diffusion and tissue displacement by the fluid.

2. Microelectroosmosis

Electroosmotic application of ethanol on single neurons can, in combination with proper neuropharmacological manipulations, help identify the neurotransmitter system, if any, by which ethanol mediates its action, and in follow-up studies, suggest which aspect of the transmitter system is involved. Thus, it is surprising that, in the face of the present "fad" over microiontophoresis, so few studies of this kind have been done with ethanol. Laboratories with this capability may be missing a golden opportunity.

A common way of applying drugs to single neurons is to use an array of micropipettes, one of which records impulse activity, and the others of which deliver charged molecules of drug in solution into the region of the recorded cell by electrophoretic (iontophoretic) action. Although this technology affords an elegant way to evaluate ethanol actions at the cell level, it is not always clear that the results obtained truly parallel what happens following normal routes of ethanol administration. For example, the one study which has compared impulse activity in the same neuron in response to microelectroosmotic ejection and to intravenous administration found opposite effects.[43] However, such results could simply mean that systemic ethanol acts in an offsetting way at sites remote to the recorded neuron.

Another problem that emerges from the iontophoretic data presented in the two previously discussed reports is that all manner of responses have been observed (i.e., some neurons were excited, some were depressed, and some seemed unaffected). Similar variations have been reported with iontophoretic applications of neurotransmitters, such as norephrine.[56] Since neurons in the ethanol studies had generally not been identified, it is possible that these differences reflect true differential effects of ethanol rather than some "pitfall" in electroosmostic technique. There are many technical problems in such a sophisticated technique as microiontophoresis, but these are beyond the scope of this review.

Among other interpretive problems is the possibility that ethanol may not be acting always on the postsynaptic neuron but rather acting presynatpically to alter transmitter release or acting on other nearby interconnected neurons whose impulse activity is not being observed.

3. Tissue Culture

For over 10 years various investigators have been developing techniques for culturing central nervous system neurons under conditions which permit development of histiotypic structural characteristics, high metabolic rates, and complex electrophysiological activity. Most recently, such methods have been used in evaluating ethanol effect on spontaneous nerve impulse activity of mouse cerebellum and cerebral neocortex.[57] Direct application of ethanol to cerebellar neurons caused initial excitation and regularization of impulse discharge, followed by depression of activity when the larger concentrations of ethanol were used. In marked contrast, neocortical neurons only responded with depression at all effective concentrations of ethanol (0.5–3.5%). Rapid washing of both kinds of culture with alcohol-free medium quickly reversed the ethanol actions, sometimes even within 15 min after discharge had been completely supressed.

4. Neuropharmacological Alterations of Transmitter Functions

Another needed approach is to treat intact animals with drugs that have very specific effects on given neurotransmitter systems. This becomes an especially powerful tool when used with neuronal pathways that can be physiologically identified. One example that is totally unexploited in this context is the selective destruction of serotonergic neurons by injection of 5,6-dihydroxytryptamine into the raphe, and the destruction of noradrenergic and terminals by the injection of 6-hydroxydopamine into the locus coeruleus. In acute rat preparations, locus coeruleus neurons are markedly depressed within about 5 min after intraperitoneal injection of ethanol (2 g/kg)[58]; this depression is consistent with previous work from this same laboratory, which showed that ethanol decreased the accumulation of labeled norepinephrine metabolites. Since the locus coeruleus noradrenergic terminals are generally known to be inhibitory, these results provide yet another example, as discussed earlier, of a disinhibition mode of ethanol action.

At a more equivocal level, but one which still could yield important insight, there are many well-accepted pharmacological ways to alter the function of most of the known neurotransmitter systems in the brain. While such strategems could not delineate specific pathways that are especially important in ethanol action, they could at the least identify the key neurotransmitter systems that should be given emphasis in future research. The use of tissue culture or other *in vitro* preparations could prove especially useful.

At the behavioral level, a great deal of this kind of research has been published. However, these kinds of data seem even more difficult to control and interpret than the relatively straightforward results that can be obtained in a properly conducted physiological study of unit activity.

5. Study of Identified Pathways

Perhaps the major problem in studying ethanol effects on impulse activity in the intact brain is the difficulty of sorting out direct from indirect effects.

One approach is to lesion known input pathways to the structure being recorded. One of the obvious procedures is to eliminate visceral and motor feedback input into the brain by transecting the spinal cord. Within the brain, the different nuclei vary enormously in the difficulty required to isolate them from indirect sources of input. The only attempt to do this was the deafferentation study of the cerebellum previously mentioned.[53] The apparent success of this study should encourage further such attempts with other brain areas where this is feasible. In isolated brain areas which show a clear, presumably direct ethanol effect on unit activity, it is feasible to achieve much more reliable and unequivocal data on the effects of pharmacological alterations of transmitter systems contained within that area (provided that blood supply is not grossly altered).

A still more powerful, but difficult to achieve, approach would be to record impulses from single cells at known points in a circuit in which the transmitters used by each cell are known. This is a very realistic possibility in molluscan systems, and even in vertebrates; although the technical problems would be formidable, the strategem is feasible on a limited scale.

6. Stimulus-Evoked Unit Activity

The other major way to distinguish specific from nonspecific effects of ethanol is with evoked-response studies. The value of this approach, recognized for many years in the context of averaged evoked EEG activity, has not been exploited in a unit-activity context. The EEG studies have led to considerable controversy, owing in part to the fact that the physiological bases of the EEG are by no means totally understood; certainly EEG data are much less interpretable than unit-activity data. Moreover, evoked EEG responses do not necessarily parallel nor correlate with evoked unit activity; for one thing EEG voltages are detectable over vastly greater volumes of tissue than are those of extracellularly recorded nerve impulses. The testing of ethanol, even if given systemically, in a paradigm that employs electrical stimulation of known, well-defined paucisynaptic anatomical pathways, seems to be an unusually promising strategem[52,54] which needs to be exploited. Such research could be especially valuable if it incorporated pharmacological alteration of the neurotransmitter systems that funtion within the involved pathways.

E. SUMMARY AND CONCLUSIONS

Research on ethanol effects on nerve impulses spans five decades, and ranges from *in vitro* studies of peripheral nerve and isolated ganglia to studies of spontaneous and evoked *in vivo* unit activity in the intact brain.

The *in vitro* research, often performed with unrealistically high concentrations, suggests that ethanol can (1) decrease impulse amplitude and ultimately block spike electrogenesis, (2) depolarize some neurons and hyperpolarize others, (3) depress excitatory and inhibitory postsynaptic potentials, (4) depress the potentiation of excitatory postsynaptic potentials by tetanic stimulation, and (5) depress cholinergic synaptic transmission.

The *in vivo* research suggests that ethanol (1) is more effective on synaptic function than on impulse propagation, (2) depresses the genesis of both excitatory and inhibitory postsynaptic potentials, (3) affects polysynaptic pathways more than monosynaptic ones, (4) can potentiate presynaptic inhibition, (5) has different actions on the various neurochemical transmitter systems, (6) can increase or decrease impulse discharge of a given neuron, (7) acts differently on neurons in different brain regions, and (8) does not affect neurons equally at the same time.

There are many stratcgems for the study of ethanol that would capitalize on the special utility of nerve impulse monitoring. Such research could provide new insight into many unresolved issues about mechanisms of ethanol action: the roles of various neurotransmitters, acetaldehyde, and biogenic aldehydes, and the processes involved in the genesis of tolerance and dependence. For complete understanding of ethanol mechanisms, biochemical and behavioral observations must be linked with electrophysiological findings. But scientists have not yet exploited the power of the nerve impulse assay for disclosing the mechanisms of ethanol actions.

F. REFERENCES

1. H. A. Davis, A. Forbes, D. Brunswick, and A. M. Hopkins, *Am. J. Physiol.* **76**:448 (1926).
2. J. Kato, *The Theory of Decrementless Conduction in Narcotised Region of Nerve*, Nankodo Co., Tokyo (1924).
3. E. B. Wright, *Am. J. Physiol.* **148**:174 (1947).
4. A. Gallego, *J. Cell. Comp. Physiol.* **31**:97 (1948).
5. J. M. Steinmann, *J. Physiol. (Paris)* **59**:175 (1967).
6. G. A. Montoya, W. K. Riker, and N. J. Russell, *J. Pharmacol. Exp. Ther.* **200**:320 (1977).
7. L. Quevedo, J. Baldeig, and J. Concha, *Pharmacology* **14**:148 (1976).

8. C. M. Armstrong and L. Binstock, *J. Gen. Physiol. (Lond.)* **48**:265 (1964).

9. J. W. Moore, W. Ulbricht, and M. Takata, *J. Gen. Physiol. (Lond.)* **48**:265 (1964).

10. D. J. Houck, *A. J. Physiol.* **216**:364 (1969).

11. J. L. Barker and H. Gainer, *Science* **182**:720 (1973).

12. J. L. Barker, *Brain Res.* **92**:35 (1975).

13. R. Chase, *Comp. Biochem. Physiol.* **50C**:37 (1975).

14. M. C. Bergmann, M. R. Klee, and D. S. Faber, *Pfluegers Arch.* **348**:139 (1974).

15. P. B. J. Woodson, M. E. Traynor, W. T. Schlapfer, and S. H. Barondes, *Nature (Lond.)* **260**:797 (1976).

16. M. F. Traynor, P. B. J. Woodson, W. T. Schlapfer, and S. H. Barondes, *Science* **193**:510 (1976).

17. R. F. Mayer and R. Garcia-Mullin in: *The Biology of Alcoholism* (B. Kissin and H. Begleiter, eds.), Vol. 2: *Physiology and Behavior*, pp. 21–66, Plenum Press, New York (1972).

18. D. S. Faber and M. R. Klee, *Brain Res.* **104**:347 (1976).

19. R. G. Grenell, in: *The Biology of Alcoholism*, (B. Kissin and H. Begleiter, eds.), Vol. 2: *Physiology and Behavior*, pp. 1–20, Plenum Press, New York (1972).

20. H. Kalant, *Fed. Proc. Symp.* **34**:1930 (1975).

21. F. Eidelberg, in: *Research Advances in Alcohol and Drug Problems* (R. J. Gibbins, Y. Isreal, H. Kalant, R. E. Popham, W. Schmidt, and R. G. Smart, eds.), pp. 147–176, John Wiley & Sons, Inc., New York (1975).

22. N. Lake, G. G. Yarbrough, and J. W. Phillis, *J. Pharm. Pharmacol.* **25**:582 (1973).

23. M. R. Klee, K. C. Lee, and M. R. Park, *Pfluegers Arch.* **355**:R85 (1975).

24. J. W. Dundee, *Anesth. Analg.* **49**:467 (1970).

25. A. Staiman and P. Seeman, *Can. J. Physiol. Pharmacol.* **52**:535 (1974).

26. H. Kalant, in: *Alcohol Intoxication and Withdrawal: Experimental Studies* (M. M. Gross, ed.), Vol. I, pp. 3–14, Plenum Press, New York (1973).

27. J. L. Story, E. Eidelberg, and J. D. French, *Archiv. Neurol.* **5** (1961).

28. C. Kornetsky, *Pharmacology: Drugs Affecting Behavior* pp. 149–163, John Wiley & Sons, Inc., New York (1976).

29. E. Eidelberg and D. F. Wooley, *Arch. Int. Pharmacodyn. Ther.* **185**:388 (1970).

30. E. K. Sauerland, T. Knauss, and C. D. Clemente, *Brain Res.* **6**:181 (1967).

31. D. S. Faber and M. R. Klee, in: *Alcohol and Opiates* (K. Blum, ed.), pp. 41–63, Academic Press, Inc. New York.

32. N. R. Banna, *Experientia* **25**:619 (1969).

33. J. T. Miyahara, D. W. Esplin, and B. Zablocka, *J. Pharmacol. Exp. Ther.* **154**:119 (1966).

34. R. A. Davidoff, *Arch. Neurol.* **28**:60 (1973).

35. E. K. Sauerland, N. Mizuno, and R. M. Harper, *Exp. Neurol.* **27**:476 (1970).

36. E. Eidelberg, M. L. Bond, and A. Kelter, *Arch. Int. Pharmacodyn. Ther.* **192**:213 (1971).

37. M. J. Wayner, T. Ono, D. Nolley, and A. De Young, in: *Recent Studies of Hypothalamic Function* (K. Cooper, K. Lederis, and W. Veale, eds.), pp. 232–250, S. Karger, Basel (1974).

38. M. J. Wayner, D. Gawronski, and C. Roubie, *Physiol. Behav.* **6**:747 (1971).

39. W. R. Klemm and R. E. Stevens, III, *Brain Res.* **70**:361 (1974).

40. W. R. Klemm, C. G. Mallari, L. R. Dreyfus, J. C. Fiske, E. Forney, and J. A. Mikeska, *Psychopharmacologia* **49**:235 (1976).

41. W. R. Klemm, L. R. Dreyfus, E. Forney, and M. A. Mayfield, *Psychopharmacologia* **50**:131 (1976).

42. J. Meyer-Lohmann, R. Hagenah, C. Hellweg, and R. Benecke, *Naunyn-Schmiedebergs Arch. Pharmakol.* **272**:131 (1972).

43. M. J. Wayner, T. Ono, and D. Nolley, *Pharmacol. Biochem. Behav.* **3**:499 (1975).

44. M. G. Larrabee and J. M. Pasternak, *J. Neurophysiol.* **15**:91 (1952).

45. P. W. Gage, *J. Pharmacol. Exp. Ther.* **150**:236 (1965).

46. K. Okada, *Jpn. J. Physiol.* **17**:245 (1967).

47. G. M. Kolmodin, *Acta Physiol. Scand.* **29**, (Suppl. 106:)530 (1953).

48. M. J. Wayner, *Ann. N.Y. Acad. Sci.* **215**:13 (1973).

49. G. G. Yarbrough, N. Lake, and J. W. Phillis, *Brain Res.* **67**:77 (1974).

50. P. Seeman, M. Chau, M. Goldberg, T. Sauks, and L. Sox, *Biochim. Biophys. Acta* **225**:185 (1971).
51. D. H. Ross, B. C. Kilber, and H. L. Cardenas, *Drug Alcohol Depend.* **2**:305 (1977).
52. A. C. Bäckstrom, *Phys. Norv.* **7**:181 (1974).
53. E. Forney and W. R. Klemm, *Res. Commun. Chem. Pathol. Pharmacol.* **15**:801 (1976).
54. W. R. Folger and W. R. Klemm, in: *Currents in Alcoholism* (F. A. Seixas, ed.), Vol. III, pp. 67–84, Grune & Stratton, Inc., New York (1978).
55. R. D. Myers, in: *Drug and Chemical Stimulation of the Brain,* 1st ed., pp. 106, 398, 401, 556, Van Nostrand Reinhold Company, New York (1974).
56. F. E. Bloom, *Life Sci.* **14**:1819 (1974).
57. F. J. Seil, A. L. Leiman, M. M. Herman, and R. A. Risk, *Exp. Neurol.* **55**:390 (1977).
58. L. A. Pohorecky and J. Brick, *Brain Res.* **131**:174 (1977).

Tolerance to a Specific Synaptic Effect of Ethanol in *Aplysia*

16

M. Elaine Traynor, Werner T. Schlapfer,
Paul B. J. Woodson, and Samuel H. Barondes

A. INTRODUCTION

In this chapter we summarize our recent experiments describing a specific neurophysiological effect of ethanol at the cellular level and the subsequent tolerance to this effect following repeated exposures to ethanol. Ethanol selectively accelerates the rate of decay of posttetanic potentiation (PTP) at an identified and presumably cholinergic synapse (called RC1-R15) in the abdominal ganglion of the marine mollusk *Aplysia californica*.

The central nervous system of *Aplysia* is particularly well suited to studies of the effects of pharmacological agents on synaptic transmission for the following reasons:

1. Many of the neuronal cell bodies are identifiable by their characteristic locations, sizes, pigmentations, and physiological properties.[1] Thus, the same individual cells and synapses can be studied in different preparations.

2. The neuronal soma are large (up to 700 μm), so prolonged intracellular recordings of synaptic and action potentials are quite feasible.

3. The nervous system is aggregated into ganglia, which can be isolated from the animal and maintained in organ culture for many days with little change in physiological or biochemical parameters.[2] Thus, the neurons may be easily exposed to pharmacological agents for prolonged periods while assaying the neurophysiological actions of these agents.

M. Elaine Traynor, Werner T. Schlapfer, Paul B. J. Woodson, and Samuel H. Barondes • Department of Psychiatry, University of California at San Diego, San Diego, California 92093, and Veterans Administration Medical Center, San Diego, California 92161.

B. MODIFIABLE SYNAPTIC TRANSMISSION IN *APLYSIA*

We have been studying a particular synapse, RC1-R15, made by an axon in the right visceropleural connective onto cell R15 in the abdominal ganglion (Fig. 1). Stimulation of the right visceropleural connective with a suction electrode within a critical range of stimulus intensities and stimulus durations evokes a large, unitary, and monosynaptic, excitatory postsynaptic potential (EPSP) in cell R15.[3] For convenience in studying this EPSP, we hyperpolarize cell R15 so as to suppress the spontaneous oscillations of the membrane potential and the concomitant bursts of action potentials. The amplitude of the EPSP produced by right connective stimulation is strongly influenced by the pattern of stimulation of the presynaptic axon. If stimuli are delivered at frequencies greater than one pulse per 2 sec, the EPSP shows facilitation (i.e., it increases in size). Subsequent to a train of stimuli, the amplitude of test EPSPs is elevated above that of isolated EPSPs for some time (Fig. 2). This phenomenon, called posttetanic

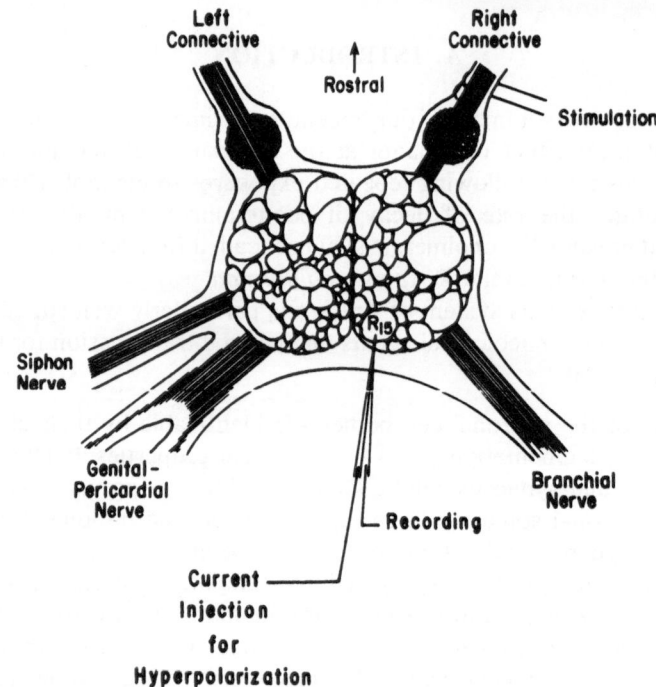

Fig. 1. Schematic representation of the abdominal ganglion of *Aplysia californica,* showing the neuronal cell bodies as seen through the transparent sheath of the ganglion. The diameter of the soma of the cell R15 is about 300 μm. The right connective was stimulated with a suction electrode within a critical range of intensities[3] to evoke unitary, monosynaptic, excitatory postsynaptic potentials (EPSPs) in cell R15. One barrel of an intracellular double-barreled microelectrode was used to record the EPSPs while the other barrel was used to hyperpolarize the cell to about −100 mV in order to suppress spontaneous action potentials in cell R15.

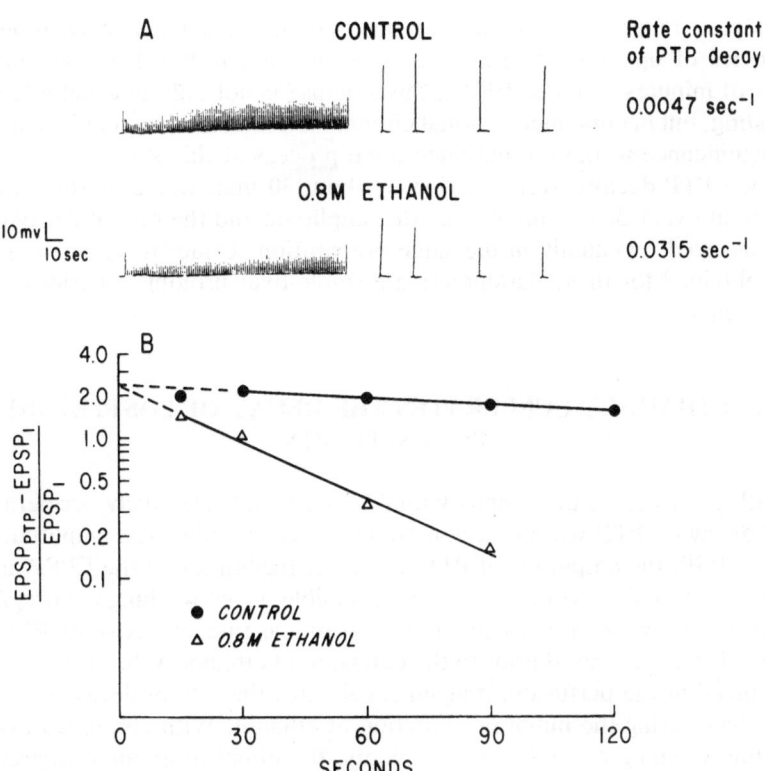

Fig. 2. Acceleration of the rate of decay of PTP by ethanol. (A) Chart recording of EPSPs during a train of stimuli (100 pulses at 1/sec) followed by test pulses ($EPSP_{PTP}$) at 15, 30, 60, and 90 sec, after the end of the train to assay the amplitude and duration of PTP. (B) Decay of $EPSP_{PTP}$ with a single exponential time course toward the size of the control EPSP ($EPSP_1$). A plot of log [($EPSP_{PTP}$ − $EPSP_1$)/$EPSP_1$] versus time gives a straight line. The zero-time intercept of this line is a measure of the amplitude of PTP, while its slope represents the rate constant of decay of PTP. In the presence of 0.8 M ethanol, the rate of PTP decay is accelerated although the amplitude of the PTP remains unchanged. [From Woodson et al.[6]; by permission of Macmillan Journals, Ltd.]

potentiation (PTP), decays with a single exponential time course such that the test EPSPs asymtotically approach the amplitude of an isolated EPSP (Fig. 2). Plots like that in Fig. 2 allow the quantification of PTP in terms of its rate of decay (defined as the slope of the plot) and its amplitude (defined as the extrapolated zero-time intercept and given as the change in the EPSP amplitude upon repetitive stimulation).[4]

As can be seen from Fig. 2, the EPSP amplitude at the RC1-R15 junction is unusually large. The amplitude and duration of the PTP at this junction, following relatively short trains of stimuli at relatively low frequencies, are also quite large compared to that observed in other systems. This PTP occurs under conditions of normal ionic medium composition and normal temperatures, and the frequency of stimulation and the length of the stimulus trains are well within the physiological range. In the intact nervous system this synapse fires in bursts at frequencies of up to seven per second, each burst lasting from seconds to a number of minutes.[5] Thus, PTP at this synapse is not only unusually large and long lasting, but occurs under normal circumstances in the intact animal and may be of significance to the normal integrative process at this synapse.

Since PTP decays over a period of about 30 min, trains of stimuli can be given about every 30–45 min so that the amplitude and the rate of decay of PTP can be assayed repeatedly in the same preparation. Under these conditions the values obtained for these parameters are stable over prolonged periods in most preparations.

C. ETHANOL ACCELERATES THE DECAY OF POSTTETANIC POTENTIATION

Bath perfusion of the ganglia with 0.8 M ethanol selectively accelerates the rate of decay of PTP without a consistent effect on either the amplitude of an isolated EPSP, the amplitude of PTP, or on the facilitation of the EPSP during a train (Fig. 2).[6] This effect of ethanol is reversible, since washing the preparation with normal seawater results in the return of the rate of decay of PTP to its control value as measured prior to the exposure to ethanol. When ethanol is then reintroduced in the perfusate, it again accelerates the rate of decay of PTP, but less so than during the initial presentation of ethanol. With continued cycles of alternating washing and ethanol treatments, the effect of ethanol progressively diminishes [i.e., the preparation becomes tolerant to the effects of ethanol (Fig. 3)].[7] This tolerance is not due to any kind of deterioration of the preparation, since in the tolerant state all parameters of synaptic transmission are identical to that of the control state before any ethanol was administered. The only difference is that now the rate of decay of PTP is unchanged upon exposure to 0.8 M ethanol (Fig. 4).

In order to formulate a working hypothesis as to the site and mechanism of action of ethanol and the development of tolerance to ethanol at the RC1-R15 synapse, it is necessary to review what is known about the synaptic site and mechanism of PTP. At the RC1-R15 synapse in *Aplysia,* as well as at various neuromuscular junctions, PTP has been shown to be a presynaptic phenomenon (i.e., PTP is due to an increase in transmitter released per presynaptic spike).[4] This increase in transmitter release is accomplished by an increase in the fraction of the available pool of transmitter released by a spike rather than by an increase in the size of the pool of available transmitter.[4] In the light of the exocytotic hypothesis of transmitter release, this implies that PTP is a tetanus-dependent increase in the efficiency of spike-induced fusion of synaptic vesicles with presynaptic terminal membrane. PTP is thus a membrane process and the decay of PTP may represent a progressive relaxation of some critical components of the presynaptic terminal membrane back to their initial unperturbed state.

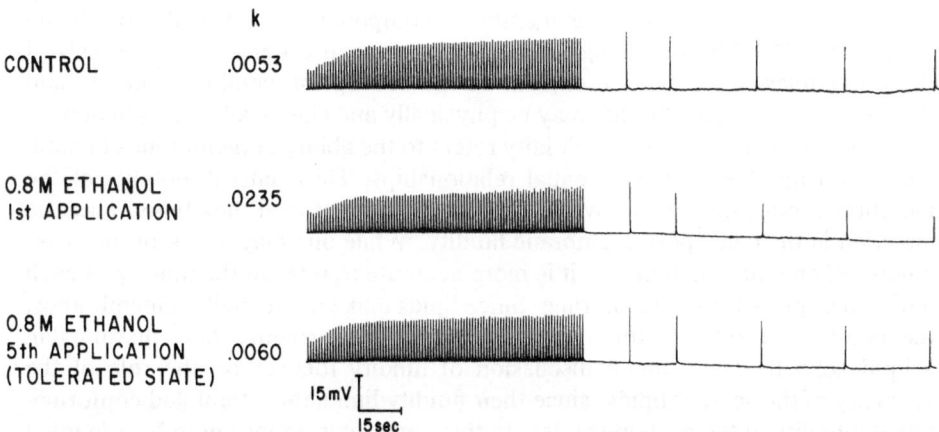

Fig. 3. Development of tolerance to repeated application of ethanol. Intracellular records of the EPSPs in cell R15 during and after repetitive stimulation similar to Fig. 2A. The rate constant of decay of PTP, *k*, was determined from the slope of plots similar to Fig. 2B. Top trace: control in artificial seawater. Middle trace: 30 min after the first application of 0.8 M ethanol in artificial seawater. The rate of decay of PTP is accelerated. Ten minutes after this train, the preparation was washed in artificial seawater for 120 min before the next application of ethanol. Bottom trace: 30 min after the fifth application of 0.8 M ethanol. The rate of decay of PTP here is similar to that in the control preparation. [From Traynor *et al.*[7]; by permission of the AAAS.]

D. MEMBRANE FLUIDITY IMPLICATED IN THE REGULATION OF DECAY OF POSTTETANIC POTENTIATION

Ethanol, through its well-documented effect of altering membrane fluidity,[8,9] may accelerate the rate of PTP decay by expediting the relaxation of the tetanus-induced state of organization of the critical presynaptic membrane components. Thus, we are proposing two hypotheses: (1) PTP decay is affected by

Fig. 4. Changes in the rate constant of PTP decay with repeated application of 0.8 M ethanol. Stimulus patterns identical with the ones shown in Fig. 3 were given every 40 min. After three control trains in artificial seawater, 0.8 M ethanol was introduced and a single train was begun 30 min later. The preparation was then washed by perfusion with artificial seawater and the procedure was repeated. The rate constant of PTP decay was determined from plots similar to those in Fig. 2B. Each data

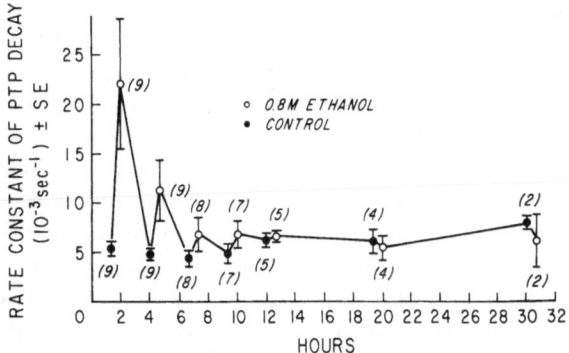

point represents the mean for the number of experiments, which is shown in the parentheses; error bars represent the standard error. [From Traynor *et al.*[7]; by permission of the AAAS.]

the fluidity of critical presynaptic membrane components, and (2) ethanol acts on the decay of PTP by fluidizing these components. In order to develop tests of these hypotheses, we need to expand on the concept of membrane fluidity and describe ways in which fluidity may be physically and chemically manipulated.

The concept of membrane fluidity refers to the ability of membrane constituents to change their relative spatial relationships. Thus, lateral mobility of the membrane constituents as well as their conformational flexibility may be included in the concept of membrane fluidity. While one may speak of the gross fluidity of an entire membrane, it is more accurate to refer to the fluidity of each individual species in the membrane. Since lipids make up the bulk of membranes, the membrane is often pictured as being composed of protein solutes dissolved in a lipid solvent. Thus, much discussion of fluidity focuses on the role of the viscosity of the solvent lipids, since their fluidity limits the lateral and conformational mobility of the protein solutes. In this regard it is important to bear in mind that each individual protein may be surrounded by unique lipids, which are likely to be different both from the general lipid composition of the membrane and from the lipids surrounding other proteins. Therefore, when agents or manipulations that alter membrane fluidity are introduced, different microenvironments may respond to such agents to different degrees and possibly even in opposite directions with respect to fluidization or rigidification.[10,11]

Ethanol has long been regarded a membrane-active agent, owing to its high solubility in both aqueous and lipid environments. It is generally thought that enthanol dissolves in the lipid portion of the membranes. Such solvation should result in alterations in the forces between membrane molecules, changing the fluidity in each patch. The change may be an increase or a decrease in the fluidity, depending on the composition of the patch. Since we postulate that ethanol accelerates the relaxation of a stimulus-induced alteration in a membrane state, we propose that ethanol fluidizes the membrane patch relevant to PTP decay.

If this is the case, then higher-chain aliphatic alcohols, which partition more readily in the lipid phase of the membrane, should be more potent in increasing the rate of decay of PTP. Our results confirm this expectation (Fig. 5). Aliphatic alcohols (C_2–C_8) increase the rate of decay of PTP with a potency correlating with their water/octanol partition coefficient.[6]

Membrane fluidity also has a strong temperature dependence (Fig. 6). In fact, many systems exhibit abrupt changes in their properties at well-defined temperatures, corresponding to temperature-induced phase transitions in the patches of membrane pertinent to the function being measured.[12] Indeed, the rate of decay of PTP decreases with decreasing temperature but shows a sharp transition when the temperature is lowered from 12 to 10°C.[13] In terms of the membrane fluidity hypothesis of the rate of decay of PTP, this would correspond to a "freezing" of the state of organization of these critical presynaptic membrane components involved in PTP. If the preparation is kept at 10°C for more than 4 hr, the temperature at which the transition will occur is shifted to a lower

Fig. 5. Potency of aliphatic alcohols in increasing the rate of PTP decay. (A) Dose dependencies of the rate constant of PTP decay in the presence of a series of aliphatic alcohols. The length of the carbon chain is indicated by the subscript of C (C_2 is ethanol; C_8 is octanol). Each point represents the first exposure of any preparation to a drug, to avoid complications owing to the development of tolerance. Alcohols with longer carbon chains are more potent. Note the steep increase in potency over a narrow range of concentrations. (B) The concentration of alcohol needed to accelerate the decay of PTP by a factor of 4 (i.e., the potency of the alcohol) is directly proportional to the octanol/water partition coefficient for that alcohol. The octanol/water partition coefficient of an agent is an index of its lipophilicity. The data points for part B were obtained by interpolating from part A the concentration of each alcohol which would give a constant rate of decay of PTP of 20×10^{-3} sec^{-1}.[6] [From Woodson et al.[6]; by permission of Macmillan Journals, Ltd.]

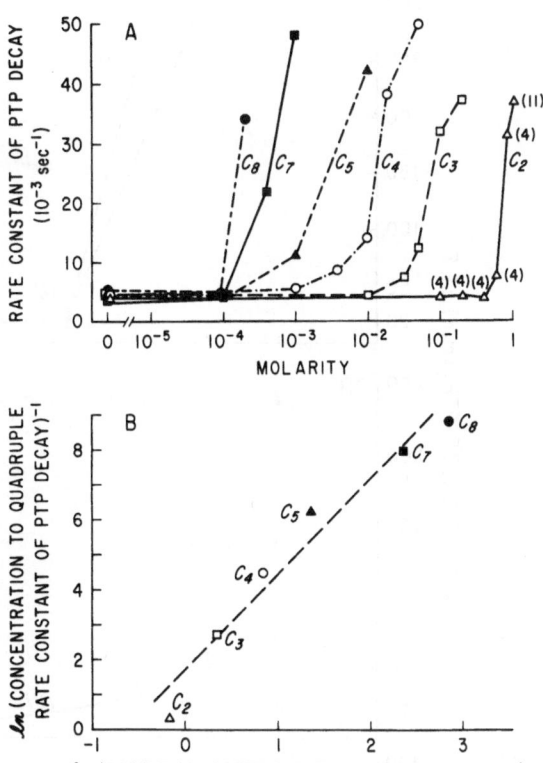

temperature. In other words, the preparation adapts to the lower ambient temperature.

Furthermore, we have found that in the presence of ethanol, no abrupt transition in the duration of the PTP is observed upon lowering the temperature from 12 to 10°C.[13] This suggests that ethanol is acting as an "antifreeze" by preventing the temperature transition from taking place.

E. TOLERANCE TO ETHANOL AS AN ADAPTIVE RESPONSE TO A PERTURBATION IN MEMBRANE FLUIDITY

If ethanol accelerates the rate of decay of PTP by increasing the fluidity of critical presynaptic membrane components, the development of tolerance to the effects of ethanol could be the result of (1) a compensatory change in the structure of the presynaptic membrane (e.g., its lipid composition) which renders it chronically less fluid, (2) a chronic change in the composition of the presynaptic membrane which renders it resistant to fluidization by ethanol, or (3) the development of a mechanism which allows much more rapid compensation to the fluidizing effects of ethanol.

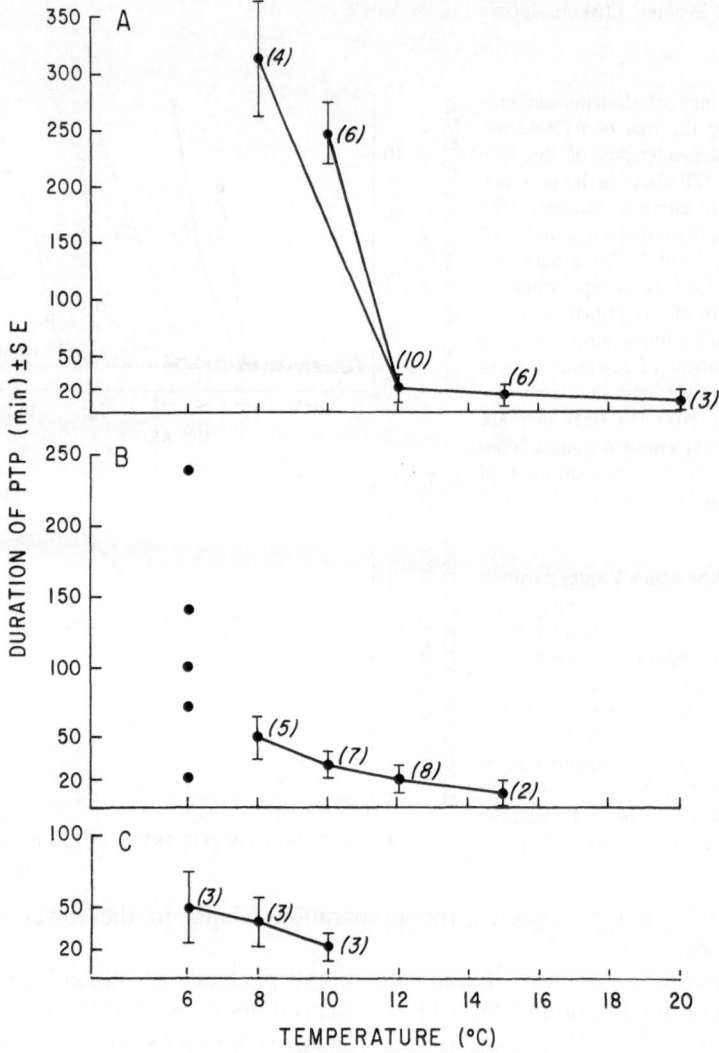

Fig. 6. Marked prolongation of PTP at a transition temperature and its adaptation. Following a train of stimuli, test pulses were administered at 10-min intervals until the size of the EPSP returned to the size of the first EPSP of the train. This time was taken as the duration of PTP. The number of preparations at each temperature is indicated in parentheses. (A) Pooled data from 10 preparations. All 10 preparations were tested at 12°C; 6 were also tested at 15°C and 3 at 20°C. In each preparation the temperature was lowered from either 12°C to 10°C (6 preparations) or from 12°C to 8°C (4 preparations). (B) Pooled data from animals which were maintained at 11°C rather than the usual 20°C for 2 days prior to the experiment. In contrast to the upper figure, the duration of PTP was only slightly prolonged when the temperature was lowered from 12°C to 10°C or 8°C, but in only some preparations did the abrupt prolongation of PTP finally occur at 6°C. The data from the individual preparations are shown at this temperature instead of the average. In no case did any of the preparations treated this way show a temperature transition between 12°C and 10°C. (C) The ganglia from three animals were removed and maintained at 10°C for 4 hr before stimulation. Under these circumstances the duration of PTP was short compared with control preparations (A) and no abrupt prolongations of PTP occurred when the temperature was lowered to 6°C. [From Schlapfer et al.[13]; by permission of Macmillan Journals, Ltd.]

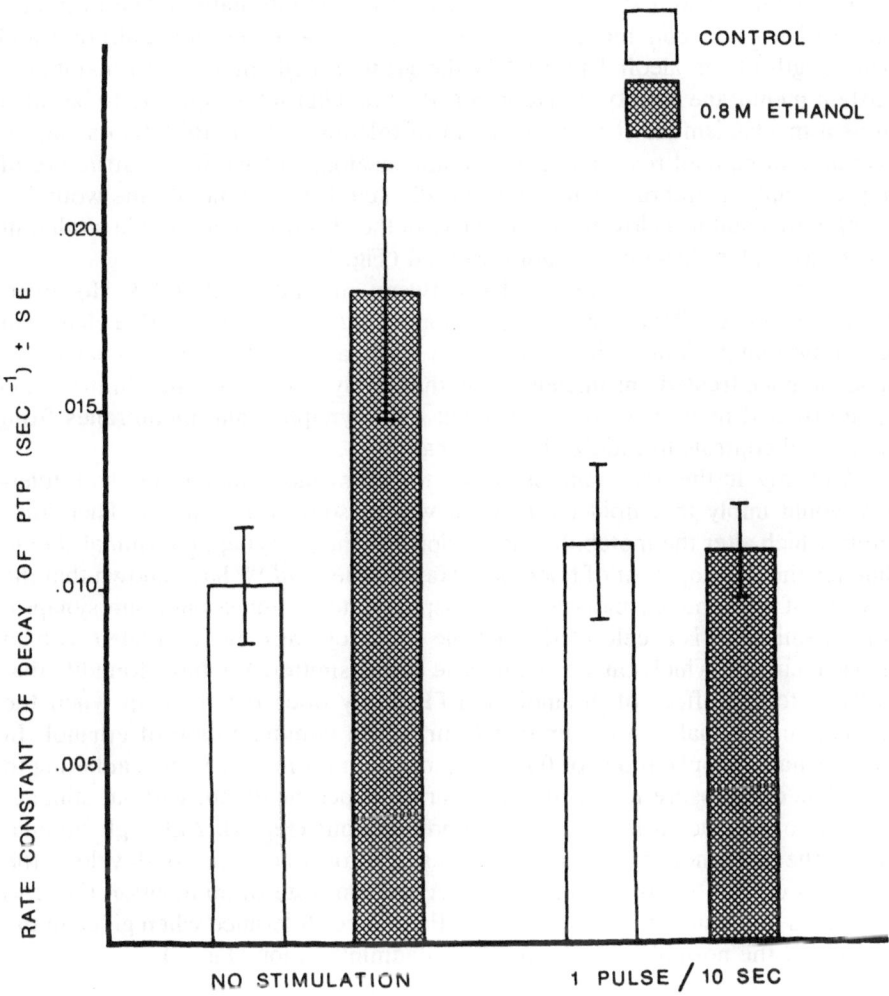

Fig. 7. Stimulation in the presence of ethanol is required for the development of tolerance. All preparations were alternately exposed to 0.8 M ethanol in artificial seawater for 45 min followed by artificial seawater alone for 45 min, such that they were each exposed to ethanol five times for 45 min at a time for a total of 225 min in the presence of ethanol. Some preparations ($n = 5$, right pair of bars) received continuous stimulation of the right connective at a rate of one pulse per 10 sec; others ($n = 10$, left pair of bars) received no stimulation. After the five alternate exposures to ethanol and ethanol-free seawater, each preparation received 100 stimuli at a rate of 1/sec and the rate constant of PTP decay was determined by administration of test pulses at 30-sec intervals. The rate constants determined in this way are indicated on the graph as CONTROL. Forty-five minutes after the control train the preparation was perfused with 0.8 M ethanol and identical train and test pulses were given again to determine the rate constant of PTP decay. Tolerance was evident only when the prior exposures to ethanol were accompanied by stimulation. Without stimulation, the administration of ethanol still significantly increased ($p < 0.05$) the rate constant of PTP decay.[17] [From Traynor *et al.*[17]; by permission of Elsevier Sequoia S.A.]

In bacteria there exists a precedent for the first alternative. The fatty acid composition of *E. coli* membranes varies according to the concentration and chain length of the alcohol present in the growth medium so as to maintain a constant membrane fluidity.[14] However, the first alternative appears to be ruled out as a mechanism for the development of tolerance at the RC1-R15 synapse. Tolerance to ethanol resulting from a compensatory change in the structure of the presynaptic membrane which chronically stabilizes the membrane would be expected to result in a slower rate of decay in the absence of ethanol in a tolerant preparation. This, however, is not observed (Fig. 2).

At present we cannot distinguish between alternatives 2 and 3. However, Chin and Goldstein[15] have provided evidence that is consistent with a change in membrane composition. They reported a "tolerance" of synaptosomal membranes in mice treated chronically with ethanol. Synaptosomal membranes from ethanol-treated mice were more resistant than synaptosomal membranes from sucrose-fed controls to fluidization by ethanol.

A change in the lipid composition of the presynaptic membrane with tolerance would imply that lipid metabolism was in some way altered. Therefore, factors which alter the metabolic rate of lipids in the presynaptic terminal should influence the development of tolerance. Hawthorne *et al.*[16] have shown that the turnover of phosphotidylinositol and phosphatidate in guinea pig brain synaptosomal membranes is accelerated when the synaptosomes are stimulated with an electrical current which causes the release of transmitter. We have found[17] that tolerance to the effect of ethanol on PTP decay does not develop when the presynaptic terminal is not stimulated during the administration of ethanol. In fact, alternating applications of 0.8 M ethanol with normal seawater, adhering to a schedule of exposure identical to our earlier experiments but without stimulation, did not produce tolerance in these preparations (Fig. 7). Although stimulation in the presence of ethanol is necessary for tolerance to develop, the generation of PTP is not a prerequisite. A uniform rate of stimulation ($1/10$ sec) which does not result in facilitation or PTP produces tolerance when given in the presence of the normal regimen of ethanol administration (Fig. 7).

F. SUMMARY AND CONCLUSIONS

In summary, our evidence suggests the following:

1. The rate of decay of PTP may be regulated by the fluidity of critical membrane components of the presynaptic terminal.

2. Ethanol accelerates the rate of PTP decay, presumably by fluidization of these membrane components.

3. Tolerance of the preparation to this effect of ethanol is either the result of a change in the membrane composition which renders the critical membrane components resistant to fluidization by ethanol or the development of a more rapid adaptive compensation to the fluidizing effects of ethanol.

G. REFERENCES

1. E. Kandel, *Cellular Basis of Behavior: An Introduction to Behavioral Neurobiology*, pp. 225–233, W. H. Freeman and Company, Publishers, San Francisco (1976).
2. S. A. Dewhurst and D. Weinreich, *J. Neurobiol.* **5:**21 (1974).
3. W. T. Schlapfer, P. B. J. Woodson, J. P. Tremblay, and S. H. Barondes, *Brain Res.* **76:**267 (1974).
4. W. T. Schlapfer, J. P. Tremblay, P. B. J. Woodson, and S. H. Barondes, *Brain Res.* **109:**1 (1976).
5. P. B. J. Woodson and W. T. Schlapfer, 7th Ann. Meet. Soc. Neurosci., Abstr. (1977).
6. P. B. J. Woodson, M. E. Traynor, W. T. Schlapfer, and S. H. Barondes, *Nature (Lond.)* **260:**797 (1976).
7. M. E. Traynor, P. B. J. Woodson, W. T. Schlapfer, and S. H. Barondes, *Science* **193:**510 (1976).
8. P. Seeman, *Pharmacol. Rev.* **24:**583 (1972).
9. P. Seeman, *Experientia* **30:**759 (1974).
10. S. H. Roth and L. Spero, *Can. J. Physiol. Pharmacol.* **54:**35 (1976).
11. K. W. Miller and K. Y. Y. Pang, *Nature (Lond.)* **263:**253 (1976).
12. S. Ohnishi, in: *Advances in Biophysics* (M. Kotani, ed.), pp. 35–82, University Park Press, Baltimore (1976).
13. W. T. Schlapfer, P. B. J. Woodson, G. A. Smith, J. P. Tremblay, and S. H. Barondes, *Nature (Lond.)* **258:**623 (1975).
14. L. O. Ingram, *J. Bacteriol.* **125:**670 (1976).
15. J. H. Chin and D. B. Goldstein, *Science* **196:**684 (1977).
16. J. N. Hawthrone, J. E. Bleasdale, and M. R. Pickard, in: *Function and Metabolism of Phospholipids in the Central and Peripheral Nervous System* (G. Porcellati, L. Amaducci, and C. Galli, eds.), pp. 199–209, Plenum Press, New York (1975).
17. M. E. Traynor, P. B. J. Woodson, W. T. Schlapfer, and S. H. Barondes, *Drug Alcohol Depend.* **2:**361 (1977).

Interaction of Ethanol with Drugs   XII

Interaction of Ethanol with Other Drugs 17

Richard A. Deitrich and Dennis R. Petersen

A. INTRODUCTION

There are two major reasons for studying the interaction of ethanol with other compounds. One is to discover and define interactions that are of therapeutic importance; the other is to use the interactions to study the mechanism of action of ethanol on one system or another. The literature in these areas is enormous. In the volume, "Interaction of Alcohol and Other Drugs", an annotated bibliography published by the Addiction Research Foundation, there are 15,000 references. The older references are largely of therapeutic importance, whereas many studies in recent years have utilized interactions as tools to dissect the intimate details of ethanol action. Our purpose here is to give an overview of the area and provide an entrance to the literature. To completely and critically review the entire area would require a very large volume indeed. We shall confine our remarks to the acute actions of ethanol.

A large number of drugs interact with ethanol in a variety of ways. There are purely metabolic interactions in which the drug interferes with the metabolism of ethanol or vice versa. There are drug effects that are additive or synergistic with those of ethanol. Drugs with one or more of these effects are found among the depressants, analgesics, antihistamines, oral antidiabetics, sedatives, anesthetics, antibiotics, and so on. Antagonism of the actions of ethanol is also possible and the search for a "sobering up" drug (amesthetic agent) has received much attention.

We shall adhere to the general principle that a compound that shares some of

Richard A. Deitrich • School of Medicine, University of Colorado, Denver, Colorado 80262. **Dennis R. Petersen** • School of Pharmacy and Insitute for Behavioral Genetics, University of Colorado, Boulder, Colorado 80309.

the actions with ethanol can be additive or supra-additive, while the effect of a compound which has a different mechanism of action can be either antagonistic or synergistic with ethanol.

Since specific receptors mediating central nervous system (CNS) effects of ethanol have not been identified, it is expected that the interaction between ethanol and other drugs in the brain will be physiological and not pharmacological in nature. The failure to find ethanol receptors may in part be a result of the reluctance to believe that they exist as identifiable entities. This, in turn, results from the fact that large concentrations of ethanol are necessary for any observable behavioral effect, and any receptor would have to have such a large dissociation constant (K_d) as to be indistinguishable from nonspecific binding. This does not rule out the possibility of finding receptors for ethanol that have K_d values in the ''respectable'' range, and these may mediate some of the effects of ethanol. Indeed, the findings of Ross *et al.*[125] that ethanol has effects on calcium metabolism at low concentrations (1–50 μM) argues for the opposite view. In fact, we know of some relatively specific receptors for ethanol with reasonable K_ds (i.e., alcohol dehydrogenase, catalase, and the cytochrome P_{450} systems that bind ethanol).

There are a number of methodological problems in investigations of ethanol–drug interactions. These have been recently reviewed by Curry.[30] The detailed critical assessment of drug interactions in lower animals is difficult at best, and it is often nearly impossible in humans. A few of the questions which must be approached in studies of this nature are:

1. What are the peak times for effects of a given dose of ethanol and drug A?
2. Does the time course or magnitude of the peak concentrations change for either ethanol or drug A in the presence of each other? These considerations could be of considerable importance for a drug like ethanol that has been reported to have different effects on gross behavior depending on whether the blood alcohol is rising or falling.
3. Do the effects being studied parallel the blood or brain concentration of each drug?
4. What are the dose–response curves for each drug? Obviously, if one is near the top of a dose–response curve for ethanol, no further increase is likely with another drug.
5. Are the interaction effects consistent throughout the dose range of both compounds? In other words, is a compound antagonist at one dose but additive at another dose?

One method to answer some of these questions for compounds with actions like those of ethanol is the use of isobolograms.[30] This is done by choosing some end point to be achieved (e.g., a certain duration of sleep). The dose of ethanol which will achieve that condition in 50% of the animals (ED_{50}) is then determined in the absence and presence of various amounts of the second drug. The resultant data are plotted as illustrated in Fig. 1. To be maximally useful such data must be accompanied by determinations of blood levels of the various agents.

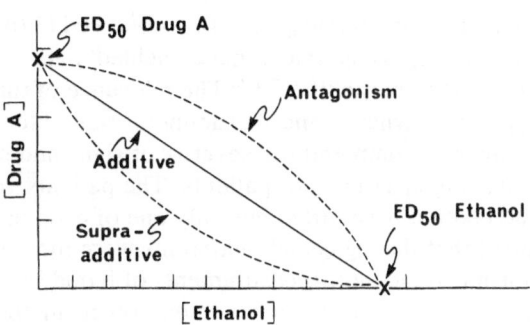

Fig. 1. Isobologram for effects of ethanol with other drugs. An end point is chosen which is shared between ethanol and the second drug. The ED_{50} for both drugs is determined. The ED_{50} for ethanol in the presence of different doses of the second drug is then determined.

B. DEPRESSANTS

Virtually all central nervous system depressants (hypnotics, narcotic analgesics, gaseous anesthetics, antihistamines, etc.) are at least additive with depressant doses of ethanol. This aspect of ethanol–drug interactions is covered in all standard textbooks of pharmacology and reviewed elsewhere.[2,76,93,96]

1. Gaseous Anesthetics

It has been known for many years that ethanol and the gaseous anesthetics have much in common since ethanol, as well as other alcohols, fits well in hydrophobicity–anesthetic potency correlations (Meyer–Overton theory). One would expect that the gaseous anesthetics and ethanol would be additive when both are given acutely, and this is a well-known phenomenon.

Ethanol has a long history of use as an anesthetic agent, although the therapeutic index for such use is small. In recent years it has again been used as an agent for induction of anesthesia.[38] Quantitative studies on the interaction of ethanol and gaseous anesthetics are not numerous in either humans or lower animals, but such studies should provide some answers as to the mechanism of the interaction.

For individuals not acutely under the influence of ethanol, but possessing tolerance to the drug, it is found that more anesthetic is required to achieve anesthesia. This is a result of cross tolerance between the gaseous anesthetics and ethanol.[2]

2. Barbiturates

The additivity or supra-additivity between ethanol and barbiturates has long been known. Among the early studies, those carried out by Dille and Ahlquist[36] are most informative. Results of these experiments indicated that, when rabbits were given both ethanol and sodium pentobarbital, a supra-additive effect was observed. Coefficients of potentiation were reportedly greater (5.8) at a dose of

10 mg pentobarbital/kg than that observed (3.0) at 20 mg pentobarbital/kg. Similar studies, using mice, have yielded data which are consistent with those obtained from rabbits.[77,130] The relevance of this drug interaction becomes more apparent when one examines the literature dealing with humans. Graham[50] compared the severity of signs and symptoms from acute barbiturate poisoning in 11 pairs of patients. The patients were paired on the basis that each had taken a barbiturate but only one of each pair had also drunk alcohol. It was noted that the signs and symptoms were most severe in those patients who had consumed alcohol. Measurements of blood barbiturate concentrations indicated that there was no significant difference in the blood levels between patients ingesting only barbiturates and those who had consumed both barbiturates and alcohol. Since blood enthanol determinations were not performed simultaneously with blood barbiturate concentrations, data from this study become only suggestive of the supra-additive nature of the interactions between these two agents.

Teare[143] has presented some very striking data concerning blood concentrations of barbiturates and ethanol in various groups of overdose patients. In 155 patients dying from an overdose of only barbiturates, the mean barbiturate concentration in blood was 3.67 μg/ml. The mean barbiturate concentration blood levels in 30 deaths resulting from a combination of alcohol and barbiturate overdose was 2.55 μg/ml. Blood ethanol levels in these 30 patients averaged 175 mg/dl. This blood alcohol content is drastically lower than the lethal levels of alcohol (500–800 mg/dl) which commonly appear in the literature.[123] These figures are in good agreement with more recent figures presented by Gupta and Kofoed[53] and Milner.[104]

It is interesting to note that, even though these two agents display such a supra-additive action, a person dependent on alcohol, who is not under the direct influence of alcohol, requires very large doses of barbiturates and other anesthetics to produce surgical anesthesia.

In an attempt to explain the supra-additive effect of these two drugs, many studies have been conducted to better elucidate the effects of ethanol on metabolism of sedative and anesthetic agents. Some evidence concerning the *in vivo* effect of ethanol on barbiturate metabolism was presented by Schüppel.[131] These experiments demonstrated that ethanol prolonged sleeping time in rats given those barbiturates which are hydroxylated by hepatic microsomes.

The ability of acute ethanol treatment to inhibit hepatic drug-metabolizing enzymes *in vitro* and decrease the rate of drug metabolism *in vivo* has been demonstrated by Rubin *et al.*[126,127] This study is of importance since acute ethanol ingestion by rats (5 g/kg) and humans (1 g/kg) significantly retarded the disappearance of pentobarbital and meprobromate from the blood of these two species. *In vitro* studies utilizing hepatic microsomes obtained from rats indicated that ethanol inhibited the activities of aniline and pentobarbital hydroxylases in addition to the demethylation of aminopyrine and ethylmorphine. Benzpyrene hydroxylase was also inhibited by ethanol. In the case of aniline hydroxylase and aminopyrine demethylase, the inhibition was competive, whereas inhibition of pentobarbital hydroxylase and ethylmorphine demethylase was a mixed type. In this study ethanol was shown to be inhibitory to NADPH cytochrome P_{450} reductase, which is essential to microsomal drug metabolism. This decreased hepatic drug metabolism resulting from acute ethanol treatment has been used to explain in part the increased sensitivity of inebriated persons to the effect of barbiturates and tranquilizers.

It is of interest that chronic ethanol ingestion in humans[128] and rats[126–129] results in an induction of the hepatic microsomal drug-metabolizing enzymes. Under such circumstances it becomes apparent why induction of this microsomal drug-metabolizing system by prior chronic ethanol ingestions results in a concomitant increase in drug clearance from the blood when ethanol is absent.[107]

Inhibition of the hepatic drug-metabolizing enzymes by acute ethanol treatment yields a partial explanation for the supra-additive effects of ethanol and barbiturates. Likewise, induction of hepatic drug-metabolizing enzymes by chronic ethanol consumption partially explains why a sober alcoholic may appear tolerant to barbiturates or other anesthetics.

Undoubtedly, in addition to the metabolic tolerance described above, cross tolerance between ethanol and barbiturates may also be due to alterations in CNS sensitivity. For instance, Kalant et al.[67] and Rubin and Lieber[128] have reported that increased barbiturate metabolism in animals chronically treated with ethanol is quite variable and unpredictable. Further studies by Khanna et al.[72] suggest that the changes in barbiturate sleeping time of rats made tolerant to ethanol are the result of alterations in CNS sensitivity and not a modification of drug dispostion. However, in determining the importance of metabolic tolerance and aberrations in CNS sensitivity in studies of the supra-additive actions of ethanol and barbiturates, additional factors should be considered. Of importance is an individual's genotype, which may be a determining factor in the development of metabolic tolerance and/or CNS sensitivity alterations to ethanol and barbiturates. As suggested by Caldwell and Sever[17], tolerance to shorter-acting barbiturates could be explained by induction of drug-metabolizing enzymes, and, as the length of action of the drug increases, CNS adaptation becomes more important.

3. Marihuana

In humans, combination of marihuana and ethanol is additive[60a] and may have metabolic as well as CNS components.[139a] In rats, ethanol at a dose of 2 g/kg increased the depressant effects of Δ^9-tetrahydrocannabinol.[119a]

C. TRANQUILIZERS

1. Phenothiazines

It is generally assumed that the interaction between phenothiazines and ethanol is at least additive or suppra-additive. Chlorpromazine has been shown to potentiate ethanol narcosis in rats[28] and increase the depressant actions of ethanol on respiration in mice.[8] The studies of Brodie *et al.*[13] and Brodie and Shore[14] also suggest that chlorpromazine–ethanol interactions are supra-additive. Several studies in humans have demonstrated that chlorpromazine potentiates ethanol impairment of motor function.[37,105,135,136,152] However, Liljequist *et al.*[83] have shown that humans treated for 2 weeks with ethanol did not suffer any deleterious effects of memory or learning.

As yet, the mechanisms that may be responsible for the supra-additive actions of phenothiazine and ethanol are not understood. However, it is believed that alterations of alcohol metabolism by chlorpromazine may be a partial explanation. Khouw *et al.*[73] also showed that chlorpromazine treatments result in higher blood alcohol levels in rabbits. The study was also extended to show that chlorpromazine inhibits alcohol dehydrogenase *in vitro*.

Studies designed to demonstrate the *in vivo* inhibition of ethanol metabolism in man by chlorpromazine are limited. Sutherland *et al.*[142] observed that chlorpromazine treatment of humans for 7 days prior to a 1-g-ethanol/kg dose resulted in greater arterial level of ethanol 15–20 min after administration than in individuals not given chlorpromazine. This difference had disappeared by 60 min following ethanol ingestion. Subjects treated with chlorpromazine also displayed significantly higher levels of arterial acetaldehyde 15–20 min after ethanol treatment as compared to the respective control groups. These data indicate that chlorpromazine may have the potential to alter both ethanol and acetaldehyde metabolism in humans.

2. Benzodiazepines

The benzodiazepines—diazepam, chlordiazpoxide, and oxazepam—are of importance when one considers their potential for reactions with ethanol. These drugs are among the most commonly prescribed medications in the United States. Greenblatt and Shader[51] have estimated that the total expenditure for these drugs surpassed $200 million in 1972.

A number of investigators have obtained conflicting results concerning actions between benzodiazepines and ethanol. A survey of the pertinent literature indicates that the benzodiazepines display additive effects with ethanol,[86,98,103] no additivity with ethanol,[12,150,151] and mild antagonist interactions with ethanol.[38,39,63]

The literature prior to 1972 clearly suggests that benzodiazepines have little or no interaction with ethanol. In fact, this literature has led Kissin[74] to suggest that, in addition to the obvious advantages of benzodiazepines, another virtue is the absence of an interaction with ethanol.

Numerous reports have appeared in the more recent literature which indicate that benzodiazepines display additive interaction when taken in combination with ethanol. Linnoila *et al.*[87] administered diazepam (15 mg daily) to human subjects for 2 weeks, or lithium carbonate (adjusted to yield serum concentrations of 0.75 meq/liter) for 2 weeks. Psychomotor skills related to driving were assessed in the absence and presence of 0.5 g ethanol/kg. The combined effects of diazepam and ethanol on psychomotor performance were particularly serious. Lithium did not appear to alter psychomotor performance significantly when taken in combination with ethanol. In fact, a slight antagonsim between lithium and ethanol became apparent in certain performance tests. In a similar experiment, Linnoila et al.[88] found that a 2-week treatment with chlordiazepoxide (10 mg daily) or flupenthixole (0.5 mg daily) impaired psychomotor skills related to driving when given in combination with 0.5 g ethanol/kg. The impairment appeared to be greatest when diazepam and ethanol were taken in combination. Results obtained in a similar study by Palva et al.[113] also indicate that chlordiazepoxide and diazepam in combination with ethanol are deleterious to psychomotor skills related to driving.

Mendelson et al.[98] administered 20 mg flurazepam, 0.25 mg triazolam or placebo, alone or in combination with 0.8 g ethanol/kg. When EEGs were obtained and analzyed 16 hr after administration, there were no residual effects from either drug alone or triazolam combined with ethanol. However, there were residual effects of flurazepam and ethanol together. These studies indicate that the interaction of flurazepam and ethanol is intensive enough to be detected by EEG analysis 16 hr after administration.

The acute effects of oxazepam (10, 20, and 40 mg), diazepam (5, 10, 20, and 40 mg), and methylperone (10, 25, and 50 mg), alone and in combination with ethanol (0.5 ml ethanol/kg), were studied by Molander and Duvhok.[108] They found that the drugs with or without ethanol affected sedation, coordination, and mood in humans. Ethanol alone reportedly had no effect on these parameters; when it was given simultaneously with the drugs, however, the effects of diazepam were markedly increased. The actions of oxazepam and methylperone appeared to be affected to a lesser extent. In this study it is interesting to note that ethanol alone did not influence the parameters studied, even though blood ethanol levels 30 min after administration were reported to be 0.555%. This must have been either an analytical or a typographical error.

A recent study by Hayes et al.[59] has focused on the effects of ethanol on oral diazepam absorption. Diazepam (0.07 mg/kg body wt) was administered alone or with ethanol (30 ml of 50% ethanol) to healthy humans. Plasma diazepam levels were significantly higher at 60, 120, and 240 min if administered with ethanol as opposed to water. These data suggest that, in addition to interactions of diazepam and ethanol in the CNS, ethanol may also interact by enhancing absorption or inhibiting metabolism of diazepam.

3. Meprobamate

It is quite clear that meprobamate and ethanol taken together results in an increased depressant effect in humans as detected by performance tasks.[49,109,152] However, it has not been determined if this action is associated with alterations in CNS sensitivity to either of the drugs or ethanol inhibition of microsomal enzyme systems involved in meprobamate disposition.

The studies of Khan et al.[71] with rats and of Seidel and Soehring[131a] with dogs indicate that meprobamate administered with ethanol does not significantly alter the ethanol metabolism rate. In these two studies, blood levels of meprobamate were not reported. Therefore, it is difficult to determine if ethanol significantly altered meprobamate metabolism rates.

Ugarte et al.[149] administered 1.2 g of meprobamate daily for 9–13 days to humans classified as alcoholics who had abstained from alcohol for a minimum of 2 weeks. Individuals receiving the same dose of meprobamate over the same period of time and not classified as alcoholics served as the control group. When subjected to a challenge dose of ethanol (1.0 g/kg), alcoholic and nonalcoholic individuals displayed nearly the same blood ethanol disappearance rates.

Studies which have focused on the influence of ethanol on meprobamate disposition have yielded interesting results. In rat liver slices, 50 mM and 100 mM ethanol have been shown to inhibit meprobamate metabolism by 40% and 60%, respectively.[127] In this same study, human volunteers were given 1 g ethanol/kg 14 hr after receiving 12–15 mg meprobamate/kg, orally. Meprobamate elimination rates from the blood were determined during the 10 hr prior to ethanol administration and up to 12 hr following ethanol administration. When compared to the pre-ethanol-treatment period, a two- to fivefold increase in blood meprobamate half-life was observed during the post-ethanol-treatment phase. This study strongly suggests that the enhanced effect when ethanol and meprobamate are administered together can be partially explained on the basis of aberrant meprobamate metabolism induced by ethanol.

It stands to reason that, if inhibition of a hepatic drug-metabolizing system results in an increased half-life of a drug, the induction of such a system should result in a decrease of drug half-life. The studies of Misra et al.[106] support this hypothesis. Humans and rats chronically exposed to ethanol were found to display enhanced rates of both meprobamate and ethanol metabolism. (The role of ethanol in inducing the microsomal drug-metabolizing enzymes is dealt with more extensively in the section on ethanol–barbiturate interactions and elsewhere in this volume.) Again, these data lend support to the hypothesis that the ethanol–meprobamate interaction may be metabolic in nature.

The most extensive studies dealing with meprobamate and ethanol interactions in humans have been carried out by Carpenter et al.[20] Ashford and Cobby,[7] and Cobby and Ashford.[22] In general, these studies demonstrated that meprobamate–ethanol pharmacokinetic and behavioral interactions are very complex and may depend on duration of treatment and dose of both ethanol and meprobamate.

D. STIMULANTS AND ANTIDEPRESSANTS

1. Tricyclic Antidepressants

The search for agents which will reverse the CNS effects of ethanol is an old one. This area has recently been reviewed[111] and will not be extensively examined here.

In the case of the tricyclic antidepressants, there are numerous reports of potentiation of the depressant effects of ethanol in laboratory animals[56,60,80, 99,100-103] and humans,[89] but no evidence of an interactive effect was reported.

It has been suggested that amitryptiline may have an antabuselike action, since it potentiated the effect of disulfiram and calcium cyanamide in precipitating the interaction of these drugs with ethanol.[90]

2. Amphetamine

Amphetamine is reported to overcome the nystagmus produced by ethanol,[9] to antagonize the depressant effect of ethanol,[54,82] and to interfere with absorption of ethanol from the intestine of humans.[122] The drug did not change the ethanol concentration necessary to cause death in rats.[55] A great many anecdotal reports or uncontrolled studies are available, but well-controlled animal studies or double-blind human studies are not numerous.

3. Caffeine

Caffeine fails to alleviate ethanol-induced psychomotor impairment in rabbits,[4] fails to antagonize ethanol effects in humans,[44,46,65,110] and fails to alter the blood level of ethanol necessary to cause death in rats.[55] In one report[62] the compound *increased* the depression present after ethanol had disappeared from the blood of rats. In cats, the two drugs were found to be antagonistic in the sense that ethanol decreased the toxicity of caffeine,[91] which is not the usual purpose for combination of the two. An early study[117] found that caffeine and alcohol were synergistic in their lethal effects at higher doses of each in cats.

The experience with other CNS stimulants, such as bemegride,[1] is similar. At doses of these compounds which cause convulsions, there is little antagonism of CNS depressant effects.

E. OTHER AGENTS

The recent discovery that thyrotropin-releasing hormone (TRH) and other peptides will antagonize ethanol depression[25,119] is of considerable interest. The action is not related to thyroid hormone action and promises to attract much attention in the future as a tool for the study of the actions of ethanol, as well as a possible therapeutic agent (see also Ref. 27).

The observation by Penn[115] that 5-hydroxymethylcytosine alone or in combination with pyridoxal would antagonize ethanol depression apparently has not been pursued further to date.

F. NEUROTRANSMITTERS

In Table I is found a list of some of the more recent work dealing with neurotransmitter agents. The work falls into two general categories: investiga-

Table I. Effect of Agents That Interact with Neurotransmitter Systems on Acute Ethanol-Induced CNS Effects

Agents	Effect	Reference
Physostigmine	Antagonize depression	41
Atropine	Increase depression	41
Sotalol	Antagonize depression	58
Sotalol	Antagonize depression	5
Propanolol	Antagonize depression	95
Propanolol	Antagonize depression	139
Propanolol	Increase depression at low doses of ethanol antagonize depression at >2 g ethanol/kg	45
Propanolol	Increase depression (μg ethanol/kg)	151a
Propanolol	No effect	97
Practolol	No effect	151a
Phentolamine	Antagonize stimulation	95
Phentolamine	Antagonize depression	45
Spiroperidol	Increase depression	97
α-Methyl-p-tyrosine	Antagonize stimulation	3
α-Methyl-p-tyrosine	Antagonize stimulation	18,19
α-Methyl-p-tyrosine	Antagonize stimulation	40
L-Dopa	Increase depression	10
L-Dopa	No effect at low dose	10
Dopamine	Increase depression	124
Dopamine	Increase depression	10
Apomorphine	Antagonize stimulation	18
Glycine	Increase depression	11
Serine	Increase depression	11
Serotonin	Increase depression	124
Tyramine	Increase depression	124
1-Tryptophane	Increase depression	66
γ-Hydroxybutyric acid	Antagonize stimulation	26
Baclophen	Antagonize stimulation	26
Aminooxyacetic acid	Antagonize stimulation	26
Nialamide	Antagonize stimulation	3

tions of agents that alter the "stimulation" action of ethanol and investigations of those that alter the "depressant" actions.

The changes produced by these drugs are obviously complicated. They depend on the dose of ethanol, the species or strain of animal used, and, undoubtedly, the effects of the agents on systems besides those which they are intended to alter. In most cases the blood or brain level of the agent was not determined and in some reports levels of ethanol were not obtained. No mechanism that is consistent with all the observations has been proposed to explain these results. As is usually the case, it probably will be found that some of the observations are not replicable, are invalid for some other reason, or require a much different interpretation.

G. METABOLIC INTERACTIONS

In addition to a great many well-documented interactions between ethanol and drugs at the level of the CNS, there are interactions at the level of metabolism of ethanol as well. In this section we shall discuss only those interactions where a clear alteration of the metabolism of ethanol or acetaldehyde can be demonstrated. We shall largely avoid discussion of the effects of ethanol on metabolism of other compounds and of agents which inhibit or accelerate the metabolism of ethanol, as these topics are left to other chapters.

1. Chloral Hydrate

Chloral hydrate may be the oldest and best known substance having a metabolic interaction with ethanol. In addition to interaction at the metabolic level in the liver, chloral hydrate also interacts with ethanol and other CNS depressants.[48] Many publications also document the metabolic interactions of chloral hydrate with other compounds.[64] To mention three, interactions have been noted with furosemide,[92] warfarin,[132,140] and tolbutamide.[44]

Marshall and Owens[94,112] observed that administration of trichloroethanol (TCE) or chloral hydrate to dogs results in appearance of trichloroacetic acid (TCA) in the plasma. The oxidation of TCE was inhibited by disulfiram and cyanamide. These workers did not attempt to detect chloral hydrate in these studies, but showed that administration of ethanol to dogs treated with disulfiram or cyanamide resulted in increased blood acetaldehyde levels determined by the Stotz procedures.[141] Similarly, infusion of acetaldehyde led to detectable amounts of ethanol in the plasma and acetaldehyde in the blood of animals treated with disulfiram or cyanamide. They also observed *in vitro* oxidation of TCE to TCA by dog and rat liver slices or by homogenates prepared from the same tissues. This oxidation was inhibited by disulfiram. Oxidation of chloral hydrate to TCA by dog and rat liver homogenates was also observed, but this reaction was not inhibited by disulfiram. Marshall and Owens also prepared a lyophilized aqueous extract of acetone-dried dog liver that had a high concentration of chloral-hydrate-oxidizing enzyme. This enzyme was inhibited neither by disulfiram nor by cyanamide. This work indicated that the route of oxidation of TCE to chloral hydrate was via the same enzyme system that oxidized ethanol. These workers tentatively suggested that acetaldehyde and chloral hydrate could share the same enzymatic pathway during conversion to their respective acids (acetic and trichloroacetic). Using purified horse liver alcohol dehydrogenase, they were able to establish that this enzyme would reduce chloral hydrate with an apparent K_m of 1.4×10^{-3} M. They were unsuccessful in demonstrating oxidation of trichloroethanol by this enzyme, however. Freidman and Cooper[47] also implicated alcohol dehydrogenase in reduction of chloral hydrate.

Sellers *et al.*[133,134] carried out studies in humans similar to those of Owens and Marshall.[112] They found that ethanol, given after chloral hydrate, resulted in higher blood levels of chloral hydrate and simultaneously depressed blood TCA levels. The reverse is also true (i.e., blood levels of ethanol are higher when chloral hydrate is given prior to ethanol and the blood acetaldehyde level is lower). These investigators could not detect any oxidation of chloral hydrate to TCA by rat liver preparations containing aldehyde dehydrogenase.[24]

No interference with acetaldehyde metabolism by chloral hydrate, TCE, or TCA was observed by Sellers et al.[133,134] in rat liver preparations. The enzyme preparation used must have contained a mixture of aldehyde dehydrogenases from mitochondria, cytoplasm, and microsomes,[146] since a detergent which solubilizes aldehyde dehydrogenases from mitochondria and microsomes, Triton X-100, was used in the homogenizing media. It is not stated how long the preparations were stored before assay, but prolonged (overnight) contact of aldehyde dehydrogenase with Triton X-100 leads to significant inhibition of the activity of this enzyme.

Cabana and Gessner[16] studied the metabolic interactions of chloral hydrate and ethanol. They concluded that ethanol accelerated both chloral hydrate disappearance and TCE appearance. Similar results in humans were obtained by Kaplan et al.[68] who found increased formation of TCE from chloral hydrate as a result of ethanol administration and inhibition of ethanol disappearance from the blood by chloral hydrate. Essentially the same results were obtained by these workers when dogs were used as experimental animals, except that no alteration of ethanol metabolism could be demonstrated.[69]

It seems clear from studies such as those cited above that treatment of humans with chloral hydrate preceding ethanol administration results in higher blood levels of ethanol but lower acetaldehyde levels. This finding was unexpected, as chloral hydrate inhibits human liver aldehyde dehydrogenase in vitro,[79] bovine brain aldehyde dehydrogenase,[43] and acetaldehyde metabolism in mice.[29]

Trichloroethylene also is metabolized to trichloroethanol and has been reported to interact with ethanol in humans.[114] Furthermore, chlorpromazine is reported to increase blood ethanol concentrations,[144] perhaps by inhibition of alcohol dehydrogenase.[138]

2. Paraldehyde

Paraldehyde is a CNS depressant and therefore expected to be additive with ethanol in this respect. Because it is also a polymer of acetaldehyde, it might be expected to depolymerize in vivo, thereby generating acetaldehyde and slowing ethanol disappearance. Evidence for this reaction is extremely scarce. A study of Keplinger and Wells[70] seems to be the most thorough, but this study, in light of more recent information, has a number of defects. Keplinger and Wells studied the effect of disulfiram on the rates of disappearance of paraldehyde and of acetaldehyde from the blood of animals treated with paraldehyde and disulfiram. Disulfiram slowed the rate of disappearance of paraldehyde from a $t_{1/2}$ of 4.79 hr to a $t_{1/2}$ of 8.87 hr. Acetaldehyde disappearance was slowed from an apparent $t_{1/2}$ of 1.94 hr to a $t_{1/2}$ of 6.35 hr. From these data they concluded that paraldehyde was in equilibrium with acetaldehyde and that disulfiram, by blocking acetaldehyde metabolism, prolonged the time that the blood level of paraldehyde remained high. They did not measure blood ethanol or show that the dose of disulfiram administered would inhibit acetaldehyde metabolism when it arose from administered ethanol or acetaldehyde. Since disulfiram has a number of actions on drug metabolism, the prolonged $t_{1/2}$ for paraldehyde may have been a result of interference with some other pathway of paraldehyde metabolism.

The tacit assumption that paraldehyde gives rise to acetaldehyde *in vivo* probably would be made less often if paraldehyde were called by its chemical name, 2,4,6-trimethyl-1,3,5-trioxane. There is, in fact, no direct evidence that acetaldehyde arises from paraldehyde in the body. Even if paraldehyde does provide some acetaldehyde *in vivo*, it is very unlikely that blood levels of about 160 μg/dl as reported by Keplinger and Wells[70] are possible. Both alcohol and aldehyde dehydrogenase would have to be severely inhibited to prevent the rapid reduction of acetaldehyde to ethanol or its oxidation to acetate.

3. Aldehyde Dehydrogenase Inhibition

a. Disulfiram

Disulfiram (tetraethylthiuramdisulfide), sold under the trade name Antabuse, is the best known of several aldehyde dehydrogenase inhibitors. Its administration to man or other animals prior to ethanol ingestion results in elevated blood acetaldehyde levels. The effect of such acetaldehyde levels, perhaps in concert with other effects of disulfiram, are collectively called the "Antabuse" reaction. This subject has recently been reviewed[75] and is covered elsewhere in this volume.

b. Cyanamide (Carbimide)

This relatively simple compound ($H_2N-C\equiv N$) was marketed as calcium cyanamide (Temposil) for use as a deterrent to ethanol ingestion. The compound inhibits aldehyde dehydrogenase when given *in vivo* but does not directly inhibit aldehyde dehydrogenase *in vitro*.[34] Incubation of cyanamide with crude preparations of liver aldehyde dehydrogenase results in inhibition of the enzyme.[108] Brain aldehyde dehydrogenase activity is less susceptible to *in vivo* inhibition than are liver enzymes, possibly because brain does not have the capacity to form the inhibitory compounds and the blood–brain barrier excludes the active form of the inhibitor.[34] The active form of cynamide is unknown, although an acidic urinary excretion product has been isolated. It is inactive as an enzyme inhibitor *in vitro*.[34]

Cyanamide is apparently responsible for the Antabuselike substance in bone charcoal.[145]

c. Coprine

It has been known for many years that ingestion of inky cap mushrooms (*Coprinus artamentarius*) prior to alcohol consumption leads to an Antabuselike reaction.[23,121] Two groups isolated and identified the active substance as coprine.[57,84] The mechanism of the reaction is not yet understood; however, it may involve breakdown of coprine to 1-aminocyclopropranol, since this compound inhibits aldehyde dehydrogenase.[145]

d. Other "Antabuselike" Compounds

Several other compounds have been reported to inhibit aldehyde dehydrogenase or to bring about an "Antabuse" reaction when ingested and followed by ethanol intake. Recently, it has been found that adminstered pargyline inhibits rather specifically the low-K_m aldehyde dehydrogenase in rat liver mitochondrial matrix[81] and brings about an increased acetaldehyde level in rats treated with pargyline and ethanol.[35,81] It is much less potent in this regard than it is as a monoamine oxidase inhibitor. Because pargyline is ineffective as an *in vitro* inhibitor of aldehyde dehydrogenase, it is assumed that some metabolic product is responsible for the inhibitory effect. It is of interest to speculate as to whether this effect of pargyline might not be responsible for some of the "beer, wine, and cheese" reaction usually associated with the use of pargyline in humans.

Other compounds reported to cause reactions to ethanol, presumably by inhibition of aldehyde dehydrogenase, are the oral hypoglycemic agents,[6,15,118,147] nitrofuran analogues, chloramphenicol, bisdichloroacetyl amines, and butyraldoxime.[32] Any of these compounds should be effective in bringing about higher than normal blood acetaldehyde levels following ethanol ingestion. In humans and in other animals this results in a decreased ethanol ingestion.

H. ACTIVATION OF ACETALDEHYDE OXIDATION BY DRUGS

1. Alteration of NAD/NADH Ratios

Since aldehyde dehydrogenase is an NAD-linked enzyme, processes which increase the rate of reoxidation of NADH to NAD would increase the rate at which acetaldehyde could be oxidized if the reoxidation of NADH were rate-limiting. Most of the metabolism of acetaldehyde occurs in the matrix fraction of liver mitochondria,[85,137] so alteration of NADH reoxidation in the mitochondrial compartment would be most likely to change acetaldehyde oxidation.

2. Induction of Acetaldehyde Metabolism

Redmond and Cohen[120] discovered that adminstration of phenobarbital to mice results in increased aldehyde dehydrogenase activity in whole liver homogenates. Deitrich et al.[31,33] and Petersen et al.[116] extended these observations to rats. In this species the effect is accomplished by genetically controlled induction of a cytoplasmic aldehyde dehydrogenase. The amount of the enzyme is increased up to 30-fold in animals treated with phenobarbital over those not so treated or those treated, but genetically incapable of responding. Qualitatively, the same induction is found in mice, although no genetic polymorphism has been detected. It is not known if such an enzyme induction occurs in humans. The induced enzyme has a relatively high K_m (about 0.2 mM) for acetaldehyde.[33] Thus, it will not function very efficiently, as liver acetaldehyde levels are rarely this high.[42,116] A second aldehyde dehydrogenase in rat liver is induced by tetrachlorodibenzo-p-dioxin (TCDD). The enzyme induction is far greater, 100- to 150-fold, but the K_m is even higher (2.6 mM) than that for the phenobarbital-induced supernatant enzyme.[33] Despite the relatively high K_m values for these enzymes, treatment of rats with phenobarbital and then ethanol results in lower blood acetaldehyde levels in those animals with induced liver aldehyde dehydrogenase.[116] Elevated blood acetaldehyde levels accomplished by administration of pargyline before ethanol could be lowered by pretreatment of the animals with either phenobarbital or TCDD.[116] The possibility exists that blood and liver acetaldehyde might be maintained at low levels by such mechanisms in humans. This would presumably prevent or ameliorate the toxic effects of ethanol due to acetaldehyde. Obviously, the use of phenobarbital (which is addicting in its own right) or TCDD (which is extremely toxic) is not possible, but less toxic inducing agents may be found.

An inducing agent for the low-K_m mitochondrial-matrix aldehyde dehydrogenase should be more effective in lowering blood acetaldehyde levels than inducing agents for the high-K_m cytosolic enzymes. Horton[61] reported that chronic ethanol adminstration to rats resulted in an increased aldehyde dehydrogenase in isolated liver mitochondria, as measured at 100-μM acetaldehyde concentration. The principal enzyme measured was presumed the low-K_m mitochondrial-matrix enzyme. This observation has not been confirmed by others; in fact, it seems to be in contrast to the results of Cederbaum and Rubin[21] and Korsten et al.,[78] who reported that ethanol ingestion leads to impaired ability of the liver to metabolize acetaldehyde.

Greenfield et al.[52] studied the effect of administered ethanol on aldehyde dehydrogenase in the mitochondria, as well as in other subcellular fractions, of livers obtained from ethanol-treated Sprague-Dawley rats. They reported a 60% increase in a high-affinity aldehyde dehydrogenase in mitochondrial membranes, but a 50% decrease in activity of a high-affinity dehydrogenase in the microsomes and a 58% decrease in a low-affinity enzyme in the mitochondrial matrix. They suggest that, as the changes found depend on the propionaldehyde concentrations used as well as on the subcellular fraction assayed, it is not surprising that conflicting reports can be found in the literature on this point.

There are several aspects of this paper worthy of comment. In order to separate high-affinity from low-affinity aldehyde dehydrogenase activity, they subtracted the rate obtained at a propionaldehyde concentration of 13.6 mM from that obtained at a concentration of 0.068 mM. The difference was labeled "low-affinity aldehyde dehydrogenase activity." A better method would have been actually to determine the K_m values for aldehyde dehydrogenase in each fraction as carried out be Tottmar et al.[146] and correct for cross contamination of the data by the method described in that paper or as described by Petersen et al.[116] Since their animals were pair-fed and the enzyme assays carried out in parallel, they used Student's t-test for correlated means. This procedure will improve the chances of finding statistically significant results. The animals are not really pairs, however, since Sprague-Dawley animals are not inbred, and each animal may well have responded to the experimental situation somewhat differently. In fact, the authors later argue that genetic differences may be important.

I. FACTORS INFLUENCING ETHANOL–DRUG INTERACTIONS

As one reviews the literature concerning ethanol–drug interactions, one very prominent feature of these studies is the extent of metabolic and/or behavioral variation observed when alcohol or other drugs are given alone or in combination. Clearly, the literature indicates that, even when such factors as dose and duration are controlled, a significant amount of variation in response is evident among individuals receiving the same treatment. Such variation in response indicates that ethanol–drug interactions are very complex and depend upon an individual's genotype and environment, as well as possible genotype–environment interactions. More meaningful data relating to the ethanol–drug interactions could result from the use of appropriate animal models which are maintained under strict genetic and environmental control. It seems logical that the mechanisms underlying both metabolic and CNS alterations resulting from ethanol and barbiturate interactions could be partially elucidated by the continued selection and maintenance of animal lines which differ in their response to different combinations of ethanol and barbiturates.

It is evident from the literature that some investigators have considered the possible interactions of specific drug metabolites and ethanol. However, few studies have considered the possibility that acetaldehyde, a metabolite of ethanol, could also play an important role in the interactions of ethanol with various drugs. Since acetaldehyde is a very reactive compound, its interactions with the various drugs and their metabolites deserve consideration. That consideration may be important is suggested by the study of Korsten et al.[78] which found that alcoholics maintain significantly higher blood acetaldehyde levels than do nonalcoholics. From the preceding discussion it appears that chronic alcoholics display both elevated blood acetaldehyde levels and induced microsomal drug-metabolizing enzymes. If a chronic alcoholic is subjected to drug therapy, the situation is optimal for an interaction between acetaldehyde and rapidly generated drug metabolites.

J. SUMMARY AND CONCLUSIONS

There are numerous drugs that interact with the metabolism of ethanol and acetaldehyde. Some of these are or may become clinically useful because of this interaction, while others are toxic and should not be combined with ethanol. In all cases, the drugs that interact metabolically with ethanol not only provide information about their mechanism of action but also shed light on the multitude of actions of ethanol as well.

ACKNOWLEDGMENTS. This work was supported in part by Grants AA00263, AA05011, and AA03527 from the National Institute on Alcohol Abuse and Alcoholism. The authors would like to thank Dr. David Jensen, Pequita Bludeau, and Rebecca Miles for their helpful comments on parts of the manuscript.

K. REFERENCES

1. A. H. Abdallah and D. M. Roby, *Proc. Soc. Exp. Biol. Med.* **148**:819 (1975).
2. J. Adriani and R. C. Morton, *Anesth. Analg.* **47**:472 (1968).
3. S. Ahlenius, R. Brown, J. Engel, T. H. Svensson, and B. Waldeck, *J. Neural Transm.* **35**:175 (1974).
4. R. L. Alstott, D. J. Brown, and R. B. Forney, *Toxicol. Appl. Pharmacol.* **17**:296 (1970).
5. A. G. Arbab and P. Turner, *J. Pharm. Pharmacol.* **23**:718 (1971).
6. M. M. Asaad and D. E. Clarke, *Eur. J. Pharmacol.* **35**:301 (1976).
7. J. R. Ashford and J. M. Cobby, *J. Stud. Alcohol,* Suppl. **7**:140 (1975).
8. H. J. Berger, *J. Stud. Alcohol* **30**:862 (1969).
9. M. E. Berstein, A. B. Richards, F. W. Hughes, and R. B. Forney, in: *Alcohol and Traffic Safety* (R. N. Harger, ed.), pp. 208–210, Indiana University Press (1966).
10. K. Blum, W. Calhoun, J. Merritt, and J. E. Wallace, *Nature (Lond)* **242**:407 (1973).
11. K. Blum, J. E. Wallace, and I. Geller, *Science* **176**:292 (1972).
12. H. A. Bowes, *Dis. Nerv. Syst.* **21**:20 (1960).
13. B. B. Brodie, P. A. Shore, and S. L. Silver, *Nature (Lond)* **175**:1133 (1955).
14. B. B. Brodie and P. A. Shore, *Ann. N. Y. Acad. Sci.* **66**:631 (1956–67).
15. H. Büttner, *Dtsch. Arch. Klin. Med.* **207**:1 (1961).
16. B. E. Cabana and P. K. Gessner, *J. Pharmacol. Exp. Ther.* **174**:260 (1970).
17. J. Caldwell and P. Sever, *Clin Pharmacol. Ther.* **16**:737 (1974).
18. A. Carlsson, J. Engel, U. Strombom, T. H. Svensson, and B. Waldeck, *Arch. Pharmacol.* **283**:117 (1974).
19. A. Carlsson, J. Engel, and T. H. Svensson, *Psychopharmacologia* **26**:307 (1972).
20. J. A. Carpenter, R. J. Gibbins, and J. A. Marshman, *J. Stud. Alcohol,* Suppl. **7**:54 (1975).
21. A. I. Cederbaum and E. Rubin, *Arch. Biochem. Biophys.* **179**:46 (1977).
22. J. M. Cobby and J. R. Ashford, *J. Stud. Alcohol,* Suppl. **7**:162 (1975).
23. B. B. Coldwell, K. Genest, and D. W. Hughes, *J. Pharm. Pharmacol.* **21**:176 (1966).
24. J. R. Cooper and P. J. Friedman, *Biochem. Pharmacol.* **1**:76 (1958).
25. J. M. Cott, G. R. Breese, B. R. Cooper, T. S. Barlow, and A. J. Prange, Jr., *J. Pharmacol. Exp. Ther.* **196**:594 (1976).
26. J. Cott, A. Carlsson, J. Engel, and M. Lindquist, *Arch. Pharmacol.* **295**:203 (1976).
27. J. Cott and J. Engel, *Psychopharmacology* **52**:145 (1977).
28. S. Courvoisier, J. Fournel, R. Sucrot, M. Kolsky, and P. Koetschet, *Arch. Int. Pharmacodyn. Ther.* **92**:305 (1953).
29. P. J. Creaven and M. K. Roach, *J. Pharm. Pharmacol.* **21**:332 (1969).

30. S. H. Curry, in: *Drug Interactions* (D. G. Grahame-Smith, ed), pp. 87–99, University Park Press, Baltimore (1977).
31. R. A. Deitrich, *Science* **173**:334 (1971).
32. R. A. Deitrich and L. Hellerman, *J. Biol. Chem.* **238**:1683 (1963).
33. R. A. Deitrich, P. Bludeau, T. Stock, and M. Roper, *J. Biol. Chem.* **252**:6169 (1977).
34. R. A. Deitrich, P. A. Troxell, and W. S. Worth, *Biochem. Pharmacol.* **24**:2733 (1976).
35. D. Dembiec, D. MacNamee, and G. Cohen, *J. Pharmacol. Exp. Ther.* **197**:332 (1976).
36. J. M. Dille and R. Ahlquist, *J. Pharmacol. Exp. Ther.* **61**:332 (1976).
37. A. Doenicke and W. Sigmund, *Arzneim.-Forsch.* **14**:907 (1964).
38. J. W. Dundee, *Anesth. Analg.* **49**:467 (1970).
39. J. W. Dundee and M. Isaac, *Med. Sci. Law (Lond.)* **10**:220 (1970).
40. J. Engel, U. Strombom, T. H. Svensson, and B. Waldeck, *Psychopharmacologia* **37**:275 (1974).
41. C. K. Erickson and W. L. Burnam, *Agents Actions* **2**:8 (1971).
42. C. J. Eriksson, *Biochem. Pharmacol.* **22**:2283 (1973).
43. V. G. Erwin and R. A. Deitrich, *J. Biol. Chem.* **241**:3533 (1966).
44. R. B. Forney and F. W. Hughes, *Combined Effects of Alcohol and Other Drugs*, C. C. Thomas, Springfield, Ill. (1968).
45. D. Frankel, H. Kalant, J. M. Khanna, and A. E. LeBlanc, *Can. J. Physiol. Pharmacol.* **54**:622 (1976).
46. H. M. Franks, H. Hagedorn, V. R. Hensley, W. J. Hensley, and G. A. Starmer, *Psychopharmacologia* **45**:177 (1975).
47. P. J. Friedman and J. R. Cooper, *J. Pharmacol. Exp. Ther.* **129**:373 (1960).
48. P. K. Gessner and B. E. Cabana, *J. Pharmacol. Exp. Ther.* **174**:247 (1970).
49. L. Goldberg, in: *Alcohol and Alcoholism* (R. E. Popham, ed.), pp. 42–56, University of Toronto Press, Toronto (1970).
50. J. P. D. Graham, *Toxicol. Appl. Pharmacol.* **2**:14 (1960).
51. D. J. Greenblatt and R. I. Shader, *N. Engl. J. Med.* **291**:1011 (1974).
52. N. J. Greenfield, R. Pietruszko, G. Lin and D. Lester, *Biochim. Biophys. Acta* **428**:627 (1976).
53. R. C. Gupta and J. Kofoed, *Can. Med. Assoc. J.* **94**:836 (1966).
54. F. Haffner, *Klin. Wochenschr.* **17**:1310 (1938).
55. H. W. Haggard, L. A. Greenberg, N. Rakieten, and L. H. Cohen, *J. Pharmacol. Exp. Ther.* **69**:266 (1940).
56. G. Halliwell. R. M. Quinton. and F. E. Williams, *Br. J. Pharmacol.* **23**:330 (1964).
57. G. M. Hatfield and J. P. Schaumberg, *Lloydia* **38**:489 (1975).
58. K. Hayashida and A. A. Smith, *J. Pharm. Pharmacol.* **23**:718 (1971).
59. S. L. Hayes, G. Pablo, T. Radomski, and R. Palmer, *N. Engl. J. Med.* **296**:186 (1977).
60. F. Herr, J. Stewart, and M. Charest, *Arch. Int. Pharmacodyn. Ther.* **134**:328 (1961).
60a. L. E. Hollister, *Ann. N.Y. Acad. Sci.* **281**:212 (1976).
61. A. A. Horton, *Biochim. Biophys. Acta* **253**:514 (1971).
62. F. W. Hughes, R. B. Forney, and A. B. Richards, *Clin. Pharmacol. Ther.* **6**:139 (1965).
63. F. W. Hughes, R. B. Forney, and A. B. Richards, *Clin. Pharmacol. Ther.* **6**:139 (1965).
64. K. S. Iyer and A. G. Kutty, *Indian J. Physiol. Pharmacol.* **18**:49 (1974).
65. G. Jansen, *Med. Exp.* **2**:209 (1960.
66. C. I. Jarowski and C. O. Ward, *Toxicol. Appl. Pharmacol.* **18**:603 (1971).
67. H. Kalant, J. M. Khanna, and J. Marshman, *J. Pharmacol. Exp. Ther.* **175**:318 (1970).
68. H. L. Kaplan, R. B. Forney, F. W. Hughes, N. C. Jain, and D. Crim, *J. Forensic Sci.* **12**:295 (1967).
69. H. L. Kaplan, N. C. Jain, R. B. Forney, and A. B. Richard, *Toxicol. Appl. Pharmacol.* **14**:127 (1969).
70. M. L. Keplinger and L. A. Wells, *J. Pharmacol. Exp. Ther.* **119**:19 (1957).
71. A. U. Khan, R. B. Forney, and F. W. Hughes, *Arch. Int. Pharmacodyn. Ther.* **150**:171 (1964).
72. J. M. Khanna, H. Kalant, and G. Lin, *Biochem. Pharmacol.* **21**:2215 (1972).
73. L. B. Khouw, T. N. Burbridge, and V. C. Sutherland, *Biochim. Biophys. Acta* **73**:173 (1963).

74. B. Kissin, in: *The Biology of Alcoholism* (B. Kissin and H. Begleiter, eds.), Vol. 3: *Clinical Pathology,* pp. 109–161, Plenum Press, New York (1974).
75. T. M. Kitson, *J. Stud. Alcohol* **38**:96 (1977).
76. K. D. Kolenda, *Med. Klin.* **70**:516 (1975).
77. E. Kopmann and F. W. Hughes, *Arch. Gen. Psychiatry* **1**:7 (1959).
78. M. A. Korsten, S. Matsuzaki, L. Feinman, and C. S. Lieber, *N. Engl. J. Med.* **292**:386 (1975).
79. R. J. Kraemer and R. A. Deitrich, *J. Biol. Chem.* **243**:6402 (1968).
80. A. A. Landauer, G. Milner, and J. Patman, *Science* **163**:1467 (1969).
81. M. E. Lebsack, D. R. Petersen, A. C. Collins, and A. D. Anderson, *Biochem. Pharmacol.* **26**:1151 (1977).
82. B. E. Leonard and B. D. Wiseman, *J. Pharm. Pharmacol.* **22**:967 (1970).
83. R. Liljequist, M. Linnoila, M. J. Mattila, I. Saario, and T. Seppälä, *Psychopharmcology* **44**:205 (1975).
84. P. Lindberg, R. Bergman, and B. Wickberg, *J. Chem. Soc. D Chem. Commun.,* 946 (1975)
85. K. O. Lindros, N. Oshino, R. Parrilla, and J. R. Williamson, *J. Biol. Chem.* **249**:7956 (1974).
86. M. Linnoila and S. Hakkinen, *Clin. Pharmacol. Ther.* **15**:368 (1974).
87. M. Linnoila, I. Saario, and M. Mäki, *Eur. J. Clin. Pharmacol.* **7**:337 (1974).
88. M. Linnoila, I. Saario, R. Liljequist, J. J. Himberg, and M. Mäki, *Arzneim.-Forsch.* **25**:1088 (1975).
89. M. F. Lockett and G. Milner, *Br. Med. J.* **1**:921 (1965).
90. W. A. G. MacCallum, *Lancet* **1**:313 (1969)
91. D. I. Macht and M. E. Davis, *Am. J. Physiol.* **100**:67 (1934).
92. M. Malach and N. Berman, *J. Am. Med. Assoc.* **232**:638 (1975).
93. S. Mallov, *Clin. Obstet. Gynecol.* **20**:483 (1977).
94. E. K. Marshall, Jr., and A. J. Owens, Jr., *Bull. Johns. Hopkins Hosp.* **95**:1 (1954).
95. J. A. Matchett and C. K. Erickson, *Psychopharmacology* **52**:201 (1977).
96. *Med. Lett.* **17**:17 (1975).
97. J. H. Mendelson, A. M. Rossi, J. G. Bernstein, and J. Kuehnle, *Clin. Pharmacol. Ther.* **15**:571 (1974).
98. W. B. Mendelson, D. W. Goodwin, S. Y. Hill, and J. D. Reichman, *Curr. Ther. Res. Clin. Exp.* **19**:155 (1976).
99. D. B. Myers, D. O. Kanyuck, and R. C. Anderson, *J. Pharm. Sci.* **55**:1317 (1966).
100. G. Milner, *Lancet* **1**:222 (1967).
101. G. Milner, *Br. J. Pharmacol.* **34**:370 (1968).
102. G. Milner, *J. Pharm. Sci.* **57**:1900 (1968).
103. G. Milner, *J. Pharm. Sci.* **57**:2005 (1968).
104. G. Milner, *Med. J. Aust.* **1**:1204 (1970).
105. G. Milner and A. A. Landauer, *Br. J. Psychiatry* **118**:351 (1971).
106. P. Misra, A. Lefevre, H. Ishii, E. Rubin, and C. S. Lieber, *Am. J. Med.* **51**:346 (1971).
107. P. S. Misra, A. Lafevre, E. Rubin, and C. S. Lieber, *Gastroenterology* **58**:308 (1970).
108. L. Molander and C. Duvhok, *Acta Pharmacol. Toxicol.* **38**:145 (1976).
109. P. Munkelt, G. A. Lienert, M. Frahm, and K. Soehring, *Arzneim.-Forsch.* **12**:1059 (1962).
110. H. W. Newman and E. J. Newman, *J. Stud. Alcohol* **17**:406 (1956).
111. E. P. Noble, R. L. Alkana, and E. S. Parker, Proc. Fourth Annu. Alcoholism Conf. Nat. Inst. Alcohol Abuse Alcoholism, 134 (1975).
112. A. H. Owens, Jr., and E. K. Marshall, Jr., *Bull. Johns Hopkins Hosp.* **97**:395 (1955).
113. E. S. Palva, M. Linnoila, and M. J. Mattila, *Mod. Probl. Pharmacopsychiatry* **11**:70 (1976).
114. S. Pardys and M. Brotman, *J. Am. Med. Assoc.* **229**:521 (1974).
115. N. W. Penn, *Life Sci.* **17**:1055 (1975).
116. D. R. Peterson, A. C. Collins, and R. A. Deitrich, *J. Pharmacol. Exp. Ther.* **201**:471 (1977).
117. J. D. Pilcher, *J. Pharmacol. Exp. Ther.* **3**:267 (1912).
118. H. Podgainy and R. Bressler, *Diabetes,* **17**:679 (1968).
119. C. C. Porter, V. J. Lotti, and M. J. DeFelice, *Life Sci.* **21**:811 (1977).
119a. G. T. Pryor, S. Usairn, C. Mitoma, and M. C. Braude, *Ann. N.Y. Acad. Sci.* **281**:171 (1976).

120. G. Redmond and G. Cohen, *Science* **171**:387 (1971).
121. W. A. Reynolds and F. H. Lowe, *N. Engl. J. Med.* **272**:630 (1965).
122. M. Rinkel and A. Myerson, *J. Pharmacol. Exp. Ther.* **71**:75 (1941).
123. J. M. Ritchie, in: *The Pharmacological Basis of Therapeutics* (L. S. Goodman and A. Gilman, eds.), 5th ed., pp. 137–146, Macmillan Publishing Co., Inc., New York (1965).
124. G. Rosenfeld, *J. Stud. Alcohol* **21**:584 (1960).
125. D. H. Ross, B. C. Kibler, and H. L. Cardenas, *Drug Alcohol Depend.* **2**:305 (1977).
126. E. Rubin, P. Bacchin, H. Gang, and C. S. Lieber, *Lab Invest.* **22**:569 (1970).
127. E. Rubin, H. Gang, P. Misral, and C. S. Lieber, *Am. J. Med.* **49**:801 (1970).
128. E. Rubin and C. S. Lieber, *Science* **162**:690 (1968).
129. E. Rubin and C. S. Lieber, *Science* **172**:1097 (1971).
130. F. Sandberg, *Acta Physiol. Scand.* **22**:113 (1951).
131. R. Schüppel, *Scand. J. Clin. Lab. Invest.* **113**:15 (1970).
131a. G. Seidel and K. Soehring, *Arzneim.-Forsch.* **15**:472 (1965).
132. E. M. Sellers and J. Koch-Weser, *N. Engl. J. Med.* **283**:827 (1970).
133. E. M. Sellers, M. Land, J. Koch-Weser, E. LeBlanc, and H. Kalant, *Clin. Pharmacol. Ther.* **13**:37 (1972).
134. E. M. Sellers, M. Lang, J. Koch-Weser, E. LeBlanc, and H. Kalant, *Clin. Pharmacol. Ther.* **13**:50 (1972).
135. T. Seppälä, *Arch. Int. Pharmacodyn. Ther.* **223**:311 (1976).
136. T. Seppälä, I. Saario, and M. J. Mattila, *Mod. Probl. Pharmacopsychiatry* **11**:85 (1976).
137. C. Siew and R. A. Deitrich, *Arch. Biochem. Biophys.* **176**:638 (1976).
138. L. Skursky, J. Kovar, and J. Michalsky, *FEBS Lett* **51**:297 (1975).
139. A. Smith, K. Hayashida, and Y. Kim, *J. Pharm. Pharmacol.* **22** (1970).
139a. C. M. Smith, *Ann. N.Y. Acad. Sci.* **281**:384 (1976).
140. K. Soehring and R. Schüppel, *Dtsch. Med. Wochenschr.* **91**:1892 (1966).
141. E. Stotz, *J. Biol. Chem.* **148**:585 (1943).
142. V. C. Sutherland, T. N. Burbridge, J. E. Adams, and A. Simon, *J. Appl. Physiol.* **15**:189 (1960).
143. R. D. Teare, in: *Drug-induced Diseases* (L. Meyler and H. M. Peck, eds.), p. 66, Exerpta Medica, Amsterdam (1956).
144. D. L. Tipton, V. C. Sutherland, T. N. Burbridge, and A. Simon, *Am. J. Physiol.* **200**:1007 (1961).
145. O. Tottmar, H. Marchner, and P. Lindberg, in: *Alcohol and Aldehyde Metabolizing Systems* (R. G. Thurman, J. R. Williamson, H. R. Drott, and B. Chance, eds.), Vol. 2, pp. 203–212, Academic Press, Inc., New York (1977).
146. S. O. C. Tottmar, H. Pettersson, and K.-H. Kiessling, *Biochem J.* **135**:577 (1973).
147. E. L. Truitt, Jr., G. Duritz, A. M. Morgan, and R. W. Prouty, *J. Stud. Alcohol* **23**:197 (1962).
148. J. A. Udall, *Curr. Ther. Res. Clin. Exp.* **17**:67 (1975).
149. G. Ugarte, T. Pereda, M. E. Pino, and H. Iturriaga, *J. Stud. Alcohol* **33**:698 (1972).
150. H. Vaapatal and H. Karppunen, *Agents Actions* **1**:43 (1969).
151. Z. Votava and H. Dyntarova, in: *Collegium Internationale Neuro-Psychopharmacologium (C.I.N.P.)*, Abstracts Second International Conference, Prague (1970).
151a. G. H. Wimbish, R. Martz, and R. B. Forney, *Life Sci.* **20**:65 (1977).
152. G. A. Zirkle, O. D. McAtee, P. D. King, and R. Van Dyke, *J. Am. Med. Assoc.* **173**:1823 (1960).

Biochemical Pharmacology of Pyrazoles

18

Frank H. Deis and David Lester

A. INTRODUCTION

Pyrazoles interfering with alcohol metabolism, the subject of this review, were used freely in the period 1970–1975. Their use has diminished recently, presumably out of concern for their toxicity and side effects. Current research, however, by describing, explaining, and differentiating these side effects may make it possible for certain pyrazoles to become not only more useful research tools, but also therapeutic agents in preventing toxicity from compounds such as ethylene glycol and methanol. Rydberg[1] reviewed this area in 1972.

B. PHARMACODYNAMICS

1. Physical Properties

Pyrazoles are stable basic compounds available in crystalline form (pyrazole, mp = 70°C, mol. wt. = 68.1) or as liquids (4-methylpyrazole, bp = 204°C, mol. wt. = 82.1); the basic structure is shown in Fig. 1. The simpler compounds are very soluble in water and relatively soluble in other solvents. They are absorbed well parenterally; because they are bound in the stomach,[2] presumably as salts of stomach acids, they may not be absorbed well when given orally. Pyrazole[2] and 4-methylpyrazole[3] appear to be distributed throughout the body water; they are neither bound to plasma protein nor taken up by fat.

Frank H. Deis • Department of Biochemistry, Rutgers University, Piscataway, New Jersey 08854.   **David Lester** • Center of Alcohol Studies, Rutgers University, Busch Campus, New Brunswick, New Jersey 08903.

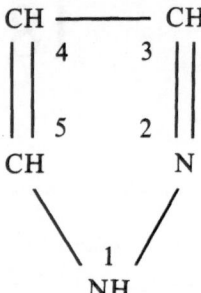

Fig. 1. Pyrazole structure.

2. Analytic Methods

Lester and Benson[6] extracted pyrazole from K_2HPO_4-saturated urine into ethylene dichloride, returned it to 0.12 N HCl, and determined its absorption at 215–217 nM; from its molar absorptivity (log e = 3.66), it may be estimated that pyrazole is detectable at about 30 μM (using 1 ml urine), although the specificity is low. Ward and Pennington's colorimetric method[5] used the colored complex, formed in the presence of sodium nitrite and acetic acid, of pyrazole and sodium pentacyanoaminoferrate. Although a simple method, the limit of detectability of pyrazole in plasma is only 500 μM. Since all pyrazoles inhibiting ADH form the colored complex, the method has been adapted as a thin-layer chromatography spray reagent.[8] Scintillation counting of radioactive pyrazole in blood and urine (after combustion to $^{14}CO_2$) has a sensitivity limited mainly by the specific activity of the [^{14}C]pyrazole, certainly as low as 10 μM.[2] This method does not, of course, discriminate between pyrazole and its metabolites unless separation is first effected. Rydberg and Buijten[4] (see also Ref. 2) determined pyrazole and 4-methylpyrazole in blood with gas–liquid chromatography (GLC) using 5% Carbowax 20M on Chromosorb W at 120°C and a flame ionization detector. Metabolite interference probably does not occur, but the lowest concentrations detected were about 400 μM pyrazole and somewhat less than 100 μM 4-methylpyrazole. Bjorkhem et al.,[7] using a 3.5% EGSS-X on Chromosorb W column at 150° with GLC/mass spectrometry and a multiple ion detector, were able to determine concentrations of 1 μM 4-methylpyrazole in blood; if the equipment is available, this appears the method of choice.

3. Kinetics

a. Pyrazole

Since the metabolism of most drugs can be described by first-order kinetics, the earlier papers dealing with the kinetics of pyrazole[3,6] and 4-methylpyrazole[3,7] were reported in these terms. Detailed studies of pyrazole in the rat[2] have shown that its "half-life" varies widely with dosage and that a better fit with the data is obtained by using zero-order kinetics, although this too falls short of accurately describing the disappearance process. With zero-order kinetics, the rate of pyrazole disappearance in rats, using the data of Rydberg et al.,[3] appears to be between -40 and -50 μmole/liter per hr, compared to the -30 μmole/liter per hr obtained by Deis et al.[2] In the presence of ethanol, pyrazole disappearance slows to about -33 μmole/liter per hr,[3] while Deis et al.[2] report that the rate decreased even more to -11 to -15 μmole/liter per hr with different doses of alcohol; however, Rydberg et al.[3] gave ethanol as a single initial injection, whereas Deis et al.[2] maintained relatively constant ethanol levels by continuous ethanol infusion. In mice, pyrazole has a half-life of 10 hr, calculated from the relative inhibition of ethanol disappearance rather than from direct determination of pyrazole in the blood.[9]

b. 4-Methylpyrazole

The data of Rydberg et al.,[3] for a dose of 0.84 mmole 4-methylpyrazole/kg in rats, appears to be first-order with a half-life of 9.1 hr; their alcohol-slowed disappearance curve also fits first-order kinetics (12.2 hr half-life after 32.6 mmole ethanol/kg). When Bjorkhem et al.[7] repeated the experiment at 1.0 mmole 4-methylpyrazole/kg, they reported a half-life of 10.8 hr, but the data fit a zero-order disappearance as well (rate $= -32.6$ μmole/liter per hr); furthermore, they state that "after administration of 0.1 mmole/kg the plasma concentration of 4-methylpyrazole decreased from about 100 μM to about 50 μM during the first 90 minutes after the injection," a rate of -33.3 μmole/liter per hr, almost identical with the rate at the tenfold greater dose. Bjorkhem et al. describe their analytic method, gas chromatography–mass spectrometry (GC-MS), as being "sensitive and accurate" in contrast to the GLC method of Rydberg et al.[3]; their data for 4-methylpyrazole disappearance in humans[7] also fit zero-order kinetics (-2.4 μmole/liter per hr) better than first-order (half-life of 1.47 hr). Using the GC-MS technique[7] to measure 4-methylpyrazole in rat blood, Blomstrand et al.[35] also report ethanol slowing of 4-methylpyrazole disappearance. The "half-life" of 4-methylpyrazole (at a 1-mmole/kg dose) was 650 min and increased to 2000 min after ethanol; the blood ethanol was not allowed to fall below 17 mM, achieved by an initial ethanol injection of 30 mmole/kg followed at about 24, 32, and 40 hr by booster doses of 10 mmole/kg. Blomstrand et al.[35] attribute the ethanol slowing of 4-methylpyrazole metabolism to interaction of its oxidation product, 4-hydroxymethylpyrazole, at the active site of alcohol dehydrogenase. While such an interaction may occur, it seems more reasonable to postulate that ethanol inhibits 4-methylpyrazole metabolism at the initial microsomal oxidation of the methyl side chain; pyrazole metabolism is also strongly inhibited by ethanol, yet pyrazole is not an ADH substrate: the common step for both pyrazole and 4-methylpyrazole would appear to be their microsomal oxidation.

c. 4-Bromopyrazole

In mice, the half-life of 4-bromopyrazole is 3 hr, calculated from the relative inhibition of ethanol disappearance.[9]

4. Metabolites

The metabolites of 4-methylpyrazole in mice[10] and rats[11] have been investigated. Only about 20% of the dose is excreted unchanged, the bulk being oxidized on the 4-methyl side chain to 4-hydroxymethylpyrazole and 4-carboxypyrazole. 4-Methylpyrazole is also conjugated as 4-methylpyrazole-N-glucosiduronic acid.[10] Similarly, Deis et al.[2] found in rats that more than 50% of administered pyrazole is excreted as 4-hydroxypyrazole in hydrolyzable conjugated forms and that ethanol can almost completely prevent formation of this metabolite; since pyrazole itself is cleared only slowly by the kidney, ethanol maintains pyrazole levels for a long time. Murphy and Clay[12] and Clay et al.[81] have characterized, in rats and mice, seven urinary metabolites and conjugates of pyrazole, including two ribosides, an interesting and unexpected conjugation; the metabolites include 3- and 4-hydroxy, 3,4-dihydroxy- and 1,3,4-trihydroxypyrazole, pyrazole-N-glucuronide, pyrazole-N-riboside, and 4-hydroxypyrazole-N-riboside. Pyrazole-N-glucuronide accounts for perhaps 10% of total radioactivity, and di- and trihydroxypyrazoles for about 20%. The findings with regard to 4-hydroxypyrazole appear consistent with those of Deis et al.[2] Some of the unidentified derivatives of 4-methylpyrazole may also prove to be ribosides. Another pyrazole-N-glucuronide was recently found in man after administration of EMD 16 923, a pyrazolo-ethyl-4-chlorophenylpiperazine.[13]

5. Toxicity

The acute LD_{50} of pyrazole in rats is reported to be as low as 12.9 mmole/kg (ip)[6] and as high as 19 or 21 mmole/kg (iv or po).[14] In mice, Magnusson et al.[14] found similar values po and iv, while Goldstein and Pal[9] reported an LD_{50} of 7.9 mmole/kg (ip). 4-Methylpyrazole's acute LD_{50} in rats or mice was 7.9 mmole/kg (po) and 3.8 mmole/kg (iv).[14] When the inhibitors were administered chronically, 4-methylpyrazole did not appear to cause harm at sublethal doses,[14] but pyrazole showed a delayed toxicity. Lelbach[15] found that 0.46 mmole/kg per day of pyrazole caused death of rats in 5–19 days with extensive liver damage; coadministration of ethanol caused death sooner. Wilson and Bottiglieri[33] reported that the LD_{50} of pyrazole in rats dropped from 13 to 2.2 mmole/kg when the doses were repeated daily for 14 days; with doses of only 0.37 or 0.18 mmole/ kg per day in dogs, they found liver damage, anorexia, diarrhea, leukopenia, and thrombocytopenia. The LD_{50} for 4-bromopyrazole in mice is 2.5 mmole/kg after a single ip injection.[9]

In none of these cases were pyrazole levels determined; the results of Deis et al.[2] imply that daily doses of higher than 0.72 mmole pyrazole/kg should result in accumulation while this dose would be lower if ethanol were also given.

C. EFFECTS OF PYRAZOLES ON ALCOHOL DEHYDROGENASE

1. In Vitro

a. Structure–Activity Relationships

Except where noted in this section, K_is mentioned are for horse liver ADH. Pyrazole[17] is a potent inhibitor of alcohol dehydrogenase ($K_i = 0.22$ μM). From the papers of Theorell et al.[17,18] and from other early papers,[19,20] it is evident that any substituent in the 1-, 3-, or 5-positions destroys the inhibitory activity of pyrazole, but that substitution in the 4-position enhances this activity if nonpolar substituents such as halo- (4-bromopyrazole, $K_i = 0.02$ μM) or alkyl- (4-methylpyrazole, $K_i = 0.013$ μM)[21] are used. The most potent ADH inhibitors known are the longer-chain 4-alkyl-pyrazoles (4-pentylpyrazole, $K_i = 0.0008$ μM; 4-butylpyrazole, $K_i = 0.0018$ μM; 4-propylpyrazole, $K_i = 0.004$ μM).[21] 4-Carboxypyrazole[22] and 4-nitropyrazole[20] are not good inhibitors, but 4-trimethylsilylpyrazole ($K_i = 0.20$ μM)[23] and 4-trifluoromethylpyrazole ($K_i = 0.08$ μM)[21] are active, demonstrating that some bulkiness is tolerable. Drott[23] found that 4-γ-halopropylpyrazoles were the most active in the series he tested (4-γ-iodopropylpyrazole, $K_i = 0.05$ μM), but 4-γ-aminopropylpyrazole ($K_i = 10.0$ μM) was less active than pyrazole.[19,23] 4-Hydroxyalkylpyrazoles similarly appear to be only weakly active ADH inhibitors (4-hydroxymethylpyrazole, $K_i = 6.6$ μM, human LADH).[22] Imidazole compounds are inactive as ADH inhibitors.[18,20] Although there have been reports that antipyrine, a pyrazolone, slows the metabolism of ethanol,[24] Lester et al.[19] found no such effect in rats.

b. Ternary Complex

The ternary complex formed by pyrazole, ADH, and NAD^+[17,25] has spectral characteristics which have been used to investigate the nature of the active site of ADH. Various perturbations of the enzyme, such as removal or exchange of its zinc ions[26–28] or use of NAD^+ analogues,[29–31] have shown that pyrazole is closely associated with zinc and with the nicotinamide ring of NAD^+ at the active site. Optical rotatory dispersion studies[25,32] have shown that pyrazole produces a substantial conformational change in the enzyme as the complex is formed.

2. *In Vivo*

a. Pyrazole versus 4-Methylpyrazole

Pyrazole has the advantages of being readily available and easily handled, but so many side effects of pyrazole are known that there is serious doubt whether pyrazole should be used at all where the intention is simply to inhibit alcohol dehydrogenase. Many of these side effects are not seen with the 4-alkylpyrazoles, and Murphy and Clay[12] and Clay et al.[81] speculate that the toxic effects of pyrazole could stem from metabolites not shared by 4-alkylpyrazoles. Thus:

1. Pyrazole causes liver damage in rats at high doses (4.4 mmole/kg)[14,16,33,34] and with ethanol at lower doses (0.18 mmole/kg per day).[15,36] 4--Methylpyrazole appears to produce no liver damage at either level.[14,37]

2. Pyrazole inhibits rat liver catalase *in vivo*[34,38,39] in the absence of alcohol.[39,40] No such inhibition is seen with 4-methylpyrazole.

3. Although pyrazole and 4-alkylpyrazoles share a CNS-depressant effect in rats and mice,[41–45] the effect of pyrazole is qualitatively different: pyrazole reduces rat brain norepinephrine[44] and 4-methylpyrazole does not.[45]

4. Pyrazole produces a feeling of illness, with nausea and vomiting in humans (from 0.09 mmole/kg per day)[16] and dogs (0.37 mmole/kg per day)[33] and aversive reactions in rats (3.0 mmole/kg).[46] No such reaction has been reported in humans given 0.12 mmole 4-methylpyrazole/kg.[47]

The use of 4-propylpyrazole is increasing[48,49] partly because it is 100 times more potent than pyrazole *in vitro*.[21] Although it might be the inhibitor of choice *in vitro*, its side effects should be investigated before it is used extensively *in vivo*, it is a CNS depressant.[42]

b. In Laboratory Animals

Soon after the early *in vitro* work, it was established that pyrazoles were effective ADH inhibitors *in vivo*[19,20,50]; *in vivo* inhibitory activities in mice[9] and rats[19] were generally close to those expected from examination of their liver homogenates. Krebs and Perkins[53] used pyrazole to produce a tenfold increase in the level of endogenous ethanol in the portal vein of rats, suggesting that the physiological substrate of liver ADH is ethanol.

c. In Human Subjects

Although toxicity studies have made it clear that pyrazole was too toxic for use in man, Blomstrand et al.[7,47,54–56] were convinced by test reports of their supplier that 4-methylpyrazole did not share this toxicity and thereupon administered low doses to human subjects with ethanol. This work indicated 4-methyl-pyrazole to be a less potent inhibitor in man, perhaps, than in the rat, the maximal dose of 0.12 mmole/kg producing a 49.6% inhibition of the rate of metabolism in one human subject, whereas 0.03 mmole/kg (po) produced 64% inhibition in rats. The inhibitor was reported to have a shorter half-life in man at the low doses given than in rats at much higher doses[7,47]; however, if the metabolic data are approached using zero-order rather than first-order kinetics, the rate of elimination appears to be much *lower* in man than in rats at the doses reported. These kinetics are discussed in detail in Section B.3.b. Salaspuro *et al.*[57] found that, in poorly nourished alcoholics, 4-methylpyrazole had practically no effect on the rate of ethanol metabolism at 0.085 mmole/kg, although this dose produced a mean inhibition of 28% in normal volunteers and in better nourished alcoholics.

Lindros[58] reported 4-methylpyrazole in low doses to prevent the ethanol-induced rise in acetaldehyde blood levels of rats, an effect not seen by him in humans, while Blomstrand[55] found 4-methylpyrazole to reduce the ethanol-induced rise in the lactate/pyruvate ratio in man and to prevent, "to a variable extent," the inhibitory effect of ethanol on fatty acid oxidation.[56]

The only human studies conducted with pyrazole were designed to investigate its antitumor activity in terminal cancer patients[16]: 0.262 mmole/kg per day produced "practically intractable nausea and vomiting, as well as the appearance of severe anemia and marked renal toxicity." Lower doses appeared to be tolerated better, but animal studies implied that liver and bone marrow damage could be taking place.[33] It should be noted that in animal experiments with pyrazole, doses of 1.0–4.4 mmole/kg are commonly used.

Cederbaum and Rubin[59] have shown that 4-methylpyrazole and 4-bromo-pyrazole share a variety of side effects with pyrazole and have cautioned against advocating their therapeutic use in man. Although Lelbach[15] found pyrazole more toxic in the presence of ethanol, Kager and Ericsson[37] found only minor toxic effects when ethanol and 4-methylpyrazole were administered together in a chronic study. However, they recommended "further study . . . before 4-methylpyrazole can be used in adequate repeated doses in man."

D. OTHER EFFECTS OF PYRAZOLES

1. Enzymes Other Than Alcohol Dehydrogenase

a. Aldehyde Dehydrogenase

Koe and Tenen[60] found that chronic ingestion of 4-bromopyrazole or 4-iodopyrazole (1.4–2.1 mmole/kg per day) "diminished the aldehyde dehydrogenase activity of livers of [ethanol-]imbibing mice and somewhat elevated the alcohol dehydrogenase activity," but acute administration had no effect on aldehyde dehydrogenase activity. Lindros *et al.*[61] showed, and Parrilla *et al.*[62] confirmed, that in acetaldehyde-perfused rat liver "the presence of 4-methylpyrazole (0.05 mM) had the effect of diminishing the rate of acetaldehyde oxidation."[62] Rydberg[1] states that pyrazole does not affect aldehyde dehydrogenase, without citing data.

b. Microsomal Ethanol-Oxidizing System

Rubin et al.[64] showed that pyrazole binds to hepatic microsomes to produce a type 2 spectrum, which could account for its inhibition of microsomal ethanol-oxidizing system (MEOS), if such inhibition exists. Some authors[65,66] have reported that pyrazole does not inhibit MEOS, while others[34] suggest that it inhibits MEOS by 10–20%. Still others[67,68] believe that the extent to which pyrazoles inhibit MEOS simply reflects the proportion of ADH in the system.

c. Liver Mitochondria

Cederbaum and Rubin[59] studied the effects of pyrazole, 4-methylpyrazole, and 4-bromopyrazole on isolated rat liver mitochondria; at 1.0–10.0 mM these pyrazoles interfere with several aspects of mitochondrial function, including oxidative phosphorylation, the ATP-^{32}P exchange reaction, and calcium uptake.

d. Catalase

Pyrazoles that are active against ADH have been shown to affect several other enzymes, 4-aminopyrazole being a better inhibitor of catalase[69] than 3-amino-1,2,4-triazole, both in vivo and in vitro. Pyrazole inhibits catalase in vivo (4.4 mmole/kg per day), but not in vitro[34,38]; the former effect occurs only after a delay[39] and is prevented in the presence of alcohols.[39,40] Pyrazole's catalase inhibition is caused by 4-hydroxypyrazole,[71] the major metabolite of pyrazole,[2] which resembles 4-aminopyrazole in having a polar group at the 4-position. 4-Methylpyrazole does not inhibit catalase.[70]

e. Tryptophan Oxygenase and Tyrosine Aminotransferase

Morland et al.[72] reported that pyrazole (10.0 mM) inhibits tryptophan oxygenase in vitro and Rouach et al.[73] found the same effect in vivo, after a delay, at a dose of 4.0 mmole/kg. These results differ from other studies[74–76] which did not detect any inhibition. Since both catalase and tryptophan oxygenase contain heme groups, it was suggested[73] that an active metabolite of pyrazole was the common inhibitor; later studies in vivo on tyrosine aminotransferase, where pyrazole appears to stimulate production of the enzyme,[77] led the same authors to suggest that pyrazole acted via an effect on protein synthesis. Murphy and Clay[12] suggest that such side effects might be mediated by one of the riboside metabolites of pyrazole.

f. Other Enzymes

Lieber et al.,[34] surveying pyrazole's effects on the liver, reported that "*in vitro,* pyrazole inhibited MEOS and microsomal aniline, pentobarbital and benzpyrene hydroxylases" (at 2.0 and 4.0 mM). *In vivo,* a single 4.4-mmole/kg dose "produced a significant decrease of ADH, catalase and MEOS activities, whereas those of benzpyrene hydroxylase and aniline hydroxylase increased." Chronic feeding at 0.95 mmole/kg per day produced decreases in ADH, MEOS, pentobarbital hydroxylase, and benzpyrene hydroxylase activities, with no change in aniline hydroxylase. It should be noted that a microsomal hydroxylase is responsible for the metabolism of pyrazole,[2,39] and these systems are easily saturated with substrate[78] to produce non-first-order kinetics.[2]

2. Central Nervous System Effects

a. In the Absence of Ethanol

It has been known for about 20 years that pyrazoles have CNS depressant effects. Owen et al.[41] showed that pyrazole and its 3-substituted derivatives are good anticonvulsants in rodents; since the 3-substitution negates ADH inhibition,[19] this implies that a non-ADH mechanism is involved. The compounds with larger substituents in the 3-position had greater anticonvulsant activity and were more toxic. Karmas[79] and Easton[80] filed patents on 4-alkyl- and 4-chloropyrazoles, respectively, as anticonvulsants for human use and Raevskii et al.[42] found 4-alkylpyrazoles to reduce "muscle tone, motor activity and body temperature." Feldstein[43] has shown that pyrazole produces "sleep" in mice at 10 mmole/kg and that it potentiates tryptophol-induced sleep at 1.5 mmole/kg, presumably by inhibiting the metabolism of tryptophol.

While brain norepinephrine in rats decreases after administration of pyrazole (0.74 mmole/kg per day, 4 days),[45] the effect is not seen with 4-methylpyrazole (0.61 mmole/kg per day) or 4-iodopyrazole (0.26 mmole/kg per day),[44,45] suggesting it may be due to a metabolite unique to pyrazole. MacDonald[45] and MacDonald et al.[44] conclude, from the fact that pyrazole's 4-substituted analogues do not share its depleting effect on brain norepinephrine, that this effect is not related to its inhibition of ADH,[45] but might be mediated through inhibition of dopamine β-hydroxylase. These ideas could be tested in two ways: (1) if pyrazole activity is mediated via 4-hydroxypyrazole, then in the presence of ethanol the inhibition of 4-hydroxypyrazole production should eliminate any brain norepinephrine decrease; and (2) 3-methylpyrazole may be expected to decrease brain norepinephrine via its possible 4-hydroxy derivative (which neither 4-methylpyrazole nor 4-iodopyrazole can form) and such an action might also be blocked in the presence of ethanol.

Pyrazole (4.3 mmole/kg, ip) more than doubled urine volume of rats from 24 to 48 hr after injection, without affecting water intake; the diuresis was neither enhanced by ethanol (1 g/kg, ip) given at 24 hr nor was the diuresis evident when the rats were treated with ethanol alone. Crow and Thompson[82] were unable to ascribe the diuresis either to changes in extracellular fluid composition or to suppression of antidiuretic hormone.

b. In the Presence of Ethanol

Goldberg and Rydberg[51] first reported ethanol-induced depression (loss of righting reflex) to last much longer after a dose of pyrazole and also to be noticeably deeper, especially with higher doses of pyrazole (1.20–8.82 mmole/kg). A later paper[83] reported that the combined impairment (tilting plane) from pyrazole (1.77 mmole/kg) and ethanol (32.6 mmole/kg) was greater than the sum of the impairments produced individually by the two drugs. Inasmuch as only single doses of each compound were employed, it appears unwarranted to call such an effect synergistic. Rydberg and Neri[84] obtained less than additive results with single doses of ethanol and 4-methylpyrazole. LeBlanc and Kalant,[85] using the moving belt test, found that doses of 0–3.5 mmole pyrazole/kg progressively increased the degree of impairment produced by ethanol (30–43 mmole/kg). They interpreted this to mean that pyrazole independently depresses the CNS, perhaps producing its own tolerance and withdrawal syndrome. Recently, Thurman and Pathman[87] reported qualitative differences in withdrawal produced with and without pyrazole that seem to support this interpretation.

Crow et al.[88], using shuttle-box avoidance and fixed-ratio operant responding, found no synergic effect of pyrazole (3.3 and 4.3 mmole/kg) with alcohol (22 and 54 mmole/kg), only a prolongation of ethanol impairment. They suggest that Goldberg's results reflect a primary motor impairment more than central depression.

Blum et al.[89] studied interactions of ethanol and pyrazole (and 3-methylpyrazole, a noninhibitor of ADH) on sleeping time in mice and rotor rod balance and lever-pressing behavior in rats. Pyrazole (3.5 mmole/kg) and 3-methylpyrazole (3.5 mmole/kg), when administered before ethanol, resulted in an increased behavioral depression, apparently caused by direct CNS depression.

More recent papers by Blum et al.[90,91] use the Goldstein[86] inhalation method to study brain amines during withdrawal from ethanol. If, as discussed above, the nature of the withdrawal reaction[87] and levels of brain catecholamines[44,45] are altered by pyrazole, its daily use during ethanol inhalation might render the results invalid—unless such side effects are mediated by a metabolite, such as 4-hydroxypyrazole; such a metabolite would be expected to appear in significant concentrations in the absence of ethanol but not be present at the levels of ethanol reached during inhalation. This possibility remains untested.

Pyrazole prevents the ethanol-induced decrease in mouse cerebral acetylcholine levels,[92] while the ethanol-induced increase in accumulation of tritiated brain catecholamines formed from [³H]tyrosine in mice was "if anything, enhanced by pretreatment with 4-methylpyrazole."[93] Since acetaldehyde also decreases brain acetylcholine,[92] ethanol *per se* apparently has no effect on this neurotransmitter, while the effect on catecholamine transmitters is in some part a result of the direct action of ethanol.

E. PYRAZOLES AS RESEARCH TOOLS

1. Nonethanol ADH Substrates

a. Outside the Central Nervous System

The metabolism of a variety of compounds is inhibited *in vivo* (Table I) and *in vitro* (Table II) by pyrazoles. The tabular information has several points of interest. Note that while Watkins and Tephly[110] found no inhibition of retinol oxidation *in vitro* at a low pyrazole concentration, Raskin *et al.*[109] found inhibition *in vivo* at about the LD_{50} of pyrazole. The large doses of ethylene glycol used are necessary to produce fatalities; where the toxicity arises from the metabolite (of ethylene glycol, methanol, trifluoroethanol, etc.), ethanol often works as well as pyrazole as a protective agent—and with fewer and lesser toxic side effects.

b. Within the Central Nervous System

1,4-Butanediol is normally a hypnotic (5.5 mmole/kg), except in the presence of pyrazole (6.6 mmole/kg),[99,115] which prevents its oxidation to γ-hydroxybutyrate (GHB). Pyrazoles prolong sleep induced by γ-butyrolactone,[99,115,116] tryptophol,[117] hexobarbital,[42,52] 2-phenylethanol,[52] and 3-methoxy-4-hydroxyphenylethanol,[118] presumably by a combination of direct CNS depression and inhibited metabolism. With GHB, only indirect evidence (prolongation of sleep by pyrazole or ethanol) points to an inhibition of its metabolism,[115] while direct evidence offers no support for GHB being an ADH substrate.[119,120]

c. Aldehyde Reduction

Pyrazole, in complexing with ADH, also prevents aldehyde reduction. Thus, Lindros *et al.*[121] have used 4-methylpyrazole to stop acetaldehyde reduction in perfused liver. This "reverse" inhibition of ADH has been used to

Table I. Substrate Metabolism Inhibited by Pyrazoles

ADH Substrate	Substrate (mmole/kg)	Pyrazole dose[a] (mmole/kg)	Species	Inhibition (%)	Reference
Methanol	32.3, ip	2.94–0.74, ip	Monkey	80–50	40
	32.3, ip	2.94, ip	Rat	56	40
	31.3, ip	6.6, ip	Rat	57	6
	65.2, po	0.61 (MP), ip	Monkey	75	94
Propanol	16.7, ip	6.6, ip	Rat	57	6
Isopropanol	16.7, ip	6.6, ip	Rat	38	6
	16.7, ip	4.4, po	Rat	75	95
Butanol	6.8, ip	6.6, ip	Rat	54	6
Isobutanol	6.8, ip	6.6, ip	Rat	49	6
Ethylene glycol	215–394,[b] po	4.4, ip	Mouse	[c]	96
	242, po	2.2, ip	Rat, dog	No deaths	97
	180, po	0.5–3.0 (P,MP)[d]	Rat	[c]	98
1,4-Butanediol	5.55, ip	6.6, ip	Rat	[c]	99
Allyl alcohol	0.73, ip	5.5, ip	Rat	[c]	100
	1.08, ip	5.5, ip	Rat	No deaths	101
Allyl formate	0.82, ip	5.3, ip	Rat	No deaths	101
Thiamine	0.0015, ip	2.94, ip	Rat	47	102
"Thiazole"[e]	0.69, ip	2.94, ip	Rat	78	102
Vinyl chloride	65 ppm, inh	4.7, ip	Rat	87[f]	103
monomer	1234 ppm, inh	4.7, ip	Rat	71[f]	103
Fluoroxene	Lethal, inh	0.015, ip	Mouse	No deaths	104
Trifluoroethanol	1.95–4.50,[b] ip	0.52 (IP), ip	Mouse	[c]	105
L-1-Deoxyfluoroglycerol	0.15–0.60, ip	1.3, ip	Mouse	50	106
Diethylnitrosamine	0.067, ip	0.59[d]	Rat	70	107
Dimethyl	0.067, ip	2.9[d]	Rat	95	108
nitrosamine	0.067, ip	0.59[d]	Rat	80	108
Nitrotoluene	3.65, po	1.47, ip	Rat	67	52
	3.65, po	0.34 (BP), ip	Rat	63	52
Retinol	Endogenous	14.7, ip	Rat	[c]	109

[a]Pyrazole, unless otherwise noted. MP, 4-methylpyrazole; BP, 4-bromopyrazole; IP, 4-iodopyrazole.
[b]LD_{50} with and without pyrazole.
[c]Percent not calculable.
[d]Route not reported.
[e]5-(2-Hydroxyethyl-)-4-methylthiazole.
[f]Measured by difference in rate of disappearance from chamber air.

differentiate pyrazole noninhibited aldehyde reductases from ADH,[122,123] especially in the CNS.[124–127]

2. Pyrazole–Alcohol Dehydrogenase Complex

a. Characterization of Alcohol Dehydrogenases

Pyrazole's inhibitory potency varies widely with known alcohol dehydrogenases. Thus, horse liver ADH ($K_i = 0.2$ μM) is 100 times more affected by pyrazole than is yeast ADH ($K_i = 29$ μM).[20] Human ($K_i = 2.6$ μM),[128] mouse

Table II. Pyrazole Inhibition of Metabolism *in Vitro*

ADH Substrate	Substrate (mM)	Pyrazole (mM)	Tissue	Inhibition (%)	References
Retinol	0.0011	0.001	Rat retina homogenate	0	110
17-DL-Hydroxystearic acid	0.055	0.1	Rat LADH	50	111
1,3-Butanediol	1.0	0.01	Rat LADH	92	112, 113
1,4-Butanediol	0.0018–0.037	0.025	Rat LADH	$K_i = 3–15\ \mu M$	99
Dimethyl nitrosamine	1.0	1.0	Rat LADH	65	114
"Thiazole"[a]	2.5–5.0	Varied	Rat LADH	$K_i = 12.5\ \mu M$	102

[a] 5-(2-Hydroxyethyl-)-4-methylthiazole.

($K_i = 0.8\ \mu M$),[9] and rat ($K_i = 4.2\ \mu M$)[20] liver ADHs are inhibited by pyrazole at intermediate concentrations. Von Wartburg[129] found that pyrazole inhibits atypical human liver ADH more strongly than normal ADH, and Raskin and Sokoloff[130] found that neural ADH closely resembles liver ADH in its interaction with pyrazole. ADHs from rat stomach and from a hepatocellular carcinoma[131] interact weakly with pyrazole ($K_i = 2000\ \mu M$).

b. Chromatography and Chemical Alteration of Alcohol Dehydrogenase

The strong pyrazole–NAD⁺–ADH complex has been used to protect the active site of ADH during chemical modifications of the enzyme.[132,133] Andersson *et al.*[134] have specifically eluted ADH from an affinity column by washing with NAD⁺ and pyrazole, and Lange and Vallee[135] have used 4-(3-aminopropyl)-pyrazole as the ADH-binding agent in an affinity chromatography column.

3. Interactions between Ethanol and Pyrazoles

a. Ethanol Effects Reversed by Pyrazoles

Ethanol produces profound effects on the redox state of the liver (see Volume 1, Chapter 10). (In the discussion that follows, 4-methylpyrazole use is indicated by a single asterisk.) When pyrazole reverses an ethanol-induced change, some authors have attributed the change to alteration of the redox state, which pyrazole prevents,[136*] and others have implicated acetaldehyde, whose concentration pyrazole reduces,[19] as the causative factor.[137,138*] Thus, pyrazole prevents the ethanol-induced *increase* in choline uptake in the liver,[139] transport of Mn in the rat jejunum,[140] activity of aminolevulinic acid synthetase,[141] half-life of galactose in humans,[57*] reduction of dehydroepiandrosterone in liver (4-bromopyrazole),[142] and activity of tryptophan oxygenase.[74]

Pyrazole reverses the ethanol-induced *decrease* in protein synthesis,[143] α-aminoisobutyric acid uptake in liver,[144] gluconeogenesis,[145] tryptamine metabolism to indoleacetic acid,[146] brain glycogen levels, [137] hepatic pyridoxal phosphate levels,[138*] and triolein oxidation in man.[56*]

Changes in various brain amines produced by ethanol are prevented by pyrazole.[92] Pyrazole reduces the incorporation of ^2H from dideuterated ethanol into bile acids[147] and abolishes the inhibitory effect of ethanol on oxidation of various substrates in rat kidney cortex slices.[148]

b. Ethanol Effects Not Reversed by Pyrazoles

Pyrazole prolongs ethanol-induced hypocalcemia[149] and potentiates the ethanol-induced slowing of amphetamine parahydroxylation.[150] It appears to increase the myocardial depressant effect of ethanol in dogs,[151] but has no effect on ethanol's decreasing ATP in intestine,[152] ethanol's inhibition of insulin secretion,[153] or ethanol mortality in hyperthyroid rats.[154]

c. Fatty Liver

Lelbach[15] administered ethanol with pyrazole to determine whether fatty liver is caused by ethanol *per se* or by metabolic changes produced by ethanol. His result, liver damage in rats at low ethanol doses, could be interpreted to mean that ethanol *per se* is the agent in liver damage, or simply that pyrazole damages the liver in the presence of ethanol. Bustos *et al.*[155] reported that pyrazole did not prevent the ethanol-induced rise in liver triglycerides, whereas Morgan and DiLuzio[156] reported that this increase was prevented by pyrazole. The two papers differed in several experimental details, subsequent papers attributing the conflicting results to differences in sex,[157] ethanol dose,[158] or time of measurement.[159,160] After some years of debate about whether pyrazole does in fact prevent the ethanol-induced fatty liver and what the implications are, certain of the participants have declared the question moot,[161,162] owing to confounding direct effects of pyrazole and 4-methylpyrazole on, for example, ethanol absorption and liver triglycerides. Others have since published on both the "ethanol *per se*"[163] and the "redox state"[164] sides of the debate. A thorough review of the points involved, from the former viewpoint, may be found in Khanna and Kalant[165] and from the latter in DiLuzio *et al.*[101]

Where Bizzi *et al.*[166,167] had shown that 3,5-dimethylpyrazole and its active metabolite, 3-carboxy-5-methylpyrazole (CMP), were potent antilipolytic agents *in vivo,* Giudicelli *et al.*[68,169] found pyrazole inactive against ACTH-stimulated lipolysis, indicating a different mechanism of action from that of CMP. Pyrazole reportedly decreases liver damage after isopropanol,[170,171] apparently implicating acetone as the causative agent.[171] It is a matter of dispute whether[172,173] or not[174] pyrazole prevents liver damage due to carbon tetrachloride.

d. Ethanol Withdrawal Syndrome (Goldstein Method)

The Goldstein method for producing ethanol dependence in mice[86,175] relies on daily injections of pyrazole to slow metabolism of inhaled ethanol (maintained at a constant concentration in the inspired air), thereby increasing the blood level of alcohol and reducing its fluctuation. The doses are low (1 mmole/kg) and the course of treatment is short, 3 days. Pyrazole's catalase inhibition would presumably be prevented by the presence of ethanol. Liver damage could begin, but probably would not affect the results. Criticism that pyrazole has a direct CNS effect[85] would not be remedied by changing to 4-methylpyrazole. Goldstein and Pal[9] considered using 4-bromopyrazole to lower toxicity but found its lower potency *in vivo* and shorter duration of action not in its favor. Recently, Thurman and Pathman[87] suggested that profound qualitative differences exist between withdrawal syndrome with and without pyrazole, stating: "Not only does pyrazole produce withdrawal itself, this reaction is markedly modified by the nutritional state of the animals. This is perhaps the most essential criticism of Goldstein and Pal."

e. Existence of Microsomal Ethanol-Oxidizing System

Pyrazoles have been employed in the MEOS controversy throughout the decade. Lieber and DeCarli[176] reported that pyrazole had little or no effect on the MEOS system. Since then, the use of pyrazoles to block the ADH pathway completely has established conclusively that other pathways of ethanol oxidation exist *in vivo*.[136,177] Controversies still exist, principally over the percentage of ethanol metabolized via non-ADH pathways in various circumstances and over the contributions of catalase or MEOS.

McCaffrey and Thurman[178] assert that the low maximal inhibition of ethanol metabolism seen in several studies[179,180] has been due to the use of pyrazole, since "others[181] have clearly and consistently demonstrated 80–90% inhibition of ethanol utilization with 4-methylpyrazole." However, there seems to be agreement that pyrazole administered to rats *in vivo* can produce 80–90% inhibition of the rate of ethanol metabolism[6,51] except when administered intragastrically,[182] perhaps because pyrazole is bound in the stomach.[2] 4-Methylpyrazole exhibits a similar maximum inhibition *in vivo*.[83,183] *In vitro,* in hepatocytes, when the two inhibitors were used in the same experiment,[184] they produce the same maximal inhibition (79%) of ethanol oxidation.

Thieden[185] and Grunnet *et al.*[179] correlated the inhibition produced by 18 mM pyrazole in rat liver cells and slices with the concentration of ethanol used and have reported that the rate, which is 90% slowed by pyrazole at 4 mM ethanol, is only about 50% slowed at 65 mM. The inhibition was measured by formation of acetate. Calculation using appropriate values for K_m and K_i will show that this difference cannot simply be the result of competitive interaction between ethanol and pyrazole; the ADH should still be almost completely inhibited at the higher ethanol level. Schulman et al.[186] criticized the method used for estimating ethanol disappearance in the papers mentioned above; measuring the disappearance of ethanol itself, they obtained 80% inhibition of ethanol metabolism by 4-methylpyrazole in hepatocytes at either 6 or 100 mM ethanol.

Other workers, using perfused liver, have found surprisingly low levels of inhibition despite low ethanol levels. Papenberg et al.,[187] with 3 mM pyrazole and 4.3 mM ethanol, found only 62% inhibition. In this case pyrazole was not added until 1 or 2 hr after ethanol perfusion was started, suggesting a possible protection by ethanol of the enzyme in vitro. McCaffrey and Thurman[178] obtained only 71% inhibition in perfused liver with 4 mM 4-methylpyrazole against 25 mM ethanol. The consequent low estimates of the proportion of ethanol oxidation due to ADH do not appear to depend on the use of pyrazole instead of 4-methylpyrazole. In cases where the inhibition falls below about 75%, it is probable that a point of technique has caused the in vitro system to cease resembling the state of affairs in vivo. This judgment is supported by recent work[63] in vivo where the rate of ethanol metabolism was invariant with doses of 11–87 mmole ethanol/kg and all doses were equally inhibited (83–92%) by 4-methylpyrazole (2.0 mmole/kg).

F. SUMMARY AND CONCLUSIONS

Pyrazoles have proven useful and powerful tools for researchers studying the biochemical pharmacology of ethanol. This review catalogs and discusses pyrazole work which is relevant to ethanol. One of the most promising aspects of this research appears to lie in the separation of ethanol's effect on the NADH/NAD$^+$ ratio from its other effects.

Knowledge of the metabolism of pyrazole, and the effects of its various metabolites on liver and brain, will be useful for interpreting early papers in which pyrazole was used without due consideration for its side effects. It has been suggested that pyrazole be used therapeutically in preventing poisoning from ethylene glycol, methanol, and the disulfiram–ethanol reaction syndrome. The elucidation of pyrazole's toxicity makes such use appear unwise, but further study of 4-methylpyrazole, 4-propylpyrazole, or 4-decylpyrazole could show that these analogues are safe and effective enough to be used in man for the purposes described. The best evidence for the safety of 4-alkylpyrazoles would be proof that pyrazole's side effects are caused by a unique metabolite. Now that the metabolites of pyrazole and 4-methylpyrazole are known, such work can proceed.

G. REFERENCES*

1. U. Rydberg, Opusc. Med. Suppl. 26:1 (1972).
2. F. H. Deis, G. W.-J. Lin, and D. Lester, in: Alcohol and Aldehyde Metabolizing Systems (R. G.

*A more complete set of references, with titles and complete pagination, may be obtained from the authors.

Thurman, J. R. Williamson, H. R. Drott, and B. Chance, eds.), Vol. 3, pp. 399–405, Academic Press, Inc., New York (1977).
3. U. Rydberg, J. C. Buijten, and A. Neri, *J. Pharm. Pharmacol.* **24**:651 (1972).
4. U. Rydberg and J. C. Buijten, *J. Chromatogr.* **64**:170 (1972).
5. T. R. Ward and S. N. Pennington, *Mikrochim. Acta* **2**:297 (1973).
6. D. Lester and G. D. Benson, *Science* **169**:282 (1970).
7. I. Bjorkhem, R. Blomstrand, O. Lantto, and L. Svensson, *Biochem. Med.* **12**:205 (1975).
8. J. P. Peyre and M. Reynier, *Ann. Pharm. Fr.* **27**:749 (1969).
9. D. B. Goldstein and N. Pal, *J. Pharmacol. Exp. Ther.* **178**:199 (1971).
10. R. C. Murphy and W. D. Watkins, *Biochem. Biophys. Res. Commun.* **49**:283 (1972).
11. R. Blomstrand and G. Ohman, *Life Sci.* **13**:107 (1973).
12. R. C. Murphy and K. L. Clay, *Fed. Proc.,* Abstr. **36**:938 (1977).
13. H. Diekmann and A. Garbe, *Naunyn Schmiedeberg's Arch. Pharmacol.,* Suppl. **287**:R77 (1975).
14. G. Magnusson, J. A. Nyberg, N. O. Bodin, and E. Hansson, *Experientia* **28**:1198 (1972).
15. W. K. Lelbach, *Experientia* **25**:816 (1969).
16. H. T. Foley, B. I. Shnider, G. L. Gold, and Y. Uzer, *Cancer Chemother. Rep.* **44**:45 (1965).
17. H. Theorell and T. Yonetani, *Biochem. Z.* **338**:537 (1963).
18. H. Theorell, T. Yonetani, and B. Sjoberg, *Acta Chem. Scand.* **23**:255 (1969).
19. D. Lester, W. Z. Keokosky, and F. Felzenberg, *Q. J. Stud. Alcohol* **29**:449 (1968).
20. M. Reynier, *Acta Chem. Scand.* **23**:1119 (1969).
21. R. Dahlbom, B. R. Tolf, A. Akeson, G. Lundquist, and H. Theorell, *Biochem. Biophys. Res. Commun.* **57**:549 (1974).
22. R. Pietruszko, *Biochem. Pharmacol.* **24**:1603 (1975).
23. H. R. Drott, in: *Alcohol and Aldehyde Metabolizing Systems* (R. G. Thurman, B. Chance, J. R. Williamson, and T. Yonetani, eds.), Vol. 2, p. 530, Academic Press, Inc., New York (1974).
24. C. Bode, *Rev. Alcool.* **18**:256 (1972).
25. H. Theorell, *Harvey Lect.* **1967**:18 (1967).
26. D. S. Sigman, *J. Biol. Chem.* **242**:3815 (1967).
27. J. M. Young and J. H. Wang, *J. Biol. Chem.* **246**:2815 (1971).
28. J. D. Shore and D. Santiago, *J. Biol. Chem.* **250**:2008 (1975).
29. H. Theorell and T. Yonetani, in: *Structure and Activity of Enzymes* (T. W. Goodwin, J. I. Harris, and B. S. Hartley, eds.), pp. 131–132, Academic Press, Inc., London (1964).
30. J. D. Shore and M. J. Gilleland, *J. Biol. Chem.* **245**:3422 (1970).
31. P. L. Luisi, A. Baici, F. J. Bonner, and A. A. Aboderin, *Biochemistry* **14**:362 (1975).
32. A. Rosenberg, H. Theorell, and T. Yonetani, *Arch. Biochem. Biophys.* **110**:413 (1965).
33. W. L. Wilson and N. G. Bottiglieri, *Cancer Chemother. Rep.* **1**:137 (1962).
34. C. S. Lieber, E. Rubin, L. M. DeCarli, P. Misra, and H. Gang, *Lab. Invest.* **22**:615 (1970).
35. R. Blomstrand, A. Lof, and H. Ostling, in: *Alcohol and Aldehyde Metabolizing Systems* (R. G. Thurman, J. R. Williamson, H. R. Drott, and B. Chance, eds.), Vol. 3, pp. 427–436, Academic Press, Inc., New York (1977).
36. M. J. Phillips, H. F. Chiu, J. M. Khanna, and H. Kalant, in: *Alcoholic Liver Pathology* (J. M. Khanna, Y. Israel, and H. Kalant, eds.), pp. 271–288, Alcoholism and Drug Addiction Research Council of Ontario, Toronto (1973).
37. L. Kager and J. L. Ericsson, *Acta Pathol. Microbiol. Scand.* **82**:534 (1974).
38. E. Feytmans and F. Leighton, *Biochem. Pharmacol.* **22**:349 (1973).
39. E. Feytmans, M. N. Morales, and F. Leighton, *Biochem. Pharmacol.* **23**:1293 (1974).
40. W. D. Watkins, J. I. Goodman, and T. R. Tephly, *Mol. Pharmacol.* **6**:567 (1970).
41. J. E. Owen, Jr., E. E. Swanson, and D. B. Meyers, *J. Am. Pharm. Assoc. Sci. Ed.* **47**:70 (1958).
42. K. S. Raevskii and Yu. M. Batulin, *Farmakol. Toksikol.* **26**:551 (1963).
43. A. Feldstein and V. DaForno, *Neuroscience,* Abstr. **2**:1085 (1976).
44. E. MacDonald, M. Marselos, and U. Nousianinen, *Acta Pharmacol. Toxicol.* **37**:106 (1975).
45. E. MacDonald, *Acta Pharmacol. Toxicol.* **39**:513 (1976).
46. M. Nachman, D. Lester, and J. LeMagnen, *Science* **168**:1244 (1970).
47. R. Blomstrand and H. Theorell, *Life Sci.* **9**:631 (1970).

48. N. Oshino, D. Jamieson, T. Sugano, and B. Chance, *Biochem. J.* **146**:67 (1975).
49. H. J. Brentzel and R. G. Thurman, in: *Alcohol and Aldehyde Metabolizing Systems* (R. G. Thurman, J. R. Williamson, H. R. Drott, and B. Chance, eds.), Vol. 2, Academic Press, Inc., New York (1977).
50. U. Rydberg, *Biochem. Pharmacol.* **18**:2424 (1969).
51. L. Goldberg and U. Rydberg, *Biochem. Pharmacol.* **18**:1749 (1969).
52. M. Reynier, *Agressologie* **11**:401, 407 (1970).
53. H. A. Krebs and J. R. Perkins, *Biochem. J.* **118**:635 (1970).
54. R. Blomstrand, in: *Metabolic Changes Induced by Alcohol* (G. Martini and C. Bode, eds.), pp. 38–52, Springer-Verlag, Berlin, New York (1971).
55. R. Blomstrand, in: *Structure and Function of Oxidation–Reduction Enzymes* (A. Akeson and A. Ehrenberg, eds.), pp. 667–679, Pergamon Press, Inc., Elmsford, N.Y. (1972).
56. R. Blomstrand and L. Kager, in: *Alcohol and Aldehyde Metabolizing Systems* (R. G. Thurman, B. Chance, J. R. Williamson, and T. Yonetani, eds.), Vol. 1, pp. 339–350, Academic Press, Inc., New York (1974).
57. M. P. Salaspuro, K. O. Lindros, and P. Pikkarainen, *Ann. Clin. Res.* **7**:269 (1972).
58. K. O. Lindros, in: *The Role of Acetaldehyde in the Actions of Ethanol* (K. O. Lindros and C. J. P. Eriksson, eds.), pp. 67–81, The Finnish Foundation for Alcohol Studies, Helsinki (1975).
59. A. I. Cederbaum and E. Rubin, *Biochem. Pharmacol.* **23**:203 (1974).
60. B. K. Koe and S. S. Tenen, *Biochem. Pharmacol.* **24**:723 (1975).
61. K. O. Lindros, R. Vihma, and O. A. Forsander, *Biochem. J.* **126**:945 (1972).
62. R. Parrilla, K. Okawa, K. O. Lindros, U. J. Zimmerman, K. Kobayashi, and J. R. Williamson, *J. Biol. Chem.* **249**:4926 (1974).
63. J. M. Khanna, K. Lindros, H. Kalant, Y. Israel, and H. Orrego, in: *Alcohol and Aldehyde Metabolizing Systems* (R. G. Thurman, J. R. Williamson, H. R. Drott, and B. Chance, eds.), Vol. 3, pp. 325–334, Academic Press, Inc., New York (1977).
64. E. Rubin, H. Gang, and C. S. Lieber, *Biochem. Biophys. Res. Commun.* **42**:1 (1971).
65. R. Teschke, Y. Hasumura, and C. S. Lieber, *Arch. Biochem. Biophys.* **163**:404 (1974).
66. J. M. Khanna, H. Kalant, and G. Lin, *Biochem. Pharmacol.* **19**:2493 (1970).
67. K. J. Isselbacher and E. A. Carter, *Biochem. Biophys. Res. Commun.* **39**:530 (1970).
68. M. P. Schulman and K. P. Vatsis, *Fed. Proc.*, Abstr. **36**:333 (1977).
69. R. N. Feinstein, J. E. Seaholm, and L. B. Ballonoff, *Enzymologia* **27**:30 (1964).
70. A. B. Makar and T. R. Tephly, *Biochem. Med.* **13**:334 (1975).
71. F. H. Deis, G. W.-J. Lin, and D. Lester, *FEBS Lett.* **78**:81 (1977).
72. J. Morland, T. Christofferson, J. B. Osnes, P. O. Seglen, and K. F. Jervell, *Biochem. Pharmacol.* **21**:1849 (1972).
73. H. Rouach, C. Ribiere, J. Nordmann, and R. Nordmann, *Life Sci.* **19**:505 (1976).
74. G. K. Kumar and V. M. Sardesai, *Biochem. Med.* **16**:143 (1976).
75. A.-A. Badawy and M. Evans, in: *Alcohol Intoxication and Withdrawal: Experimental Studies* (M. M. Gross, ed.), Vol. I, pp. 105–123, Plenum Press, New York (1973).
76. A.-A. Badawy and M. Evans, in: *Alcohol Intoxication and Withdrawal: Experimental Studies* (M. M. Gross, ed.), Vol. II, pp. 229–251, Plenum Press, New York (1975).
77. C. Ribiere, H. Rouach, R. Nordmann, and J. Nordmann, *FEBS Lett.* **64**:419 (1976).
78. T. Koizumi, M. Ueda, M. Kakemi, J. Shibasaki, S. Matsumoto, and R. Shinagawa, *Chem. Pharm. Bull. (Tokyo)* **22**:988 (1974).
79. G. Karmas, U.S. Patent 2,931,814 (1960).
80. N. R. Easton, U.S. Patent 2,992,163 (1961).
81. Clay, K. L., Watkins, W. D. and Murphy, R. C., *Drug Metab. Dispos.* **5**:149 (1977).
82. L. T. Crow and J. R. Thompson, *Psychol. Rep.* **32**:1209 (1973).
83. L. Goldberg, C. Hollstedt, A. Neri, and U. Rydberg, *J. Pharm. Pharmacol.* **24**:593 (1972).
84. U. Rydberg and A. Neri, *Acta Pharmacol. Toxicol.* **31**:421 (1972).
85. A. E. LeBlanc and H. Kalant, *Can. J. Physiol. Pharmacol.* **51**:612 (1973).
86. D. B. Goldstein and N. Pal, *Science* **172**:288 (1971).

87. R. G. Thurman and D. E. Pathman, in: *The Role of Acetaldehyde in the Actions of Ethanol* (K. O. Lindros and C. J. P. Eriksson, eds.), pp. 217–231, The Finnish Foundation for Alcohol Studies, Helsinki (1975).

88. L. T. Crow, C. S. Edelbrock, and P. A. Martin, *Physiol. Psychol.* **3**:56 (1975).

89. K. Blum, I. Geller, and J. E. Wallace, *Br. J. Pharmacol.* **43**:67 (1971).

90. K. Blum and J. E. Wallace, *Br. J. Pharmacol.* **51**:109 (1974).

91. K. Blum, J. D. Eubanks, J. E. Wallace, and H. A. Schwertner, *Experientia* **32**:493 (1976).

92. A. K. Rawat, *J. Neurochem.* **22**:915 (1974).

93. T. H. Svensson and B. Waldeck, *Psychopharmacologia* **31**:229 (1973).

94. K. E. McMartin, A. B. Makar, G. Martin-Amat, M. Palese, and T. R. Tephly, *Biochem. Med.* **13**:319 (1975).

95. R. Nordmann, C. Ribiere, H. Rouach, and J. Nordmann, *Life Sci.* **13**:919 (1973).

96. R. L. Mundy, L. M. Hall, and R. S. Teague, *Toxicol. Appl. Pharmacol.* **28**:320 (1974).

97. E. W. Van Stee, A. M. Harris, M. L. Horton, and C. K. Back, *J. Pharmacol. Exp. Ther.* **192**:251 (1975).

98. K. E. Richardson and J. Chou, *Fed. Proc., Abstr.* **36**:938 (1977).

99. S. P. Bessman and E. McCabe, *Biochem. Pharmacol.* **21**:1135 (1972).

100. W. D. Reid, *Experientia* **28**:1058 (1972).

101. N. R. DiLuzio, T. Stege, and E. O. Hoffman, in: *Alcoholic Liver Pathology* (J. M. Khanna, Y. Israel, and H. Kalant, eds.), pp. 245–260, Alcoholism and Drug Addiction Council of Ontario, Toronto (1973).

102. R. R. Dalvi, H. E. Sauberlich, and R. A. Neal, *J. Nutr.* **104**:1476 (1974).

103. R. E. Hefner, Jr., P. G. Watanabe, and P. J. Gehring, *Environ. Health Perspect.* **11**:85 (1975).

104. H. F. Cascorbi, *Int. Anesthesiol. Clin.* **12**:107 (1974).

105. M. M. Airaksinen, P. H. Rosenberg, and T. Tammisto, *Acta Pharmacol. Toxicol.* **28**:299 (1970).

106. T. P. Fondy, R. W. Pero, K. L. Karker, G. S. Ghangas, and F. H. Batzold, *J. Med. Chem.* **17**:697 (1974).

107. J. C. Phillips, B. G. Lake, M. J. Minski, S. D. Gangolli, and A. G. Lloyd, *Biochem. Soc. Trans.* **3**:285 (1975).

108. J. C. Phillips, C. E. Heading, B. G. Lake, S. D. Gangolli, and A. G. Lloyd, *Biochem. Soc. Trans.* **2**:885 (1974).

109. N. H. Raskin, K. P. Sligar, and R. H. Steinberg, *Brain Res.* **50**:496 (1973).

110. W. D. Watkins and T. R. Tephly, *J. Neurochem.* **18**:2397 (1971).

111. I. Bjorkhem, *Eur. J. Biochem.* **30**:441 (1972).

112. R. L. Tate, M. A. Mehlman, and R. B. Tobin, *J. Nutr.* **101**:1719 (1971).

113. M. A. Mehlman, R. B. Tobin, and C. R. Mackerer, *Fed. Proc. Symp.* **34**:2182 (1975).

114. B. G. Lake, M. J. Minski, J. C. Phillips, C. E. Heading, S. D. Gangolli, and A. G. Lloyd, *Biochem. Soc. Trans.* **3**:183 (1975).

115. P. V. Taberner, J. T. Rick, and G. A. Kerkut, *Life Sci.* **11**:335 (1972).

116. D. Benton, C. P. Kyriacou, J. T. Rick, and P. V. Taberner, *Eur. J. Pharmacol.* **27**:288 (1974).

117. A. Feldstein and J. M. Kucharski, *Life Sci.* **10**:961 (1971).

118. L. D. Waterbury, O. T. Wendell, and L. A. Pearch, *Res. Commun. Chem. Pathol. Pharmacol.* **6**:855 (1973).

119. J. D. Doherty, R. W. Stout, and R. H. Roth, *Biochem. Pharmacol.* **24**:469 (1975).

120. R. H. Roth and N. J. Giarman, *Biochem. Pharmacol.* **15**:1333 (1966).

121. K. O. Lindros, N. Oshino, R. Parrilla, and J. R. Williamson, *J. Biol. Chem.* **249**:7956 (1974).

122. I. Bjorkhem, H. Danielsson, and K. Wikvall, *Eur. J. Biochem.* **36**:8 (1973).

123. J. B. Silverman, P. S. Babiarz, K. P. Mahajan, J. Buschek, and T. P. Fondy, *Biochemistry* **14**:2252 (1975).

124. R. A. Deitrich, A. C. Collins, and V. G. Erwin, *Biochem. Pharmacol.* **20**:2663 (1971).

125. M. M. Ris and J. P. von Wartburg, *Res. Commun. Chem. Pathol. Pharmacol.* **7**:217 (1974).

126. B. Tabakoff, C. Vugrincic, R. Anderson, and S. G. Alivatsos, *Biochem. Pharmacol.* **23**:455 (1974).

127. J. P. von Wartburg, D. Berger, M. M. Ris, and B. Tabakoff, in: *Alcohol Intoxication and Withdrawal: Experimental Studies* (M. M. Gross, ed.), Vol. II, pp. 119–138, Plenum Press, New York (1975).

128. T. K. Li and H. Theorell, *Acta Chem. Scand.* **23**:892 (1969).

129. J. P. von Wartburg and P. M. Schurch, *Ann. N.Y. Acad. Sci.* **151**:936 (1968).

130. N. H. Raskin and L. Sokoloff, *J. Neurochem.* **19**:273 (1972).

131. A. I. Cederbaum, R. Pietruszko, J. Hempel, F. F. Becker, and E. Rubin, *Arch. Biochem. Biophys.* **171**:348 (1975).

132. R. Dworschak, G. Tarr, and B. V. Plapp, *Biochemistry* **14**:200 (1975).

133. D. C. Sogin and B. V. Plapp, *J. Biol. Chem.* **250**:206 (1975).

134. L. Andersson, H. Jornvall, A. Akeson, and K. Mosbach, *Biochim. Biophys. Acta* **364**:1 (1974).

135. L. G. Lange and B. L. Vallee, *Biochemistry* **15**:4681 (1976); L. G. Lange, A. J. Sytkowski, and B. L. Vallee, *Biochemistry* **15**:4687 (1976).

136. R. G. Thurman and W. R. McKenna, in: *Biochemical Pharmacology of Ethanol* (E. Majchrowicz, ed.), pp. 57–76, Plenum Press, New York (1975).

137. C.-J. Estler and V. Lachmann, *J. Neurochem.* **26**:653 (1976).

138. R. L. Veitch, L. Lumeng, and T. K. Li, *J. Clin. Invest.* **55**:1026 (1975).

139. D. J. Tuma, A. J. Barak, D. F. Schafer, and M. F. Sorrell, *Can. J. Biochem.* **51**:117 (1973).

140. D. F. Schafer, D. V. Stephenson, A. J. Barak, and M. F. Sorrell, *J. Nutr.* **104**:101 (1974).

141. D. S. Beattie, G. M. Patton, and E. Rubin, *Enzyme* **16**:252 (1973).

142. W. H. Admirand, T. Cronholm, and J. Sjovall, *Biochim. Biophys. Acta* **202**:343 (1970).

143. A. Perin, G. Scalabrino, A. Sessa, and A. Arnaboldi, *Biochim. Biophys. Acta* **366**:101 (1974).

144. V. J. Piccirillo and J. W. Chambers, *Res. Commun. Chem. Pathol. Pharmacol.* **13**:297 (1976).

145. H. A. Krebs, R. A. Freedland, R. Hems, and M. Stubbs, *Biochem. J.* **112**:117 (1969).

146. M. M. Asaad, H. Barry, III, D. E. Clarke, and N. Dixit, *Br. J. Pharmacol.* **50**:277 (1974).

147. T. Cronholm, I. Makino, and J. Sjovall, *Eur. J. Biochem.* **24**:507 (1972).

148. A. K. Rawat, *Arch. Biochem. Biophys.* **151**:93 (1972).

149. T. C. Peng, C. W. Cooper, and P. L. Munson, *Endocrinology* **91**:586 (1972).

150. P. J. Creaven, T. Barbee, and M. K. Roach, *J. Pharm. Pharmacol.* **22**:828 (1970).

151. J. Nakano and A. V. Prancan, *Arch. Int. Pharmacodyn. Ther.* **196**:259 (1972).

152. E. A. Carter and K. J. Isselbacher, *Proc. Soc. Exp. Biol. Med.* **142**:1171 (1973).

153. C. H. Bivens and J. M. Feldman, *Q. J. Stud. Alcohol* **35**:635 (1974).

154. M. E. Hillbom and A. R. Poso, *Toxicol. Appl. Pharmacol.* **32**:168 (1975).

155. G. O. Bustos, H. Kalant, J. M. Khanna, and J. Loth, *Science* **168**:1598 (1970).

156. J. C. Morgan and N. R. DiLuzio, *Proc. Soc. Exp. Biol. Med.* **134**:462 (1970).

157. R. Domanski, D. Riferenberick, F. Stearns, R. M. Scorpio, and S. A. Narrod, *Proc. Soc. Exp. Biol. Med.* **138**:18 (1971).

158. R. Nordmann, C. Ribiere, H. Rouach, and J. Nordmann, *Rev. Eur. Etud. Clin. Biol.* **17**:592 (1972).

159. O. Johnson, O. Hernell, G. Fex, and T. Olivecrona, *Life Sci.* **10**:553 (1971).

160. E. S. Higgins and W. H. Friend, *Proc. Soc. Exp. Biol. Med.* **141**:944 (1972).

161. J. M. Khanna, H. Kalant, J. Loth, and F. Seymour, *Biochem. Pharmacol.* **23**:3037 (1974).

162. R. Beuge, M. Clement, J. Nordmann, and R. Nordmann, *Biomed. Express (Paris)* **25**:70 (1976).

163. C.-J. Estler, *Res. Exp. Med.* **163**:95 (1974).

164. R. Blomstrand, L. Kager, R. Eklof, and O. Lantto, *Acta Chir. Scand. Suppl.* **446**:1 (1974).

165. J. M. Khanna and H. Kalant, in: *Alcoholic Liver Pathology* (J. M. Khanna, Y. Israel, and H. Kalant, eds.), pp. 261–269, Alcoholism and Drug Addiction Research Council of Ontario, Toronto (1973).

166. A. Bizzi, M. T. Tacconi, E. Veneroni, and S. Garattini, *Nature (Lond.)* **209**:1025 (1966).

167. A. Bizzi, E. Veneroni, and S. Garattini, *J. Pharm. Pharmacol.* **18**:611 (1966).

168. Y. Giudicelli, J. Nordmann, and R. Nordmann, *Biochem. Pharmacol.* **21**:2095 (1972).

169. Y. Giudicelli, J. Nordmann, and R. Nordmann, *Biomed. Express (Paris)* **19**:48 (1973).

170. G. J. Traiger and G. L. Plaa, *J. Pharmacol. Exp. Ther.* **183**:481 (1972).

171. F. Beauge, M. Clement, J. Nordmann, and R. Nordmann, *Biomed. Express (Paris)* **23**:137 (1975).
172. N. D'Acosta, J. A. Castro, F. C. de Ferreyra, M. I. Diaz Gomez, and C. R. de Castro, *Res. Commun. Chem. Pathol. Pharmacol.* **4**:641 (1972).
173. E. G. de Toranzo, M. I. Diaz Gomez, and J. A. Castro, *Biochem. Biophys. Res. Commun.* **64**:823 (1975).
174. N. R. DiLuzio, *Fed. Proc. Symp.* **32**:1875 (1973).
175. D. B. Goldstein, *Fed. Proc. Symp.* **34**:1953 (1975).
176. C. S. Lieber and L. M. DeCarli, *J. Biol. Chem.* **245**:2505 (1970).
177. C. S. Lieber, L. M. DeCarli, L. Feinman, Y. Hasumura, M. Korsten, S. Matsuzaki, and R. Teschke, in: *Alcohol Intoxication and Withdrawal: Experimental Studies* (M. M. Gross, ed.), Vol. II, pp. 185–227, Plenum Press, New York (1975).
178. T. B. McCaffrey and R. G. Thurman, in: *Alcohol and Aldehyde Metabolizing Systems* (R. G. Thurman, T. Yonetani, J. R. Williamson, and B. Chance, eds.), Vol. 1, pp. 483–492, Academic Press, Inc., New York (1974).
179. N. Grunnet, B. Quistorff, and H. I. Thieden, *Eur. J. Biochem.* **40**:275 (1973).
180. K. J. Isselbacher and E. A. Carter, in: *Alcohol and Aldehyde Metabolizing Systems* (R. G. Thurman, T. Yonetani, J. R. Williamson, and B. Chance, eds.), Vol. 1, pp. 271–286, Academic Press, Inc., New York (1974).
181. J. R. Williamson, K. Ohkawa, and A. J. Meijer, in: *Alcohol and Aldehyde Metabolizing Systems* (R. G. Thurman, T. Yonetani, J. R. Williamson, and B. Chance, eds.), Vol. 1, pp. 365–381, Academic Press, Inc., New York (1974).
182. C. S. Lieber and L. M. DeCarli, *J. Pharmacol. Exp. Ther.* **181**:279 (1972).
183. K. O. Lindros, M. Salaspuro, and P. Pikkarainen, in: *Alcohol and Aldehyde Metabolizing Systems* (R. G. Thurman, J. R. Williamson, H. R. Drott, and B. Chance, eds.), Vol. 3, pp. 343–354, Academic Press, Inc., New York (1977).
184. R. Rognstad, *Arch. Biochem. Biophys.* **163**:544 (1974).
185. H. I. D. Thieden, *Acta Chem. Scand.* **25**:3421 (1971).
186. M. P. Schulman, B. Andersson, and S. Orrenius, in: *Alcohol and Aldehyde Metabolizing Systems* (R. G. Thurman, J. R. Williamson, H. R. Drott, and B. Chance, eds.), Vol. 3, pp. 355–366, Academic Press, Inc., New York (1977).
187. J. Papenberg, J. P. von Wartburg, and H. Aebi, *Enzymol. Biol. Clin.* **11**:237 (1970).

Biochemical Pharmacology of Disulfiram

19

Morris D. Faiman

A. INTRODUCTION

Hanzlik and Irvine,[1] investigating the toxicity of several thioureas and thiuram disulfides, reported on the toxicity of tetraethylthiuram disulfide. However, it was Williams,[2] a physician working in a chemical company which manufactured rubber accelerators, who observed that workers coming into contact with either tetramethylthiuram disulfide or tetraethylthiuram monosulfide, developed a sensitivity to ethanol. This sensitivity was manifested by hypotension, accelerated pulse rate, flushed skin, and headache. He subsequently suggested that these agents may be useful in the treatment of alcoholism.

Hald *et al.*[3] began extensive investigations on the sensitizing effect of disulfiram to ethanol, subsequently leading to its clinical use in the treatment of alcoholism. A series of studies on the early work with disulfiram, its use in the treatment of alcoholism, and its possible mechanism of action first were reported by a group of investigators working in Sweden.[3-7]

B. CHEMISTRY

Disulfiram [(tetraethylthioperoxydicarbonic diamide), bis(diethylthiocarbamyl) disulfide, tetraethylthiuram disulfide, bis(diethylthiocarbamoyl) disulfide] is a disulfide with the formula

$$(C_2H_5)_2{-}N{-}\overset{\overset{\displaystyle S}{\|}}{C}{-}S{-}S{-}\overset{\overset{\displaystyle S}{\|}}{C}{-}N{-}(C_2H_5)_2$$

Morris D. Faiman • Department of Pharmacology and Toxicology, University of Kansas, Lawrence, Kansas 66045.

and a molecular weight of 296.54. Its composition is C 40.50%, H 6.80%, N 9.45% and S 43.25%. Disulfiram is a white to light gray crystalline powder with a melting point of 70°C and density of 1.30. Its solubility in various solvents is: water, 0.02%; ethanol, 3.82%; ether, 7.14%. It also is soluble in acetone, benzene, chloroform, and carbon disulfide. The preparation of disulfiram and its use as an acceletator in the rubber industry has been discussed by Cummings and Simmons.[8]

C. ANALYTICAL METHODS

A number of different analytical methods for the determination of disulfiram and its metabolites have been employed. These include colorimetric procedures in which a copper complex is formed between the metal and diethyldithiocarbamate, a reduction product of disulfiram, and its subsequent spectrophotometric determination.[9-17] Polarographic methods also have been employed in the analysis of disulfiram,[18-22] as well as proton magnetic resonance.[23] More recently analytical methods employing gas chromatography have been used to determine disulfiram or some of its metabolites.[24-26] The use of radioactive disulfiram in man and animals also has been used to study the metabolism and excretion characteristics of disulfiram.[27-32] More recently a high-pressure liquid chromatographic method has been developed which can simultaneously determine disulfiram, diethyldithiocarbamate, the diethyldithiocarbamate methyl ester, carbon disulfide, and diethylamine.[33]

The colorimetric procedures, although simple to use, are considered to lack specificity and sensitivity. Furthermore, the spectrophotometric methods appear to be inaccurate in the presence of proteins, probably the result of incomplete and varying extraction of the compounds to be measured.[34] The gas chromatographic methods are time consuming, while the polarographic methods and the proton magnetic resonance methods, although appropriate for analysis of pharmaceutical preparations, may not be suitable for biological fluids and tissues. Because of the protein-binding characteristics of disulfiram, the use of radioactive disulfiram in man probably has limited applicability. The high-performance liquid chromatography method of Faiman et al.[33] is a new method, with advantages of simplicity, sensitivity, and specificity. Furthermore, disulfiram, diethyldithiocarbamate, the diethyldithiocarbamate methyl ester, carbon disulfide, and diethylamine can all be determined simultaneously.

D. ABSORPTION, DISTRIBUTION, METABOLISM, EXCRETION

1. Absorption

Disulfiram appears to be rapidly but incompletely absorbed. After the intraperitoneal administration of [^{35}S]disulfiram to rats, Eldjarn[27] found that approximately 20% and 65% of the ^{35}S radioactivity appeared in feces and urine respectively over a 5-day period. De Saint-Blanquat et al.[35] administered disulfiram orally to rats and found that 70–90% of the administered dose was absorbed. Two hours after dosing, both disulfiram and diethyldithiocarbamate were found in blood, liver, kidney, and muscle. Studies also have been carried out in man. After the administration of 2 g of disulfiram to volunteers, approximately 20% of the unchanged drug appeared in the feces.[3] Eldjarn[28] administered [35S]disulfiram to two patients and detected radioactive sulfur in plasma within half an hour. Cobby et al.[26] found diethyldithiocarbamate in blood 30 min after the administration of 250 mg disulfiram to alcoholics. Iber et al.[29] studied the excretion of radioactive sulfur following [35S]disulfiram administration to alcoholics. It was observed that excretion in feces varied from 1% to 15% of the ingested dose. No differences in the degree of absorption between alcoholics with liver disease and those with normal liver function were apparent.

2. Distribution

Following absorption, disulfiram and its metabolites appear to be readily distributed to various tissues. De Saint-Blanquat et al.[35] found that after the oral administration of disulfiram to rats, disulfiram and diethyldithiocarbamate were detected in the stomach, both the proximal and distal portions of the intestine, cecum, colon, liver, kidney, and muscle. Blood and brain contained little or no disulfiram at low doses, while only trace amounts of either disulfiram or diethyldithiocarbamate were found at larger doses. Eldjarn[27] administered [35S]disulfiram intraperitoneally to rats and found radioactive sulfur in the adrenals, liver, spleen, tibia plus marrow, muscle, blood, and kidney. However, the metabolites contributing to this radioactivity were not identified. In studies with two patients, Eldjarn[28] found that after the oral ingestion of [35S]disulfiram, plasma contained both disulfiram and diethyldithiocarbamate. Radioactivity also was detected in plasma but not in red blood cells.

Strömme[31] studied the tissue distribution of disulfiram after the subcutaneous administration to mice. Four hours after dosing, [35S]disulfiram was detected in plasma, liver, kidney, lung, spleen, and intestinal mucosa. Only trace amounts were found in heart, brain, and testes. After the intraperitoneal administration of [35S]diethyldithiocarbamate, highest amounts of thiol were found in plasma, liver, and kidney, whereas lowest amounts were found in brain. When [35S]disulfiram was administered subcutaneously to mice, the plasma and supernatant of liver homogenates contained free diethyldithiocarbamate, the glucuronide, inorganic sulfate, and carbon disulfide. In detailed studies by Faiman et al.[36] the distribution of [35S]disulfiram and its metabolites was investigated in mice. [35S]disulfiram was found in plasma, brain, lung, liver, and kidney. Furthermore, [35S]disulfiram was found in plasma only during the first 45 min after disulfiram administration. This could explain why others have been unable to detect free disulfiram in plasma when plasma was analyzed for disulfiram 1 or more hours after drug dosing. In addition to disulfiram, diethyldithiocarbamate, the methyl ester, the diethyldithiocarbamic acid glucuronide, and inorganic sulfate were also found in plasma, brain, liver, lung, and kidney. Generally, the

Table I. Distribution of [35S]Disulfiram and 35S Metabolites in Mice[a]

| | Time (hr) after [35S]disulfiram dosing | | | | | | | | | | | | | | |
| | Plasma | | | Brain | | | Lung | | | Liver | | | Kidney | | |
	1	3	6	1	3	6	1	3	6	1	3	6	1	3	6
[35S]DSF	0	0	0	0.5* ± 0.1	0.1* ± 0.02	0	4.5* ± 4	0.7 ± 0.2	0	5.7 ± 3.9	1.5* ± 0.4	2.0* ± 0.5	7.2* ± 4.9	1.0* ± 0.2	0.3* ± 0.2
[35S]DDTC	8.6* ± 3.8	2.1† ± 0.8	0.5† ± 0.3	1.3† ± 0.6	0	0	6.7 ± 6.0	1.4† ± 0.9	0	11.5 ± 3.9	6.4* ± 1.0	14 ± 3.9	13.6 ± 5.4	3.8 ± 1.6	2.0 ± 1.6
[35S]DDTC-Me	0	0	0	3.1 ± 0.7	0.7 ± 0.2	0	2.6 ± 0.9	1.3 ± 0.4	0	10.4 ± 2.6	3.1* ± 1.3	1.9 ± 0.2	10 ± 1.1	4.6 ± 1.4	2.5 ± 0.9
[35S]Water-soluble fraction	434* ± 152	625 ± 187	614† ± 237	26 ± 4	43 ± 10	:5 ± 6.7	362 ± 98	531 ± 152	709 ± 121	841 ± 339	779 ± 214	1115 ± 117	3612 ± 1422	3162 ± 905	2984 ± 630

[a] 70 μCi of [35S]disulfiram ([35S]DSF) was mixed with nonradioactive DSF and administered intraperitoneally to mice in a dose of 200 mg/kg. Each value is the mean ± SEM for four mice, except in some instances where the means reflect three (*) and two (†) mice. Means expressed as DPM × 10^{-3}/g tissue or ml/plasma.

Source: Faiman et al.[28]; with permission from PJD Publications Limited, New York.

brain contained the smallest amount of disulfiram and metabolites (Table I). In other studies, Faiman et al.[37] investigated the tissue distribution of total radioactivity in rats given [^{35}S]disulfiram either intraperitoneally or orally. One-half hour after either intraperitoneal or oral [^{35}S]disulfiram, radioactivity was found in the thyroid, adrenals, pancreas, stomach, small and large intestine, muscle, brain, liver, testes, kidney, lung, spleen, heart, and plasma. Further investigation as to the source of this radioactivity indicated that diethyldithiocarbamate glucuronide and inorganic sulfate accounted for approximately 75% of the radioactive sulfur, while lesser amounts were the result of the protein-bound fraction, the thiol, the methyl ester, and disulfiram. Furthermore, it was observed that the percentage of each metabolite differed for the various organs. In bile, approximately 75% of the radioactivity was caused by the diethyldithiocarbamate glucuronide, while inorganic sulfate accounted for the remaining radioactivity.

3. Protein Binding

Disulfiram has been found to exhibit protein binding, forming mixed disulfides with protein SH groups in vitro.[34] When administered to rats intraperitoneally, [^{35}S]disulfiram was bound to plasma proteins and to soluble proteins in the liver. Furthermore, a significant fraction of protein binding also has been found after the administration of diethyldithiocarbamate.[30] The ^{35}S bound to proteins was released by addition of reduced glutathione. This appears to suggest that both disulfiram and diethyldithiocarbamate form mixed disulfides. Strömme[30] estimated that after the intraperitoneal administration of disulfiram to rats, approximately 8% of the titratable SH groups of plasma proteins were blocked.

4. Metabolism

Studies with rat liver homogenates by Johnston and Prickett[38] have suggested that disulfiram metabolism follows two steps: (1) enzymatic reduction to diethyldithiocarbamate, and (2) formation of diethylamine and carbon disulfide. From buffer experiants the rate of carbon disulfide formation at pH 7.3, although quite slow, is directly proportional to the concentration of diethyldithiocarbamate. The pH dependancy upon the rate of thiol decomposition has been detailed by Aspila et al.[39] Evidence suggests that the glutathione–glutathione reductase system mediates the rapid reduction of disulfiram to diethyldithiocarbamate.[40] In in vitro studies, Cobby et al.[26] found that when disulfiram was added to blood, reduction to the thiol occurred within 4 min. This rapid reduction also occurs in vivo, as diethyldithiocarbamate is found in plasma within 5 min after the administration of disulfiram to mice or dogs.[36]

The main pathway for disulfiram metabolism appears to be as shown in Fig. 1.[41] The formation of diethyldithiocarbamate methyl ester as a metabolite of disulfiram also has been observed by others in both animals and man.[26,36]

Methylation is mediated by microsomal S-methyltransferase,[42] with the enzyme found in both kidneys and liver. The greatest activity is associated with

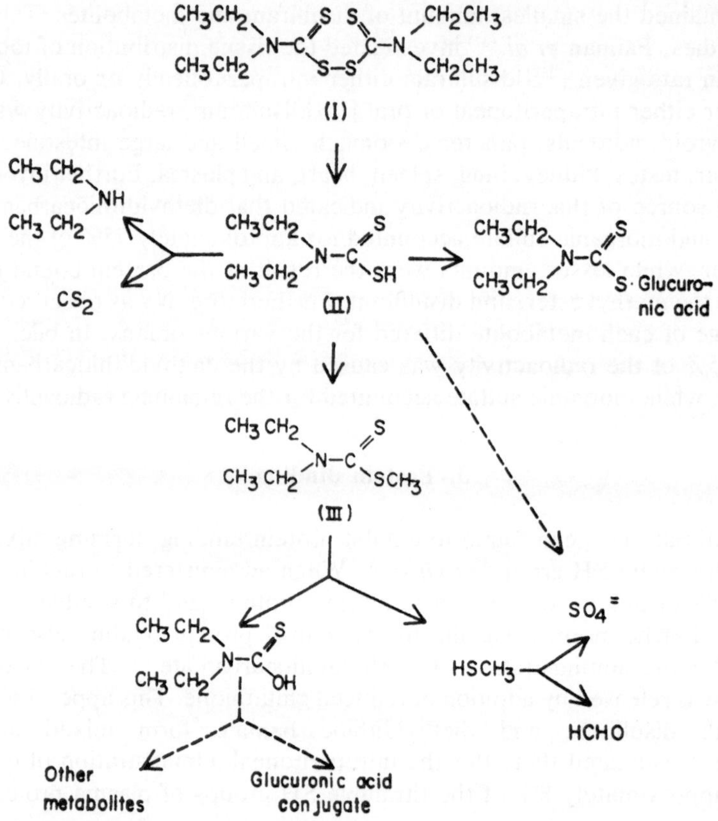

Fig. 1. Metabolic pathways *in vivo* for the sulfur of disulfiram (I), diethyldithiocarbamic acid (II), and diethyldithiocarbamic acid methyl ester (III). [From Gessner and Jakubowski[41]; with permission from Pergamon Press, Inc., Elmsford, N.Y.]

the 100,000g microsomal fraction. The distribution of S-adenosyl methionine transferase in organs other than the liver could explain why the methyl ester also has been found in lung, kidney, brain, and plasma.[36] When urinary metabolites of both diethyldithiocarbamate and diethyldithiocarbamate methyl ester were compared,[41] it was found that over 60% of the dose of the methyl ester administered to rats was excreted as sulfate the first 2 days, whereas when diethyldithiocarbamate was given, only 16% of the dose was excreted as sulfate. No glucuronide was found in the urine of the methyl-ester-treated rats, while diethyldithiocarbamate glucuronide was the major metabolite of the thiol-treated animals.

5. Excretion

Disulfiram is rapidly metabolized and the parent compound and its metabolites excreted via the kidney, feces, and lung. Approximately 5–20% of an administered dose of disulfiram is excreted unchanged in the feces.[3,27,28] It has been shown from animal studies that diethyldithiocarbamate glucuronide and inorganic sulfate are the major metabolites of disulfiram metabolism. Both these water-soluble metabolites are excreted by the kidney. Rats receiving disulfiram either intraperitoneally or orally, excreted approximately 37% and 29% of the administered dose of disulfiram as the glucuronide and inorganic sulfate, respectively, within 48 hr after disulfiram administration. Only trace amounts of disulfiram and diethyldithiocarbamate-methyl ester have been found in urine,[37] while approximately 0.6% of the dose of disulfiram administered has been identified as the thiol.[36] Strömme,[30] similarily, could only find traces of thiol in the urine of rats. Disulfiram also is metabolized to diethylamine which is excreted in the urine. However, to date no estimate of the percentage of the dose of disulfiram excreted as diethylamine is available.

After the administration of [35S]disulfiram to patients, Eldjarn[28] found that one-third of the plasma radioactivity was the result of the thiol 6 hr after dosing. The diethyldithiocarbamate is rapidly conjugated with glucuronic acid, with the glucuronide found in both man[43] and animals.[34,36] Although Kaslander[43] could only account for 0.75% of the disulfiram dose as glucuronide, the low value may be attributed to loss during the extraction procedure. Larger concentrations have been found in animals. For example Strömme,[30] found that 4 hr after the subcutaneous administration of [35S]disulfiram, approximately 57% of the 35S administered was recovered in the urine as the glucuronide. Similar findings were made by Faiman et al.[37] after the intraperitoneal administration of [35S]disulfiram to rats.

Carbon disulfide is an important metabolite of disulfiram. Although small amounts are found in urine of alcoholics receiving disulfiram, the major route of excretion is the lung. Strömme[30] found approximately 2% of the administered dose of disulfiram eliminated as carbon disulfide 4 hr after [35S]disulfiram administration, while Prickett and Johnston[44] found approximately 3% of the dose as carbon disulfide 2 hr after intravenous disulfiram. In recent studies with rats, it has been found that approximately 4% of the dose is eliminated from breath as carbon disulfide 4 hr after intraperitoneal disulfiram, while after 48 hr almost all carbon disulfide, about 12% of the dose, was eliminated.[37] Merlevede and Casier[45] estimated that 46–53% of a dose of disulfiram is eliminated in patients as carbon disulfide. In studies with alcoholic volunteers by Faiman et al.,[46] only about 12% of the disulfiram dose was found to be eliminated as carbon disulfide. The reason for the difference between the studies of Merlevede and Casier[45] and Faiman et al.[46] is not apparent at this time.

E. BIOCHEMICAL AND PHARMACOLOGICAL EFFECTS

Disulfiram has been found to be an inhibitor of numerous enzymes both *in vitro* and *in vivo*. Because of its nonspecific general inhibitory action, disulfiram can affect carbohydrate metabolism, mitochondrial oxidations, neurotransmission, and drug metabolism. These actions, however, do not appear to produce major physiological effects when disulfiram is taken alone, although they may contribute to several of the side effects attributed to disulfiram.

1. Intermediary Metabolism

Glycolysis, the tricarboxylic acid cycle, and the pentose phosphate shunt all have been found to be inhibited by disulfiram as a result of enzyme inhibition. In *in vitro* studies, 10^{-3}–10^{-6} M disulfiram has been shown to inhibit rabbit muscle glyceraldehyde-3-phosphate dehydrogenase,[47] heart muscle succinic dehydrogenase,[48] rat liver xanthine oxidase,[49] calf brain hexokinase,[50] and liver nicotinamide adenine dinucleotide dehydrogenase.[51] Brewer's yeast fructose 1,6-diphosphate dehydrogenase also has been shown to be inhibited by disulfiram,[52] as has glucose-6-phosphate dehydrogenase in insects.[53] In *in vivo* studies by Dodd and Faiman,[54] disulfiram administered intraperitoneally in a dose of 200 mg/kg to mice inhibited cortical glucose-6-phosphate dehydrogenase. It is generally believed that binding occurs between the sulfur of the disulfide and that of the SH enzyme, resulting in mixed disulfide formation. This would explain the lack of specificity for disulfiram and its general interaction with SH-containing enzymes.

2. Aldehyde Dehydrogenase

The enzyme which has received the most attention has been aldehyde dehydrogenase.[51,55,56] Detailed *in vitro* studies by Graham[51] showed that concentrations of 10^{-5} M of disulfiram inhibited liver aldehyde dehydrogenase, and that disulfiram acted as a competitive inhibitor, competing with NAD for the active site of the enzyme. The affinity of the liver aldehyde dehydrogenase for disulfiram was found to be about 46 times greater than that for NAD, and approximately 350 times greater than that for acetaldehyde. Deitrich and Hellerman,[55] using bovine liver homogenates, also found disulfiram in a concentration of 10^{-7} M to be a competitive inhibitor of NAD. The effect of orally administered disulfiram to rats on mitochondrial and tissue homogenate supernatant aldehyde dehydrogenase activity was investigated by Deitrich and Erwin.[56] These investigators found that aldehyde dehydrogenase activity declined within 2 hr, reached a maximum reduction within 16–40 hr, and returned to control levels within 140 hr. They proposed that aldehyde dehydrogenase inhibition occurred as a result of the formation of mixed disulfides between the SH groups of the reduced form of disulfiram and the sulfhydryl groups of the enzyme. Furthermore, the administration of cyclohexamide, a protein synthesis inhibitor, blocked the return of aldehyde dehydrogenase activity in both the mitochondrial and supernatant fraction. This appears to suggest that enzyme activity is dependent upon new protein synthesis.

Recently, several physically distinct isoenzymes of aldehyde dehydrogenase in various subcellular fractions of rat liver have been identified.[57–61] In *in vitro* studies with horse liver, Eckfeldt *et al.*[61] identified two isoenzymes, with both isoenzymes utilizing NAD as a coenzyme. One isoenzyme exhibited a very low K_m for NAD (3 μM) and a higher K_m for acetaldehyde (70 μM), while the other isoenzyme was found to have a higher K_m for NAD (30 μM) and low K_m for acetaldehyde (0.2 μM). The isoenzyme exhibiting a low K_m for NAD was extremely sensitive to disulfiram inhibition. In *in vivo* studies, Tottmar and Marchner[60] administered disulfiram orally to rats. They found that high doses of disulfiram (150–600 mg/kg) markedly decreased the low K_m mitochondrial acetaldehyde dehydrogenase. No significant effects were found on the high K_m enzyme activity present in either the mitochondrial, microsomal, and cytosolic fractions. Similar isoenyzme patterns have been reported for human[62] and sheep liver.[63]

Koivula and Koivusalo[64] found that in rat liver the cytoplasmic, mitochondrial, and microsomal fractions contained 10–15%, 45–50%, and 35–45%, respectively, of the total aldehyde dehydrogenase activity. Furthermore, the cytoplasmic and mitochondrial fractions contained two separable aldehyde dehydrogenases, one with a low K_m and the other with a high K_m. The isoenzyme in each fraction was inhibited by disulfiram. These different isoenzyme patterns could play an important role in disulfiram therapy, and perhaps explain the different degree of sensitivity found in patients receiving the same dose of disulfiram.

3. Dopamine-β-hydroxylase

Effects on carbohydrate metabolism can also occur indirectly as a result of inhibition of dopamine β-hydroxylase.[65] Dopamine-β-hydroxylase, a metalloenzyme, catalyzes the conversion of dopamine to norepinephrine. Several studies have shown that disulfiram is an effective inhibitor of dopamine-β-hydroxylase both *in vitro* and *in vivo*, in several tissues including heart, spleen, and various regions of the brain, such as the caudate nucleus, hypothalmus, and brain stem.[65–68] Binding or uptake of norepinephrine is not affected by disulfiram. In addition, octopamine formation from tyramine, which also is mediated by dopamine-β-hydroxylase, is inhibited by disulfiram in heart and spleen.

In *in vitro* studies with purified enzyme from adrenal medulla, Goldstein *et al.*[67] found that 10^{-6} M disulfiram completely inhibited the conversion of dopamine to norepinephrine. These investigators also demonstrated that disulfiram was completely and quantitatively reduced by ascorbate to diethyldithiocarbamate, and that diethyldithiocarbamate was the moiety responsible for the inhibition of dopamine-β-hydroxylase and subsequent inhibition of norepinephrine synthesis. The effect of disulfiram on plasma dopamine-β-hydroxylase in alcoholics has also been investigated. Lake *et al.*[69] administered 250–500 mg disulfiram per day to alcoholics for 3 weeks but were unable to find any effect on plasma dopamine-β-hydroxylase. Ewing *et al.*,[70] similarily, were unable to find changes in plasma dopamine-β-hydroxylase after the administration of disulfiram to alcoholics.

4. Superoxide Dismutase

Recently, Heikkila et al.[71] reported that the administration of diethyldithio-carbamate intraperitoneally to mice inhibited superoxide dismutase in brain, liver, and erythrocytes. The enzyme superoxide dismutase is a copper/zinc enzyme, which has been suggested to play an important biological role as a defense mechanism against endogenously generated superoxide radicals.[72] Since disulfiram is immediately reduced to diethyldithiocarbamate in vivo, disulfiram may also inhibit superoxide dismutase, although this has not yet been reported.

5. Inhibition of Drug Metabolism

Further evidence of disulfiram's nonspecificity is its inhibitory action of enzyme systems involved in drug metabolism. The microsomal drug-metaboliz-ing system is thought of as a mixed-function oxidase system. The rate of drug biotransformation by the mixed-function oxidase system is determined by several parameters. These include the concentration of cytochrome P_{450}, the proportions of the various forms of cytochrome P_{450} and their affinities for the substrate, the concentration of cytochrome c reductase, and the rate of reduction of the drug–cytochrome P_{450} complex. Glucuronide formation is catalyzed by various microsomal glucuronyltransferases, with uridine diphosphate glucuronic acid as the donor of glucuronic acid. Uridine diphosphate glucuronic acid is generated from glucose by enzymes in the cytosol.

Disulfiram has been shown to affect a number of enzymes associated with drug metabolism. For example, the nitrogen oxidation in vivo of azomethane to azoxymethane in rats, an important pathway in the metabolic activation of the carcinogen 1,2-dimethylhydrazine, has been found to be inhibited by disulfiram.[73] Lake et al.[74] found that when disulfiram was given intraperitoneally to rats for 7 days, liver weight increased, and ethylmorphine N-demethylase activity and cytochrome P_{450} decreased. Lang et al.[75] similarly found that after the oral administration of disulfiram to rats, the liver/body weight ratio, as well as the total hepatic protein content, increased. Disulfiram also increased the microsomal protein and phospholipid content. Disulfiram and diethyldithiocarbamate decreased both cytochrome P_{450} and cytochrome P_{420}, as well as nitroanisole O-demethylase activity. NADPH-cytochrome c reductase is also enhanced by disulfiram.

Zemaitis and Greene[76] have shown that the activities of both plasma and microsomal carboxylesterase and plasma cholinesterase were decreased by disulfiram. The plasma esterases recovered more rapidly than the microsomal enzymes (Fig. 2), suggesting that the extent to which hydrolysis of other drugs is impaired by disulfiram may depend upon whether the drug is hydrolyzed by plasma or microsomal esterases. Results from electrophoretic analysis of the microsomal esterases suggested that disulfiram inhibited these enzymes by interacting with essential SH groups of the enzyme. This would be consistent with

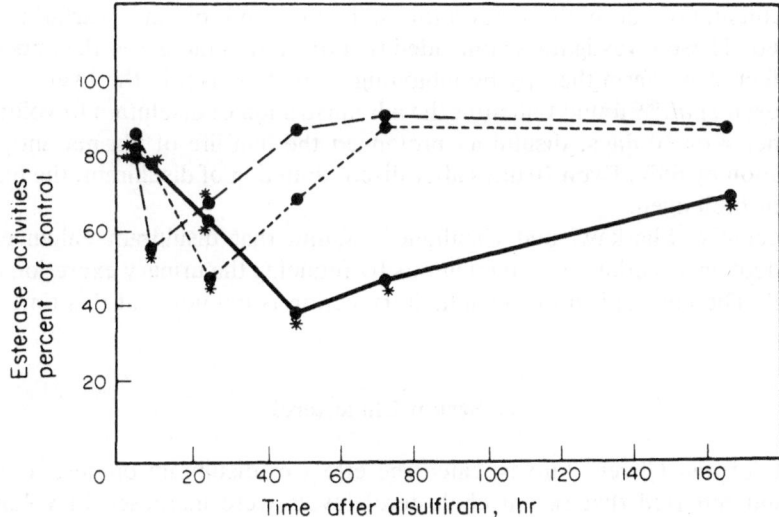

Fig. 2. Time course of esterase impairment after a single oral dose of disulfiram (2.0 g/kg). At the indicated times, control and treated animals ($N = 4$) were sacrificed and esterase activities were determined. Values are expressed as percentage of control activity. Microsomal carboxylesterase (————); plasma carboxylesterase (— —); plasma cholinesterase (- - - -). An asterisk indicates different from control ($p < 0.05$). [From Zemaitis and Greene[76]; with permission from Pergamon Press, Inc., Elmsford, N.Y.]

the suggested mechanism for inhibition of hexokinase,[40] aldehyde dehydrogenase,[55,56] and D-amino acid oxidase.[77]

Disulfiram can also affect drug conjugation reactions. Marselos *et al.*[78] found that the intragastric administration of either disulfiram or diethyldithiocarbamate to rats induced changes in enzyme activity of the D-glucuronic acid pathway. For example, disulfiram inhibited glucose-6-phosphate dehydrogenase and β-glucuronidase, while the activities of UDP glucose dehydrogenase, UDP glucuronic acid pyrophosphatase, UDP glucuronyltransferase, and L-gulonate dehydrogenase were increased.

6. Drug Interactions

Changes in activity of enzymes associated with drug metabolism have important considerations when disulfiram is administered in combination with other drugs. It has been found that after disulfiram administration, the half-life of diphenylhydantoin is increased from 11 to 19 hr, while the metabolic clearance, which is related to the apparent volume of distribution of the drug and the rate constant of elimination, decreased from 51 to 34 hr.[79] O'Reilly[80] administered disulfiram orally to volunteers for 12–20 days and on the fourth day of disulfiram therapy a single dose of warfarin was administered, while in another study both disulfiram and warfarin were given simultaneously for 3 weeks. In both studies an augmentation of both hypoprothrombinemia and plasma warfarin levels occurred. These investigators concluded that disulfiram increased the anticoagulant effect of warfarin therapy by inhibiting its metabolism in the liver.

Vesell et al.[81] found that after the administration of disulfiram to volunteers for either 4 or 10 days, disulfiram prolonged the half-life of plasma antipyrine elimination by 68%. Even 10 days after discontinuation of disulfiram, the half-life was still prolonged.

Recently, Sharkawi and Cianflone[82] found that disulfiram enhanced the pharmacological action of barbital in rats by reducing the urinary excretion of the barbital. The mechanism of this inhibitory action is unknown at this time.

7. Serum Cholesterol

Major and Goyer[83] investigated the effect of disulfiram on serum cholesterol and reported that serum cholesterol levels were increased in volunteers receiving 500 mg disulfiram/day, but not when the dose was reduced to 250 mg/day. The increase in cholesterol was prevented when disulfiram was taken in conjunction with pyridoxine. The mechanism by which serum cholesterol is increased or prevented by pyridoxine is not clear. It has been reported that pyridoxine deficiency increases serum cholesterol in animals.[84,85] Since disulfiram is metabolized to carbon disulfide, and carbon disulfide reacts with pyridoxine causing a pyridoxine deficiency, this could explain the relationship between disulfiram and the increase in serum cholesterol.

8. Cerebral Blood Flow

The effect of disulfiram on cerebral blood flow in patient volunteers has also been studied.[86] After the administration of approximately 1 g of disulfiram for 4 days, cerebral blood flow was decreased by 30% in 9 of 10 patients studied. This was accompanied by a slight decrease in the cerebral utilization of oxygen. No effects were found on the respiratory quotient, the arterial and venous glucose values, and the blood pressure. It was suggested disulfiram had no effect on cerebral glucose and cerebral glucose metabolism. This would appear inconsistent with studies in mice in which disulfiram increased both cerebral and blood glucose.[87] Since disulfiram has been shown to inhibit hexokinase,[40] an increase in brain and blood glucose would be expected.

9. Catecholamines

The effect of disulfiram on putative neurotransmitters has not been studied in as much detail as its action on liver aldehyde dehydrogenase. Of the neurotransmitters investigated, the catecholamines have received the most attention. Biogenic amines such as norepinephrine and dopamine are found in various tissues, with synthesis and catabolism essentially similar for the liver, heart, and brain. Disulfiram is known to inhibit dopamine-β-hydroxylase and aldehyde dehydrogenase, two enzymes which play an important role in norepinephrine metabolism. Dopamine-β-hydroxylase is the enzyme responsible for the conversion of dopamine to norepinephrine, while aldehyde dehydrogenase catalyzes several steps in dopamine and norepinephrine catabolism. These include the formation of 3,4-dihydroxyphenylacetic acid from its corresponding aldehyde, 3,4-dihydroxyphenylacetaldehyde, the formation of homovanillic acid from 3-methoxy-4 hydroxyphenylacetaldehyde, the formation of 3,4-dihydroxymandelic acid from its corresponding aldehyde, and 3-methoxy-4-hydroxymandelic acid from 3-methoxy-4-hydroxyphenylglycol aldehyde.

The formation of these metabolites vary depending upon the endogenous substrate and species. The major products of dopamine catabolism are 3,4-dihydroxyphenylacetic acid and homovanillic acid, while the glycols 3,4-dihydroxyphenylglycol appear to be the major products of newly formed norepinephrine, with 3,4-dihydroxyphenylglycol occurring in approximately twice the quantities of 3-methoxy-4-hydroxyphenylglycol in rabbit cerebral cortex slices.[88]

Major et al.[89] administered 250–500 mg disulfiram/day to a group of volunteers and found that cerebrospinal fluid homovanillic acid decreased approximately 30%. This would be consistent with the findings of Rutledge and Jonason[88] that homovanillic acid represents a major pathway of dopamine metabolism.

Waterbury et al.[90] administered disulfiram orally to rats in two doses, 500 mg/kg 24 hr before sacrifice and 1 g/kg 4 hr before sacrifice, and found an increase in 3-methoxy-4-hydroxyphenylethanol in both cortex and brain stem. This increase was explained on the basis of aldehyde dehydrogenase inhibition, thus preventing normal homovanillic acid formation.

Berger and Weiner[91] studied the effect of disulfiram on the metabolism of catecholamines in rat liver and brain employing both tissue slices and perfusion techniques. Disulfiram inhibited 3,4-dihydroxyphenylacetic acid approximately 16% in liver slices, whereas about a 48% inhibition was found in brain when the tissues were incubated with dopamine. Homovanillic acid was not affected in either liver or brain by disulfiram. When norepinephrine was used as the substrate, disulfiram had no effect on the rate of norepinephrine disappearance. In perfusion studies of the caudate nucleus with dopamine, disulfiram inhibited the formation of 3,4-dihydroxyphenylacetic acid. The observation that disulfiram only inhibited liver metabolism of dopamine 15%, while markedly inhibiting liver metabolism of acetaldehyde, appears to suggest the occurrence of different isoenzymes of aldehyde dehydrogenase in the oxidation of these two aldehydes. Furthermore, the finding that 3,4-dihydroxymandelic acid formation is not inhibited would seem to suggest that the aldehyde derived from norepinephrine is utilized by still another isoenzyme of aldehyde dehydrogenase which is insensitive to disulfiram.

10. Serotonin

The effect of disulfiram on serotonin metabolism has not been as extensively investigated as the catecholamines. Since serotonin is metabolized to 5-hydrox-yindoleacetic acid via 5-hydroxyindoleacetaldehyde which is mediated by alde-hyde dehydrogenase, it seems probable that disulfiram would inhibit this reaction and elevate serotonin levels. Indeed, Faiman et al.[92] have shown that 1 hr after the intraperitoneal administration of disulfiram to mice, serotonin levels are elevated approximately 20% above controls.

11. γ-Aminobutyric Acid

The effect of disulfiram on other neurotransmitters such as γ-aminobutyric acid (GABA) also have not been studied in detail. In a limited study, however, the intraperitoneal administration of disulfiram had no effect on cortical GABA.[93]

F. DISULFIRAM–ETHANOL REACTION

The disulfiram–ethanol reaction is a term which refers to the symptoms which result from the ingestion of ethanol during disulfiram therapy. Symptoms reported include blurred vision, nausea, dizziness, hypotension, palpitations, tachycardia, and flushing of the face. The symptoms range from minor effects to severe reactions. Several deaths as a result of the disulfiram–ethanol reaction have been reported.[94–96]

At the present time the mechanism by which the disulfiram–ethanol reaction occurs is not too well understood. Casier and Polet[97] suggested that the reaction was the result of a complex mechanism producing a synergistic toxic action in the body between disulfiram or one of its metabolites and ethanol. Casier and Merlevede[98] proposed that the disulfiram–ethanol reaction was due to the formation of a toxic quaternary ammonium compound formed *in vivo* between disulfiram and ethanol. However, a recent study by Kitson[99] has indicated that based upon chemical considerations, this mechanism was not feasible. Strömme[30] hypothesized that conjugation of disulfiram with glucuronic acid was inhibited by the combination of ethanol with disulfiram, which resulted in a decreased detoxification of disulfiram.

Lamboeuf et al.[100] found that disulfiram enhanced rat gastric alcohol dehy-drogenase activity, which they suggested could explain the hyperacetaldehyde-mia in patients exhibiting the disulfiram–ethanol reaction. De Saint-Blanquat et al.[101] observed that ethanol and disulfiram increased gastric secretion in the rat, producing hyperacidity, which they suggested could account for the symptoms of vomiting and vertigo found in patients. Others have proposed that the symp-toms of the disulfiram–ethanol reaction were caused by a mechanism involving serotonin. Since disulfiram inhibits aldehyde dehydrogenase, it has been sug-gested that when ethanol is ingested, increased acetaldehyde releases serotonin from storage vesicles, producing the symptoms of the disulfiram–ethanol reaction. This seems unlikely, however, since other drugs which affect serotonin metabolism and release are not known to have a disulfiramlike action.

It is generally believed that the mechanism of the disulfiram–ethanol reaction can be explained on the basis of liver aldehyde dehydrogenase inhibition. This theory has been supported by the observation that many of the symptoms associated with the disulfiram–ethanol reaction can be duplicated by the administration of acetaldehyde. Unfortunately, acetaldehyde administration does not produce all the symptoms associated with the disulfiram–ethanol reaction. Furthermore, it is agreed that acetaldehyde has a sympathomimetic action and should not produce the flushing, dilitation of blood vessels, and hypotension associated with the disulfiram–ethanol reaction.[102] This anomaly could be explained by acetaldehyde's effect on norepinephrine metabolism. Kitson[103] has suggested that after the administration of disulfiram, the release of acetaldehyde and vasoconstriction are temporary, since norepinephrine synthesis is blocked as a result of dopamine-β-hydroxylase inhibition. The elevated acetaldehyde then acts directly on the heart and smooth muscle producing hypotension.

Sauter et al.[104] studied the disulfiram–ethanol reaction in man in order to assess the importance of acetaldehyde on various pharmacological parameters, and also attempted to determine the various parameters that could predict the severity of the reaction. In these studies disulfiram alone did not increase blood acetaldehyde. When ethanol was ingested, significant increases were found in blood acetaldehyde (Fig. 3). Disulfiram, both at low and high doses, increased

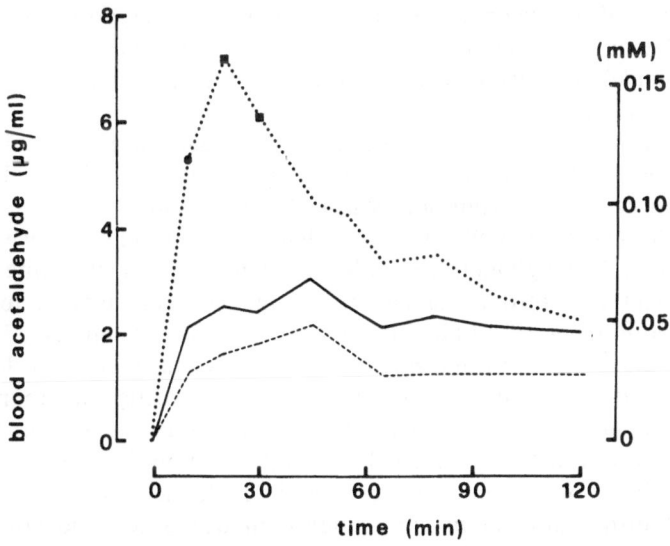

Fig. 3. Mean blood acetaldehyde concentration. Disulfiram was given to volunteers at a dose of either 100 mg/day for 3 days (- - - -) or 800 mg/day for 3 days (· · · · · ·). Controls (———) did not receive any disulfiram. Statistically significant from control at $p < 0.05$ (●) and $p < 0.01$ (■). [From Sauter et al.[104]; with permission from the publisher.]

blood pH, while only the high dose of disulfiram (800 mg/day for 3 days) increased pulse rate and decreased plasma potassium. No significant changes in the systolic and diastolic blood pressure, blood P_{O_2}, and respiration were seen after disulfiram administration. Ethanol, however, produced marked decreases in both the systolic and diastolic blood pressure in patients on the larger dose of disulfiram. Of the various parameters studied, these investigators suggested that the determination of pH before and after disulfiram may be used as a predictor of the severity of the disulfiram–ethanol reaction.

Reports on blood acetaldehyde are variable, probably because of the limitation of the analytical method employed. These problems have been reviewed by Majchrowicz and Mendelson[105] and Majchrowicz.[106] Apparently, nonenzymatic formation of acetaldehyde that occurs during the analysis produces high acetaldehyde levels. In addition, there may be uneven distribution of acetaldehyde in blood as a result of binding. However, new methods employing gas chromatography have confirmed the increased acetaldehyde levels reported in earlier studies.[107,108]

At the present time the mechanism by which the disulfiram–ethanol reaction occurs is not well understood. Although the prevailing theory appears to involve increased acetaldehyde, many uncertainties remain.

G. DRUGS REPORTED TO INDUCE A DISULFIRAMLIKE ETHANOL REACTION

Numerous drugs when taken with ethanol have been reported to produce an effect similar to the disulfiram–ethanol reaction. These include calcium cyanamide (carbimide), hypoglycemic agents such as tolbutamide and chlorpropamide, analgesic and antiarthritic agents such as phenylbutazone, as well as agents such as isonicotinyl hydrazine, cyclothiazide, metronidazole, pyrogallol, coprinus atramentarius (edible mushroom), and animal charcoal. These have been reviewed previously by Truitt and Walsh[108] and Kitson.[103]

Bonfiglio and Donadio[109] reported that metronidazole produced a disulfiramlike action. Although metronidazole has been used as a substitute for disulfiram, its effect is controversial and the drug has been difficult to evaluate. Goodwin and Reinhard[110] studied Flunidazole in patient volunteers. This drug, a chemical analogue of metronidazole, lacks the bitter taste and can therefore be better assessed in a blind control study. These investigators found that no disulfiram–ethanol reaction was experienced when Flunidazole was taken with ethanol. In view of metronidazole's disulfiramlike action, Kalant et al.[111] investigated the effects of metronidazole on acetaldehyde metabolism. No effect was observed. Furthermore, in in vitro studies, metronidazole did not show any effect on liver alcohol dehydrogenase.

Since the oral hypoglycemics have been reported to produce symptoms similar to the disulfiram–ethanol reaction when taken with ethanol, several groups of investigators have investigated their effect on aldehyde dehydrogenase. Truitt et al.[112] found tolbutamide and chlorpropamide increased blood acetaldehyde. However, large doses of these oral hypoglycemics were required to produce this effect. Podgainy and Bressler[113] found that tolbutamide and chlorpropamide inhibited aldehyde dehydrogenase, with the inhibition exhibiting noncompetative kinetics. These findings differed from those of Asaad and Clarke,[114] who found that tolbutamide and chlorpropamide had no effect on liver aldehyde dehydrogenase. These investigators explained the differences between their findings and those of Podgainy and Bressler[113] by suggesting that those investigators failed to establish a clear linear relationship between the concentration of acetaldehyde and NADH production, thus leading to erroneous results.

Of interest is the observation that the mushroom *Coprinus atramentarius* when taken with ethanol produces symptoms similar to the disulfiram–ethanol reaction. Hatfield and Schaumberg[115] found that in mice hyperacetaldehydemia is produced when *Coprinus atramentarius* is combined with ethanol. However, coprine, the active constituent, has no effect on liver aldehyde dehydrogenase. It was suggested that coprine might be metabolized *in vivo* to an active aldehyde dehydrogenase inhibitor.

A number of drugs, such as calcium cyanamide (citrated calcium carbimide), metronidazole, and the oral hypoglycemics, have been reported to produce disulfiramlike symptoms when combined with ethanol. However, well-controlled studies, including proper experimental design, have not been carried out. Thus, it is very difficult to evaluate the efficacy not of only disulfiram but also of drugs producing disulfiramlike effects when combined with ethanol. Questions surrounding the efficacy of disulfiram and other antialcohol-type drugs have been reviewed by Mottin.[116]

H. DISULFIRAM TOXICITY AND SIDE EFFECTS

Early studies by Hanzlik and Irvine[1] in rabbits and puppies found disulfiram to be relatively nontoxic. The minimal fatal dose of intragastrically administered disulfiram was approximately 3 g/kg. No marked symptoms were apparent until the next day, with the animals generally recovered by the end of the second day. In these studies disulfiram was mixed with water and administered. Signs of disulfiram toxicity after oral administration include diarrhea, anorexia, emesis in dogs, lethargy, ataxia, hypothermia, and respiratory arrest.[117] The toxic dose reported by Hanzlik and Irvine[1] appears frequently in the literature and gives the impression that disulfiram is relatively nontoxic. Because disulfiram was suspended in water and administered intragastrically, the drug may not have been completely absorbed, and the values reported may be erroneous. For example, when disulfiram is administered intraperitoneally to mice, the LD_{50} is approximately 400 mg/kg.[118] Furthermore, rats appear to be even more sensitive to disulfiram. Species differences, as well as differences in the route of administration, markedly affect the LD_{50} of disulfiram. In view of the numerous side effects reported clinically, disulfiram may be more toxic than first believed.

Toxic reactions for disulfiram reported clinically include skin eruptions,[119] myocardial arrhythmias, ischemia and infarction,[120] optic neuritis, peripheral neuropathy,[121] drowsiness, headache, tiredness, vertigo, and reduced sexual potency.[122-124]

Price and Silberfarb[125] reported the occurence of a grand mal seizure in a patient taking 500 mg disulfiram/day for approximately 9 months. Two months before admission to a hospital the patient had marked changes in personality, increased lethargy, periodic confusion, and paranoid delusions. Disulfiram also has been implicated in seizures by Liddon and Satran[126] and Hotson and Langston.[127] There also have been reports of acute hepatitis due to disulfiram. Patients receiving 500 mg each day for 5 days, then 250 mg/day, developed fatigue, malaise, and jaundice after 10 days of disulfiram therapy. Liver function tests returned to normal within 6 weeks after disulfiram was terminated. When disulfiram was again started, acute hepatities developed.[128]

Rainey[129] recently reviewed disulfiram neurotoxicity. He attempted to relate the various side effects and toxicity to carbon disulfide, a metabolite of disulfiram. Unfortunately, the amount of carbon disulfide theoretically available based upon the molecular structure of the compound is not found clinically. For example, Faiman et al.[46] could only detect approximately 4% of the dose of disulfiram eliminated in the breath as carbon disulfide. Although carbon disulfide may produce some of the side effects attributed to disulfiram, these side effects are difficult to interpret in view of the disease state and the psychological makeup of the patient being treated.

1. DOSAGE

The dose of disulfiram used is variable. It is generally recommended that patients initiating treatment receive 500 mg daily as a single dose for 1–2 weeks and then 250 mg daily as maintenance. However, various dosage patterns seem to be employed. These range from 125 mg daily to 250 mg two to three times a week. It is recommended that after disulfiram therapy has been discontinued, ethanol should be avoided for at least 10 days. Recently, Iber and Chowdhury[130] found that the reaction to disulfiram and ethanol does not persist beyond 24 hr in normal patients but may continue for as long as 96 hr in patients with liver disease. Whenever side effects with disulfiram occur, the dosage should be reduced or the drug stopped.

A major problem with disulfiram dosage regimens seems to be that the same dosage schedule is employed for most patients regardless of sex, weight, and liver or kidney function. Because of differences in rates of metabolism apparent with disulfiram in animals and man, and the importance of the liver and kidney in disulfiram elimination, it appears that the disulfiram dosage should be more individualized. This probably would eliminate many of the reported side effects.

J. DISULFIRAM IMPLANTATION

Disulfiram implantation has been employed in an attempt to prevent patients from discontinuing their medication. Although this mode of administration has been used in Europe, this route has not been used extensively in North America.

Implantation generally involves making an incision in the lower abdomen, inserting 100-mg sterile disulfiram trocars, and then suturing the incision. About eight of the disulfiram trocars are used. Disulfiram implantation has been recently reviewed by Malcolm et al.[131] and Wilson.[132]

At the present time, implantation as a route for disulfiram administration has not been overwhelmingly accepted. The fact that surgery and anesthesia are required has produced some resistance by clinicians. More important is the question of efficacy. Malcolm et al.[131] suggested that for disulfiram to be effective, blood levels of disulfiram in excess of 0.1 mg% are required. These investigators believe that implantation does not give the patient a pharmacologically active dose of disulfiram, and suggest that any benefits derived are only psychological. Sauter et al.[104] studied the disulfiram–ethanol reaction in patients receiving several doses of disulfiram. At a dose of 100 mg disulfiram/day for 2 days, a suggested model for implantation, no disulfiram–ethanol reaction occurred in patients given ethanol.

Another problem with implantation methods is the continuous dose of disulfiram required. Leaching of a drug from a matrix is a very complex pharmaceutical problem. If disulfiram could be implanted in specially designed matrices that allowed for continuous and effective blood levels, the results might be more favorable.

K. PATIENT COMPLIANCE

Although disulfiram is widely used, patient compliance is difficult to monitor. Since noninvasive techniques must be used to assess compliance, only breath or urine sampling is feasible. Two procedures employed to determine compliance are breath and urine analysis. Determination of breath carbon disulfide involves the use of McKees solution,[133,134] a solution containing diethylamine, triethanolamine, cupric acetate solution, and 95% ethanol. Breath from the patient is trapped in the McKees solution, and the carbon disulfide combines with the cupric acetate to form a yellow color. The absorbance can then be measured spectrophotometrically at 430 nm to determine carbon disulfide quantitatively, or the yellow color used qualitatively. Kraml[133] found that at a concentration of 1.0 μg carbon disulfide, 90% of the tests were correctly identified, while at 0.5 μg carbon disulfide, only 25% of the tests were identified correctly.

Another test employs the determination of urinary diethylamine. Gordis and Peterson,[135] added a copper–carbon disulfide reagent mix to 2 ml of urine, converting the secondary amine to the copper salt of diethyldithiocarbamate, which was then analyzed by thin-layer chromatography. These investigators found that tests for diethylamine in urine remain positive even after 2 months if the urine is refrigerated.

Neiderhiser et al.[136] also have used the determination of diethylamine in the urine to monitor patient compliance. The method they employed involves the conversion of the diethylamine to a derivative, N,N-diethyl-3,5-dinitrobenzamide, which is soluble in organic solvents. The derivative is readily extracted with diethyl ether and then quantified using thin-layer chromatography.

Since alcoholics may not always be relied upon to continue their disulfiram therapy program, it is important that physicians have a method for accurately determining compliance. Unfortunately, to date, rapid screening methods are not available. Those described can be employed to detect disulfiram compliance; however, the methods are not simple and cannot be readily employed in a physician's office. The methods can, however, be used to follow the excretion characteristics of disulfiram metabolites in breath and urine.

L. RISK/BENEFIT

Although disulfiram has been used for almost 30 years, its effectiveness is controversial. This is because most of the clinical trials evaluating disulfiram have been uncontrolled or were poorly designed.[137] Armor et al.[138] recently reviewed various antialcohol treatment programs. They concluded that disulfiram met with some degree of therapeutic success.

Recently, a well-designed preliminary study employing 128 males was undertaken by Fuller[139] to evaluate the effectiveness of disulfiram. The three treatment groups were comparable in sex, age, race, marital status, employment status, and duration of alcohol abuse. The results of the study suggested that disulfiram may be a useful adjunct to becoming abstinent. The percentage of patients that remained abstinent was twice as large as those not receiving disulfiram (23% versus 12%). Also, patients on disulfiram drank fewer days, worked more days, and kept appointments more regularly than non-disulfiram-treated patients.

Side effects and occasional reports of drug toxicity make disulfiram a drug with some risk. However, it appears that if disulfiram is effective, the risk/benefit ratio is small. Unfortunately, a large-scale study properly designed with appropriate controls has never been carried out.

M. SUMMARY AND CONCLUSIONS

Disulfiram, after oral administration, is rapidly absorbed and metabolized. In animals, the diethyldithiocarbamate-glucuronide and inorganic sulfate are the major metabolites, and these as well as disulfiram, the corresponding thiol, and the methyl ester are found in numerous organs, including the lung, liver, kidney, brain, adrenals, and thyroid. Diethylamine and carbon disulfide also are found in urine, while only carbon disulfide is found in breath. Organ distribution of disulfiram and its metabolites in man are unknown, although the metabolites found in the plasma and urine of animals are also found in man. Disulfiram inhibits numerous enzymes, including aldehyde dehydrogenase, dopamine-β-hydroxylase, superoxide dismutase, enzymes important in drug metabolism, as well as those enzymes involved in intermediary metabolism.

The ingestion of ethanol by patients on disulfiram therapy produces an array of unpleasant symptoms which is referred to as the disulfiram–ethanol reaction. The mechanism by which this reaction occurs is unknown, although increased acetaldehyde is the theory which has received the most support. In addition to disulfiram, a number of other drugs have been reported to produce disulfiramlike reactions when taken with ethanol. Most frequently reported are metronidazole and the oral hypoglycemics. Many side effects, such as depression, anorexia, gastrointestinal irritation, hepatitis, and convulsions, have been attributed to disulfiram. Unfortunately, the causes of these side effects are difficult to evaluate because of the disease state being treated and the physiological effects of ethanol withdrawal.

Disulfiram is usually administered in a dose of 500 mg/day for 1 to 2 weeks, and then 250 mg/day. Many clinicians, however, employ modifications of this dosing schedule. Because the disulfiram dosage is fairly standardized, this could account for some of the side effects reported. If the dosage could be more individualized, taking into account such factors as sex, weight, and liver and kidney function, some of the undesirable effects might not occur. Disulfiram implantation, although used in Europe, has not met with acceptance in North America. No evidence presently exists supporting the effectiveness of disulfiram implants. Attempts have been made to monitor patient compliance, measuring either carbon disulfide in the breath or urinary diethylamine. These methods, however, cannot be readily used in an office environment.

Although disulfiram has been used as a deterrent in the treatment of alcoholism for almost 30 years, well-designed studies with suitable controls to test its effectiveness have never been carried out. Until the efficacy of disulfiram as an ethanol deterrent has been proven, the risk/benefit ratio remains unknown.

N. REFERENCES

1. P. J. Hanzlik and A. Irvine, *J. Pharmacol. Exp. Ther.* **17**:349 (1921).
2. E. E. Williams, *J. Am. Med. Assoc.* **109**:1472 (1937).
3. J. Hald, E. Jacobsen, and V. Larsen, *Acta Pharmacol. Toxicol.* **4**:258 (1948).
4. E. Asmussen, J. Hald, E. Jacobsen, and G. Jorgensen, *Acta Pharmacol. Toxicol.* **4**:297 (1948).
5. J. Hald and E. Jacobsen, *Acta Pharmacol. Toxicol.* **4**:305 (1948).
6. E. Asmussen, J. Hald, and V. Larsen, *Acta Pharmacol. Toxicol.* **4**:311 (1948).
7. V. Larsen, *Acta Pharmacol. Toxicol.* **4**:321 (1948).
8. A. D. Cummings and H. E. Simmons, *Ind. Eng. Chem.* **20**:1173 (1928).
9. G. Domar, A. Fredga, and H. Linderholm, *Acta Chem. Scand.* **3**:1441 (1949).
10. H. Linderholm and K. Berg, *Scand. J. Clin. Lab. Invest.* **3**:96 (1951).
11. K. J. Divatia, C. H. Hine, and T. N. Burbridge, *J. Lab. Clin. Med.* **39**:974 (1952).

12. S. L. Tompsett, *Acta Pharmacol. Toxicol.* **21**:20 (1964).
13. A Farago, *Arch. Toxicol.* **22**:396 (1967).
14. R. Fried, A. N. Masoud, and F. M. Klein, *J. Pharm. Sci.* **62**:1368 (1973).
15. R. Fried, *Ann. N.Y. Acad. Sci.* **273**:212 (1976).
16. A. M. Sauter, W. Wiegrebe, and J. P. von Wartburg, *Arzneim.-Forsch.* **26**:173 (1976).
17. H. Yoshida, M. Toga, and S. Hikime, *Bunseki Kagaku* **16**:605 (1967).
18. M. W. Brown, G. S. Porter, and A. E. Williams, *J. Pharm. Pharmacol.*, Suppl. **26**:95P (1974).
19. E. C. Gregg and W. P. Tyler, *J. Am. Chem. Soc.* **72**:4561 (1950).
20. A. F. Taylor, *Talanta* **11**:894 (1964).
21. M. J. D. Brand and B. Fleet, *Analyst (London)* **95**:1023 (1970).
22. D. G. Prue, C. R. Warner, and B. T. Kho, *Drug Stand.* **61**:249 (1972).
23. E. B. Sheinin and W. R. Benson, *J. Assoc. Off. Anal. Chem.* **61**:55 (1978).
24. J. Wells and E. Koves, *J. Chromatogr.* **92**:442 (1974).
25. A. M. Sauter and J. P. von Wartburg, *J. Chromatogr.* **133**:167 (1977).
26. J. Cobby, M. Mayersohn, and S. Selliah, *J. Pharmacol. Exp. Ther.* **202**:724 (1977).
27. L. Eldjarn, *Scand. J. Clin. Lab. Invest.* **2**:198 (1950).
28. L. Eldjarn, *Scand. J. Clin. Lab. Invest.* **2**:202 (1950).
29. F. L. Iber, S. Dutta, M. Shamszad, and S. Krause, *Alcohol. Clin. Exp. Res.* **1**:359 (1977).
30. J. H. Strömme, *Biochem. Pharmacol.* **14**:393 (1965).
31. J. H. Strömme, *Biochem. Pharmacol.* **15**:287 (1966).
32. M. D. Faiman, D. E. Dodd, R. J. Nolan, L. Artman, and R. E. Hanzlik, *Res. Commun. Chem. Pathol. Pharmacol.* **17**:481 (1977).
33. M. D. Faiman, D. E. Dodd, S. S. Minor, and R. E. Hanzlik, *Alcohol. Clin. Exp. Res.*, **2**:366 (1978).
34. J. H. Strömme, *Biochem. Pharmacol.* **14**:381 (1965).
35. G. de Saint-Blanquat, G. Vidaillac, A. Lindenbaum, and R. Derache, *Arch. Int. Pharmacodyn. Ther.* **223**:339 (1976).
36. M. D. Faiman, D. E. Dodd, and R. Hanzlik, *Res. Commun. Chem. Pathol. Pharmacol.* **21**:543 (1978).
37. M. D. Faiman, K. Haya, and L. Artman, *Alcohol. Clin. Exp. Res.* **2**:217 (1978).
38. C. D. Johnston and C. S. Prickett, *Biochim. Biophys. Acta* **9**:219 (1952).
39. K. I. Aspila, V. S. Sastri, and C. L. Chakrabarti, *Talanta* **16**:1099 (1969).
40. J. H. Strömme, *Biochem. Pharmacol.* **12**:705 (1963).
41. T. Gessner and M. Jakubowski, *Biochem. Pharmacol.* **21**:219 (1972).
42. M. Jakubowski and T. Gessner, *Biochem. Pharmacol.* **21**:3073 (1972).
43. J. Kaslander, *Biochim. Biophys. Acta* **71**:730 (1963).
44. C. S. Prickett and C. D. Johnston, *Biochim. Biophys. Acta* **12**:542 (1953).
45. E. Merlevede and H. Casier, *Arch. Int. Pharmacodyn. Ther.* **132**:427 (1961).
46. M. D. Faiman, K. Haya, and J. A. Ewing, *Am. J. Psychiatry* **135**:623 (1978).
47. A. P. Nygaard and J. B. Sumner, *Arch. Biochem. Biophys.* **39**:119 (1952).
48. D. Keilin and E. F. Hartree, *Proc. R. Soc. B Biol. Sci. Lond.* **129**:277 (1940).
49. D. A. Richert, R. Vanderlinde, and W. W. Westerfield, *J. Biol. Chem.* **186**:261 (1950).
50. J. H. Strömme, *Biochem. Pharmacol.* **12**:157 (1963).
51. W. D. Graham, *J. Pharm. Pharmacol.* **3**:160 (1951).
52. H. D. Sisler and C. E. Cox, *Am. J. Bot.* **42**:351 (1955).
53. W. Chefurka, *Enzymologia* **18**:209 (1957).
54. D. E. Dodd and M. D. Faiman, *Biochem. J.* **174**:769 (1978).
55. R. A. Deitrich and L. Hellerman, *J. Biol. Chem.* **238**:1683 (1963).
56. R. A. Deitrich and V. G. Erwin, *Mol. Pharmacol.* **7**:301 (1970).
57. R. A. Deitrich, A. C. Collins, and V. G. Erwin, *J. Biol. Chem.* **247**:7232 (1972).
58. S. O. C. Tottmar, H. Pettersson, and K. H. Kiessling, *Biochem. J.* **135**:577 (1973).
59. A. Horton and M. Barrett, *Arch. Biochem. Biophys.* **167**:426 (1975).
60. O. Tottmar and H. Marchner, *Acta Pharmacol. Toxicol.* **38**:366 (1976).
61. J. Eckfeldt, L. Mope, K. Takio, and T. Yonetani, *J. Biol. Chem.* **251**:236 (1976).

62. N. J. Greenfield and R. Pietruszko, *Biochim. Biophys. Acta* **483**:35 (1977).
63. K. E. Crow, T. M. Kitson, A. K. H. MacGibbon, and R. D. Batt, *Biochim. Biophys. Acta* **350**:121 (1974.)
64. T. Koivula and M. Koivusalo, *Biochim. Biophys. Acta* **397**:9 (1975).
65. M. Goldstein, B. Anagnoste, E. Lauber, and M. R. McKereghan, *Life Sci.* **3**:763 (1964).
66. J. Musacchio, I. J. Kopin, and S. Snyder, *Life Sci.* **3**:769 (1964).
67. M. Goldstein, E. Lauber, and M. R. McKereghan, *J. Biol. Chem.* **240**:2066 (1965).
68. M. Goldstein and K. Nakajima, *J. Pharmacol. Exp. Ther.* **157**:96 (1967).
69. C. R. Lake, L. F. Major, M. G. Ziegler, and I. J. Kopin, *Am. J. Psychiatry* **134**:1411 (1977).
70. J. A. Ewing, B. A. Rouse, R. A. Mueller, and D. Silver, *Alcohol. Clin. Exp. Res.* **2**:93 (1978).
71. R. E. Heikkila, F. S. Cabbat, and G. Cohen, *J. Biol. Chem.* **251**:2182 (1976).
72. I. Fridovich, *Acc. Chem. Res.* **5**:321 (1972).
73. E. S. Fiala, G. Bobotas, C. Kulakis, L. W. Wattenberg, and J. H. Weisburger, *Biochem. Pharmacol.* **26**:1763 (1977).
74. B. G. Lake, R. C. Longland, R. A. Hodgson, B. J. Severn, and S. D. Gangolli, *Biochem. Soc. Trans.* **5**:308 (1977).
75. M. Lang, M. Marselos, and R. Törrönen, *Chem.-Biol. Interact.* **15**:267 (1976).
76. M. A. Zemaitis and F. E. Greene, *Biochem. Pharmacol.* **25**:453 (1976).
77. A. H. Neims, D. A. Coffey, and L. Hellerman, *J. Biol. Chem.* **241**:5941 (1966).
78. M. Marselos, M. Lang, and R. Törrönen, *Chem.-Biol. Interact.* **15**:277 (1976).
79. T. L. Svendsen, M. B. Kristensen, J. M. Hansen, and L. Skovsted, *Eur. J. Clin. Pharmacol.* **9**:439 (1976).
80. R. A. O'Reilly, *Ann. Intern. Med.* **78**:73 (1973).
81. E. S. Vesell, G. T. Passananti, and C. H. Lee, *Clin. Pharmacol. Ther.* **12**:785 (1971).
82. M. Sharkawi and D. Cianflone, *Science* **201**:543 (1978).
83. L. F. Major and P. F. Goyer, *Ann. Intern. Med.* **88**:53 (1978).
84. P. E. Dussault and M. Lepage, *J. Nutr.* **105**:1371 (1975).
85. C. Hinse and P. Lupien, *Can. J. Biochem.* **49**:933 (1971).
86. C. H. Hine, A. F. Shick, L. Margolis, T. N. Burbridge, and A. Simon, *J. Pharmacol. Exp. Ther.* **106**:253 (1952).
87. R. J. Nolan, D. E. Dodd, and M. D. Faiman, *Undersea Biomed. Res.* **5**:87 (1978).
88. C. O. Rutledge and J. Jonason, *J. Pharmacol. Exp. Ther.* **157**:493 (1967).
89. L. F. Major, J. C. Ballenger, F. K. Goodwin, and G. L. Brown, *Biol. Psychiatry* **12**:635 (1977).
90. L. D. Waterbury, O. T. Wendel, and L. A. Pearce, *Biochem. Med.* **8**:61 (1973).
91. D. Berger and H. Weiner, *Biochem. Pharmacol.* **26**:1741 (1977).
92. M. D. Faiman, R. G. Mehl, and M. B. Myers, *Life Sci.* **10**:21 (1971).
93. M. D. Faiman, R. J. Nolan, C. F. Baxter, and D. E. Dodd, *J. Neurochem.* **28**:861 (1977).
94. E. Amador and A. Gazdar, *Q. J. Stud. Alcohol* **28**:649 (1967).
95. C. A. Cahill, *N. Engl. J. Med.* **287**:935 (1972).
96. D. Fernandez, *N. Engl. J. Med.* **286**:610 (1972).
97. H. Casier and H. Polet, *Arch. Int. Pharmacodyn. Ther.* **113**:439 (1958).
98. H. Casier and E. Merlevede, *Arch. Int. Pharmacodyn. Ther.* **139**:165 (1962).
99. T. M. Kitson, *J. Stud. Alcohol* **38**:1771 (1977).
100. Y. Lamboeuf, G. de Saint-Blanquat, and R. Derache, *Clin. Exp. Pharmacol. Physiol.* **1**:361 (1974).
101. G. de Saint-Blanquat, O. Öztürkcan, and R. Derache, *Arch. Int. Pharmacodyn. Ther.* **208**:343 (1974).
102. E. S. Perman, *Acta Physiol. Scand. Suppl.* **55**(Suppl. 190):1 (1962).
103. T. M. Kitson, *J. Stud. Alcohol* **38**:96 (1977).
104. A. M. Sauter, D. Boss, and J. P. von Wartburg, *J. Stud. Alcohol* **38**:1680 (1977).
105. E. Majchrowicz and J. H. Mendelson, *Science* **168**:1100 (1970).
106. E. Majchrowicz, in: *Biochemical Pharmacology of Ethanol* (E. Majchrowicz, ed.), pp. 111–140, Plenum Press, New York (1975).

107. E. B. Truit, Jr., and M. J. Walsh, in: *Biology of Alcoholism* (B. Kissin and H. Begleiter, eds.), Vol. 1: *Biochemistry*, pp. 161–195, Plenum Press, New York (1971).
108. E. B. Truitt, Jr., and M. J. Walsh, in: *Proc. First Annu. Alcoholism Conf., Natl. Inst. Alcohol Abuse Alcoholism* (M. E. Chafetz, ed.), DHEW Publ. No. 74-675 (1973).
109. G. Bonfiglio and G. Donadio, *Br. J. Addict.* **62:**249 (1967).
110. D. W. Goodwin and J. Reinhard, *Q. J. Stud. Alcohol* **33:**734 (1972).
111. H. Kalant, A. E. LeBlanc, M. Guttman, and J. M. Khanna, *Can. J. Physiol. Pharmacol.* **50:**476 (1972).
112. E. B. Truitt, Jr., G. Duritz, A. M. Morgan, and R. W. Prouty, *Q. J. Stud. Alcohol* **23:**197 (1962).
113. H. Podgainy and R. Bressler, *Diabetes* **17:**679 (1968).
114. M. M. Asaad and D. E. Clarke, *Eur. J. Pharmacol.* **35:**301 (1976).
115. G. M. Hatfield and J. P. Schaumberg, *Lloydia* **38:**489 (1975).
116. J. L. Mottin, *Q. J. Stud. Alcohol* **34:**444 (1973).
117. G. P. Child and M. Crump, *Acta Pharmacol. Toxicol.* **8:**305 (1952).
118. J. Maj and E. Przegalinski, *Diss. Pharm. Pharmacol. XIX* **5:**505 (1967).
119. H. M. Lewis and H. H. Bremers, *J. Am. Med. Assoc.* **142:**1141 (1950).
120. E. A. Macklin, M. Koklow, A. Simon, and W. Schottstaedt, *J. Am. Med. Assoc.* **146:**1377 (1951).
121. D. J. Dalessic, *Bull. L.A. Neurol. Soc.* **33:**136 (1968).
122. B. Knutsen, *Norw. Laeger* **69:**436 (1949).
123. O. Martensen-Larsen, *Q. J. Stud. Alcohol* **14:**406 (1953).
124. N. Rathod, *Q. J. Stud. Alcohol* **19:**418 (1958).
125. T. R. P. Price and P. M. Silberfarb, *J. Stud. Alcohol* **37:**980 (1976).
126. S. C. Liddon and R. Satran, *Am. J. Psychiatry* **123:**1284 (1967).
127. J. R. Hotson and J. W. Langston, *Arch. Neurol.* **33:**141 (1976).
128. E. B. Keeffe and F. W. Smith, *J. Am. Med. Assoc.* **230:**435 (1974).
129. J. M. Rainey, Jr., *Am. J. Psychiatry* **134:**371 (1977).
130. F. L. Iber and B. Chowdhury, *Alcohol. Clin. Exp. Res.* **1:**365 (1977).
131. M. T. Malcolm, J. S. Madden, and A. E. Williams, *J. Psychiatry* **125:**485 (1974).
132. A. Wilson, *J. Stud. Alcohol* **36:**555 (1975).
133. M. Kraml, *Can. Med. Assoc. J.* **6:**578 (1973).
134. S. M. Paulson, S. Krause, and F. L. Iber, *Johns Hopkins Med. J.* **141:**119 (1977).
135. E. Gordis and K. Peterson, *Alcohol. Clin. Exp. Res.* **1:**213 (1977).
136. D. H. Neiderhiser, R. K. Fuller, L. J. Hejduk, and H. P. Roth, *J. Chromatogr.* **117:**187 (1976).
137. L. Lundwall and F. Baekeland, *J. Nerv. Ment. Dis.* **153:**381 (1971).
138. D. J. Armor, J. M. Polich, and H. B. Stambul, Alcoholism Treatment No. R-1739-NIAA, pp. 120–121, Rand Corporation, Santa Monica, Calif. (1976).
139. R. K. Fuller, *Clin. Res.* **24:**568A (1976).

Amethystic Agents 20

Reversal of Acute Ethanol Intoxication in Humans

Ronald L. Alkana and Ernest P. Noble

A. INTRODUCTION

Ethanol, like other depressant drugs, produces both excitatory and depressant effects on behavioral and physiological parameters. However, its predominant acute effect is depression of brain function. Unfortunately, the resultant impairment of motor coordination, judgment, attention, and other aspects of behavior represents a serious threat to the health and safety of both the drinker and the nondrinker. In fact, the estimated cost associated with ethanol-related motor vehicle accidents has been placed at $4.7 billion annually. More importantly, evidence suggests that excessive ethanol use is associated with approximately one-half of all automobile fatalities (about 25,000 in one year).[1]

The identification of a means of quickly and safely antagonizing acute ethanol intoxication could provide a mechanism for reducing this toll and could have additional benefits for the fields of alcoholism and alcohol research. For example, such an amethystic agent might be useful in the treatment of acute ethanol overdose, a condition for which there is no specific remedy at this time. It may provide information leading to a better understanding of, as well as a tool with which to investigate, ethanol's mechanism of action. In addition, the discovery of methods to reverse ethanol's acute effects may contribute valuable insights into the adaptive processes that lead to drug dependence. Finally, since central depression, tolerance, and withdrawal may represent a continuum of

---

Amethystic: From the Greek *amethystos* (not drunken). The Greeks believed that the amethyst prevented intoxication. Defined here as sobering.

Ronald L. Alkana • School of Pharmacy, University of Southern California, Los Angeles, California 90033. **Ernest P. Noble** • University of California, College of Medicine, Irvine, California 92717.

ethanol's effects over time, it is not unreasonable to speculate that aborting or mitigating one of the initial steps in this sequence could interfere with the subsequent changes induced by ethanol.

A critical issue in the search for amethystic agents is the identification of drugs with high potential as ethanol antagonists. Although numerous studies report antagonistic interactions between ethanol and other drugs and several folk remedies purportedly hasten sobriety, no chemicals or techniques are known which, as with the opiate antagonists, unequivocally reverse the central depressant effect of ethanol or other general anesthetics on humans. Furthermore, with the exception of dialysis,[2-5] there are no effective methods for rapidly eliminating ethanol from the body.

This paper identifies and evaluates several potential amethystic agents via a review and synthesis of available information regarding ethanol's pharmacokinetics, its mechanism(s) of action in the brain, its effects on physiological and behavioral systems, and its interactions with other drugs. Since many of these topics are treated in detail elsewhere in this volume and in an earlier review of amethystic agents,[6] the following discussion is selective and is limited to a presentation of the rationale and pertinent evidence leading to the identification or elimination of specific agents or classes of agents as potentially safe, effective, and useful ethanol antagonists. In the hope of facilitating the formation of a clearly defined and quantitative picture of the effects of potential amethystic agents discussed in the text, the most typical results and representative experimental values from the literature are presented in tabular form (Tables I–III).

B. CHARACTERISTICS OF AN IDEAL AMETHYSTIC AGENT

Several characteristics are necessary for a prospective amethystic agent to be safe and practicable in humans as a means of reducing ethanol-related vehicular accidents. For example, in addition to antagonistic efficacy, the ideal agent should:

1. Restore behavior to predrinking levels without causing marked stimulation or inducing disturbance of natural sleep.

2. Have a rapid onset of action. It should be effective within 15–30 min after it is taken.

3. Have a long duration of action so that blood ethanol concentrations can fall below 50 mg/dl.

4. Have a convenient method of administration with minimal discomfort (oral, sublingual, or inhalation).

5. Have no toxic or cumulative adverse effects either in the presence or in the absence of ethanol.

6. Have a low abuse liability.

7. Be inexpensive and possess a long shelf life.

The criteria of potential efficacy, safety, and practicality are utilized in the following discussion as means of evaluating possible amethystic agents.

C. REVERSING ETHANOL'S EFFECTS

Once ethanol is administered, there are essentially two methods available for reversing its central effects: (1) reducing the amount of ethanol in the brain by decreasing its absorption, altering its distribution, or enhancing its removal (pharmacokinetic antagonism); and (2) counteracting the effects of ethanol by antagonizing its actions or effects on the brain (pharmacodynamic antagonism). A given drug or environmental manipulation may act to reduce intoxication by one or both of these methods.

1. Pharmacokinetic Antagonism (Table I)

a. Reduced Ethanol Absorption

Several means are available for reducing the rate of ethanol absorption. These generally involve substances or techniques which delay gastric emptying time or reduce gastrointestinal circulation, such as the ingestion of food or the administration of sympathomimetic and anticholinergic drugs.[7] However, as discussed below, once an intoxicating blood ethanol concentration has been reached (100 mg/dl or greater), the limiting factor in reducing the blood ethanol concentration is the maximum rate at which ethanol can be eliminated from the body. Since this rate is approximately 15 mg/dl per hr (Ref. 8, p. 47), it would take several hours in order to appreciably reduce intoxicating blood ethanol concentrations, even with ethanol absorption blocked. Therefore, absorption blockade alone cannot achieve a *rapid reversal* of intoxication. Such a reversal must depend upon altered ethanol distribution, increased ethanol elimination, or centrally mediated antagonism.

b. Altered Ethanol Distribution

Recently, Eskelson et al.[9] found that rats given 15 mg/kg of DH-524 [2(3,4-dichlorophenoxy) methyl-2-imidazoline] were partially protected from the hypnotic effect of 3.0 g ethanol/kg (30%) given either orally or intraperitoneally (ip). The authors attribute the protection to a DH-524-induced reduction in plasma ethanol concentration, resulting in part from shifts in body water compartments. These findings suggest the possibility of reducing intoxication by lowering the concentration of ethanol at its sites of action via manipulating its distribution within the body. However, McNamee and co-workers[10] report that pretreat-

Table I. Putative Pharmacokinetic Ethanol Antagonists: Representative Experiments

Treatment	Dose	Protocol	Ethanol dose	Species	Measure	Effect	Reference
Fenmetazole	100–200 mg	Pretreat	0.8 g/kg, oral	Human	Mood, memory, psychomotor	No change	10
NAD	2 mg/kg, iv	Posttreat	0.5–2.0 g/kg, ip	Rat	Ethanol elimination	No change	21
	500 mg/kg, ip (3 days)	Pretreat	0.5–2.0 g/kg, ip	Rat	Ethanol elimination	No change	21
Dinitrophenol	34 mg/kg, ip	Pretreat	1.1–3.0 g/kg, ip	Rat	Ethanol elimination	Increase (50%)	25
	500 mg, oral	Posttreat	1.5 ml/kg, iv; 20%	Human	Ethanol elimination	Increase (5%)	27
	100 mg, oral (tid[a] × 2 weeks)	Pretreat	0.5 ml, iv	Human	Ethanol elimination	No change	27
Fructose	198 mg/dl (11 mM)	In vitro	20 ng/dl (4 nM)	Human liver	Ethanol oxidation rate	Increase (76%)	29
	75 g, iv; 10%	Concurrent	105 g, oral, bourbon	Human	BEC[b]	Decrease (43%)	30
	100 g, iv; 10%	Posttreat	150 mg/dl minimum	Human	Ethanol elimination	No change	31
	125 g, oral	Pretreat	0.8 g/kg, oral; 20%	Human	Serum uric acid, lactate; BEC[b]	Increase; Decrease (60%) (block absorption)	32

[a] Three times a day.
[b] Blood ethanol concentration.

ment with this drug (fenmetazole 100 mg, 200 mg) failed to antagonize or significantly modify the acute effects of ethanol (100 mg/dl) on mood, psychomotor
performance, memory, and degree of intoxication in humans. There is no explanation at present for the discrepancy between the animal and human studies, but
in view of the absence of antagonism at the tested doses, it is unlikely that fen-
metozole represents an effective ethanol antagonist in humans.

c. Enhanced Ethanol Elimination

It is now acknowledged that approximately 95% of the ethanol entering an
organism is eliminated via oxidation, primarily in the liver. The remainder is
excreted via the respiratory and urinary tracts and through the sweat glands.[11]
For an agent or method to be effective in enhancing ethanol removal, it must
accelerate hepatic ethanol metabolism, unless a radical change in the normal
routes of elimination can be affected. It is evident, therefore, why methods
designed to increase ethanol excretion, such as the induction of hypothermia[12,13]
or hyperthermia,[14] increasing respiration,[15] or utilizing diuretics[16] have generally been unsuccessful in greatly increasing ethanol elimination. Dialysis has
been used successfully to reduce blood ethanol concentrations in human emergency situations[2,3] and experimentally in dogs.[4,5] However, this extreme
approach is of little practical value in the treatment of ordinary intoxication.

A detailed discussion of the enzymatic pathways of ethanol metabolism is
presented in Volume 1, Chapter 4. Briefly, the rate of ethanol elimination is
primarily determined by the rate of hepatic oxidation of ethanol to acetaldehyde
via the NAD dependent enzyme alcohol dehydrogenase (ADH) with the coenzyme NAD acting as the hydrogen acceptor. Theorell and Chance[17] were the
first to propose the mechanism of this reaction. ADH reacts with NAD^+ to form
an $ADH \cdot NAD^+$ complex. The $ADH \cdot NAD^+$ complex reacts with ethanol to
produce $ADH \cdot NADH + H^+$ + acetaldehyde. The acetaldehyde is normally
oxidized rapidly by an NAD-dependent aldehyde dehydrogenase to acetate,
which in turn enters the tricarboxylic acid cycle as well as other metabolic reactions via acetyl-CoA.

Liver ADH is saturated with ethanol at low concentrations (ca. 9 mg/dl).[18]
Thus, the enzyme usually catalyzes dehydrogenation of ethanol at a constant rate
until the ethanol concentration falls below the saturation level. In humans, the
rate of ethanol elimination at saturation blood ethanol concentrations is approximately 100 mg/kg per hr, which corresponds to a decline in blood ethanol of 15
mg/dl per hr (Ref. 8, p. 47). Accordingly, it would take approximately 3–4 hr to
metabolize a sufficient amount of ethanol to drop human blood ethanol concentrations from 100 mg/dl to 50 mg/dl, if ethanol absorption was blocked, and longer
if it was not blocked. Therefore, a minimum increase of 10 times the normal rate
of ethanol elimination must be achieved in order to initiate a rapid reversal of
intoxication.

Early studies with purified horse liver ADH indicated that the rate of oxidation of ethanol to acetaldehyde is controlled by the rate of dissociation of the ADH · NADH complex.[17] More recent investigations suggest that the rate of reoxidation of NADH formed during the metabolism of ethanol and acetaldehyde may be the rate-limiting factor.[19] Theoretically, therefore, ethanol metabolism may be enhanced by augmenting the oxidation of NADH to NAD, or perhaps by increasing the NAD/NADH ratio by adding exogenous NAD.

Attempts at increasing ethanol metabolism by administering exogenous NAD or related substances have been generally unsuccessful. For example, after a report by O'Hollaren[20] that NAD benefited the treatment of acute and chronic states of alcoholism and withdrawal syndrome in humans, Majchrowicz *et al.*[21] investigated the effects of NAD (2 mg/kg, iv; 500 mg/kg, ip) on ethanol-treated (0.5–2.0 g/kg, ip) rats. They found no difference in the rate of ethanol metabolism between controls and NAD-injected rats using [1-^{14}C]ethanol. Similar negative findings were reported by Wilson,[22] who found that nicotinamide raised the concentration of NAD in mouse liver but did not alter the rate of ethanol elimination.

On the other hand, it appears that agents which increase oxidation of NADH can increase ethanol metabolism. In early studies, Ewing[14] and others (reviewed in detail by Newman)[23] reported that dinitrophenol, which uncouples oxidative phophorylation and thus accelerates the mitochondrial oxidation of NADH, increased ethanol elimination in dogs at doses as low as 7.5 mg/kg. Newman and Tainter[24] found that dinitrophenol (6–10 mg/kg, im) increased the respiratory loss of ethanol in dogs and suggested that this route rather than enhanced metabolism explained dinitrophenol-stimulated ethanol elimination. But subsequent experiments by Israel and his group[25] demonstrated that respiratory and evaporatory losses only explained 7–25% of the enhanced elimination, indicating that dinitrophenol does stimulate ethanol metabolism. However, the *in vivo* enhancement in animals is small (50%) at near toxic doses (34 mg/kg, ip).[25] The same appears true for arsenate, another uncoupler.[26] In fact, at safe levels of dinitrophenol, a single oral dose of 500 mg or 100 mg three times daily for 2 weeks, did not appreciably accelerate ethanol elimination in humans.[27] Therefore, oxidative uncouplers may moderately increase elimination in animals, but their toxicity limits their usefulness in humans.[27]

Although numerous other agents have been tested, including carbohydrates, insulin, proteins, amino acids, glucagon, and glucocorticoids, no known substance is capable of producing a tenfold acceleration in the rate of ethanol elimination. The reader is referred to the reviews of this topic by Kalant[7] and by Wallgren and Barry (pp. 100–104)[8] for details. It appears that fructose is the most effective stimulator of ethanol elimination available. Reported accelerations vary from 15% to over 100%,[28–30] but the effect is inconsistent.[31] The mechanism of the increased rate of elimination is unknown. It may involve a fructose induced increase in oxidation of NADH to NAD[29] coupled with inhibited ethanol absorption.[32] In view of the reports indicating that fructose may induce dangerous adverse effects[31,33] and the relative small degree of acceleration accomplished compared to the required tenfold increase, it is clear that fructose does not represent a practical amethystic agent.

In conclusion, drugs which act primarily to reduce blood ethanol concentrations, either by blockade of absorption, or by acceleration of metabolism or excretion are poor candidates as rapid acting amethystic agents. On the other hand, nontoxic substances which reduce blood ethanol concentrations might be useful as adjuncts to centrally acting sobering agents.

2. Pharmacodynamic Antagonism (Table II)

In the following presentation, possible ethanol antagonists are identified and evaluated in light of their reported interactions with ethanol and/or their theoretical antagonistic potential as suggested by available information regarding their mechanisms of action and effects.

a. Catecholamine and Related Substances

Considerable attention has focused on elucidating a possible link between ethanol's behavioral effects and changes in central catecholamine systems. These experiments have taken the form of investigations of ethanol's effects on central catecholamine levels and turnover and investigations of the effects of catecholamine-altering drugs on ethanol intoxication. At present, precise relationships are unknown, but the available evidence, as outlined below, suggests that pharmacological manipulation of central catecholamine systems represents a promising method of reversing ethanol's effects.

Recent neurochemical studies, although still controversial,[34] suggest that ethanol or its metabolite acetaldehyde may increase release,[35,36] turnover,[37] and synthesis[38] of brain catecholamines while reducing their endogenous levels.[39,40] Furthermore, ethanol may block catecholamine reuptake.[41,42] These findings suggest that ethanol may deplete releasable central catecholamine stores via alteration in the function and/or metabolism of these monoamines. Since reserpine and α-methyl-p-tyrosine reportedly cause behaviorial depression via central catecholamine depletion,[43,44] it is possible that part of ethanol's depressant effects may be mediated via a similar reduction in central catecholamines. This hypothesis predicts that drugs which block or otherwise attenuate catecholamine systems should increase ethanol depression, whereas drugs which augment catecholamine systems should antagonize ethanol depression. Drug-interaction studies support this hypothesis.

In general, attenuators of central catecholamine function markedly increase ethanol's depressant effects. For example, Blum et al.[45] found that pretreatment with α-methyl-p-tyrosine (80 mg/kg, ip), an inhibitor of catecholamine synthesis, prolonged ethanol sleep time (4.0–7.0 g/kg, ip) in mice. Ahlenius et al.[46] demonstrated that α-methyl-p-tyrosine pretreatment (4 × 0.5 g; 4 × 1.0 g) increased ethanol-induced (100 g, orally) impairment of psychomotor performance and reaction time in humans. Similar increases in ethanol depression have been shown with the catecholamine-depleting agent, reserpine.[47]

Table II. Putative Pharmacodynamic Ethanol Antagonists: Representative Experiments

Treatment	Dose	Protocol	Ethanol dose	Species	Measure	Effect	Reference
Catechol-related							
Propranolol	2.5–5.0 mg/kg, ip	Pretreat	4.0 g/kg, ip; 25% w/v	Mouse	Sleep time	Decrease	49
	10.0–20.0 mg/kg, ip	Pretreat	4.0 g/kg, ip; 25% w/v	Mouse	Sleep time	Increase	49
	40 mg, oral	Posttreat	1.1 g/kg, oral	Human	EEG, divided attention, inebriation rating	Increase	52
Amphetamine	5 mg/kg, ip	Pretreat	2 g/kg, ip; 10%	Rat	Tilt plane	Decrease	55
	5 mg/kg, ip	Pretreat	3 g/kg, ip; 10%	Rat	Tilt plane	No change	55
	15 mg, oral	Pretreat	1.2 g/kg, oral	Human	Addition Psychomotor	Decrease No change	62
Pipradrol	5–20 mg/kg	Pretreat	2.0 g/kg, ip	Mouse	Locomotor depression	Decrease	67
L-Dopa	1.5 g, oral	Posttreat	0.8 g/kg, oral	Human	Balance, divided attention, EEG Memory, mood, other	Decrease No change	74
Caffeine	400 mg/kg	Pretreat	4.0–7.0 g/kg, ip; 25% v/v	Mouse	Sleep time	Increase	54
	300 mg, oral	Pretreat	0.2 g/kg, oral repeated (BEC[a] 55–152 mg/dl)	Human	Balance Steadiness, EEG, flicker fusion	Decrease No change	85
Aminophylline	200 mg, oral	Posttreat	0.8 g/kg	Human	Balance, EEG Divided attention, other	Decrease No change	74
Ephedrine	50 mg, oral	Posttreat	0.8 g/kg	Human	Balance, EEG	Decrease	74
TRH[b]	1–10 mg/kg, ip	Pretreat	5.0 g/kg, ip; 25% w/v	Mouse	Sleep time Hypothermia	Decrease Decrease	89
	5–20 µg, ic[c]	Posttreat	5.0 g/kg, ip; 25% v/v	Mouse	Sleep time Hypothermia	Decrease Decrease	89

Dibutyryl cyclic AMP	30–90 µg/rat, icv[d]	Posttreat	5.0 g/kg, ip; 25% v/v	Rat	Sleep time	Decrease	90
	30–90 µg/rat, icv[d]	Pretreat	5.0 g/kg, ip; 25% v/v	Rat	Sleep time	No change	90
Apomorphine	5 mg, oral	Posttreat	0.8 g/kg	Human	Divided attention inebriation Memory, other	Increase No change	92
Acetylcholine-related							
Physostigmine	0.005–0.3 mg/kg, ip	Pretreat	4.5 g/kg, ip; 25% w/v	Mouse	Sleep time	Decrease	97
	0.005–0.3 mg/kg, ip	Posttreat	4.5 g/kg, ip; 25% w/v	Mouse	Sleep time	Decrease	97
	0.2 mg/kg, ip	Pre- or posttreat	1.5–2.5 g/kg, ip; 20% v/v	Rat	EEG synchrony Behavior	Decrease No change	103
Atropine	0.5 mg, oral	Pretreat (?)	0.5 g/kg, oral	Human	Attention	Increase	99
GABA-related							
Aminooxyacetic acid	50 mg/kg, ip	Pretreat	2.0–2.25 g/kg, ip; 10% v/v	Rat	Tilt plane	Increase	114
Bicuculline	2 mg/kg, sc[e]	Pretreat	2.0–2.25 g/kg, ip; 10% v/v	Rat	Tilt plane	Decrease	114
Analeptics							
Pentylenetetrazol	9 ml; 10%, injected	Posttreat	BEC[a] 439 mg/dl	Human	Coma	Decrease	119
	4.4–7.0 mg/kg, iv	Posttreat	2.5–5.5 g/kg, iv	Dog	Respiratory depression, coma	Decrease	118
Strychnine	0.05–2.0 mg, iv	Posttreat	LD[f], ip or stomach tube	Dog, cat	Respiration, ataxia Spinal reflex	No change Decrease	122
Cations							
Potassium chloride	0.5 mmole/kg, ip	Simultaneous	2.0 g/kg, ip; 20% w/v	Rat	Tilt plane	Decrease	124
Rubidium chloride	5.0 mg/kg, ip (7 days)	Pretreat	5.0 g/kg, ip; 25% v/v	Mouse	Sleep time	Decrease	131

(Continued)

Table II. *(Continued)*

Treatment	Dose	Protocol	Ethanol dose	Species	Measure	Effect	Reference
Cesium chloride	5.0 mg/kg, ip (7 days)	Pretreat	5.0 g/kg, ip; 25% v/v	Mouse	Sleep time	Decrease	131
Lithium chloride	5.0 mg/kg, ip (7 days)	Pretreat	5.0 g/kg, ip; 25% v/v	Mouse	Sleep time	Increase	131
Calcium chloride	2.0 g, oral	Pretreat	1.0 g/kg, oral	Human	Intoxication	Decrease	137
	10–20 μmole/kg, icv[d]	Pretreat	4.5 g/kg, ip; 20% w/v	Mouse	Sleep time	Increase	136
Ethylenediamine-tetraacetic acid	4 μmole/kg, icv[d]	Pretreat	4.5 g/kg, ip; 20% w/v	Mouse	Sleep time	Decrease	136

[a] Blood ethanol concentration. [d] Intracerebroventricularly.
[b] Thyrotropin-releasing hormone. [e] Subcutaneously.
[c] Intracisternally. [f] Lethal dose.

In contrast to the findings with α-methyl-p-tyrosine and reserpine, pretreatment with propranolol, a centrally active β-adrenergic receptor blocking agent,[48] antagonized ethanol depression in mice. Smith et al.[49] found that pretreatment with low (1–5 mg/kg) but not high (10–20 mg/kg) doses of propranolol diminished ethanol's (4 g/kg, 25% w/v, ip) respiratory and hypnotic effect. Matchett and Erickson[50] demonstrated that propranolol pretreatment (5, 10, 20 mg/kg) also significantly antagonized ethanol's (2.0 g/kg, ip, 20% w/v) depressant effect on locomotor activity in mice. The antagonism was more evident at low doses versus high doses of propranolol. Despite this evidence of antagonism in animals, Mendelson et al.[51] failed to demonstrate a significant effect of sustained propranolol pretreatment (10 mg, four times daily × 3 days, orally) on ethanol intoxication (75 mg/dl; 100 mg/dl) in male alcoholics. More importantly, Alkana et al.,[52] using healthy male human subjects, found that a single dose of propranolol (40 mg) ingested 15 min following ethanol (1.1 g/kg) significantly increased ethanol's depressant effect on the electroencephalogram, auditory divided attention performance, and inebriation assessments without increasing blood ethanol concentrations. Propranolol also showed additive, rather than antagonistic, trends with ethanol on other tests in their battery. It is not evident why acute pretreatment with propranolol in mice reduces whereas postethanol treatment in humans increases ethanol depression. It has been suggested that this contradiction may reflect species differences or cross-tolerance mechanisms involving brain catecholamines.[52] Further research is necessary to clarify this issue, but the data[51,52] clearly indicate that propranolol is not a sobering agent in humans.

Despite some contradictory findings at high drug doses [ca. 0.2–1.0 g/kg[53,54]], there is a growing body of evidence suggesting that catecholamine-augmenting drugs antagonize ethanol's depressant effect. Wallgren and Tirri[55] found that epinephrine reduced ethanol performance impairment on the tilt-plane task in rats. Allen et al.[56] demonstrated that α-methyl-m-tyrosine, which mediates the release of endogenous norepinephrine centrally,[57] also antagonized ethanol depression of behavioral performance in rats.

Pre- or concurrent treatment with amphetamine, which enhances release and/or blocks reuptake of catecholamines by presynaptic terminals,[58] antagonizes ethanol's effect as measured by sleep time (4.5 g ethanol/kg, ip; 3 and 4 mg amphetamine/kg, ip)[40,59]; tilt plane (2 g ethanol/kg, ip, 10% w/v; 5 mg amphetamine/kg, ip)[55]; and cortical EEG (blood ethanol 300 mg/dl, ethanol iv; 6 mg amphetamine/kg, iv)[60] in rats, mice, and rabbits. On the other hand, pretreatment with amphetamine (2–5 mg/kg, ip) did not antagonize an ip dose of ethanol sufficient to inhibit 75–95% of responses on a shock termination task.[56] Haggard et al.[61] also found that posttreatment with amphetamine (3.0 mg/kg, injected) did not reduce ethanol-induced (blood ethanol ca. 930 mg/dl) respiratory failure in rats. Interestingly, Wallgren and Tirri[55] found that amphetamine (5 mg/kg, ip) antagonized ethanol impairment (10% w/v, ip) on the tilted-plane test in rats at 2 g/kg but not at 3 g/kg, suggesting that amphetamine's antagonist effect on intoxication might be limited to low doses of ethanol.

Amphetamine–ethanol antagonism has also been demonstrated in humans. Wilson et al.[62] found that amphetamine (15 mg, orally) significantly blocked ethanol's effect (1.2 g/kg, orally) on an addition task and learning of new material, but not its effect on several psychomotor tasks. However, in other experiments, amphetamine pretreatment (20 mg, orally, four times daily) failed to alter ethanol's (0.5 g/kg, orally) effect on arithmetic and verbal tests under the stress of delayed auditory feedback[63]; nor did amphetamine pretreatment (15 mg, orally) significantly change the blood ethanol concentration at which test failure occurred on a test battery, including balance, hand steadiness, EEG (drop in dominant frequency by 1 Hz), and flicker fusion frequency.[64]

In summary, amphetamine's amethystic action is limited and inconsistent in humans and animals. Therefore, particularly in view of its high misuse potential,[65] amphetamine is not a good candidate for further testing. The same holds true for methamphetamine.[65,66]

Amphetamine-related drugs have also been tested with ethanol. Allen et al.[56] found that methylphenidate (10 mg/kg, ip), phenmetrazine (20 mg/kg, ip), and pipradrol (10 mg/kg, ip), but not amphetamine (2–5 mg/kg, ip), pretreatment antagonized ethanol's effects on a shock-termination paradigm in rats. Erickson and Matchett[67] demonstrated that pretreatment with pipradrol HCl (5, 10, 20 mg/kg) also blocked ethanol-induced (2.0 g/kg) depression of spontaneous locomotor activity in mice. This finding suggests that amphetamine-related drugs might have greater amethystic efficacy than amphetamine and should be investigated further. Such studies could yield information regarding structure–activity relationships for amethystic agents and thus provide clues aiding in the design of new agents with greater amethystic effect and lower misuse potential.

L-Dopa, the biosynthetic precursor of dopamine used clinically to treat Parkinson's disease,[68] increases brain catecholamine levels[69] and has central stimulant properties.[70,71] Systemically administered L-dopa reversed behavioral depression resulting from reserpine-induced catecholamine depletion,[72] and, in contrast to amphetamine, it reversed the effect of α-methyl-p-tyrosine on neuronal firing rates in rat brain.[73] Taken with the evidence discussed above suggesting that pharmacological stimulation of brain catecholamine systems antagonizes ethanol depression, these findings suggest that L-dopa may antagonize ethanol intoxication. Alkana et al.[74] investigated L-dopa's effectiveness as an amethystic agent in humans. They demonstrated that L-dopa (1.5 g), ingested 15 min following ethanol (0.8 g/kg), significantly improved performance on several measures, including platform balance and divided attention, and significantly reduced ethanol's depressant effect on the electroencephalogram. These results are promising and suggest that L-dopa alone or in combination with a peripheral decarboxylase inhibitor, which would reduce L-dopa's dose-limiting peripheral adverse effects while increasing its central effects,[75] may be useful sobering agents. These drugs need to be tested further in human amethystic studies.

Recent evidence suggests a role for 3',5'-adenosine monophosphate (cyclic AMP) in mediating the postsynaptic effects of catecholamine transmitters in several brain regions.[76,77] Moreover, Volicer and Gold[78] found that ethanol induces a dose-dependent decrease in rat cerebellar cyclic AMP beginning 1 hr after ethanol ingestion. Taken together, these results suggest that increasing central cyclic AMP by blocking its metabolism with inhibitors of phosphodiesterase, the enzyme catalyzing the conversion of cyclic AMP to AMP, and/or increasing synthesis of cyclic AMP via stimulation of adenylate cyclase, the enzyme promoting the conversion of ATP to cyclic AMP,[79] may reverse some of ethanol's depressant effects. Several lines of evidence support this hypothesis.

Caffeine, a methylated xanthine with phosphodiesterase-inhibiting properties,[80] and caffeine-containing beverages have long been regarded as antagonists to ethanol.[81] Research on the interaction of these drugs has confirmed caffeine's capacity to antagonize ethanol's depressant effects as measured by driving simulator performance,[66] balance and motor coordination,[64,82] cognitive capacity,[83] and respiratory depression[80,84] in humans as well as narcosis in cats.[85] However, as with amphetamine, the antagonism appears to be concentration-dependent and inconsistent.[66,85,86] Therefore, despite several indications of antagonism, caffeine does not represent a reliable nor effective amethystic agent.

Caffeine is not the only drug available that increases central cyclic AMP concentration. For example, theophylline, a potent phosphodiesterase inhibitor resembling caffeine structurally, has pronounced central nervous system-stimulating properties,[87] and ephedrine, which also acts as a central stimulant, reportedly activates adenylate cyclase via direct action on catecholamine receptors as well as by inducing release of central catecholamines.[88] Alkana et al.[74] tested these drugs as amethystic agents in humans. They found that aminophylline (200 mg), ephedrine (50 mg), or a combination of aminophylline (200 mg) plus ephedrine (50 mg) ingested 15 min after ethanol (0.8 g/kg) significantly reduced ethanol's effects on platform-balance performance and EEG without apppreciably altering blood ethanol concentrations. These drugs did not reduce ethanol's effect on divided attention and memory performance, nor did they influence mood or intoxication ratings. Despite the limited antagonism observed, these results are encouraging. Higher doses of aminophylline and ephedrine should be tested to see if they can produce a stronger or more complete antagonism.

Other drugs which increase cyclic AMP also interact antagonistically with ethanol. For example, thyrotropin-releasing hormone (1-10 mg/kg, ip; 5–20 μg, intracisternal) both blocked and reversed ethanol's hypnotic and hypothermic effects (5.0 g/kg, ip, 25% v/v) in mice and rats.[89] Cohn et al.[90] supply more direct evidence that stimulating cyclic AMP function antagonizes ethanol depression. They demonstrated that postethanol, but not preethanol, treatment with intracerebroventricularly administered dibutyryl cyclic AMP (30–90 μg/rat) induced a significant dose-related reduction in ethanol sleep time (ca. 5.0 g/kg, ip, 25% v/v) in rats. Under the same conditions Cohn et al.[90] also found that dibutyryl cyclic AMP induced a dose-related reduction in sleep time with seven other central depressants—amobarbital, paraldehyde, chloral hydrate, halothane, methanol, ketamine, and diazepam. These findings prompted the authors to suggest that cyclic AMP may be the common pathway and regulator of narcosis through which many, if not all, hypnotic agents may function. These exciting findings should be replicated and extended. Furthermore, in view of the proposed cyclic-AMP-stimulating effect of some prostaglandins,[91] these agents may hold some promise as amethystic agents.

Several reports of ethanol's interaction with central monoamines indirectly suggest that dopamine and norepinephrine may play different, perhaps opposite, roles in modulating ethanol's effects. Blum et al.[54] found that pretreatment of mice with large doses of L-dopa (400 mg/kg) enhanced ethanol sleep time (4.0–7.0 g/kg, 25% v/v, ip). This observation was correlated with increased concentrations of brain dopamine, but not with increased brain norepinephrine, leading the authors to speculate a causal relationship between dopaminergic stimulation and ethanol's soporific effect. Recent human studies support this finding. Alkana et al.[92] observed that postethanol (0.8 g/kg) ingestion of a subemetic dose of apomorphine (5 mg), a central dopaminergic agonist that does not stimulate noradrenergic systems,[93,94] significantly increased ethanol's effect on auditory divided attention performance and inebriation ratings without altering blood ethanol concentrations. There was no indication of antagonism on other measures employed. These data suggest that specific dopaminergic stimulation augments rather than reduces ethanol depression.

In contrast, norepinephrine stimulation appears to antagonize ethanol depression. Amphetamine, which reduces some aspects of ethanol depression, reportedly depends on norepinephrine release for its stimulatory behavioral effects.[58] Moreover, Ewing et al.[95] report that subjects with high plasma activity of dopamine-β-hydroxylase, the enzyme catalyzing the biosynthesis of norepinephrine from dopamine, felt less intoxicated than subjects with the same blood ethanol concentrations (66 mg/dl) but with low dopamine-β-hydroxylase activity. Although highly speculative, these findings support the notion that high norepinephrine availability may counteract or block some aspects of ethanol depression.

Experiments with noradrenergic inhibitors further support an ethanol antagonist role for norepinephrine. As discussed previously, propranolol, which purportedly blocks brain norepinephrine but not dopamine receptors,[77] enhances ethanol depression in humans[52] and in mice at high doses (20 mg/kg).[49] Furthermore, Blum et al.[54] found that the dopamine-β-hydroxylase inhibitor disulfiram had a similar augmenting effect on sleep time in mice, which apparently did not result from disulfiram-induced inhibition of aldehyde dehydrogenase.

These findings indirectly suggest that dopaminergic stimulation may augument, whereas noradrenergic stimulation may reduce, ethanol depression. This conclusion is consistent with Antelman and Caggiula's[96] hypothesis that norepinephrine and dopamine play balanced but opposing roles in behavioral and physiological functions. Such differential roles in altering ethanol intoxication could explain the failure of drugs which stimulate both noradrenergic and dopaminergic systems, such as amphetamine, L-dopa, aminophylline, and ephedrine, to completely or consistently reverse intoxication. That is, any dopaminergic stimulation induced by these drugs might tend to increase ethanol depression and thus offset any antagonism resulting from noradrenergic stimulation. According to this perspective, future studies of amethystic agents should investigate the possible differential roles of dopaminergic and noradrenergic systems in modulating and perhaps mediating ethanol's effects. Selective pharmacological stimulation of norepinephrine systems and/or blockade of dopamine systems may exert a more comprehensive reversal of ethanol depression than that seen following treatment with drugs that stimulate both systems.

b. Acetylcholine and Related Substances

Neurochemical and behavioral studies suggest a possible role for central cholinergic drugs in modulating ethanol intoxication. Erickson and Burnam[97] found that pre- or concurrent administration of the centrally active cholinesterase inhibitor physostigmine (0.005–0.3 mg/kg, ip), but not the peripherally limited neostigmine (0.01–0.2 mg/kg, ip), significantly shortened sleep time in mice when compared to controls given saline and ethanol (4.5 g/kg, ip, 25% w/v). Atropine in doses greater than 3 mg/kg (ip), but not atropine methyl nitrate (1 mg/kg, ip), blocked physostigmine's antagonistic effect. Erickson and Chai[98] extended these findings to show that physostigmine (0.05–0.3 mg/kg, iv) reversed ethanol-induced (1.0 g/kg, iv, 20% v/v) EEG synchrony in rats without altering blood ethanol concentrations. Moreover, Linnoila[99] demonstrated that atropine (0.5 mg, oral) enhances ethanol's (0.5 g/kg, oral) depressant effect on attention in humans. Taken with the growing body of evidence indicating that ethanol attenuates central cholinergic function,[100–102] these findings suggest that part of ethanol depression may be mediated via inhibition of central cholinergic systems, and that stimulation of these systems antagonizes this depression. Therefore, physostigmine and other centrally active cholinergics may be useful amethystic agents.

On the other hand, recent findings indicate that cholinergic stimulants have a limited spectrum of antagonistic interaction with ethanol. Klemm,[103] investigating the interaction of ethanol and physostigmine in rats, found that pre- and posttreatment with physostigmine (0.2 mg/kg, ip) reduced ethanol-induced (1.5, 2.5 g/kg, ip, 20% v/v) deactivation (synchrony) of the EEG but did not prevent behavioral signs of intoxication. Based on these findings, the author suggests that intoxication does not primarily involve impairment of cholinergic systems. Graham and Erickson[104] also investigated the effects of acetylcholine-altering drugs on behavior in rats. In these experiments, parenterally administered atropine (5 mg/kg, ip) and intraventricularly administered hemicholinium (0.05, 0.10 mg/kg) or acetylcholine (4.5 or 9.0 μg/min $\times$ 60 min) failed to alter ethanol's (4.5 g/kg, orally) effect on an active avoidance task. These authors conclude that the reported reduction in free acetylcholine during ethanol intoxication may reflect decreased neuronal activity rather than representing an important causative factor in ethanol-induced depression.

The paucity of studies in humans and the mixed findings in animals preclude a decision as to the merits of cholinergic drugs as amethystic agents. There is sufficient indication of antagonism, however, to warrant human studies. These studies should proceed with caution, though, in view of the report that peripheral cholinergic stimulation augments ethanol toxicity.[97]

c. γ-Aminobutyric Acid and Related Substances

The role played by γ-aminobutyric acid (GABA) in the acute and chronic effects of ethanol has drawn considerable interest. This interest, in part, has been generated by evidence indicating several possible biological functions of GABA, including its putative role as an inhibitory central transmitter,[105] its implication in neural excitability including convulsive disorders,[106,107] and its biosynthetic pathway with its unique connection with the tricarboxylic acid cycle via the GABA "shunt."

The fairly extensive literature regarding ethanol's effect on central GABA systems is inconsistent. Measurement of GABA levels after ethanol administration have shown increase,[108,109] decrease,[110,111] or no change[112,113] in brain concentrations. Nevertheless, the available behavioral pharmacological studies suggest a possible role for GABA systems in modulating intoxication.[114-116] Häkkinen and Kulonen[114] recently demonstrated that pretreatment with aminooxyacetic acid (50 mg/kg), ip), an inhibitor of GABA breakdown, increased ethanol's (2.0, 2.25 g/kg, ip, 10% v/v) detrimental effect on a tilt-plane test in rats without increasing blood ethanol concentrations. Cott et al.[116] observed that intraperitoneal pretreatment with aminooxyacetic acid (40 mg/kg), or the GABA derivatives baclophen (5 mg/kg) and γ-hydroxybutyric acid (200 mg/kg), eliminated or reduced ethanol's (2.4 g/kg, ip, 15% v/v) stimulatory effect on locomotor activity in mice. This can be taken as a further indication that GABA augmentors enhance ethanol's depressant effects. More importantly, Häkkinen and Kulonen found that pretreatment with bicuculline (2 mg/kg, sc), which purportedly blocks GABA receptors, antagonized ethanol's (2.0, 2.25 g/kg, ip, 10% v/v) effect on rat tilt-plane performance. These findings tentatively suggest that ethanol depression may involve stimulation of GABA systems and that GABA blockers may be useful amethystic drugs. The efficacy of bicuculline and other GABA antagonists in reversing ethanol's acute effects remains to be explored.

d. Analeptics

The analeptics are a group of structurally unrelated drugs known for their powerful central stimulant effect. They have been used as respiratory stimulants to treat overdose with depressant drugs such as barbiturates and ethanol, but this use is not indicated at present.[117] This group includes strychnine, picrotoxin, pentylenetetrazol, bemegride, nikethamide, ethamivan, and flurothyl. The ability of these drugs to induce seizures represents a potential hazard but also suggests that their excitatory capacity may be more potent and/or more pervasive than that of other drugs.

Pentylenetetrazol is used clinically as a diagnostic aid in epilepsy and may be useful in the management of regressed geriatric patients.[117] McCrea and Taylor[118] reported that pentylenetetrazol (4.4–7.0 mg/kg, iv) reversed the respiratory depressant effect of lethal ethanol doses (2.4–5.5 g/kg, iv) in dogs within a few seconds of its administration without altering blood ethanol concentrations. At ethanol doses between 2.6 and 4.0 g/kg (iv), pentylenetetrazol treatment revived dogs from their comatose state to the point where they could support their front quarters and head for 10–15 sec at a time. This behavioral antagonism lasted 30–45 min, a duration which corresponded to the end of maximum respiratory stimulation. In human alcoholics, Rosenbaum[119] found that slow injection of pentylenetetrazol (9 ml, 10%) revived comatose patients with blood ethanol concentrations up to 439 mg/dl to a responsive state without lowering blood ethanol concentrations. Pentylenetetrazol has not been tested in humans after moderately intoxicating doses of ethanol. The evidence above, although sparse, suggests this drug may be a potent amethystic agent.

Strychnine and picrotoxin have also been studied for their capacity to antagonize ethanol depression. Marshall et al.[120] and Ramsey and Haag[121] present evidence suggesting that picrotoxin is ineffective as an ethanol antagonist. On the other hand, strychnine may have some efficacy. Gold and Travell[122] found that strychnine (0.05–2.0 mg) reversed several aspects of intoxication induced by lethal ethanol doses (5–10 ml, ip or stomach tube, 24% v/v) in dogs and cats, but the animals typically showed a different form of abnormal behavior. Gold and Travell[122] conclude that it is not possible to abolish all of alcohol's effects even with toxic doses of strychnine. The effectiveness of strychnine in reversing moderate intoxication in humans has not been investigated, although it has enjoyed some popularity as a treatment for human ethanol overdose.[84] The results of Gold and Travell[122] suggest that the doses of strychnine required to initiate reversal may be too toxic for human use in non-life-threatening situations.

In conclusion, analeptic drugs may be of value in antagonizing the behavioral and cognitive aspects of ethanol intoxication. Pentylenetetrazol stands out as the most promising agent tested. This and other analeptics with less toxicity than strychnine should be further investigated.

e. Cations

The intimate involvement of ion metabolism in the electrical properties of the neuron[123] has prompted a multitude of studies of ethanol's effects on sodium and potassium ions and the processes in which they are involved. The bulk of these studies suggest that ethanol inhibits the activity of Na^+,K^+-ATPase,[124] reduces the membrane resting potential,[125] depresses the stimulus-evoked increase in sodium conductance presently held as the electrochemical basis of the rising phase of the action potential,[126-128] and reduces the peak and rate of rise of the action potential.[126,128,129]

There have been few experimental attempts at modifying ethanol's effects via the manipulation of monovalent cation levels. Israel et al.[124,130] demonstrated that the ethanol-induced (1 mg/dl) inhibition of Na^+,K^+,Mg^+-activated ATPase in rat and guinea pig brain and eel electroplaque tissue was antagonized by increasing the in vitro concentration of potassium (1.0–5.0 mM), whereas sodium enhanced the inhibition. In addition, they found that potassium chloride (0.5 mM/kg) mixed with ethanol and injected ip antagonized ethanol-induced (2.0 g/kg, 20% w/v) impairment of tilt-plane performance in rats. The antagonism was not due to a reduction in blood ethanol concentrations and there was no evidence that sodium chloride had a similar antagonistic effect. Based on their findings, Israel et al.[124] suggest that inhibition of active potassium transport plays an important role in ethanol intoxication. More recently, Messiha[131] found that pretreatment with 5 meq/kg (ip) daily for 7 days with rubidium chloride and cesium chloride, but not lithium chloride, antagonized ethanol-induced (5.0 g/kg, ip, 25% v/v) sleep time in mice without altering the specific activity of liver alcohol or aldehyde dehydrogenases. Blood ethanol concentrations were not monitored. While far from conclusive, these results suggest that some monovalent cations, particularly potassium, rubidium, and cesium, may antagonize ethanol's effects. These agents should be tested in animals at moderately intoxicating doses of ethanol on a variety of measures.

Several lines of indirect evidence suggest that calcium ions may be involved in mediating ethanol intoxication. Seeman et al.[132] found that ethanol increased the concentration of membrane-bound calcium, while Ross and co-workers[133,134] reported that ethanol (1.5 mg/kg, ip, 50% w/v; 1.5–2.5 g/kg, ip) significantly reduced brain calcium content in the corpus striatum, cortex, hippocampus, hypothalamus, cerebellum, medulla pons, midbrain, and thalamus. In addition, Peng et al.[135] found that oral narcotic doses of ethanol (2–8 g/kg) produced a rapid, sustained, and dose-related hypocalcemia in rats and dogs. The duration of hypocalcemia was coincident with the duration of anesthesia, suggesting an interrelation between the anesthetic and hypocalcemic effects of ethanol. Further evidence comes from the work of Erickson et al.,[136] who found that pretreatment with intracerebroventricularly injected calcium chloride (10–20 μmole/kg) significantly increased ethanol sleep time (4.5 g/kg, ip, 20% w/v) in mice, while similar treatment with the calcium chelators, ethylenediaminetetraacetic acid or ethylenebis(oxyethylenenitrilo) tetraacetic acid (4 μmole/kg), significantly reduced ethanol sleep time by approximately 26%. Blood ethanol concentrations were not measured. Based on these findings and other evidence in their report, Erickson et al.[136] suggest that a central calcium pool is involved in ethanol intoxication in rodents. Furthermore, their results suggest that decreasing the availability of endogenous extracellular or loosely bound membrane calcium decreases ethanol intoxication. In contrast, Alha[137] reported that 2 g of calcium chloride ingested immediately before 1.0 g ethanol/kg slightly reduced the symptoms of intoxication in humans without altering blood ethanol concentrations. Although possibly due to species differences or differences between the effects of intraventricular and oral administration of calcium chloride, it is difficult to reconcile these conflicting findings. Further studies are necessary to elucidate the potential of calcium ions or calcium chelators as amethystic agents.

f. Other Drugs

A number of substances not discussed in the present paper purportedly act as ethanol antagonists or have actions or effects suggesting that they may act as ethanol antagonists. Many of these agents, including alcohols, amino acids, antianxiety medications, carbohydrates, diethanolamine, food, hallucinogens/ psychotogens, hormones, isoniazid, maleate, monoamine oxidase inhibitors, pyrithioxine, tricyclic antidepressants, and vitamins were considered in a previous review.[6] The available evidence indicates that these agents tend to be ineffective, impractical, or toxic and do not represent strong candidates as amethystic agents.

3. Environmental Manipulations (Table III)

Several environmental manipulations reportedly modify ethanol's effects. For example, Leikola[138] demonstrated that the stress of forced swimming reduced the degree of intoxication (2.0 g/kg, ip) in rats on the tilted plane task by 60% without significantly altering blood ethanol concentrations. However, Wallgren and Tirri[55] found that the antagonism induced by forced swimming in rats on this task diminished at higher ethanol doses (3.0 g/kg, ip), whereas less exercise-oriented stressors such as loud noises (110 phone) and electric shock (500 μA, 1100 V) induced a low level of improvement at the 2.0 g/kg dose (ip), which lasted only 15 minutes after removal of the stress. Therefore, it appears that the induction of stressful states is of doubtful value as a strong and long-lasting countermeasure for intoxication. Likewise, other manipulations including the induction of hyperthermic states,[14] hypothermic states[12,13] and the administration of raised partial pressures of oxygen at normal atmospheric pressures[139,140] do not appear to represent methods of rapidly reversing intoxication.

On the other hand, hydrostatic and hyperbaric environments appear to have a strong amethystic effect. After observing that the application of high hydrostatic pressures could restore the luminosity of luminous bacteria exposed to halothane and ethanol,[141] Johnson and Flagler[142] found that hydrostatic pressure also reversed the narcosis induced in tadpoles by a solution containing 3–6% ethanol. The narcotized tadpoles resumed normal swimming in the presence of ethanol when the pressure was raised to between 2000–5000 psi. These experiments indicate that increased environmental pressure quickly reverses ethanol narcosis, but the use of such high pressures in humans would be dangerous and

Table III. Putative Environmental Ethanol Antagonists: Representative Experiments

Treatment	Dose	Protocol	Ethanol dose	Species	Measure	Effect	Reference
Stress							
Forced swimming	15 g/kg tail weight	Pretreat	2.0 g/kg, ip; 10% w/v	Rat	Tilt plane	Decrease (24%)	55
	15 g/kg tail weight	Pretreat	3.0 g/kg, ip; 10% w/v	Rat	Tilt plane	Decrease (15%)	55
Electric shock	500 μA, 1100 V	Pretreat	2.0 g/kg, ip; 10% w/v	Rat	Tilt plane	Decrease (11%)	55
Temperature							
Hyperthermia	4°C increase in body temp.	Posttreat	0.8 g/kg, iv; 20% w/v	Dog	Ethanol elimination BEC[a]	No change Increase	14
Hypothermia	Body temp. at 24°C	Pretreat	1.3, 2.6 g/kg, iv; 14% w/v	Dog	Ethanol elimination	No change	12
	Environment at 2°C for 5 days	Pretreat	0.8–1.6 g/kg, ip; 20% v/v	Rat	Ethanol elimination	Increase (15%)	13
Oxygen							
Breathing pure O_2	2–3 liters/min	Posttreat	93 $\mu mole$/ min, iv	Cat	Ethanol elimination	No change	139
Hydrostatic/hyperbaric pressure							
Hydrostatic	2000–5000 psi	Posttreat	3.0–6.0 mg/dl	Tadpole	Depression of swimming	Decrease	142
Hyperbaric (100% O_2)	30–60 psi gas	Posttreat	3.2 g/kg, ip; 16% w/v	Mouse	Sleep time	Decrease	143

[a]Blood ethanol concentration.

impractical. However, Alkana *et al.*[143] recently demonstrated that low level hyperbaric treatment with 100% oxygen at 30 and 60 psi of gas reduced mean ethanol sleep time (3.2 g/kg, ip, 16% w/v) in mice by 33 and 60%, respectively. Further studies with hyperbaric oxygen–helium mixtures at 6–8 atm absolute (90–120 psi) indicate that the antagonism does not result from the direct effects of oxygen nor from enhanced ethanol elimination but appears to involve a direct effect of pressure *per se*.[144,145] Although the interaction of ethanol with increased atmospheric pressure has not been studied in humans, these findings suggest that this method should be explored further, since it may prove useful in the treatment of ethanol overdose.

D. SUMMARY AND CONCLUSIONS

The goal of this review was to identify potential amethystic agents—agents which may be capable of rapidly reversing ethanol intoxication in humans. To this end, drugs and techniques with theoretical and/or purported value as ethanol antagonists were reviewed and evaluated using the criteria of efficacy, safety, and practicability.

Although no known chemicals or techniques were found to unequivocally reverse acute ethanol intoxication, several agents and techniques warrant further investigation. Further studies should thoroughly explore the efficacy of central catecholamine stimulants as amethystic agents, including the possibility that noradrenergic and dopaminergic systems may play opposing roles in modulating intoxication. In particular, the amethystic efficacy of L-dopa, aminophylline, and ephedrine need to be explored further in human studies. The recent animal studies with thyrotropin-releasing hormone and dibutyryl cyclic AMP require replication and extension and preliminary studies in animals should be initiated investigating prostaglandin–ethanol interactions. In addition, synthesis of new centrally active adrenergic substances with powerful receptor-stimulating properties and low misuse potential could be attempted. On the other hand, investigations of propranolol, amphetamine, and caffeine can be deemphasized.

Future studies should also focus on elucidating the scope and efficacy in reversing intoxication of centrally active cholinergic stimulants and γ-aminobutyric acid antagonists. Moreover, the studies demonstrating that cationic substances, particularly potassium, rubidium, and cesium chloride, antagonize ethanol require systematic investigation, as do the conflicting reports that both calcium chloride and calcium chelators antagonize intoxication.

Even though strychnine appears to be too toxic for use in humans and picrotoxin ineffective, other analeptics show promise as sobering agents and should be explored further. These studies should include tests of the amethystic efficacy of pentylenetetrazol at moderately intoxicating blood ethanol concentrations in humans. Since the practicality of analeptics may be limited by their tendency to produce seizures, attempts to synthesize agents with similar central stimulant properties but with lowered convulsive potency need to be encouraged.

The effectiveness of low-level hyperbaric treatments in antagonizing acute intoxication in mice suggests that this procedure may be useful in treating overdose in emergency room situations. Therefore, although probably not practical for the treatment of ordinary inebriation, this technique should be tested further beginning with animal toxicity studies.

Agents which act primarily to reduce blood ethanol concentrations, either by blocking ethanol absorption or accelerating its metabolism or excretion, cannot rapidly reduce blood ethanol. Therefore, they appear to be poor candidates as amethystic agents. They may, however, be useful as adjuncts to centrally acting sobering drugs. Other substances that do not appear to be strong candidates include alcohols, amino acids, antianxiety medications, carbohydrates, diethanolamine, dinitrophenol, fenmetazol, food, hallucinogens/psychotogens, hormones, isoniazid, maleate, monoamine oxidase inhibitors, pyrithioxine, tricyclic antidepressants, and vitamins.

As evidenced in this review, the majority of ethanol–drug interaction studies measure the effect of pretreatment with a given drug on the level of intoxication resulting from subsequent ethanol administration. This pretreatment protocol makes it difficult to interpret the effectiveness of the drug in reversing intoxication. For example, preethanol treatment may lead to the development of cross-tolerance, absorption inhibition, or other changes which can falsely indicate antagonism, while postethanol administration of the same agent could have no effect on or increase intoxication. Accordingly, amethystic studies should employ a protocol in which treatment is initiated after, rather than before, ethanol is administered. Careful attention to this and other variables in the design and reporting of amethystic studies will facilitate the interpretation of results and may reduce the number of inconsistent findings.

Ethanol intoxication probably represents a characteristic succession of physical, biochemical, physiological, and behavioral changes. Unless an antagonist intervenes at the first step(s) in this sequence, it may reverse some, but not all, of ethanol's effects. Therefore, precise information regarding ethanol's primary site(s) of action in the brain would promote the development of sobering agents by identifying specific mechanistic targets for antagonists. Since it appears that ethanol, like other general anesthetics, does not act via a specific receptor, it is possible that no single agent, analogous to the opiate antagonists, will be discovered which reverses all aspects of intoxication. Ultimately, a useful amethystic agent may require the combination of several antagonists selected for their abilities to reverse different aspects of intoxication.

The 25,000 lives lost each year in ethanol-related vehicular accidents underscore the value of developing a safe, effective means of reversing human intoxication. It is hoped that the present review will help draw attention to this need and stimulate vigorous and systematic efforts to develop amethystic agents.

ACKNOWLEDGMENTS. R. Alkana is grateful for financial support from the School of Pharmacy, University of Southern California. The authors wish to thank Linda Alkana, Haydee Bermúdes, Dorthea Douglas, Micheal Eckerdt,

Kyle Lovett, Ila MacLeod, Richard Malcolm, Elizabeth Parker, and Dena Sadamune for their comments and assistance.

E. REFERENCES

1. M. Keller (ed.), Second Special Report to the U.S. Congress on Alcohol and Health (1974).
2. B. Perey, S. Helle, and L. MacLean, *Can. J. Surg.* **8**:194 (1965).
3. J. Wieth and H. Jorgensen, *Dan. Med. Bull.* **8**:103 (1961).
4. J. Marc-Aurele and G. Schreiner, *J. Clin. Invest.* **39**:802 (1960).
5. T. Koppanyi, J. Canary, and G. Maengwyn-Davies, *Q. J. Stud. Alcohol,* Suppl. **1**:724 (1961).
6. E. Noble, R. Alkana, and E. Parker, in: *Proc. 4th Annu. Alcoholism Conf., Natl. Inst. Alcohol Abuse Alcoholism,* pp. 134–170 (1975).
7. H. Kalant, in: *The Biology of Alcoholism* (B. Kissin and H. Begleiter, eds.), Vol. 1: *Biochemistry,* pp. 1–62, Plenum Press, New York (1971).
8. H. Wallgren and H. Barry, III, *Actions of Alcohol,* Elsevier Publishing Company, Amsterdam (1970).
9. C. D. Eskelson, L. E. Myers, C. M. Calkins, and C. R. Cazee, *Life Sci.* **18**:1149 (1976).
10. H. McNamee, J. Mendelson, and J. Korn, *Clin. Pharmacol. Ther.* **17**:735 (1975).
11. R. Harger and R. Forney, in: *Progress in Chemical Toxicology* (A. Stolman, ed.), Vol. 1, pp. 53–134, Academic Press, Inc., New York (1963).
12. D. MacGregor, E. Schönbaum, and W. Bigelow, *Am. J. Physiol.* **208**:1016 (1965).
13. N. Platonow, B. Coldwell, and L. Dugal, *Q. J. Stud. Alcohol* **24**:385 (1963).
14. P. Ewing, *Q. J. Stud. Alcohol* **1**:483 (1940).
15. H. Haggard, L. Greenberg, N. Rakieten, and H. Cohen, *J. Pharmacol. Exp. Ther.* **69**:266 (1940).
16. R. Dixon and D. Rall, *Proc. Soc. Exp. Biol. Med.* **118**:970 (1965).
17. H. Theorell and B. Chance, *Acta Chem. Scand.* **5**:1127 (1951).
18. F. Lundquist and H. Wolthers, *Acta Pharmacol. Toxicol.* **14**:265 (1958).
19. K. Lindros, R. Vihma, and O. Forsander, *Biochem. J.* **126**:945 (1972).
20. P. O'Hollaren, *West. J. Surg. Obstet. Gynecol.* **69**:101 (1961).
21. E. Majchrowicz, B. Bercaw, W. Cole, and D. Gregory, *Q. J. Stud. Alcohol* **28**:213 (1967).
22. E. C. Wilson, in: *Biochemical Factors in Alcoholism* (R. P. Maickel, ed.), pp. 115–124, Pergamon Press, Inc., Elmsford, N.Y. (1967).
23. H. Newman, *Acute Alcoholic Intoxication: A Critical Review,* Stanford University Press, Stanford, Calif. (1941).
24. H. Newman and M. Tainter, *J. Pharmacol. Exp. Ther.* **57**:67 (1936).
25. Y. Israel, J. Khanna, and R. Lin, *Biochem. J.* **120**:447 (1970).
26. L. Videla and Y. Israel, *Biochem. J.* **118**:275 (1970).
27. H. Newman and W. Cutting, *J. Clin. Invest.* **14**:945 (1935).
28. S. Damgaard, F. Lundquist, K. Tonnesen, F. Hansen, and L. Sestoft, *Eur. J. Biochem.* **33**:87 (1973).
29. H. Thieden, N. Grunnet, S. Damgaard, and L. Sestoft, *Eur. J. Biochem.* **30**:250 (1972).
30. L. Lowenstein, R. Simone, P. Boulter, and P. Nathan, *J. Am. Med. Assoc.* **213**:1899 (1970).
31. R. Levy, T. Elo, and I. Hanenson, *Arch. Intern. Med.* **137**:1175 (1977).
32. E. Clark, I. Hughes, E. Letley, *J. Pharm. Pharmacol.* **25**:319 (1973).
33. R. Cohen, *Lancet* **2**:1086 (1972).
34. A. Sun, *Res. Commun. Chem. Pathol. Pharmacol.* **15**:705 (1976).
35. H. Corrodi, K. Fuxe, and T. Hökfelt, *J. Pharm. Pharmacol.* **18**:821 (1966).
36. J. Darden and W. Hunt, *J. Neurochem.* **29**:1143 (1977).
37. W. Hunt and E. Majchrowicz, *J. Neurochem.* **23**:549 (1974).
38. A. Carlsson and M. Lindqvist, *J. Pharm. Pharmacol.* **25**:437 (1973).
39. A. Carlsson, T. Magnusson, T. Svensson, and B. Waldeck, *Psychopharmacologia* **30**:27 (1973).

40. C. Erickson and J. Matchett, in: *Alcohol Intoxication and Withdrawal: Experimental Studies* (M. M. Gross, ed.), Vol. II, pp. 419–430, Plenum Press, New York (1975).
41. Y. Israel, F. Carmichael, and J. MacDonald, *Ann. N.Y. Acad. Sci.* **215**:38 (1973).
42. M. Roach, D. Davis, W. Pennington, and E. Nordyke, *Life Sci.* **12**:433 (1973).
43. H. Blaschko and T. Chruściel, *J. Physiol. (Lond.)* **151**:272 (1960).
44. S. Spector, A. Sjoerdsma, and S. Udenfriend, *J. Pharmacol. Exp. Ther.* **147**:86 (1965).
45. K. Blum, J. Merritt, J. Wallace, R. Owen, J. Hahn, and I. Geller, *Curr. Ther. Res.* **14**:324 (1972).
46. S. Ahlenius, A. Carlsson, J. Engle, and T. Svensson, *Clin. Pharmacol. Ther.* **14**:586 (1973).
47. R. Forney, H. Hulpieu, and F. Hughes, *Experientia* **18**:468 (1962).
48. D. Masuoka and E. Hansson, *Acta Pharmacol. Toxicol.* **25**:447 (1967).
49. A. Smith, K. Hayashida, and Y. Kim, *J. Pharm. Pharmacol.* **22**:644 (1970).
50. J. Matchett and C. Erickson, *Psychopharmacology* **52**:201 (1977).
51. J. Mendelson, A. Rossi, J. Bernstein, and J. Kuehnle, *Clin. Pharmacol. Ther.* **15**:571 (1974).
52. R. Alkana, E. Parker, H. Cohen, H. Birch, and E. Noble, *Psychopharmacology* **51**:29 (1976).
53. G. Rosenfeld, *Q. J. Stud. Alcohol* **21**:584 (1960).
54. K. Blum, W. Calhoun, J. Merritt, and J. Wallace, *Nature (Lond.)* **242**:407 (1973).
55. H. Wallgren and R. Tirri, *Acta Pharmacol. Toxicol.* **20**:27 (1963).
56. L. Allen, H. Ferguson, and G. McKinney, *Eur. J. Pharmacol.* **15**:371 (1971).
57. J. Van Rossum, *Psychopharmacologia* **4**:271 (1963).
58. S. Snyder, K. Tayler, J. Coyle, and J. Meyerhoff, *Am. J. Psychiatry* **127**:199 (1970).
59. J. Matchett and C. Erickson, *Fed. Proc.,* Abstr. **32**:697 (1973).
60. R. Greenberg, *Q. J. Stud. Alcohol* **28**:1 (1967).
61. H. Haggard, L. Greenberg, N. Rakieten, and L. Cohen, *J. Pharmacol. Exp. Ther.* **69**:266 (1940).
62. L. Wilson, J. Taylor, C. Nash, and D. Cameron, *Can. Med. Assoc. J.* **94**:478 (1966).
63. F. Hughes and R. Forney, *Psychopharmacologia* **6**:234 (1964).
64. H. W. Newman and E. J. Newman, *Q. J. Stud. Alcohol* **17**:406 (1956).
65. J. Innes and M. Nickerson, in: *The Pharmacological Basis of Therapeutics* (L. Goodman and A. Gilman, eds.), pp. 478–523, Macmillan Publishing Co., Inc., New York (1970).
66. J. Rutenfranz and G. Jansen, *Int. Z. Angew. Physiol. Einschl. Arbeitsphysiol.* **18**:62 (1959).
67. C. Erickson and J. Matchett, *Fed. Proc.,* Abstr. **35**:814 (1976).
68. D. Franz, in: *The Pharmacological Basis of Therapeutics* (L. Goodman and A. Gilman, eds.), pp. 227–244, Macmillan Publishing Co., Inc., New York (1975).
69. D. Reis, D. Moorhead, and N. Merlino, *Arch. Neurol.* **22**:31 (1970).
70. J. Hanig and J. Seifter, *Experientia* **27**:168 (1971).
71. B. Angrist, G. Sathananthan, and S. Gershon, *Psychopharmacologia* **31**:1 (1973).
72. C. Creveling, J. Daly, T. Tokuyama, and B. Witkop, *Biochem. Pharmacol.* **17**:65 (1968).
73. B. Bunney, G. Aghajanian, and R. Roth, *Nature (Lond.)* **245**:123 (1973).
74. R. Alkana, E. Parker, H. Cohen, H. Birch, and E. Noble, *Psychopharmacology* **55**:203 (1977).
75. T. Chase and A. Watanabe, *Neurology* **22**:384 (1972).
76. F. Bloom, *Life Sci.* **14**:1819 (1974).
77. L. Iversen, *Science* **188**:1084 (1975).
78. L. Volicer and B. Gold, in: *Biochemical Pharmacology of Ethanol* (E. Majchrowicz, ed.), pp. 211–237, Plenum Press, New York (1975).
79. R. Butcher and E. Sutherland, *J. Biol. Chem.* **237**:1244–1250 (1962).
80. J. Ritchie, in: *The Pharmacological Basis of Therapeutics* (L. Goodman and A. Gilman, eds.), pp. 358–370, Macmillan Publishing Co., Inc., New York (1970).
81. H. Nash, *Q. J. Stud. Alcohol* **27**:727 (1966).
82. E. Strongin and A. Winsor, *J. Abnorm. Soc. Psychol.* **30**:301 (1935).
83. O. Graf, in: *German Aviation Medicine: World War II,* Vol. 2, pp. 1080–1103, Department of the Air Force, Washington, D.C. (1950).
84. M. Victor, *Psychosomat. Med.* **28**:636 (1966).
85. J. Pilcher, *J. Pharmacol. Exp. Ther.* **3**:267 (1912).

86. R. Forney and F. Hughes, *Q.J. Stud. Alcohol* **26**:206 (1965).
87. J. Ritchie, in: *The Pharmacological Basis of Therapeutics* (L. Goodman and A. Gilman, eds.), pp. 367–378, Macmillan Publishing Co., Inc., New York (1975).
88. I. Innes and M. Nickerson, in: *The Pharmacological Basis of Therapeutics* (L. Goodman and A. Gilman, eds.), pp. 514–532, Macmillan Publishing Co., Inc., New York (1975).
89. G. Breese, J. Cott, B. Cooper, A. Prange, Jr., and M. Lipton, *Life Sci.* **14**:1053 (1974).
90. M. L. Cohn, M. Cohn, F. Taylor, and F. Scattaregia, *Neuropharmacology* **14**:483 (1975).
91. A. Cooper, F. Bloom, and R. Roth, *The Biochemical Basis of Neuropharmacology*, Oxford University Press, New York (1974).
92. R. Alkana, T. Willingham, H. Cohen, E. Parker, and E. Noble, *Fed. Proc.*, Abstr. **36**:331 (1977).
93. A. Ernst, *Psychopharmacologia* **10**:316 (1967).
94. N. Andén, A. Rubenson, K. Fuxe, and T. Hökfelt, *J. Pharm. Pharmacol.* **19**:627 (1967).
95. J. Ewing, B. Rouse, and R. Mueller, *Res. Commun. Chem. Pathol. Pharmacol.* **8**:551 (1974).
96. S. Antelman and A. Caggiula, *Science* **195**:646 (1977).
97. C. Erickson and W. Burnam, *Agents Actions* **2**:8 (1971).
98. C. Erickson and K. Chai, *Neuropharmacology* **15**:39 (1976).
99. M. Linnoila, *Eur. J. Clin. Pharmacol.* **6**:107 (1973).
100. C. Erickson and D. Graham, *J. Pharmacol. Exp. Ther.* **185**:583 (1973).
101. Y. Israel, F. Carmichael, and J. MacDonald, in: *Alcohol Intoxication and Withdrawal: Experimental Studies* (M. M. Gross, ed.), Vol. II, pp. 55–64, Plenum Press, New York (1975).
102. W. Hunt and T. Dalton, *Brain Res.* **109**:628 (1976).
103. W. Klemm, *Nature (Lond.)* **251**:234 (1974).
104. D. Graham and C. Erickson, *Psychopharmacologia* **34**:173 (1974).
105. K. Krnjević, *Nature (Lond.)* **228**:119 (1971).
106. J. Wood and S. Peesker, *J. Neurochem.* **20**:379 (1973).
107. S. Simler, I. Ciesielski, M. Maitre, H. Randrianarisoa, and P. Mandel, *Biochem. Pharmacol.* **22**:1701 (1973).
108. H. Häkkinen and E. Kulonen, *J. Neurochem.* **10**:489 (1963).
109. A. Rawat, *J. Neurochem.* **22**:915 (1974).
110. R. Ferrari and A. Arnold, *Biochim. Biophys. Acta* **52**:361 (1961).
111. E. Higgins, *Biochem. Pharmacol.* **11**:394 (1962).
112. D. Hagen, *Q. J. Stud. Alcohol* **28**:613 (1967).
113. U. Sutton and M. Simmonds, *Biochem. Pharmacol.* **22**:1685 (1973).
114. H. Häkkinen and E. Kulonen, *J. Neurochem.* **27**:631 (1976).
115. E. McCabe, E. Layne, D. Sayler, N. Slusher, and S. Bessman, *Science* **171**:404 (1971).
116. J. Cott, A. Carlsson, J. Engel, and M. Lindqvist, *Naunyn-Schmiedeberg's Arch. Pharmacol.* **295**:203 (1976).
117. D. Franz, in: *The Pharmacological Basis of Therapeutics* (L. Goodman and A. Gilman, eds.), pp. 359–366, Macmillan Publishing Co., Inc., New York (1975).
118. F. McCrea and H. Taylor, *J. Pharmacol. Exp. Ther.* **68**:41 (1940).
119. M. Rosenbaum, *Am. Med. Assoc. Arch. Neurol. Psychiatry* **48**:1010 (1942).
120. E. Marshall, Jr., E. Walzl, and D. LeMessurier, *J. Pharmacol. Exp. Ther.* **60**:472 (1937).
121. H. Ramsey and H. Haag, *J. Pharmacol. Exp. Ther.* **88**:313 (1946).
122. H. Gold and J. Travell, *J. Pharmacol. Exp. Ther.* **52**:345 (1934).
123. B. Katz, *Nerve, Muscle, Synapse,* McGraw-Hill Book Company, New York (1966).
124. Y. Israel, H. Kalant, and I. Laufer, *Biochem. Pharmacol.* **14**:1803 (1965).
125. E. Knutsson and A. Katz, *Acta Pharmacol. Toxicol.* **25**:54 (1967).
126. C. Armstrong and L. Binstock, *J. Gen. Physiol.* **48**:265 (1964).
127. J. Moore, *Psychosomat. Med.* **28**:450 (1966).
128. F. Inoue and G. Frank, *Br. J. Pharmacol. Chemother.* **30**:186 (1967).
129. D. Houck, *Am. J. Physiol.* **216**:364 (1969).
130. Y. Israel and I. Salazar, *Arch. Biochem. Biophys.* **122**:310 (1967).
131. F. Messiha, *Pharmacology* **14**:153 (1976).

132. P. Seeman, M. Chau, M. Goldberg, T. Sauks, and L. Sax, *Biochim. Biophys. Acta* **225**:185 (1971).
133. D. Ross, *Ann. N.Y. Acad. Sci.* **273**:280 (1976).
134. D. Ross, M. Median, and H. Cardenas, *Science* **186**:63 (1974).
135. T. Peng, C. Cooper, and P. Munson, *Endocrinology* **91**:586 (1972).
136. C. Erickson, T. Tyler, and R. Harris, *Science* **199**:1219 (1978).
137. A. Alha, *Ann. Med. Exp. Biol. Fenn.* **29**:125 (1951).
138. A. Leikola, *Q. J. Stud. Alcohol.* **23**:369 (1962).
139. J. A. Larsen, *Acta Physiol. Scand.* **73**:186 (1968).
140. R. Fleming and D. Reynolds, *J. Pharmacol. Exp. Ther.* **54**:236 (1935).
141. F. Johnson, D. Brown, and D. Marsland, *J. Cell. Comp. Physiol.* **20**:269 (1942).
142. F. Johnson and E. Flagler, *Science* **112**:91 (1950).
143. R. Alkana, P. Syapin, and E. Noble, *Alcohol. Clin. Exp. Res.* **2**:190 (1978).
144. R. L. Alkana and R. D. Malcolm, in: *Proc. 4th Bienn. Int. Symp. Biomed. Res. Alcohol.*, (H. Begleiter and B. Kissin, eds.), Plenum Press, New York, in press (1979).
145. R. L. Alkana and R. D. Malcolm, *Neurosci. Abst.* **4**:485 (1978).

Drug Therapy of the Alcohol Withdrawal Syndrome

21

Peter K. Gessner

A. INTRODUCTION

The pharmacological treatment of the alcohol withdrawal and delirium tremens syndromes is one of the few instances in which drug therapy has proved eminently effective in alleviating the consequences of excessive ethanol consumption. As the rationale for such treatment has become better understood, mortality from delirium tremens has decreased from the 33% reported at the turn of the century[28] to 13–15% in the 1950s[253,270] and to an even lower incidence more recently (Table II). Although this is not as easy to document, there are indications that there has also been a parallel and concurrent decrease in the morbidity of alcohol withdrawal.[29]

In tracing the development of the successful drug treatment of these syndromes, it is apparent that drugs now known to be effective in their treatment have long been employed in the therapy of the alcohol withdrawal and delirium tremens syndromes. A lack of an understanding of the etiology of the syndromes involved, however, led to these drugs not being employed either consistently or effectively. The major strides made in the 1950s in the understanding of the etiology of these syndromes led to an increased interest in potentially effective drug therapies for these syndromes. As a result, controlled clinical trials of the effectiveness of drugs in the treatment of the alcohol withdrawal and delirium tremens syndromes were vigorously pursued during the 1960s and much was learned. Even though such trials have continued unabated during the 1970s, much less has been learned from them. In some part this has been because the diversity of patient populations, dosage regimens, and outcome criteria

Peter K. Gessner • Department of Pharmacology and Therapeutics, School of Medicine, State University of New York at Buffalo, Buffalo, New York 14214.

Table I. Structures and Names of Drugs Discussed [a,b]

ANTIPSYCHOTICS

Phenothiazines

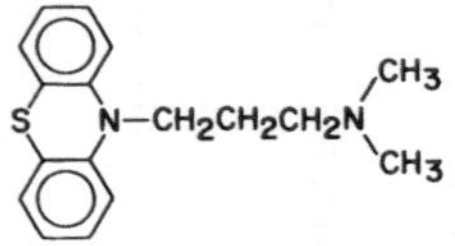

Promazine
INN, NF, BAN, DCF, NFN
N,N-Dimethyl-10H-phenothiazine
-10-propanamine

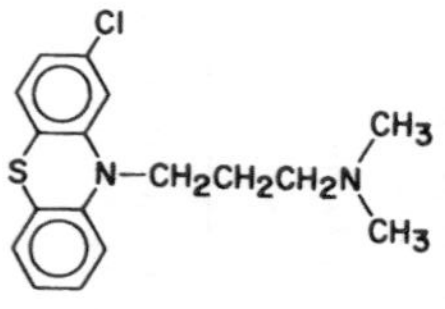

Chlorpromazine
INN, USP, BAN, DCF, NFN
Clopromazina
DCIT
2-Chloro-N,N-dimethyl-10H-phenothiazine
-10-propanamine

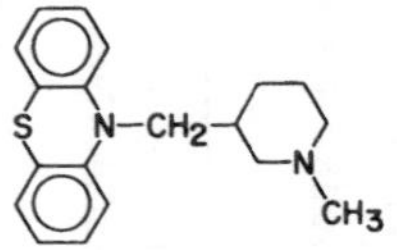

Mepazine
NND64
Pecazine
INN, BAN, NFN
10-[(1-methyl-3-piperidinyl)methyl]
10H-phenothiazine

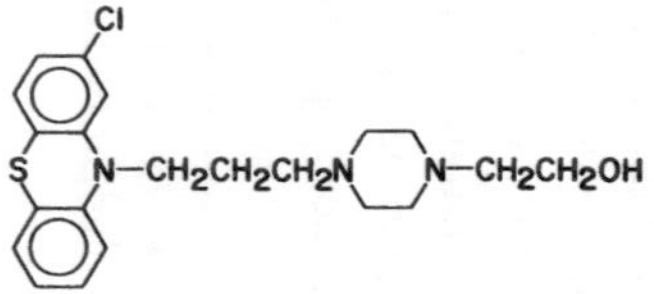

Perphenazine
INN, NF, BAN, NFN
Perfenazina
DCIT
4-[3-(2-Chlorophenothiazin-10-yl)
propyl]-1-piperazinethanol

Buterophenones

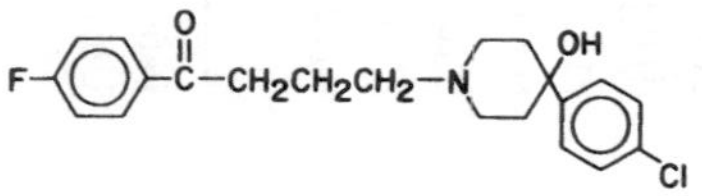

Haloperidol
INN, USAN, BAN, DCF, NFN
4-[4-(4-Chlorophenyl)-4-hydroxy-1
-piperidinyl)]-1-(4-fluorophenyl)-1-
butanone

ALCOHOL–BARBITURATE-TYPE AGENTS

Benzodiazepines

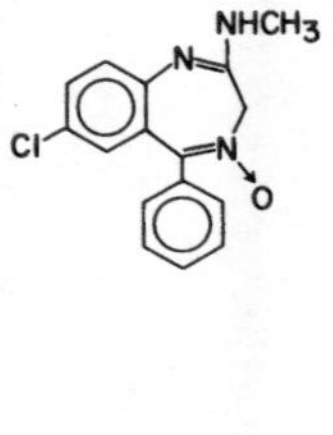

Chlordiazepoxide
INN, BAN, DCF, NFN
Chlordiazepoxide hydrochloride
USAN
Cloridazepossido
DCIT
Methaminodiazepoxide†
Dizepin†
7-Chloro-N-methyl-5-phenyl-3H-
1,4-benzodiazepin-2-amine
4-oxide

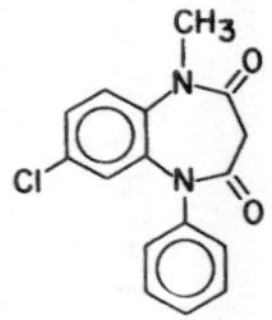

Diazepam
INN, USAN, BAN, DCF, NFN
7-Chloro-1,3-dihydro-1-methyl-5-phenyl-
2H-1,4-benzodiazepin-2-one

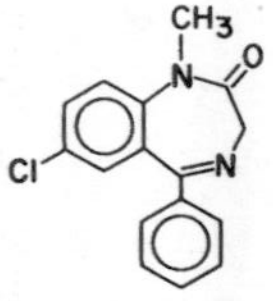

Clobazam
INN, USAN, BAN, DCF, NFN
7-Chloro-1-methyl-5-phenyl-1H-
1,5-benzodiazepine-2,4(3H,5H)-dione

Aliphatic agents

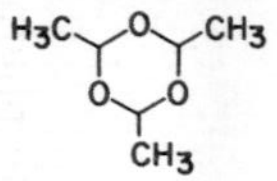

Paraldehyde
USP, BP, PN
Paracetaldehyde

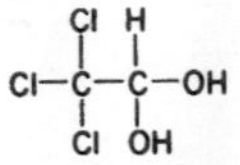

Chloral hydrate
USP, BP, FP, PN
2,2,2-Trichloro-1,1-ethanediol

ALCOHOL–BARBITURATE-TYPE AGENTS (*Continued*)

Barbiturates

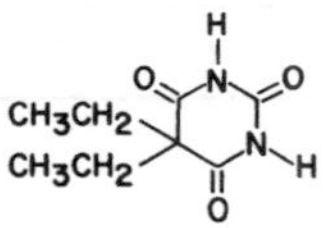

Barbital
INN, DCF, NF60
Barbitone sodium
BP
Diemalum
NFN
5,5-Diethyl-2,4,6(1H,3H,5H)-pyrimidinetrione

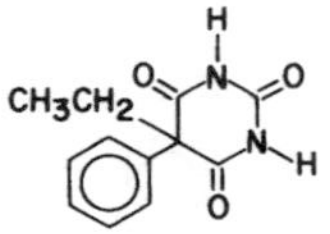

Phenobarbital
INN, USP, DCF
Phenobarbitone sodium
BP
Phenemalum
NFN
5-Ethyl-5-phenyl-2,4,6-(1H,3H,5H)-pyrimidinetrione

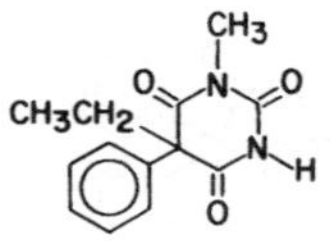

Mephobarbital
NF
Methylphenobarbital
INN, DCF
Enphenemalum
NFN
5-Ethyl-1-methyl-5-phenyl-2,4,6(1H,3H,5H)-pyrimidinetrione

ALCOHOL–BARBITURATE-TYPE AGENTS (*Continued*)

Thiazoles

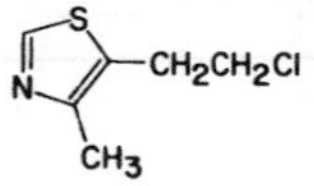

Clomethiazole
INN, DCF, NFN
Chlormethiazole
BAN
SCTZ†
5-(2-Chloroethyl)-4-methylthiazole

OTHER SEDATIVES

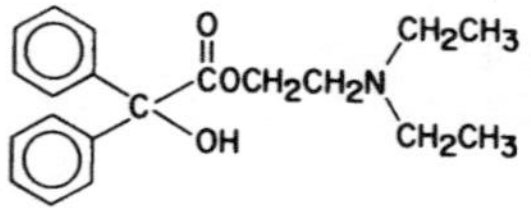

Benactyzine
INN, DE71, BAN, DCF, NFN
α-Hydroxy-α-phenylbenzeneacetic acid-
2-(diethylamino)ethyl ester

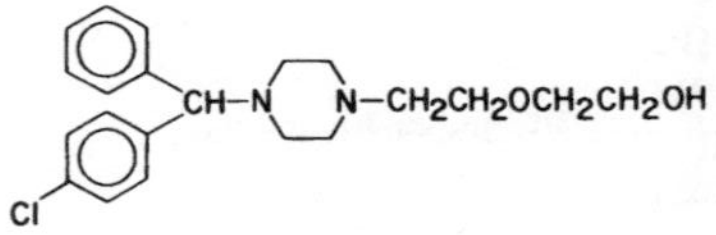

Hydroxyzine
INN, NF, BAN, DCF, NFN
Idrossizina
DCIT
2-[2-[4-[(4-Chlorophenyl)phenyl-
methyl]1-piperazinyl]ethoxy]
ethanol

(*Continued*)

Table I. (Continued)

ANTICONVULSANTS

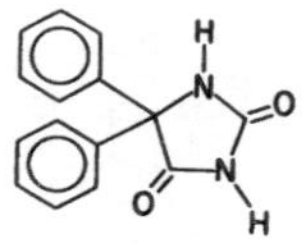

Phenytoin
INN, USAN, BP, DCF, NFN
Fenitoina sodica
DCIT
Diphenylhydantoin†
5,5-Diphenyl-2,4-imidazolidinedione

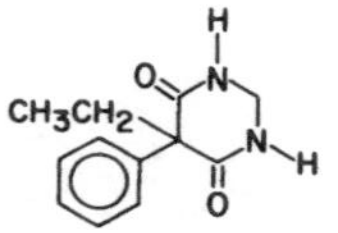

Primidone
INN, USP, BAN, NFN
Primaclone
DCF
5-ethyldihydro-5-phenyl-4,6(1H,5H)-primidinedione

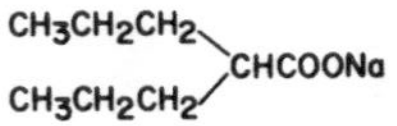

Valproate sodium
USAN
Valproic acid
INN, DCF, BAN, NFN
Dipropylacetate†
2-Propylpentanoic acid

CH₃CH₂CH₂ and CH₃CH₂CH₂ —CHCOONa

BETA-BLOCKING AGENTS

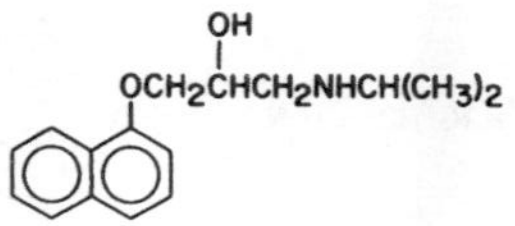

Propranolol
INN, USP, BAN, DCF, NFN
Propranolol hydrochloride USAN
1-(Isopropylamino)-3-(1-naphthyloxy)-2-propanol

OCH₂CHCH₂NHCH(CH₃)₂ with OH

CATIONS

Li₂CO₃ Lithium carbonate
MgSO₄ Magnesium sulfate

---

aIn listing drug name synonyms, preference was given to the English language. Where American usage, the English language version of the International Nonproprietary Name, and the British name differ significantly, they are listed separately and in that order. Other national formulations and previously used generic names (†) are also provided if they differ from those above by more than a change in suffix (i.e., -e or -um). Sources are indicated by abbreviations.[292–295] The last entry for each drug is its chemical name.[296]

bAbbreviations: BAN, British Approved Name; BP, British Pharmacopoeia, 1973; DCF, Denomination commune francaise; DCIT, Denominazione comune italiana; DE, Drug Evaluations, American Medical Association, 1977; DE71, Drug Evaluations, 1971; FP, Pharmacopee francaise; INN, International Nonproprietary Name, World Health Organization; NF, National Formulary, XIV, 1975; NF60, National Formulary, XI, 1960; NFN, name approved by the Nordic Pharmacopoeia Council (Nordiska Farmakopenamnden); NND64, New and Nonofficial Drugs, American Medical Association, 1964; PN, Pharmacopoea Nordica, editio svecica; USAN, United States Approved Name; USP, United States Pharmacopeia, XIX, 1975.

employed has hindered comparative discussion, leaving the field inchoate.[69] To make comparisons easier a major effort was made, in writing this review, to provide a more uniform presentation of the salient details of these studies (Table III). The resulting comparisons are discussed under the individual drugs or drug groups (see Table I).

An additional confounding characteristic of the field is the plethora of names applied to the conditions under discussion. As will be made clear, "delirium tremens," though the oldest term, is not encompassing. "Alcohol withdrawal syndrome," probably the most commonly used term, and that embodied in the title of this review, is encompassing, but has diagnostic overtones and is suggestive of a disease as opposed to a medical condition. As such its use becomes questionable when it is applied to a condition brought about experimentally, particularly in animals. Contraction to "alcohol withdrawal" gives a term devoid of such overtones and frequently used[59,66,94,95,125,128,168,216,219,229,234,252,257,261,290] but syntactically ambiguous. Alternate qualifiers of "alcohol withdrawal" that have been used include "symptoms,"[127,131,169,199,269,286] "manifestations,"[145] "states,"[48,190] and "reaction."[111,112,228] The latter term has perhaps the greatest utility, and "alcohol withdrawal reaction" is used synonymously with "alcohol withdrawal syndrome" in this review. The use and utility of yet other terms is discussed in Section F.1.a.

B. ETIOLOGY

1. Clinical Observations

Sutton, who introduced the term "delirium tremens" in 1813, firmly believed that all cases of the disease were connected with excessive drinking.[248] In the same year Armstrong[9] noted that, in some cases, the syndrome did not develop until 2–3 days after cessation of drinking. By 1886, Wilson was stressing[283] in Pepper's *System of Practical Medicine* that alcoholic delirium was never the result of direct primary actions of ethanol upon the nervous system, ascribing it instead to nervous system changes brought on by prolonged alcoholic saturation. Thus, it long has been evident that delirium tremens was not a form of alcoholic inebriation; the concept that it was a withdrawal phenomenon, however, was not convincingly espoused until the 1940s.[112]

We owe a comprehensive description of the natural history of the alcohol withdrawal and delirium tremens syndromes to Victor and Adams,[270] who based it on their clinical observation of a large number of hospitalized patients with obvious alcoholic complications. These workers forcefully advanced the concept that withdrawal of ethanol from an individual previously consuming it in large quantities led to the clinical symptoms or syndromes of tremulousness, hallucinations, seizures, and delirium tremens and that these, insofar as they occurred in a given patient, tended to do so in a definite time sequence. They defined delirium tremens as a distinct clinical state characterized by psychomotor, speech and autonomic overactivity, disorientation, confusion, and a disordered sense perception. They also reported that, defined in this way, delirium tremens developed in 5% of patients hospitalized with obvious alcoholic complications, and that it followed the development of the other symptoms. In particular, they found that in patients developing both seizures and delirium tremens, the seizures preceded the development of the delirium in 97% of the cases.

2. Experimental Approach

As classical as this 1953 study by Victor and Adams[270] was, it only suggested that the symptomatology in question had as a proximate cause cessation of, or a reduction in, ethanol ingestion. In 1955, Isbell and his co-workers[142] tested this hypothesis experimentally with six individuals with a history of morphine addiction, three of whom also had a history of excessive drinking. All had been abstinent from both agents in the preceding 3 months. All received ethanol in divided doses around the clock for a mean of 66 ± 7 days. The doses were incremented as increases in their tolerance permitted. During the phase of maximum intake (average duration 52 ± 8 days) they consumed an average of 442 g ethanol/day. Upon withdrawal of the ethanol, but at no time previously, all of these subjects experienced the minor signs associated with the alcohol withdrawal syndrome: tremor, weakness, perspiration, elevation of blood pressure, nausea, and anorexia. All six also developed the more serious signs and symptoms associated with this syndrome. Thus, two had grand mal convulsions, four had visual and auditory hallucinations, three became disoriented, and all had fever. Only two, however, experienced delirium tremens (Table II).

A parallel but shorter study (24 days) is that of Mendelson and LaDou.[181] It involved 10 alcoholics abstinent for the preceding 23 ± 9 days. During the phase of maximum intake (5 days) they consumed an average of 343 g ethanol/day. Withdrawal of ethanol at the end of this period resulted in tremors in 8 of the 10 subjects, hallucinations in 5 and disorientation in 3; no convulsions were observed. None of the three episodes of disorientation could be equated with delirium tremens.

These experimental studies went a long way toward causally linking the manifestations of tremor, nausea, insomnia, hallucinations, convulsions, and disorientation to the cessation of ethanol ingestion. Because so few of the individuals in the abovementioned studies[142,181] experienced delirium tremens, however, the proximal cause of this latter syndrome must be considered as less well established. Consequently, although most hold that delirium tremens is in some way related to the cessation of ethanol ingestion, there is no unanimity about this.[74,166,188,193,205,220,243] It is of interest that it has not proved possible to link the occurrence of convulsions with the subsequent development of delirium tremens.[204]

The experiments of Isbell and co-workers[142] and those of Mendelson and LaDou[181] do not lend themselves to precise quantitation of the withdrawal

Table II. Withdrawal Reaction Severity as a Function of Preceding Exposure to a Depressant

Study	Depressant	Period of maximum ingestion		n	Incidence of withdrawal syndrome signs and symptoms						Reference
		Daily amount (g)	Duration (days)		Minor		Major				
					Tremor (%)	Other (%)	Hallucinations (%)	Seizures (%)	Fever (%)	Delirium tremens (%)	
Isbell et al., 1955	Ethanol	442	52	6	100	100[c]	67	33	100	33	142
Mendelson and LaDue, 1964		343	23	10	80	80[d]	50	—	10	—	181
Isbell et al., 1955		377	10	4	100	100[e]	—	—	a	—	142
Sellers et al., 1977		280	5	6	a	a	a	a	a	a	229
Sellers et al., 1976		280	3	6	100	a	a	a	a	a	227
Isbell et al., 1950	Barbiturate[b]	1.5[f]	90	18	100	100[g]	67[h]	78	i	39[j]	85,87,141
Fraser et al., 1957		0.8	90	8	a	100[k]	26	13	a	—	83,87
Fraser et al., 1957		0.6	90	18	a	50[k]	—	14	a	—	83,87
Fraser et al., 1957		0.4	90	18	a	6[k]	—	—	a	—	83,87

[a] Information not available.
[b] Seco and pentobarbital.
[c] All had weakness, hyperreflexia, insomnia, nausea, vomiting, diarrhea, elevated blood pressure, anorexia, and perspiration.
[d] Eighty percent had hyperreflexia and laterial nystagmus.
[e] All had weakness, 3/4 had anorexia and perspiration, 1/2 had nausea, and 1/4 had elevated blood pressure.
[f] SE ± 0.1, range 0.9–2.2 g.
[g] All had weakness, anxiety, insomnia, and anorexia.
[h] Sum of those identified as having intermittent hallucinations and/or disorientation and those with hallucinations and loss of insight constantly present.
[i] "Usually" for the 67% identified as having hallucinations.
[j] Disorientation and loss of insight constantly present.
[k] Not further differentiated.

reaction's severity as a function of the amount and duration of ethanol ingestion; they are, however, suggestive of such a relationship (Table II). Thus, individuals in the Mendelson and LaDou study[181] consumed less ethanol and for a shorter period of time than individuals in the Isbell *et al.* study and, as a group, experienced a less severe withdrawal reaction. Moreover, the study of Isbell *et al.* involved an additional 4 individuals who stopped drinking after an average of only 18 days of drinking, and who consumed an average of 377 g ethanol/day during a maximum intake period of 10 ± 5 days. Although, upon withdrawal, all four experienced tremor, weakness, and other minor symptoms, none experienced fever, hallucinations, convulsions, or disorientation. More recently Sellers *et al.*[227,229] have reported that individuals maintained on 280 g ethanol/70 kg per day for 3–5 days all had tremor during the withdrawal period but did not have seizures or hallucinations.

Work with experimental animals has lent itself better to this type of quantitation. In particular Goldstein,[110] working with mice, obtained a linear relationship between seizure scores, on the one hand, and the product of blood ethanol and duration of exposure, on the other. Majchrowicz,[171] working with rats, has also noted a relationship between duration of treatment with ethanol and severity of the withdrawal reaction.

3. Withdrawal from Other Depressants

Human studies of barbiturate withdrawal provide quantitative evidence that severity of the withdrawal reaction is a function of the dosage employed in the prewithdrawal period.[83,85,87,141] In these studies individuals were maintained on various dosages, ranging from 0.2 to 2.2 g/day of secobarbital or pentobarbital, for at least 32 days. The severity of the symptomatology observed upon withdrawal appeared to be correlated to the amount of barbiturate consumed daily (Table II). Thus, although 39% of those receiving 0.9 g/day or more (average 1.5 ± 0.1 g) developed delirium, none of those receiving daily doses of 0.8, 0.6, or 0.4 g/day did (26% of those receiving 0.8 g/day, however, did develop hallucinations). Similarly, convulsions were experienced by 78% of those receiving doses of 0.9 g/day or more, but only by a minority of those receiving 0.8 and 0.6 g/day (13 and 14%, respectively) and by none of those receiving 0.4 or 0.2 g/day.[84,85]

When the descriptions of the withdrawal reactions from barbiturates and ethanol are considered side by side, their great similarity is apparent; Isbell *et al.*[141] called it striking. This equivalence is further underlined by the mutual ability of barbiturates and ethanol to terminate each others' withdrawal syndrome in humans.[4,86,104] Admittedly, ethanol does not fully control the barbiturate withdrawal syndrome; neither, however, does ethanol always fully control the alcohol withdrawal syndrome.[108] The reasons for this are not clear but are likely to be related to the dosage regimens employed.

Similar withdrawal symptomatology has been noted with such older hypnotics as chloral hydrate and paraldehyde,[150] as well as with more recently introduced agents, such as meprobamate, glutethimide, methyprylon, ethinamate, ethchlorvynol, chlordiazepoxide, and diazepam.[61,63,68,208,214,242] Consideration of these similarities and the other pharmacological effects of the drugs in question led Isbell[139] to define this group of agents (class IB) as of the "alcohol-barbiturate type." Under this heading he included the following central nervous system depressants: (1) ethyl alcohol: beer, wine, whiskey, etc.; (2) barbiturates: all those sold in the United States for sedation and hypnosis; (3) paraldehyde; (4) chloral hydrate; (5) meprobamate (Miltown, Equanil); (6) piperidinediones: glutethemide (Doriden), methprylon (Nodular); (7) benzodiazepines: chlordiazepoxide (Librium), diazepam (Valium); and (8) tertiary carbonols: methparafynol (Dormison), ethchlorvynol (Placidyl), ethinamate (Valmid).

Although not used as sedatives or hypnotics, another group of agents that possesses the sedative–hypnotic–anesthetic spectrum of action of the alcohol-barbiturate types of drugs are the gases and volatile liquids used clinically for the induction and maintenance of general anesthesia. The inebriant properties of diethyl ether, for instance, were known before it was discovered that this agent could be used to render surgery painless.[76] Moreover, individuals who used diethyl ether for its inebriant properties were known to be difficult to anesthetize with it.[176] To further explore the similarity between diethyl ether and other agents of the alcohol–barbiturate type, Gessner[96] exposed mice for a period of 3 days to subanesthetic levels of diethyl ether in the respired air. He observed, upon withdrawal, a syndrome analogous to that seen following exposure to ethanol. This syndrome, moreover, was controlled by ethanol and by phenobarbital administration. These facts strongly suggest that diethyl ether, and possibly, by extension, other anesthetic agents, can be appropriately listed as an alcohol–barbiturate type of drug. The etiology of the alcohol withdrawal reaction and delirium tremens can then be seen as a specific subset of a pharmacologically more widespread phenomenon: withdrawal reactions from depressant drugs of the alcohol–barbiturate type.

C. TREATMENT STRATEGIES

1. Substitution and Replacement Therapy

Most of the drugs shown to be clinically useful in the treatment of the alcohol withdrawal and delirium tremens syndromes are of the alcohol–barbiturate type. The use of drugs of this type for the treatment of these syndromes goes back to the 1880s when first chloral hydrate[283] and then, within a year of its introduction, paraldehyde[126] were thus employed. The latter agent remained a mainstay of alcohol withdrawal treatment for many years,[21,206] although the majority of alcohol withdrawal treatments described in the literature of that period did not involve the use of alcohol–barbiturate types of drugs.[31] Moreover, it was not until 1942 that it was suggested that the reason treatment of the alcohol withdrawal syndrome with paraldehyde and other alcohol–barbiturate types of agents was successful was that it was based on the substitution of one central nervous system depressant for another of the same type.[150]

It was left to Isbell[140] to explicitly and forcefully advance the substitution rationale for the treatment of the alcohol withdrawal syndrome. Taking the position that both the early phase of the alcohol withdrawal syndrome and delirium tremens are reactions to abstinence from ethanol, he stated that the principle used in treating the condition is (1) to substitute one of the hypnotic drugs of the alcohol–barbiturate type, thus reintoxicating the individual and, then, (2) to gradually reduce the dose of the substitute. In those instances where serious symptoms of ethanol abstinence (i.e., convulsions, hallucination, delirium) have already occurred, the reintoxication, Isbell maintained, should be rapid, using parenteral administration if necessary. As pointed out by Gessner,[94] this substitution method of treatment is analogous to that used in the treatment of the barbiturate withdrawal syndrome[138] a procedure elaborated into a careful titration of the patient with pentobarbital[19,68,222a,282] or phenobarbital.[242]

The titration method of treatment is not fully applicable to delirium tremens because, unlike the early symptomatology of the alcohol withdrawal reaction, delirium tremens, once started, cannot be readily controlled by any treatment.[130a] This is also a characteristic of the analogous withdrawal psychosis seen in severe cases of the barbiturate withdrawal reaction.[138] Thomas and Freedman[257] saw in these facts support for the belief that withdrawal from ethanol, although a major and necessary factor for the development of delirium tremens, is not the only one. This led them to suggest that most therapies of delirium tremens act by suppressing the condition rather than as antagonists of whatever process brings the condition about. They further suggested[257] that, since delirium tremens does not respond readily to treatment, titration of patients in this condition with potential therapeutic agents raises questions of whether adjustment of dose to "clinical judgment" might introduce extraneous variables that could bias the results of controlled trials. Indeed, Kramp et al.[157b] have found that titration of delirium tremens leads to higher drug blood levels in patients who fail to respond satisfactorily than in those who do. Thomas and Freedman[256,257] decided, in a controlled trial of paraldehyde and promazine, to use fixed-dose schedules, with the therapy being terminated suddenly once the patient was free of symptoms. Clearly, even if Thomas and Freedman[257] were correct in that treatment of delirium tremens is suppressive rather than antagonistic, their experimental paradigm is not problem-free. Thus, should a treatment not prove effective, it would be hard to know whether to consider the ineffectiveness to be an inherent property of the drug or of the dosage regimen employed.[222a] Although it is possible to partially obviate this difficulty by comparing more than one dosage regimen for each drug, as was done by Sereny and Kalant,[232] logistically this becomes necessarily very difficult.

2. Blockage of Noradrenergic Hyperactivity

The alcohol withdrawal reaction has been observed to be associated with hyperactivity of the noradrenergic system. Clinically, this is evidenced by such biochemical findings as elevated blood[43a] and urinary[10,100,196] norepinephrine levels, increased serum dopamine-β-hydroxylase,[10] and increased cerebrospinal fluid[10] and urinary[196] 3-methoxy-4-hydroxyphenylglycol levels. Animal studies have provided similar evidence.[6,137,207,292] Inhibiting catecholamine synthesis with α-methyl-p-tyrosine[22,112,119] or otherwise interfering with catecholamine pathways[112] has been found to aggravate the alcohol withdrawal syndrome in experimental animals. This led Goldstein and others[113,137,207] to suggest that the increased norepinephrine turnover activity is the result rather than the cause of the withdrawal syndrome. Athen *et al.*[10] have taken this a step further, speculating that the increased norepinephrinergic activity may have a protective function. If the alcohol withdrawal reaction represents a hyperactivity of all those functions which were previously inhibited by ethanol, then blocking the hyperactivity of only one of these, the noradrenergic one, may further disequilibrate a system already under stress.

The clinical effects on the course of the alcohol withdrawal syndrome of agents able to interfere more or less specifically with noradrenergic function—namely, phenothiazines, haloperidol, and propranolol[227]—are discussed under the individual specific treatment headings; apomorphine, a dopamine agonist, has been reported to have no significant effect on postintoxication symptoms.[274a] The evidence to date leaves open the question of whether such agents are useful in the treatment of medically significant alcohol withdrawal reactions.

It is of interest that norepinephrine levels remain normal in patients being withdrawn from ethanol if these patients are treated with clomethiazole[292] and that administration of ethanol to experimental animals at the start of withdrawal delays the rise in brain catecholamine concentrations.[119] This suggests that if a depressant substitution–replacement therapy is properly instituted, noradrenergic hypersensitivity may never become significant.

3. Other Treatments

Factors other than drug administration are also likely to influence the course and severity of the alcohol withdrawal syndrome. As pointed out by various writers,[54,101,159,165,170,239,267] nondrug clinical support conditions, that is, good vigilant nursing and medical care, absence of restraint, and correction of dehydration, electrolyte depletion, and, if present, hypoglycemia, are important in reducing the morbidity of the withdrawal syndrome.

Additionally, the hypothesis has been advanced[239,280] that development of the hallucinations associated with the alcohol withdrawal reaction and of the delirium tremens which frequently follows are, to some degree, a function of the sensory deprivation to which the individual undergoing withdrawal in an institutional setting may be subjected. These authors also suggested that administration of sedatives can further increase the isolation of the patients from their surroundings.

In keeping with this hypothesis it has been suggested that the hallucinations can be alleviated and the delirium prevented by providing patients with "reality orientation," that is, doing all that is possible to restore the patient's link with reality.[280] The concept parallels the suggested role of sensory deprivation in the genesis of postoperative and intensive-care delirium.[266,284] The reports of success attributed to the use of the "reality orientation" approach in the treatment of the alcohol withdrawal syndrome come, however, from uncontrolled studies performed at a location that made self-selection of treatment facilities possible.[280] More extensive discussion of this area, as well as that of the use of antibiotics in cases of alcohol withdrawal with coexistent infection and that of vasopressor drugs for circulatory collapse, falls outside the scope of the present review.

D. PATIENT POPULATIONS

When comparing the results of controlled clinical studies of the efficacy of various agents in controlling the alcohol withdrawal and delirium tremens syndromes, it is important to bear in mind that the severity of the presenting condition can vary markedly from study to study. This is likely to be the main reason why the course of illness in patients reported as admitted with acute alcohol withdrawal and administered placebo therapy is quite variable. The incidence of seizures in such patients, for instance, varies from 2 out of 10[127] to 0 out of 103.[146] A confounding factor is the possible variation in the nondrug clinical support conditions characterizing various studies, a variable that is very difficult to quantify.

Some effort at quantifying the severity of the presenting condition is made by authors who note the percentage of patients with delirium tremens at admission. Even then, however, the problem of how this syndrome is defined arises. The most explicit and classical definition of delirium tremens is that of Victor and Adams,[270] here quoted as paraphrased by Victor[268]:

> Delirium tremens . . . in its fully developed form . . . is characterized by gross confusion, constant agitation and sleeplessness, crude generalized tremors, garbled speech, vivid hallucinations and delusions, and, most typically, by signs of overactivity of the autonomic nervous system, such as fever, tachycardia, excessive sweating and dilated pupils. The confusional state in patients with delirium tremens is not simply one of slight temporal disorientation or impaired ability to recall events which occured at the height of the period of alcohol intoxication or in the first day or two of abstinence. Confusion of this mild degree is a common occurrence in the early stage of the withdrawal period, in patients with tremulousness and hallucinosis. The mental derangement which characterized delirium tremens is of an entirely different order; these patients are profoundly confused, being unable to identify people or to interpret properly the meaning of what they see or what they hear, or for that matter, any sensory impression.

A vivid and graphic account of an individual in fully developed delirium, given by Zola in *L'Assommoir*,[291] brings this definition to life.

A shorter definition is that of the Diagnostic and Statistical Manual (DSM-II)[58] of the American Psychiatric Association as

> a variety of acute brain syndrome characterized by delirium, coarse tremor and frightening visual hallucinations, usually becoming more intense in the dark. It is

distinguished from 'other alcoholic hallucinosis' by the tremors and the disordered sensorium.

It is obvious, when considering the studies in Table III, that not all of them adhere to such strict criteria, and the variability in the definition of delirium tremens by different workers is quite considerable.[174]

Gross *et al.*,[123] pointing out that many alcoholics do not fit readily into the classifications of Victor and Adams,[270] proposed more complex classifications, some based on factor analysis of as many as 30 variables.[120-125] However, these and other similar classifications[73,131,220] have yet to be shown to have diagnostic utility in controlled clinical trials of treatments of the alcohol withdrawal syndrome.

E. SPECIFIC TREATMENTS

Although over the years many treatments of the alcohol withdrawal and delirium tremens syndromes had been tried, it was only in the late 1950s that the technique of double-blind studies began to be applied to the comparative evaluation of such treatments. These studies ushered in an era of rigorous evaluation of the effectiveness and safety of drugs in such treatment, which is the principal subject of this review. At the time, two groups of drugs were vying for primacy in the field. These were (1) the sedative/hypnotics in general and paraldehyde in particular, that is, the treatments most consistently employed in the first half of this century; and (2) the then recently introduced phenothiazines, which had proved dramatically effective in controlling a variety of psychoses. As will be shown, double-blind clinical trials of these two groups of agents made it apparent that the ability of a drug, such as a phenothiazine, to control the early and minor signs and symptoms of the alcohol withdrawal syndrome does not necessarily presage that it will have the ability to forestall the triad of more serious consequences associated with this syndrome: convulsions, delirium tremens, and death. Because this concept is an important one, the antipsychotic agents (phenothiazines and haloperidol) will be discussed first. This will be followed by a discussion of alcohol–barbiturate types of agents in the order of the popularity they currently enjoy (benzodiazepines, paraldehyde, barbiturates, ethanol, and chlormethiazole), of sedatives of other than the alcohol–barbiturate type (benactyzine and hydroxyzine), then of the anticonvulsants, special attention being given to phenytoin and dipropylacetate, and finally of other agents regarding whose effectiveness it is harder to draw any conclusions at this time.

1. Antipsychotic Agents

a. Phenothiazines

With the introduction of the major tranquilizers into medicine reports began to appear in the literature, suggesting phenothiazines to be very effective in controlling the alcohol withdrawal syndrome.[7,71,74,115,186,219,225] Even though

Table III. Comparison of Treatments of the Alcohol Withdrawal and Delirium Tremens Syndromes

Study	Patients	Treatments compared			Findings[a]			Other medication[b]
		n	Drugs	Dosage regimens[b]	Outcome measures		Outcomes[c]	
1. Laties et al., 1953[162]	Agitated, with acute brain syndrome associated with alcohol intoxication (DT's)	16	Chlorpromazine	50 mg im stat/50 mg po qid; mean daily dose 225 mg	On awakening from first dose percent: (1) less anxious, (2) less tremulous	36	21	
		16	Promazine	50 mg im stat/50 mg po qid; mean daily dose 270 mg		85*	23	
2. Friedhoff and Zitrin, 1959[88]	Male patients with DT's	15	Chlorpromazine	Average daily dose: 252 mg	No. of days for: (1) return to normal temperature, (2) recovery	8	8	Vitamin B complex, vitamin C and 100 mg thiamine parenterally per day; iv fluids prn
		16	Paraldehyde	Average daily dose: 30 ml		5	6*	
3. Ewing, 1960[67]	Drinking last few days, still shaky and nervous; DT's not sought; 95% drunk on admission	29	Benactyzine	2 mg q4–6h prn	(1) No. developing DT's; (2) No. failures; (3) days to reach "comfort"; (4) reaching "comfort" in 3 days	1 19*	2 2.9	Parenteral vit; iv fluids prn
		26	Paraldehyde	Im stat/po max: 30 ml/24 hr		2 17*	2 2.6	
		29	Pyridoxine	100 mg po prn max: 500 mg/24 hr		4	5 3.7	
4. Hart, 1961[128]	Drinking more than 5 of previous 10 days, 22% with DT's	56	Paraldehyde	6 ml po q3h prn.	(1) No. developing seizures; (2) percent of DT's developing seizures; (3) days hospitalized; relative significant change in: (4) anxiety duration, (5) insomnia control, (6) anorexia duration	2 → 0 →	5.2 →	Barbiturates for seizures if on promazine; vitamin B complex im for 3 days; po multivitamins
		57	Promazine	100 mg im stat/100 mg po q3h prn.		5 ↑ 38	4.1* ↑	
		62	Promazine + Paraldehyde	100 mg PR im stat/100 mg PR po qid + 6 ml PA po q4h prn		5 29		

No.	Author	N	Treatment	Clinical condition	Criteria	Dose/route	(1)	(2)	(3)	Comments
5.	Rosenfeld and Bizzoco, 1961[216]	30	Chlordiazepoxide	Significant blood alcohol level in 73%, ≥0.1% in 30%	No. developing (1) DT's, (2) hallucinosis; (3) no. with global improvement at 5 days	100 mg iv stat, q1h sos sos∫po in diminishing doses	2	2	22*	Glutethimide for insomnia 0.5 g; phenytoin for seizures; vitamins
		30	Placebo				2	0	8	
6.	Beroz et al., 1962[17]	22	MgSO₄	With acute brain syndrome due to alcohol, but not hallucinating on admission	No. hallucinating on: (1) day 2, (2) day 3	1g im q4h for 4–6 doses∫1 g im qid for 24 hr	5*	3*		Promazine: 100 mg im q4h for 4–6 doses/100 mg po qid for 24 hr/50 mg po qid; vit. B complex po tid
		16	Placebo		See text for comments		10	9		
7.	Koutsky and Sletten, 1963[157]	7	Chlordiazepoxide	Chronic alcoholics undergoing alcohol withdrawal; 3 with DT's	(1) Duration acute symptoms (days)	100 mg im q8h for 24 hr ∫ placebo q8h for 24 hr	4.6			Paraldehyde for sedation prn
		5	Placebo			Placebo q8h for 24 hr ∫ 100 mg C im q8h for 24 hr	7.8			
8.	Thomas and Freedman, 1964[256]	33	Paraldehyde	In alcohol withdrawal	(1) No. deaths; (2) no. developing DT's; (3) duration of symptoms (days)	10 ml po q4–6h	0	0	2.9	Fluids po 3 liter per day; vit po and im; aspirin and im; aspirin and alc sponging if fever > 37.7°C
		34	Promazine			200 mg po q4–6h	1	4	2.6	
		22	Paraldehyde	In delirium tremens	(1) No. deaths; (2) duration of symptoms (days)	10 ml po q4–6h	1	3.1		
		17	Promazine			200 mg po q4–6h	6*	4.0*		
9.	Chambers and Schultz, 1965[48]	34	Chlordiazepoxide	In active alcohol withdrawal; 53% with active DT's, 35% with impending DT's, 68% hallucinating	No. developing (1) DT's, (2) uncontrolled seizures; (3) no. with good to excellent improvement in 4 days	Day 1: 75–150 mg po tid; day 2:50–100 mg po tid; day 3: 25–75 mg po tid; day 4: 25–75 mg po	0	0	24	Chloral hydrate 'nightly; phenytoin for seizures 250 mg im stat/ (100 mg phenytoin + 32 mg phenobarbital) qid prn

(Continued)

Table III. *(Continued)*

Study	Patients	n	Treatments compared — Drugs	Dosage regimens[b]	Findings[a] — Outcome measures	Outcomes[c]	Other medication[b]
9. *(Continued)*		35	Diazepam	Day 1: 30–60 mg po tid; day 2:20–40 mg po tid; day 3: 10–30 mg po tid; day 4: 10–30 mg po		0 0 17	
		34	Promazine	Day 1: 75–150 mg po tid; day 2: 50–100 mg po tid; day 3: 25–75 mg po tid; day 4: 25–75 mg po		2 0 24	
10. Glatt *et al.*, 1965[105]	In alcohol intoxication; major-ity with ≥5 years alcohol-ism, 58% with tremor, 26% with nausea/vomiting	49	Chlormethiazole	2 g po stat; days 1–6: 5g, 3.5 g, 2.5 g, 1.5 g, 1 g, and 0.5 g po, respectively	No. developing (1) DT's, (2) depression; (3) percent incidence seda-tion; percent assessed "improved" on (4) day 1, (5) day 2, (6) day 3	0 2* 58* 50* 67* 71*	Phenytoin
		48	Placebo			1 10 35 22 27 41	
11. Sereny and Kalant, 1965[232]	Male alcoholics heavily drink-ing (5 days to several months) up to admission (71%) or up to 10 hr previ-ously (29%)	∑58 {	Chlordiazepoxide	50 mg qid	No. developing (1) DT's, (2) seizures; percent time slept on: (3) day 1, (4) day 2, (5) day 3; (6) change in systemic b.p. after ½ day; (7) percent with increased tremor after ½ day	0 0 23 16 13 6 90	
			Chlordiazepoxide	100 mg qid		0 0 36 27 28* –13 89	
			Promazine	100 mg qid		1 2 38 20 13 –32* 72	

	Clinical state	n	Drug	Dose				Outcome measures	Comments
			Promazine	200 mg qid	0, 32*, 44*	2, 16	60, −32*		
			Placebo	qid	*100*	0, 6	*11*, *2*		
12. Motto, 1966[189]	In alcohol withdrawal syndrome; simple intoxication not sufficient for admission	38	Chlorpromazine	50–100 mg po or im stat, q2h prn	1	−	−	(1) No. developing seizures; relative significant remission of: (2) hallucinations at 48 hr; (3) agitation at 24 hr; (4) agitation at 48 hr; (5) GI upset at 24 hr	Chloral hydrate 6–10 ml q6h prn; vitamin B complex and vitamin C; phenytoin 100 mg tid if history of seizures
		35	Mepazine	50 mg stat, q2h prn	0	↓	→		
					↓	↑			
		32	Perphenazine	8 mg po or 5 mg im stat/(8 mg po q4h or 5 mg im q3h) prn	0	↓	→		
					↑	−	↑		
		35	Promazine	100 mg po or im stat, q2h prn	0	↑	→		
					−	−			
		32	Placebo		0	↓	0		
					↑	→			
13. Golbert et al., 1967[108]	In tremulous and agitated states and acute hallucinosis of alcohol withdrawal	12	Chlordiazepoxide	100 mg im stat/100 mg (q1h im + q3h po) prn/100 mg po q4h: max 3200 mg/24 hr	0	6	8	(1) No. deaths; (2) no. developing DT's; (3) no. with complications	
		12	Ethanol	200–600 ml/24 hr: ½ po as whiskey and ½ iv in 5% glucose	0	5	7		

(Continued)

Table III. (Continued)

Study	Patients	n	Drugs	Dosage regimens[b]	Outcome measures	Outcomes[c]	Other medication[b]
13. (Continued)		12	Paraldehyde + chloral hydrate	10 ml P po or im + 0.5–1.0 g CH po statʃ(10 ml P q2h + 1.0 g CH q4h) prnʃ10 ml P q4h + 0.5–1.0 g CH q6h		0 1* 1*	
		13	Promazine	100 mg im statʃ 100 mg (q1h + q2h poʃ prn 100 mg po q4h: max. 3600 mg/24 hr		2 7 8	
	In delirium tremens	13	Paraldehyde + chloral hydrate	10 ml P po or im + 0.5–1.0 g CH po statʃ(10 ml P q2h + 1.0 g CH q4h) prnʃ10 ml P q4h + 0.5–1.0 g CH q6h	(1) No. deaths; (2) no. with serious complications	0 1	
		12	Promazine	100 mg im statʃ100 mg (q1h im + q2h po) prnʃ100 mg po q4h: max. 3600 mg/24 hr		2 11*	
14. Harfst et al., 1967[127]	Males in acute alcohol withdrawal; 91% drinking up to time of admission	10	Amobarbital	Day 1–2: 600 mg po qid; day 3–4: 400 mg po qid; day 5–6: 200 mg po qid	(1) No. developing seizures; (2) self-assessment of night sleep; (3) hr day sleep; (4) change in b.p.; (5) change pulse rate	0 5.3* 1.2* −13* −11.7*	

No.	n	Drug	Condition	Dosage	Measure	Results				
	10	Chlormethiazole		Day 1–2: 1500 mg po qid; day 3–4: 1000 mg po qid; day 5–6: 500 mg po qid		0	−7.7*	5.5* 0.9	2.2*	Orphenedrine for parkinsonian side effects
	10	Placebo				2* 10	−12.4*	3.3	0.2	
15. Greenberg and Rosenfeld, 1969[118]	44	Haloperidol	Alcoholics in the "acute phase" of their illness	5 mg im stat, q2h sos∫2 mg po bid or tid	No. responding at 24 hr: (1) well, (2) poorly; percent remission at 24 hr of (3) anxiety, (4) tremor, (5) insomnia, (6) anorexia. (7) nausea	66* 28* 100*	11 93*	82* 99*		
	45	Placebo				29 2.5 58	49* 73	31 79		
16. Kaim et al., 1969[146]	103	Chlordiazepoxide	Recently drinking and showing at least 4 of following: GI distress, sweating or flushing, insomnia, tremulousness, irritability, apprehension, depression, clouded sensorium or confusion	Day 1: 50 mg im q4h; day 2–9: po. flexible and decreasing	(1) No. deaths: no. developing: (2) DT's, (3) seizures	0	1*	1*		
	103	Chlorpromazine		Day 1: 100 mg po q4h; day 2–9: po. flexible and decreasing		0	7	12		
	98	Hydroxyzine		Day 1: 100 mg po q4h; day 2–9: po. flexible and decreasing		1	4	8		

(Continued)

Table III. (*Continued*)

Study	Patients	n	Drugs	Dosage regimens[b]	Outcome measures	Outcomes[c]			Other medication[b]
					Treatments compared → **Findings**[a]				
16. (*Continued*)		103	Thiamine	Day 1: 100 mg im q4h; day 2–9: po, flexible and decreasing		0	4	7	
		130	Placebo			0	8	9	
17. Madden et al., 1969[168]	In alcohol withdrawal	50	Chlormethiazole	Days 1–7: 1.60 g, 3.52 g, 2.56 g, 1.60 g, 1.28 g, 0.64 g, and 0.32 g, respectively	No. developing (1) DT's; (2) seizures; (3) no. with daytime sleep or drowsiness; relative significant change in (4) anxiety, (5) headache, (6) GI distress	0 ←	2 ←	28* ←	Methaqualone at night 300 mg for 70–74% of each group; phenytoin 100 mg tid for those receiving trifluoperazine
		50	Trifluoperazine	Days 1–7: 4.40 mg, 9.68 mg, 7.04 mg, 4.4 mg, 3.52 mg, 1.76 mg, and 0.88 mg, respectively		1 →	2 →	9 →	
18. Muller, 1969[190]	With alcohol withdrawal tremulousness	20	Chlordiazepoxide	125–500 mg/24 hr.	No. developing (1) DT's, (2) seizures, (3) fever	1	0	0	Phenytoin 100 mg tid if history of alc. withdrawal seizures; secobarbital 100 mg at night prn
		20	Chlorpromazine	200–500 mg/24 hr		0	1	0	
		20	Paraldehyde + chloral hydrate	16–28 ml P + 2–5 g CH/24 hr		0	0	1	
	With delirium tremens	9	Chloridazepoxide	200–400 mg/24 hr	No. developing (1) fever, (2) prolonged confusion, (3) hypotension; (4) hospital stay (days)	3 3	1	0	
		9	Chlorpromazine	300–800 mg/24 hr		0 4	0	1	
		9	Paraldehyde + chloral hydrate	20–25 ml P + 2–6 g CH/24 hr		0 4	0	0	

Ref	Characteristics	n	Drug	Dose	Measures				Notes
19. Schwarz and Fjeld, 1969[226]	Drinking for ≥21 days	16	Hydroxyzine HCl	50 mg im stat, day 1: 100 mg po qid + 50 mg im bid prn; day 2: 100 mg po qid; day 3–8: 50 mg po qid	No. developing (1) DT's, (2) seizures, (3) severe hallucinosis; relative significant change in: (3) time asleep, (5) anorexia	0 →	2 ↑	4	Vits. im and po; phenytoin 250 mg iv/100 mg po qid for seizures
		14	Placebo			1 →	0	1 ↑	
20. Friend et al., 1971[89]	With delirium tremens	13	Fructose	100 g infused iv as a 10% sol. over 6 hrs	No. (1) deaths, (2) continuing delirium, (3) seizures, (4) unimproved hallucinations, (5) unimproved tremor	3	5	2	1 patient: 100 mg chlordiazepoxide 6 hr prior; 1 patient: 10 mg diazepam
						0	10		
		11	Dextrose	100 g infused iv as a 10% sol. over 6 hr		0	1	1	
						7	6		
21. Ritter and Davidson, 1971[215]	In alcohol withdrawal with tremor, anxiety, insomnia, nausea, hallucinations, depression	42	Haloperidol	5 mg im stat/q6h prn	(1) No. developing hallucinations; (2) percent improved at 48 hr; no. with worse: (3) insomnia, (4) tremor, (5) anxiety, (6) all symptoms	12	95	5	
						2	1	9*	
		38	Perphenazine	5 mg im stat/q6h prn		9	80	5	
						2	3	26	
22. Brown et al., 1972[35]	Recently drinking alcoholics with average blood alc. 112 mg%, disorientation, hallucinations, tremor	7	Chlordiazepoxide	3.1 mg/min iv till calm; max. 150 mg, q8h sos sos	(1) No. recovered at 72 hr; (2) no. improved at 72 hr; (3) min to onset of maximum effect	1	5	10–15	
		7	Diazepam	1 mg/min iv till calm: max. 60 mg, q8h sos sos		2	5	1–2	
23. Kaim and Klett, 1972[147]	With history of alcoholism and manifestations of disorientation, tremor and hallucinations	46	Chlordiazepoxide	100 mg im stat/50–100 mg im q3h prn/flexible: max 300 mg/24 hr	No. developing: (1) seizures; (2) hyperthermia, (3) severity ratings (low rating = mild)	1	0	65	

(Continued)

Table III. (Continued)

| | Treatments compared | | | | Findings[a] | | | |
Study	Patients	n	Drugs	Dosage regimens[b]	Outcome measures	Outcomes[c]		Other medication[b]
23. (Continued)		55	Paraldehyde	8 ml po or 5 ml im staf/flexible: day 1: max 60 ml po or 30 ml im; day 2 on: max 40 ml po or 20 ml im		1	1 59	
		41	Pentobarbital	150 mg im staf/50–150 mg im q3h prn/flexible: max 450 mg im or 1200 mg po/24 hr		0	1 71	
		46	Perphenazine	10 mg im staf/5–10 mg im q3h prn/flexible: max 30 mg im or 48 mg po/24 hr		1	0 70	
24. Palestine, 1973[(198)]	Alcoholics with organic brain syndrome secondary to alcohol ingestion	17	Haloperidol	2 mg im stat, q1h sos sos sos	(1) No. therapeutic success at 4 hr on basis of global rating; (2) BPRS scores at 4 hr	14*	21.3*	
		14	Mesoridazine	25 mg im stat, q1h sos sos sos		3	35.0	
	Same as above	25	Haloperidol	2 mg im stat, q1h sos sos sos	Same as above	18*	22.7*	
		25	Hydroxyzine HCl	100 mg im stat. q1h sos sos sos		10	28.8	

Study	Population	N	Treatment	Dose	Outcome			Other treatment
25. Rothstein, 1973[217]	Consuming ≥480 ml liquor/day for ≥5 days; those on phenytoin or with seizures in prior 2 weeks excluded	100	Phenytoin	200 mg po bid	No. developing: (1) DT's, (2) seizures	5	0	Chlordiazepoxide day 1: 360 mg po or im/prn; thiamine 100 mg daily
		100	Controls			4	0	
26. Sampliner and Iber, 1974[221]	Alcoholics with heavy alc. intake in prior 4 weeks and a history of seizures	70	Phenytoin	100 mg po tid	(1) No. developing seizures	2		Chlordiazepoxide day 1: 400 mg, day 2: 200 mg
		66	Placebo			11*		
27. Martin, 1975[173]	"Dangerous" alcoholics undergoing detoxification	10	Clobazam	15 mg/day	At 12 days no. of markedly improved-physician rating of (1) anxiety, (2) psychosomatic difficulties; patient rating of (3) anxiety, (4) psychosomatic difficulties	8	8	Nonanxiolytic hypnotics; antidepressants
						8	8	
		10	Clobazam	30 mg/day		10*	10*	
						10*	10*	
		10	Diazepam	15 mg/day		4	5	
						5	4	
		10	Placebo			0	0	
						0	0	
28. McGrath, 1975[175]	Admitted for treatment of alcoholism; 3/4 with GI disturbances and apprehension, 2/5 with disorientation	41	Chlordiazepoxide	50 mg po stat ∫50 mg po 1 hr later∫ 1: 100 mg q6h; day 2 on: decreasing	No. developing (1) DT's, (2) hallucinations on days 2–8	4*	3	Vit. B complex + vit. C; methaqualone for insomnia
		46	Chlormethiazole	384 mg po stat ∫384 mg po 1 hr later∫ day 1: 768 mg q6h; day 2 on: decreasing		0	0	

(Continued)

Table III. (Continued)

Study	Patients	n	Drugs	Dosage regimens[b]	Outcome measures	Outcomes[c]	Other medication[b]
29. Thompson et al., 1975[258]	In DT's; all had fever ≥ 37.9°C, tachycardia >90/min hallucinations, agitation, disorientation	17	Diazepam	10 mg iv stat/5 mg iv q5m prn∫5–10 mg im q1–4h	(1) No. deaths; (2) no. developing apnea; (3) time (hr) to calm; (4) duration (hr) of DT's	0 0 1.1* 56.6	Fluids iv prn
		17	Paraldehyde	10 ml pr in 20 ml oil q30m prn∫5–10 ml pr q2–4h		2 3 3.0 52.9	
30. Palestine and Alatorre, 1976[199]	Chronic alcoholics in alcohol withdrawal	25	Chlordiazepoxide	50 mg im stat, q1h sos sos sos	Withdrawal symptoms severity score reduction at: (1) 1 hr, (2) 2 hr; (3) no. controlled at 4 hr	2 8 11	
		24	Haloperidol	5 mg im stat, q1h sos sos sos 300 mg q8h		8* 14* 17	
31. Sellers et al., 1976[227]	With alc. intake of ≥160 g/day for 7 days + 4 g/kg/day for 3 days	6	Lithium carbonate		No. withdrawal symptoms on (1) day 1, (2) day 2, (3) day 3: (4) rel. sig. Δ in tracking errors	23⋄ ↑ 18* 17*	
		6	Placebo			28 ↓ 37 31	
32. Dilts et al., 1977[59]	In alcohol withdrawal	144	Clorazepate	15 mg q4h prn/45 mg sf 1st night∫22.5 mg sf 2nd night	No. developing (1) DT's, (2) seizures	0 0	Thiamine 100 mg on day 1–3/day 4-on po. Rx for insomnia; for DT antipsychotics
		121	Hydroxyzine	Day 1: 100 mg qid∫ day 2: 50 mg qid∫ day 3: 25 mg qid		6* 3	
33. Pena-Ramos, 1977[203]	In mild to moderate withdrawal and no history of seizures	≥70	Chlordiazepoxide	25–200 mg/24 hr for 4 weeks	See text for comments		
			Thioridazine	25–200 mg/24 hr for 4 weeks			

				Tremor size × 10^{-4} G^2 on (1) day 2, (2) day 2; percent decrease in: (3) systolic pressure, (4) heart rate; (5) hours sleep per day		
34. Sellers et al., 1977[229]	With alcohol intake of ≥ 160 g for ≥7 days + 4 g/kg per day for 5 days	6	Propanolol	10 mg q6h	8* 16	15 7.9	5
		6	Propanolol	40 mg q6h	4* 20	6* 7.6	10
		6	Propanolol + chlordiazepoxide	(10 mg P + 25 mg C) q6h	5* 15	25 8.7*	10
		6	Chlordiazepoxide	25 mg qh6	8* 10 31 16	15 8.6* 16 73	3 0.3
		6	Placebo				
35. Shulsinger et al., 1977[235]	In alcohol detox; 55% in moderate withdrawal; 9% in severe withdrawal	34	Diazepam	10–20 mg po or im q4h prn	0	(1) No. developing seizures; no significant relative change with respect to: insomnia, hallucinations, tremor, b.p., h.r., temperature	
		21	MgSO₄ + diazepam	2 g MgSO₄ q6h for 48 hr + 10–20 mg D po or im q4h prn	0		
		20	MgSO₄	2 g im q6h for 48 hr	1		
		19	MgSO₄	3 g im q6h for 48 hr	0		

(Continued)

Table III. (*Continued*)

Study	Patients	n	Drugs	Dosage regimens[b]	Outcome measures	Outcomes[c]	Other medication[b]
36. Dencker *et al.*, 1978[(57a)]	Chronic alcoholics with 10 yrs alc abuse history admitted evening before following ≥ 7 days drinking	28	Chlormethiazole	0.6 g po qid except 0.9 g at night, reduced after 3 days	No. developing: (1) hallucinosis, (2) seizures; sleep disturbance scores on (3) day 2, (4) day 3, (5) day 4	1 0 1.4 / 1.0 0.8	Up to 1 g chloral hydrate in first night following admission
		32	Piracetam	1600 mg po tid		2 1 2.1* / 1.9* 1.7*	
37. Funderburk *et al.*, 1978[(89a)]	Consuming a mean of 280 ml absolute alcohol per day for ≥ 3 days with mean 0.1% blood alcohol level at admission	18	Chlordiazepoxide	50 to 200 mg po per day for 2 to 4 days/reduced dosage on last day	In 6-day posttreatment period mean nightly: (1) No. REM episodes, (2) min δ sleep	11.1 0.4	
			Ethanol	60 ml of 47.5% soln po q2h from 10.00 to 20.00 hr daily for 2 to 5 days/30 ml of 47.5% solution po q2h from 10.00 to 20.00 on last day		19.0* 19.8*	
38. Kramp and Rafaelsen, 1978[(157a)]	In alcohol withdrawal with tremor and hallucinosis. Alcohol free at admission.	11	Barbital	500 mg po q30 m prn for sleep to max of 5 g per 24 hr	(1) No. showing satisfactory effect, (2) hr to sleep onset	8 5	Poly-B-vitamin preparation, fluids prm for dehydration, antibiotics prm for infections

	8	Diazepam	20 mg im q30 m prm for sleep to max of 200 mg per 24 hr	7	14*
Frank DTs: same as above with disorientation	17	Barbital	500 mg po q30 m prm for sleep to max of 5 g per 24 hr	15*	8
	13	Diazepam	20 mg im q30 m prm for sleep to max of 200 mg per 24 hr	6	11

[a]Convention employed in analysis of findings: outcome measures are numbered sequentially; the corresponding numerical values are listed serially, separated by commas, under "Outcomes." To the extent that it was reported the information regarding the number dying, developing delirium tremens (DT's), seizures, and hallucinations is presented first.

[b]Convention and nomenclature employed in dosage regimen description: im, intramuscular; iv, intravenous; po, orally; pr, rectally; q_nh, every n hours; q_nm, every n minutes; bid, twice a day; tid, three times a day; qid, four times a day; prm, repeated as required (any number of times); sos, repeated if required (once); +, together with; $\int$, followed by.

[c]Convention used to indicate significant differences: *value given is significantly ($p < 0.05$) different from italic values obtained with the other treatments. Where it was reported that there was a significant difference between results obtained with different treatments, but no values were given, $\uparrow$ and $\downarrow$ are used to indicate this.

properly controlled experiments demonstrating the superiority of these agents over paraldehyde, barbiturates, and other sedative drugs were lacking, by 1958 many clinics considered phenothiazines as drugs of choice in the management of agitated alcoholics suffering from delirium tremens and related states.[162] Controlled comparisons among the various phenothiazines suggested promazine to be more effective than chlorpromazine[162] (Table III: study 1), although the reverse holds for their antipsychotic activity. Such comparisons also suggest both these agents, as well as perphenazine, to be more effective than mepazine[189] (Table III: study 12).

As early as 1956 it became apparent, however, that the phenothiazines have the capacity to reduce the threshold to seizures.[72,75,224] Moreover, starting in 1959, and in contrast to the early clinical reports mentioned above, more controlled studies have consistently suggested treatment with phenothiazines to be a less successful therapy of the alcohol withdrawal syndrome and delirium tremens than treatment with one of the alcohol–barbiturate types of agents. The most compelling evidence in this respect is the excess mortality observed in a 1964 study by Thomas and Freedman[256] and a 1967 study of Golbert et al.[108] among patients treated with phenothiazines relative to that observed in those treated with alcohol–barbiturate types of drugs. Specifically, Thomas and Freedman[256] reported that among patients admitted with delirium tremens 6 of 17 treated with promazine died as compared to only one among the 22 treated with paraldehyde ($p < 0.05$). Additionally, of the patients admitted with a mild alcohol withdrawal syndrome, one of the 34 treated with promazine died, although there were no mortalities among the 33 treated with paraldehyde.

Similarly, Golbert et al.[108] reported that among patients treated with promazine, 2 of the 12 admitted with delirium tremens and 2 of 13 admitted in mild withdrawal died, although there was no mortality among 12 patients admitted with delirium tremens and treated with a paraldehyde–chloral hydrate combination or among three groups of 12 admitted in mild withdrawal and treated, respectively, with the paraldehyde–chloral hydrate combination, chlordiazepoxide, or ethanol.

Additional evidence for the conclusion that phenothiazines are inferior to alcohol–barbiturate types of drugs in the treatment of alcohol withdrawal and delirium tremens syndromes comes from eight studies where it is possible to compare the outcome of treatment with these two types of agents. In all these studies phenothiazine treatment was consistently associated with a higher combined incidence of seizures and development of delirium tremens. Specifically,

1. It results from a 1965 report by Chambers and Schultz[48] that promazine treatment was associated with significantly higher incidence of uncontrolled seizures than either chlordiazepoxide ($p = 0.01$) or diazepam ($p = 0.01$) treatment.

2. It is apparent from a 1965 study by Sereny and Kalant[232] that the combined incidence of seizures and delirium tremens development was significantly higher ($p < 0.05$) in patients treated with promazine than in those treated with chlordiazepoxide.

3. Reference to the 1967 study by Golbert *et al.*[108] shows a significantly higher ($p < 0.05$) incidence of delirium tremens development among patients receiving promazine than those receiving a combination of paraldehyde and chloral hydrate.

4. It is evident from the 1964 data of Thomas and Freedman[256] that promazine treatment was associated with an almost significantly higher ($p = 0.06$) incidence of development of delirium tremens than treatment with paraldehyde.

5. Kaim *et al.*[146] reported in 1969 that chlorpromazine treatment was associated with a significantly higher incidence of seizures ($p < 0.05$) and delirium tremens development ($p < 0.01$) than treatment with chlordiazepoxide.

The same trend was observed in the remaining three studies[128,147,190] (Table III: studies 4, 18, and 23), although it was not a statistically significant one.

Comparison of phenothiazine treatment with placebo also reveals a similar nonsignificant trend of phenothiazine therapy being associated with a higher incidence of seizures and delirium tremens development[146,189,232] (Table III: studies 11, 12, and 16).

Animal studies also suggest phenothiazines to be ineffective in controlling the alcohol withdrawal reactions. Thus, Goldstein[111] found neither promazine nor chlorpromazine to reduce withdrawal scores of mice rendered dependent on ethanol, but, on the contrary, to significantly enhance them. Similar results have been reported by Essig and Fraser,[65] who observed chlorpromazine administration to dogs in the analogous barbiturate withdrawal reaction to double, on the average, the number of convulsions experienced by these animals.

Given the evident inferiority of phenothiazines in the treatment of the alcohol withdrawal and delirium tremens syndromes, it is pertinent to ask why these agents achieved such wide popularity at one time in the treatment of these conditions. The results of objective studies suggest that one of the reasons was that patients experiencing a mild alcohol withdrawal syndrome improved more rapidly on promazine than paraldehyde. Hart[128] reported in 1961, for instance, that promazine-treated patients have significantly shorter durations of anxiety, insomnia, anorexia, and hospitalization ($p < 0.01$ in each instance) than those receiving paraldehyde. Likewise, in the 1964 study of Thomas and Freedman,[256] the duration of symptoms among the 30 patients in mild withdrawal for whom promazine therapy was effective (i.e., those who did not develop delirium tremens) was significantly ($p < 0.001$) shorter (2.1 days) than among those receiving paraldehyde (2.9 days). However, the reverse was observed to be true for patients in delirium tremens. Already in 1959, Friedhoff and Zitrin,[88] for instance, reported such patients to recover significantly faster ($p < 0.05$) if treated with paraldehyde rather than with chlorpromazine. Likewise, Thomas and Freedman[256] observed significantly ($p < 0.05$) shorter durations of signs and symptoms among delirium tremens patients receiving paraldehyde rather than promazine.

From the above data it becomes evident that for patients with a serious dependency, and thereby likely to develop a severe withdrawal syndrome, phenothiazine therapy is dangerous and clearly contraindicated. For patients experiencing a mild withdrawal syndrome, however, phenothiazines could have some utility. Nevertheless, since controlled studies to date do not provide evidence that such patients can be identified *a priori* with any degree of confidence, phenothiazine therapy of the alcohol withdrawal syndrome is contraindicated.

As an addendum reference is made to a 1977 study by Pena-Ramos,[203] who compared, using three psychiatric rating scales, long-term therapy (4 weeks) with the phenothiazine thioridazine to that with chlordiazepoxide in patients originally presenting with a mild to moderate degree of alcohol withdrawal symptoms. The primary intent of this study appears to have been to determine the efficacy of thioridazine in the therapy of the depression to which the alcoholic is prone. As such, it would not have warranted mention in this review. The title of the report speaks, however, of "controlling symptoms attributable to alcohol withdrawal" and in its conclusions claims to have shown thioridazine "to be as efficacious and as safe as chlordiazepoxide in controlling symptoms attributable to alcohol withdrawal."[203] On the basis of the experimental design of the study and the results presented, these claims appear unwarranted. As discussed elsewhere, thioridazine's pharmacological properties and the pathophysiological manifestations of the alcohol withdrawal syndrome evidence incompatibilities.[252,261]

b. Haloperidol

Haloperidol is a butyrophenone drug which has proven to be an effective alternative to phenothiazines as an antipsychotic. Pharmacologically, its properties are similar, in kind if not degree, to those of the phenothiazines.[39] Since 1969, the results of five controlled clinical trials of the effectiveness of this agent in the treatment of the alcohol withdrawal syndrome have been published (Table III: studies 15, 21, 24, and 30). Given haloperidol's pharmacological similarity to phenothiazines, it is valid on the basis of experience accumulated with the phenothiazines to raise two questions of particular interest and concern: (1) What is the prognosis of patients who are on haloperidol therapy and do not promptly improve? (2) Are patients receiving haloperidol more prone to seizures than those receiving other treatments? Unfortunately, the published studies provide no information in this regard. Paradoxically, the experimental design of most or all of them would appear to be such as to preclude any possibility of these questions being answered.

This is evident if one considers the two 1973 studies of Palestine[198] [also published by Maerz[169]], comparing haloperidol to mesoridazine (a phenothiazine) and to hydroxyzine, respectively, and the 1976 study of Palestine and Alatorre[199] comparing haloperidol to chlordiazepoxide. In each of these studies haloperidol was reported to be the significantly more effective treatment. These outcomes have to be considered, however, in the context of both the experimental design of the studies and the criteria of success applied to them. The criteria of success were improvement in perambulation, in the eating of solid food, in the drinking of liquids, and in socialization. While these are desirable outcomes, they cannot be equated to a drug's ability to prevent and control the more severe manifestations of the alcohol withdrawal and delirium tremens syndromes. Moreover, the success of the treatment was evaluated after only 4 hr, that is, at the end of a period during which the drugs were administered hourly by intramuscular injection until the withdrawal syndrome was controlled or five injections had been given. Patients successfully controlled in this period of time were followed for another 24[199] to 48[198] hr, but the basic conclusions of the studies were based on the evaluation at 4 hr. The brevity of this experimental period is noteworthy, since, as previously discussed, the course of the alcohol withdrawal syndrome spans a much longer period of time. This raises questions regarding the presenting condition of the patients and the objectives of the treatment. The patients were stated as having been diagnosed to have "an organic brain syndrome secondary to alcohol ingestion." and exhibiting "symptoms of alcohol withdrawal with impending or frank delirium."[198] Nonetheless, the symptoms of delirium and hyperpyrexia were scored as absent in all patients, both initially and finally. Regarding objectives, Palestine[198] notes that the rapid control of agitation, combativeness, and hallucinatory behavior enabled the staff to better provide further medical and psychiatric treatment (unspecified), and Palestine and Alatorre[199] note further that 4 hr would be the maximum period afforded most primary care and emergency room physicians to make decisions to either admit the patient to the hospital or release him to home care. Given the severity and expected duration of a syndrome such as delirium tremens, the strategy implied by these statements has to be seriously questioned, particularly in view of the unavailability of information relevant to the two questions posed about haloperidol therapy earlier. In this respect it is also important to consider the work of Blum et al.[23] These investigators rendered mice dependent on ethanol by the Goldstein inhalation technique. Using a range of haloperidol doses from 0.5 to 10 mg/kg, they found that at no dose did haloperidol decrease seizure scores. On the contrary, the 2- and 10-mg/kg doses of haloperidol significantly enhanced convulsive scores. Under the same experimental conditions 20 mg chlordiazepoxide/kg markedly reduced seizure scores for many hours.

The remaining published studies report haloperidol therapy as significantly superior to placebo[118] and perphenazine.[215] They are based on 24- and 48-hr observation periods, respectively, but do not provide additional insights regarding the issues raised above. Finally, reference should be made to a study by Gross et al., the experimental details of which remain unpublished, but which the authors state to have shown chlordiazepoxide and paraldehyde to be more effective than haloperidol, although not in the first 48 hr.[124]

2. Agents of the Alcohol–Barbiturate Type

a. Benzodiazepines

In the last 25 years a number of benzodiazepines have been introduced into medicine and have become very extensively used; they have become the most frequent drug choice of American physicians for the treatment of the alcohol withdrawal and delirium tremens syndromes.[70] They are agents of the alcohol–barbiturate type as defined by Isbell[139] notwithstanding the presence in the brain of benzodiazepine-specific receptors,[186a,245a,246a] because benzodiazepines, if taken in sufficient dosage and withdrawn, lead to the appearance of a syndrome with convulsions and confusion.[61,65,208,242] This syndrome develops more slowly than that associated with alcohol withdrawal, owing probably to the longer *in vivo* half-life of the commonly used benzodiazepines and their active metabolites. In considering studies of the therapeutic effectiveness of benzodiazepines in the treatment of the alcohol withdrawal syndrome the studies will be grouped on the basis of the nature of comparisons made: whether the comparison made was with (1) phenothiazines, (2) placebos, or (3) other agents of the alcohol–barbiturate type.

The first of these comparisons, which shows treatment with the benzodiazepines to result in a significantly lower combined incidence of serious complications (seizures and delirium tremens development) than treatment with phenothiazines,[48,146,232] is mostly of historical interest. This is so because, as already discussed, phenothiazines are known to lower seizure threshold and their use appears to be associated with a greater incidence of complications than are placebos, even if only marginally so.

Double-blind comparative evaluations of benzodiazepine, or more specifically chlordiazepoxide and placebo, can be found in four studies, all of which suggest chlordiazepoxide to be significantly more effective than placebo in controlling the alcohol withdrawal syndrome. The most important of these is a 1969 study of Kaim *et al.*,[146] which showed treatment with chlordiazepoxide to be associated with a significantly lower incidence of seizures ($p < 0.05$) and development of delirium tremens ($p < 0.05$) than was placebo administration. The other three studies are (1) a 1965 study by Sereny and Kalant,[232] who found that on each of the first days of treatment, patients receiving chlordiazepoxide spent a significantly ($p < 0.02$) higher percentage of time sleeping than did the placebo controls; (2) a 1961 study by Rosenfeld and Bizzoco,[216] who found that after 5 days of therapy a significantly ($p < 0.001$) greater proportion of patients receiving chlordiazepoxide were rated by the hospital staff as improved than were those receiving placebos; and (3) a 1963 study of Koutsky and Sletten,[157] who, using a 24-hr crossover design (i.e., one in which patients received chlordiazepoxide in either the first 24 hr or the second 24 hr of treatment only), found those receiving chlordiazepoxide in the first 24 hr to have an overall shorter duration of syndrome, this approaching significance ($p < 0.07$). It should be added that in both of the latter studies patients were administered additional alcohol–barbiturate types of drugs (Table III: studies 5 and 7), a factor which could have biased the outcomes of these studies.

The third comparison, that of benzodiazepines with other drugs of the alcohol-barbiturate type involves five studies (Table III: studies 13, 18, 23, 29, and 38).[108,147,157a,157b,190,258] These not only fail to provide evidence of chlordiazepoxide's superiority over therapy with ethanol[108] or a barbiturate (pentobarbital) but provide evidence that such therapy is significantly less effective than a combination of paraldehyde and chloral hydrate. Evidence for the latter comes primarily from the 1967 study of Golbert *et al.*,[108] who found chlordiazepoxide therapy to be associated with a significantly ($p < 0.05$) higher incidence of delirium tremens development than that observed in patients receiving the combination of paraldehyde and chloral hydrate. This outcome cannot be ascribed to inadequate dosing with chlordiazepoxide, since the dosage paradigm used by Golbert et al.,[108] which was similar to that of Isbell,[139] was such that patients received increasing doses of chlordiazepoxide until they either became calm or received a maximum of 3.2 g/day. Additional evidence that chlordiazepoxide is less effective than the combination of paraldehyde and chloral hydrate comes from a 1969 study by Muller,[190] whose results indicate that patients with delirium tremens had a significantly ($p = 0.04$) higher combined incidence of complications (fever or prolonged confusion) if treated with chlordiazepoxide rather than with the combination of paraldehyde and chloral hydrate. Kramp and Rafaelsen[157a] reported similar findings in a comparison of diazepam and barbital, therapy with barbital being assessed as significantly more satisfactory on the basis of onset of action, course of clinical complications, and patient ability to cooperate. The results of two studies comparing benzodiazepines with paraldehyde are much less clear cut. In a generally inconclusive 1972 study, Kaim and Klett[147] compared chlordiazepoxide to paraldehyde and found no significant differences, although the withdrawal syndromes of patients treated with paraldehyde was slightly less severe than the one experienced by patients receiving chlordiazepoxide. More problematical is a study (Table III: study 29) comparing diazepam with paraldehyde in patients with delirium tremens.[258] No significant differences were observed in this study in the duration of the delirium tremens in patients receiving the two treatments, nor were any other of the concomitants of alcohol withdrawal syndrome (seizures, fever, etc.) reported on. Nevertheless, 2 of the 17 patients receiving paraldehyde, administered rectally in oil, died within hours of the start of therapy, while others suffered incidents of apnea. These events are most extensively discussed in the section on paraldehyde. They are likely to be less a function of the delirium tremens process or of the characteristics of paraldehyde than to be related to the route and rate of administration chosen for paraldehyde.

Animal studies appear to confirm the ability of benzodiazepines to control the alcohol withdrawal reaction. Goldstein[111] reported in 1972 that diazepam in doses of 50, 20, and 5 mg/kg and chlordiazepoxide in doses of 100, 50, and 20 mg/kg suppressed withdrawal scores of mice rendered dependent on ethanol. Gessner and Hu[99] extended the chlordiazepoxide dosage series under similar experimental conditions to 10, 5, 2, and 1 mg/kg and found all but the lowest dose to significantly reduce withdrawal seizure scores of ethanol-dependent mice.

Benzodiazepines have been compared to each other in only three controlled studies (Table III: studies 9, 22, and 27). In one of these the comparison was between clobazam and diazepam.[173] The vague description of the criteria for admission, the long duration of treatment (12 days), the limitation of effectiveness assessment to relief of anxiety, and the coadministration of other medication preclude any conclusions being reached on the basis of this study. The other two studies dealt with chlordiazepoxide and diazepam, administered orally in the one[48] and intravenously in the other.[35] No differences in the outcome of the therapy were found in either study. Onset of the effects was stated by Brown et *al.*,[35] however, to be much faster following intravenous diazepam than intravenous chlordiazepoxide. Although they did not quantitate this precisely, Brown *et al.*[35] offered estimates of 1–2 min and 10–15 min for the respective onset times. If a paradigm similar to that of Isbell[139] is used in the treatment of the alcohol withdrawal syndrome, the short onset time of diazepam would appear, on the face of it, to offer distinct advantages, since it should allow more accurate titration of the dosage required to reintoxicate the patient. Unfortunately, such use of diazepam is complicated by (1) the accumulation of an active diazepam metabolite,[90] and (2) the unusual pharmacokinetics seen following diazepam administration, diazepam blood levels surging several hours after administration,[11] a phenomenon which some believe to be related to food intake.[155]

The long half-life of chlordiazepoxide, diazepam, and their active metabolites poses a potential problem even when they are administered orally. As pointed out by Sellers and Kalant,[228] repeated daily dosage results in the accumulation of the parent compound or metabolites (or both) in the body. Desired therapeutic and unwanted toxic effects may therefore not appear until after several days of continuous therapy. It is possible that the slow dissipation of this accumulation may be responsible for the reduction in REM sleep and suppression of δ sleep that is a residual effect of chlordiazepoxide but not ethanol treatment of the alcohol withdrawal syndrome.[89a]

Finally, it must be added that there is increasing concern and evidence regarding the abuse potential of benzodiazepines and particularly diazepam. [33,52,61,62,80,103,114,157a,201,208,213,213a,230,241,250,281]

b. Paraldehyde

Paraldehyde has been for many years the mainstay in the treatment of alcohol withdrawal syndrome and delirium tremens. With only one exception, in which patients were administered exceptionally high doses, controlled studies have consistently indicated that therapy of these two conditions with paraldehyde, either alone or in combination with chloral hydrate, is very effective. Nevertheless, this agent is now used less often than the benzodiazepines,[70] having received more than its fair share of unfavorable publicity. On analysis, much of this publicity is directly attributable to misconceptions and a general unawareness of paraldehyde's properties. These aspects of paraldehyde's pharmacology will be discussed subsequently to a more detailed account of the results of the controlled studies.

The effectiveness of paraldehyde or combinations of paraldehyde and chloral hydrate in the treatment of the alcohol withdrawal and delirium tremens syndromes has been compared to that of other agents in eight clinical studies. In seven of these the paraldehyde was administered orally with intramuscular administration being resorted to occasionally for patients not in condition to accept oral medication. In all seven of these studies, paraldehyde, or its combination with chloral hydrate, was found to be either more or as effective as the other agents employed. The most dramatic of these results was that, in the therapy of patients in delirium tremens, paraldehyde,[256] alone or in combination with chloral hydrate,[108] was associated with a significantly ($p < 0.001$) lower mortality than the use of promazine. Of more central concern, however, given that promazine is no longer considered a satisfactory agent for the control of this condition, are comparisons of the effectiveness of paraldehyde, with or without chloral hydrate, to that of other drugs of the alcohol–barbiturate type in general and benzodiazepines in particular. Unfortunately, few such studies have been done, possibly because of the known reluctance of drug manufacturers to have their products tested against well-established agents.[53] The most important of these studies is the 1967 one by Golbert et al.,[108] in which paraldehyde was used in combination with chloral hydrate. The paraldehyde combination proved significantly ($p < 0.05$) more effective in preventing development of delirium tremens than either chlordiazepoxide or promazine, and it did so even though chlordiazepoxide proved no more effective than promazine. In this same study 5 of 12 patients treated with ethanol developed delirium tremens as opposed to 1 of 12 treated with the paraldehyde combination; while this difference is not significant ($p = 0.08$), the difference between the total number developing complications while treated with ethanol (7/12) and those treated with the paraldehyde combination (1/12) is significant ($p < 0.05$). In another study involving the use of a combination of paraldehyde and chloral hydrate (Table III: study 18) patients in delirium tremens who were treated with the combination experienced significantly ($p < 0.05$) fewer complications than did ones receiving chlordiazepoxide.[190] Surprisingly, and in spite of the apparent effectiveness of the combination of paraldehyde and chloral hydrate, no reports of studies or even use of this combination have been published in the 1970s. Orally administered paraldehyde has been compared to chlordiazepoxide in only one study (Table III: study 23), and although treatment with paraldehyde was associated with withdrawal episodes which were on the average less severe than those observed in patients receiving chlordiazepoxide or pentobarbital, differences observed were not statistically significant.[147]

In the studies reviewed above, paraldehyde, or combinations of paraldehyde and chloral hydrate, have been consistently found to be better or at least as good as any other drug used in the treatment of the alcohol withdrawal syndrome and delirium tremens. In contrast, the 1975 study by Thompson et al.[258] suggested diazepam to be significantly better for the treatment of severe delirium tremens than paraldehyde, for although the duration of delirium tremens in the patients receiving the two treatments was comparable, diazepam was associated with a significantly ($p < 0.05$) shorter induction time and a significantly ($p < 0.001$) smaller incidence of untoward reactions. There are several major difficulties with this study, however, some of which have been discussed elsewhere.[117,218,259,260] These difficulties stem primarily from the route and rate of paraldehyde administration used in the study. In contrast to the studies previously reviewed in which paraldehyde was administered orally, or in uncooperative patients intramuscularly, Thompson et al.[258] administered the paraldehyde rectally in two volumes of cottonseed oil. The amounts of paraldehyde used for induction of a calm state (88 ml on average with a maximum of 175 ml) and the rate at which it was administered (10 ml every 30 min) by far exceeded those used in the other studies in which the maximum dose in the first 24 hr of treatment varied from 30 ml[67] to 65 ml,[88] at rates varying from 6 ml every 3 hr[128] to 10 ml every 4 hr.[108,256]

As the 1940 work of Gardner *et al.*[91] and more recently that of Anthony *et al.*[8] make clear, paraldehyde absorption is slower if it is administered rectally rather than orally, particularly so if administered in oil. In the presence of a slow absorption rate, a higher rate of rectal administration could be expected to result in the formation of a reservoir of paraldehyde in the lower gastrointestinal tract, the size of the reservoir increasing with each additional dose. That the rectum can act as such a reservoir of unabsorbed paraldehyde is clear from two case reports.[14,102] The sequestering of paraldehyde in the rectum would make the agent appear as less potent than expected, necessitating additional administration of paraldehyde. When eventually fully absorbed, on the other hand, the additional paraldehyde could make the agent appear more potent than expected, in some cases leading to marked central nervous system depression. It is instructive in this context to consider the nature of the untoward reactions Thompson *et al.*[258] observed in patients administered paraldehyde. They reported that

> Two patients . . . died . . . 17 and 24 hours after onset of therapy. Their deaths were not expected or explained, and postmortem examinations were refused. One other patient with delirium tremens alone, treated with paraldehyde, bit his nurse severely. . . . Two . . . patients had sudden apnea develop 20 and 80 minutes after the last dose of paraldehyde during induction, and the first patient had apnea again 90 minutes after a maintenance dose of paraldehyde. Both patients were promptly resuscitated and survived. Two other patients . . . sustained bilateral brachial plexus injuries and sheet burns from episodic violent agitation during induction. One patient seriously wounded his intern by biting, and one patient broke from restraints and was caught as he jumped from a third-floor window.

The occurrence of apnea in two of the patients very strongly suggests an overdose effect and leads to the conclusion, subscribed to implicitly by Thompson *et al.*[258] and explicitly by Thompson and Maddrey,[260] that their finding that paraldehyde was less predictable than diazepam can be ascribed to the mode of administration employed. These conclusions, together with the evidence that duration of delirium tremens was similar for the two treatments, preclude use of this study in any comparative assessment of the inherent value of the two drugs.

Animal studies confirm the ability of both paraldehyde[111] and chloral hydrate[97] to control the alcohol withdrawal reaction in mice rendered dependent on ethanol. In spite of the apparent clinical effectiveness of a combination of paraldehyde and chloral hydrate, however, no investigation has been made of the combined effects of paraldehyde and chloral hydrate in an animal model.

It is interesting to note that Thompson *et al.*[258] considered intravenous administration of paraldehyde, only to reject it on the basis that intravenous injection of oils leads to pulmonary edema. Introduced into medicine in 1882 as a hypnotic,[47] paraldehyde is a clear, colorless liquid with a pungent smell and a density similar to that of water. It is not an oil; it is not lighter than water and it is volatile and partially soluble in water. This solubility is greatest (12.8%) at 12°C and decreases as temperature rises or falls below this point. At 37°C, in particular, the solubility is 7.8%.[200] Unawareness of these facts has led to the practice, widespread at one time, of injecting paraldehyde intravenously in its pure form[40,41,277] or as a 10% solution.[240] In either event the solubility of paraldehyde at 37°C being exceeded, one might expect that droplets of pure paraldehyde would be formed in the blood and pulmonary embolism might be induced. Description of the events (coughing, tachycardia, tachypnea) occurring after intravenous administration of pure paraldehyde or 10% solutions of it in water[15,38,240] suggest this is indeed the case. On the other hand, solutions of paraldehyde that do not exceed its solubility at 37°C can be safely administered intravenously.[8,194]

Another concern with respect to paraldehyde is the reputation it has acquired for "idiosyncratic" toxicity of small doses [listings[12,38,156]]. No evidence of this, however, has been reported in any of the controlled studies involving paraldehyde. It is likely that the vast majority of reported toxic reactions to small amounts of paraldehyde were due to the use of samples which had decomposed. Paraldehyde is a trimer of acetaldehyde and it has a slight tendency to depolymerize back to acetaldehyde. In the presence of air, acetaldehyde oxidizes to acetic acid, which then acts as a catalyst for further depolymerization.[262] Accordingly, the once common practice of keeping open-stock paraldehyde in hospitals and pharmacies resulted in some samples containing as much as 40–98% acetic acid,[30,129] yet as little as 7 ml of paraldehyde containing 40% acetic acid has proved fatal.[1,2] The 1965 and 1970 U.S. Pharmacopea specifications that paraldehyde must be preserved "in well-filled, tight, light-resistant containers . . . not exceeding 30 ml" and that "the user . . . discard the unused contents of any container that has been opened for more than 24 hours" appear to have completely eliminated this problem. No instances of idiosyncratic toxicity have been reported since the institution of the new packaging.

One further comment regarding paraldehyde concerns its metabolism. It is widely held that paraldehyde is metabolized to acetaldehyde *in vivo*. A careful review of the literature suggests, however, that not only is there no unambiguous evidence to support this hypothesis, but evidence exists that is incompatible with it. The two studies most frequently cited as indicative that acetaldehyde is a metabolite of paraldehyde are a 1943 study by Hitchcock and Nelson[132] and a 1957 one by Keplinger and Wells.[152] Hitchcock and Nelson[132] considered the results of their study to be ambiguous. A similar conclusion appears appropriate with regard to the observation by Keplinger and Wells[152] that disulfiram pretreatment of dogs led to higher acetaldehyde levels following paraldehyde administration. Analysis of the original data[151] reveals that at all times acetaldehyde blood levels were a function of the paraldehyde levels and the latter were 1000-fold greater than the acetaldehyde ones. This suggests that the observed acetaldehyde may have been formed artifactually by a slight depolymerization of paraldehyde in the samples at the time of assay. Conversely, the 1956 report by Christie[51] on the condition of patients administered paraldehyde while on disulfiram therapy is incompatible with the *in vivo* formation of acetaldehyde from paraldehyde, since the toxic syndrome normally associated with ethanol ingestion by disulfiram-treated individuals appeared to be entirely absent following the ingestion of paraldehyde.

c. Barbiturates

As prototypical drugs of the alcohol–barbiturate type, these agents would be expected to effectively control the alcohol withdrawal syndrome. Although some American physicians do use barbiturates in the therapy of the alcohol withdrawal syndrome,[70,140,222,222a] the use of these agents has been limited by the concern that it would lead the alcoholic to become habituated to barbiturates as well as ethanol.[21,56,82] In Denmark, on the other hand, barbital has been considered the drug of choice for the treatment of the alcohol withdrawal syndrome and delirium tremens for more than 50 years,[130a,157a,192,245] and although since 1954 other sedative drugs have replaced barbital in the treatment of delirium tremens to some extent, both a 1965 analysis by Nielsen[192] and a 1978 controlled comparison to diazepam by Kramp and Refaelsen[157a] suggest that barbital continues to constitute the best treatment. Finally, barbiturate therapy is considered the treatment of choice in the management of dependence to other depressant drugs of the alcohol–barbiturate type.[19,68,242,282]

In considering the dichotomies this presents, it should be remembered that barbiturates vary markedly in their onset and durations of action and that, as pointed out by Smith,[243] alcoholics are prone to enjoy the euphoria produced by short-acting barbiturates, but are less likely to enjoy that produced by long-acting barbiturates. Accordingly, Smith[243] advanced mephobarbital as a drug which is effective in the treatment of delirium tremens and does not produce euphoria or habituation. Barbital, likewise, is a very long-acting barbiturate. Phenobarbital, used by Isbell[139] in the treatment of the alcohol withdrawal syndrome and more recently in the treatment of withdrawal syndromes from other depressants of the alcohol-barbiturate type,[242] is only slightly shorter acting than barbital. Moreover, it is interesting to note that according to one recent study[250] no barbiturate is responsible for as many such hospitalizations for prescription drug abuse as diazepam. Phenobarbital, in particular, was responsible during the period considered (1966–1972) for 0.38 times as many hospitalizations[250] and 0.36 times as many prescriptions[20] as diazepam, suggesting that the addictive potential of the two agents may be quite similar.

There are only three studies in Table III (studies 14, 23, and 38) in which barbiturates were compared to other agents. In the first of these, Harfst et al.[127] found that in patients with the alcohol withdrawal syndrome amobarbital led to significantly better self-assessment of night sleep and a significantly higher nurse assessment of day sleep than did placebos. The second study involved pentobarbital[147] and was generally inconclusive. It should be noted that both of these agents are intermediate-acting barbiturates. Of the long-acting barbiturates, that is mephobarbital, phenobarbital, and barbital, only the latter has been tested in a controlled trial. Kramp and Rafaelsen[157a] compared it with diazepam. They found that of patients with frank delirium tremens a significantly greater number ($p < 0.05$) showed a satisfactory effect with barbital therapy than with diazepam therapy in terms of onset of effect, course of clinical condition, and the patient's ability to cooperate.

Animal studies confirm the ability of barbiturates to control the alcohol withdrawal reaction. Essig et al.[66] found pentobarbital to significantly reduce the incidence of both withdrawal mortality and seizures in dogs rendered dependent on ethanol. Similarly, Goldstein[111] found withdrawal scores in mice rendered dependent on ethanol to be significantly reduced by pentobarbital, phenobarbital, and barbital.

d. Ethanol

The 1955 study of Isbell et al.,[142] in which individuals were maintained consuming an average of 366 g ethanol daily for 48 days without the appearance of any signs and symptoms of the alcohol withdrawal syndrome as long as ethanol consumption was continued, makes it clear that ethanol ingestion can forestall withdrawal under these conditions. Whether ethanol ingestion would continue to forestall withdrawal symptomatology over longer periods of time or when still higher daily amounts are ingested remains unresolved. Obviously, many alcoholics fail to forestall the withdrawal syndrome through continued drinking, but overnight abstinence may be sufficient for the appearance of some of the symptoms[144] and, in any event, the drinking of alcoholics tends to be cyclic.[179] Also, the question arises whether ethanol would control the withdrawal syndrome once it had gone beyond a certain point.

Ethanol has been used in the treatment of the alcohol withdrawal syndrome, being administered both orally[82,89a,108,236,243] and intravenously.[164] In a 1967 controlled study Golbert et al.[108] compared ethanol to promazine, chlordiazepoxide, and a combination of paraldehyde–chloral hydrate and found it to have an effectiveness comparable to chlordiazepoxide, being, like the latter, significantly less effective than the paraldehyde–chloral hydrate combination with respect to the general incidence of side effects, but superior, although not significantly so, to promazine insofar as the latter was associated with a 15% mortality. More recently Funderburk et al.[89a] have compared the posttreatment sleep patterns of alcoholics treated during the withdrawal period with either ethanol or chlordiazepoxide. Ethanol, unlike chlordiazepoxide, induced little suppression of REM or δ sleep, a finding that led the authors to consider the possibility that an "ethanol detoxification regimen may prove more beneficial to the healthy alcoholic patient than current regimens which employ other psychoactive medication."[89a] The effectiveness of ethanol in controlling the alcohol withdrawal reaction in animals has been demonstrated,[111] as has that of the higher polyhydric homologue, 1,3-butanediol.[172] Additionally, it might be noted that 1,2-pro-panediol[2] is known to have marked anticonvulsant activity,[210,289] although its effectiveness in controlling the alcohol withdrawal syndrome has not yet been determined.

e. Chlormethiazole (Clomethiazole)

Chlormethiazole is an agent chemically very similar to the thiazole part of thiamine (vitamin B_1). It was found by Charonnat et al. in 1957 to possess anticonvulsive and sedative properties,[49] and in the same year it was introduced as a treatment for delirium.[160] It has achieved significant popularity in Europe as an agent for the treatment of the alcohol withdrawal syndrome. It is not available in the United States and, for various nonmedical reasons, is likely to remain so. As discussed below, there is some evidence that it may be more effective in the treatment of the alcohol withdrawal syndrome than is chlordiazepoxide. Because of its relatively short duration of action[145] and its effectiveness as a suicidal agent,[136] its use should be restricted to hospitalized patients.[32,104]

Lundquist[167] described chlormethiazole not only as an excellent hypnotic but stated that it had been employed as a general anesthetic for minor operations. This suggests that it may properly belong to the alcohol–barbiturate class of compounds, a possibility enhanced by the demonstration[249] that in its anesthetic effects it is simply additive with ethanol and with pentobarbital. Additionally, there have been reports of patients experiencing grand mal convulsions 2–3 days following chlormethiazole withdrawal[209] and, generally, of a withdrawal syndrome involving both convulsions and "confusional delirium or psychoses."[158,212] Clinical experience indicates, therefore, that chlormethiazole may be added to the group of hypnotic and sedative drugs resembling ethanol,[212,275] that is, that it is an alcohol–barbiturate type of drug.[139]

In view of the fact that chlormethiazole not only resembles the thiazole portion of vitamin B_1 but is also a metabolite of this vitamin,[16] its use in the therapy of the alcohol withdrawal syndrome may have been occasioned, to some degree, by a theory that B_1 deficiency plays a part in the pathogenesis of alcoholic delirium. Schultz has been quoted by Gastager[92] as having found chlormethiazole to have, however, $1/300$ of the activity of vitamin B_1, and thus it is likely that the therapeutic effects of chlormethiazole in the treatment of the alcohol withdrawal syndrome are due primarily to its being an agent pharmacologically similar to ethanol. This is evidenced by the mode in which it is employed. Bergener,[16] for example, suggested that the drug should be given in increasing doses until the therapeutic response is obtained, the therapeutic response in question being defined as superficial sleep. Similarly, Madden et al.[168] and McGrath[175] made it clear that the patients are very heavily sedated during the first 2 days of chlormethiazole therapy. Finally, Glatt et al.[105,106] found, in a placebo-controlled study of chlormethiazole, that among patients receiving this agent the success rate was much greater ($p < 0.001$) among those who were sedated as opposed to nonsedated individuals.

Of the five controlled studies of chlormethiazole in the treatment of the alcohol withdrawal syndrome (Table III: studies 10, 14, 17, 28, and 36) the most important is the 1975 one by McGrath,[175] who compared it to chlordiazepoxide and found that, although none of the 46 patients receiving chlormethiazole developed delirium tremens, 4 of the 41 receiving chlordiazepoxide did. Although McGrath did not report this difference as statistically significant, calculation of the exact binomial probabilities[109] indicates that it is ($p < 0.05$), suggesting that chlormethiazole is indeed an excellent agent for the control of the alcohol withdrawal syndrome. This analysis of McGrath's study[175] is somewhat confounded, however, by the patients having been administered methaqualone for insomnia, the chlormethiazole patients having received this medication slightly more frequently on the average. In the remaining studies, both Glatt *et al.*[105] and Harfst *et al.*[127] found chlormethiazole therapy to be significantly more effective than placebo, Glatt *et al.* with respect to the number developing depression, the incidence of sedation, and global improvement on each of the first 3 days of treatment, and Harfst *et al.* with respect to self-assessment of night sleep and nurse-assessed hours of day sleep. The latter authors also compared chlormethiazole to amobarbital and found no significant differences. Finally, chlormethiazole was found to be superior to a phenothiazine (trifluoperazine) by Madden *et al.*[168] and to piracetam by Dencker *et al.*[26] The administration of additional medications confounds the finding reported by Madden *et al.*[168]

A number of other and early studies of chlormethiazole use in the treatment of the alcohol withdrawal syndrome have been reviewed by Gershon.[93] None of these were controlled, however, and they provide no further information regarding the usefulness of this agent. It is appropriate, nevertheless, to add some remarks regarding the form in which the drug was administered. In the studies quoted so far, the drug was administered in capsule form. In other studies it had been introduced intravenously as either an injection or an infusion. Two percent solutions of chlormethiazole cause thrombophlebitis at the injection site.[249] Moreover, chlormethiazole at concentrations greater than 0.6% has been found to cause hemolysis as a function of concentration. This probably accounts for two unusual phenomena: (1) the large variations in the value of the LD_{50} of chlormethiazole in mice when administered by different routes (intravenously: 200 mg/kg, intraperitoneal: 380 mg/kg, per os: 800 mg/kg), and (2) the minimum lethal dose of slowly intravenously infused chlormethiazole being inversely proportional to the chlormethiazole concentration [see Lechat [163]]. Because of these considerations it has become customary, when intravenous chlormethiazole is used, to administer it in concentrations no greater than 0.8%.

Goldstein,[111] working with mice rendered dependent on ethanol, reported 80 and 160 mg/kg intraperitoneal to be ineffective in reducing withdrawal scores. This result is a surprising one since Ross [quoted by Svedin[249]] found the hypnotic ED_{50} of chlormethiazole in the same species to be 123 mg/kg.

3. Other Sedatives

a. Benactyzine

Benactyzine is an agent which has sedative properties, although it is not a drug of the alcohol–barbiturate type. It also has muscarinic cholinergic blocking activity and is about one-fifth as potent as atropine in this respect.[60] The effectiveness of benactyzine in controlling the alcohol withdrawal syndrome was compared in 1960 to that of paraldehyde and that of pyridoxine by Ewing,[67] who found that a significantly ($p < 0.01$) larger proportion of the patients receiving benactyzine than those receiving pyridoxine reached "comfort" in 3 days. Otherwise, the outcomes obtained were similar to those obtained with paraldehyde.

b. Hydroxyzine

Hydroxyzine is an antihistaminic agent with sedative actions, which, like benactyzine, does not fall into the category of alcohol–barbiturate types of drugs. In 1969, Schwartz and Fjeld[226] evaluated its effectiveness relative to placebo in the therapy of the alcohol withdrawal syndrome and found that the patients receiving hydroxyzine slept significantly ($p < 0.001$) less, had significantly ($p < 0.05$) poorer appetites, and had a greater incidence of seizures and severe hallucinosis than did the placebo group, although not significantly so. One placebo patient developed delirium tremens. Schwartz and Fjeld concluded that the placebo group exhibited more favorable characteristics than the hydroxyzine group. Similar conclusions can be reached from the 1969 study of Kaim et al.[146] in which one of the patients in the hydroxyzine group died, although none did in the placebo group. Additionally, Kaim et al. found patients receiving hydroxyzine to experience a significantly ($p < 0.05$) greater incidence of seizures than those treated with chlordiazepoxide.[146] Likewise, the 1977 data of Dilts et al.[59] indicate that patients treated with hydroxyzine by one therapeutic team developed delirium tremens significantly ($p < 0.01$) more frequently than those treated with clorazepate, a benzodiazepine, by a second therapeutic team. Thus, the results of Dilts et al.,[59] although possibly biased by the use of two treatment teams, fit into the trend established by previous studies in suggesting that hydroxyzine is less than an optimal agent for the treatment of the alcohol withdrawal syndrome.

4. Anticonvulsants

The seizures associated clinically with the alcohol withdrawal syndrome are of the grand mal type,[142,267,270,271] that is, they are major, generalized, nonfocal seizures with loss of consciousness; they are an inherent part of the alcohol withdrawal syndrome.[108,142,146,256,270] Analysis of mortality figures of patients with delirium tremens has revealed that mortality among patients who manifest seizures at the onset or during delirium tremens is twice as high as in those who do not.[253] Control of alcohol withdrawal seizures can, therefore, be viewed as an important therapeutic goal in itself. Drugs of the alcohol–barbiturate type generally, and benzodiazepines and paraldehyde in particular, are effective in controlling and preventing such seizures. These drugs can induce dependence, however, and there is, therefore, some reluctance to use them because of the concern that the alcoholic may become addicted to them.[153,177,276] The possibility that such seizures might be controlled or prevented by the use of specific nonaddictive anticonvulsants is, therefore, an attractive one.

a. Phenytoin

Of the clinically employed, nonbarbiturate anticonvulsants, the most widely used is phenytoin (diphenylhydantoin). Over the years, and in the absence of any experimental investigation of its ability to prevent alcohol withdrawal seizures,[95] phenytoin has been repeatedly advanced as a drug to be used in the therapy of the alcohol withdrawal syndrome. It should be noted, in this context, that the term "anticonvulsant" is a somewhat misleading one, since it suggests a universality of anticonvulsant action for the drug thus named which in reality may not exist. For instance, phenytoin does control the seizures of grand mal epilepsy[263] and, both clinically[263a] and experimentally,[50,251] it prevents, or at least modifies, those consequent on maximal electroshock. On the other hand, it is ineffective in controlling febrile seizures in the young,[178] experimentally induced pentylenetetrazole (Metrazole) seizures[50,251] in animals, and barbiturate-withdrawal seizures in dogs,[64] cats,[197] and purportedly in humans.[63]

Two clinical studies of phenytoin therapy in the alcohol withdrawal syndrome have been published since 1973 (Table III: studies 25 and 26), the first by Rothstein[217] the second by Sampliner and Iber.[221] The two studies are similar in that in both all patients received chlordiazepoxide with an average dose in the first 24 hr of 360 and 400 mg, respectively. They differ diametrically, however, in both patient selection criteria and in reported outcomes. Rothstein[217] excluded patients who had a history of long-term phenytoin use or had had seizures in the previous 2 weeks, a total of 9.5% of the patients being excluded on this basis. He reported that none of the 200 patients admitted to the study experienced seizures during the alcohol withdrawal period, although 20% of the patients had had seizures under analogous conditions previously.

In contrast, Sampliner and Iber[221] reported that in their investigation "patients assigned to the study met the following criteria: They had been consecutively admitted to our alcohol detoxication unit and had a history of seizures in adulthood *(without regard to whether seizures were related to alcohol withdrawal)*" (italics added). On this basis 93.8% of patients admitted for treatment of the alcohol withdrawal syndrome were excluded from the study. Of the patients admitted to the study, only 2 of the 70 treated with phenytoin experienced seizures, whereas 11 of 66 controls did ($p < 0.005$).

In considering the reason for the contrasting findings reported in these two studies, one possibility could be that Sampliner and Iber's patients had presented a more severe withdrawal syndrome on admission. This seems unlikely, since Rothstein[217] reported disorientation combined with hallucination in 9 individuals (4.5%), whereas Sampliner and Iber[221] reported no patients with delusions and only 4 with hallucinations (2.9%). Another possibility is that a high proportion of the patients experiencing seizures in Sampliner and Iber's study had idiopathic epilepsy. The information provided in their study regarding this is fragmentary, but it suggests that 7 of 13 of those who experienced seizures had a history of seizures unrelated to drinking. Individuals with idiopathic epilepsy are more prone to develop seizures when withdrawing from ethanol.[271] This might then provide a partial explanation for Sampliner and Iber's findings. The matter remains unresolved, however, particularly so since Sampliner and Iber also showed that, with the phenytoin regimen employed, blood phenytoin levels in alcoholics rose to a maximum of 5.6 μg/ml after 2.6 days of therapy, whereas the accepted therapeutic level is 10 μg/ml. Other questions regarding Sampliner and Iber's study have been raised by Sapira and Cherubin.[222b]

A relevant clinical experience is that of Smith,[244] who reported in 1976 on changes in seizure incidence occurring upon sequential changes in medication. This incidence, which was 21/500 when a combination of phenytoin and a phenothiazine was employed, decreased to 14/500 upon replacement of the phenothiazine by chlordiazepoxide, then to 9/500 upon replacement of the phenytoin by primidone, a barbituratelike anticonvulsant, and finally to 0/500 upon the primidone being gradually, rather than abruptly, withdrawn. Although these results are subject to bias, as are all studies using historical controls, they do suggest that it may be worthwhile to undertake a controlled clinical trial of combination therapy with primidone and an alcohol–barbiturate type of agent.

Gessner[97] reported in 1974 that, in mice rendered dependent on ethanol, phenytoin, administered either orally or intraperitoneally in doses ranging from 12 to 100 mg/kg, was ineffective in decreasing seizure scores during the withdrawal period, although under the same circumstances chloral hydrate lowered seizure scores promptly and in a dose-related manner. Later, Gessner and Hu[99] explored, in the same species, the possibility of the potentiation of the effects of chlordiazepoxide by phenytoin. Using a range of four chlordiazepoxide doses (1–10 mg/kg) they found seizure scores in animals coadministered phenytoin to be slightly, but not significantly, lower than the saline-coadministered controls. Thus, phenytoin appears to have no effect on alcohol withdrawal seizures in mice.

Mention should be made of a 1976 claim by Sprague and Craigmill[246] that administration of 20 mg phenytoin/kg in 10 ml 40% propylene glycol/kg and 10% ethanol to mice dependent on ethanol significantly decreased seizure scores during the withdrawal period. Methodologically, however, the relevance of Sprague and Craigmill's data may be an artifact of the use of groups unmatched for the severity of their convulsive scores prior to drug administration. Additionally, it should be borne in mind that propylene glycol in a dose of 4 g/kg and ethanol in a dose of 1 g/kg will have powerful anticonvulsant effects of their own,[210,289] a matter of some importance with respect to clinical studies, since phenytoin for injection is supplied as a solution in a mixture of propylene glycol, ethanol, and water.

b. Dipropylacetate

Dipropylacetate (2-n-propyl-pentanoic acid, di-n-propylacetic acid, sodium valporate) is a new anticonvulsant agent found to be useful in the clinical treatment of epilepsy[143,238,272] and one which has been thus used in Europe for several years. It is a remarkable agent, in that it appears to be able to control a wide spectrum of epileptic disorders[161] and in that it is an inhibitor of γ-aminobutyric acid transaminase and that it increases brain γ-aminobutyric acid levels.[107,237] Animal studies suggest that this agent may also control the convulsions associated with the alcohol withdrawal syndrome. Hillbom,[133] for example, reported in 1975 that rats rendered dependent on alcohol and administered dipropylacetate orally at the start of the withdrawal and 12 hr later (100 and 200 mg/kg, respectively) had significantly fewer audiogenic seizures at 15 hr than did water-administered controls. Similarly, Noble et al.[195] reported in 1976 that rats rendered dependent on ethanol and given dipropylacetate intraperitoneally (200 mg/kg) 24 hr after withdrawal had significantly fewer audiogenic seizures 1 hr later than did saline controls. Interesting as these results are, they raise the question of whether it is appropriate to equate alcohol withdrawal seizures audiogenically induced with spontaneous ones.

Dipropylacetate was routinely used in the treatment of the alcohol withdrawal and delirium tremens syndromes in Italy by Bonfiglio et al.[25] In 1972, these workers reported on favorable results of dipropylacetate treatment of 23 cases of the alcohol withdrawal syndrome.[24] Since then they reported[25] on a 2-year period during which dipropylacetate (300–400 mg three times a day) was used without concurrent administration of sedatives in the treatment of 141 alcoholic patients who presented alcohol withdrawal symptomatology, 87 of whom had delirium tremens. The average duration of the withdrawal syndrome during this period was shorter, according to Bonfiglio et al., than that in the preceding 2 years, when 109 patients presenting withdrawal symptomatology (including 88 with delirium tremens) received more conventional therapy. Similarly, in a group of 22 patients with previous alcohol withdrawal syndrome hospitalizations, the average duration of the syndrome was shorter when treated with dipropylacetate and in those experiencing frank delirium on previous hospitalizations; dipropylacetate therapy appeared to prevent its recurrence. Finally, whereas in the previous 2 years three deaths had occurred that could not be attributed to causes extraneous to the delirium tremens, no such death occurred since dipropylacetate treatment was begun. Since the only controls for these studies[24,25] were historical ones, however, it is not possible to make definitive conclusions regarding the effectiveness of dipropylacetate in the treatment of the alcohol withdrawal and delirium tremens syndromes. Since this agent has purportedly no sedative properties and hence is not of the alcohol–barbiturate type, it will be most interesting if its effectiveness can be confirmed. Further controlled studies of dipropylacetate should, therefore, be undertaken.

c. Other Anticonvulsants

With the exception of Smith's 1976 report[244] suggesting that primidone was more effective than phenytoin, there has been no reported use of other anticonvulsants in the treatment of the alcohol withdrawal syndrome.

Gessner[98] has investigated, using mice, the ability of anticonvulsants other than phenytoin to control the seizures associated with the alcohol withdrawal reaction. Choosing agents representative of different types of anticonvulsants, he reported mephynetoin, phenacemide, and paramethadione, but not ethotoin, to significantly decrease seizure scores. The only pharmacological property that distinguished the three effective agents from ethotoin and phenytoin is that the former possess some sedative properties, with somnolence being a known clinical side effect. Okamoto et al.,[197] working with cats in barbiturate withdrawal, reported that phenytoin was ineffective for most withdrawal signs and actually intensified some. Trimethodione and dimethadione, however, significantly reversed most of these signs, even though dimethadione was less effective and more toxic than trimethodione.

5. Cations

a. Magnesium Sulfate

The possibility that therapy with magnesium might be effective in alleviating the alcohol withdrawal and delirium tremens syndromes is a particularly intriguing one since serum magnesium levels are abnormally low in these syndromes.[78,79] Moreover, low serum magnesium is generally associated with increased seizure susceptibility[3,231,274] and with increased susceptibility to photomyoclonus during the period of withdrawal from ethanol.[269] Furthermore, low serum magnesium in delirium tremens is associated with an increased severity of the syndrome[235] and a higher incidence of alcohol encephalopathies, this being a common diagnostic denominator used to designate (1) Korsakoff's psychosis, (2) Wernicke's disease, (3) dementia alcoholica, and (4) intellectual reduction to a degree incompatible with social readjustment.[247] No correlation has been found, on the other hand, between low magnesium levels and the duration of the delirium tremens.[34]

Flink[77] claimed in 1956 that parenteral administration of magnesium shortens the duration of delirium tremens, but this was challenged[273] on statistical grounds. Later, in 1962, Beroz et al.[17] reported, on the basis of a controlled trial involving 50 cases of the alcohol withdrawal syndrome, that parenteral magnesium treatment led to a lower incidence of hallucinations than that observed in the control placebo group. Evaluation of an additional 100 cases, however, led them to retract this conclusion.[18] In 1977, Shulsinger et al.[235] reported parenteral magnesium sulfate to be as effective as diazepam in controlling the alcohol withdrawal syndrome. This conclusion was based on the absence of any significant differences in several outcomes between groups of patients treated with diazepam, two dosage levels of magnesium sulfate, and a combination of magnesium sulfate and diazepam. This conclusion is weakened by the only outcome reported quantitatively having a very low incidence (1 seizure among the 94 patients), which raises questions regarding the severity of the patient's presenting condition, a question which in the absence of a placebo group becomes difficult to answer.

A new dimension has been added to studies of magnesium metabolism by two recent reports. The first of these is the 1975 study of Brooks and Adams,[34] who found that cerebrospinal fluid (CSF) magnesium levels are significantly lower in individuals who have experienced alcohol withdrawal seizures 5–12 hr previously than (1) in controls without seizure disorders, (2) in individuals who have experienced an idiopathic seizure in the previous 6 hr, and (3) in individuals with delirium tremens not preceded by seizures. The second is the 1977 report of Buck et al.[37] who, working with magnesium-depleted rats, showed that the increased seizure susceptibility observed in such animals was countered when CSF magnesium levels were replenished, although replenishment of blood magnesium levels alone was ineffective.

Finally, it should be noted that although alcoholics have an overall magnesium deficit[182] perhaps due to enhanced urinary magnesium excretion[149] or nutritional in origin,[223] the hypomagnesemia observed during withdrawal is considered to be largely a redistributive phenomenon,[183] since as the withdrawal syndrome wanes, serum magnesium levels tend to return to normal in spite of the absence of efforts to replenish them.[286,287]

b. Lithium Carbonate

Lithium is known to be effective in the treatment of manic-depressive illness. Two controlled studies have suggested that it may also be effective in the chronic treatment of depressive alcoholics.[154,184] Sellers et al.[227] reported in 1976 that in individuals experiencing a mild withdrawal syndrome lithium therapy reduced the patients subjective scores on a 34-item check list of "symptoms commonly associated with alcohol intoxication and withdrawal" by 18, 51, and 45% relative to placebo controls on the first 3 days of treatment ($p < 0.05$, $p < 0.001$, and $p < 0.05$, respectively). Performance on a complex tracking task was also significantly improved by lithium relative to placebo. It is difficult to determine the relevance of these findings to the clinical treatment of the alcohol withdrawal syndrome in the absence of more explicit information on the identity of the 34 items on the check list and in view of the subjective manner in which they were rated. In contrast, Ho and Tsai[134] reported in 1976 that in animal studies lithium did not alleviate the severity of the alcohol withdrawal symptomatology but rather, if given in sufficient dosage, aggravated it.

6. Miscellaneous

a. Propranolol

Propranolol (Inderal) is a β-blocking agent used to control cardiac activity in the treatment of angina pectoris, hypertension, and cardiac arrhythmias.[191] It also possesses some sedative properties and consideration has been given to its use in the treatment of anxiety.[27,116,264,279] The use of propranolol in the treatment of this condition is still being evaluated and it has not been approved by the U.S. Food and Drug Administration.[130] There also has been some interest with respect to its use in chronic alcoholics.[45] Finally, there has been some interest as to the effectiveness of this agent in the treatment of the alcohol withdrawal syndrome.[42–44,211,229]

There have been three comparative studies of the effect of propranolol on the alcohol withdrawal syndrome. In the first of these, Carlsson and Johansson[44] claimed that propranolol relieved significantly the tension and depression associated with the alcohol withdrawal syndrome. The patients involved were also receiving four other drugs, however, and although the study was later stated to have been double-blind,[43] patients treated with propranolol received 2.6 times as much chlormethiazole as the placebo group. Given the effectiveness of chlormethiazole alone in controlling the alcohol withdrawal syndrome, it is not clear what conclusions, if any, can be drawn from this study regarding the effectiveness of propranolol in the treatment of the alcohol withdrawal syndrome. In a 1976 crossover study, Terävänen and Larsen[254] were not able to show propranolol to be more effective than placebo in reducing either the frequency or amplitude of positional tremor associated with the alcohol withdrawal syndrome. Sellers et al.[229] reported in 1977, on the other hand, that propranolol significantly reduced the tremor associated with cessation of ethanol ingestion in a group of men maintained on 280 g ethanol/day for the previous 7 days. One of the differences between these two studies was the severity of the withdrawal syndrome. Thus, in spite of the fact that all of Terävänen and Larsen's patients received chlormethiazole and some also diazepam, 25% experienced convulsions.[254] None of the individuals in the study of Sellers et al. experienced convulsions, even though the controls in this study received no medication at all.[229] Reference to the study of Isbell et al.[142] in which four subjects consumed an average of 377 g ethanol/day for 10 ± 5 days and experienced, upon withdrawal, tremor, weakness, and other minor signs but no fever, hallucinations, convulsions, or disorientation suggests that the alcohol withdrawal syndrome incurred by the subjects in the study of Sellers et al.[229] following consumption of 280 g ethanol daily for 5 days must have been a rather mild one. This point is an important one, since in an animal model of the alcohol withdrawal syndrome, propranolol administration resulted in a significant intensification of the withdrawal syndrome.[112] It is surprising how little exploration of the ability of this agent to control alcohol withdrawal was done in animal models prior to clinical experiments being undertaken.

An interesting observation reported by Sellers et al.[229] was that individuals receiving propranolol slept significantly longer than controls. It has been shown that both d- and l-propranolol will prolong ethanol sleep times in mice and that this effect is not due to an alteration in brain ethanol levels.[285] Since the dextro isomer possesses about $\frac{1}{60}$ of the β-blocking activity of the racemic mixture,[13] it would be interesting to find out whether the results of Sellers et al. could be replicated using d-propranolol. Such a finding with respect to the effect of propranolol on tremors associated with mild alcohol withdrawal syndrome would resolve the present controversy regarding whether the effect is mediated centrally and via β-receptor blockade.[288,290]

Finally, chronic excessive ethanol consumption is associated with the development of alcoholic cardiomyopathy, a condition difficult to diagnose except at postmortem.[278] Given that sudden withdrawal of propranolol from individuals with, or even without, heart disease can be followed by myocardial infarction,[5,57,185,233] the usefulness of the agent in the treatment of the alcohol withdrawal syndrome remains questionable.

b. Fructose

Two uncontrolled studies have suggested intravenous fructose to be beneficial in the treatment of delirium tremens.[55,187] A 1971 double-blind study by Friend et al. has failed, however, to substantiate these findings.[89]

The rationale for using fructose in the treatment of delirium tremens is not clear. Fructose increases the rate of human ethanol clearance.[36,46,202,255,265] Hastening the ethanol disappearance would be expected, however, to result in a more acute and severe withdrawal syndrome. Moreover, by the time the individual undergoing withdrawal from ethanol develops delirium tremens, it can be assumed that blood ethanol has all been cleared. Thus, the use of fructose in the treatment of delirium tremens is not supported by any published empirical observations or theoretical considerations.

F. PERSPECTIVES FOR FUTURE RESEARCH

1. Diagnostic Problems

a. Admission Criteria

The diversity of the presenting conditions of the patients admitted to the various studies in Table III underscores the existing differences in what is viewed as requiring therapeutic intervention and what is encompassed by the term "alcohol withdrawal," be it used alone or in conjunction with such qualifiers as syndrome, reaction, etc. Thus, whereas in some studies frank delirium was a requirement for inclusion, in others[67,105,216] inebriation at the time of admission appears to have constituted sufficient reason for admission to the study. In still others, the terms used in describing the patient's presenting condition were sufficiently ambiguous to leave open the question of whether they encompassed acute inebriation. Examples of such terms are "acute brain syndrome secondary to alcohol ingestion,"[198] "acute alcoholic state,"[48] "acute alcoholism,"[118] and "acute alcoholic psychosis."[215] Inebriation, unless preceded by chronic and excessive drinking, does not presage anything more severe by way of withdrawal than a hangover and the question of whether it is a condition requiring therapeutic intervention is an open one.[39,105] Moreover, although by stretching syntax, inebriated patients denied further access to ethanol could be considered as being withdrawn from ethanol, administration of drugs normally used in the control of alcohol withdrawal to inebriated patients would likely only deepen the central nervous sytem depression induced by the ethanol.

It is important in the evaluation of agents for the treatment of the postalcoholic phase to identify the condition being treated, be this uncomplicated drying out, a mild or marked alcohol withdrawal syndrome, or delirium tremens. Drugs which may be effective in the treatment of one of these conditions may prove ineffective or inappropriate as treatment for another. Publication of enthusiastic reports regarding the effectiveness of an agent for "the detoxication of dangerous alcoholics,"[173] for instance, only adds to the confusion. It follows that criteria for admission to a study should be carefully elaborated and clearly reported. However, essential as the use of such criteria is, it is not sufficient, since at the time of admission it may not be possible to differentiate, on the basis of the presenting condition, a mild withdrawal syndrome from the early stages of a severe one. Other means of securing the diagnosis should, therefore, also be employed.

b. Placebo Groups

In the context of experimental studies, the most straightforward way to secure the diagnosis, if only on a group basis, is to randomly select individuals who are then administered placebos and in whom the syndrome is allowed to take its natural course. Given, however, that the effectiveness of some agents vis-à-vis placebos has now been established, this approach is ethically problematical. As a result, the incidence of placebo-controlled studies has declined markedly. Thus, for instance, placebos were used in 11 of the 19 studies published during the 1960s but in only four of those published in the 1970s (Table III). Moreover, in two of the later studies, either all[221] or such patients as needed[173] were receiving effective medication, while in the other two the subjects were all known to have only a moderate degree of dependence.[227,229] An alternative strategy for securing the diagnosis, that of withholding sedative medication and allowing patients, initially presenting inebriation, to dry out and develop withdrawal symptomatology, may also present ethical problems, since once started, delirium tremens is not readily controlled by drug therapy (see Section C) and consequently it is felt, strongly by some, that therapy should be started without waiting for the development of classical withdrawal symptomatology.[245]

c. Diagnostic Procedures

Given that withholding treatment is no longer a viable diagnostic strategy, there is a need for an alternative procedure which would provide both a diagnosis of the presenting condition and its prognosis if not treated. Although hitherto such a procedure has not been worked out, the potential for it exists. First, much could be learned, with regard to the inebriated patient, from a comparison of the blood ethanol level to the degree of impairment presented. The discrepancy between the observed blood ethanol level and that associated with a similar degree of impairment in nontolerant individuals[81,148] could be used as a quantitative measure of tolerance and the coexisting dependence. Similarly, blood ethanol levels in apparently sober individuals, or ones in mild withdrawal, would also be indicative of a history of chronic, excessive ethanol ingestion.[142] The tolerance of patients not inebriated at admission could be investigated using a test dose of a short-acting alcohol–barbiturate type of agent. This is the recommended standard practice in the treatment of withdrawal illness from depressants of the alcohol–barbiturate type[68] and in the detection of barbiturate dependence in narcotic addicts,[126a] lack of tolerance being inferred from a patient's response to the test drug and tolerance from the continued absence of such a response.[210a]

Alternatively, as proposed by Gessner,[94] the patient could be administered increasing amounts of ethanol or, because of concern that it could create an even greater emotional dependency on ethanol, of another relatively short-acting drug of the alcohol–barbiturate type, until evidence of CNS depression is evident. The amount administered would then constitute a measure of the patient's degree of tolerance. This approach has been used by Quinn,[209] who infused chlormethiazole solution until the stage of light sleep was reached and estimated the oral maintenance dose on this basis, and by Sapira,[222] who used secobarbital. It has been elaborated into a step-by-step procedure by Sapira and Cherubin.[222a] As Quinn[209] points out, it is essential, when this approach is used, to take into consideration the patient's blood ethanol level, since ethanol will act additively with the administered agent.

The diagnostic approach discussed above is not applicable in patients who have already developed delirium tremens, since that syndrome, once started, is not easily revised by administration of alcohol–barbiturate types of drugs. For such patients the severity of their symptoms, including body temperature, pulse rate, and blood pressure, should be recorded. Additionally, Salum[220] has shown that pseudoopisthotonos (i.e., retraction of the head and hyperextension of the back) is a useful index of the severity of delirium tremens.

Regardless of whether at admission patients are diagnosed as experiencing a mild and uncomplicated alcohol withdrawal syndrome or delirium tremens, an effort should be made to record and publish what are the values of a range of physiological variables and how these are changed by treatment. Martin and Jasinski's work on physical dependence on morphine and withdrawal therefrom could be used as a model.[173a]

Finally, there have been reports that the severity[235] and sequelae,[247] although not the duration,[34] of delirium tremens are correlated to the degree of presenting hypomagnesemia, that the duration of delirium tremens is correlated to the presenting lactic acid levels in the cerebrospinal fluid,[34] and that the ratio of plasma γ-amino-n-butyric acid to leucine is elevated following heavy drinking.[234] These relationships, if confirmed, could be used diagnostically. Even if it were not possible to determine these levels quickly enough to influence the initial diagnosis and treatment decisions, such information could secure the diagnosis *ex post facto* and would be most useful in the evaluation of the outcomes of controlled studies.

2. Dosage Regimens

The majority of studies listed in Table III employed flexible dosage regimens. The merits of using, respectively, fixed and flexible dosage regimens in testing agents for effectiveness in the treatment of the alcohol withdrawal syndrome and delirium tremens has been discussed in Section C.1. When effective agents are tested against placebos or clearly ineffective drugs, definitive results have been obtained using both fixed[232,257] and flexible regimens.[108,146] Investigation of more subtle differences will probably require, however, a more sophisticated approach, because of the shortcomings of both these regimen paradigms. Thus, a fixed schedule will necessarily lead to less than optimal individualization of treatment and consequently a greatly varied response to it.[127] Individualization of dose on the basis of the attending physician's perception of the patient's needs, on the other hand, is not only open to the objections made by Thomas and Freedman[257] and reviewed earlier, but to the degree that such perception is subjective, it precludes the establishment of treatment paradigms suitable for more general adoption. It follows that dosage regime formulations should incorporate explicit description of the therapeutic decision-making process and the diagnostic criteria employed at the time such decisions are made. Two other items of information would greatly aid the interpretation of reported study outcomes, namely (1) the actual amounts of drug the patients received,[222b] subdivided by whether outcome was successful or not, and (2) the blood levels of the drug administered, similarly subdivided.[157b]

Table IV.   Relative Effectiveness of Drugs in the Treatment of the Alcohol Withdrawal Syndrome[a]

More effective				Less effective
		Paraldehyde	>	Promazine
Paraldehyde + chloral hydrate	>	Chlordiazepoxide	>	Promazine
		Diazepam	>[b]	Promazine
		Chlordiazepoxide	>	Chlorpromazine
		Chlordiazepoxide	>	Placebo
		Chlordiazepoxide	>	Thiamine
Chlormethiazole	>[b]	Chlordiazepoxide		
		Chlordiazepoxide	>	Hydroxyzine
		Clorazepate	>[b]	Hydroxyzine

[a]All inequalities represent significant differences in effectiveness in indicated direction.
[b]Possibly confounded by concomitant use of additional medication.

G. SUMMARY AND CONCLUSIONS

1. Comparative Evaluation

The variation in the outcome criteria reported by the authors of the various studies in Table III is probably mainly a reflection of the variation, from study to study, in the seriousness of the patient's presenting condition. Given the present state of the art, however, secure *a priori* prognoses of how the patients would have fared without effective treatment could not be arrived at. Accordingly, in evaluating the studies in Table III and their outcomes, all patients were considered as potentially at risk with respect to the three most serious or catastrophic concomitants of the alcohol withdrawal syndrome: seizures, development of delirium tremens, and death. The studies reviewed suggest that, with respect to protecting patients against this triad of outcomes, drugs differ significantly and consistently in their effectiveness and can be ranked accordingly as shown in Table IV.

2. Treatment versus Diagnostic Category

As pointed out in the introduction, although clinical trials of therapies for the control of the alcohol withdrawal and delirium tremens syndromes have continued unabated in the 1970s, much less has been learned from them than from those published in the 1960s. This suggests that, for whatever reason, the 1970s have witnessed what Melmon termed a "suboptimal use of human subjects in this area." The most immediate cause of this has been the virtual evanescence of placebo controls and the absence of a concurrent development of other procedures to secure the diagnosis. As a consequence, the interaction between diagnostic category and treatment has not been explicitly investigated. Were this possible, it might prove, for instance, that high degrees of dependence are best treated with relatively short acting agents [such as paraldehyde (with or without chloral hydrate) or chlormethiazole], that intermediate degrees of dependence are best treated with long-acting agents that form active and equally persistent metabolites (chlordiazepoxide or diazepam), and that low degrees of dependence may be optimally treated with agents providing symptomatic relief (possibly haloperidol). Unfortunately, at present, the above must necessarily remain just speculation. Moreoever, there exists the real danger that drugs potentially useful for one of these indications may prove ineffective when applied to an undifferentiated population of alcoholic patients and that, thereby, future testing of such agents for the treatment of specific alcohol dependence conditions may become ethically difficult.

The drugs that, on the basis of available studies, result as the most effective for the treatment of severe alcohol withdrawal syndrome are paraldehyde (particularly in combination with chloral hydrate), chlormethiazole, and the benzodiazepines.

Anomalously, although the efficacy and safety of the combination of paraldehyde and chloral hydrate was reported in the late 1960s, the combination does not appear to have been used since that time. Similarly, although chlormethiazole has been widely used in Europe and elsewhere for a number of years, it has been neither available nor tested in the United States.

These drugs differ from the benzodiazepines in that they have generally shorter half-lives, which should permit a better adjustment of dose to the patient's condition, and that in overdose they are more readily fatal, necessitating greater clinical acuity in their use. They are clearly drugs that should be used only in hospitalized patients. It is, of course, open to serious question whether any potentially addictive drug, including the benzodiazepines, should ever be used in alcoholics on an outpatient basis.

It should be noted that the agents shown to be more effective were, in all instances, drugs of the alcohol–barbiturate type. Based on this evidence it must be concluded that the substitution and replacement rationale advanced by Isbell remains the most viable strategy for the treatment of severe cases of the alcohol withdrawal syndrome. This conclusion is strengthened by the fact that when the benzodiazepines and chlormethiazole were first employed for the treatment of the alcohol withdrawal syndrome, it was not generally appreciated that they belonged to this group of drugs.

ACKNOWLEDGMENTS. I wish to thank Peter Van Oot for help in the formulation of this review and his editorial assistance. The work on this review was supported in part by PHS grant 5 RO1 AA 00699.

H. REFERENCES

1. Anonymous, *Br. Med. J.* **2:**1114 (1954).
2. Anonymous, *Lancet* **2:**912 (1954).

3. Anonymous, *Br. Med. J.* **2**:195 (1967).
4. Anonymous, *Med. Lett.* **13**:19 (1971).
5. Anonymous, *FDA Drug Bull.* **5**:6 (1975).
6. L. Ahtee and M. Svartström, *Acta Pharmacol. Toxicol.* **36**:289 (1975).
7. S. N. Albert, E. L. Rea, C. A. Duverney, J. Shea, and J. F. Fazekas, *Med. Ann. D.C.* **23**:245 (1954).
8. R. M. Anthony, A. C. Andorn, I. Sunshine, and W. L. Thompson, *Fed. Proc.*, Abstr. **36**:285 (1977).
9. J. Armstrong, *Edinburgh Med. Surg. J.* **9**:58 (1813).
10. D. Athen, H. Beckmann, M. Ackenheil, and M. Markianos, *Arch. Psychiatr. Nervenkr.* **224**:129 (1977).
11. E. S. Baird and D. M. Hailey, *Br. J. Anaesth.* **44**:803 (1972).
12. G. Baratham and L. F. Tinckler, *Med. J. Aust.* **2**:877 (1964).
13. A. M. Barrett and V. A. Cullum, *Br. J. Pharmacol.* **34**:43 (1968).
14. R. A. Bartholomew, *J. Am. Med. Assoc.* **104**:367 (1935).
15. J. A. Beauchemin, R. G. Springer, and G. A. Elliot, *Med. Times* **63**:179 (1935).
16. M. Bergener, *Acta Psychiatr. Scand. Suppl.* **192**:65 (1966).
17. E. Beroz, P. Conran, and R. W. Blanchard, *Am. J. Psychiatry* **118**:1042 (1962).
18. E. Beroz, P. Conran, and R. W. Blanchard, *Am. J. Psychiatry* **119**:482 (1962).
19. P. H. Blachly, *Am. J. Psychiatry* **120**:894 (1964).
20. B. Blackwell, *J. Am. Med. Assoc.* **225**:1637 (1973).
21. M. A. Block, *J. Am. Med. Assoc.* **162**:1610 (1956).
22. K. Blum and J. E. Wallace, *Br. J. Pharmacol.* **51**:109 (1974).
23. K. Blum, J. D. Eubanks, J. E. Wallace, and H. Hamilton, *Clin. Toxicol.* **9**:427 (1976).
24. G. Bonfiglio, S. Falli, and A. Pacini, *Lav. Neuropsichiatr.* **50**:115 (1972).
25. G. Bonfiglio, S. Falli, and A. Pacini, *Minerva Med.* **68**:4233 (1977).
26. S. J. Dencker, G. Wilhelmson, E. Carlsson, and F. J. Bereen, *J. Int. Med. Res.* **6**:395 (1978).
27. J. A. Bonn, P. Turner, and D. C. Hicks, *Lancet* **1**:814 (1972).
28. L. N. Boston, *Lancet* **1**:18 (1908).
29. J. Boucharlat, J. Ledru, A. Maitre, M. Ratel, and R. Wolf, *Ann. Med.-Psychol.* **134**:557 (1976).
30. G. C. Bowles, Jr., *Mod. Hosp.* **103**:130 (1964).
31. K. M. Bowman and E. M. Jellineck, in: *Effects of Alcohol on the Individual: A Critical Exposition of Present Knowledge* (E. M. Jellineck, ed.), Vol. 1, pp. 81–169, Yale University Press, New Haven, Conn. (1942).
32. J. J. Bradley, *Br. Med. J.* **2**:774 (1977).
33. S. M. Bramson, *J. Am. Med. Assoc.* **225**:749 (1973).
34. B. R. Brooks and R. D. Adams, *Neurology* **25**:943 (1975).
35. J. H. Brown, D. E. Moggey, and F. H. Shane, *Scott. Med. J.* **17**:9 (1972).
36. S. S. Brown, *Lancet* **2**:898 (1972).
37. D. R. Buck, A. W. Mahoney, and D. G. Hendricks, *Pharmacol. Biochem. Behav.* **5**:529 (1976).
38. C. L. Burstein, *J. Am. Med. Assoc.* **121**:187 (1943).
39. R. Byck, in: *The Pharmacological Basis of Therapeutics* (L. S. Goodman and A. Gilman, eds.), pp. 152–200, MacMillan, New York (1975).
40. W. E. Caldwell, *Am. J. Obstet. Gynecol.* **4**:313 (1922).
41. W. E. Caldwell, *N.Y. State J. Med.* **26**:57 (1926).
42. C. Carlsson, *Int. J. Clin. Pharmacol. Ther. Toxicol.*, Suppl. **3**:61 (1969).
43. C. Carlsson, *Postgrad. Med. J.* **52**, Suppl. (4):166 (1976).
43a. C. Carlsson and J. Häggendal, *Lancet* **2**:889 (1967).
44. C. Carlsson and T. Johansson, *Br. J. Psychiatr.* **119**:605 (1971).
45. C. Carlsson and B. G. Fasth, *Br. J. Addict.* **71**:321 (1976).
46. T. M. Carpenter and R. C. Lee, *J. Pharmacol. Exp. Ther.* **60**:286 (1937).
47. V. Cervello, *Arch. Sci. Med.* **6**:177 (1882).
48. J. F. Chambers and J. D. Schultz, *Q. J. Stud. Alcohol* **26**:10 (1965).

49. R. Charonnat, P. Lechat, and F. Chareton, *Therapie* **12**:68 (1957).
50. G. Chen and C. R. Ensor, *Arch. Neurol. Psychiatry* **63**:56 (1950).
51. G. L. Christie, *Med. J. Aust.* **1**:789 (1956).
52. A.W. Clare, *Br. Med. J.* **4**:340 (1971).
53. Competitive Problems in the Drug Industry: Hearings on Present Status of Competition in the Pharmaceutical Industry before the Subcommittee on Monopoly of the Senate Select Committee on Small Business, Part 2, 90th Congress, 2nd Session (1976), p. 461, testimony of Dr. Harry Williams.
54. B. J. Cutshall, *Q. J. Stud. Alcohol* **26**:423 (1965).
55. M. S. Dalton and D. W. Duncan, *Med. J. Aust.* **1**:659 (1970).
56. F. Damrau, *Med. Rec.* **147**:557 (1938).
57. Z. Danilevicus, *J. Am. Med. Assoc.* **237**:53 (1977).
58. Diagnostic and Statistical Manual of Mental Disorders (DSM-II), 2nd ed., American Psychiatric Association, Washington, D.C. (1968).
59. S. L. Dilts, D. L. Keleher, G. Hoge, and B. Haglund, *Am. J. Psychiatry* **134**:92 (1977).
60. E. F. Domino, in: *Drill's Pharmacology in Medicine* (J. R. DiPalma, ed.), 3rd ed., pp. 356–364, McGraw-Hill Book Company, New York (1965).
61. M. W. Dysken and C. H. Chan, *Am. J. Psychiatry* **134**:573 (1977).
62. R. Edgley, *Med. J. Aust.* **1**:186 (1970).
63. C. F. Essig, *J. Am. Med. Assoc.* **196**:714 (1966).
64. C. F. Essig and W. W. Carter, *Neurology* **12**:481 (1962).
65. C. F. Essig and H. F. Fraser, *Clin. Pharmacol.* **7**:466 (1966).
66. C. F. Essig, E. Jones, and R. C. Lam, *Arch. Neurol.* **20**:554 (1969).
67. J. A. Ewing, *Q. J. Stud. Alcohol* **21**:68 (1960).
68. J. A. Ewing and W. E. Bakewell, *Am. J. Psychiatry* **123**:909 (1967).
69. A. R. Favazza and P. Martin, *J. Am. Med. Assoc.* **29**:219 (1974).
70. A. R. Favazza and P. Martin, *Am. J. Psychiatry* **131**:1031 (1974).
71. J. F. Fazekas, J. D. Schultz, P. D. Sullivan, and J. G. Shea, *J. Am. Med. Assoc.* **161**:46 (1956).
72. J. F. Fazekas, J. G. Shea, W. R. Ehrmantraut, and R. W. Alman, *J. Am. Med. Assoc.* **165**:1241 (1957).
73. W. Feuerlein, *Br. J. Addict.* **69**:141 (1974).
74. F. A. Figurelli, *J. Am. Med. Assoc.* **166**:747 (1958).
75. G. B. Fink and E. A. Swinyard, *J. Am. Pharm. Assoc. Sci. Ed.* **49**:510 (1960).
76. J. T. Flexner, *Doctors on Horseback, Pioneers in American Medicine,* Garden City Publishing, New York (1939).
77. E. B. Flink, *J. Am. Med. Assoc.* **160**:1406 (1956).
78. E. B. Flink, in: *The Biology of Alcoholism* (B. Kissin and H. Begleiter, eds.), Vol. 1: *Biochemistry,* pp. 377–395, Plenum Press, New York (1971).
79. E. B. Flink, F. L. Stutzman, A. R. Anderson, T. Konig, and R. Fraser, *J. Lab. Clin. Med.* **43**:169 (1954).
80. J. B. Floyd and C. M. Murphy, *Ky. Med. J.* **74**:549 (1976).
81. R. B. Forney and R. N. Harger, in: *Drill's Pharmacology in Medicine* (J. R. DiPalma, ed.), pp. 275–302, McGraw-Hill Book Company, New York (1971).
82. V. Fox, *Am. Pract. Dig. Treat.* **7**:1461 (1956).
83. H. F. Fraser, *Annu. Rev. Med.* **8**:427 (1957).
84. H. F. Fraser, *Am. J. Public. Health* **48**:561 (1958).
85. H. F. Fraser, H. Isbell, A. J. Eisenman, A. Wikler, and F. T. Pescor, *Arch. Intern. Med.* **94**:34 (1954).
86. H. F. Fraser, A. Wikler, H. Isbell, and N. K. Johnson, *Q. J. Stud. Alcohol* **18**:541 (1957).
87. H. F. Fraser, A. Wikler, C. F. Essig, and H. Isbell, *J. Am. Med. Assoc.* **166**:126 (1958).
88. A. J. Friedhoff and A. Zitrin, *N.Y. State J. Med.* **59**:1060 (1959).
89. W. G. Friend, M. S. Vishwanath, J. J. Pitt, and J. C. Hale, *J. Am. Med. Assoc.* **217**:474 (1971).
89a. F. A. Funderburk, R. P. Allen, and A. M. I. Wagman, *J. Nerv. Ment. Dis.* **166**:195 (1978).
90. J. A. S. Gamble, J. W. Dundee, and R. C. Gray, *Br. J. Anaesth.* **48**:1087 (1976).
91. H. L. Gardner, H. Levine, and M. Bodansky, *Am. J. Obstet. Gynecol.* **40**:435 (1940).

92. H. Gastager, *Acta Psychiatr. Scand. Suppl.* **192**:185 (1966).
93. S. Gershon, *Psychol. Res. Rep. Am. Psychiatr. Assoc.* **24**:166 (1968).
94. P. K. Gessner, *J. Am. Med. Assoc.* **193**:165 (1965).
95. P. K. Gessner, *J. Am. Med. Assoc.* **219**:1072 (1972).
96. P. K. Gessner, *Pharmacologist,* Abstr. **16**:304 (1974).
97. P. K. Gessner, *Eur. J. Pharmacol.* **27**:120 (1974).
98. P. K. Gessner, *Fed. Proc.,* Abstr. **34**:719 (1975).
99. P. K. Gessner and E. H. Hu, *Pharmacologist,* Abstr. **18**:237 (1976).
100. E. Giacobini, S. Isikowitz, and A. Wegmann, *Arch. Gen. Psychiatry* **3**:289 (1960).
101. E. Giacobini and I. Salum, *Acta Psychiatr. Scand.* **37**:198 (1961).
102. A. J. Gilbert, *Ohio State Med. J.* **52**:826 (1956).
103. M. M. Glatt, *Br. Med. J.* **2**:444 (1967).
104. M. M. Glatt, *Br. Med. J.* **2**:1088 (1977).
105. M. M. Glatt, H. R. George, and E. P. Frisch, *Br. J. Med.* **2**:401 (1965).
106. M. M. Glatt, H. R. George, and E. P. Frisch, *Acta Psychiatr. Scand. Suppl.* **192**:121 (1966).
107. Y. Godin, L. Heiner, J. Mark, and P. Mandel, *J. Neurochem.* **16**:869 (1969).
108. T. M. Golbert, C. J. Sanz, H. D. Rose, and T. H. Leitschuh, *J. Am. Med. Assoc.* **201**:113 (1967).
109. A. Goldstein, *Biostatistics: An Introductory Text,* Macmillan Publishing Co., Inc., New York (1964).
110. D. B. Goldstein, *J. Pharmacol. Exp. Ther.* **180**:203 (1972).
111. D. B. Goldstein, *J. Pharmacol. Exp. Ther.* **183**:14 (1972).
112. D. B. Goldstein, *J. Pharmacol. Exp. Ther.* **186**:1 (1973).
113. D. B. Goldstein, *Life Sci.* **18**:553 (1976).
114. E. B. Gordon, *Br. Med. J.* **1**:112 (1967).
115. R. C. Grandon, W. Heffley, T. Hensel, and S. Bashore, *Am. Pract. Dig. Treat.* **7**:231 (1956).
116. K. L. Granville-Grossman and P. Turner, *Lancet* **1**:788 (1966).
117. L. W. Gray, *Ann. Intern. Med.* **82**:852 (1975).
118. L. A. Greenberg and J. E. Rosenfeld, *Psychosomatics* **10**:172 (1969).
119. P. J. Griffiths, J. M. Littleton, and A. Ortiz, *Br. J. Pharmacol.* **50**:489 (1974).
120. M. M. Gross, E. Halpert, and L. Sabot, *J. Nerv. Ment. Dis.* **145**:500 (1968).
121. M. M. Gross, S. M. Rosenblatt, S. Chartoff, A. Hermann, M. Schachter, D. Sheinkin, and M. Broman, *Q. J. Stud. Alcohol* **32**:611 (1971).
122. M. M. Gross, S. M. Rosenblatt, B. Malenowski, M. Broman, and E. Lewis, *Q. J. Stud. Alcohol* **33**:400 (1972).
123. M. M. Gross, E. Lewis, and M. Nagarajan, in: *Alcohol Intoxication and Withdrawal: Experimental Studies* (M. M. Gross, ed.), Vol. 1, pp. 365–376, Plenum Press, New York (1973).
124. M. M. Gross, E. Lewis, and J. Hastey, in: *The Biology of Alcoholism* (B. Kissin and H. Begleiter, eds.), Vol. 3: *Clinical Pathology,* pp. 191–263, Plenum Press, New York (1974).
125. M. M. Gross, E. Lewis, S. Best, N. Young, and L. Feuer, in: *Alcohol Intoxication and Withdrawal: Experimental Studies* (H. M. Gross, ed.), Vol. II, pp. 315–331, Plenum Press, New York (1975).
126. Gugl, *Z. Ther.* **1**:153 (1883).
126a. E. Hamburger, *J. Am. Med. Assoc.* **193**:143 (1965).
127. M. J. Harfst, J. G. Greene, and F. G. Lassalle, *Q. J. Stud. Alcohol* **28**:641 (1967).
128. W. T. Hart, *Am. J. Psychiatry* **118**:323 (1961).
129. J. N. Hayward and B. R. Boshell, *Am. J. Med.* **23**:965 (1957).
130. J. F. Heiser and D. DeFrancisco, *Am. J. Psychiatry* **133**:1389 (1976).
130a. R. Hemmington, P. Kramp, and O. J. Rafaelsen, *Acta Psychiatr. Scand.,* **59**:337 (1979).
131. H. I. Hershon, *Br. J. Addict.* **68**:295 (1973).
132. P. Hitchcock and E. E. Nelson, *J. Pharmacol. Exp. Ther.* **79**:286 (1943).
133. M. E. Hillbom, *Neuropharmacology* **14**:755 (1975).
134. A. K. S. Ho and C. S. Tsai, *Ann. N.Y. Acad. Sci.* **273**:371 (1976).
135. L. E. Hollister, *J. Am. Med. Assoc.* **237**:1432 (1977).
136. J. M. Horder, *Br. Med. J.* **2**:614 (1977).

137. W. A. Hunt and E. Majchrowicz, *J. Neurochem.* **23**:549 (1974).
138. H. Isbell, *Med. Clin. N. Am.* **34**:425 (1950).
139. H. Isbell, in: *Loeb's Textbook of Medicine* (P. B. Beeson and W. McDermott, eds.), 12th ed., pp. 1494–1500, W. B. Saunders Company, Philadelphia (1967).
140. H. Isbell, in: *Loeb's Textbook of Medicine* (P. B. Beeson and W. McDermott, eds.), 12th ed., pp. 1500–1502, W. B. Saunders Company, Philadelphia (1967).
141. H. Isbell, S. Altschul, C. H. Kornetsky, A. J. Eisenman, H. G. Flanary, and H. F. Fraser, *Arch. Neurol. Psychiatry* **64**:1 (1950).
142. H. Isbell, H. F. Fraser, A. Wikler, R. E. Belleville, and A. J. Eisenman, *Q. J. Stud. Alcohol* **16**:1 (1955).
143. P. M. Jeavons and J. E. Clark, *Br. Med. J.* **2**:584 (1974).
144. R. B. Johnson, *Q. J. Stud. Alcohol,* Suppl. **1**:66 (1961).
145. K. G. Jostell, S. Agurell, L. E. Hollister, and B. Wermuth, *Clin. Pharmacol. Ther.* **23**:181 (1978).
146. S. C. Kaim, C. J. Klett, and B. Rothfeld, *Am. J. Psychiatry* **125**:1640 (1969).
147. S. C. Kaim and C. J. Klett, *Q. J. Stud. Alcohol* **33**:1065 (1972).
148. H. Kalant, A. E. LaBlanc, and R. J. Gibbins, *Pharmacol. Rev.* **23**:135 (1971).
149. J. M. Kalbfleisch, R. D. Lindeman, H. E. Ginn, and W. O. Smith, *J. Clin. Invest.* **42**:1471 (1963).
150. L. B. Kalinowsky, *Arch. Neurol. Psychiatry* **48**:946 (1942).
151. M. L. Keplinger, Ph.D. thesis, Northwestern University, Chicago (1956).
152. M. L. Keplinger and J. A. Wells, *J. Pharmacol. Exp. Ther.* **119**:19 (1957).
153. N. J. Khoury, *Postgrad. Med.* **43**:119 (1968).
154. N. S. Kline, J. C. Wren, T. B. Cooper, E. Varga, and O. Canal, *Am. J. Med. Sci.* **268**:15 (1974).
155. K. Kortilla, M. J. Mattila, and M. Linnoila, *Br. J. Anaesth.* **48**:333 (1976).
156. J. Kotz, G. B. Roth, and W. A. Ryon, *J. Am. Med. Assoc.* **110**:2145 (1938).
157. C. D. Koutsky and I. W. Sletten, *Minn. Med.* **46**:354 (1963).
157a. P. Kramp and O. J. Rafaelsen, *Acta Psychiatr. Scand.* **58**:174 (1978).
157b. P. Kramp, R. Rønsted, and T. Hansen, *Acta Psychiatr. Scand.,* **59**:263 (1979).
158. V. K. Krypsin-Exner and R. Mader, *Wien. Med. Wochenschr.* **121**:811 (1971).
159. H. Krystal, *Am. J. Psychiatry* **116**:137 (1959).
160. H. Laborit, R. Coirault, R. Damasio, R. Gaviard, G. Laborit, and P. Fabrizy, *Presse Med.* **65**:1051 (1957).
161. J. W. Lance and M. Anthony, *Arch. Neurol.* **34**:14 (1977).
162. V. G. Laties, L. Lasagna, G. M. Gross, I. L. Hitchman, and J. Flores, *Q. J. Stud. Alcohol* **19**:238 (1958).
163. P. Lechat, *Acta Psychiatr. Scand. Suppl.* **192**:15 (1966).
164. J. Lereboullet, P. Benda, and M. Poisson, *Presse Med.* **68**:473 (1960).
165. J. Lereboullet, *Rev. Pract.* **12**:3293 (1962).
166. G. Lundquist, *Acta Psychiatr. Neurol. Scand.* **36**:443 (1961).
167. G. Lundquist, *Acta Psychiatr. Scand. Suppl.* **192**:113 (1966).
168. J. S. Madden, D. Jones and E. P. Frisch, *Br. J. Psychiatry* **115**:1191 (1969).
169. J. C. Maerz, *Clin. Toxicol.* **7**:277 (1974).
170. V. Magnan, *Etude Experimentale et Clinique sur l'Alcoolisme, Alcool et Absinthe; Epilepsie Absinthique,* Renou and Moulde, Paris (1871).
171. E. Majchrowicz, *Psychopharmacologia* **43**:245(1975).
172. E. Majchrowicz, W. A. Hunt, and C. Piantadosi, *Science* **194**:1181 (1976).
173. J. C. Martin, *J. Pharmacol. Clin.* **2**:21 (1975).
173a. W. R. Martin and D. R. Jasinski, *J. Psychiatr. Res.* **7**:9 (1969).
174. S. D. McGrath *Acta Psychiatr. Scand. Suppl.* **192**:165 (1966).
175. S. D. McGrath, *Br. J. Addict.* **70**:81 (1975).
176. F. H. McMechan, *The Lancet-Clinic* **116**:313 (1916).
177. R. W. McNichol, W. J. Cirksena, J. T. Payne, and M. C. Glasgow, *South. Med. J.* **60**:7 (1967).

178. J. D. Melchior, F. Buchtal, and M. Lennox-Buchthal, *Epilepsia* **12**:55 (1971).
179. N. K. Mello and J. H. Mendelson, in: *Recent Advances in Studies on Alcoholism* (N. K. Mello and J. H. Mendelson, eds.), pp. 647–686,U.S. Government Printing Office, Washington, D.C. (1971).
180. K. L. Melmon, *Clin. Pharmacol. Ther.* **20**:125 (1976).
181. J. H. Mendelson and J. LaDou, *Q. J. Stud. Alcohol,* Suppl. **2**:1 (1964).
182. J. H. Mendelson, B. Barnes, C. Mayman, and M. Victor, *Metabolism* **14**:88 (1965).
183. J. H. Mendelson and N. K. Mello, *Annu. Rev. Med.* **27**:321 (1976).
184. J. Merry, C. M. Reynolds, J. Bailey, and A. Coppen, *Lancet* **2**:481 (1976).
185. R. R. Miller, H. G. Olson, E. A. Amsterdam, and D. T. Mason, *N. Engl. J. Med.* **293**:416 (1975).
186. E. H. Mitchell, *Am. J. Med. Sci.* **229**:363 (1955).
186a. H. Möhler and T. Okada, *Science* **198**:849 (1977).
187. F. A. Montgomery and R. Solowan, *Hosp. Commun. Psychiatry* **24**:808 (1973).
188. R. Moroz and E. Rechter, *Psychiatr. Q.* **38**:619 (1964).
189. J. A. Motto, *Q. J. Stud. Alcohol* **29**:917 (1966).
190. D. J. Muller, *South. Med. J.* **62**:495 (1969).
191. M. Nickerson and B. Collier, in: *The Pharmacological Basis of Therapeutics* (L. S. Goodman, A. Gilman, A. G. Gilman, and G. B. Koelle, eds.), pp. 533–564, Macmillan Publishing Co., Inc., New York (1975).
192. J. Nielsen, *Acta. Psychiatr. Scand. Suppl.* **187**:5 (1965).
193. J. Nielsen and E. Strömgren, *Bibl. Psychiatr. Neurol.* **1969**:165 (1969).
194. I. Nitzescu and J. Iacobovici, *Presse Med.* **42**:331 (1934).
195. E. P. Noble, R. Gillies, R. Vigran, and P. Mandel, *Psychopharmacologia* **46**:127 (1976).
196. M. Ogata, J. H. Mendelson, N. K. Mello, and E. Majchrowicz, *Psychosomatics* **33**:159 (1971).
197. M. Okamoto, H. C. Rosenberg, and N. R. Boisse, *J. Pharmacol. Exp. Ther.* **202**:479 (1977).
198. M. L. Palestine, *Q. J. Stud. Alcohol* **34**:185 (1973).
199. M. L. Palestine and E. Alatorre, *Curr. Ther. Res.* **20**:289 (1976).
200. P. Pascal and Dupuy, *Bull. Soc. Chim. Fr.* **4** (27):353 (1920).
201. V. D. Patch, *N. Engl. J. Med.* **290**:807 (1974).
202. G. L. S. Pawan, *Nature (Lond.)* **220**:374 (1968).
203. A. Pena-Ramos, *Dis. Nerv. Sys.* **38**:144 (1977).
204. M. Philipp, N. Seyfeddinipur, and A. Marneros, *Nervenarzt* **47**:192 (1976).
205. P. Piker, *Am. J. Psychiatry* **93**:1387 (1937).
206. P. Piker and J. V. Cohn, *J. Am. Med. Assoc.* **108**:345 (1937).
207. L. A. Pohorecky, *J. Pharmacol. Exp. Ther.* **189**:380 (1974).
208. S. H. Preskorn and L. J. Denner, *J. Am. Med. Assoc.* **237**:36 (1977).
209. J. T. Quinn, *Acta. Psychiatr. Scand. Suppl.* **192**:161 (1966).
210. M. S. Radomski, W. J. Watson, and L. J. McBurney, *Int. Hyperbaric Congr. Proc., 5th, 1973,* **1**:142 (1974).
210a. D. Raft, R. Gomez, and J. A. Ewing, *J. Nerv. Ment. Dis.* **159**:366 (1974).
211. R. T. Rappolt, *Clin. Toxicol.* **6**:293 (1973).
212. T. M. Reilly, *Br. J. Psychiatry* **128**:375 (1976).
213. J. L. Rementeria and K. Bhatt, *Pediatrics* **90**:123 (1977).
213a. T. A. Rejent and K. C. Whal, *Clin. Chem.* **22**:889 (1976).
214. A. Rifkin, F. Quitkin, and D. F. Klein, *J. Am. Med. Assoc.* **236**:2172 (1976).
215. R. M. Ritter and D. E. Davidson, *South. Med. J.* **64**:249 (1971).
216. J. E. Rosenfeld and D. H. Bizzoco, *Q. J. Stud. Alcohol,* Suppl. **1**:77 (1961).
217. E. Rothstein, *Am. J. Psychiatry* **130**:1381 (1973).
218. M. C. Ruddy, *Ann. Intern. Med.* **83**:279 (1975).
219. A. A. Sainz, *Psychiatr. Q.* **31**:275 (1957).
220. I. Salum, II, *Acta Psychiatr. Scand. Suppl.* **235**:17 (1972).
221. R. Sampliner and F. L. Iber, *J. Am. Med. Assoc.* **230**:1430 (1974).
222. J. D. Sapira, *South. Med. J.* **66**:1417 (1973).

222a. J. D. Sapira and C. E. Cherubin, *Drug Abuse—A Guide for the Clinician*, pp. 337–356, American Elsevier, New York (1975).

222b. J. D. Sapira and C. E. Cherubin, *Drug Abuse—A Guide for the Clinician*, p. 435, American Elsevier, New York (1975).

223. P. D. Saville and C. S. Lieber, *J. Nutr.* **87**:477 (1965).

224. W. Schlichter, M. E. Bristow, S. Schultz, and A. L. Henderson, *Can. Med. Assoc. J.* **74**:364 (1956).

225. J. D. Schultz, E. L. Rea, J. F. Fazekas, and J. C. Shea, *Q. J. Stud. Alcohol* **16**:245 (1955).

226. L. Schwartz and S. P. Fjeld, *Behav. Neuropsychiatry* **1**:7 (1969).

227. E. M. Sellers, S. D. Cooper, D. H. Zilm, M. Phil, and C. Shanks, *Clin. Pharmacol. Ther.* **20**:199 (1976).

228. E. M. Sellers and H. Kalant, *N. Engl. J. Med.* **294**:757 (1976).

229. E. M. Sellers, D. H. Zilm, and N. C. Degani, *J. Stud. Alcohol* **38**:2096 (1977).

230. J. W. Selig, *J. Am. Med. Assoc.* **198**:951 (1966).

231. M. S. Sellig, A. R. Berger, and N. Spielholz, *Dis. Nerv. Sys.* **36**:461 (1975).

232. G. Sereny and H. Kalant, *Br. Med. J.* **1**:92 (1965).

233. D. G. Shand, *N. Engl. J. Med.* **93**:449 (1975).

234. S. Shaw, B. Stimmel, and C. S. Lieber, *Science* **194**:1057 (1976).

235. O. Z. Shulsinger, P. J. Forni, and B. B. Clyman, in: *Currents in Alcoholism* (F. A. Seixas, ed.), Vol. I, pp. 319–327, Grune & Stratton, Inc., New York (1977).

236. W. D. Silkworth, *Med. Rec.* **145**:321 (1937).

237. S. Simler, L. Ciesielski, M. Maitre, H. Randrianariso, and P. Mandel, *Biochem. Pharmacol.* **22**:1701 (1973).

238. D. Simon and J. K. Penry, *Epilepsia* **16**:549 (1975).

239. R. K. Simpson, E. Fitz, B. Scott, and L. Walker, *J. Am. Osteopath. Assoc.* **68**:123 (1968).

240. S. H. Sinal and J. E. Crowe, *Pediatrics* **57**:158 (1976).

241. C. M. Smith, in: *Handbuch der Experimentellen Pharmakologie* (new series), Vol 45/1 (W. R. Martin, ed.), p. 524, Springer-Verlag, Berlin (1977).

242. D. E. Smith and D. R. Wesson, *J. Am. Med. Assoc.* **213**:294 (1970).

243. J. A. Smith, *J. Am. Med. Assoc.* **152**:384 (1953).

244. R. F. Smith, *Ann. N.Y. Acad. Sci.* **273**:378 (1976).

245. B. F. Sørensen, *Dan. Med. Bull.* **6**:261 (1959).

245a. R. C. Speth, G. J. Wastek, P. C. Johnson, and H. I. Yamamura, *Life Sci.* **22**:859 (1978).

246. G. L. Sprague and A. L. Craigmill, *Res. Commun. Chem. Pathol. Pharmacol.* **15**:721 (1976).

246a. R. F. Squires and C. Braestrup, *Nature* **266**:732 (1977).

247. G. Stendig-Lindberg, *Acta Psychiatr. Scand.* **50**:465 (1974).

248. T. Sutton, *Tracts on Delirium Tremens, on Peritonitis, and on Some Other Internal Inflammatory Affections*, pp. 1–77, Thomas Underwood, London (1813).

249. C. O. Svedin, *Acta Psychiatr. Scand. Suppl.* **42**:27 (1966).

250. D. W. Swanson, R. L. Weddige, and R. M. Morse, *Mayo Clin. Proc.* **48**:359 (1973).

251. E. A. Swinyard, W. C. Brown, and L. S. Goodman, *J. Pharmacol. Exp. Ther.* **106**:319 (1952).

252. M. A. Sydney, *Br. Med. J.* **4**:467 (1974).

253. M. E. Tavel, W. Davidson, and T. D. Batterton, *Am. J. Med. Sci.* **242**:18 (1961).

254. H. Teräväinen and H. Larsen, *J. Neurol. Neurosurg. Psychiatry* **39**:607 (1976).

255. H. I. D. Thieden, N. Grunnet, S. E. Damgaard, and L. Sestoft, *Eur. J. Biochem.* **30**:250 (1972).

256. D. W. Thomas and D. X. Freedman, *J. Am. Med. Assoc.* **188**:316 (1964).

257. D. W. Thomas and D. X. Freedman, *J. Am. Med. Assoc.* **193**:166 (1965).

258. W. L. Thompson, A. D. Johnson, W. L. Maddrey, and Osler Medical Housestaff, *Ann. Intern. Med.* **82**:175 (1975).

259. W. L. Thompson and W. C. Maddrey, *Ann. Intern. Med.* **82**:852 (1975).

260. W. L. Thompson and W. C. Maddrey, *Ann. Intern. Med.* **83**:279 (1975).

261. W. E. Thornton, *Dis. Nerv. Syst.* **38**:1024 (1977).

262. J. S. Toal, *Q. J. Pharm. Pharmacol.* **10**:439 (1937).

263. J. E. P. Toman, in: *The Pharmacological Basis of Therapeutics* (L. S. Goodman and A. Gilman, eds.), 4th ed., p. 212, Macmillan Publishing Co., Inc., New York (1970).

263a. J. E. P. Toman, S. Loewe, and L. S. Goodman, *Arch. Neurol. Psychiatry* **58**:312 (1947).

264. P. Turner, *Proc. R. Soc. Med.* **69**:375 (1976).

265. N. Tygstrup, K. Winkler, and F. Lundqvist, *J. Clin. Invest.* **44**:817 (1965).

266. S. Vaisrub, *Arch. Intern. Med.* **130**:297 (1972).

267. M. Victor, *Psychosomat. Med.* **28**:636 (1966).

268. M. Victor, in: *Psychopharmacology: A Review of Progress, 1957–1967* (D. H. Efron, ed.), U.S. Govt. Printing Office, Washington, D.C. (1968).

269. M. Victor, *Ann. N.Y. Acad. Sci.* **215**:235 (1973).

270. M. Victor and R. D. Adams, *Res. Publ. Assoc. Nerv. Ment. Dis.* **32**:526 (1953).

271. M. Victor and C. Brausch, *Epilepsia* **8**:1 (1967).

272. E. Völzke and H. Doose, *Epilepsia* **14**:185 (1973).

273. W. E. C. Wacker and B. L. Vallee, *N. Engl. J. Med.* **259**:475 (1958).

274. W. E. C. Wacker and A. F. Parisi, *N. Engl. J. Med.* **278**:658, 712, 772, 776 (1968).

274a. J. Wadstein, H. Öhlin, and P. Stenberg, *Drug Alcohol Depend.* **3**:281 (1978).

275. H. Wallgren and H. Barry, III, *Actions of Alcohol*, Vol. 2, p. 665, Elsevier Publishing Company, Amsterdam (1970).

276. P. J. F. Walsh, *Am. J. Psychiatry* **119**:262 (1962).

277. I. S. Wechsler, *J. Am. Med. Assoc.* **114**:2198 (1940).

278. C. C. Welch, *Postgrad. Med.* **61**(5):138 (1977).

279. D. Wheatley, *Br. J. Psychiatry* **115**:1411 (1969).

280. C. L. Whitfield, G. Thompson, A. Lamb, V. Spencer, M. Pfeifer, and M. Browning-Ferrando, *J. Am. Med. Assoc.* **239**:1409 (1978).

281. J. Wiersum, *J. Am. Med. Assoc.* **227**:79 (1974).

282. A. Wikler, *Am. J. Psychiatry* **125**:758 (1968).

283. J. C. Wilson, in: *A System of Practical Medicine by American Authors* (W. Pepper, ed.), Vol. 5: *Diseases of the Nervous System*, pp. 573–646, Lea Brothers, Philadelphia (1886).

284. L. M. Wilson, *Arch. Intern. Med.* **130**:225 (1972).

285. G. M. Wimbish, R. Martz, and R. B. Forney, *Life Sci.* **20**:65 (1977).

286. S. M. Wolfe and M. Victor, *Ann. N.Y. Acad. Sci.* **162**:973 (1969).

287. S. M. Wolfe and M. Victor, in: *Recent Advances in Studies of Alcoholism* (M. K. Mello and J. H. Mendelson, eds.), U.S. Government Printing Office, Washington, D. C. (1971).

288. R. R. Young, *N. Engl. J. Med.* **294**:785 (1976).

289. J. F. Zaroslinki, R. K. Browne, and L. H. Possley, *Toxicol. Appl. Pharmacol.* **19**:573 (1971).

290. D. H. Zilm, E. M. Sellers, and S. M. MacLeod, *N. Engl. J. Med.* **294**:785 (1976).

291. E. Zola, *L'Assommoir*, Chap. 13, Paris (1877).

292. International Nonproprietary Names (INN) for Pharmaceutical Substances Cumulative List #4, World Health Organization, Geneva (1976).

293. International Nonproprietary Names (INN) for Pharmaceutical Substances Cumulative List #5, World Health Organization, Geneva (1977).

294. *Index Nominum 1975/76* (Pharmaceutical Society of Switzerland, ed.), Zurich (1975).

295. M. Griffiths, J. J. Dickerman, and L. C. Miller (eds.), *USAN and the USP Dictionary of Drug Names*, United States Pharmacopeial Convention, Inc., Rockville, Md. (1976).

296. M. Windholz, S. Budavari, L. Y. Stroumtsos, and M. E. Fertig (eds.), *The Merck Index*, 9th ed., Merck & Company, Inc., Rahway, N.J. (1976).

Physiology, Behavior, and Animal Models of Alcohol Dependence

XIII

Ethanol-Induced Changes in Body Temperature and Their Neurochemical Consequences

22

Gerhard Freund

A. INTRODUCTION

Many drugs that affect the central nervous system, including ethanol, anesthetics, opioids, cannabis, monoamine oxidase inhibitors, and phenothiazines, also impair the temperature-regulating centers. If the environmental temperature is low, the body temperature may be lowered, and high ambient temperatures may induce hyperthermia. When rodents are treated with drugs that impair temperature regulation, the standard laboratory temperature of 25°C usually causes their body temperature of 37°C to be lowered by several degrees. The great practical importance of this phenomenon is that various metabolic, physiological, or behavioral changes attributed to the drug are in reality a result of body temperature changes and are only indirectly related to the drug. For example, the slowing of central nervous system (CNS) protein synthesis after the administration of chlorpromazine can be completely prevented by maintaining the body at a constant temperature.[1] Raising the environmental temperature to keep the animal's body temperature constant when drugs cause hypothermia is a simple method to control and separate the effects of the drug itself from those effects caused by an altered body temperature.

There are three phases of loss of temperature control related to ethanol and other drugs: (1) acute drug administration causes loss of temperature control; (2) continued administration results in adaptation, restoration of temperature control, or "tolerance"; and (3) abrupt cessation of drug administration again causes temporary loss of temperature control.

Gerhard Freund • Medical Service, Veterans Administration Medical Center, and Department of Medicine, College of Medicine, University of Florida, Gainesville, Florida 32602.

B. PHYSIOLOGICAL PRINCIPLES OF TEMPERATURE REGULATION

In homoiothermic animals the core body temperature remains constant within narrow limits in the presence of widely fluctuating environmental temperatures. In humans, the evening body temperatures are normally from 0.5 to 1.2°C higher than are temperatures in the morning. Food consumption, physical activity, stage of menstrual cycle, and other physiological variables may cause minor (1°C), slow changes in core body temperature. This precise control is possible because a central "thermostat" integrates various temperature measurements (sensory inputs) to regulate the rates of various chemical and physiological mechanisms of heat production and dissipation. According to engineering principles of feedback loops,[2] the thermostat may be conceptualized as a comparator that compares an integrated input with an intrinsic reference standard input (Fig. 1). At "set point" conditions the input is such that the output is zero. Regulatory output responses are triggered when the body temperature exceeds a certain range. Drugs may completely abolish thermostat function or raise or lower this set point, but they may also broaden it. When the temperature input range is broad before a response occurs (e.g., in decerebrate animals), it is termed a "broad set point" or "wider neutral zone." This implies a less efficient response to changes in environmental temperatures.

The various reflex outputs, such as vasoconstriction and dilation, muscle shivering, sweating, and piloerection, appear to be controlled by temperature-sensitive neurons in the preoptic area of the anterior hypothalamus. Lesions in this area abolish these regulatory outputs. Local temperature changes evoke responses as if the entire body temperature had been changed. The result of a reflex response (e.g., muscle tremor) is a compensatory change in the thermal

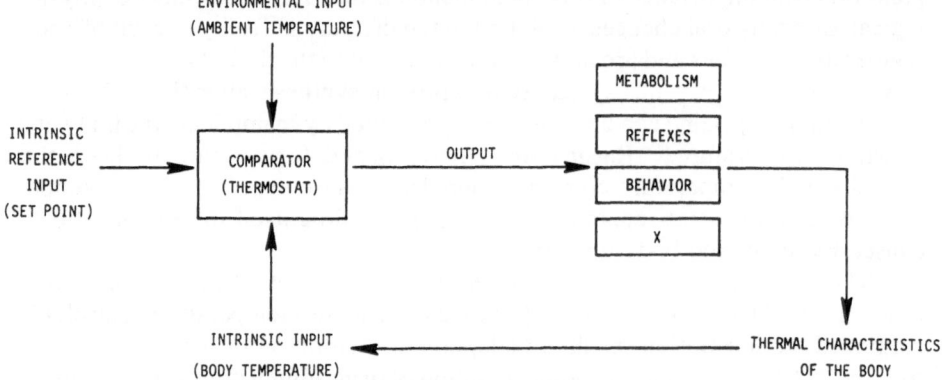

Fig. 1. Basic elements of temperature regulation by feedback loop control. Output signals from the hypothalamic thermostat control various physiological and behavioral mechanisms for heat generation, preservation, and dissipation. These appropriate responses result in maintenance of constant body temperature, while environmental temperature fluctuates. The administration of drugs may result in loss of temperature control by interfering with cellular functions at any one or several sites of the body temperature control system. The resultant biological effects of altered body temperature may be erroneously attributed to the drugs.

characteristics of the body (e.g., increased heat production). The report of this change to the thermostat completes the feedback loop. Animals have another set of responses besides reflex outputs, namely behavior.[3] Animals can move to environments with temperatures more compatible with their needs, or they can be trained to push levers to turn heating lamps or cooling fans off and on. Such thermally motivated operant behavior appears to be controlled primarily in the posterior lateral hypothalamus, where lesions may abolish behavioral responses and leave reflex responses intact. Both reflexive and operant responses can compensate for each other when one type is impaired. Both types are also affected independently by various drugs.

In summary, it is possible that a drug may impair temperature regulation at any one point of the aforementioned series of events: thermostat (raised, lowered, broadened, or abolished set point), various effector reflexes with their responses (including changes in metabolism), and behavior (free or operant).

C. EFFECTS OF DRUGS ON BODY TEMPERATURE REGULATION

The mechanisms whereby ethanol may affect temperature control must be seen in the broader context of effects of drugs in general.[4,5] Drugs may interfere with peripheral effector mechanisms (e.g., quinine-induced hypothermia is a result of prevention of muscle tremors).

To attribute the action of a drug to effects on the set point of the thermostat, two requirements must be met[6]: (1) The drug must change body temperature in the same direction under various environmental conditions. If a drug shifts the set point downward, the body temperature should fall within the limits of the effector capabilities, regardless of the ambient temperature. (2) All physiological and behavioral responses should be coordinated to create the change in temperature. If a drug shifts the set point upward, the animal should shiver, increase metabolic rate, vasoconstrict, and perform behavioral acts to obtain heat. For example, endogenous pyrogens released from leukocytes increase the body temperature by shifting the set point upward. This causes increased heat production and decreased heat loss until the new set-point temperature is reached. In spite of a rising body temperature, animals will work and depress bars to obtain more heat, the more so the colder the ambient temperature is. Physiological and behavioral responses are complementary, indicating a raised set point. In contrast, chlorpromazine induces hypothermia, which has been attributed to general CNS depression. If chlorpromazine shifted the set point downward, one would expect a decrease in operant work for heat in a warm environment and not in an already cold environment. But, in fact, chlorpromazine inappropriately reduces operant work to obtain heat in the cold environment as well, which is indicative of general CNS depression.[6]

Virtually all general CNS depressants, including general anesthetics, barbiturates, and reserpine, impair temperature control by suppressing neuronal activity in the thermoregulatory cells.[7] The usual laboratory temperatures of 23–25°C, cause hypothermia in animals that normally maintain their body temperature at 37°C. The hypothermic effect of anesthetics is enhanced by monoamine oxidase inhibitors.[8]

Morphine has long been known in rodents to cause hyperthermia in small doses[9] and hypothermia in larger doses. The hypothermic effect may be abolished by the depletion of brain serotonin and enhanced by catecholamine depletion.[10] However, pharmacological manipulation of brain monoamine transmitters are difficult to interpret, because some monoamine oxidase inhibitors cause sedation and hypothermia, whereas others do not,[11] even though they all presumably have the same effect on transmitters.

Cannabinols cause hypothermia in men, monkeys, and mice, the extent varying proportionately with ambient temperatures.[12] Interestingly, tolerance to the hypothermic effect of marihuana and chlorpromazine begin to develop after one dose and are evident by the second day of drug administration.[13] The significance of these observations lies in the fact that both of these drugs do not induce physical dependence or behavioral tolerance and yet they induce tolerance to the hypothermic effect of the drugs.

Before the results of studies in animals can be extrapolated to humans, it is necessary to recognize that humans, unlike rodents, develop hypothermia in response to CNS depressants only under unusual circumstances. Therefore, animal studies with CNS depressants that prevent hypothermia by increasing the ambient temperature are more relevant to human physiology than are uncontrolled studies that permit hypothermia to develop. Humans have a greater capacity to maintain normal body temperature than do rodents. Humans develop hypothermia only if the CNS depression is severe (coma, deep anesthesia) and if heat dissipation is extensive (immersion into very cold water and very low ambient temperatures). Hypoglycemia,[14] hypothyroidism,[15] Wernicke's encephalopathy,[16] and severe general diseases may occasionally cause hypothermia.[17] These instances probably represent temporary damage to the thermostat, with loss of control rather than specific changes in set point.

The neurons in the regions involved in temperature regulation appear to be affected by the intraventricular or intrahypothalamic injection of all the three major transmitters: norepinephrine, serotonin, and acetylcholine. Unfortunately, the results obtained by many investigators are conflicting. There seem to be real species differences.[3,5] No universally accepted coherent concept has emerged as to how these transmitter systems are involved in temperature regulation.

D. EFFECTS OF ETHANOL ON BODY TEMPERATURE REGULATION

1. Studies in Humans

The first report of ethanol-induced hypothermia in man was published over 100 years ago, when it was observed in people who fell or jumped into the cold harbor water of Hamburg.[18] Weyman and co-workers[19] reviewed the few clinical reports since then and added observations on 39 skid row patients admitted to the hospital. Body temperatures as low as 22.5°C were observed. It was concluded that the underlying disease, rather than the degree of hypothermia or rate of rewarming, determined the prognosis. Mortality was 6.25%, and the most common cause of death was cardiac arrhythmia. Acute hypothermia in normal healthy adults was induced by hypoglycemia, which in turn was caused by acute alcohol ingestion.[20] No increase in metabolic rate and no shivering occurred during the cold exposure. These results are difficult to interpret because ethanol effect is compounded by hypoglycemia.

The effects of ethanol intoxication and withdrawal on body temperature of patients admitted to a hospital are difficult to evaluate because it is usually a surgical or medical disease that leads to the discontinuation of ethanol consumption and hospitalization. Febrile illnesses are common and usually attributable to infections, such as pneumonia, and to dehydration.[21] Sometimes these infections are acquired during the hospitalization. Hypothermia may be attributable to associated diseases, such as Wernicke's encephalopathy, in which it is a rare phenomenon.[16] The treatment of ethanol withdrawal syndromes and associated diseases with a variety of drugs, which also may alter temperature control, further compounds the problem. For these reasons it is necessary to evaluate the effects of alcohol intoxication on body temperature of humans under controlled laboratory conditions.

One study[22] with 10 male alcoholic volunteers who received ethanol, 3.2 g/kg body wt per day, for 5 or 7 days reported a decrease of 22:00-hr rectal temperatures during intoxication and an increase during withdrawal. In view of the small magnitude (0.25°C) and great variability, these changes are of doubtful statistical and biological significance. The environmental temperature, clothing, and other conditions were not described.

In the classical study of Isbell and colleagues[23] of healthy former morphine addicts, there was no significant effect of ethanol intoxication on body temperature. There was no change during sudden withdrawal from alcohol, if the withdrawal signs were mild and consisted only of mild tremors, nausea, and weakness. In subjects who developed a more severe withdrawal reaction, including seizures, disorientation, and hallucinations, fevers of up to 41.4°C rectally were recorded. This was associated with severe tremors, profuse perspiration, and tachycardia. It is therefore possible that the hyperthermia was a result of intense muscle tremors and hypermetabolism exceeding the peripheral heat dissipation capacity. A central change of thermostat setting need not be postulated under these conditions. If a raised set point of the thermostat were involved, one would expect a dry, cold skin rather than perspiration.

2. Studies in Animals

a. Physiological Effects

Over 100 years ago, Walther[24] observed hypothermia induced by ethanol in experimental animals. He reported that one rabbit received 35 ml of brandy by stomach tube. During the rabbit's exposure to an ambient temperature of 12.5°C, the body temperature gradually decreased from 38.8°C to 25°C in 2 hr. After 2½ hr, the rabbit died when the body temperature had reached 20°C. The temperature of a control rabbit under identical conditions varied by only 2°C. Walther also described the hypothermic effects of morphine.

Since that time the hypothermic effect of ethanol had been virtually ignored in the experimental literature and in the design and interpretation of experiments. This is surprising, considering the fact that ethanol is one of the general CNS depressants that is expected to share many pharmacological properties with them. In 1973[25] a dose-dependent hypothermia of up to 4.5°C after ethanol administration (from 2 to 6 g/kg) was described in mice maintained at an ambient temperature of 25°C. The hypothermia was independent of the route of administration (gavage or intraperitoneal). This hypothermia could be prevented by adjusting ambient temperatures. Ambient temperatures of 37°C induced hyperthermia of up to 5.5°C. Again, the magnitude of effect depended on the amount of ethanol administered. The fact that the body temperature could be abnormally lowered and raised by alterations in environmental temperature strongly suggested that ethanol did not simply change the set point but that it nonspecifically depressed the neurons of the thermostat system in a manner similar to that caused by general anesthetics. Mice fed liquid diets containing ethanol for 10 days were found to have no hypothermia, and "tolerance" could be inferred from this. Benzyl alcohol, in concentrations used as a preservative in commercial sterile saline solutions, also caused hypothermia. During acute ethanol intoxication of the mice, lowered body temperature decreased the rate of blood–brain transport of α-aminoisobutyric acid, a nonmetabolizable amino acid. This rate was restored to normal when hypothermia was prevented. It was pointed out that many biochemical changes that were previously attributed to ethanol *per se* could have been caused by hypothermia instead. Differences in experimental results in different laboratories may be a result of different degrees of hypothermia being caused by different environmental laboratory temperatures.

Pohorecky and associates[26] confirmed the hypothermic effects of a single dose of ethanol (4 g/kg intraperitoneally) in adult rats (Table I). They also found a 10–15% increase of brain serotonin concentration associated with acute intoxication, but did not determine whether the elevation of serotonin concentrations could be prevented by avoiding hypothermia with higher ambient temperatures. Subsequent studies by this group demonstrated that pharmacological blockade of prostaglandin, norepinephrine, and dopamine systems did not alter ethanol-induced hypothermia. In contrast, 5-hydroxytryptamine blockade enhanced ethanol-induced hypothermia. They hypothesized that "reduced stimulation of receptors for 5-HT" is the mechanism whereby ethanol induces hypothermia.[27]

The induction of hypothermia in mice by a single dose of ethanol (3 g/kg, intraperitoneally), its prevention by modestly raised ambient temperatures, and hyperthermia by further increases in ambient temperatures were confirmed by Tabakoff and colleagues.[28] Confirmed also was the development of tolerance to ethanol-induced hypothermia after several days of feeding ethanol in liquid diet, in that hypothermia was no longer demonstrable in the continued presence of ethanol.[29,30] Tolerance that developed to the hypothermic effect of ethanol was

Table I. Ethanol-Induced Hypothermia

Ethanol				Maximal blood ethanol (mg/dl)	Species	Ambient temp. (°C)	Maximal hypothermia (°C)	Comments	References
Form	Dose	Route	Duration						
10% (v/v) in 0.9% NaCl	1.9 g/kg 3.8 g/kg 5.7 g/kg	Oral, ip	Single dose Single dose Single dose		Mouse	25 25 25	1.3 3.1 4.0	Ataxia Light anesthesia, responding to stimuli, deep anesthesia	25
	3.8 g/kg		Single dose			35	0	Prevention of hypothermia, tolerance	
7% in liquid diet	10 ml/day	Oral	10 days	160–280		25	0		29
2% in 0.9% NaCl	2.0 g/kg 3.0 g/kg 4.0 g/kg	ip	Single dose Single dose Single dose		Mouse	22 22 22	2.3 2.6 4.4		
7.5% in liquid diet	9.5–11 ml/ day	Oral	7 days	350		22–24	1.0	Day 5, maximum	
Sucrose	Ad libitum		7 days	350		22–24	0.4	Day 7, tolerance	
			—	0		22–24	4.5	6–8 hr postwithdrawal	
Sucrose	Ad libitum		—	0		22–24	0	25 hr postwithdrawal	
20% in 0.9% NaCl	3.0 g/kg	ip	Single dose		Rat	21	1.3		26, 27
7% in liquid diet		Oral	16 days			21	1.1		
Sucrose	Ad libitum	Oral	—	0		21	0	24 hr postwithdrawal	

demonstrated to extend also to pentobarbital administration; that is, cross tolerance developed.

In contrast to the findings of Pohorecky and co-workers, in acute intoxication no changes in 5-HT concentrations were found, but this may have been
because of differences in animal species or conditions. It was observed, however, that turnover rates of 5-hydroxyindoleacetic acid were increased and of
serotonin decreased. The former was not prevented by maintaining body temperatures at 37°C.[30] In rats a single intragastric dose of ethanol, 5 g/kg, and
continued ethanol gavage for several days resulted in an increase of over 50% of
dopamine metabolites. When the rats were withdrawn from ethanol, these values
reverted to normal. Noradrenaline metabolite concentrations were similarly elevated but, in contrast, remained high during ethanol withdrawal.[31] In addition, it
has been reported that after the administration for up to 7 days of liquid diets
containing ethanol, hypothermia occurred 6–8 hr after ethanol was withdrawn
from the diet.[29,30] The lowering of body temperature at an ambient temperature
of 22°C was positively correlated with the degree of observable manifestations of
withdrawal. Decreases of as much as 6.1°C were noted in the most severely
affected animals. Others have found only a 1°C decline of body temperature of
rats during intoxication and no change during withdrawal at unspecified ambient
temperatures.[32] This finding may simply be a result of the method of ethanol
administration, which was intermittent gavage for 7–10 days, and the timing of
temperature determinations, which were not specified in relation to stage of
intoxication and were determined only once 3 hr after ethanol was withdrawn.

b. Behavioral Effects

The behavioral aspects of ethanol-induced hypothermia were studied by
Brick and Pohorecky.[33] They reported that rats during ethanol withdrawal
preferred a warm (30°C) alley of a T-maze to the normally preferred neutral
(19°C) or the usually avoided cold (2°C) compartment. This observation is in
accord with the concept that hypothermia during withdrawal is not a downward
shift in set point but a generally inhibited thermostat.[3] What is not easily
explainable is that the selection of the warm compartment persisted for 6 days
after ethanol consumption ceased and when hypothermia had long subsided.
Possibly this is a result of "learned preference"; that is, the animals associated
the normally avoided warm compartment with the positive reinforcement of a
normal body temperature.

c. Biochemical Correlations

Because neurons are constantly communicating with each other by the
release of transmitters into synaptic clefts, it would be expected that many
physiological and pathological states alter transmitter release and metabolism in

various pools. Such changes may be causes, effects, or irrelevant correlates of either hypothermia or ethanol depression. Although changes in body temperature have been correlated with changes in transmitter concentration, or turnover, it is often not clear whether the transmitter changes cause hypothermia or hypothermia causes the transmitter changes. The question naturally arises of whether ethanol-induced alterations in body temperature are mediated by ethanol affecting specific neurotransmitter systems. Alternatively, changes in transmitters correlated with ethanol-induced hypothermia could be a result of either hypothermia or exposure to ethanol as such. Injections of transmitters into the ventricle adjacent to the hypothalamus, injections into specific areas within the hypothalamus, and pharmacologically induced depletion or accumulation of transmitters indicate that norepinephrine, 5-hydroxytryptamine, and acetylcholine are all somehow involved in temperature regulation.[3,5] Specifically how they are involved in a particular animal species is controversial because of a large body of contradictory data.

Pohorecky and associates[27] and Tabakoff and co-workers[28] have suggested that ethanol-induced hypothermia may be mediated by serotonergic mechanisms. This hypothesis is based essentially on the demonstrated increase of serotonin concentrations and turnover in acute ethanol intoxication. Pohorecky and colleagues[27] base this conclusion further on the potentiation of ethanol-induced hypothermia by serotonin, blockade with p-chloroamphetamine, and a lack of effect by norepinephrine and prostaglandin blockade under the conditions of this experiment.

Hoffman and Tabakoff[34] suggested that "dopaminergic systems may be involved in expression of tolerance to the hypothermia produced by ethanol." This suggestion is based on the following experimental evidence in mice: An acute dose of the dopamine agonist piribedil (like ethanol or the monoamine oxidase-inhibitor pargyline) induced hypothermia. During a 7-day period of drinking ethanol in liquid diet, mice developed tolerance to the hypothermic effects of single doses of ethanol and piribedil. The tolerance in the form of absence or lessening of hypothermic effects is demonstrable beginning 24 hr after withdrawal and lasts for up to 5 days. This cross tolerance and a parallel decrease in activities of dopamine-sensitive adenylate cyclase lead to the suggestion that the tolerance to ethanol-induced hypothermia may be mediated by a decrease in dopamine receptor densities in the brain. Such decreased density in dopamine receptors is thought to be a compensatory reaction to the ethanol-induced increase in turnover and release of dopamine.

Tabakoff and Ritzmann[35] suggested that the noradrenergic system may be necessary for the "consolidation" but not the expression of tolerance. When the noradrenergic system in mice was partially destroyed by the intraventricular injection of 6-hydroxydopamine *before* they consumed the liquid diet containing ethanol, no tolerance developed to the hypothermic and behavioral effects of single doses of ethanol or barbiturates. The injection of 6-hydroxydopamine, *after* tolerance had developed, did not interfere with the expression of tolerance. Interestingly, the 6-hydroxydopamine treatment did not interfere with manifestations of physical dependence. Frye and Ellis[36] also found no effect of long-term depletion of norepinephrine or dopamine, or both, on the development of physical dependence on ethanol. This could imply that tolerance to and physical dependence on ethanol are separate phenomena resulting from different pathogenetic mechanisms.

The afferent, integrating, and efferent circuits of the thermoregulatory centers probably involve several transmitter systems that interact in very complex ways.[5] Therefore, many pharmacological and morphological manipulations may interfere in this delicate system at one or several sites. The result may be complete loss of control of body temperature or various changes in set point width or direction. Quite possibly ethanol causes loss of temperature control primarily by its generalized depressant effects on the CNS. The primary effect of ethanol, like other general anesthetics, may be at the membrane level of biological organization, with resulting secondary changes in transmitter systems. Possibly the sensitivity of membranes to the expanding or fluidizing effects of ethanol differs between cells belonging to different transmitter systems. Finally, some changes in transmitter turnover may not be the cause of hypothermia but may be a result of hypothermia or other effects of ethanol on the CNS.

As discussed in Section C, the development of tolerance to drug-induced hypothermia is not restricted to drugs evoking physical dependence or behavioral tolerance.[13] Chlorpromazine, for instance, does not cause significant behavioral tolerance to increasing dosages, yet continued administration induces tolerance to the induction of hypothermia. This suggests that the development of behavioral and temperature tolerance may be related phenomena, sometimes parallel but not identical. The dopamine agonist piribedil[34] and the dopamine-receptor antagonist chlorpromazine[1] both cause acute hypothermia and induce reversible tolerance to hypothermia upon their continued administration.[13] Apparently, adaptation occurs when drug administration continues, whether the drug is an agonist or an antagonist. This suggests that the cellular adaptation may involve changes in receptor numbers or affinity. At the cellular level there may be many different mechanisms of adaptation to the prolonged presence of a drug. In the case of hypothalamic temperature-regulating neurons, the end result of any kind of adaptation means resumption of normal regulatory function irrespective of what kind of cellular adaptive mechanism was operative.

Of course, cells can adapt to drug effects by many possible mechanisms. Such adaptations may range from stiffening of membranes[37,38] to enzymatic changes. As reported by Knapp and Mandell,[39] the effects of acute (single-dose) drug administration on neurotransmitter metabolism are often different from or even opposite to effects resulting from continual administration. If acute administration changes serotonergic transmitter turnover, continued administration consistently changes transportable enzyme activity in a direction compensatory for the acute effects. The end result of continued drug administration is a return of transmitter turnover to predrug exposure conditions.[39,40] In the case of the neurons of the thermostat, this would simply mean resumption of normal regulatory functions. Again, when drug administration was abruptly ceased, temporary disruption would ensue because of the enzyme adaptation that had occurred while the drug was present. Whether or not such cellular adaptations are the basis for behavioral or temperature tolerance remains to be established.

E. NEUROCHEMICAL CONSEQUENCES OF HYPOTHERMIA

Hypothermia *per se,* irrespective of the mechanism whereby it is induced, results in certain neurochemical changes.[7] These effects are distinguished from the separate effects of drugs themselves by being reversible or preventable by raising the environmental temperature. Furthermore, one may suspect that a particular effect is caused by hypothermia by considering physiological changes caused by hypothermia not induced by drugs. Moderate hypothermia causes a decreased incorporation of ^{32}P into various brain phospholipids, as well as decreased transport[25] and incorporation[7,41] of amino acids into proteins.[7,41] Concentrations of cyclic AMP were reported to be decreased in the cerebral cortex of hypothermic rats and mice; the metabolic rate was decreased to 65% of control values, and the interconversion of the active and inactive forms of glycogen synthetase was retarded.[42] These animals were made hypothermic to body temperatures of 18–22°C by having them swim in water of room temperature, then exposing them to a "cold chamber." The effect of more modestly lowered body temperatures (in the 32–34°C range), as is relevant to ethanol-treated rodents, was not determined in this study. Drastic hypothermia (~21.5°C) in rats immersed in 2–3°C water causes brain seizures and retrograde amnesia[43] and is not further considered in this context. But lowering of body temperatures in poikilotherms, which would appear to be less drastic, may impair the formation of consolidation of memory in fish.[44]

Secondary and tertiary changes from hypothermia can, in turn, affect CNS metabolism and function. Stress with its concomitant hormonal alterations, including increased hypophyseal, adrenal corticoid, and medullary hormone secretions may be induced by hypothermia,[45] as well as by ethanol itself. Nonspecific stress alters serotonin and γ-aminobutyric acid metabolism in the CNS.[46,47]

It is well established that certain environmental alterations of *in vivo* conditions change the metabolism of the *in vitro* tissue preparations taken from such animals. Even behavioral manipulations *in vivo* may carry over into subcellular *in vitro* systems. For instance, sleep deprivation in rats may depress subsequent *in vitro* incorporation of amino acids into brain proteins.[48] *In vivo* ethanol intoxication and withdrawal alter protein synthesis *in vitro* in subcellular fractions of brains derived from ethanol-treated animals.[49] The effects of *in vivo* hypothermia on *in vitro* subcellular systems have not been studied. But it would seem evident that it is as important to control *in vivo* conditions for subsequent *in vitro* experiments as it is for studies confined to *in vivo* systems.

F. SUMMARY AND CONCLUSIONS

Acute ethanol intoxication impairs body temperature regulation in man and rodents. Instead of remaining constant, the core body temperature fluctuates with the environmental temperature. There are three phases of loss of temperature control: (1) acute intoxication causes loss of temperature control; (2) continued intoxication results in restoration of temperature control, adaptation, and "tolerance"; and (3) abrupt drug withdrawal again causes temporary loss of temperature control. Because CNS depressants other than ethanol have the same effect, the mechanism is probably impaired function of the hypothalamic neurons involved in the thermostat control.

Human subjects suffer ethanol-induced hypothermia only under unusually severe environmental conditions. In contrast, rodents at usual laboratory temperatures develop considerable hypothermia after receiving intoxicating doses of ethanol. Therefore, many biochemical and physiological observations in rodents may be attributable to hypothermia rather than to ethanol itself. This extends to *in vitro* studies with subcellular fractions derived from animals that are hypothermic during acute ethanol intoxication or withdrawal. Because humans usually do not develop significant temperature changes during acute ethanol intoxication or withdrawal, the results of experiments with rodents in the absence of temperature controls may not be relevant to humans. The body temperature changes in rodents can be prevented, however, by appropriately altering the environmental temperatures.

With continued drug administration, rodents develop tolerance to the intoxicating and hypothermic effects of ethanol within a few days. It is uncertain whether or not the tolerance to the hypothermic and the sedative effects of ethanol are different or are the expressions of the same cellular adaptations. Tolerance to hypothermia but not to the behavioral effects develops with nonaddicting drugs such as chlorpromazine and marihuana. This dichotomy between the development of behavioral and temperature tolerances suggests that they may have different cellular mechanisms. Initial hypothermia and then tolerance with continued administration of drugs develop with dopamine agonists (piribedil) as well as antagonists (chlorpromazine). This suggests the possibility that cells adapt to chronic drug exposure by changes in drug receptor number or affinity. Whatever the exact mechanism of cellular adaptation may be, "tolerance" (i.e., restoration of normal temperature control in the continued presence of the drug) means resumption of normal function of the neurons in the temperature-regulating center of the hypothalamus.

Once tolerance and physical dependence on ethanol have developed, the acute withdrawal of ethanol may under some conditions cause loss of temperature control (i.e., hypothermia). This has been considered by some investigators to represent another manifestation of physical dependence. If the same phenomenon were demonstrable with nonaddicting hypothermic drugs, this interpretation could be questionable. If the cellular mechanism for the development of behavioral and temperature tolerance were the same, all drugs that induce temperature tolerance should also induce behavioral tolerance and possibly physical dependence.

ACKNOWLEDGMENT. Part of this work was supported by the Medical Research Service of the Veterans Administration.

G. REFERENCES

1. L. Shuster and R. V. Hannam, *J. Biol. Chem.* **239**:3401 (1964).
2. J. Bligh, *Front. Biol.* **30**:1 (1973).
3. E. Satinoff and R. Hendersen, in: *Handbook of Operant Behavior* (W. K. Honig and J. E. R. Staddon, eds.), pp. 153–173, Prentice-Hall, Inc., Englewood Cliffs, N.J. (1977).
4. P. Lomax and E. Schönbaum (eds.), *Drugs and Body Temperature,* Marcel Dekker, Inc., New York (1979).
5. R. F. Hellon, *Pharmacol. Rev.* **26**:289 (1975).
6. E. Satinoff and E. R. Hackett, in: *Drugs, Biogenic Amines and Body Temperature* (K. E. Cooper, P. Lomax, and E. Schönbaum, eds.), pp. 87–95, S. Karger, Basel (1977).
7. J. E. Cremer, in: *Handbook of Neurochemistry* (A. Lajtha, ed.), Vol. 6, pp. 311–323, Plenum Press, New York (1971).
8. J. Baird and W. J. Lang, *Eur. J. Pharmacol.* **22**:1 (1973).
9. M. Sharkawi, *Br. J. Pharmacol.* **44**:544 (1972).
10. D. R. Haubrich and D. E. Blake, *Life Sci.* **10**:175 (1971).
11. B. Tabakoff and F. Moses, *Biochem. Pharmacol.* **25**:2555 (1976).
12. C. O. Haavik and H. F. Hardman, *J. Pharmacol. Exp. Ther.* **187**:568 (1973).
13. R. K. Liu, *Res. Commun. Chem. Pathol. Pharmacol.* **9**:215 (1974).
14. G. W. Molnar and R. C. Read, *J. Am. Med. Assoc.* **227**:916 (1974).
15. J. L. Verbov, *Lancet* **1**:194 (1964).
16. G. Philip and J. F. Smith, *Lancet* **2**:122 (1973).
17. L. D. Hudson and R. D. Conn, *J. Am. Med. Assoc.* **227**:37 (1974).
18. J. J. Reincke, *Dtsch. Arch. Klin. Med.* **16**:12 (1875).
19. A. E. Weyman, D. M. Greenbaum, and W. J. Grace, *Am. J. Med.* **56**:13 (1974).
20. J. S. J. Haight and W. R. Keatinge, *J. Physiol.* **229**:87 (1973).
21. H. D. Rose, T. M. Golbert, C. J. Sanz, and T. H. Leitschuh, *Am. J. Med. Sci.* **260**:112 (1970).
22. M. M. Gross, E. Lewis, S. Best, N. Young, and L. Feuer, in: *Alcohol Intoxication and Withdrawal: Experimental Studies* (M. M. Gross, ed.), Vol. II, pp. 477–493, Plenum Press, New York (1975).
23. H. Isbell, H. F. Fraser, A. Wikler, R. E. Belleville, and A. J. Eisenman, *Q. J. Stud. Alcohol* **16**:1 (1955).
24. A. Walther, *Arch. Anat. Physiol. Wiss. Med.* **25**:51 (1865).
25. G. Freund, *Life Sci.* **13**:345 (1973).
26. L. A. Pohorecky, L. S. Jaffe, and H. A. Berkeley, *Res. Commun. Chem. Pathol. Pharmacol.* **8**:1 (1974).
27. L. A. Pohorecky, J. Brick, and J. Y. Sun, *J. Pharm. Pharmacol.* **28**:157 (1976).
28. B. Tabakoff, R. F. Ritzmann, and W. O. Boggan, *J. Neurochem.* **24**:1043 (1975).
29. R. F. Ritzmann and B. Tabakoff, *Ann. N.Y. Acad. Sci.* **273**:247 (1976).
30. R. F. Ritzmann and B. Tabakoff, *J. Pharmacol. Exp. Ther.* **199**:158 (1976).
31. F. Karoum, R. J. Wyatt, and E. Majchrowicz, *Br. J. Pharmacol.* **56**:403 (1976).
32. L. Ahtee and M. Svartström-Fraser, *Acta Pharmacol. Toxicol.* **36**:289 (1975).
33. J. Brick and L. A. Pohorecky, *Alcohol. Clin. Exp. Res.* **1**:207 (1977).

34. P. L. Hoffman and B. Tabakoff, *Nature (Lond.)* **268**:551 (1977).
35. B. Tabakoff and R. F. Ritzmann, *J. Pharmacol. Exp. Ther.* **203**:319 (1977).
36. G. D. Frye and F. W. Ellis, *Drug Alcohol Depend.* **2**:349 (1977).
37. J. H. Chin and D. B. Goldstein, *Science* **196**:684 (1976).
38. M. Curran and P. Seeman, *Science* **197**:910 (1977).
39. S. Knapp and A. J. Mandell, *Science* **177**:1209 (1972).
40. A. J. Mandell, S. Knapp, and L. L. Hsu, *Life Sci.* **14**:1 (1974).
41. A. Lajtha and H. Sershen, *Life Sci.* **17**:1861 (1975).
42. W. D. Lust and J. V. Passonneau, *J. Neurochem.* **26**:11 (1976).
43. R. M. Vardaris, C. Gaebelein, and D. C. Riccio, *Physiol. Psychol.* **3**:204 (1973).
44. J. H. Neale, P. D. Klinger, and B. W. Agranoff, *Behav. Biol.* **9**:267 (1973).
45. J. G. Sprunt, D. Maclean, and M. C. K. Browning, *Lancet* **1**:324 (1970).
46. A. Yuwiler, in: *Handbook of Neurochemistry* (A. Lajtha, ed.), Vol. 6, pp. 103–169, Plenum Press, New York (1971).
47. D. M. Woodbury, in: *Handbook of Neurochemistry* (A. Lajtha, ed.), Vol. 7, pp. 255–287, Plenum Press, New York (1972).
48. P. Bobillier, F. Sakai, S. Seguin, and M. Jouvet, *J. Neurochem.* **22**:23 (1974).
49. S. Tewari, M. A. Goldstein, and E. P. Noble, *Brain Res.* **126**:509 (1977).

Interrelationship between Ethanol Consumption and Circadian Rhythm

23

Irving Geller and Robert H. Purdy

A. INTRODUCTION

In 1970, Freund[1] reported that mice, given aqueous ethanol solution as their sole drinking fluid, consumed more than 70% of their total daily intake of ethanol during the 12-hr dark cycle. The possible interrelationship between ethanol consumption and circadian rhythm derives further support from the observation[2] that rats kept in total darkness drink significantly more of an ethanol solution than they drink when maintained in a photoperiod consisting of 9 hr of light and 15 hr of darkness. Several other reports[3,4] substantiated this observation. In a critique of this work, Sinclair[3] agreed that the evidence for a darkness-induced increase of ethanol drinking by rats was rather strong but that its applicability was limited only to young rats. Burke and Kramer[4] confirmed that the amount of 4% (v/v) ethanol solution consumed by rats was influenced by environmental lighting and increased during constant darkness.

B. ETHANOL PREFERENCE IN THE RAT AS A FUNCTION OF PHOTOPERIOD: EFFECTS OF ALTERATION OF LIGHT–DARK CYCLE, MELATONIN ADMINISTRATION, AND PINEALECTOMY

Since we first reported that rats drink more ethanol solution in darkness than in light, this observation has been repeated many times. The data in Fig. 1 illustrate the development of ethanol preference in a typical laboratory rat. Initially, the rat's daily intake of water approximated 40 ml while intake of the ethanol solution was minimal. Nine days after the rat was placed in total

Irving Geller and Robert H. Purdy • Southwest Foundation for Research and Education, San Antonio, Texas 78284.

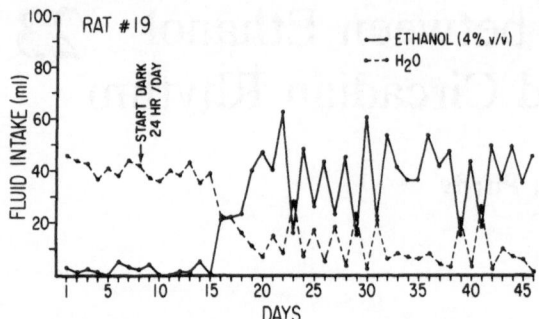

Fig. 1. Effect of darkness on ethanol intake in the rat. Solid lines indicate ethanol intake and broken lines, water intake. Arrow shows onset of continuous darkness.

darkness, water intake was reduced and a concomitant increased intake of a 4% (v/v) ethanol solution occurred. After 11 days of exposure to darkness, the rat showed a definite preference for the ethanol solution.

The possibility of pineal involvement accounting for this phenomenon derives from the work of other investigators, who found that rats kept in darkness have larger pineals than rats kept in constant illumination.[5-7] This observation as well as the reported darkness-induced increased activity of the pineal melatonin-forming enzymes, hydroxyindole-O-methyltransferase and N-acetyltransferase, prompted us to evaluate the effects of melatonin administration on the drinking of ethanol solution by the rat. The drinking curves of Fig. 2, illustrating the effects of melatonin administration, are very similar to the curves shown in Fig. 1. Prior to melatonin administration, there was a definite preference for water. After 7 days of melatonin treatment at 1.0 mg/kg, the curves cross one another and a preference for the ethanol solution is evident. Similarly, Burke and Kramer[4] reported that for those rats showing the least preference for a 4% ethanol solution, administration of melatonin resulted in significant increases of ethanol drinking with concomitant reductions of water intake. However, Blum et al.[8] reported that melatonin had no significant effect on the consumption of 5% ethanol solution by the rat or the hamster. In the earlier investigations[2,4] melatonin was administered subcutaneously, whereas in these studies[8] it was administered in subcutaneous wax implants. Then, too, Blum's animals were pinealectomized and were already drinking large quantities of ethanol solution prior to treatment with melatonin. One might justifiably expect difficulty in demonstrating increases in fluid consumption in animals already drinking at maximal levels.

Data obtained on the role of the pineal in ethanol preference are as conflicting and controversial as those cited above for melatonin. On the one hand, Burke and Kramer[4] reported that pinealectomy had no significant effects on ethanol consumption of rats subjected to various environmental lighting conditions. On the other hand, Reiter et al.[9] reported a significant decrease of ethanol drinking in sighted or blinded rats or hamsters. These differences in findings of pineal involvement on ethanol drinking may be due to procedural differences between experiments, and therefore a definitive role for the pineal will have to await future experimentation.

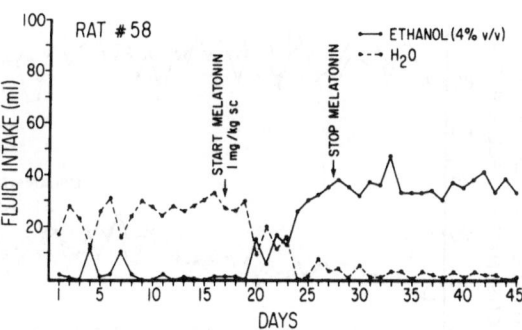

Fig. 2. Effect of chronic subcutaneous administration of 1 mg melatonin/kg on ethanol intake in the rat. Solid lines indicate ethanol intake and broken lines, water intake. Arrows show beginning and termination of chronic administration of 1 mg melatonin/kg per day.

C. ALTERATION OF ETHANOL PREFERENCE IN HAMSTERS: EFFECTS OF CHANGES IN PHOTOPERIOD AND PINEALECTOMY

The observation of a darkness-induced increase of ethanol drinking in the laboratory rat has also been extended to the golden hamster. The hamster, unlike the Sprague-Dawley rat, shows a marked preference for ethanol in free-choice experiments. Given a choice of water or a 10% ethanol solution, hamsters will drink 88% of their total fluid as the ethanol solution.[10] Furthermore, the most preferred concentration of ethanol for the male hamster is 15%, in contrast to the rat, which shows a preference only for much lower concentrations.[11] Geller and Hartmann[12] reported an increased intake of ethanol solution in hamsters maintained in total darkness. Figure 3, taken from their study,[12] illustrates the development of preference for ethanol solution by hamsters given a choice of water or ethanol solutions. The curves show milliliters of ethanol solution consumed and amount of ethanol as a percent of total fluid intake. The points on the curves represent data averaged for six hamsters for each 24-hr period. The groups did not differ as a function of increasing ethanol concentrations from 2 to 25%. Furthermore, placing group A (solid lines) in total darkness while maintaining ethanol concentrations at 25% (v/v) yielded no differences between groups over the following 28-day period. Averaged daily intake of ethanol solutions for each group ranged between 10 and 20 ml. Subsequent reduction of the ethanol concentration to 6% (lower portion of Fig. 3) resulted in an increased average intake of ethanol to the 20- to 30-ml range for group A kept in total darkness. Ethanol intake for group B remained unchanged under these new conditions. Switching group B to total darkness and group A to a "normal" photoperiod produced, after 14 days (day 54 of the curve), an increased ethanol intake in group B and a reduced ethanol intake in group A.

Thus, ethanol drinking by the hamster is affected by darkness in a manner similar to that of the rat. A greater intake of ethanol has also been reported for blinded hamsters, compared to controls.[9] The possibility of pineal involvement in this study is suggested by the fact that the exaggerated ethanol preference was reduced by pinealectomy or decentralization of the superior cervical ganglion, which causes the pineal to become nonfunctional.

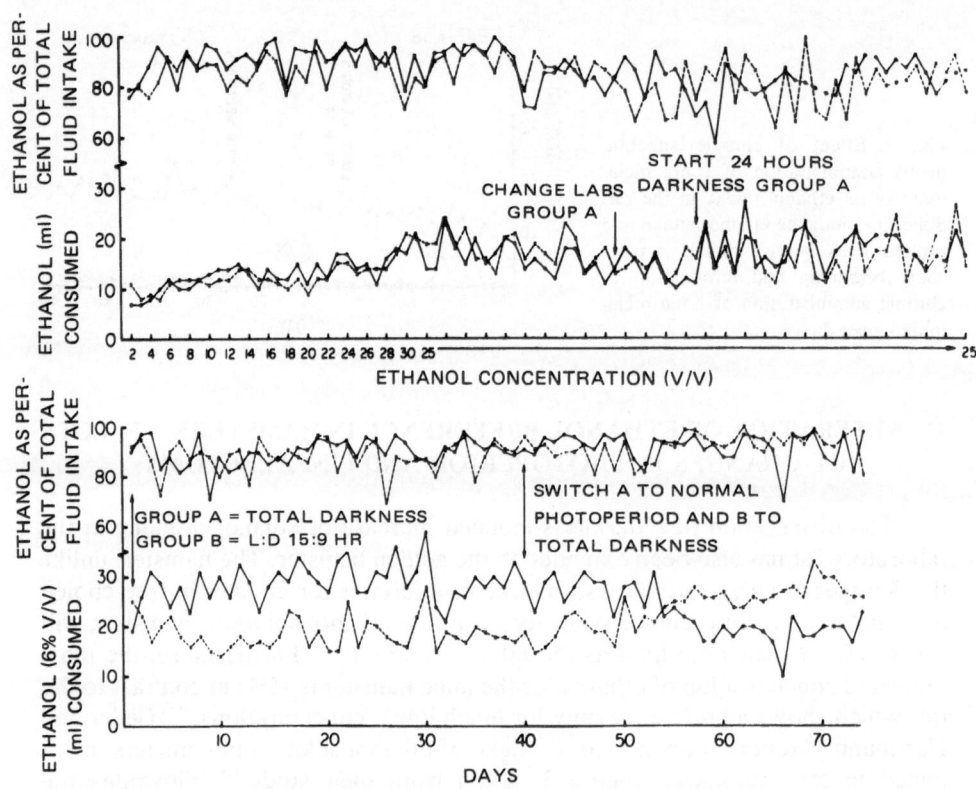

Fig. 3. Ethanol intake in the hamster: effects of photoperiods and concentrations of ethanol used. Solid lines indicate group A. Broken lines indicate group B. [From Geller and Hartmann.[12]]

Another parameter may also be relevant to this animal model system. The growth curves of Fig. 4 demonstrate that the average weight for the group in darkness was generally lower than the group in the light–dark environment. Two weeks after group A was returned to a normal photoperiod and group B was placed in total darkness, the trend was reversed so that the average weight for group B was lower than for group A. Perhaps the calories derived from ethanol led to a reduced food intake and the concomitant lowering of body weight that was observed.

D. EXTENSION OF PRECEDING STUDIES TO IMPLICATE NEUROTRANSMITTER INVOLVEMENT

The increased drinking of ethanol solutions by rats and hamsters kept in darkness has given rise to an interesting speculation regarding a possible mechanism accounting for ethanol preference in rodents. This speculation is based on the report of a darkness-induced increase of pineal N-acetyltransferase and hydroxyindole-O-methyltransferase activity in rats and a concomitant increased conversion of serotonin to melatonin via N-acetyl serotonin.[5,6] If brain serotonin is lowest in periods of darkness, it seems reasonable to assume that lowering brain serotonin might increase ethanol preference, while raising brain serotonin might decrease ethanol preference of rats. The data in Fig. 5 taken from Geller[13] show the effects of 5-hydroxytryptophan, the serotonin precursor, on ethanol preference of rats. Rats administered 5-hydroxytryptophan showed a diminished preference for ethanol solution while chronic or acute administration of equivalent volumes of saline administered for the same length of time had no effect on ethanol preference. The reduction of ethanol preference by 5-hydroxytryptophan may be attributable to increased brain serotonin, since administration of 5-hydroxytryptophan to rats significantly raises brain serotonin.[14,15] Further support for this mechanism is provided by another study, in which intraventricular administration of serotonin also decreased ethanol drinking of laboratory rats.[16]

If ethanol intake of rats is decreased through an increase of brain serotonin, one might expect that decreasing brain serotonin should produce an increase of ethanol drinking. The availability of p-chlorophenylalanine, a tryptophan hydroxylase inhibitor reported to selectively deplete brain serotonin,[17,18]

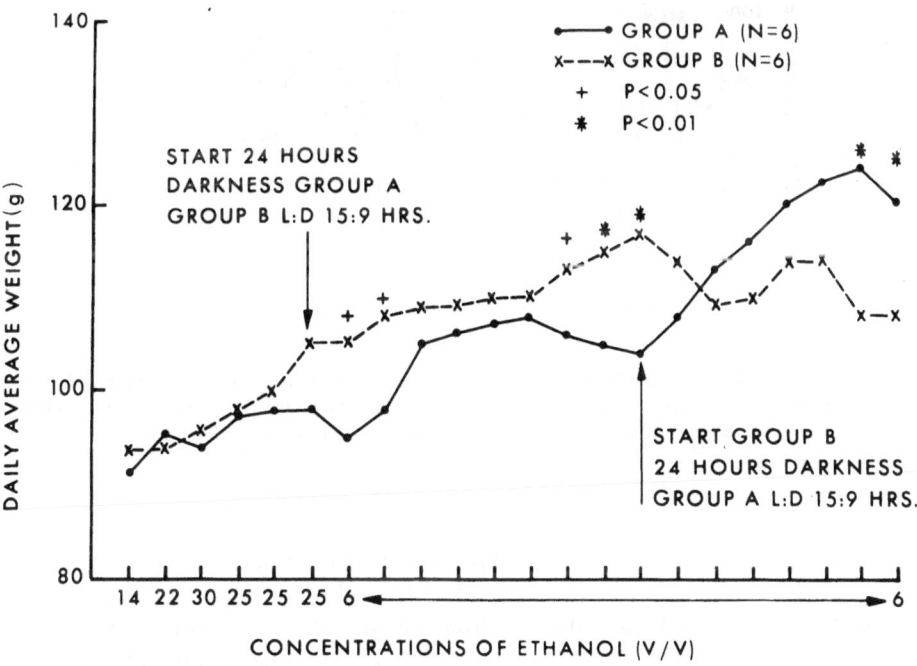

Fig. 4. Weekly average weights of hamsters kept in darkness or normal photoperiod. L/D 15:9 hr. Solid lines indicate group A. Broken lines indicate group B. [From Geller and Hartmann.[12]]

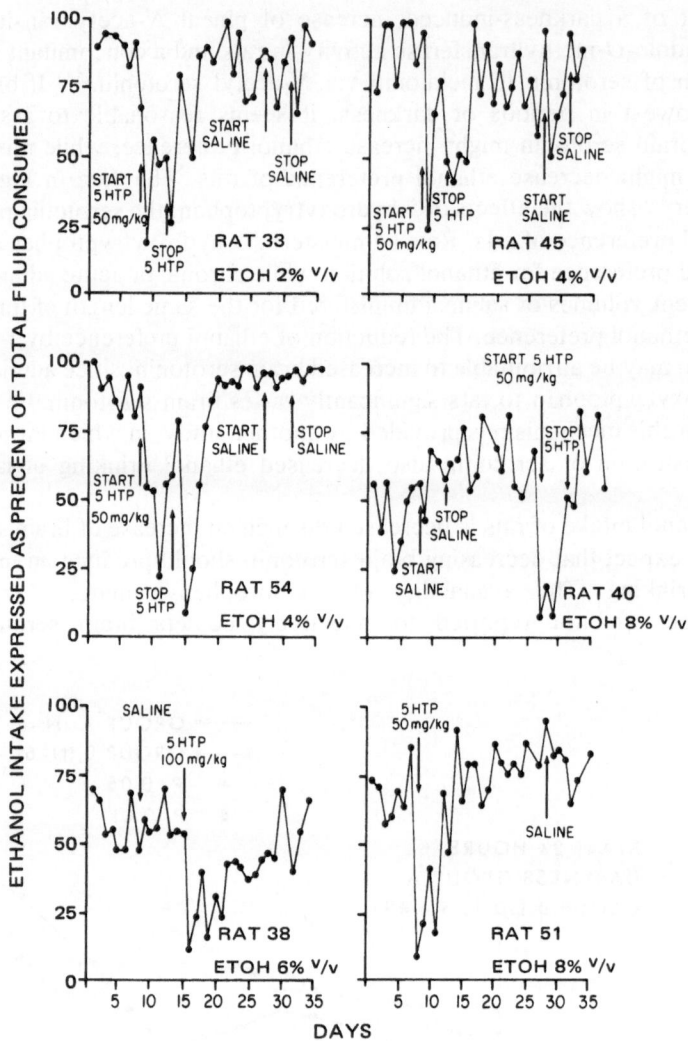

Fig. 5. Effects of chronic or acute administration of 5-hydroxytryptophan or saline on ethanol intake of rats. Ethanol concentrations for each rat are as indicated. The points on the graphs show ethanol intake expressed as a percent of total fluid consumed during a 24-hr period. The arrows show 24-hr readings obtained after the first, last, or individual treatments. [From Geller.[13]]

allowed this hypothesis to be tested. Myers and Veale[19] and Veale and Myers[20] reported an aversion to ethanol in rats administered p-chlorophenylalanine and a marked rejection of ethanol after the drug was discontinued. On the other hand, the data in Fig. 6[13] show an increased ethanol preference in rats administered chronic doses of p-chlorophenylalanine ranging from 75 to 150 mg/kg. Similarly, Hill, who was unable to produce a p-chlorophenylalanine-induced rejection of

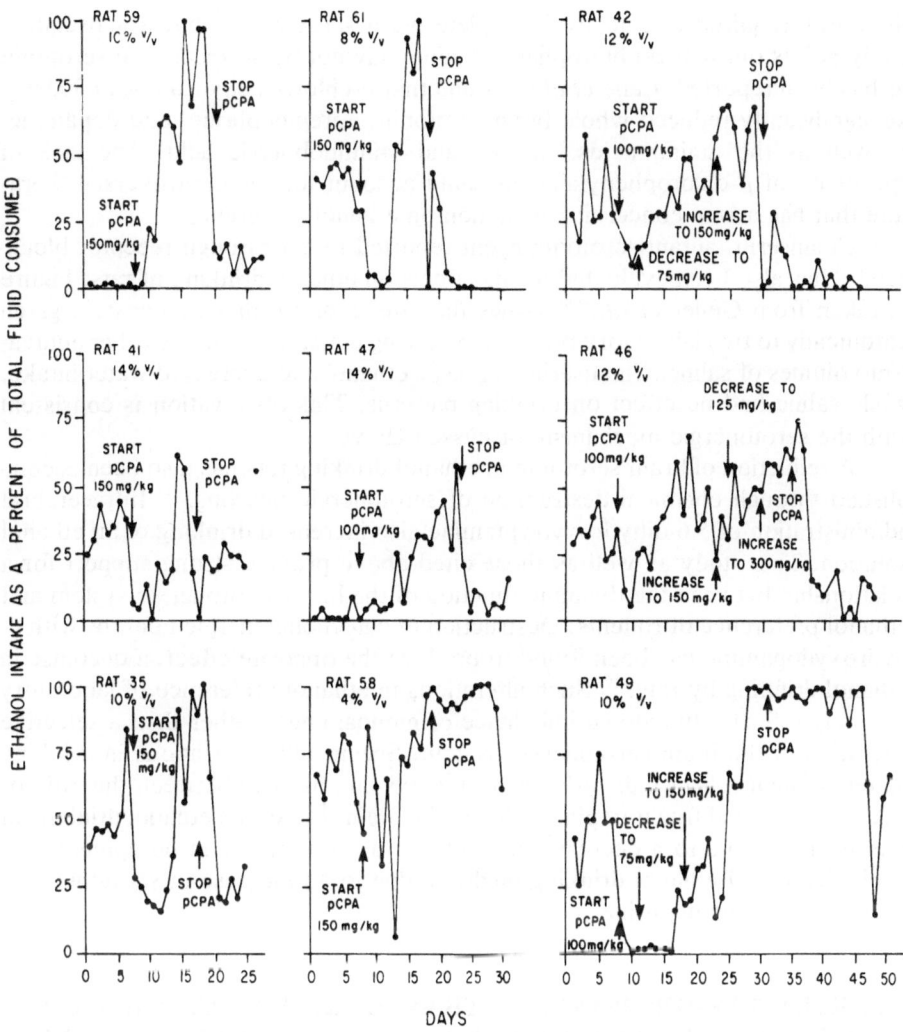

Fig. 6. Effects of chronic oral administration of *p*-chlorophenylalanine on ethanol intake of rats. Ethanol concentrations for each rat are as indicated. The points on the graphs show ethanol intake expressed as a percent of total fluid consumed during a 24-hr period. The first and last arrow on each graph indicates the 24-hr intake readings taken on the days following the first and last *p*-chlorophenyl-alanine administrations, respectively. [From Geller.[13]]

ethanol, found that *p*-chlorophenylalanine increased intake of 3 and 5% ethanol concentrations.[21] The work of other investigators suggests that the demonstrated *p*-chlorophenylalanine-induced aversion to ethanol was probably not mediated through a serotonergic involvement but rather represented a conditioned aversion established through pairing of a noxious agent (*p*-chlorophenyl-alanine) with ethanol drinking.[22,23] Not only is *p*-chlorophenylalanine toxic in the doses required to selectively deplete brain serotonin,[18] but a subsequent study points out that p-chlorophenylalanine may not be as specific to serotonin as has been reported. Lane et al.[24] found that p-chlorophenylalanine at 300 mg/kg significantly reduced whole brain serotonin, norepinephrine, and dopamine, as well as the amino acids, alanine and γ-aminobutyric acid. The lack of specificity of p-chlorophenylalanine could account for the controversial literature that has arisen concerning its action on ethanol preference.

Cinanserin, an antiserotonin agent reported to act through receptor blockade,[25] has also been evaluated for its effects on ethanol drinking of rats. Figure 7, taken from Geller et al.,[26] shows the effects of 50 mg/kg cinanserin given chronically to two laboratory rats, one receiving cinanserin and the other equivalent volumes of saline. Cinanserin increased ethanol and decreased water intake, while saline had no effect on drinking patterns. This observation is consistent with the serotonergic mechanism discussed above.

A reduction of brain serotonin in ethanol-drinking rats has also been accomplished through chemical destruction of serotonergic neurons.[27] Intracerebral administration of 5,6-dihydroxytryptamine also increased drinking of an ethanol solution. This study as well as those cited above provide strong support for a relationship between the dynamic function of the brain serotonergic system and ethanol preference of rodents. Destruction of catecholaminergic neurons with 6-hydroxydopamine has been found to produce the opposite effect, a decrease in ethanol drinking by rats.[27] Such alterations in ethanol preference of laboratory rats may reflect a functional imbalance of monoamines, rather than a selective alteration in the brain serotonergic system. Specifically, alterations in drinking patterns might reflect a disturbance of the existing balance between the ratio of free serotonin and free norepinephrine in the brain. Increased ethanol drinking in rats might be due to a decrease in serotonin or an increase in norepinephrine, while decreased ethanol drinking might be due to an increase in serotonin or a decrease in norepinephrine.

E. RELATIONSHIP BETWEEN ETHANOL PREFERENCE AND BRAIN CHEMISTRY: ALTERATIONS PRODUCED BY β-CARBOLINES AND RELATED INDOLE DERIVATIVES

Evaluation of β-carbolines for effects on ethanol preference in rats seems justified for several reasons. First, β-carbolines have been reported to be inhibitors of monoamine oxidase and therefore should allow a buildup of brain serotonin. Ho et al.[28] reported a preferential rise of whole brain serotonin over norepinephrine in mice administered 6-methoxy-1,2,3,4-tetrahydro-β-carboline. Essman[29] has also reported that in mice given an intraperitoneal injection of the carboline compound, brain serotonin was increased more than 50% higher than in saline-treated mice. The work of Buckholtz and Boggan[30] suggest that brain serotonin might be increased in carboline-treated animals by a mechanism other than monoamine oxidase inhibition. They reported that 6-methoxytetrahydrohar-

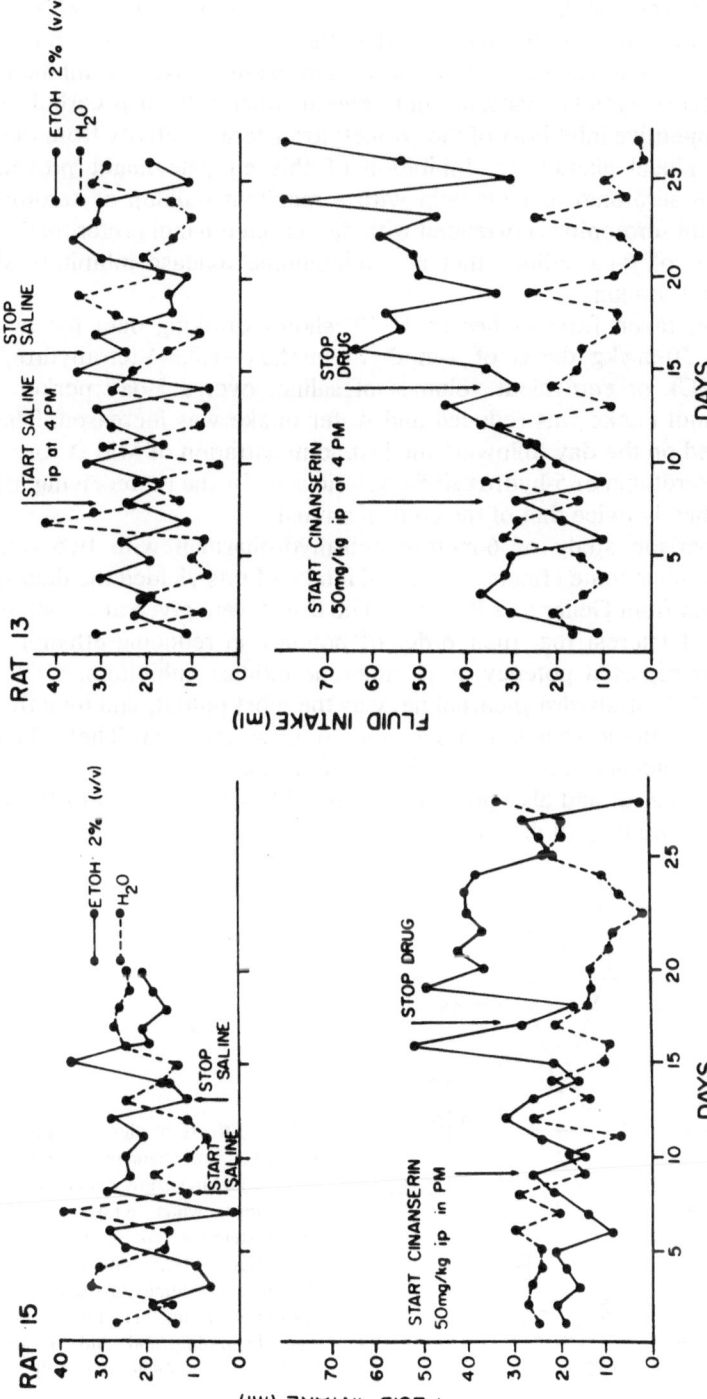

Fig. 7. Effects of chronic administration of cinanserin or saline on ethanol intake in the rat. Solid lines indicate ethanol intake and broken lines indicate water intake. The arrows show 24-hr readings obtained after the first or last treatment. [From Geller et al.[26]]

man and noreleagnine inhibited monoamine oxidase *in vitro* and serotonin uptake both *in vitro* and *in vivo* and that the inhibition of serotonin uptake occurred at doses lower than required for monoamine oxidase inhibition. In addition to those effects, harmine and several other related β-carbolines are effective competitive inhibitors of the N-acetyltransferase activity from the brain but not the pineal gland.[31–33] Inhibition of this enzyme might prevent the conversion of serotonin to melatonin with a resultant buildup of serotonin. If increased brain serotonin is correlated with decreased ethanol preference in rats, administration of β-carbolines that are monoamine oxidase inhibitors should reduce ethanol drinking.

Figure 8, taken from Geller *et al.*,[34] shows drinking data for two rats administered 50-mg/kg doses of 1-methyl-6-methoxy-1,2,3,4-tetrahydro-β-carboline (MMTC), or equivalent volumes of saline, over a 3-day period. With MMTC, ethanol intake was reduced and water intake was increased. The rats were sacrificed on the day following the last administration of MMTC or saline. Whole brain serotonin or 5-hydroxyindoleacetic acid for the rat receiving MMTC was approximately twice that of the control animal.

A comparison study of 6-methoxytetrahydroharman with two other β-carbolines for their acute effects on ethanol intake of rats yielded the data shown in Fig. 9, taken from Geller and Purdy.[35] The drugs were given at 25, 40, and 50 mg/kg. It is of interest that their order of potency in reducing ethanol intake parallels their reported potency as monoamine oxidase inhibitors.[36–38] Noreleagnine (1,2,3,4-tetrahydro-β-carboline) was the most potent, and 6-methoxytetrahydroharman the least potent monoamine oxidase inhibitor. These data provide further evidence for a possible serotonergic involvement in ethanol preference by the rat and also provide a reasonable interpretation of the induction of ethanol drinking in darkness.

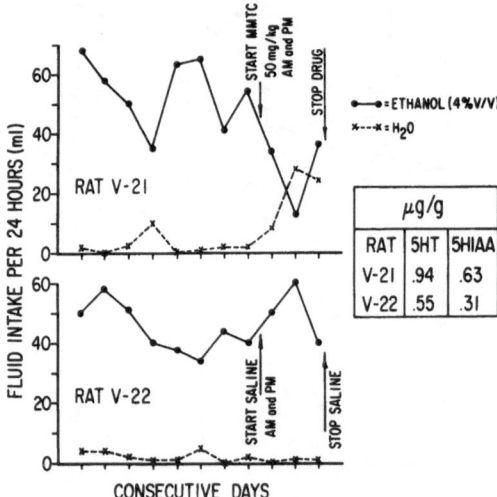

Fig. 8. Effects of chronic administration of MMTC and saline on ethanol intake in rats. Solid lines show ethanol intake; broken lines, water intake. MMTC was administered intraperitoneally in saline at 50 mg/kg at 10 A.M. and 4 P.M. each day. Equivalent volumes of saline were given on the same time schedule. Arrows indicate beginning and end of treatment periods. [From Geller *et al.*[34]]

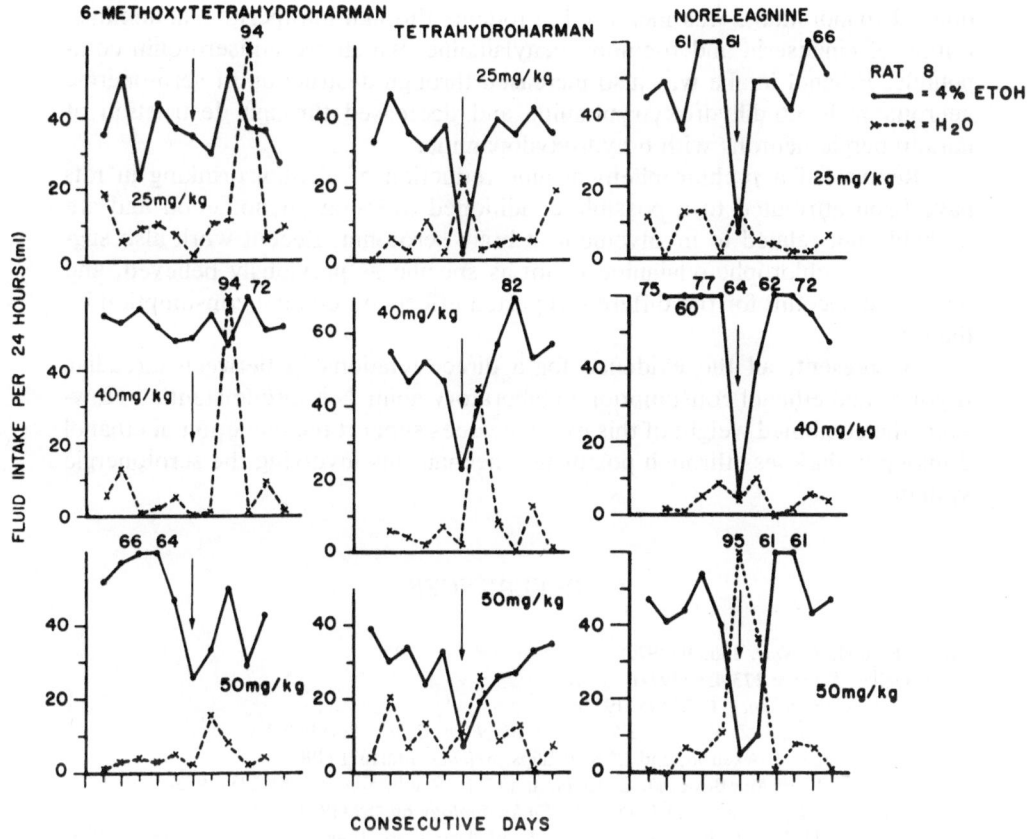

Fig. 9. Effects of acute administration of tetrahydroharman, 6-methoxytetrahydroharman, and nore-leagnine on ethanol intake in the rat. Solid lines show ethanol intake; broken lines, water intake. Drugs were injected intraperitoneally at 4 P.M. at 25, 40, or 50 mg/kg. Arrows represent readings taken 18 hr after injection. [From Geller and Purdy.[35]]

F. SUMMARY AND CONCLUSIONS

Several investigators have reported that laboratory rodents drink more aqueous ethanol solution in darkness than they do in light. The possibility of a brain serotonergic involvement accounting for this observation derives in part from the well-established increase in activity of pineal melatonin-forming enzymes of rats kept in darkness. The resulting increased formation of melatonin and concomitant reduction of brain serotonin served as a basis for subsequent studies in which ethanol intake was changed through indirect manipulation of brain serotonin. Ethanol intake was decreased in rodents through intraperitoneal administration of 5-hydroxytryptophan and β-carbolines and intraventricular administration of serotonin, substances which raise levels of serotonin in the

brain. Ethanol intake was increased in rodents through intraperitoneal administration of cinanserin and p-chlorophenylalanine, which are antiserotonin compounds. Ethanol intake was also increased through destruction of serotonergic
neurons with 5,6-dihydroxytryptamine and decreased through destruction of
noradrenergic neurons with 6-hydroxydopamine.

Reports of a p-chlorophenylalanine reduction of alcohol drinking in rats
have been attributed to a possible conditioned aversion phenomenon and are
probably not related to involvement of brain serotonin. Recent work also suggests that p-chlorophenylalanine is not as specific as previously believed, and
this could account for the different reported effects on ethanol consumption by
the rat.

At present, all the evidence for a direct relationship between circadian
rhythms and ethanol consumption in laboratory animals is circumstantial. However, the combined weight of this evidence does support the induction of ethanol
drinking in darkness through postulated mechanisms involving the serotonergic
system.

G. REFERENCES

1. G. Freund, *J. Nutr.* **100**:30 (1970).
2. I. Geller, *Science* **173**:456 (1971).
3. J. D. Sinclair, *Science* **175**:1143 (1972).
4. L. P. Burke and S. Z. Kramer, *Pharmacol. Biochem. Behav.* **2**:459 (1974).
5. R. J. Wurtman, J. Axelrod, and L. S. Phillips, *Science* **142**:1071 (1963).
6. S. Binkley, S. E. MacBride, D. C. Klein, and C. L. Ralph, *Science* **181**:273 (1973).
7. V. M. Fiske, K. Bryant, and J. Putnam, *Endocrinology* **66**:489 (1960).
8. K. Blum, J. H. Merritt, R. S. Reiter, and J. E. Wallace, *Curr. Ther. Res. Clin. Exp.* **15**:25 (1973).
9. R. J. Reiter, K. Blum, J. E. Wallace, and J. H. Merrit, *Comp. Biochem. Physiol.* **47**:11 (1974).
10. A. Arvola and O. Forsander, *Nature (Lond.)* **191**:819 (1961).
11. A. Arvola and O. Forsander, *Q. J. Stud. Alcohol* **24**:591 (1963).
12. I. Geller and R. J. Hartmann, in: *Alcohol Intoxication and Withdrawal: Experimental Studies* (M. M. Gross, ed.), Vol. IIIb, pp. 223–233, Plenum Press, New York (1977).
13. I. Geller, *Pharmacol. Biochem. Behav.* **1**:361 (1973).
14. H. Green and J. L. Sawyer, *Prog. Brain Res.* **8**:150 (1964).
15. A. Pellagrino de Iraldi, L. M. Zieher, and E. DeRobertis, *Life Sci.* **2**:691 (1963).
16. S. Y. Hill, *Biol. Psychiatry* **8**:151 (1974).
17. E. Jequier, W. Lovenberg, and A. Sjoerdsma, *Mol. Pharmacol.* **3**:274 (1967).
18. B. K. Koe and A. Weissman, *J. Pharmacol. Exp. Ther.* **154**:499 (1966).
19. R. D. Myers and W. L. Veale, *Science* **160**:1469 (1968).
20. W. L. Veale and R. D. Myers, *Neuropharmacology* **9**:317 (1970)
21. S. Y. Hill, *Q. J. Stud. Alcohol* **35**:34 (1974).
22. M. Nachman, D. Lester, and J. LeMagnen, *Science* **168**:1244 (1970).
23. L. F. Parker and B. L. Radow, *Pharmacol. Biochem. Behav.* **4**:535 (1976).
24. J. D. Lane, J. E. Smith, P. A. Shea, and W. J. McBride, *Life Sci.* **19**:1663 (1976).
25. H. L. Metcalfe and P. Turner, *Arch. Int. Pharmacodyn. Ther.* **183**:148 (1970).
26. I. Geller, R. J. Hartmann, and F. S. Messiha, *Proc. West. Pharmacol. Soc.* **18**:141 (1975).
27. R. D. Myers and C. L. Melchior, *Res. Commun. Chem. Pathol. Pharmacol.* **10**:363 (1975).
28. B. T. Ho, D. Taylor, and W. M. McIsaac, in: *Brain Chemistry and Mental Disease* (B. T. Ho and W. McIsaac, eds.), pp. 97–112, Plenum Press, New York (1971).

29. W. B. Essman, *J. Pharmacol.* **6**:313 (1975).

30. N. S. Buckholtz and W. O. Boggan, *Biochem. Pharmacol.* **25**:2319 (1976).

31. H.-Y. T. Yang and N. H. Neff, *Mol. Pharmacol.* **12**:69 (1976).

32. W. Schloot, F. J. Tigges, H. Blaesner, and H. W. Goedde, *Hoppe-Seyler's Z. Physiol. Chem.* **350**:1353 (1969).

33. H.-Y. T. Yang and N. H. Neff, *Neuropharmacology* **15**:561 (1976).

34. I. Geller, R. Purdy, and J. H. Merritt, *Ann. N.Y. Acad. Sci.* **215**:54 (1973).

35. I. Geller and R. Purdy, in: *Alcohol Intoxication and Withdrawal: Experimental Studies* (M. M. Gross, ed.), Vol. II, pp. 295–301, Plenum Press, New York (1975).

36. W. M. McIsaac and V. Estevez, *Biochem. Pharmacol.* **15**:1625 (1966).

37. B. T. Ho, W. M. McIsaac, and L. W. Tansey, *J. Pharm. Sci.* **58**:998 (1969).

38. B. T. Ho, W. M. McIsaac, K. R. Walker, and V. Estevez, *J. Pharm. Sci.* **57**:269 (1968).

Pharmacologic and Electrophysiologic Effects of Ethanol in Relation to Sleep

24

Wallace B. Mendelson

A. INTRODUCTION

Sleep research and the study of ethanol dependence intersect on several common grounds. Clinically, the soporific effects of acutely ingested ethanol and the chronic sleep disturbance of the "dry" alcoholic are well recognized. The hallucinations which occur during delirium tremens have been postulated by some authors to be associated with the pathological daytime occurrence of certain sleep phenomena. Moreover, the metabolism of the neurotransmitters thought to play a major role in sleep regulation—the biogenic amines and acetylcholine—have been found to be altered by the ingestion of ethanol. These interrelationships will be explored in this review, following a brief summary of current understanding of the regulation of sleep.

B. PHYSIOLOGY OF SLEEP

Since the first electrophysiologic recordings of sleep in the 1930s, a fundamental principle of research in this area has been that sleep can usefully be divided into several discrete stages, based on the interpretation of simultaneously recorded electroencephalogram (EEG), electrooculogram (EOG), and electro-

Wallace B. Mendelson • Laboratory of Clinical Psychopharmacology, Division of Special Mental Health Research, Intramural Research Program, National Institute of Mental Health, Saint Elizabeth's Hospital, Washington, D.C. 20032, and Unit on Sleep Studies, Biological Psychiatry Branch, Division of Clinical and Behavioral Research, Intramural Research Program, National Institute of Mental Health, Bethesda, Maryland 20014.

myogram (EMG). Following the observation of rapid-eye-movement (REM) sleep in 1953, this basic classification has been thought to include REM sleep and four stages of non-REM sleep. (REM sleep, non-REM sleep, and the waking state are often referred to as the three "states of consciousness.") The components of non-REM sleep are: stage 1, in which the EEG shows a low-amplitude, mixed-frequency signal; stage 2, characterized by the appearance of sleep spindles and K complexes; and stages 3 and 4, in which moderate or large amounts of delta waves appear. Stages 3 and 4 are also referred to collectively as "slow-wave sleep" (SWS) or "delta sleep." REM sleep is characterized by an activated EEG consisting of low-voltage, mixed-frequency signals, conjugate eye movements, and atonia of the major muscles. Much interest has focused on it because of its frequent association to dreaming, although it should be noted that some kinds of mentation also occur in non-REM sleep. Also characteristic of REM sleep are monophasic waves, which can be recorded in the pons, geniculate body, and occipital cortex, referred to as "PGO waves."

The sleep stages occur cyclically, in a non-REM/REM rhythm which occurs approximately every 90 min in the young adult. The relative content of these cycles also changes as the night progresses, with more delta sleep occurring early at night, and relatively more REM sleep appearing toward the following morning.

C. NEUROTRANSMITTERS AND SLEEP

Theories of the regulation of sleep must necessarily account not only for the appearance of each sleep stage, each with its attendant electrophysiologic properties, but also must explain the cyclic occurrence of REM and non-REM sleep. Theories of sleep regulation, couched in purely electrical terms ("dry neurophysiology"), which suggested that sleep is a passive condition occurring whenever there is inadequate discharge of the reticular activating system, have been less adequate in explaining these cyclic phenomena. Perhaps a more fruitful approach has emphasized "wet neurophysiology," the study of the actions of neurotransmitters. The seminal studies of Jouvet have emphasized the role of biogenic amines (norepinephrine, serotonin, and dopamine), and of acetylcholine. Central to this theory were observations that lesions of serotonergic neurons in the raphe nuclei of the midbrain and pons acutely led to insomnia, while destruction of the noradrenergic neurons of the locus coeruleus in the pons led to reduction of REM sleep. On the basis of such lesion studies and similar pharmacologic manipulations of the biogenic amines, Jouvet[1] speculated that non-REM sleep is initiated and maintained by the firing of serotonergic neurons of the rostral raphe nuclei. REM sleep is "primed" by activity of the serotonergic neurons of the caudal raphe, while the "executive" control of some aspects of REM sleep is regulated by catecholamines and acetylcholine. Thus, EMG suppression during REM sleep might be regulated by noradrenergic neurons located in the caudal third of the locus coeruleus of the pons, while noradrenergic neurons in the medial third regulate the occurrence of PGO waves and other components of

REM sleep. Wakefulness and cortical arousal are thought to be regulated by noradrenergic neurons in the locus coeruleus, dopaminergic neurons in the mesencephalic reticular formation, and cholinergic cortical neurons. The utility of this theory and its relation to newer data is a complex subject which has recently been reviewed in detail.[2] In general, one difficulty has been that many of the acute effects of altering the biogenic amines—such as the insomnia induced by lesions of the raphe or by administration of parachlorophenylalanine—have tended to disappear in more chronic studies. Another has been that several studies have not agreed with the original conception that even acute depletion of brain serotonin reduces sleep time. Nor have single-unit recording studies found evidence suggesting that serotonergic neurons regulate the initiation or maintenance of non-REM sleep. There are, however, some lines of evidence to suggest that serotonergic neurons in the raphe play an important role in the regulation of PGO spikes, confining their occurrence to REM sleep.[3-6] In general, it seems more useful to think of these groups of aminergic neurons not as a series of relatively isolated systems, each controlling a given aspect of sleep, but rather as a homeostatic mechanism achieved by a balance of activity among several different neurotransmitters.[2]

Another theory, which emphasizes the interaction of neurotransmitter systems and which has been derived largely from single-unit studies in head-restrained cats, has been described by Hobson et al.[3] The basic observation leading to this was that noradrenergic cells of the locus coeruleus fall into two groups. "D-ON" cells fire more rapidly in waking, non-REM, and REM sleep, in ascending order, while the opposite pattern obtains for "D-OFF" cells. In the Hobson–McCarley model, "D-OFF" cells fire in a reciprocal relationship to cholinergic neurons of the gigantocellular tegmental fields (or FTG of Berman), which fire more rapidly during REM sleep. The "D-OFF" cells are thought to be inhibitory both to themselves and the FTG cells, whereas the firing of the latter are thought to be facilitory to themselves and the locus coeruleus. This reciprocal relationship is postulated to be the genesis of the REM/non-REM sleep cycle. Unlike Jouvet's theory, this model suggests that some noradrenergic neurons are actually inhibitory to REM sleep. Several aspects of the Hobson–McCarley model require further elucidation. There is some question as to the relative amounts of "D-OFF" and "D-ON" cells in the locus coeruleus. Hobson and McCarley suggest that the former represent perhaps 60% of the cells, whereas Chu and Bloom[7] found that the fraction may be more like one-fourth. It has also not yet been clearly demonstrated that the FTG cells in question are cholinergic, although it does appear likely that they are cholinoceptive. Finally, Siegel and McGinty[8] suggest that in freely moving cats the firing of FTG units is not specific for REM sleep, but rather occurs in several types of specific motor activation situations. In perspective, these objections are perhaps of less concern than is the useful conception that the sleep cycle may be regulated by the interaction of reciprocally innervated neurotransmitter groups, which can be profitably explored by further single-unit studies.

A third approach to the regulation of sleep has been the suggestion that a circulating "hypnotoxin," which accumulates during waking, may initiate or maintain sleep. Such substances have been derived by obtaining material from the brain or cerebrospinal fluid of sleep-deprived animals and then observing the behavioral or electrophysiologic effects when it is administered to naive animals.[9,10] Other studies have described sleep-inducing properties of perfusates from the reticular formation of sleeping cats[11] or dialysates from the blood of rabbits with cortical EEG slow waves induced by electrical stimulation of the thalamus.[12] These substances are thought to be peptides, and in one case the amino acid sequencing has been performed.[13] One problem with this area of research has been that human Siamese twins[14] and animal cross-circulation studies[15] have often (but not always)[16] shown independent sleep patterns. Nonetheless, there is a very real possibility that circulating peptides may play some form of modulating role in the occurrence of sleep and waking.

D. ETHANOL AND SLEEP

Given the complex balance of neurotransmitter systems (aminergic, cholinergic, and possibly others) that may regulate sleep, and the multiplicity of effects of ethanol on the metabolism of these substances, it would have been quite surprising if ethanol did not influence electroencephalographically monitored sleep. Ethanol administration does, of course, have a variety of influences on sleep; these will now be considered, first in animals and then in humans.

1. Animal Studies

In one of the earliest studies in this area, Yules et al.[17] administered 1 g ethanol/kg by gastric tube to cats and recorded 7-hr EEGs. Percentage REM sleep time was found to be decreased relative to preethanol control nights, particularly in the first half of the recording. This resulted from a decrease in the mean length of each REM episode. (Interpretation of these data must be made with caution because of an additional methodological variable—the authors deprived these animals of REM sleep by nonpharmacologic means before performing sleep recordings.) Ethanol-induced changes in sleep were characterized more fully by subsequent dose–response studies. Branchey et al.,[18] who administered 0.5 and 1.5 g ethanol/kg intraperitoneally to 400- to 450-g rats, reported that the lower dose increased the amount of non-REM sleep (expressed as a percentage of total recording time). At the higher dose, non-REM sleep again rose, while the percentage of REM sleep decreased. Mendelson and Hill[19] administered several doses (1.1–2.5 g/kg) intraperitoneally to 200- to 250-g rats and found the same pattern, although at a somewhat higher dose range: REM sleep (minutes and percentage of total sleep) decreased, and the percentage of non-REM sleep increased with 2.5 g/kg when compared to animals who received saline or 1.1 g ethanol/kg. The decrease in REM sleep was related to a decrease in the number of episodes of REM sleep and an increase in the length of the REM/non-REM cycle (the mean time from the beginning of one REM sleep episode until the next). Other data, which (perhaps indirectly) provide some information about short-term administration of ethanol and sleep, are derived from studies by (1) Hill and Reyes,[20] whose study was designed to assess the effects of chronic self-administration of ethanol but includes data on ethanol-naive groups who received an intraperitoneal injection of 2.0 g ethanol/kg (SE group) or saline (SS group); (2) Hill and Reyes,[21] who compared the effects of orally administered tryptophan (600 mg/kg), ethanol (2.0 g/kg), and placebo; and (3) Mendelson et al.,[22] in whose study approximately 10 g ethanol/kg per 24 hours was given in multiple doses by nasogastric tube during the 4-day induction period in a model of ethanol dependence. The sleep during this period may be compared to preinduction days, in which animals received saline. As the blood ethanol concentrations (BEC) on the third induction day were substantially higher than on the first day, a comparison of sleep on these days also provides some indirect data on the relation of blood ethanol concentrations to sleep changes.

The studies described here cannot be compared with each other in excessive detail because they vary in definitions of terms and amounts of sleep data presented. Still, some patterns do seem to emerge. First, doses in the range 1.5–2.5 g/kg have no effect on total sleep time over a 7- or 8-hr recording in the rat,[19-22] although there may be some increase in the first 3.5 hr.[21,22] The latter phenomenon may be related to the consistently reported decrease in sleep latency[19,21,22] and the increase in non-REM sleep[18,20-22] previously discussed. The increase in non-REM sleep may reach significance in only the first 3.0–3.5 hr[20,22] or it may be significant for the entire 7 or 8 hr, although the effect is primarily in the first 3.5 hr.[21] This phenomenon may at least partially relate to blood ethanol concentrations, as well as duration of treatment. Mendelson et al.[22] found an increase in percentage non-REM sleep in only the first 3.5 hr on the first induction day, when blood ethanol concentrations were 164 ± 15 mg/dl; on the third day when they had risen to 291 ± 17 mg/dl, non-REM sleep was increased for the entire recording (see also Ref. 18). The duration of increase in non-REM sleep may also be influenced by route of administration. When Hill and Reyes[21] gave 2.0 g/kg by oral intubation, there were about 20 min more non-REM sleep in both 3.5-hr periods of the recording compared to saline controls; when they administered the same dose intraperitoneally,[20] there were about 3 min more in the first 3.5 hr, and no difference in the second 3.5 hr.

In response to acute administration of high doses of ethanol, REM sleep (minutes and percentage of total sleep) decreases over 7 or 8 hr, a change of relatively greater significance in the second half of the recording.[19,21,22] Mendelson and Hill,[22] for instance, found that during the first half of the recording REM time was 7.3 ± 1.9 min for saline animals and 4.4 ± 1.8 min for animals receiving 2.5 g ethanol/kg (difference not significant). During the second half the values were 29.4 ± 2.9 and 18.6 ± 2.0 min, respectively ($p < 0.05$). Thus, although the relative decrease in REM time induced by ethanol was about the same in each half (about 40%), the decrease in minutes was greater in the second half, and was reflected as a statistically significant change (also see Ref. 21). It should be noted that this effect occurs only at higher doses, and was not seen with 1.5 g/kg in the rat[18] or 1.0 g/kg in the cat.[17]

The mechanism whereby REM sleep is reduced is not entirely clear. Mendelson and Hill[19] and Mendelson et al.[22] found a decrease in the number of REM episodes, and the Hill and Reyes[21] study may have reflected a nonsignificant trend in that direction (a decrease from 34.1 ± 15.5 to 18.6 ± 6.1 episodes on placebo and 2.0 g ethanol/kg, respectively). Mendelson and Hill[19] found this to be caused by an increase in the length of the REM/non-REM cycle, and there appeared to be a nonsignificant change in that direction in Mendelson et al.[22] In a study of substantially lower doses (1.0 g/kg) administered orally in cats,[17] the change was thought to be in the length of REM episodes. There is general agreement[17,19,21,22] that REM latency is not affected by acute doses of ethanol.

A summary interpretation of these data on acute effects of relatively high doses of ethanol on the sleep of rats is as follows: Sleep latency is decreased and total sleep is increased, primarily in the first 3.5 hr. This is due not only to the decreased sleep latency but also to an increase in non-REM sleep during the first 3.5 hr. REM sleep time is decreased over 7 or 8 hr. In contrast to non-REM sleep, the decrease in REM sleep (in the higher-dose studies) tends to be as great or greater in the second 3.5 hr of a 7-hr daytime recording. Although REM latency is relatively unaffected by ethanol, there is a decrease in the number of REM episodes. This may reflect an increase in the REM/non-REM cycle length.

Several studies have examined changes in sleep following repeated administration of ethanol. Yules et al.[17] whose work with cats given 1.0 g ethanol/kg was described earlier, continued to administer ethanol for four nights. Their data, which consisted only of REM sleep measures, showed that after the initial drop of REM percentage on the first two nights, values returned to baseline levels on the third and fourth nights. As will be seen later, a similar phenomenon occurs in nonalcoholic human volunteers to whom similar doses of ethanol are given for several days. There has been little available animal data, however, on sleep during withdrawal from repeated administration of ethanol. Mendelson et al.[22] presented such information when characterizing the sleep of rats in a previously developed model of ethanol dependence based on behavioral observations.[23] The EEG findings during the 4-day induction period described earlier consisted of an initial increase in non-REM sleep, followed on the third day by a greater increase in non-REM and a decrease in REM. (This presumably reflected the increased blood ethanol concentrations achieved due to accumulation from multiple doses.) When ethanol was discontinued, there were a variety of changes in sleep. On the first day of withdrawal, sleep time was reduced by 65% compared to the baseline values, primarily due to loss of non-REM sleep. By the third day total sleep time returned to baseline values, and in fact slightly exceeded them. At this time, REM sleep increased to twice the baseline values, primarily due to decreased REM latency and an increase in the number of REM sleep episodes. REM sleep percentage (and all other sleep parameters) returned to baseline levels on the fourth withdrawal day.

Mendelson et al.[22] also compared the preceding sleep data with blood ethanol concentration values and numerical scores of withdrawal.[23] It was found that the EEG was showing profound withdrawal changes—greatly reduced sleep time at the beginning of the recording, which started 7.5 hr from the last dose of ethanol—well before the earliest behavioral signs of withdrawal were occurring at 10 hr. EEG changes also persisted longer than the behavioral signs of withdrawal; by the third withdrawal day the marked "REM rebound" was at its highest, yet all behavioral signs were greatly diminished and approaching normality. The blood ethanol concentration was found to be significantly related in an inverse manner to both the withdrawal behavioral score and to the EEG sleep latency.

Given these data that EEG sleep disturbances persist for several days after cessation of ethanol administration, and the well-recognized clinical sleep disturbances of the withdrawing and dry alcoholic, one might wonder whether chronic ethanol consumption alters CNS functioning even after ethanol levels are greatly declining or gone. Two studies have examined the possibility that previous chronic ethanol consumption may alter the response to a rechallenge with ethanol. Gitlow et al.[24] administered 4–9 g ethanol/kg orally to rats for 4–6 weeks, then fed them a more conventional diet without ethanol for 6–8 months. At this time a sleep recording revealed no differences in REM sleep time or number of REM sleep episodes compared to saline-fed controls. When a test dose of 4 g ethanol/kg was administered, however, the two groups responded differently: in both groups REM sleep time and number of REM sleep episodes declined, but this effect was much more pronounced in the animals which had previously received ethanol. Hill and Reyes[20] examined this same phenomenon by placing rats on a 12-day diet in which the only source of fluid was 10% ethanol (w/v) in a 1% saccharin solution. Animals who drank minimally were eliminated from the study, and those who consumed 7.1–9.8 g/kg daily were examined. This amount of consumption was thought to be nonintoxicating, and previous data suggested that it might produce a BEC of 60 mg/dl at 9:00 hr, the presumed time of maximum consumption. Following 2 days in which the animals received distilled water, they were given 2 g ethanol/kg intraperitoneally (these are referred to as the EE group), and their sleep was compared to that of rats who had been on an ethanol-free diet and were then challenged with either ethanol (SE) or saline injections (SS). The comparison of the SS and SE groups has been described earlier in the discussion of the effects of ethanol on ethanol-naive rats. When all three groups were compared, it was found that there was no significant difference in decrease in REM sleep among the three groups, in contrast to the Gitlow et al.[24] study, in which the rats previously exposed to ethanol had an exaggerated decrease in REM time in response to ethanol. The three groups did respond differently in terms of waking time, however. A two-way analysis of variance (treatment × half of recording) was significant, indicating that in the SS and SE groups there was a relative reduction in waking time from the first to second half, whereas waking time remained about the same across the recording in the EE group. The implication, then, is that previous ethanol consumption may alter the sleep response to rechallenge with ethanol 2 days after the previous ethanol consumption has been discontinued. Whether this differential response would have occurred had there been a longer period between cessation of ethanol consumption and rechallenge with ethanol is unknown.

2. Nonalcoholic Human Volunteers

In contrast to the relative paucity of data on the effects of ethanol on the sleep of animals, there are a variety of studies on ethanol and human sleep. The interested reader is referred to the extensive reviews on this subject.[2,25,26] Human sleep and ethanol will be discussed here, first in terms of studies on normal subjects and then regarding alcoholics.

There has been a relatively consistent literature developed on the acute effects of ethanol administration to nonalcoholic humans.[27-31] In general, the expected sedative effect has been demonstrated; as in the animal studies, this is reflected more in a decreased sleep latency than in changes in total sleep time. The effects on REM, slow-wave sleep, and waking time are summarized in Table I. Decreased REM sleep is evident, primarily in the first half of the night. This may occur even when ethanol was administered several hours before sleep,[29] resulting in bedtime blood ethanol concentrations of 50 mg/dl, below that associated with obvious inebriation. In some studies there is an actual increase in percentage REM in the second half of the night.[25,30,31] Thus, in some cases the decrease in REM percentage is "washed out," with no resultant change over the entire recording,[25] and in others it is manifest as a decrease.[27-29] The explanation for the inconsistency in whether REM sleep is suppressed for the whole night may be that it is a dose–response problem. Knowles et al.[30] for instance, found that when a subject was given 3.5 oz of ethanol, there was the familiar pattern of decreased REM percentage in the first half and increases in the second half. When 6.0 oz was given, the REM percentage decreased in both halves. The inconsistency in the data may be due to the historical phenomenon that virtually all human studies on normal subjects have been performed on the clinically substantial dose of 1.0 g/kg [in the range of the higher dose of Knowles et al.[30]]. It may be that this dose is somewhat less than that needed to consistently suppress REM throughout the night.

The available rat studies, described previously, have not shown the biphasic effect on REM percentage. In the data of Mendelson and Hill,[19] for instance, lower doses (1.1–2.0 g/kg intraperitoneally) had no effect on REM percentage, while a higher dose (2.5 g/kg) decreased REM percentage over the entire 7-hr recording. The reason this effect has not been seen in rats is unclear; there are, of course, a variety of issues in difference in rate of metabolism of ethanol, dose range, and method of administration which may obscure this effect.

The human studies do not suggest a clear mechanism by which REM sleep is decreased. Most of the data indicate that there is probably no change in REM

Table I. Effects of Ethanol on Sleep of Nonalcoholic Humans[a]

Subjects	Dose (g/kg)	Duration of administration (days)	Sleep stages on first night of administration[b] % REM	% SWS	% Waking	Sleep stages on subsequent nights of administration[b] % REM	% SWS	% Waking	Sleep stages during withdrawal[b] % REM	% SWS	% Waking	Comment	Reference
7	1	5	→			←							27
3	1	5	→	↑	↑	↑	↑		←	↑	↑ ↑		28
4	1	3	→ (Stage 4)	←		↑			←	→ (Stage 4)			29
1	0.5–1.0	10	↓[c]		↑	↑		↑	↑		↑	Administration 4 hr before sleep	30
6	0.87	3	→[d]	↑		↑	↑	↑	↑	↑	↑	Administration for 10 nonconsecutive days	25
7	0.9	3	→[d]	↑		↑	↑	↑	↑	↑	↑		31
10	0.87	1	→[d]	←	→				↑	↑	↑		25
10	0.9	1	↑	↑	↑				↑	↑	↑		31

[a] Modified from Mendelson et al.[2]; reprinted with permission.
[b] ↑, Increase; ↓, decrease; →, no change.
[c] Decrease in % REM only in first half of night at lower dose.
[d] Decrease in % REM only in first half of night.

cycle length, although a decrease was described by Rundell et al.[31] in a multiple-night (but not a single-dose) study. One study[29] found a decrease in the mean length of REM episodes, and two[25,31] found a decrease in the length of the first REM episode.

When ethanol is given for 3–5 days to normal volunteers in doses of 0.9–1.0 g/kg nightly, the initial decrease in REM percentage returns to baseline (or slightly above baseline) values by the second or third night.[25,28,31] In two studies in which ethanol was given for three nights, there was no increase in REM percentage (the "rebound" phenomenon) upon discontinuation of ethanol[25,31]; after five nights of ethanol, a REM rebound was observed.[28]

In summary, ethanol when given to nonalcoholic human volunteers results in a decreased percentage REM sleep in the first few hours of sleep. In the second half of the night, REM sleep may return toward normal or may actually increase above normal, demonstrating a biphasic effect. The REM-suppressing effects of ethanol may occur at declining blood levels lower than those necessary to produce intoxication.[29] The human studies, once again similar to the animal studies, do not make clear which REM sleep parameter accounts for decreased REM; REM latency is probably the least likely candidate, never having been reported changed. When ethanol is given for several days to normal humans, REM sleep percentage returns to normal levels after several days, and there may be a "rebound" increase starting immediately after ethanol is discontinued.

The REM rebound phenomenon has been the basis of two themes often seen in the literature on ethanol and sleep. The first of these is that ethanol induces a "self-sustaining disregulation" of sleep even when it is declining or entirely gone from the body.[29] The second is that the REM rebound is particularly characteristic of drugs which induce dependence.[32,33] These concepts will be explored in the forthcoming discussion of sleep in chronic alcoholics.

3. Sleep in Chronic Alcoholics

The sleep of abstinent alcoholics has been compared with normal values,[34,35] or with values during drinking and withdrawal,[35–39] as summarized in Table II. In the study by Lester et al.,[35] alcoholics abstinent for at least 3 weeks tended to have more REM and stage 1 sleep and less stage 3 sleep than age-matched normal volunteers. Younger alcoholics (ages 24–39) also had less stage 4 sleep. (This was not apparent in comparing older alcoholics with controls, presumably because of the normal age-related decrease in stage 4 seen in the control subjects.) The increased REM was associated with a shortening of the REM/non-REM cycle. The sleep of the alcoholics also contained more arousals and changes of sleep stages (which is thought to be a rough measure of disturbance of sleep). The decreased slow-wave sleep and increased number of stage changes persist for a long time in the abstinent alcoholic—perhaps 1 or 2 years.[34] Wagman and Allen[40] presented data implying that the recovery of slow-wave sleep might require 200 weeks; Gross et al.,[38] who examined

Table II. Effects of Ethanol on Sleep of Chronic Alcoholics[a]

Subjects	Dose[c]	Duration of administration (days)	Sleep stages when drinking compared to abstinence[b]			Sleep stages during withdrawal compared to abstinence[b]			Comment	Reference
			% REM	% SWS	% Awake	% REM	% SWS	% Awake		
A. Studies with data on abstinence, drinking, and withdrawal										
3	30–150 ml of 43% q. 4 hr	4–10	↓			↑				36
12	4.4 oz	14–32			↓			↑	Behavioral data only	37
4	3.1 g/kg	15	↓	↑	↓	↑	↓	↓		38
17	150 mg/dl[d]	2	↓	↑	→	↑	↑	↑	Changes in first half of night	35
14	50–300 mg/dl[d]	12	↓ in 36%			↑ in 29%	→ (Stage 3)		Data estimated from graphs of individual points	39
B. Studies comparing sleep during drinking with withdrawal										
14	150 mg/dl[d]	2				↑	↑	↑	Decreased intermittent waking	42
6	8 oz	3–7				↓	↑	↓		44
C. Studies comparing abstinent alcoholics with normal values										
	Duration of abstinence					Sleep stages compared to normal values				
10	1–2 years					↑	↑	↑	Uncontrolled	34
17	3 weeks minimum					→	→	↑		35
D. Studies comparing sleep during acute withdrawal to normal values										
	Duration since drinking (days)					Sleep stages compared to normal values				
4	0–4					↓	→	↓	Uncontrolled	43
14	0					↓		↓	Uncontrolled	36

[a] Modified from Mendelson et al.[a]; reprinted with permission.
[b] ↑, Increase; ↓, decrease; →, no change.
[c] Method of presentation of dosage varies in different studies. When sufficient information was provided, this was converted to g/kg; otherwise, it is listed as presented in the papers.
[d] Blood ethanol concentration.

decreased slow-wave sleep in acute withdrawal, have suggested that its return to normal may be a useful marker of the return to health. Smith *et al.*[41] have remarked that the disturbed sleep and decreased slow-wave sleep of the alcoholic is similar to that of the elderly.

When abstinent alcoholics do drink, their sleep response has some similarities to that of normal subjects. There is a decrease in REM percentage,[35,36,38] and there may be an increase in percentage slow-wave sleep.[35,38] Analogously to the Rundell *et al.*[31] study in normals, Lester *et al.*[35] found that ethanol induced a decrease in the interval between REM episodes. Mello and Mendelson,[37] observing behaviorally defined sleep in alcoholics spontaneously consuming ethanol, reported that when drinking the alcoholics tended to sleep in multiple brief episodes, although there might be an increase in total sleep time over a 24-hr period.

When ethanol is discontinued in an alcoholic, there is often, but not always,[42] a decrease in total sleep.[36,38,43,44] Usually after four to six nights the amount of REM sleep increases, often greatly exceeding normal amounts and sometimes comprising more than 90% of sleep time.[38,42,44] The mechanism of the increased REM appears to be an increased number of episodes and decreased time between episodes.[42] In alcoholics who are hallucinating, REM (which in some cases does start to increase on the first or second withdrawal night) may occupy over 90% of sleep time.[36,43] Greenberg and Pearlman[36] reported that the increase in REM was greater in those alcoholics who developed delirium tremens than in those who had nonpsychotic abstinence syndromes. Wolin and Mello[45] noted that although the association of a REM rebound and the development of hallucinations was not inevitable, there was some positive association between the two phenomena.

During the hallucinations of alcohol withdrawal syndromes, patients may have "non-emergent stage-1 REM," in which the eyes are closed, there is prominent alpha, and rapid eye movements occur[43]; similarly "pre-sleep REM," with a low-voltage fast EEG and active eye movements, has been described in patients with delirium tremens.[36] Observations such as these have led to two hypotheses relating REM sleep to ethanol dependence. The first suggests that the REM rebound contributes to the manifestation of hallucinations during withdrawal; the second concludes that the REM rebound is characteristic of drugs with a large addiction potential. These hypotheses will be discussed in turn.

E. POSSIBLE RELATIONSHIPS OF REM SLEEP PHENOMENA TO ETHANOL DEPENDENCE

1. The REM Intrusion Hypothesis

The notion that hallucinations during acute withdrawal reflect intrusions of REM sleep into waking, based on studies such as those described above, might have been anticipated as early as 1881, when Lasegue wrote a paper entitled

"Alcoholic Delirium Is Not a Delirium, but Is a Dream."[46] In more modern form, this is part of the "REM intrusion hypothesis," which has been reviewed by Vogel.[47] There are, of course, a variety of possible relationships between the REM rebound and withdrawal-related hallucinations: (1) it may be that hallucinations do represent an intrusion of REM sleep phenomena into waking, (2) there may be no relationship between the two phenomena, or (3) both may be reflections of some more fundamental central nervous system event. Any hypothesis linking the two would have to allow for the occurrence of hallucinations in alcoholics in whom amounts of REM sleep are normal,[39] and for the occurrence of REM rebounds without hallucinations in alcoholics.[39] Part of the explanation may be that there is a quanititative difference in the amount of REM in those having hallucinations and those who do not.[36] Another possibility is that some additional change in nervous system function must be present before hallucinations occur. Feinberg[48] has suggested that the second factor may be defective regulation of stage 4 sleep.

Another difficulty with the REM intrusion hypothesis is that REM rebounds may occur without hallucinations or nightmares after cessation of moderate doses given to normal volunteers for only a few days (e.g., Ref. 28). Similar REM rebounds without psychotic symptoms may occur following withdrawal of other agents such as Δ^9-tetrahydrocannabinol from volunteers.[49] The question that arises is whether there is some difference in the nature of the rebound in an alcoholic and in a normal volunteer with only a few day's consumption. Two differences come to mind. First, in an alcoholic the rebound is usually, but not always, delayed for a few days after cessation of drinking.[38,42,44] This is seen also in an animal model as reported by Mendelson et al.,[22] in which the REM rebound does not occur until the third day of ethanol withdrawal [the same time, incidentally, at which the REM rebound occurs during morphine withdrawal in rats[50]]. The second quality that seems to differentiate the REM rebound in alcoholics is that, unlike that of normal volunteers, it is preceded by seriously disrupted sleep. The clinical sensation of insomnia[36] or behavioral fragmentation of sleep time over the 24 hr[37] have been well described. EEG studies have reported initial nocturnal sleep loss[38,43,44] in most, but not all[42] cases. Once again, this is also seen in an animal model,[22] where the first 2 days after withdrawal are characterized by greatly reduced sleep. In sum, the REM rebound of the addicted individual may differ from that of a normal volunteer in that there is an initial period of severely disrupted sleep, followed by a delayed REM rebound (which, one might speculate, can begin only when the disrupted architecture of sleep returns close enough to normal to "allow" REM sleep to occur).

2. The REM Rebound–Addictive Potential Hypothesis

The hypothesis that we have been considering stated that an REM rebound is characteristic of patients with either withdrawal hallucinations, or at least perhaps of less dramatic withdrawal syndromes in addicted individuals. A second, related, hypothesis is that even in short-term use in nonalcoholic subjects, the REM rebound is an indication of the addictive potential of a drug. Oswald[32] and Kales et al.[33] have noted a similar pattern—REM suppression, followed by a return to normal levels during repeated administration, then a rebound upon withdrawal—in a variety of other addicting agents, including amphetamines, morphine, and barbiturates. In contrast, other psychoactive agents which are not thought to have much potential for dependence—such as lithium and chlorpromazine—may acutely suppress REM but have little REM rebound upon withdrawal. Once again, the same range of possible relationships exists as in the REM intrusion hypothesis. The pattern of initial REM suppression followed by a withdrawal rebound and a potential for inducing dependence may be etiologically related, they both may be manifestations of some other process, or they may, of course, be unrelated. A hypothesis relating these processes would, however, have to account for such phenomena as these: (1) The REM rebound is not invariable. Feinberg et al.,[51] for instance, found that several barbiturates given to normal volunteers for 5–8 days did not produce a withdrawal rebound. (2) There may be no REM suppression during administration, but an increase after cessation of administration of amphetamines.[52] (3) The latter pattern may also occur with clearly nonaddicting agents, such as chlorpheniramine maleate.[33] Thus, as in the case of the REM intrusion hypothesis, the linkage of the REM rebound and addiction potential is a challenging notion that is suggestive, but does not explain a number of observations.

F. ETHANOL, NEUROTRANSMITTERS, AND SLEEP

This review has first summarized data suggesting that the biogenic amines and acetylcholine play an important role in the regulation of sleep, and then described the effects of ethanol on sleep. The question is whether the known effects of ethanol on these or other neurotransmitters might mediate the observed effects on sleep. The actions of ethanol on these transmitters are reviewed elsewhere in this volume. It seems clear that ethanol has a variety of effects on several different amines. Among the more consistently found actions are a decrease in steady-state levels of brain norepinephrine with chronic administration, and an increase in norepinephrine turnover with both acute and chronic administration.[53] Serotonin turnover may be reduced during acute or chronic administration, with no change in steady-state levels.[54] The effects on the cholinergic system are complex, including an initial increase in concentrations of acetylcholine in the brainstem and striatum, followed by a decrease as the animal becomes intoxicated (see Chapter 14). These actions are so varied and complex that there seems little profit in speculating on which specific biochemical action has which effect on sleep. One strategy, however, by which such mechanisms may be examined is to observe the effects of a drug with relatively specific neurotransmitter actions on a nervous system influenced by ethanol. At present there are at least two such sleep studies available, both involving the serotonergic system. Hill and Reyes[21] administered 2.0 g ethanol/kg and 600 mg L-tryptophan/kg by oral intubation to rats. When given separately, ethanol had no effect on REM latency, while L-tryptophan administration produced a small but significant decrease. When given in combination, however, the decrease in REM latency induced by L-tryptophan was greatly potentiated. The authors took this to mean that L-tryptophan and ethanol share a common mechanism of action, which they speculated might be an increased rate of serotonin synthesis. Several considerations come to mind in evaluating this possibility. One problem is that, although ethanol has been reported to increase serotonin turnover in mice,[55] it may decrease turnover in rats.[54] The latter process would less easily be reconciled with ethanol potentiating EEG effects of L-tryptophan. Another issue is the possibility that tryptophan produces its actions on sleep by some other mechanism than enhancing serotonin synthesis (see discussion in Ref. 2).

A second study exploring the interaction of pharmacologic manipulations of serotonergic activity and ethanol was performed by Zarcone and Hoddes.[56] In view of the possibility that abnormalities of REM sleep in "dry" alcoholics might be due to a deficit in serotonergic function, they administered 300 mg of the serotonin precursor 5-hydroxytryptophan nightly to alcoholics who had been abstinent for 23–133 days. It was found that the marked fragmentation of the REM episodes (low "REM efficiency") of the alcoholics was substantially decreased by 5-hydroxytryptophan. The authors concluded that there may be abnormal regulation of sleep by serotonergic neurons in alcoholics and that this defect persists during abstinence.

G. SUMMARY AND CONCLUSIONS

Acute and chronic administration of ethanol produces a variety of effects on sleep. Acutely, there is a sedative effect in the sense that there is decreased sleep latency, but there is usually not an increase in total sleep time over an 8-hr recording. Changes in REM sleep are very sensitive to ethanol; in rat studies, for instance, a decrease in REM sleep may persist into the second half of an 8-hr recording, when BECs are greatly decreased, and a decrease in REM occurs in humans even when ethanol was taken several hours before bedtime, so that blood ethanol concentrations are below levels associated with behavioral intoxication. The specific REM sleep parameter whose change leads to a decrease in overall amounts of REM sleep is not clear. Rat studies tend to indicate that an important factor is the decrease in number of REM sleep episodes; studies in nonalcoholic humans emphasize a decrease in the length of the first (or all) REM episodes.

Multiple-dose administration of ethanol produces initial decreases in REM sleep, followed by a gradual return to baseline amounts, and a rebound increase when ethanol is discontinued. There is some evidence that chronic administration of ethanol may induce long-lasting changes in sleep regulation. This is implied by animal data showing exaggerated decreases in REM sleep in rats who had been previously treated with chronic ethanol administration, then had been "dry" for 6–8 months, and later rechallenged with ethanol. Similarly, alcoholics who have been "dry" for 1 or 2 years continue to have decreased slow-wave sleep and an increased number of stage changes. When alcoholics do drink, there may be a decrease in percentage REM sleep and an increase in percentage slow-wave sleep. Upon withdrawal, there is usually a large "rebound" increase in REM sleep which often occurs following an initial period of very disturbed sleep, described both in human alcoholics and in an animal model of physical dependence.

The REM rebound has been the basis of two hypotheses relating sleep and alcoholism. The first states that the hallucinations which occur during ethanol withdrawal syndromes may represent the intrusion of REM sleep phenomena into the waking state. The second suggests that an initial REM suppression followed by a withdrawal rebound may be characteristic of drugs with high addiction potential when given for a few days to normal subjects. Both hypotheses have been of heuristic value, but as presently formulated cannot account for a variety of reported observations.

Among the many biochemical effects of ethanol are alterations in the metabolism of the biogenic amines and acetylcholine—the neurotransmitters which are thought to play a major role in the regulation of sleep. Whether the observed effects of ethanol on sleep are mediated by changes in these compounds or by some other means is not clear. In addition, the actions on the biogenic amines and acetylcholine are so complex that it is not possible at this time to relate a specific metabolic effect to a particular change in sleep. One strategy which might help to elucidate this relationship is to administer agents with relatively specific neurotransmitter effects to animals given ethanol or to human alcoholics. Some preliminary work in this area suggests that some of the actions of ethanol on REM sleep may be mediated by changes in serotonergic function.

H. REFERENCES

1. M. Jouvet, *Ergeb. Physiol. Biol. Chem. Exp. Pharmakol.* **64:**166 (1972).
2. W. B. Mendelson, J. C. Gillin, and R. J. Wyatt, *Human Sleep and Its Disorders*, pp. 1–260, Plenum Press, New York (1977).
3. J. A. Hobson, R. W. McCarley, and T. M. McKenna, *Prog. Neurobiol.* **6:**280 (1976).
4. W. Dement, V. Zarcone, J. Ferguson, H. Cohen, T. Pivik, and J. Barchas, in: *Schizophrenia—Current Concepts and Research* (D. V. Seva, ed.), pp. 775–811, PJD Publications, New York (1969).
5. D. C. Brooks, M. D. Gershon, and R. P. Simon, *Neuropharmacology* **11:**511 (1972).
6. D. J. McGinty, R. M. Harper, and M. K. Fairbanks, in: *Serotonin and Behavior* (J. Barchas and E. Usdin, eds.), pp. 267–279, Academic Press, Inc., New York (1973).
7. N. S. Chu and F. Bloom, *J. Neurobiol.* **5:**527 (1974).
8. J. M. Siegel and D. J. McGinty, *Science* **196:**678 (1977).
9. J. R. Pappenheimer, G. Koski, V. Fencl, I. Karnovsky, and J. Krueger, *J. Neurophysiol.* **38:**1299 (1975).

10. H. Nagasaki, M. Iriki, S. Inoue, and K. Uchizaro, *Proc. Jpn. Acad.* **50**:241 (1974).
11. R. R. Drucker-Colin and C. W. Spanis, *Prog. Neurobiol.* **6**:1 (1976).
12. M. Monnier, G. A. Schoenenberger, A. Glatt, L. Dudler, W. Mehlose, R. Gachter, and L. Knappova, 2nd Int. Sleep Res. Congr., Edinburgh, June 30–July 4 (1975).
13. M. Monnier, L. Dudler, R. Gachter, P. F. Maier, H. J. Tobler, and G. A. Schoenenberger, *Experientia* **33**:548 (1977).
14. H. G. Lenard and F. J. Schulte, *J. Neurol. Neurosurg. Psychiatry* **35**:756 (1972).
15. I. DeAndres, E. Gutierrez-Rivas, E. Nava, and F. Reinoso-Suarez, *Neurosci. Lett.* **2**:13 (1976).
16. J. Matsumoto, J. Sogabe, and Y. Hori-Santiago. *Experientia* **15**:1043 (1972).
17. R. B. Yules, J. A. Ogden, F. P. Gault, and D. X. Freedman, *Psychon. Sci.* **5**:97 (1966).
18. M. H. Branchey, H. Begleiter, and B. Kissin, *Commun. Behav. Biol.* **5**:75 (1970).
19. W. B. Mendelson and S. Y. Hill, *Pharmacol. Biochem. Behav.* **8**:723 (1978).
20. S. Y. Hill and R. B. Reyes, *J. Stud. Alcohol* **39**:47 (1978).
21. S. Y. Hill and R. B. Reyes, *Psychopharmacology,* **58**:229 (1978).
22. W. B. Mendelson, E. Majchrowicz, N. Mirmirani, S. Dawson, J. C. Gillin, and R. J. Wyatt, *J. Stud. Alcohol,* **39**:1213 (1978).
23. E. Majchrowicz, *Psychopharmacologia* **43**:245 (1975).
24. S. E. Gitlow, S. H. Bentkover, S. W. Dziedzic, and N. Khazan, *Psychopharmacologia* **33**:135 (1973).
25. H. L. Williams and A. Salamy, in: *The Biology of Alcoholism* (B. Kissin and H. Begleiter, eds.), Vol. 2, pp. 435–483, Plenum Press, New York (1972).
26. F. R. Freemon, *Sleep Research,* pp. 111–114, Charles C Thomas, Publisher, Springfield, Ill. (1972).
27. S. C. Gresham, W. B. Webb, and R. L. Williams, *Science* **140**:1226 (1963).
28. R. B. Yules, D. X. Freedman, and K. A. Chandler, *Electroencephalogr. Clin. Neurophysiol.* **20**:109 (1966).
29. R. B. Yules, M. E. Lippman, and D. X. Freedman, *Arch. Gen. Psychiatry* **16**:94 (1967).
30. J. B. Knowles, S. G. Laverty, and H. A. Kuechler, *Q. J. Stud. Alcohol* **29**:342 (1968).
31. O. H. Rundell, B. K. Lester, W. J. Griffiths, and H. L. Williams, *Psychopharmacologia* **26**:201 (1972).
32. I. Oswald, in: *Sleep: Physiology and Pathology* (A. Kales, ed.), pp. 317–330, J. B. Lippincott Company, Philadelphia (1969).
33. A. Kales, G. Heuser, J. D. Kales, W. H. Rickles, Jr., R. T. Rubin, M. B. Scharf, J. T. Ungerleider, and M. D. Winters, *Ann. Intern. Med.* **70**:591 (1969).
34. J. Adamson and J. A. Burdick, *Arch. Gen. Psychiatry* **28**:146 (1973).
35. B. K. Lester, O. H. Rundell, L. C. Crowden, and H. I. Williams, in: *Alcohol Intoxication and Withdrawal: Experimental Studies* (M. M. Gross, ed.), Vol. I, p. 261, Plenum Press, New York (1973).
36. R. Greenberg and C. Pearlman, *Am. J. Psychiatry* **124**:133 (1967).
37. N. K. Mello and J. H. Mendelson, *J. Pharmacol. Exp. Ther.* **175**:94 (1970).
38. M. M. Gross, D. R. Goodenough, J. Hastey, and E. Lewis, *Ann. N.Y. Acad. Sci.* **215**:254 (1973).
39. S. J. Wolin and N. K. Mello, *Ann. N.Y. Acad. Sci.* **215**:266 (1973).
40. A. M. I. Wagman and R. P. Allen, in: *Alcohol Intoxication and Withdrawal: Experimental Studies* (M. M. Gross, ed.), Vol. II, pp. 453–466, Plenum Press, New York (1975).
41. J. W. Smith, L. C. Johnson, and J. A. Burdick, *Q. J. Stud. Alcohol* **32**:982 (1971).
42. L. C. Johnson, A. Burdick, and J. Smith, *Arch. Gen. Psychiatry* **22**:406 (1970).
43. M. M. Gross, D. Goodenough, M. Tobin, E. Halpert, D. Lepore, A. Perlstein, M. Sirota, J. Dibianco, R. Fuller, and I. Kishner, *J. Nerv. Ment. Dis.* **142**:493 (1966).
44. R. P. Allen, A. Wagman, L. A. Faillace, and M. McIntosh, *J. Nerv. Ment. Dis.* **153**:424 (1971).
45. S. J. Wolin and N. K. Mello, *Ann. N.Y. Acad. Sci.* **215**:266 (1973).
46. C. Lasegue, *Arch. Gen. Med.* **88**:513 (1881).
47. G. W. Vogel, *Arch. Gen. Psychiatry* **18**:312 (1968).
48. I. Feinberg, P. H. Wender, R. L. Koresko, F. Gottlieb, and J. A. Piehuta, *J. Psychiatr. Res.* **7**:101 (1969).

49. I. Feinberg, R. Jones, J. M. Walker, B. A. Cavness, and J. March, *Clin. Pharmacol. Ther.* **17**:458 (1975).
50. N. Khazan and B. Colasanti, *J. Pharmacol. Exp. Ther.* **183**:23 (1972).
51. I. Feinberg, S. Hibi, C. Cavness, and J. March, *Science* **185**:534 (1974).
52. I. Feinberg, S. Hibi, M. Braun, C. Cavness, G. Westerman, and A. Small, *Arch. Gen. Psychiatry* **31**:723 (1974).
53. R. A. Lahti, in: *Biochemical Pharmacology of Alcohol* (E. Majchrowicz, ed.), pp. 239–253, Plenum Press, New York (1975).
54. W. A. Hunt and E. Majchrowicz, *Brain Res.* **72**:181 (1974).
55. K. Kuriyama, G. E. Rauscher, and P. Y. Sze, *Brain Res.* **26**:450 (1971).
56. V. P. Zarcone and E. Hoddes, *Am. J. Psychiatry* **132**:74 (1975).

Clinical Pharmacology of the Fetal Alcohol Syndrome

25

Henry L. Rosett

A. INTRODUCTION

Over the past 250 years, a vast body of clinical observations, together with experimental research of varying degrees of sophistication, suggested that ethanol during pregnancy had an adverse effect on offspring. In America, during the four decades between 1930 and 1970, these data were generally ignored or ridiculed. With the description of the specific morphologic characteristics of the fetal alcohol syndrome (FAS), the importance of investigating the various effects of ethanol on mother and offspring became evident.

Few teratogens have been as thoroughly investigated as ethanol. Multiple biochemical and pathophysiologic effects have been attributed to ethanol and its metabolites. Ethanol has the potential to alter the development of the embryo and fetus differently at various stages of pregnancy.

Prospective clinical studies as well as experimental animal research now support the view that there may be a wide range of fetal ethanol effects in the absence of the full syndrome. Fetal ethanol effects have been observed in offspring of women who drank heavily during pregnancy, but with a lower level and a more sporadic pattern than those of typical chronic alcoholic women.

Ethanol is our most widely used drug. The greatest increase in ethanol consumption during the past decade has occurred among young women of high school age. Some of this group probably will continue to drink heavily throughout their reproductive years. Protection of their offspring should become a public health goal of high priority.

Henry L. Rosett • Boston University School of Medicine, and Maternal Health and Child Development Program, Boston City Hospital, Boston, Massachusetts 02165.

B. HISTORICAL SURVEY AND CLINICAL DESCRIPTION OF SYNDROME

Concern about the adverse effects on offspring of ethanol consumed during pregnancy has been recorded since the time of Aristotle.[1] Observations of increased infant morbidity and mortality during England's gin epidemic (1720–1750) are substantiated by subsequent medical literature.[2] In 1899, Sullivan found higher rates of stillbirth and early mortality among children of alcoholic women as compared with offspring of nondrinking relatives.[3] He also observed that several alcoholic women who had infants with severe and often fatal complications later bore healthy children when abstinent during pregnancy.

In 1968, Lemoine et al.[4] described 126 offspring from 69 French families in which there was chronic alcoholism. A pattern of prenatal and postnatal growth deficiency, developmental delay, mental retardation, microcephaly, and a number of facial, limb, and cardiac defects were attributed to the effects of ethanol in utero. Five mothers who had given birth to abnormal children subsequently bore normal infants when abstinent during pregnancy.

Awareness of the problem was stimulated in 1970 with a report from the University of Washington in Seattle.[5] Chronic alcoholism was identified retrospectively in 11 women. Ten of their 12 infants (83%) were born undergrown for gestational age; their subsequent development was retarded. Among nonalcoholic mothers, only 2.3% of offspring were undergrown. In 1973, Jones et al.[6,7] described the fetal alcohol syndrome—a pattern of abnormalities in children born to alcoholic women almost identical to that which had been independently reported from France. The Seattle group has published their experience with 41 patients,[8] as well as psychological test findings.[9] Additional cases have been described by clinicians from around the world.[10–29] These severely affected infants were all born to mothers whose chronic alcoholism was identified retrospectively.

A recent review of 245 published cases of the fetal alcohol syndrome classified the characteristic features into four categories: central nervous system dysfunctions, growth deficiencies, a typical facial appearance, and various malformations.[30] The manifestations in the central nervous system include mild to moderate retardation, microcephaly, poor coordination, and hypotemia, as well as irritability and hyperactive behavior. Growth deficiency in both weight and length is usually present at birth and frequently persists during the postnatal period. The facial features observed most frequently are short palpebral fissures often associated with myopic small eyes, ptosis and strabismus, a hypoplastic upper lip with a thin vermilian and flattened philtrum, and midfacial and mandibular growth deficiency. Malformations of various other organ systems, particularly cardiovascular, renogenital, and skeletal, occur in the affected children with greater than the expected frequency.

Variability in the pattern of defects is probably related to differences at critical stages of pregnancy in terms of maximum blood ethanol concentrations, binge drinking versus relatively steady-state ethanol levels, types of beverages, general nutritional status, and synergistic effects of cigarette smoking. Genetic differences in susceptibility to the dysmorphogenic effects of ethanol probably are also important. A pair of fraternal twins born to a mother who consumed at least 4.8 dl of red wine and an unspecified amount of hard liquor daily throughout pregnancy were compared.[16] One boy was markedly affected, with weight and length below the tenth percentile and head circumference at the tenth percentile, while the weight of his twin brother was at the 30th percentile, length at the 15th percentile, and head circumference at the 60th percentile. Further clarification requires prospective studies in which quantity, frequency, and variability of ethanol consumption during pregnancy are investigated and the offspring are examined in the newborn period and then studied as they mature to identify later effects on central nervous functions.

C. PROSPECTIVE STUDIES

The largest prospective study in which ethanol consumption was systematically studied during the first trimester of pregnancy was conducted in Paris between 1963 and 1969.[31] Data on more than 9000 births at 12 maternity hospitals revealed that offspring of women who consumed the equivalent of over 4 dl wine/day had birth weights which were lower than those born to women who drank less than 4 dl/day. Women who drank the equivalent of 44 ml absolute ethanol or less daily bore infants who were no smaller than those born to women who abstained from ethanol. Those who consumed over 6 dl/day (66 ml absolute ethanol) had the smallest infants, but they were only 106 g lighter than those born to women who did not drink at all. No differences were found in the incidence of malformations.

The National Institute of Neurological Diseases and Stroke's Perinatal Project study of 55,000 pregnancies omitted direct questions about ethanol consumption.[32] However, a review of 69 charts in which alcoholism was mentioned revealed 23 in which there was clear evidence of chronic alcoholism before and during pregnancy.[33] Each of these cases was matched with two control cases and charts of the offspring were reviewed by an investigator who had no information about maternal drinking habits. Among the 23 offspring of the alcoholic women, four died in the perinatal period and six had physical characteristics consistent with those of the fetal alcohol syndrome. Offspring of the alcoholic women were smaller as newborns and at age seven, and their IQ at age seven also was significantly lower than that of the controls. The methodology and conclusions of this study have been questioned.[34,35]

A prospective study of women registering for prenatal care at Boston City Hospital revealed that about 10% were heavy drinkers; they consumed 5 or 6 drinks on some occasions and averaged at least 45 drinks per month.[36] Heavy drinkers consumed an average of 174 ml ethanol daily (2.2 g/kg per day).[37] Among offspring born to 42 heavy drinkers, as compared with those born to 280 women who abstained or drank moderately, the frequency of congenital anomalies, growth retardation, and functional abnormalities was more than doubled

($p < 0.001$). Major congenital abnormalities were found in 17% of the offspring of the heavy drinkers as compared with 3% in the other infants. While the heavy drinkers' babies often had more than one anomaly, the full fetal alcohol syndrome was not diagnosed.

A subsample of infants was also studied in terms of sleep–awake state regulation over a 24-hr period on the third day of life.[38] A continuous nonintrusive bassinet sleep monitor provided data suggesting disordered development of basic regulatory mechanisms in offspring of heavy-drinking women who continued to drink throughout pregnancy. Infants born to heavy drinkers who abstained during the third trimester had cyclical patterns of sleep and awake states more like those of control infants born to mothers who never drank heavily.

Within the group of 42 heavy-drinking women, 15 were able to abstain or moderate drinking for the third trimester as a result of counseling.[39] Infants born to the women who abstained or reduced drinking demonstrated fewer abnormalities than those born to mothers who continued heavy drinking ($p < 0.001$).

The only prospective study in which the complete fetal alcohol syndrome has been identified was conducted in Seattle on 163 selected offspring of 1529 mothers who had participated in a larger project.[40] When an infant was born to a mother who had reported consumption of 30 ml or more of ethanol daily, or of occasions when she had five or more drinks, a control infant was randomly selected from the offspring born the same day to mothers who were abstainers or infrequent drinkers. The two infants were then examined by a dysmorphologist who had no knowledge of the maternal drinking history. Of the 163 infants examined, 11 had features compatible with the fetal alcohol syndrome. Two of these infants were found to have enough features to be judged as clearly having the full syndrome; both were born to mothers diagnosed as alcoholic. Of the nine other infants with signs compatible with prenatal effects of ethanol, seven were born to women in the heavier-drinking group. While these results are suggestive and also statistically significant, the experimental design would have been improved if a larger number of control infants had been examined.

Naturalistic observations were conducted on a subsample of 124 one-day-old neonates from the University of Washington prospective study and recorded by means of an electronic data acquisition system.[41] Offspring of moderate to heavy drinkers as compared with infants born to abstainers or infrequent drinkers exhibited decreased vigorous limb activity and more time with eyes open and head positioned to the left ($p < 0.05$). On the second day of life 225 neonates were tested with one of two operant learning tables.[42] Moderate maternal ethanol intake coupled with moderate to heavy cigarette smoking exerted a deleterious effect on learning which did not occur when either substance had been used separately ($p < 0.05$). There is no evidence indicating that the performance measured in the neonatal period will predict later development. However, the fact that relatively small amounts of ethanol consumed early in pregnancy seem to influence neonatal behavior indicates the need for follow-up studies to determine if there are also effects on more advanced motor and cognitive development. Persistence of effects would suggest either long-lasting changes in the developing nervous system or else late behavior developing secondary to some earlier effects of ethanol on infant–caretaker interaction. Pharmacologic effects of ethanol on the fetus and neonate may be one determinant of later behavioral problems in the offspring of alcoholics.

The fetal alcohol syndrome (FAS) has been identified in at least 245 infants of differing racial backgrounds. Typically, FAS dysmorphology has been identified first and maternal alcoholism documented subsequently. Retrospectively, accurate information about alcohol consumption and other risk factors during pregnancy is often difficult to obtain. Prospective studies are necessary to determine the incidence of the fetal alcohol syndrome, the range of symptoms, and the relationship between anomalies and amounts of alcohol consumed at different stages of pregnancy and the other risk factors present.

Few infants clearly manifesting the morphologic characteristics of the complete fetal alcohol syndrome have been identified in prospective studies. There are many more cases with only parts of the syndrome; a single malformation, retarded growth and development, or behavioral patterns such as jitteriness and abnormal state regulation. These features, compatible with exposure to ethanol *in utero*, may result from various blood ethanol concentrations at critical developmental stages. Malformations probably are produced by high concentrations at specific periods during the first trimester when embryonic development of the central nervous system is taking place. Growth may be most vulnerable to heavy drinking during the second and third trimesters. Behavioral disturbances and intellectual impairment may be related to chemical and structural disruption of the central nervous system early in pregnancy or during the later period of cell division associated with rapid brain growth and functional organization.

D. EXPERIMENTAL MODELS OF THE FETAL ALCOHOL SYNDROME

1. Morphologic Effects

Separation of the multiple compounding variables which affect outcome of clinical studies is difficult. Ethical issues preclude human experimentation. Therefore, there is a long history of animal models. In 1888, Combemale reported mating two dogs and exposing the bitch to ethanol for the first 23 days of gestation.[43] Of the six pups, three were stillborn and three were of "weak intelligence." When they were mated with normal studs, the young were "defective."

In 1894, Féré exposed hens' eggs to ethanol vapors before incubation. Ethanol which penetrated the eggshell produced a broad range of abnormalities.[44] Similar effects were obtained by Stockard using fish and chicken eggs.[45] Anomalies of the eyes were most common, followed by abnormal development of the central nervous system. Experiments with mammals were stimulated by the demonstration by Nicloux that ethanol crossed the placenta and could be demonstrated in the fetus of the guinea pig, dog, and woman at concentrations close to that in the maternal circulation.[46,47] Hodge treated pregnant cocker spaniels with ethanol from 1895 to 1897, and observed in 1903 that there was both a greater percentage of deformity as well as less vigor and vitality in the offspring of the ethanolized dogs.[48]

Pearl treated hens by inhalation and found that ethanolized chicks were superior in viability and growth and suggested that ethanol "acts as a selective agent upon the germ cells of the alcoholized animals, eliminating the weak and permitting the survival of the vigorous and highly resistant."[49] MacDowell studied maze behavior in ethanolized white rats and found inferior learning ability in treated animals and their offspring.[50] For a review of the effects of ethanol on animal growth and reproduction, including a tabular comparison of 33 experimental regimens, see Wallgren and Barry.[51]

The many conflicting observations in the early literature can be related to species differences, as well as to the amount of ethanol administered, route of administration, stage of pregnancy when ethanol was introduced, and lack of paired feeding (matching food intake of the control animal with the amounts consumed by the experimental animal). Some of these problems have been overcome in the experimental design of recent animal research on the effects of prenatal ethanol exposure.

Sandor and Elias studied the effects of ethanol, simulating the blood ethanol levels found in human alcoholics in the early stages of development in chick embryos.[52] Early maldevelopment and mortality occurred in a considerable portion of the embryos, and the remaining specimens showed a significant loss of weight toward the end of the incubation. The central nervous system was the most sensitive. Sandor and Amels investigated the effects of ethanol on the prenatal development of albino rats.[53] Ethanol diluted by distilled water was administered intravenously in order to obtain dosages of 1.5 g/kg and 2 g/kg. Some embryos were removed at 9.5 days. Fetuses were examined at 19.5 days. Injections were made at 6, 7, and 8 days of pregnancy with one group, and at 6 and 7 days gestation with another group. Ethanol in the concentration of 2 g/kg body wt induced twice the number of malformations that 1.5 g/kg induced. Two heavy intoxications were more injurious than three lighter concentrations. Effects on the bones seemed to be most apparent in the extremities and facial areas, the same regions observed in humans showing signs of the fetal alcohol syndrome.

Chernoff administered ethanol orally via a liquid diet to two highly inbred strains of mice which differed in ethanol preference, ethanol dehydrogenase activity, and ethanol sleep times.[54] Blood ethanol levels were maintained in a range from 73 to 398 mg/dl for at least 30 days before mating and throughout gestation. Females were sacrificed on day 18 and the uterine contents examined. Fetal resorptions increased with increasing ethanol concentrations. There was a definite growth deficiency due to ethanol; beyond a point this deficiency was fatal. Fetuses were sectioned and anomalies affecting the skeleton, brain, heart, and eyes were observed. The pattern of growth deficiency, together with ocular, neural, cardiac, and skeletal anomalies, was similar to that of the fetal alcohol syndrome in humans.

Randall also administered a liquid diet with 25% of total daily calories supplied by ethanol to mice, but only administered it from gestation day 5 through gestation day 10.[55] Blood ethanol levels ranged between 70 and 120 mg/dl. A pair-fed control group was treated similarly except that sucrose substituted isocalorically for ethanol. Gravid females were sacrificed on gestation day 19. The ethanol-fed group implanted a larger number of ova, but twice as many of their fetuses were resorbed compared to the control group. The increase in resorptions decreased litter size so that the average number of fetuses and fetal weight were similar between groups. Fifteen of the 16 experimental litters had at least one malformed fetus; of the 29 control litters, only 5 had a malformed fetus. In both the ethanol and sucrose groups, defective fetuses weighed significantly less than their normal litter mates. In the ethanol-fed group, anomalies involved the limbs, including syndactyly, adactyly, and ectrodactyly of the forelimbs. No limb anomalies were observed in the control fetuses. Cardiovascular anomalies included abnormalities of both the major branches of the aorta and vena caval system, as well as intracardiac anomalies such as atresia of the mitral valve and intraventricular septal defects. Urogenital anomalies included hydronephrosis and/or hydroureter. Head anomalies included exencephaly, hydrocephalus, anopthalmia, and microopthalmia. These anomalies are analogous to those observed in the children of alcoholic mothers.

Ellis and Pick have reported efforts to develop an experimental model of the fetal alcohol syndrome in the beagle.[56] A high-dose group received 5.8 g ethanol/kg. The moderate-dose group received 4.5 g/kg, and a low-dose group received 3.0 g/kg. The high dose of ethanol induced severe physical dependence and completely suppressed intrauterine tissue differentiation and development of the fertilized and implanted ovum. The moderate dose produced dependence of a milder degree and permitted a more advanced intrauterine development with spontaneous abortion at 6–7 weeks of gestation, or in retention within the uterus of abnormal dead fetuses. A low dose produced no detectable dependence and was followed by birth of normal pups at term. A threshold dose of ethanol which will consistently produce malformations in viable fetuses has been found.

2. Behavioral Effects on Animal Offspring

There are many inconsistencies and contradictions in the literature on the effects of ethanol during pregnancy on the behavior of rat offspring. Several reasons for this become clear when the methodologies of different researchers are compared.[57–64]

Although animal experiments offer the potential for controlling many of the uncertainties which necessarily exist in collecting data from humans, such as the precise amount of ethanol, date of ingestion, blood ethanol concentration, and genetic characteristics, the literature contains many inconsistencies. Often these are due to differences in technique. The route of administration may be oral by means of a liquid diet with a proportion of the total calories supplied by ethanol, or by ethanol and water as the only source of liquid. In dogs, gastric intubation has been used. Intraperitoneal injection of ethanol may have direct effects on the ovaries and uterus. There also is considerable variation between researchers in terms of the time period selected for administration of ethanol. Some feed it to the mother throughout her pregnancy; others select a few critical days during embryonic development. Some continue the ethanol through lactation; others stop it at birth. Timing is critical in terms of producing morphologic abnormalities during specific growth periods. In rats, pups fail to grow if ethanol is administered to the mother while nursing. This effect is probably secondary to its inhibiting action on the release of oxytocin, essential for milk ejection. Poor growth also may be related to altered maternal behavior or aversion of the offspring to ethanol-contaminated milk. Collard and Chen found that mouse pups weighed less than their controls at weaning when ethanol was injected into the nursing mother.[65] When litters were raised in ethanol vapor chambers, pups weighed more at weaning and at 7 weeks of age than did controls, presumably owing to additional calories from the inhaled ethanol. Control animals should be pair-fed and receive the same number of calories as the experimental animals. The liquid diet, with the experimental animals receiving carbohydrate in the form of ethanol and the pair-fed control animals in the form of sucrose, seems to be the model closest to human alcoholism.

Despite methodologic differences, offspring exposed to ethanol *in utero* and during lactation generally show hyperactivity, impaired learning, and greater ethanol preferences. It has been suggested[57–59] that hyperactivity which abates with maturation and cognitive difficulties which persist also characterize hyperactive and minimal brain-damaged children.

E. MECHANISMS OF ETHANOL'S EFFECTS ON THE MATERNAL–PLACENTAL–FETAL SYSTEM

While the mother, placenta, and fetus interact as a dynamic system, it is helpful to consider the effects of ethanol in terms of mechanisms which directly affect the fetus, mechanisms which have their primary effects on maternal metabolism and physiology, and the associated maternal risk factors.

1. Direct Effects of Ethanol on Fetal Physiology and Metabolism

In 1900, Nicloux demonstrated that ethanol ingested by the mother crossed the placenta and reached the fetus in concentrations similar to that found in the maternal circulation.[47] Placental transfer and tissue distribution have recently been studied by employing ethanol labeled with radioactive carbon in the pregnant mouse, hamster, and monkey.[66,67] The radioactive ethanol and its metabolites distributed quickly throughout the bodies of the mother and fetus.

a. Acute Effects

The use of intravenous ethanol to prevent premature labor served to stimulate research on the acute effects of ethanol on the fetus and the mother close to term.[68] Ethanol inhibits release of oxytocin by the pituitary at concentrations equivalent to blood levels used to arrest premature labor (100–160 mg/dl).[69] There is no effect on the uterine muscle itself.[70,71] Because of the immaturity of fetal hepatic enzymes, the fetal blood ethanol concentration falls at only half the rate of the mother.[72,73]

The effect of ethanol infusion on maternal and fetal acid–base balance was investigated in pregnant ewes by Mann et al.[74] Ewes close to term were administered 15 ml/kg body wt of 9.75% solution of ethanol for 1–2 hr. The peak concentration in the maternal blood was 240 mg/dl at 90 min and in the fetal blood 220 at 120 min. A significant maternal hyperlactacidemia and hyperglycemia were noted, but this did not result in signficant alteration of the maternal acid–base balance. An initial fetal metabolic acidosis and later a mixed acidosis were observed during the ethanol infusion; this worsened during the postinfusion period. They also found that the fetal EEG showed a decrease in amplitude and a slowing of the dominant rhythm as the blood ethanol concentration increased.[75] The EEG became isoelectric on occasion during the postinfusion period associated with severe fetal acidosis. Fetal cerebral uptake of oxygen was unaffected, while the cerebral uptake of glucose and the glucose–oxygen utilization ratio was significantly increased. Horiguchi et al. carried out similar investigations with 13 pregnant rhesus monkeys, with fetal ages ranging from 120 to 160 days (term is about 168 days).[76] They were infused during 60 min with 2–4 g ethanol/kg body wt after a spontaneous onset of labor or following its induction by infusion of oxytocin. Maximum BEC were 237 mg/dl. The maternal respiratory rate was decreased and there was increase in the fetal heart rate. They also observed a fetal acidosis and concluded that intravenous infusion of ethanol in doses sufficient to suppress labor may be hazardous because the fetus becomes progressively asphyxiated.

These experimental studies demonstrate the probable physiologic changes in the human fetus when the mother engages in binge drinking. Repeated episodes of severe acidosis and hypoxia may be important factors in the impaired neurologic functioning of the fetal alcohol syndrome babies. Other dangers are related to the acidity and volume of maternal gastric secretion stimulated by the ethanol administration. Subsequent anesthesia has been associated with aspirations of a highly acid secretion followed by pneumonitis.[77]

b. Chronic Effects on Fetal Metabolism

The most fundamental effects of ethanol are on cellular metabolism. The concentration of ethanol and the duration of exposure determine whether it is a rapidly metabolized nutrient or a toxic agent. At low concentrations, passive permeability of normal resting cell and capillary membranes are not influenced by ethanol. However, at higher concentrations of 0.1 M, there is a significant effect on membrane enzyme systems which employ energy derived from the cleavage of ATP to transport Na^+ to the outside of the cell and K^+ to the inside, against their respective concentration gradients. The variable effects of ethanol on mitochondrial membranes with alteration of permeability and swelling have been reviewed by Kalant.[78] Cederbaum *et al.* demonstrated that chronic ethanol ingestion is associated with striking ultrastructural changes in the mitochondria, as well as by persistent impairment of mitochondrial oxidation of fatty acids to carbon dioxide.[79] Electron miscroscopy confirms the structural changes in the mitochondria.[80]

Sze *et al.* found that ethanol exposure *in utero* induced increased activity of the alcohol dehydrogenase as well as the microsomal alcohol-oxidizing system in mice.[81] Pikkarainen and Räihä showed that ethanol dehydrogenase activity is present in 2-month-old fetal livers at an activity level of 3–4% of that of adult livers.[82] Alcohol dehydrogenase activity increases with maturation and reaches 18% of the adult capacity at birth and by age five is comparable to that found in the adult liver.

Rawat studied the effect of maternal ethanol consumption on fetal hepatic metabolism in the rat.[83] The first measurable changes were observed after 18 days of pregnancy, when there was a steady linear increase in the cytoplasmic and mitochondrial NAD reduction upon ethanol administration; adult levels were reached 12 days after birth. Alcohol dehydrogenase seems to control hepatic ethanol oxidation capacity and is not affected by the sudden changes observed in neonatal hepatic redox state immediately upon birth.

Kesäniemi found that the elimination rate of ethanol was equal in pregnant and nonpregnant rats, but that the acetaldehyde content of the peripheral blood after ethanol administration was higher in pregnant than in nonpregnant animals.[84] Since no differences were found in the liver ethanol and aldehyde dehydrogenase activity, a difference in the extrahepatic metabolism of acetaldehyde was suggested. Kesäniemi and Sippel subsequently demonstrated that following administration of ethanol to pregnant rats, the ethanol content of the maternal aortic blood was comparable with that of the intact placenta and the whole fetus.[85] However, the acetaldehyde content of the placenta was only 25% of that present in the maternal blood and no acetaldehyde was found in the intact fetal tissue. Since both ethanol and paraldehyde cross the placenta freely, it is improbable that the acetaldehyde should be unable to cross the placenta because of physiochemical properties. It is more likely that it is oxidized as it crosses the placenta. No acetaldehyde is found in the milk of lactating women who have consumed ethanol even when the maternal peripheral blood demonstrates a considerable amount of acetaldehyde.[86] Thus, offspring seem to be protected from this toxic by-product of ethanol oxidation.

Rawat studied rates of protein synthesis by livers of fetal and neonatal rats born to mothers who had consumed ethanol during pregnancy.[87] Rates of incorporation of labelled leucine into hepatic proteins were significantly lower in the offspring of the ethanol-fed rats compared with the control group. Maternal ethanol consumption resulted in a decrease in hepatic total RNA content, the RNA/DNA ratio, and the ribosomal protein content of the fetal liver. Inhibition of protein synthesis could be directly related to the retarded growth of the offspring of alcoholic mothers.

Stoewsand and Anderson exposed weaning mice to wine, 12% ethanol, or distilled water.[88] After 10 weeks of gradually increased intake, the experimental mice were receiving exclusively wine or ethanol as their sole dietary liquid. They then were mated and their offspring studied in terms of cholesterol and triglyceride in blood plasma and liver. After weaning, offspring were exposed to either wine, ethanol, or water. Liver cholesterols of the offspring from wine-fed parents were significantly higher than liver cholesterols of the offspring of water-drinking parents ($p < 0.05$). Data indicated that male mice born from water-fed parents and later fed wine had liver triglyceride and cholesterol levels which were 60% of the levels of mice fed a solution containing the equivalent amount of ethanol ($p < 0.05$). No comparable effect was seen in female offspring. A complex interaction was demonstrated between the influence of parental treatment and sex on cholesterol and triglyceride metabolism. It also was shown that the physiologic effects of wine are related to both its ethanol and nonethanol components.

c. Effects on the Developing Central Nervous System

Exposure of the fetal central nervous system to moderate or high concentrations of ethanol probably has different effects at different stages of development. Malformations at the earliest stages of embryonic growth are probably incompatible with life. Clarren et al. presented neuropathological data on brains of four human neonates exposed in utero to high peak concentrations of ethanol.[89] The most frequent finding was a sheet of aberrant neural and glial tissue covering part of the brain surface, termed leptomeningeal neuroglial heterotopia. Brain lesions were found in two infants who showed no external dysmorphic features of the FAS. In some infants, abnormal brain structure and function may be the only abnormality caused by ethanol in utero.

Bauer-Moffett and Altman studied morphologic effects on the developing cerebellar cortex of rat pups who inhaled ethanol vapor from days 3 to 20.[90] Ethanol exposure during this period of cerebellar neurogenesis produced great reductions of cerebellar tissue, nearly twice that observed in parts of the brain which developed at an earlier time.

Rosman and Malone maintained pregnant rats on isocaloric liquid diets containing 10%, 21%, and 36% ethanol.[91] Fetal loss was 100% when the mother was on the highest level diet, and 60% on the diets containing less ethanol. Pups were sacrificed between postnatal days 12 and 28. Starting at day 24, brains of the experimental animals, as compared with pair-fed controls, showed delayed myelination affecting all fiber tracts.

Ethanol affects the central nervous system via multiple mechanisms. In mature animals ethanol in physiologic concentrations has its primary effect on the neuronal membrane; active transport of sodium and potassium across the membrane is impaired.[92] Thiamine, pyridoxine, and folic acid as well as calcium, magnesium, and zinc, all essential for central nervous system enzymes, frequently are depleted due to the malnutrition and diureses associated with chronic ethanol ingestion.[93,94]

Tewari et al. investigated the effects of chronic ethanol ingestion on brain RNA metabolism in mature mice.[95,96] Changes in RNA metabolism were due to an alteration in the transcription and/or the processing of RNA in the nucleus. Pilstrom and Kiessling administered ethanol to pregnant rats and studied the effects on liver and brain mitochondria of offspring.[97] No significant effect could be found in the capacity to oxidize pyruvate, glutamate, and β-hydroxybutyrate or in oxidative phosphorylation.

Rawat studied effects of long-term ethanol consumption by pregnant rats on the incorporation of [^{14}C]leucine into fetal and neonatal brain ribosomes.[98] He found a 30% decrease in the rate of [^{14}C]leucine incorporation by the fetal cerebral ribosomes while neonatal rats suckling on ethanol-fed mothers showed about 60% decrease as compared to the control group.

Impaired learning, observed in animal behavior experiments as well as in psychological evaluation of FAS children, may be related to altered RNA metabolism. Increased motor activity, another behavioral manifestation observed in animals exposed to ethanol in utero, could be due to changes in neurohumoral amine metabolism. Ethanol affects uptake, storage, and release of serotonin, catecholamines, acetylcholine, and γ-aminobutyric acid.[99] Branchey and Friedhoff found in the caudates of rat pups exposed to ethanol in utero that the activity of tyrosine hydroxylase, a rate-limiting enzyme in catecholamine biosynthesis, was increased at 1, 2, and 3 weeks of age to levels 12% greater than in controls.[100] Increased activity in litter mates at 23 days was also observed. Rawat studied the influence of prolonged ethanol consumption by pregnant and lactating rats on the activities of several neurotransmitters together with the activities of the enzymes in fetal and neonatal brains.[101] The cerebral content of γ-aminobutyric acid in the 21-day-old fetus was 250% greater than in the controls, while in the 6-day-old suckling neonate it was 140% greater than in the controls. Serotonin and norepinephrine levels in the 6-day-old suckling were 120% greater than in the controls, while acetylcholine was only 75% of the controls' level. No significant differences in levels of neurotransmitters in the brains of experimental and control fetal brains were found.

At the cellular level, alterations of the membrane of the neuron and of neurotransmitters interfere with proper synaptic transduction. Heavy ethanol consumption alters physiologic functions in many areas of the central nervous system, and sleep disturbances are common.[102] EEG studies have revealed marked disruption of the quantitative composition of rapid-eye-movement (REM) and slow-wave sleep (SWS), together with instability or fragmentation of the circadian rhythm of 24-hr cycles and the ultradian rhythms (shorter cycles occurring periodically). In adult alcoholics, continued disturbance of sleep rhythms persists months after abstinence has been attained. Data from the study of sleep–awake state regulation on the third day utilizing a monitoring bassinet is consistent with EEG studies of adult alcoholics.[36,37]

2. Effects of Ethanol on Maternal Metabolism and Physiology

Chronic heavy ethanol consumption adversely affects almost every organ system in the body. Nonspecific risk factors for the pregnancy are associated with disease of the liver and other parts of the gastrointestinal system, the cardiovascular system, the hematopoietic system, as well as the body's defense mechanisms against infectious disease.[103] Ethanol-induced metabolic disturbances in the mother, such as ethanol hypoglycemia, ethanol ketoacidosis, alterations in lactate, uric acid, and lipid metabolism, or changes in the metabolism of individual amino acids, probably all have effects on the fetus.

a. Alcoholic Ketoacidosis

Cooperman et al. described seven episodes of severe ketoacidosis in six nondiabetic patients.[104] All were women who indulged both in heavy chronic ethanol use and binges. One patient was admitted because of metabolic acidosis four times during two pregnancies. During her first episode in the 30th week of pregnancy she delivered a premature fetus. During her second pregnancy, she was admitted at 8 weeks, 28 weeks, and 32 weeks of gestation. Each episode followed excessive drinking and each time the metabolic acidosis was progressively more severe. The second pregnancy she delivered a normal-term infant. It was suggested that ovarian and placental hormones, together with the fetal drain on carbohydrate reserves, may play a part in the pathogenesis of alcoholic ketoacidosis during pregnancy.

b. Withdrawal Syndrome

Acute withdrawal syndromes have been described in newborn infants delivered by severely alcoholic mothers.[31,105,106] If a woman undergoes an acute alcohol withdrawal syndrome in midpregnancy, it seems probable that the fetus would be subject to major metabolic and physiologic disturbances. Two patients who had been consuming over a quart of vodka per day had been withdrawn during the second trimester on the Boston City Hospital Obstetrical Service.[107] Progressively reduced doses of ethanol as well as chlordiazepoxide were utilized for 1 week to modify withdrawal signs and symptoms. No adverse effects were detected by monitoring fetal heart rate or on careful neurologic examination in the newborn nursery.

c. Mineral Deficiencies

Deficiencies of magnesium, zinc, and calcium occur in chronic alcoholism as a consequence of increased urinary excretion, loss due to vomiting and diarrhea, and inadequate intake.[94] A 100% increase in urinary excretion of calcium and a 167% increase in magnesium excretion have been observed to begin within 20 min of the ingestion of 30 ml of ethanol by normal volunteers and to continue for about 2 hr. Normally, calcium loss is replaced from vast reserves in the skeletal system. The Ca^{2+} ion bound to the synaptic membrane is a regulator of nervous system excitability. The possibility of depletion of Ca^{2+} ion from the brain of FAS neonates has not been investigated. Magnesium has an important role in fetal development since it stabilizes DNA, RNA and binds sRNA to the ribosome, and also is involved in the activating and transfer systems of all amino acids. Zinc is needed for a number of enzymes, including carbonic anhydrase and alcohol dehydrogenase, and is also necessary for RNA metabolism and DNA synthesis. Female rats, maintained on a zinc-deficient diet from the time of weaning[108] or only for the period of pregnancy,[109] bore fetuses with abnormalities of the skull, limb, heart, eyes, and urogenital system. Preschool children consuming diets deficient in zinc had many more height, weight, and head circumference measurements below the third percentile.[110] Recently, the low zinc content of certain infant diets has been shown to be growth-limiting.[111]

d. Vitamin Metabolism

Chronic alcoholics have multiple disturbances of vitamin metabolism; dietary intake often is poor, intestinal absorption impaired, storage limited by hepatic damage, and urinary loss may be elevated.[93] In a study of 120 indigent or low-income adult patients admitted to a general hospital, 59% had a significant reduction in circulating levels of two or more vitamins. The most common deficiency was of folate, measured by the serum folate level. Ethanol causes impaired intestinal absorption of folic acid. Sullivan and Herbert demonstrated that hematologic response to folic acid therapy was repeatedly prevented by the concomitant administration of whisky, wine, or ethanol.[112] When body stores are decreased and dietary intake is poor, ethanol may act as a weak folate antagonist. Folate antagonists have been shown to cause fetal resorption, stillbirths, and congenital malformations in the rat, and there have been reports suggesting that human fetal malformations may result from dietary deficiency of folate.[113,114]

Thiamine deficiency, frequently seen in chronic alcoholics, is caused by decreased dietary intake during drinking episodes, impaired absorption, and possibly also acute liver injury, which lowers the reponse to administered thiamine.[93] Magnesium deficiency concomitant with thiamine deficiency may interfere with response to thiamine. Central nervous system lesions produced by experimental thiamine deficiency are comparable to those observed in the Wernicke–Korsakoff syndrome. The effects of thiamine pyridoxine and folate deficiency on the developing fetal nervous system should be investigated.

e. Effects on Endocrine Function

Stokes reviewed alcohol–endocrine interrelationships and stressed the importance of differentiating effects of ethanol on the anterior pituitary, its trophic hormones, their hypothalmic releasing factors, and the target glands.[115] Stokes found evidence of effects of ethanol on adrenomedullary function and thyroid function, but little systematic data on the effects of ethanol on gonadal function. Recently, information about the effects of ethanol administration on sex hormone metabolism in normal men has been acquired.[116] These include an augmented conversion of androgenic precursors to estrogens, a higher plasma concentration and production rate of estradiol, and an increased plasma concentration of estrone. The alteration of metabolism of androgens has been shown to be coupled with direct effects of ethanol on hypothalmic–pituitary function, with changes in plasma luteinizing hormone. No data on corresponding changes in women have been published. This may be an important factor in the fetal alcohol syndrome, since a link has been demonstrated between exposure to female sex hormones early in pregnancy and cardiovascular malformations.[117] Heavy-drinking women with chronic liver disease might endogenously produce hormones which alter fetal cardiovascular and genital development.

Hypothalmic–pituitary function in four children ages 9–15, born to an alcoholic woman, were studied to determine if hormonal abnormalities account for FAS aberrant growth patterns.[15] Biochemical and endocrine studies were all within normal limits. Tze et al. studied severe postnatal growth deficiency in five cases of FAS.[26] Following insulin-induced hypoglycemia and arginine infusion there was a slight hyperresponse to growth hormone and normal somatomedin activity. Elevation of growth hormone level in the presence of growth failure has been previously reported in cases of severe malnutrition and in rare cases of pituitary dwarfism in which peripheral tissue insensitivity to growth hormone has been postulated.

If these findings can be replicated, one could speculate that exposure to ethanol in utero may have caused reduction of the number of cells in the peripheral tissue of FAS children.

f. Effects on the Placenta

Placental growth and function may be affected by chronic ethanol use, but this problem has received little direct study. Kaminski et al. compared the mean weight of the placentas from 236 women who consumed over 44 ml absolute alcohol/day with 4074 who drank less than that amount.[31] Heavier drinkers had placentas that weighed 22 g less than the lighter drinkers. This difference attained statistical difference ($p < 0.01$). However, the clinical significance of this small difference in placental size is unclear. Ethanol may have direct detrimental effects on active transport mechanisms which carry nutrients and electrolytes across placental membranes.[78]

3. Associated Risk Factors

Heavy drinking is often associated with other variables that contribute to reproductive risk, such as nutritional deficiency, heavy smoking, use of other drugs, and emotional stress. Each of these associated factors has been extensively studied; only brief references to some of the more recent reviews will be cited here.

a. Nutrition

The relationship between alcoholism and malnutrition has been reviewed by Hillman.[119] Ethanol and vitamin metabolism have been reviewed by Vitale and Coffey.[93] The clinical stereotype of the malnourished alcoholic has been questioned by Neville et al.[120] Thirty-four alcoholics were admitted to a research ward. A careful evaluation of nutrient intake for the month prior to admission was made and excretion levels of various vitamins were studied. There was no significant difference between the mean excretion of vitamin metabolites by alcoholics and normal controls. These results did not support the view that the nutritional status of alcoholics is markedly inferior to that of nonalcoholics, particularly those of similar economic and health histories. At the Boston City Hospital Prenatal Clinic, less than 10% of patients' diets of all women registering met the minimum daily requirement of the National Research Council.[107] The heavy-drinking women did not report diets significantly poorer than those reported by the abstinent women. However, utilization may be impaired by the effects of alcohol on intestinal absorption, liver function, urinary excretion of vitamins and trace minerals, disturbances of intermediate metabolism, and gastrointestinal disturbances such as vomiting and diarrhea.

b. Smoking

The association between heavy drinking and heavy cigarette smoking, which has been repeatedly observed,[121] is confirmed by current epidemiologic data coming from Boston, Seattle, and Loma Linda. Smoking mothers have more low-birth-weight infants. However, no increase in neonatal mortality rate or congenital anomaly rates was found. Among the many studies relating smoking and low birth weight in these offspring, none took into account that smoking itself was associated with heavier alcohol use which could have contributed a significant portion of the variance. The review by Yerushalmy on the relationship of cigarette smoking to outcome of pregnancy points out some of the problems of inferring causation from observed associations.[122]

c. Other Drugs

The use of other drugs, such as heroin, methadone, LSD, barbiturates, and dilantin, has also been associated with a higher incidence of low-birth-weight infants and increased perinatal mortality.[123] Heavier drinkers are more likely to have tried other drugs in the past; however, in the Boston City Hospital Prenatal Clinic study, only 4% of the total group reported currently using drugs other than ethanol.[35] In the Loma Linda study, 6% reported illicit drug use.[42] In both populations, drugs were used by women with low, moderate, and high ethanol intake. Total abstainers from ethanol also avoided illicit drugs.

d. Caffeine

Heavy coffee consumption is often associated with heavy ethanol use and smoking.[32] While there have been animal studies of caffeine as a teratogen, the dosages employed range from the equivalent of 40 to 100 cups of coffee per day.[124] No human malformations have been attributed to caffeine. This may be because man is protected by a rapid metabolism of caffeine, with only 1% excreted unchanged.

e. Disulfiram

Disulfiram (Antabuse) which may be useful as an adjunct to the treatment of alcoholism should only be prescribed during pregnancy when probable benefits outweigh risks, since it inhibits several enzymes, including aldehyde dehydrogenase, needed for oxidation of acetaldehyde, and dopamine-β-hydroxylase, which catalyzes the conversion of dopamine to norepinephrine.[125]

f. Paternal Drinking

The possibility that heavy ethanol consumption by the male can cause genetic damage has been the subject of speculation for over 200 years. The mutagenic role of ethanol in male mice was studied by administering ethanol in a dose of 1.24 g/kg by gastric tube to male mice for three consecutive days.[126] Each mouse was then caged with a virgin untreated female of the same strain. At weekly intervals, the males were transferred to cages with different virgins. Pregnant females were sacrificed 13–15 days after conception, and dead and live implants were scored for each pregnancy. Females mated 4–13 days after treatment of the male with ethanol had a two- to fourfold increase in the number of dead implants found. The conclusion was that ethanol in the doses used induced dominant lethal mutations in several different spermatogenic stages. Efficiency of induction was most pronounced in the epididymal spermatozoa and the late spermatid stage.

Klassen and Persaud fed a 10% ethanol Metrecal diet to 12 male rats for 35 days, while 12 controls received Metrecal with an isocaloric amount of sucrose substituted for the ethanol.[127] They were not pair-fed. The ethanol-fed males lost weight, became lethargic, and had ruffled hair and small pale eyes. Tests revealed significantly lower blood sugar and testosterone levels among the ethanol-fed males. The male rats were mated with normally fed females. The number of successful matings and the number of offspring per litter were significantly greater in females mated to the control rats than in females mated to ethanol-fed rats. Pups of ethanol-fed fathers demonstrated significantly lower weight, length, and placental index ($p < 0.01$).

The effects of ethanol on spermatogonal cells in the rat have been directly investigated by giving rats a nutritionally adequate diet containing 10% ethanol as the only supply of liquid for 70 days.[128] Following this experiment, testicular tissue was directly examined for the frequency of aberrations such as chromosome breaks, chromatid breaks, and chromatid gaps. There was no significant difference between the experimental and control groups. Subsequently, the experiment was repeated with male rats who received a thiamine-deficient diet as well as ethanol.[129] Combining abnormal thiamine supply with ethanol feeding did not lead to any significant increase in the frequency of chromosomal aberrations.

Ethanol has no damaging effects on human chromosomes *in vitro*.[130] However, chromosomes of alcoholics show a significant increase in aberrations. One possible reason may be an inhibition of cellular RNA synthesis which impedes cellular repair *in vivo*.

g. Maternal Psychological Stress

Emotional stress has been demonstrated to be related to perinatal complications.[131] Differences have not been found between various psychiatric diagnostic groups; however, when psychiatric patients are divided on a dimension of chronicity, those with the greatest number of psychiatric contacts and hospitalizations had infants with the most perinatal complications. Heavy-drinking women frequently use ethanol to relieve symptoms of chronic anxiety and depression. Emotional stress data should be reexamined in terms of ethanol and other drug use, including tranquilizers, by the patients with the more severe and chronic diagnoses.

Sameroff and Chandler emphasize the importance of viewing the continuum between prenatal and newborn care.[131] A number of studies of failure to thrive describe the infants as being irritable, difficult to manage, unappealing, having fussy eating habits, poor food intake, and frequent regurgitations. Traditionally, these symptoms were interpreted as the result of parental neglect. These characteristics may be part of the etiology of parental neglect rather than the consequence of poor mothering.

There are almost no case studies of the management of the pregnant chronic alcoholic patient in literature. One report of the successful cooperation of the psychiatrist and obstetrician following a patient jointly seemed to have been successful in enabling an alcoholic mother who had a potential for psychosis to deal successfully with the stress of pregnancy and also control her ethanol use.[132] The dire consequences of fragmentation of medical care are illustrated in another case report of a 37-year-old alcoholic woman who died of complications following a ruptured uterus at 32 weeks that was not managed properly.[133]

Offspring of alcoholics are a high-risk group for child abuse and neglect,[134] hyperactivity, delinquency and other behavioral disorders,[135] and, as they mature, alcoholism. This pessimistic prognosis undoubtedly is multidetermined. The family of an alcoholic has many crises in which the childrens' loyalties to each parent become tools in the struggle between mother and father.[136] Genetic predispositions to affective disorders or alcoholism itself have also been investigated.[137] Several studies of the families of hyperactive children have indicated that the parents have a higher incidence of psychiatric illness, and particularly alcoholism.[138,139] Goodwin et al. reported a higher occurrence of ethanol problems in adoptees raised apart from their alcoholic biological parents, and found a high incidence of hyperactive behavior in the childhood of a sample of these alcoholic adoptees.[140,141] Waldrop and Halverson described minor physical anomalies associated with hyperactive behavior in young children, but did not investigate parental ethanol use.[142] Rapoport et al. studied a similar group and also found a higher than normal mean plasma dopamine-β-hydroxylase activity.[143] These observations resemble findings in rat offspring exposed to ethanol *in utero*. Intrauterine exposure of the developing central nervous system to ethanol may contribute to hyperactive behavior in childhood.[57-59] This may contribute to subsequent behavioral disturbances and development of alcoholism.

F. RESEARCH ON PREVENTION

Epidemiologic research should be designed to provide data needed for development of prevention programs. Careful attention to information on changes in individual drinking patterns during pregnancy may help explain why some offspring of alcoholic women escape damage. Binges should be described as carefully as possible, in terms of the types and amounts of ethanol ingested, duration of drinking, and precise stage of the pregnancy. Development of affected offspring should be followed as long as possible so that parents can have more facts about prognosis for different components of the fetal alcohol syndrome. Educational programs for teenagers should present facts about the fetal alcohol syndrome together with information on the changes in drinking patterns among adolescents.[144]

Animal models of the fetal alcohol syndrome have been produced in a number of species.[145] These models could be modified to examine effects of abstinence occurring at different stages of pregnancy. Some morphologic and functional damage may be reversible if ethanol is discontinued while the fetus is not fully developed. Other possible prevention techniques that should be studied in animals include potential benefits of supplementary diets with essential vitamins and trace elements to facilitate catch-up growth after ethanol is no longer present.

The prognosis for FAS infants has been pessimistic. However, this has developed from observations of the most severely affected offspring of chronic alcoholic women. Frequently, multiple compounding risk factors exist. Initial experience with infants born to women about to abstain or significantly reduce ethanol use in midpregnancy suggests that rebound growth and physiologic adaptation can occur in some instances and compensate for early disruption.[35]

Long-term follow-up of children impaired by perinatal risk factors such as anoxia and prematurity suggests that in the less severely damaged infants, a facilitating environment can effect relatively normal developmental outcomes.[131] Deviant development may result from severe insult to the organism's integrative mechanism or severe familial or social abnormalities which can prevent the restoration of normal growth processes. Prevention approaches must be developed to effect early identification and reduction of ethanol use by heavy-drinking pregnant women. Particular techniques may be more effective in different settings with various ethnic and socioeconomic groups. Improvement of social supports during pregnancy and assistance with care giving should be provided to help parents with children who overtax their emotional resources.

Research on the fetal effects of ethanol may serve as a model for investigation of teratogenicity of other substances in common use. Most pharmacologic agents, especially prescription drugs, are ingested in relatively standardized dosage over a well-defined time period. In contrast, patterns of consumption of ethanol by humans have great variability in terms of both quantity and frequency. Marked differences between individuals in rate of metabolism of ethanol and CNS tolerance also have been demonstrated. Teratogenic effects probably are related to the blood ethanol concentration at critical stages of pregnancy, acting synergistically with associated risk factors. Experimental research employing animal models should facilitate identification of these critical variables and permit development of more specific strategies for prevention.

G. SUMMARY AND CONCLUSIONS

The full fetal alcohol syndrome has been reported in 245 offspring of chronic alcoholic mothers. Components of the fetal alcohol syndrome—growth retardation, malformations, and abnormalities of the central nervous system—have been seen in many other offspring of heavy-drinking women. Prospective studies have demonstrated that pregnant women who consume over 90 ml of ethanol on some occasions and at least 45 drinks a month as compared with abstainers or moderate drinkers are two to three times more likely to bear a child with features suggestive of a prenatal effect of ethanol on growth and morphogenesis. Mild growth retardation has been reported in women who averaged 30 ml ethanol/day; however, the developmental consequences of this have not been ascertained. The effects of ethanol in utero on later growth, health, and psychological development should be studied by follow-up studies of FAS children as well as those exposed to low concentrations of ethanol. Prospective developmental evaluation also is needed to determine whether reduction of heavy drinking in midpregnancy improves the prognosis for the baby.

Animal models of the fetal alcohol syndrome have been developed using several species. Animals with short gestational periods facilitate experiments with many subjects and controls at lower costs. Species with longer gestational periods permit evaluation of the effects of single binge doses, or modification of ethanol administration in midpregnancy. Many clinical questions which are difficult to answer from observations in humans could be tested in animal experiments:

1. Which of the many pharmacologic effects of ethanol with the potential to effect embryonic and fetal growth and development are the most important?

2. What are the specific biochemical and physiological systems involved?

3. Can growth retardation be modified by supplementary vitamins, minerals, or other nutrients? Will they have any effect if ethanol is present?

4. How do the effects of ethanol interact with other drugs ingested during pregnancy (e.g. nicotine, caffeine, benzodiazopines, opiates, etc.)?

5. Can direct biochemical measures be developed to predict which fetuses are at high risk?

6. Is there an effect of ethanol consumption before pregnancy if none is ingested after conception?

7. Does ethanol administered to the male before impregnation affect fetal development?

8. How do genetic differences in ethanol metabolism affect the outcome of the pregnancy?

Basic research and clinical investigations in these areas should provide significant benefits for children and their parents and should also contribute to our fundamental understanding of the pharmacology of ethanol and the pathophysiology of alcoholism.

ACKNOWLEDGMENTS. This review is based on an earlier paper prepared under Contract NIA 76-25 (P). Initial research was supported by ADAMHA Career Teacher Awards T01DA00031 (NIDA) and PHSAA07008 (NIAAA). Further research was supported in part by NIAAA Grant AA02446-01, the National Council on Alcoholism, the Massachusetts Developmental Disabilities Council, the United States Brewers Association, Inc., and both University Hospital and

Boston City Hospital General Research Support Awards. I also wish to acknowledge the editorial assistance of Jill Marcus.

H. REFERENCES

1. R. Burton, *The Anatomy of Melancholy*, Vol. 1, Part I, Section 2: *Causes of Melancholy*, William Tegg, London (1906); originally 1621.
2. R. H. Warner and H. L. Rosett, *J. Stud. Alcohol* **36**:1395 (1975).
3. W. C. Sullivan, *J. Ment. Sci.* **45**:489 (1899).
4. P. Lemoine, H. Haronsseau, J.-P. Borteryu, and J. C. Menuet, *Ouest Med.* **25**:476 (1968).
5. C. Ulleland, R. P. Wennberg, R. P. Igo, and N. J. Smith, *Pediatr. Res.* **4**:474 (1970).
6. K. L. Jones, D. W. Smith, C. Ulleland, and A. P. Streissguth, *Lancet* **1**:1267 (1973).
7. K. L. Jones and D. W. Smith, *Lancet* **2**:999 (1973).
8. J. W. Hanson, K. L. Jones, and D. W. Smith, *J. Am. Med. Assoc.* **235**:1458 (1976).
9. A. P. Streissguth, *N.Y. Acad. Sci.* **273**:140 (1976).
10. P. E. Ferrier, I. Nicod, and S. Ferrier, *Lancet* **2**:1496 (1973).
11. H. P. Palmer, E. M. Ouellette, L. Warner, and S. R. Leichtman, *Pediatrics* **53**:490 (1975).
12. J. Saule, *Klin. Pediatr.* **186**:452 (1974).
13. M. S. Tenbrinck and S. Y. Buchin, *J. Am. Med. Assoc.* **232**:1144 (1975).
14. R. G. G. Barry and S. O'Nuallain, *Ir. J. Med. Sci.* **144**:286 (1975).
15. A. W. Root, E. O. Reiter, M. Andriola, and G. Duckett, *J. Pediatr.* **87**:585 (1975).
16. K. K. Christoffel and I. Salafsky, *J. Pediatr.* **87**:963 (1975).
17. J. Manzke and F. R. Grosse, *Med. Welt* **26**:709 (1975).
18. L. Reinhold, H. Hütteroth, and H. Schulte-Wisserman, *Muench. Med. Wochenschr.* **117**:1731 (1975).
19. G. Loiodice, G. Fortuna, A. Guidetti, N. Ria, and R. D'Elia, *Minerva Pediatr.* **27**:1891 (1975).
20. J. J. Mulvihill, J. T. Klimas, D. C. Stokes, and H. M. Risemberg, *Am. J. Obstet. Gynecol.* **125**:937 (1976).
21. J. J. Mulvihill and A. M. Yeager, *Teratology* **13**:345 (1976).
22. J. R. Bierich, F. Majewski, R. Michaelis, and I. Tillner, *Eur. J. Pediatr.* **121**:155 (1976).
23. K. Ijaiya, A. Schwenk, and E. Gladtke, *Dtsch. Med. Wochenschr.* **101**:1563 (1976).
24. B. D. Hall and W. A. Orenstein, *Lancet* **1**:680 (1974).
25. J. A. Noonan, *Am. J. Cardiol.* **37**:160 (1976).
26. W. J. Tze, H. G. Friesen, and P. M. MacLeod, *Arch. Dis. Child.* **51**:703 (1976).
27. S. Pierog, O. Chandavasu, and I. Wexler, *J. Pediatr.* **90**:630 (1977).
28. F. Majewski, J. R. Bierich, H. Löeser, R. Michaelis, B. Lieber, and F. Bettecken, *Muench. Med. Wochenschr.* **118**:1635 (1976).
29. H. Löeser, F. Majewski, J. Apitz, and J. R. Bierich, *Klin. Pediatr.* **188**:233 (1976).
30. S. K. Clarren and D. W. Smith, *N. Engl. J. Med.* **298**:1063 (1978).
31. M. Kaminski, C. Rumeau-Rouquette, and D. Schwartz, *Rev. Epidemiol. Sante Publique* **24**:27 (1976); English translation, *Alcohol. Clin. Exp. Res.* **2**:155 (1978).
32. K. R. Niswander and M. Gordon (eds.), The Collaborative Perinatal Study of the National Institute of Neurological Diseases and Stroke: The Women and Their Pregnancies, pp. 500–524, DHEW Publ. No. (NIH) 73-379.
33. K. L. Jones, D. W. Smith, A. P. Streissguth, and N. C. Myrianthopoulos, *Lancet* **1**:1076 (1974).
34. H. L. Rosett, *Lancet* **2**:218 (1974).
35. R. Sturdevant, *Lancet* **2**:349 (1974).
36. H. L. Rosett, E. M. Ouellette, and L. Weiner, *Ann. N.Y. Acad. Sci.* **273**:118 (1976).
37. E. M. Ouellette, H. L. Rosett, N. P. Rosman, and L. Weiner, *N. Engl. J. Med.* **297**:528 (1977).
38. L. W. Sander, P. A. Snyder, H. L. Rosett, A. Lee, J. B. Gould, and E. M. Ouellette, *Alcohol. Clin. Exp. Res.* **1**:233 (1977).
39. H. L. Rosett, E. M. Ouellette, L. Weiner, and E. Owens, *Obstet. Gynecol.* **51**:41 (1978).

40. J. W. Hanson, A. P. Streissguth, and D. W. Smith, *J. Pediatr.* **92:**457 (1978).
41. S. Landesman-Dwyer, L. S. Keller, and A. P. Streissguth, *Alcohol. Clin. Exp. Res.* **2:**171 (1978).
42. J. Martin, D. C. Martin, C. A. Lund, and A. P. Streissguth, *Alcohol. Clin. Exp. Res.* **1:**243 (1977).
43. F. Combemale, *Misc. Nerv. Syst.* **82:**14 (1888).
44. C. Féré, *C. R. Seances Soc. Biol. Fil.* **46:**646 (1894).
45. C. R. Stockard, *Am. J. Anat.* **10:**369 (1910).
46. M. Nicloux, *C. R. Seances Soc. Biol. Fil.* **51:**980 (1899).
47. M. Nicloux, *Obstetrique* **5:**97 (1900).
48. C. F. Hodge, in: *Physiological Aspects of the Liquor Problem* (W. O. Atwater, J. S. Billings, H. P. Bowditch, R. H. Chittenden, and W. H. Welch, eds.), pp. 359–375, Houghton Mifflin Company, Boston (1903).
49. R. Pearl, *Proc. Natl. Acad. Sci. USA* **2:**380 (1916).
50. E. C. MacDowell, *J. Exp. Zool.* **37:**417–456 (1923).
51. H. Wallgren and H. Barry, III, *Actions of Alcohol*, Vol. I, pp. 482–489, Elsevier Publishing Company, Amsterdam (1970).
52. S. Sandor and S. Elias, *Rev. Roum. Embryol. Cytol. Ser. Embryol.* **5:**51 (1968).
53. S. Sandor and D. Amels, *Rev. Roum. Embryol. Cytol. Ser. Embryol.* **8:**105 (1971).
54. G. F. Chernoff, *Teratology* **15:**223 (1977).
55. C. L. Randall, W. J. Taylor, and D. W. Walker, *Alcohol. Clin. Exp. Res.* **1:**219 (1977).
56. F. W. Ellis and J. R. Pick, *Pharmacologist*, Abstr. **18:**190 (1976).
57. N. W. Bond and E. L. Digiusto, *Psychopharmacologia* **46:**163 (1976).
58. L. Branchey and A. J. Friedhoff, *Ann. N.Y. Acad. Sci.* **273:**328 (1976).
59. B. A. Shaywitz, J. H. Klopper, and J. W. Gordon, *Pediatr. Res.* **10:**451 (1976).
60. E. L. Abel, *Arch. Int. Pharmacodyn. Ther.* **210:**121 (1974).
61. E. L. Abel, *J. Stud. Alcohol* **36:**654 (1975).
62. M. Auroux and M. Dehaupas, *C. R. Seances Soc. Biol. Fil.* **164:**1432 (1970).
63. M. Auroux, *C. R. Seances Soc. Biol. Fil.* **167:**626 (1973).
64. J. C. Martin, D. C. Martin, P. Sigman, and B. Redow, *Dev. Psychobiol.* **10:**435 (1977).
65. M. E. Collard and C. S. Chen, *Q. J. Stud. Alcohol* **34:**1323 (1973).
66. B. T. Ho, G. E. Fritchie, J. E. Idanpaan-Heikkila, and W. M. McIsaac, *Q. J. Stud. Alcohol* **33:**485 (1972).
67. C. Akesson, *Arch. Int. Pharmacodyn. Ther.* **209:**296 (1974).
68. F. Fuchs, A. R. Fuchs, V. F. Poblete, Jr., and A. Risk, *Am. J. Obstet. Gynecol.* **99:**627 (1967).
69. A. R. Fuchs, *J. Endocrinol.* **35:**125 (1966).
70. G. Wagner and A. R. Fuchs, *Acta Endocrinol.* **58:**133 (1968).
71. K. H. Wilson, R. Landesman, A. R. Fuchs, and F. Fuchs, *Am. J. Obstet. Gynecol.* **104:**436 (1969).
72. M. Seppälä, N. C. R. Räihä, and V. Tamminen, *Lancet* **1:**1188 (1971).
73. R. Waltman and E. S. Iniquez, *Obstet. Gynecol.* **40:**180 (1972).
74. L. I. Mann, A. Bhakthavathsalan, M. Lui, and P. Makowski, *Am. J. Obstet. Gynecol.* **122:**837 (1975).
75. L. I. Mann, A. Bhakthavathsalan, M. Lui, and P. Makowski, *Am. J. Obstet. Gynecol.* **122:**845 (1975).
76. T. Horiguchi, K. Suzuki, A. C. Comas-Urrutia, E. Mueller-Huebach, A. M. Boyer-Milic, R. A. Baratz, H. O. Morishima, L. S. James, and K. Adamsons, *Am. J. Obstet. Gynecol.* **122:**910 (1975).
77. B. S. Greenhouse, R. Hook, and F. W. Hehre, *J. Am. Med. Assoc.* **210:**2393 (1969).
78. H. Kalant, in: *The Biology of Alcoholism* (B. Kissin and H. Begleiter, eds.), Vol. 1: *Biochemistry*, pp. 1–62, Plenum Press, New York (1971).
79. A. I. Cederbaum, C. S. Lieber, D. S. Beattie, and E. Rubin, *J. Biol. Chem.* **250:**5122 (1975).
80. M. Beskid, T. Majdecki, and J. Skladzinski, *Folia Histochem. Cytochem.* **13:**175 (1975).
81. P. Y. Sze, J. Yanai, and P. E. Ginsburg, *Biochem. Pharmacol.* **25:**215 (1976).

82. P. H. Pikkarainen and N. C. R. Räihä, *Pediatr. Res.* **1**:165 (1967).
83. A. K. Rawat, *Ann. N.Y. Acad. Sci.* **273**:175 (1976).
84. A. K. Kesäniemi, *Biochem. Pharmacol.* **23**:1157 (1974).
85. A. K. Kesäniemi and H. W. Sippel, *Acta Pharmacol. Toxicol.* **37**:43 (1975).
86. A. K. Kesäniemi, *J. Obstet. Gynaecol. Br. Commonw.* **81**:84 (1974).
87. A. K. Rawat, *Biochem. J.* **160**:653 (1977).
88. G. S. Stoewsand and J. L. Anderson, *J. Food Sci.* **39**:957 (1974).
89. S. K. Clarren, E. C. Alvord, S. M. Sumi, A. P. Streissguth, and D. W. Smith, *J. Pediatr.* **92**:64 (1978).
90. C. Bauer-Moffett and J. Altman, *Exp. Neurol.* **48**:378 (1975).
91. N. P. Rosman and M. J. Malone, *Neurology,* Abstr. **24**:377 (1974).
92. R. G. Grennell, in: *The Biology of Alcoholism* (B. Kissin and H. Begleiter, eds.), Vol. 2: *Physiology and Behavior,* pp. 1–19, Plenum Press, New York (1972).
93. J. J. Vitale and J. Coffey, in: *The Biology of Alcoholism* (B. Kissin and H. Begleiter, eds.), Vol 1: *Biochemistry,* pp. 327–352, Plenum Press, New York (1971).
94. E. B. Flink, in: *The Biology of Alcoholism* (B. Kissin and H. Begleiter, eds.), Vol. 1: *Biochemistry,* pp. 377–395, Plenum Press, New York (1971).
95. S. Tewari, E. W. Fleming, and E. P. Noble, *J. Neurochem.* **24**:561 (1975).
96. S. Tewari and E. P. Noble, in: *Alcohol and Abnormal Protein Biosynthesis: Biochemical and Clinical* (M. A. Rothschild, M. Oratz, and S. S. Schreiber, eds.), pp. 421–448, Pergamon Press, Inc., Elmsford, N.Y. (1975).
97. L. Pilstrom and K.-H. Kiessling, *Acta Pharmacol. Toxicol.* **25**:225 (1967).
98. A. Rawat, *Res. Commun. Chem. Pathol. Pharmacol.* **12**:723 (1975).
99. A. Feldstein, in: *The Biology of Alcoholism* (B. Kissin and H. Begleiter, eds.), Vol. 1: *Biochemistry,* pp. 127–159, Plenum Press, New York (1971).
100. L. Branchey and A. J. Friedhoff, *Psychopharmacologia* **32**:151 (1973).
101. A. Rawat, in: *The Role of Acetaldehyde in the Actions of Ethanol* (K. O. Lindros and C. J. P. Erikkson, eds.), pp. 159–176, The Finnish Foundation for Alcohol Studies, Helsinki (1975).
102. B. K. Lester, O. H. Rundell, L. C. Cowden, and H. L. Williams, in: *Alcohol Intoxication and Withdrawal: Experimental Studies* (M. M. Gross, ed.), Vol. I, pp. 261–279, Plenum Press, New York (1973).
103. F. A. Seixas, K. Williams, and S. Eggleston, *Ann. N.Y. Acad. Sci.* **252**:399 (1975).
104. M. T. Cooperman, F. Davidoff, R. Spark, and J. Pallotta, *Diabetes* **23**:433 (1974).
105. O. Schaefer, *Can. Med. Assoc. J.* **87**:1333 (1962).
106. M. M. Nichols, *Am. J. Dis. Child.* **113**:714 (1967).
107. H. L. Rosett, E. M. Ouellette, L. Weiner, and E. Owens, in: *Currents in Alcoholism* (F. A. Seixas, ed.), Vol. I, pp. 419–430, Grune & Stratton, Inc., New York (1977).
108. L. S. Hurley and H. Swenerton, *Proc. Soc. Exp. Biol. Med.* **123**:692 (1966).
109. L. S. Hurley, J. Gowan, and H. Swenerton, *Teratology* **4**:199 (1971).
110. K. M. Hambridge, P. A. Walravens, R. M. Brown, J. Webster, S. White, M. Anthony, and M. L. Roth, *Am. J. Clin. Nutr.* **29**:734 (1976).
111. P. A. Walravens and K. M. Hambridge, *Pediatr. Res.* **9**:310 (1975).
112. L. W. Sullivan and V. Herbert, *J. Clin. Invest.* **43**:2048 (1964).
113. L. W. Sullivan, in: *Newer Methods of Nutritional Biochemistry* (A. A. Albanese, ed.), Vol. III, pp. 365–406, Academic Press, Inc., New York (1967).
114. J. Lindenbaum, in: *The Biology of Alcohol* (B. Kissin and H. Begleiter, eds.), Vol. 3: *Clinical Pathology,* pp. 461–480, Plenum Press, New York (1974).
115. P. E. Stokes, in: *The Biology of Alcohol* (B. Kissin and H. Begleiter, eds.), Vol. 1: *Biochemistry,* pp. 397–436, Plenum Press, New York (1971).
116. G. G. Gordon, K. Altman, A. L. Southren, E. Rubin, and C. S. Lieber, *N. Engl. J. Med.* **295**:793 (1976).
117. O. P. Heinonen, D. Slone, R. R. Monson, E. B. Hook, and S. Shapiro, *N. Engl. J. Med.* **296**:67 (1977).

118. L. D. Longo, in: *The Pathophysiology of Gestation* (N. S. Assali eds.), Vol. II: *Fetal–Placental Disorders*, pp. 1–76, Academic Press, Inc., New York (1972).

119. R. W. Hillman, in: *The Biology of Alcoholism* (B. Kissin and H. Begleiter, eds.), Vol. 3: *Clinical Pathology* pp. 513–581, Plenum Press, New York (1974).

120. J. N. Neville, J. A. Eagles, G. Samson, and R. E. Olson. *Am. J. Clin. Nutr.* **21**:1329 (1968).

121. U.S. Department of Health, Education and Welfare, The Health Consequences of Smoking, DHEW Publ. No. (HSM) 73-8704, pp. 99–149 (1973).

122. J. Yerushalmy, *Am. J. Epidemiol.* **93**:443 (1971).

123. P. Rothstein and J. B. Gould, *Pediatr. Clin. North Am.* **21**:307 (1974).

124. J. J. Mulvihill, *Teratology* **8**:69 (1973).

125. E. B. Truitt and M. J. Walsh, in: *The Biology of Alcoholism* (B. Kissin and H. Begleiter, eds.), Vol. 1: *Biochemistry*, pp. 161–195, Plenum Press, New York (1974).

126. F. M. Badr and R. S. Badr, *Nature Lond.* **253**:134 (1975).

127. R. W. Klassen and T. V. N. Persaud, *Exp. Pathol.* **12**:38 (1976).

128. T. Kohila, K. Erikkson, and O. Halkka, *Med. Biol.* **54**:150 (1976).

129. O. Halkka and K. Eriksson, in: *Alcohol Intoxication and Withdrawal: Experimental Studies* (M. M. Gross, ed.), Vol. IIIa, pp. 1–6, Plenum Press, New York (1977).

130. G. Obe, H. J. Ristow, and J. Herha, in: *Alcohol Intoxication and Withdrawal: Experimental Studies* (M. M. Gross, ed.), Vol. IIIa, pp. 47–70, Plenum Press, New York (1977).

131. A. J. Sameroff and M. J. Chandler, in: *Review of Child Development Research* (F. D. Horowitz, ed.), Vol. IV, pp. 187–243, University of Chicago Press, Chicago (1975).

132. A. Silber, W. Gottschalk, and C. Sarnoff, *Psychiatr. Q.* **34**:461 (1960).

133. J. F. Jewett, *N. Engl. J. Med.* **294**:335 (1976).

134. J. Mayer and R. Black, in: *Currents in Alcoholism* (F. A. Seixas, ed.), Vol. II, pp. 429–444, Grune & Stratton, Inc., New York (1977).

135. M. E. Chafetz, H. T. Blane, and M. J. Hill, *Q. J. Stud. Alcohol* **32**:687 (1971).

136. J. K. Jackson, *Q. J. Stud. Alcohol* **15**:562 (1954).

137. D. W. Goodwin, *Arch. Gen. Psychiatry* **25**:545 (1971).

138. D. P. Cantwell, *Arch. Gen. Psychiatry* **27**:414 (1972).

139. J. R. Morrison and M. A. Stewart, *Biol. Psychiatry* **3**:189 (1971).

140. D. W. Goodwin, F. Schuslinger, L. Hermansen, S. B. Guze, and G. Winokur, *Arch. Gen. Psychiatry* **28**:238 (1973).

141. D. W. Goodwin, F. Schuslinger, L. Hermansen, S. B. Guze, and G. Winokur, *J. Nerv. Ment. Dis.* **160**:349 (1975).

142. M. F. Waldrop and C. F. Halverson, in: *The Exceptional Infant* (J. Hellmuth, ed.), Vol. II, pp. 343–380, Bruner Mogel, New York (1971).

143. J. L. Rapoport, P. O. Quinn, and F. Lamprecht, *Am. J. Psychiatry* **131**:386 (1974).

144. J. W. Demone, Jr., and H. Wechsler, in: *Alcoholism Problems in Women and Children* (M. Greenblatt and M. Schuckit, eds.), pp. 197–210, Grune & Stratton, Inc., New York (1976).

145. C. Randall, in: *Alcohol and Opiates: Neurochemical and Behavioral Mechanisms* (K. Blum, D. L. Bord, and M. G. Hamilton, eds.), pp. 91–107, Academic Press, New York (1977).

Behavioral Manifestations of 26
Ethanol Intoxication and Physical Dependence

A. INTRODUCTION

Ethanol intoxication drastically changes many physiological functions. The magnitude of the changes is generally underestimated because the behavioral responses include compensatory reactions, counteracting the direct drug effects.

Studies of behavior reveal both the direct actions of ethanol and the compensatory reactions. Behavior constitutes the total, organized responses to the conditions of the external and internal environment. The central nervous system is the focus of behavior because this is where the motor responses are integrated with each other and with the stimuli from the external and internal environment.

Section B of this chapter summarizes the behavioral responses to ethanol intoxication. These are primarily depressant effects, including stupor and anesthesia at high doses, in accordance with the pharmacological actions of the drug. The behavioral responses also often include feelings of exhilaration and behavioral arousal, which may partly express the compensatory reactions, counteracting the depressant effect.

Section C describes two different types of learning involved in the behavioral responses to ethanol. One type is the compensatory response, counteracting the drug effect. This is seen as tolerance, which develops within several hours after a single ethanol dose and more prominently after repeated doses or during prolonged maintenance of intoxication. The other type of learning is to establish the drug effect as a classically conditioned, discriminative signal. Alternative

Herbert Barry, III ● Department of Pharmacology, University of Pittsburgh School of Pharmacy, Pittsburgh, Pennsylvania 15261.

responses are learned, depending on whether the drug effect is present or absent. Contrary to tolerance, this type of learning emphasizes the difference of the drug effect from the nondrug condition.

Section D describes the behavioral manifestations of withdrawal illness, after the organism has become physically dependent on ethanol. Contrary to the depressant effect of intoxication, the withdrawal signs and symptoms constitute excessive stimulation. Compensatory responses are seen in lethargic behavior and avoidance of sensory stimulation. Hangover is a brief, mild form of withdrawal illness after a single episode of intoxication. The withdrawal illness can be very severe, including convulsions, hallucinations, disorientation, and occasionally death, after prolonged intoxication.

Each topic begins with an account of behavioral responses by human beings, the same species as all the readers. The responses by various species of laboratory animals show the same general patterns, but different techniques and test situations make it convenient to describe their behavior separately from the corresponding behavior of human beings.

B. INTOXICATION

Effects of ethanol on behavior have been studied extensively both in humans and in laboratory animals. The present section is limited to a brief, general survey of the principal effects. A more detailed review, published in 1970, appears in Volume 1, Chapters 6 and 7, of a book by Wallgren and Barry.[1] This includes studies both on humans and on laboratory animals. Effects of alcohol on several aspects of behavior, mostly in humans, are surveyed in several chapters in a book edited by Kissin and Begleiter[2] and in subsequent reviews by Barry,[3-4] Moskowitz,[5] and Levine et al.[6]

1. General Effects

Ethanol is a depressant agent, pervasively impairing, retarding, and disorganizing the functions of the central nervous system. Nevertheless, many of the behavioral effects of ethanol are stimulant. These contrasting effects are explainable by some general attributes of behavior.

Normal behavior consists of highly organized responses to the sensory signals transmitting changes in the external environment, such as visual, auditory, tactual, olfactory, and gustatory stimuli, and to the motivating forces related to physiological needs, such as hunger, thirst, pain, and fatigue. Inhibitory functions of the central nervous system are important mediators of the controlled behavior. The multitudinous rapid changes in sensory stimuli would be overwhelming if they elicited continuous attention. Instead the responses to the stimuli are inhibited by adaptation, both at the receptors and in the integrating functions of the central nervous system.

The powerful emotions aroused by the physiological needs can potentially occupy the individual's full attention and prevent adaptive responses to external stimuli. Instead the responses to these motivating forces are often suppressed or modified. One of the mechanisms for focusing on sensory signals, and obtaining relief from the pressure of physiological needs, is the classically conditioned response to conditioned stimuli that are signals for relief from hunger, pain, or other physical discomforts.

The general depressant effect of ethanol impairs the controls over the responses to the sensory stimuli and physical needs. Since these controls are predominantly inhibitory, ethanol thereby gives rise to behavioral arousal. Simultaneous or rapidly shifting attention to sensory stimuli expresses the disinhibitory effect of ethanol. Suppressed emotions are displayed, such as anger or dependency.

The behaviorally arousing effect of ethanol is partly attributable to a general tendency for the inhibitory functions to be more susceptible to disruption than are the excitatory functions. The depressant drug effect, by impairing the inhibitory functions more strongly than the excitatory functions, thereby has a behaviorally stimulant effect. A further contribution to the behavioral stimulation caused by ethanol is a compensatory response, counteracting the depressant drug effect.

The prominent disinhibiting or arousing effects of a pharmacologically depressant agent are not limited to ethanol. Anesthetic agents, such as barbiturates and ether, likewise give rise to disorganized, excited behavior at subanesthetic doses. The "minor" tranquilizers, such as benzodiazepines and meprobamate, show disinhibitory effects under some conditions.

Ethanol is often classified with the barbiturates as a sedative/hypnotic. These drugs have in common a drastic depressant effect on behavior, indicated by stupor or anesthesia, at doses only slightly higher than those with arousing effects. The "minor" tranquilizers have a weaker and more selective arousing effect at low doses, but they also alleviate anxiety, with a "taming," muscle relaxant effect at doses far below the amount that causes anesthesia. These drug effects are shared to a greater degree by ethanol than by barbiturates.

In general, ethanol has a distinctive pattern of behavioral effects. Barbiturates are the most closely similar drugs, but in some respects ethanol resembles benzodiazepines more than barbiturates.

2. Behavior of Humans

Ethanol affects mood and selectively impairs some types of performance at a blood concentration of 0.05%. The footnote to Table I shows correspondences with this measure. In comparison with percentage ethanol in the blood, the measure per mille is a multiple of 10 and the measure mg/dl is a multiple of 1000. This blood ethanol concentration usually occurs during the span of 30–60 min after fairly rapid drinking of a dose of 0.5 g/kg, which is 0.05% of the total body

**Table I.   Measures of Ethanol Quantity
Corresponding to 0.5 g/kg (0.6 ml/kg) Consumed by a
Man or Woman of Typical Body Weight**[a]

	Man	Woman
Body weight of drinker		
Kilograms	70	50
Pounds	154	110
Ethanol consumed		
Grams	35	25
Milliliters	44	31
Ounces	1.5	1.0

[a] Blood ethanol concentration: %, 0.05; per mille, 0.5; mg/dl, 50.

weight. The effects of incomplete absorption and of partial metabolism during this time span are approximately equally balanced by the fact that ethanol is absorbed very poorly in fat and thus has a higher concentration in body water, which is the predominant component of blood.

Table I shows the corresponding amounts of consumption of absolute ethanol by a man and a woman of typical body weight. The amount of 1.5 oz, shown for a man, is contained in three typical drinks, in the form of shots of hard liquor (each 1.0 oz of 40–50% ethanol) or glasses of wine (each 2.5 oz of 15% ethanol) or bottles or cans of beer (each 12 oz of 4% ethanol). The amount 1.0 oz, shown for a woman, is contained in two instead of three typical drinks.

These equivalences are approximations, and the blood ethanol levels are affected by variations in rate of drinking, absorption, and metabolism. The estimates in Table I do not take into account a sex difference in body fat. Since women have 10–20% more body fat, which excludes most of the ethanol, the same ethanol intake in terms of body weight (g/kg) results in higher blood ethanol concentrations in women than in men. Therefore, the same quantity of liquor consumption by a typical woman results in appreciably higher blood ethanol concentration, not only because of her lower body weight but also because of her higher proportion of body fat.

The rate of ethanol metabolism is about 0.01%/hr in the typical person. Thus, 5 hr is required to eliminate a blood ethanol concentration of 0.05%. A steady concentration would be maintained by a rate of one typical drink in slightly less than 2 hr for men and in more than 2 hr for women.

Table II shows a comparison among types of performance with respect to the blood ethanol concentration usually required to cause substantial impairment. Three types of functions are designated as being impaired at 0.05% blood ethanol: divided attention, vigilance, and standing steadiness.

Attention is divided between two simultaneous tasks when the person responds to visual signals presented in two different locations or does arithmetical calculations while responding to visual signals. Two reviews[5,6] emphasize the detrimental effects of low ethanol doses on tasks of divided attention.

Perceptual vigilance is tested by presenting stimuli infrequently at irregular intervals. The increase in failures to respond to the signals might be a special instance of impairment in divided attention, because inattention to the infrequently presented signals might indicate the effects of competing attention to other sensory stimuli or to fantasies or other internal stimuli.

Standing steadiness is measured by swaying of the body (Romberg test). Usually, the person is blindfolded during the test. If the eyes are kept open, the swaying is greatly decreased, but ethanol causes a greater percentage increase in swaying under this condition than when the eyes are closed. Therefore, the visual environment is more effective in stabilizing the body position in the sober condition than during intoxication.

The performances impaired at 0.10% blood alcohol (Table II) are functions of motor control and complex intellectual performances. Higher blood ethanol concentration (0.15%) is required for large, consistent effects on well-practiced motor functions, on response speed, and on sensory acuity. The ethanol concentration that impairs these functions is close to the concentrations that cause stupor (0.20%) and anesthesia (0.25%). The lethal dose, which fatally depresses the respiratory reflex, is approximately twice the anesthetic dose.

Self-reported moods and emotions are predominantly pleasant and exhilarated after doses that give rise to blood ethanol concentrations of 0.05–0.15%. Social behavior is typically cheerful, friendly, and boisterous. Depressant effects are sometimes reported, however. Some drinkers feel sleepy, calm, and contented; others feel hazy, confused, withdrawn, and tired. There is suggestive evidence that the stimulant reaction is more prevalent among men, the depressant reaction more frequent among woman.[3] A sufficiently high dose causes all drinkers to show the depressed response of stupor or unconsciousness.

Table II.   Blood Ethanol Concentrations That Usually Cause
Substantial Impairment in the Specified Types of Performance

Concentration	Nature of impairment
0.05%	Impaired attention to two simultaneous tasks
	More failures to perceive infrequent stimuli
	Swaying of body when standing
0.10%	More failures in sensorimotor tracking
	Staggering when blindfolded
	More errors in arithmetical calculations
	Impaired short-term memory
0.15%	Staggering when eyes are open
	Slower speed in arithmetical calculations
	Slower response to a signal
	Slurred speech
	Decreased sensitivity to sensory stimuli
0.20%	Responsiveness to stimuli abolished
0.25%	Consciousness abolished
0.50%	Breathing abolished

A sex difference in effects of ethanol on emotions is apparent in studies of the themes of stories told in response to a standard series of pictures (Thematic Apperception Test). Intoxication in men increases the frequency of themes of power and of sexual and aggressive behavior, whereas intoxication in women increases the frequency of themes of femininity and maternal behavior. Ethanol thus appears to intensify expression of the characteristics socially approved for the drinker's sex. This effect is consistent with the previous conclusion[4] that the disinhibitory effect of ethanol is at a superficial level, arousing responses that are culturally approved or weakly restrained. The more deeply repressed emotions and desires continue to be excluded from conscious awareness or direct expression.

Complex performance, such as driving an automobile, is affected both by performance capability and by mood. Ethanol increases willingness to make risky decisions and also causes the objective hazard to be underestimated. Many experimental studies of ethanol effects include reports that the subjects believed they were performing normally or better than usual, after a dose that substantially impaired their performance. Accordingly, frequency of accidents by automobile drivers shows evidence for being increased by blood ethanol concentrations well below 0.05%, when most tests of performance capabilities show no perceptible effect. Depressant effects of high ethanol doses might increase the frequency of accidents not only by impairing performance but also by diminishing alertness. The extreme case is when the intoxicated driver falls asleep, sometimes while driving at excessive speed due to the concurrent disinhibitory effect of ethanol.

Reports of improved performance under the influence of ethanol are rare and limited to low blood concentrations, about 0.05%, under unusual conditions that inhibit or impair normal performance. In a series of difficult problems of logic, a low dose of ethanol increased the number of solutions while causing a greater increase in the number of incorrect attempted solutions.[7] In a task of speed and accuracy of response to visual stimuli, a continuous noise increased the number of errors. Doses of ethanol that increased number of errors in the quiet condition counteracted this detrimental effect of noise.[8]

Studies of effects of ethanol on aggressive and sexual behavior appear to have been limited to men. Ethanol intoxication increases aggressiveness[9-10] and decreases sexual arousal.[11-12] Enhancement of sexual arousal under some conditions, however, was indicated by greater maximum penile diameter, although with a slower rate of tumescence, in a test with a low blood ethanol concentration of 0.025%[11] and by a tendency for a medium ethanol dose (1.0 g/kg) to impair ability to suppress penile tumescence.[12]

3. Behavior of Laboratory Animals

Effects of ethanol are most often tested by intraperitoneal injection in rats or mice. The studies in which blood ethanol concentration was measured show that performance is substantially impaired above 0.20%, anesthesia measured by loss

**Table III.   Effects of Ethanol Doses[a] on Blood
Ethanol Concentration and Percentage of Maximum
Time That Rats Successfully Avoided Shock on the
Moving Belt[b]**

Dose (g/kg)	Concentration (%)	Percentage of maximum time
1.6	0.20	88
1.9	0.17	90
2.2	0.26	46
2.5	0.30	28

[a]Injected intraperitoneally 23–29 min earlier.
[b]This initial ethanol test for five rats tested with each dose was reported by Gibbins *et al.*[13]

of righting reflex occurs at 0.30%, and the lethal dose is about 0.80%. Table II shows that the corresponding effects in humans are found at lower doses, especially for the lethal effect.

Table III summarizes data obtained at 23–29 min after the first time when rats were administered ethanol.[13] The blood ethanol concentrations exceed somewhat this amount of alcohol injected as a percentage of body weight. Thus, 2.5 g ethanol constitutes 0.25% of a kilogram. In Table III this dose is shown to give rise to 0.30% blood ethanol concentration. The concentration decreases rapidly after its peak at 10–20 min after injection because of the rapid rate of metabolism (0.03%/hr in rats, 0.05–0.06%/hr in mice).

Oral administration might appear to be more similar to the usual method of ethanol intake by humans, but the greater stressfulness and difficulty of oral intubation of rodents cause most experimenters to prefer the intraperitoneal route. An oral dose of 4.0 g/kg gives rise to a peak blood ethanol concentration of only about 0.20% 30–60 min afterward, as a result of rapid metabolism while the ethanol is slowly absorbed from the gastrointestinal tract.

Blood ethanol concentrations above 0.20% were necessary to cause substantial impairment of the performance shown in Table III, whereas most performance tests in humans show substantial decrement at a blood ethanol concentration of 0.15% (Table II). The moving-belt test in rats (Table III) consisted of the simple, well-practiced response of running at a moderate speed on a flat surface, and failures were punished by painful electric shock. These characteristics may have enhanced resistance to the effect of ethanol.

Most other tests of learning, performance, and sensory capabilities likewise show substantial impairments only with doses sufficient to give rise to blood ethanol concentrations above 0.20%. An example is the tilted-plane test, which does not require active movement by the animal and does not punish failure. The front end of a flat surface on which the rat crouches is gradually lifted off the horizontal plane and the angle is measured at which the rat starts to slide backward. This apparatus, which is simple to construct and operate, is often used for testing ethanol intoxication in rats.

Some reports indicate that blood ethanol concentrations of 0.15% or less substantially impair the ability of rats and mice to cling with their forepaws to a wire or to run on a curved, rotating surface. In common with humans, different types of motor functions are affected differentially by ethanol.

Measures of exploratory behavior and other spontaneous motor activity are sometimes affected by low ethanol doses. In rats, decreased grooming and rearing have been reported at blood ethanol concentrations below 0.10%. In mice, motor activity may be increased by a wide range of doses, but the increase in motor activity is accompanied by impairment of motor control, measured by increased frequency with which the feet slip through the openings in the wire mesh floor. In one study,[14] motor activity of mice was increased by low ethanol doses (0.5 or 1.0 g/kg), whereas a higher dose (2.0 g/kg) decreased activity in the first hour but increased it above the control level in the second hour.

Experiments on cats and rats have demonstrated the effects of ethanol on motivations and emotions. Ethanol can induce a food-approach response that has been inhibited by avoidance of a punishing shock. The drug has been shown to diminish the strength of the approach response but to diminish even more the strength of the avoidance. Similarly, a diminished avoidance of the punishing experience of frustration appears to explain findings that rats persist longer in running to a food cup (show greater resistance to extinction of the approach response) when food is no longer presented. These effects occur at doses that give rise to blood ethanol concentrations of about 0.15–0.20%, substantially below the level required to impair performance capabilities.

Ethanol depresses rather than disinhibits approach behavior in some approach–avoidance conflict situations, notably when a rat is required to press a lever in a chamber rather than to run down an alley. The situation or type of response appears to be an important influence on the disinhibitory effect of ethanol.

Effects of ethanol on aggressive behavior have been studied in male animals of various species. A review[15] concluded that attack and defensive reactions can be enhanced by low ethanol doses but are consistently suppressed by higher doses. A subsequent study[16] showed that in a fight between two rats, injection of ethanol in the subordinate rat increased the frequency of attacks and wounds if it was naive but not if it had previous fighting experience. Therefore, lack of experience with the test situation impaired the ability of the subordinate animal to counteract the detrimental effect of ethanol on defensive behavior.

Ethanol depresses sexual behavior in laboratory animals as in humans. Studies on male and female rats and on male dogs and monkeys indicate that ethanol inhibits sexual responsiveness, with a stronger depressant effect on penile erection than on ejaculation.[5]

C. DIFFERENT TYPES OF LEARNING

Learning in general constitutes modification of behavior, adjusting to environmental or internal forces. If the individual is comfortable and satisfied, there is no incentive for the effort and stress involved in learning. Ethanol intoxication is a strong internal force, which thereby fulfills a basic requirement for learning. This section describes two contrasting types of learned responses to ethanol intoxication.

One type of learned response is compensatory, counteracting the effect of ethanol. This response is commonly described as the development of tolerance to the drug effect. It is an example of the learned response of sensory adaptation to a strong stimulus, thereby minimizing the disturbance caused by the stimulus. Tolerance to ethanol is reviewed in Volume 2, Chapter 9, of a book by Wallgren and Barry[1] and in an article by Kalant et al.[17]

The other type of learned response has the opposite effect, of emphasizing and maintaining differential responses to the drug. The drug effect becomes a classically conditioned stimulus, signaling the requirement for differential responses. Studies in rats on ethanol as a discriminative stimulus have indicated some attributes of the drug effect and especially have identified other drugs with similar discriminative stimulus attributes. This type of learned response to ethanol has been reviewed by Overton[18] and by Barry and Krimmer.[19]

1. Compensatory Response of Tolerance

After rapid ethanol consumption, performance of humans is impaired more greatly while the blood ethanol concentration is rising than at the same concentration after the peak level. This indicates acute tolerance in a single session. Therefore, effects of ethanol on performance are already diminishing at the time of maximum blood ethanol concentration.

The compensatory response is enhanced by the opportunity to perform it under the influence of ethanol. This is indicated by progressive improvement in performance, tested continuously or repeatedly while the blood ethanol concentration is maintained at a steady level. Acute tolerance may be especially effective during continuous performance beginning immediately after ethanol intake, so that the compensatory response is learned while the blood ethanol concentration gradually increases. This might account for a report that the same blood ethanol concentration had less detrimental effect on performance in studies in which the testing began within 30 min rather than 1 hr or more after drinking.[6]

In addition to acute tolerance, within a single session, chronic tolerance develops when ethanol intoxication is not continuously maintained but is repeated on successive days. The acute tolerance begins sooner after the onset of drinking and develops to a greater degree. Chronic tolerance usually can be observed within 1–2 weeks after daily ethanol intake. Ethical considerations have severely limited controlled experiments on chronic tolerance in humans. In many studies, however, the same ethanol dose has been shown to have less detrimental effect on performance of people with a history of heavier or more frequent drinking. Alcoholics have often been observed to show little or no behavioral effect of blood ethanol concentrations above 0.20%, which severely impair performance of most people.

Tolerance has been studied more thoroughly in laboratory animals. The results of these controlled experiments agree well with the observations on humans.

Acute tolerance in rats has been measured accurately by correlating brain ethanol levels with shock avoidance on the moving belt[13] in groups of animals tested for 2 min at three time intervals after various ethanol doses.[20] The brain ethanol concentration required to cause failure to avoid for 50% of the time was 0.15% at 9–11 min, 0.20% at 29–31 min, and 0.30% at 59–61 min. These differences suggest that the magnitude of acute tolerance is larger than has generally been recognized.

When high blood ethanol concentrations in rats are maintained continuously for several days,[21] concentrations above 0.30% are generally required to cause noticeable staggering and above 0.50% to cause loss of righting reflex. These are higher than the levels sufficient to cause the same effects after a single dose in rats. A general debilitating effect of continuous intoxication, however, is indicated by a lower lethal concentration (0.70%) than the level of 0.80% after a single dose.

When ethanol intoxication is repeated but not continuously maintained on successive days, measures of performance after a subsequent test dose can distinguish between two types of tolerance. Physiological tolerance is due to a general compensatory response by the central nervous system, causing a decreased effect of ethanol on a response that has not been practiced previously during ethanol intoxication. The usual experimental procedure is to train the animal in the performance shortly before each daily ethanol administration. Psychological tolerance, also called behaviorally augmented tolerance, is the enhancement of physiological tolerance when the animal has been trained shortly after each daily ethanol administration, thereby practicing the response while intoxicated.

Experiments on two types of performance tests have shown that behaviorally augmented tolerance develops more rapidly but to the same final level as physiological tolerance.[22,23] An ethanol dose of 1.4 g/kg, giving rise to a blood ethanol concentration of 0.17%, caused a larger detrimental effect on a difficult food-rewarded maze-running performance[22] than a higher ethanol dose (2.5 g/kg) on shock-avoidance performance on the moving belt.[23] Behaviorally augmented tolerance reached its maximum after 3 days in both situations. The same level of physiological tolerance was reached after 5 days in the maze-running test and after 9 days in the moving-belt test. This difference gives evidence that during repeated ethanol doses, the general compensatory response by the central nervous system (physiological tolerance) affects most rapidly the functions that are most severely impaired by ethanol. This is in accordance with the adaptive purpose of the compensatory response.

2. Discriminative Response to Drug Stimulus

Since ethanol intoxication is accompanied by pervasive differences from the normal state of the central nervous system, responses learned in the ethanol or placebo state should be more likely to occur when tested subsequently in the same state than in the alternative state. Several studies on humans have demonstrated superior memory for verbal material when tested in the same ethanol or placebo state as the prior learning.[18] The detrimental effect of the change in drug state is greater if the task requires more general or difficult retrieval of the memory. Thus, the effect of the change in drug state is greater for free recall than when the categories of words are given as cues[24] and for words that have low rather than high imagery.[24,25]

The occurrence of state-dependent amnesia is indicated by anecdotal accounts that alcoholics sometimes hide a liquor bottle while drunk, forget the location when sober, but remember when drunk again. Under most circumstances, however, memories transfer from one state to the other. The principal effect of the differential drug state is to facilitate the learning of different responses, such as bold or dependent behavior when intoxicated but cautious or self-reliant behavior when sober. This differential learning can develop because the intoxicated and sober states evoke different types of responses. A response that is reinforced only under one state will occur only in the presence of that state.

The importance of ethanol intoxication as a discriminative stimulus for differential behavior is usually not recognized by humans. Most behavior is controlled by sensory stimuli, such as the clock or perception of responses by other people. The effects of internal stimuli, such as hunger, fatigue, and drug effects, are not fully accessible to conscious awareness. Differences in learned behavior in the intoxicated and sober states are commonly attributed to differences in the social situation or physical environment.

Nevertheless, the ethanol state can be an important discriminative stimulus. This is demonstrated by experiments in which the differential ethanol or placebo state serves as the only available signal that a desired reinforcement is obtained by making alternative responses. In a preliminary experiment,[26] two women learned to earn money by pressing a different lever, depending on whether they had drunk a beverage containing ethanol (approximately 0.6 g/kg) or placebo.

This type of discrimination learning has been studied extensively in rats. Verbal labeling of the differential treatments is thereby precluded, and the predominant intraperitoneal route of administration prevents perception of differential tastes. Rats differentiate the ethanol state from placebo at doses substantially less than those sufficient to cause other behavioral changes. After training to discriminate 1.0–1.2 g ethanol/kg from placebo, at 5–20 min after intraperitoneal injection, the dose eliciting 50% choice of the ethanol response (ED_{50}) is 0.5–0.6 g/kg.[19] Rats have been trained to discriminate 0.63 g ethanol/kg from placebo at 20 min after intraperitoneal injection[27] and 0.1 g ethanol/kg from placebo at 5 min after intravenous infusion.[28]

Ethanol is a strong unconditioned stimulus with rewarding effects indicated by voluntary consumption of ethanol solutions[29] and punishing effects indicated by avoidance of a conditioned stimulus associated with ethanol.[30] When ethanol is established as a discriminative stimulus, however, it becomes a conditioned stimulus for alternative responses, associated with a different unconditioned stimulus, such as food or escape from shock. The discriminative ethanol stimulus thus functions as a signal, as do sensory stimuli, and the unconditioned rewarding or punishing effects are probably minimized in this situation.[31] Accordingly, rats differing in amount of ethanol voluntarily consumed have been reported to show no difference in rate of learning alternative responses in the ethanol and placebo states.[32]

Unlike visual, auditory, and other sensory stimuli, the conditioned drug stimulus is effective, although a delay of 5 min or more intervenes between the onset of the drug effect and the opportunity to make the discriminative response. This persistence of the conditioned drug stimulus might be attributable to the pervasive changes in the central nervous system caused by the drug or to the slow, gradual nature of the change in drug state.[31]

The discriminative stimulus attributes of ethanol resemble those of sedative, hypnotic, anesthetic agents, such as barbiturates and chloral hydrate, and "minor" tranquilizers, such as chlordiazepoxide, diazepam, and meprobamate.[33,34] Rats trained to discriminate ethanol from placebo choose the ethanol response in tests with sufficiently high doses of pentobarbital or chlordiazepoxide but not in tests with morphine, Δ^9-tetrahydrocannabinol, amphetamine, or chlorpromazine. Under some conditions, however, rats trained to discriminate pentobarbital or chlordiazepoxide from placebo do not select the drug response in tests with ethanol.[34] Also, rats have been trained to discriminate between an ethanol and pentobarbital dose on the basis of a qualitative rather than merely quantitative difference between the two drugs.[19,35]

The discriminative stimulus attributes of ethanol thus partially resemble and partially differ from those of pentobarbital and chlordiazepoxide. These relationships among drugs might be explained by a model[19] according to which they cause both sedation and arousal. The predominant effect of ethanol is sedation, and the discriminative ethanol response readily transfers to drugs that also cause sedation. Arousal is an additional important component of the effect of pentobarbital and to a lesser degree of chlordiazepoxide. The discriminative drug response fails to transfer to ethanol if the training conditions have caused arousal rather than sedation to be the predominant component of the discriminative drug stimulus.[31] This model therefore explains findings that the pentobarbital or chlordiazepoxide response consistently transfers to ethanol in a shock–escape situation, in which sedation is the principal effect of all three drugs, but not in a food-rewarded situation, in which arousal is an important effect of pentobarbital and chlordiazepoxide.

D. PHYSICAL DEPENDENCE

Ethanol intoxication refers to the disturbances of functioning during high blood ethanol concentrations. After the individual has adjusted to intoxication, with the help of physiological and behavioral compensatory responses, withdrawal of the ethanol can cause illness and other disturbances of functioning.

Physical dependence on ethanol is inferred from the occurrence of withdrawal illness.

The hangover is a form of withdrawal illness following a single episode of intoxication. Research on this topic has been meager and limited to humans. The available information was summarized in Volume 2, Chapter 9, of a review by Wallgren and Barry.[1] An article by Chapman[36] reviewed previous studies and listed the various signs and symptoms of hangover.

Much more research has been done on the potentially fatal withdrawal illness following prolonged, heavy ethanol intake. Studies on humans and laboratory animals have been reviewed in Volume 1, Chapter 9, of Wallgren and Barry[1] and by Kalant et al.[17] Subsequently, the withdrawal signs and symptoms in humans have been reviewed by Mello and Mendelson.[37] Also, Mello[38] summarized techniques for sustaining intoxication in laboratory animals and thus eliciting the withdrawal illness within a few days.

1. Hangover

Exhilaration and pleasurable effects are important attributes of ethanol intoxication, but they are absent from the hangover. The withdrawal illness is completely unpleasant, although it is brief and does not threaten life when it occurs as a hangover after a single episode of intoxication.

Table IV lists the signs and symptoms of hangover in three general categories. The first category shows effects of excessive stimulation of the central nervous system or autonomic nervous system. The second category lists direct behavioral expressions of excessive stimulation. The third category lists compensatory responses, which attempt to minimize or counteract the effects of excessive stimulation.

People during hangover feel sick and incapable of their maximum physical or mental performance. Nevertheless, studies of motor and intellectual functions

Table IV. Signs and Symptoms of Hangover Reported in Humans

Signs of physical overstimulation
  Dizziness, vertigo
  Headache
  Nausea, vomiting
  Thirst
  Sweating
Behavioral expressions of overstimulation
  Tremor
  Irritability
  Tense feelings
Compensatory responses
  Fatigue
  Apathy
  Avoidance of stimulation

indicate that the detrimental effects of hangover are small and inconsistent.[39,40] Contrary to intoxication, when the person feels well but performs badly, during hangover the person feels badly but usually performs as well as normally.

A study of the time course of hangover[41] indicated that the signs and symptoms typically last about 12 hr, beginning shortly after the blood ethanol concentration begins to decline and reaching the peak of severity at about the same time as the blood ethanol concentration reaches zero. The self-rated hangover symptoms began to decrease while the physical signs seen by an observer were still increasing. A compensatory response thereby appears to mitigate and abbreviate the subjectively perceived symptoms.

In laboratory mice, susceptibility to convulsions was increased following intraperitoneal injection of a single high dose of ethanol (5 g/kg). The peak effect coincided approximately with the disappearance of ethanol from the blood, but the effect began while the blood ethanol concentration was still high enough to impair motor performance.[42]

2. Withdrawal Illness in Humans

A severe and sometimes fatal withdrawal illness occurs in some alcoholics. The signs and symptoms have been reproduced in experiments in which ethanol was abruptly withdrawn following several days of sufficiently steady drinking to maintain continuously high blood ethanol concentrations.

Table V summarizes the most prominent signs and symptoms, separately for muscular reactions, sensory effects, and changes in mood. The compensatory responses and the mild direct responses correspond closely to signs and symptoms of hangover (Table IV). The more severe direct responses, however, are far beyond the range of hangover symptoms.

The duration of the withdrawal illness is typically about 4 days. Convulsions usually constitute the most severe sign in the first and second days. Hallucinations and disorientation occur later, in the third and fourth days after cessation of drinking. Delirium tremens constitutes profound disorientation, agitation, and vivid hallucinations, accompanied by symptoms of intense overstimulation of the autonomic nervous system. This condition is sometimes fatal, owing to severely accelerated heart rate, resulting in cardiac overload and cardiovascular collapse.[43]

In a study of more than 1000 patients in acute ethanol withdrawal, Gross et al.[44] reported on the frequency of hallucinations and the degree of disorientation. Hallucinations were often both auditory and visual, but if they were limited to one modality, it was visual more often than auditory. The preponderance of visual hallucinations is in accordance with effects of hallucinogenic drugs but contrary to schizophrenic hallucinations. The most frequent category of disorientation was more than two calendar days in time but with intact orientation for place and person. The majority of the patients were tested while experiencing the signs and symptoms of delirium tremens.

**Table V.   Types of Ethanol Withdrawal Signs and Symptoms in Humans Listed
Separately for Each of Three Modalities of Response**

Type of signs and symptoms	Modality of response		
	Muscular	Sensory	Mood
Compensatory	Weakness	Unresponsiveness	Apathy
Direct—mild	Tremors	Disorientation	Irritability
Direct—medium	Shaking	Hallucinations	Anxiety
Direct—severe	Convulsions	Delirium tremens	Terror

The severe physical illness of ethanol withdrawal results in various behavioral responses. Gross et al.[45] have summarized patterns of physical and behavioral signs and symptoms. Gross and Hastey[46] have reviewed studies on sleep disturbances.

3. Withdrawal Illness in Laboratory Animals

Various techniques have been developed for maintaining continuous ethanol intoxication in laboratory animals for several days.[38] When the ethanol is withdrawn, the signs and reactions appear to correspond closely to those observed in humans (Table V).

Withdrawal signs and reactions observed in rats after 4 days of continuous intoxication[21] are summarized in Table VI. All four types of signs and reactions indicate a syndrome of excessive stimulation. Two of the types (tremors and convulsions) correspond directly to prominent signs in humans. The other two types (behavioral stimulation and rigidity) might indicate disorientation and hallucinations, which are important symptoms in humans (Table V) but not directly measured in rats.

Additional signs and reactions, reported in some rats, include stereotyped movement and aggressiveness. These also express excessive stimulation. Stereo-

**Table VI.   Four Types of Typical Withdrawal Signs and Reactions in
Rats, with Specific Behaviors[a]**

Sign/reaction	Behaviors
1. Behavioral stimulation	Restlessness; frenzied exploration
	Agitation; enhanced startle reflex
2. Tremors	Tail; rear part of body
	Body; wet shakes
	Head; chattering of teeth
3. Rigidity	Tail stiffly extended or curled around a rod
	Rigid body posture
4. Convulsions	Audiogenic
	Spontaneous

[a]From Majchrowicz.[21]

typy is a response of high doses of amphetamine, and fighting is a typical behavioral response during withdrawal from morphine or other narcotic analgesics. The withdrawal signs and reactions ended in death for a few of the animals. Convulsions preceded the mortality in most but not all of these cases.[21]

In accordance with humans, the withdrawal signs and reactions in rats[21] began before the blood ethanol concentrations reached zero, and convulsions often occurred while the blood ethanol was 0.10%. The withdrawal syndrome gradually intensified, usually reaching the peak at 2–6 hr after disappearance of ethanol from the blood. Occasionally, perceptible signs and responses (tremors and tail stiffness) persisted 2–3 days later.

The four types of withdrawal signs and reactions shown in Table VI were used for a rating scale of severity of withdrawal reactions in rats after 1–7 days of continuous ethanol intoxication.[47] The severity of withdrawal increased progressively after longer duration of intoxication up to 4 days but not beyond that number, although further tolerance after the fourth day was indicated by increased blood ethanol concentrations necessary to cause the behaviorally depressant effects of ethanol.[47]

A similar pattern of withdrawal signs and reactions has been reported after lower blood ethanol concentrations maintained for longer durations (10–30 days) by consumption of a liquid diet containing ethanol.[48] Criteria for withdrawal included piloerection, broad-based gait, sprawling episodes, and spontaneous squealing, in addition to the signs and reactions listed in Table VI. Observations of behavioral activity indicated an increase at 1–2 hr before the blood ethanol concentration reached zero, after onset of the withdrawal syndrome, and depressed activity for several hours afterward, while the withdrawal signs and responses were most severe.

The same pattern of withdrawal signs and reactions has been found, regardless of whether rats had experienced several days of intoxication with t-butanol or with ethanol.[49] Since t-butanol is not oxidized to the corresponding aldehyde, the withdrawal signs and reactions are thereby attributable to ethanol rather than the metabolite, acetaldehyde.

Two different stages of the withdrawal syndrome have been identified in rats after consumption of a liquid diet containing ethanol for the extraordinarily long duration of 270 days.[50] Within 12 hr after ethanol withdrawal they showed tremors, rigidity, and convulsions (the last three types of signs and reactions in Table VI). Their spontaneous activity was depressed but their reactions to sensory stimulation were abnormally strong. On 3–5 days after ethanol withdrawal, however, their coordinated, exploratory activity was stimulated, indicated by increases in forward locomotion, rearing, grooming, and sniffing in the open field.

Other signs and reactions of ethanol withdrawal syndrome in rats are a decrease in the intensity of electric shock sufficient to cause flinching or jumping,[51] suppression of a food-rewarded lever-pressing response,[52] decreased locomotor and exploratory behavior,[53] and preference for a warmer environment.[54] When water and an ethanol solution are both available during withdrawal, rats can learn to drink a sufficient amount of ethanol to counteract the withdrawal reactions. This learning occurred only in some of the animals, after one or more previous experiences with withdrawal.[55] The conditions were unfavorable for this learning, because the withdrawal illness caused a decrease in fluid intake, and when water and ethanol were both available the ethanol consumption was negligible, not only during ethanol withdrawal but also in prior choice tests and by a control group.

In mice, after continuous inhalation of ethanol vapor for several days, a quantitative scale for withdrawal syndrome was based on the signs of lethargy, tremor, tail lift, startle to noise, convulsions on handling, spontaneous convulsions, and death.[56] A quantitative scale of susceptibility to convulsions on handling[42] showed onset of the withdrawal signs and reactions before the blood ethanol concentration reached zero, the peak intensity several hours after disappearance of the ethanol, and persistence of the signs and reactions for 1–3 days. Increasing durations of exposure to ethanol (1–9 days) progressively increased the duration of the withdrawal syndrome. Similar results were found after 2–12 days of continuous ethanol intoxication by consumption of a liquid diet containing ethanol.[57]

Ethanol withdrawal syndrome has been studied most extensively in rats and mice because these species of laboratory animals are relatively inexpensive to purchase and maintain and are most frequently used for behavioral studies. Similar withdrawal signs and reactions have been observed in dogs, rhesus macaque monkeys, and chimpanzees. The experiments on these species have been summarized by Mello,[38] Freund,[58] Pieper,[59] and Cicero (Chapter 27).

E. SUMMARY AND CONCLUSIONS

The predominantly depressant effects of ethanol are evident with doses sufficient to cause stupor or anesthesia. Lower doses impair motor and intellectual performance. In humans, low ethanol doses have the greatest detrimental effect on divided attention, perceptual vigilance, and standing steadiness. Laboratory rats and mice are less sensitive to the effects of ethanol, but tests of motor coordination seem to show the greatest detrimental effect. In humans and laboratory animals, moderate ethanol doses increase aggressive behavior and decrease sexual behavior.

Behavior is modulated by complex, interrelated elements and by mechanisms for adapting to changes of the internal and external environment. These factors account for exhilarated moods and increased motor activity during ethanol intoxication. Increased boldness is attributable to the general depressant effect, because ethanol acts more strongly on inhibitory functions, such as avoidance of punishment, than on excitatory functions, such as approach of a desired goal. A stimulant response, counteracting the prevalently depressant effect of ethanol, may also contribute to feelings and expressions of excitement.

Ethanol intoxication results in two contrasting types of learning. One type counteracts the effects of ethanol, thereby stabilizing the internal environment. This is the adaptive response of tolerance to the drug. The other type of learning establishes the drug effect as a discriminative stimulus or as a state of the internal environment dissociated from the nondrug state. Differential responses are thereby elicited and rewarded in the intoxicated and sober states. The magnitude of both types of learning is generally underestimated. The adjustive response of tolerance develops rapidly, during the initial exposure to the drug, so that the intensity of the drug effect is not recognized as being diminished by this mechanism. Differences in behavior between the intoxicated and sober state are ascribed to the physiological changes caused by ethanol or to a distinctive situation associated with drinking rather than to the discriminative or dissociative aspect of the drug effect.

The signs and symptoms of the withdrawal illness are opposite to the effects of ethanol intoxication. During withdrawal, generalized stimulation is accompanied by sickness. During intoxication, generalized depressant effects are accompanied by exhilarated feelings. Since the signs and symptoms of withdrawal illness are terminated by ethanol administration, they indicate physical dependence on ethanol. Since the signs and symptoms are also reversed by other depressant agents, they are due to a general effect of this type of drug rather than to a specific attribute of ethanol. Since the signs and symptoms are predominantly attributable to excessive stimulation of the central nervous system, it is a reasonable although unproven inference that they constitute persistence of a compensatory, stimulant response. This compensatory response is seen as tolerance while ethanol exerts its depressant effects and as the withdrawal illness immediately afterward.

The withdrawal signs and symptoms begin soon after the blood ethanol concentration starts declining; the peak signs and symptoms occur within a few hours after the blood ethanol reaches zero. The severity and duration of the illness appear to depend mostly on the duration of continuous intoxication. Hangover is the mild brief withdrawal illness after a single, brief drinking episode. The signs and symptoms include behavioral expressions of overstimulation (tremor, irritability, tense feelings) and compensatory responses (fatigue, apathy, avoidance of stimulation). The illness disappears about 12 hr after the blood ethanol concentration reaches zero. Performance capabilities show no consistent impairment during hangover in spite of the unpleasant symptoms.

Experiments on rats and mice show that a few days of continuous ethanol intoxication are sufficient to cause a much more severe and prolonged withdrawal illness, including convulsions and sometimes death. In humans, the occurrence of severe withdrawal illness only in some alcoholics, and only after many months or years of excessive drinking, may be due to the fact that even alcoholics usually do not maintain high blood ethanol concentrations continuously for several days. Two stages of the withdrawal illness have been distinguished in humans. Convulsions constitute the principal signs in the first 1 or 2 days. Hallucinations and disorientation are characteristic on days 3–5. The drastic illness of delirium tremens, which is sometimes fatal, consists of hallucinations and disorientation, together with severe overstimulation of the autonomic nervous system.

Ethanol has been compared with other drugs by tests with these drugs in rats previously trained to respond to the discriminative stimulus attributes of ethanol or another drug. Ethanol shows substantial resemblance to sedative/hypnotic drugs, such as barbiturates and chloral hydrate. In common with ethanol, these drugs cause consistently large effects on behavior only at doses close to the anesthetic level, and behavioral excitement or disinhibition is seen at lower doses. Ethanol shows equivalent resemblance to "minor" tranquilizers, such as benzodiazepines and meprobamate. The basis for this resemblance might be a muscle relaxant, taming effect, so that with low doses the depressant drug effect tends to predominate over the exciting or disinhibiting effect.

In some respects, however, ethanol is differentiated from all other types of drugs tested. Ethanol is differentiated to a larger degree from other types of depressants, such as narcotic analgesics, major tranquilizers, and Δ^9-tetrahydrocannabinol, and also from stimulants and hallucinogens. The unique attribute of ethanol may consist of a unique combination of effects, even if each one of the effects of ethanol is duplicated by other drugs.

ACKNOWLEDGMENTS. Preparation of this chapter was supported by U.S. Public Health Service Research Grant MH-13595 from the National Institute of Mental Health.

F. REFERENCES

1. H. Wallgren and H. Barry, III, *Actions of Alcohol,* Vols. 1 and 2, Elsevier Publishing Company, Amsterdam (1970).
2. B. Kissin and H. Begleiter (eds.), *The Biology of Alcoholism,* Vol. 2: *Physiology and Behavior,* Plenum Press, New York (1972).
3. H. Barry, *J. Saf. Res.* **5**:200 (1973).
4. H. Barry, III, in: *Drug Abuse: Clinical and Basic Aspects* (S. N. Pradhan and S. N. Dutta, eds.), pp. 78–101, The C. V. Mosby Company, St. Louis (1977).
5. H. Moskowitz, *J. Saf. Res.* **5**:185 (1973).
6. J. M. Levine, G. G. Kramer, and E. M. Levine, *J. Appl. Psychol.* **60**:285 (1975).
7. J. A. Carpenter and B. M. Ross, *Q. J. Stud. Alcohol* **26**:561 (1965).
8. W. P. Colquhoun and R. S. Edwards, *Ergonomics* **18**:81 (1975).
9. R. E. Boyatzis, *Q. J. Stud. Alcohol* **35**:959 (1974).
10. S. P. Taylor and C. B. Gammon, *J. Stud. Alcohol* **37**:917 (1976).
11. G. M. Farkas and R. C. Rosen, *J. Stud. Alcohol* **37**:265 (1976).
12. H. B. Rubin and D. E. Henson, *Psychopharmacology* **47**:123 (1976).
13. R. J. Gibbins, H. Kalant, and A. E. LeBlanc, *J. Pharmacol. Exp. Ther.* **159**:236 (1968).
14. J. A. Matchett and C. K. Erickson, *Psychopharmacology* **52**:201 (1977).
15. K. A. Miczek and H. Barry, III, in: *Behavioral Pharmacology* (S. D. Glick and J. Goldfarb, eds.), The C. V. Mosby Company, St. Louis (1976).
16. K. A. Miczek and H. Barry, III, *Psychopharmacology* **52**:231 (1977).
17. H. Kalant, A. E. LeBlanc, and R. J. Gibbins, *Pharmacol. Rev.* **23**:135 (1971).
18. D. A. Overton, in: *The Biology of Alcoholism* (B. Kissin and H. Begleiter, eds.), Vol. 2: *Physiology and Behavior,* pp. 193–217, Plenum Press, New York (1972).

19. H. Barry, III, and E. C. Krimmer, in: *Discriminative Stimulus Properties of Drugs* (H. Lal, ed.), pp. 72–93, Plenum Press, New York (1977).
20. A. E. LeBlanc, H. Kalant, and R. J. Gibbins, *Psychopharmacologia* **41**:43 (1975).
21. E. Majchrowicz, *Psychopharmacologia* **43**:245 (1975).
22. A. E. LeBlanc, R. J. Gibbins, and H. Kalant, *Psychopharmacologia* **30**:117 (1973).
23. A. E. LeBlanc, H. Kalant, and R. J. Gibbins, *Psychopharmacology* **48**:153 (1976).
24. R. C. Petersen, *Psychopharmacology* **55**:141 (1977).
25. H. Weingartner, W. Adefris, J. E. Eich, and D. L. Murphy, *J. Exp. Psychol. Hum. Learn. Mem.* **2**:83 (1976).
26. J. L. Altman, J.-M. Albert, S. L. Milstein, and I. Greenberg, in: *Discriminative Stimulus Properties of Drugs* (H. Lal, ed.), pp. 187–206, Plenum Press, New York (1977).
27. J. C. Winter, *Psychopharmacologia* **44**:209 (1975).
28. K. Ando, *Psychopharmacologia* **45**:47 (1975).
29. J. D. Sinclair and K. Kiianmaa (eds.), *The Effects of Centrally Active Drugs on Voluntary Alcohol Consumption,* Vol. 24, The Finnish Foundation for Alcohol Studies, Helsinki (1975).
30. L. M. Barker and T. Johns, *J. Stud. Alcohol* **39**:39 (1978).
31. H. Barry, III, in: *Psychopharmacology of Aversively Motivated Behavior* (H. Anisman and G. Bignami, eds.), pp. 455–485, Plenum Press, New York (1978).
32. J. L. York, *Experientia* **34**:224 (1978).
33. H. Barry, III, *Fed. Proc. Symp.* **33**:1814 (1974).
34. H. Barry, III, and E. C. Krimmer, in: *Drug Discrimination and State Dependent Learning* (B. T. Ho, D. W. Richards, and D. L. Chute, eds.), pp. 3–32, Academic Press, Inc., New York (1978).
35. D. A. Overton, *Psychopharmacology* **53**:195 (1977).
36. L. F. Chapman, *Q. J. Stud. Alcohol,* Suppl. **5**:67 (1970).
37. N. K. Mello and J. H. Mendelson, in: *American Handbook of Psychiatry* (S. Arieti, general ed.), Vol. 4: *Organic Conditions and Psychosomatic Medicine* (M. F. Reiser, ed.), 2nd ed., pages 371–403, Basic Books, Inc., New York (1975).
38. N. K. Mello, *Pharmacol. Biochem. Behav.* **1**:89 (1973).
39. T. Seppälä, T. Leino, M. Linnoila, M. Huttunen, and Y. Ylikahri, *Acta Pharmacol. Toxicol.* **38**:209 (1976).
40. U. Rydberg, in: *Recent Advances in the Study of Alcoholism* (C.-M. Ideström, ed.), pp. 32–40, Excerpta Medica, Amsterdam (1977).
41. R. H. Ylikahri, M. O. Huttunen, C. J. P. Eriksson, and E. A. Nikkilä, *Eur. J. Clin. Invest.* **4**:93 (1974).
42. D. B. Goldstein, *J. Pharmacol. Exp. Ther.* **180**:203 (1972).
43. H. Kalant, in: *Alcohol Intoxication and Withdrawal* (M. M. Gross, ed.), Vol. IIIb, pp. 57–64, Plenum Press, New York (1977).
44. M. M. Gross, S. M. Rosenblatt, B. Malenowski, M. Broman, and E. Lewis, *Q. J. Stud. Alcohol* **33**:400 (1972).
45. M. M. Gross, E. Lewis, and J. Hastey, in: *The Biology of Alcoholism* (B. Kissin and H. Begleiter, eds.), Vol. 3: *Clinical Pathology,* pp. 191–263, Plenum Press, New York (1974).
46. M. M. Gross and J. M. Hastey, in: *Alcoholism: Interdisciplinary Approaches to an Enduring Problem* (R. E. Tarter and A. A. Sugerman, eds.), pp. 257–307, Addison-Wesley Publishing Company, Inc., Reading, Mass. (1976).
47. E. Majchrowicz and W. A. Hunt, *Psychopharmacology* **50**:107 (1976).
48. B. E. Hunter, J. N. Riley, D. W. Walker, and G. Freund, *Pharmacol. Biochem. Behav.* **3**:619 (1975).
49. H. Wallgren, A.-L. Kosunen, and L. Ahtee, *Isr. J. Med. Sci.,* Suppl., **9**:63 (1973).
50. S. Liljequist and J. Engel, in: *Alcohol Intoxication and Withdrawal* (M. M. Gross, ed.), Vol. IIIb, pp. 235–250, Plenum Press, New York (1977).
51. R. J. Gibbins, H. Kalant, A. E. LeBlanc, and J. W. Clark, *Psychopharmacologia* **19**:95 (1971).
52. V. DeNoble and H. Begleiter, *Pharmacol. Biochem. Behav.* **5**:227 (1976).
53. L. A. Pohorecky, *Psychopharmacology* **50**:125 (1976).
54. J. Brick and L. A. Pohorecky, *Alcohol. Clin. Exp. Res.* **1**:207 (1977).

55. B. E. Hunter, D. W. Walker, and J. N. Riley, *Pharmacol. Biochem. Behav.* **2:**523 (1974).
56. D. B. Goldstein and N. Pal, *Science* **172:**288 (1971).
57. D. B. Goldstein and V. W. Arnold, *J. Pharmacol. Exp. Ther.* **199:**408 (1976).
58. G. Freund, in: *Biochemical Pharmacology of Ethanol* (E. Majchrowicz, ed.), pp. 311–325, Plenum Press, New York (1975).
59. W. A. Pieper, in: *Biochemical Pharmacology of Ethanol* (E. Majchrowicz, ed.), pp. 327–337, Plenum Press, New York (1975).

A Critique of Animal Analogues of Alcoholism 27

Theodore J. Cicero

A. INTRODUCTION

In this chapter recent attempts to generate animal analogues of human alcoholism will be reviewed. To properly assess the relative advantages and disadvantages of the various animal analogues of addiction to ethanol, however, three important factors must be considered: What are the objectives of animal analogues? Why are they necessary? What criteria must they meet to be valid?

1. Objectives of Animal Analogues

In general, there are four research objectives which have prompted the formulation of animal analogues: (1) an examination of the biomedical complications associated with chronic ethanol administration, such as liver damage, brain dysfunction, or other organ pathology; (2) an evaluation of possible therapeutic devices useful in the management of the ethanol withdrawal syndrome; (3) an examination of the neurobiological mechanisms underlying tolerance to and physical dependence on ethanol; and (4) an examination of the factors which give rise to and maintain the abnormal consumption of ethanol. There are, of course, other ethanol-related problems to which animal analogues can be applied, but those listed above represent the ones which have provided the greatest incentive to develop animal analogues. There are several issues, however, which clearly cannot be explored with animals. For example, the complex psychosocial variables involved in why a human elects to maladaptively consume ethanol cannot be

Theodore J. Cicero • Department of Psychiatry, Washington University School of Medicine, St. Louis, Missouri 63110.

reasonably examined in an animal.[1,2] In addition, the psychological characteristics of the alcoholic are also not amenable to study with infrahuman analogues.

2. Why Are Animal Models Required?

The reasons why animal models are required are twofold. First, many ethanol-related problems cannot be rigorously examined in the human for a number of obvious technical and ethical reasons. Second, human alcoholics have numerous biomedical and psychosocial problems in addition to their alcoholism, such as poor nutrition, liver damage, psychiatric illnesses or multidrug abuse.[3,4] Any of these variables could interfere with studies of the consequences or causes of abnormal ethanol ingestion or to an identification of the neurobiological mechanisms underlying the adaptive changes produced by chronic ethanol administration. An animal analogue is the only way in which the effects of ethanol *per se* can be examined free from the contaminating influences of other factors.

3. Criteria for an Animal Analogue of Human Alcoholism

A number of criteria for animal analogues of alcoholism have been proposed.[5–9] Ideally, a true model of alcoholism should encompass all the key features of the human condition, with the obvious exception that the constellation of psychosocial variables which give rise to and maintain excessive ethanol intake in the human probably cannot be incorporated into an animal analogue. With this in mind, the following criteria should ideally be satisfied.

1. The animal must self-administer ethanol in pharmacologically significant amounts. Specifically, the following conditions must be met:
    a. Ethanol should be self-administered by the oral route of administration.
    b. Ethanol should be preferentially consumed when there is a choice between it and other solutions. The alternative solutions should consist of water and solutions whose caloric value or palatability match or exceed those of the ethanol solution.
    c. Ethanol must be consumed as a drug. That is, the amount of ethanol volitionally consumed must give rise to pharmacologically meaningful blood ethanol concentrations (BECs).
    d. The voluntary intake of ethanol should be based solely on its pharmacologic properties and not be related to some other characteristic, such as the calories it provides or its gustatory or olfactory properties.
2. Tolerance to ethanol should be demonstrable following a period of continuous consumption. Specifically, it must be demonstrated that animals are

markedly less affected by the same dose of ethanol and, more importantly, at the same BEC after a period of chronic exposure.

3. Dependence on ethanol should also develop after a period of continuous consumption. Dependence is characterized by two components, "psychological dependence" and "physical dependence" in humans. Although attempts are often made to differentiate the two in some rigid sense, there seems to be little need to do so. Presumably all the behavioral and physiological responses observed upon abrupt ethanol withdrawal are reflections of certain basic, adaptive changes in key biological substrates produced by chronic exposure of the brain to ethanol. In this scheme, a "need" for ethanol may simply be viewed as a behavioral or subjective expression of these adaptive changes, whereas convulsions or tremors are biological expressions.

     a. Psychological dependence on ethanol cannot really be examined in an animal (see below for a detailed discussion), but the ability of ethanol to act as a reinforcer can and should be measured.

     b. Physical dependence is defined by those behavioral and biological responses which are overtly expressed during acute ethanol withdrawal. These signs and reactions should be objectively, rather than subjectively, defined and preferably quantified (see Section B.1.c).

At present, no single analogue meets the criteria outlined above, and thus in a strict sense there is no animal analogue of alcoholism. However, it should be apparent that many research objectives do not require that all of the criteria for an animal analogue be met. For example, if one is interested in peripheral neuropathies induced by chronic ethanol treatment or the neurobiological mechanisms underlying tolerance and dependence, it does not really matter if the experimental animal voluntarily consumes ethanol. Rather, if a means can be found to simply force enough ethanol into the animal to produce very high BEC for periods sufficient to produce the adaptive changes associated with dependence and tolerance, this should suffice. Consequently, a number of animal analogues have been devised which meet only some of the criteria outlined above (e.g., tolerance and dependence). Animal analogues of alcoholism must, therefore, be considered in light of what they were designed to do and not necessarily simply on the basis of whether or not they meet all the criteria outlined above. These criteria simply outline the conditions which should be met for a true analogue of human alcoholism.

In the following sections, animal analogues of alcoholism have been arbitrarily divided into two categories: self-administration and forced-administration analogues. In most self-administration studies attempts have been made to generate the volitional administration of ethanol in amounts sufficient to produce dependence and tolerance. In the forced-administration studies, on the other hand, the goal has simply been to produce dependence and tolerance without worrying about volition. To aid in an examination of the various analogues of alcoholism, Table I has been devised to permit easy access to the methods, their salient features, and where a discussion of them appears in the text.

Table I. Animal Analogues of Alcoholism

	Species	Route of administration	BEC[a]	Tolerance[b]	Dependence[b]	Induction time[c]	Description, advantages, disadvantages	References
Self-administration								
Preference	Mouse, rat, guinea pig, monkey	Oral	Nil	Nil	Nil	Long	pp. 539–540	40–55
Intravenous	Rat, monkey	Intravenous	High	Substantial	Substantial	Moderate	pp. 541–542	58–64
Intragastric	Rat, monkey	Intragastric	High	N.S.E.	N.S.E.	Unknown	pp. 542–543	68–71
Operant paradigm	Rat, monkey	Oral	High	N.S.E.	N.S.E.	Unknown	pp. 543–544	72–81
Forced consumption								
Forced intake	Rat, monkey	Oral	Low	Slight	Slight	Long	pp. 545–546	38, 82–84
Inhalation	Mouse, rat	Inhalation	High	Substantial	Substantial	Brief	pp. 546–547	87–99
Polydipsia	Mouse, rat, monkey	Oral	High	Substantial	Substantial	Long	pp. 547–549	114–123
Diet	Mouse, rat, monkey	Oral	High	Substantial	Substantial	Moderate	pp. 549–552	8, 126–144
Intubation	Mouse, rat	Intragastric	High	N.S.E.	Substantial	Brief	pp. 552–553	19, 21, 23 148–152
Intragastric	Dog, rat	Intragastric	High	N.S.E.	Substantial	Moderate	pp. 553–554	153–158
Intraventricular	Rat, monkey	Intraventricular	High	N.S.E.	N.S.E.	Unknown	pp. 554–555	159–163 166, 167

[a]BECs have been classified as nil (< 50 mg/dl), low (50–100 mg/dl), moderate (100–200 mg/dl), or high (> 200 mg/dl).
[b]Tolerance and dependence have been classified as nil (none), slight (modest signs and reactions which are difficult to quantify or a slight degree of tolerance if measured or inferred), or substantial (marked withdrawal signs and reactions or tolerance if measured or inferred). N.S.E., not systematically examined.
[c]Induction time has been categorized as brief (less than 1 week), moderate (1–3 weeks), or long (3 weeks plus).

B. EVALUATION OF ANIMAL ANALOGUES

1. Technical and Theoretical Considerations

The determination that ethanol is consumed as a drug and the definition and measurement of tolerance and physical dependence represent significant theoretical and technical problems in the development of valid animal analogues of alcoholism. In this section, some of these issues will be discussed.

a. Ethanol as a Drug

In any animal analogue it must be shown that ethanol is administered in amounts sufficient to produce a pharmacologic effect. The only certain way to demonstrate this is to measure BEC. Since there must be ethanol in the blood before it can be safely concluded that a pharmacologically significant effect can occur, BECs provide the most direct evidence that ethanol is consumed as a drug. Unfortunately, however, BECs have not been measured routinely in most studies, and it must be inferred whether meaningful amounts of ethanol have been consumed. Such inferences are, of course, highly speculative and are often incorrect. Since there are numerous methods available to measure BEC,[10-13] which are relatively free of technical problems and can be routinely carried out in virtually any laboratory, such needless speculation is difficult to understand or tolerate.

b. Measurement of Tolerance

There has been an unfortunate tendency to assume rather than demonstrate that tolerance has developed. This is due to two factors. First, tolerance to ethanol is rather subtle.[14-16] In the tolerant human or animal, the ED_{50} or LD_{50} for ethanol shifts only slightly (less than twofold), whereas in the subject tolerant to the narcotics, for example, shifts of six- to tenfold are not uncommon. Second, the tests used to measure tolerance (usually behavioral) are relatively insensitive and imprecise, hence it is difficult to accurately state that tolerance has developed.[15-17] Because of these problems, tolerance has not routinely been examined in most studies. Since it is widely assumed that tolerance and dependence develop simultaneously,[15,18] most investigators infer that if the animal is dependent on ethanol, it is also tolerant. This is unfortunate and clearly much better means must be devised to measure the development of tolerance.

c. Measurement of Dependence on Ethanol

I. Physical Dependence. The characteristics of the ethanol withdrawal syndrome in animals have been the subject of several excellent reviews[8,9,19-21] and thus will not be described here. However, in general, it appears that the severity of the withdrawal syndrome, both in terms of the signs and reactions which develop and their intensity, is a function of the species examined, the length of exposure to ethanol, and the dose to which the animal is exposed.[18-22]

There has been an unfortunate tendency to subjectively rate the intensity of withdrawal behavior rather than to quantify it. Thus, such descriptions as "irritability," "tremulous behavior," "hyperactivity," and "hyperreactivity" are frequently used in the literature to characterize the ethanol withdrawal syndrome in animals. Such subjective observations are of questionable value for fairly obvious reasons, not the least of which is that investigators seldom look at an animal's behavior for extended periods of time. Under these circumstances, one can generally see behavioral abnormalities if one expects to see them. A blind study should theoretically overcome this bias but, in practical terms, chronic ethanol studies cannot be carried out blind. It is very difficult to disguise the fact that a rat, for example, has been subjected to a chronic period of ethanol intoxication. Thus, the experienced observer can frequently recognize the difference between an ethanol-treated and control rat and be influenced in his rating of the severity of its withdrawal behavior.

There are many objective measures of physical dependence which avoid the need to rate withdrawal behavior. Perhaps the most widely used method is the induction of seizure activity during withdrawal, either by the use of a chemical or auditory stimulus.[23-25] Since susceptibility to seizures reflects the heightened neuroexcitability present during acute ethanol withdrawal, this technique provides a useful and fairly reliable quantitative index of the severity of the withdrawal syndrome. Two methods of quantifying induced-seizure activity have been used: recording the percentage of ethanol-treated and control animals that convulse in response to the stimulus or, preferably, by determining the intensity of the stimulus needed to induce seizures in the two groups. The latter is particularly valuable since it permits a grading of the ethanol withdrawal syndrome, whereas the former provides a more or less all-or-none answer. In addition to seizure susceptibility, a number of investigators have described other quantifiable indices of withdrawal behavior in both the human and animal, such as changes in sleep and other $\overline{\text{EEG}}$ profiles,[26-33] evoked responses,[34-36] or startle thresholds[37,38] to name only several, which seem to be quite useful.

II. Psychological Dependence. A "need" or "psychological dependence" on ethanol, in the human sense of these words, cannot be established in an animal. The terms "need" and "psychological dependence" are uniquely human constructs and simply cannot be applied to the ethanol-choice behavior of an animal. That is, it has often been inferred that if an animal prefers ethanol to water or will work to obtain it, it must therefore "need" ethanol or otherwise it would not choose to consume it. However, since this choice may be based on certain reinforcing properties of ethanol, which are unrelated to its characteristics as a drug (e.g., its taste), the use of the terms "need" or, worse yet, "psychological dependence" to describe this behavior is unacceptable, since they both imply something much more than what the behavior itself allows. Mello[39] has advanced a similar argument regarding "craving" and other semantic aspects of alcoholism in the human and the reader is referred to this excellent paper.

III. Nutritional Variables. The primary problem in ethanol-related studies in humans is the ability to differentiate specific effects of ethanol from nonspecific ones generated by a number of confounding variables. Indeed, the primary motivation for using an animal analogue is to circumvent the many biomedical and psychosocial problems which are inherent characteristics of the human alcoholic, such as polydrug abuse, liver damage, brain dysfunction, psychiatric disturbances, and, particularly, poor nutrition. While virtually any animal analogue can circumvent a number of these problems, particularly psychosocial variables, controlling for nutritional variables remains a problem in any chronic ethanol study. The basic difficulty in dealing with ethanol, which is unique with respect to other drugs, is that it serves as a source of calories—usually at the expense of calories derived from other nutrients. As will be evident in this review, assuring a good nutritional profile in chronic ethanol studies in animals has been a very difficult task to accomplish. That is, if one succeeds in getting an animal to drink ethanol, it generally restricts its consumption of food-derived calories, which may lead to nutritional difficulties. At the very least a pair-feeding design must be employed in all chronic ethanol studies. Even when these necessary controls are employed (which is sometimes difficult), however, there is some question about whether the effects of ethanol in an animal deprived of nutrient-derived calories—especially in a long-term study—are the same as those in a normal preparation. This seems unlikely but, given the present state of the art, the best that can be done in terms of examining ethanol-specific effects is to utilize pair-fed controls in long-term ethanol studies. This design does not necessarily remove nutrition as a concern, however, and such problems should be kept in mind in examining any ethanol-related issue.

2. Self-Administration Models

a. Ethanol Preference Studies

Richter's observation several decades ago[40-42] that a rat would select a weak ethanol solution when offered a choice between it and water triggered an immediate explosion of interest in the possibility that rats, and subsequently other animals, might serve as analogues for human alcoholism. Indeed, Richter, and many others after him, termed an animal's volitional selection of ethanol as "alcoholic" behavior.[41] The ethanol-preference paradigm, needless to say, has not lived up to these high expectations and, moreover, has become a major theoretical stumbling block in the development of animal analogues of human alcoholism. That is, many investigators continue to treat ethanol preference in animals as if it were analogous to the consumption of ethanol by the human. However, after literally thousands of papers it can be concluded that animals frequently prefer ethanol to water for a variety of reasons, such as the calories it provides, its taste, smell, or perhaps novel aspects.[43-52] Unfortunately, however, animals, unlike people, do not appear to select ethanol for its pharmacologic properties. When BECs have been measured in ethanol preference studies (which has been very rare), negligible levels are generally found. Investigators usually state that their animals drank a given level of ethanol which, depending upon the distribution of drinking, could have exceeded the animals ability to metabolize it. This is totally unacceptable; such studies are essentially worthless and contribute little except confusion to an already overburdened literature. If the animal drinks ethanol for its pharmacologic properties, it will have measurable BEC. BEC should be employed as the only acceptable basis for concluding that ethanol is self-selected for its drug-related properties.

An apparent exception to the conclusion that preference studies rarely generate significant BEC is the report of Fitz-Gerald et al.[53] that a few chimpanzees and orangutans consumed ethanol solutions, dissolved in orange or other fruit juices, to the point of intoxication (BECs were not measured, unfortunately). Whether this behavior was drug-related or occurred simply as an accidental correlate of drinking the orange juice was not established nor was any evidence of tolerance or physical dependence provided. Moreover, this 10-year-old paper stands alone and has apparently never been replicated. There are other reports in the literature that animals appeared to be intoxicated occasionally during a prolonged period of free access to ethanol, but these infrequent episodes have not been rigorously supported by BEC and seem to be accidental or random occurrences.

Although it seems clear from the foregoing discussion that ethanol is not generally consumed as a drug in preference studies, it seems absurd to question the well-documented observation that an animal often chooses ethanol in preference to water. On the basis of this observation, there is little reason to quarrel with the conclusion that animals find ethanol "reinforcing" in some way. What the animal finds reinforcing about ethanol is unclear, however, but it seems doubtful that it is consumed as a drug. Therefore, studies of ethanol preference in animals, particularly those in which attempts to modify preference by, for example, drug treatments or alterations in brain chemistry, would seem to have very little relevance to the human alcoholic, whose ethanol consumption is unquestionably based exclusively on the pharmacologic impact of the drug. It matters very little whether an investigator can modify the preference for ethanol in an animal if he has no idea what this preference implies. The still prevalent assumption that ethanol preference in animals will tell us something about abnormal ethanol consumption in humans is difficult to understand and, moreover, is naive, has no conceptual or theoretical basis, and has in the final analysis impeded the development of other, more suitable, animal analogues of alcoholism. A number of recent reviews[6,9,54-57] are available on this subject.

b. Other Self-Administration Studies

The inability to produce ethanol addiction via the oral route of administration may be due to one of two factors: animals appear to avoid ethanol solutions because of their aversive taste or smell, particularly at concentrations likely to produce pharmacologic effects[55]; and the delay between the oral ingestion of ethanol and its impact on the nervous system (5–15 min depending upon the animal) is quite long,[5,56] thereby producing a very poor learning situation. A number of self-administration techniques have been devised to overcome one or both of these limitations and will be discussed in the following subsections.

1. Intravenous Self-Administration. Description. Several groups of workers, notably Woods, Winger, and their colleagues, have demonstrated that primates will self-infuse ethanol into their bloodstreams via permanent indwelling venous (jugular) catheters.[58–61] This method has also been applied to other species (e.g., the rat) but with less success.[62] The use of the intravenous route of administration, rather than the oral route, has two advantages: (1) this route of administration considerably shortens the delay between the self-administration of ethanol and its pharmacologic effect, and (2) it has been shown that many drugs with significant addictive liability are readily administered via indwelling cannulae.[63,64]

In the intravenous self-administration analogue, monkeys are trained to press a bar to infuse ethanol (usually 0.1 g/kg per infusion) into their jugular veins. The schedules of reinforcement can be manipulated to determine the strength of the monkey's "drive" to self-administer ethanol. In early results of studies utilizing the intravenous self-administration paradigm, monkeys frequently infused enough ethanol to become grossly intoxicated and reach very high BECs (up to 400 mg/dl) if they were allowed free access to ethanol throughout the day. However, under these conditions, they would spontaneously terminate their self-infusion of ethanol at episodic intervals, resulting in a withdrawal syndrome.[61] This pattern of self-administration resembles in a crude way the "binge" human drinker and has attracted some interest in this regard.[65] However, this propensity to self-terminate ethanol administration is not desirable from a pharmacologic point of view. In subsequent studies, therefore, Woods, Winger, and their colleagues[60,61] showed that by limiting the monkeys access to ethanol to just one session each day, this tendency to terminate self-administration could be eliminated. Under these conditions, stable ethanol responding could be maintained for long periods of time such that the monkeys routinely infused enough ethanol (7–8 g/kg per day) to become grossly intoxicated, reaching BEC of between 200 and 300 mg/dl.

The self-infusion of ethanol via the intravenous route of administration culminates in the development of tolerance and physical dependence. Within a relatively brief period of time (around 10 days), tolerance appears to develop,[61] as indicated by an increased clearance of ethanol from the body (i.e., "metabolic" tolerance) and an increase in the response rate for ethanol over time. Physical dependence can be readily demonstrated by replacing the ethanol with sucrose or saline. Under these conditions responding on the previously reinforced ethanol lever increases dramatically and is difficult to extinguish, giving some idea of the strength of ethanol as a reinforcer. Moreover, a full-blown withdrawal syndrome, consisting of tremors, hyperactivity, convulsions (which occasionally terminate in death), and several bizarre behavioral patterns, which have been interpreted as hallucinatory in nature, can also be demonstrated.

Advantages and Disadvantages. The advantages of the self-administration of ethanol via indwelling intravenous cannulae are numerous. First, this technique provides a means of assessing the volitional aspects of ethanol administration (although not by the oral route). Thus, a number of psychopharmacological studies, such as determining the factors which give rise to and maintain the self-administration of ethanol or attempts to modify its administration, are possible with this method which could not be conducted with other currently available animal analogues of alcoholism. In addition, this technique does not involve weight reduction, either by design or generated by the procedure, and hence nutritional variables seem well under control. In addition, the monkey is a good subject for a number of experiments which are poorly conducted in rodents. For example, cirrhosis may only occur in the human or subhuman primate.[66,67] Moreover, the complexity of the monkey's behavior, relative to that of other species, permits substantially more sophisticated manipulations than would otherwise be possible. Finally, it seems probable that studies with monkeys or other subhuman primates (e.g., chimpanzees) are much more likely to provide data with direct relevance to the human than would experiments with other species.

The primary disadvantages of the technique of intravenous self-administration, other than the fact that it is not based on oral intake, which is a distinct drawback, are that not all laboratories are equipped to handle the physical requirements of maintaining monkeys or have the relatively sophisticated surgical and technical expertise available to utilize this analogue. In addition, it should also be noted that not all monkeys can be trained to self-administer ethanol.[58] Although a number of procedures are useful in ultimately eliciting self-administration,[61] they are not always successful and can be quite laborious and time-consuming. For this reason, and the costs and physical requirements involved in housing and maintaining monkeys in experimental chambers, only a few monkeys can be treated simultaneously. Consequently, once animals have been trained to self-infuse ethanol they are extremely valuable preparations, requiring that the choice of experiments to be conducted be very judiciously made. The technique seems to be most suitable for studies involving the specification and manipulation of those factors which give rise to and maintain the self-infusion of ethanol, whereas it appears to be rather unsuitable for experiments in which the neurobiological correlates of tolerance and dependence or the biomedical complications of alcoholism, either of which require killing the animals or the use of a very large subject number, are examined (unless, of course, no other species will do).

II. Intragastric Self-Administration. Description. This analogue is quite similar to the intravenous self-administration analogue in design and rationale. The objectives are to shorten the time between the administration of ethanol and its pharmacologic effects and to use the normal route of absorption of ethanol but, at the same time, circumvent the possibly aversive gustatory and olfactory properties of ethanol. In this model,[68–71] monkeys and rats are surgically implanted with chronic indwelling intragastric cannulae into which ethanol solutions can be infused. By first passively infusing the animals with ethanol and then gradually "shaping" them to press a bar to obtain their infusions, it has been possible to demonstrate that ethanol will be infused intragastrically in pharmacologically significant amounts resulting in high BECs and gross intoxication.[68-71] In addition, some indication of ethanol withdrawal behavior has also been noted in these animals, although dependence and tolerance development have not been as systematically examined as would be desirable. This analogue is relatively new and hence its full potential has yet to be explored.

Advantages and Disadvantages. The advantages of the intragastric self-administration procedure are virtually the same as those described above for the intravenous analogue (e.g., assessment of the volitional aspects of alcoholism) as are the general disadvantages. However, the principal advantage of this procedure over the intravenous model is that the route of ethanol administration is the normal one and, hence, this analogue incorporates an important feature of the human alcoholic. There is an additional problem with the intragastric technique, relative to the intravenous procedure, however. Specifically, there is an increased incidence of postsurgical complications, including the development of ulceration or irritations of the linings of the stomach and small intestine subsequent to the infusion of high ethanol concentrations. These gastrointestinal disturbances result in reduced food intake and could result in poor nutritional status. Nevertheless, this analogue seems to offer great promise as a potentially useful animal analogue.

III. Ethanol as a Reinforcer in an Operant Conditioning Paradigm. Description. It has been demonstrated for many years that rats or monkeys will press a bar or perform some other work to obtain ethanol solutions.[72-75] These observations indicate, therefore, that an ethanol reward can increase the frequency of an operant response (i.e., it is reinforcing). For this observation to have any meaning, however, it must be shown that the rat or monkey selects ethanol for its pharmacologic properties rather than for its caloric, olfactory, taste, or other characteristics. Until several years ago this had not been demonstrated and, hence, one of the non-drug-related properties of ethanol probably accounted for its reinforcing properties in most operant paradigms.

The recent work of Meisch, Thompson, and their colleagues at the University of Minnesota has, however, altered the direction of this research.[76-81] These investigators have shown that animals will press a bar to obtain ethanol solutions, ranging from 8 to 32%, during an operant session in which food is also intermittently available (somewhat analogous to the psychogenic–polydipsia paradigm described below). Moreover, the intake of ethanol is sufficient to produce significant BECs under these conditions. When food-access sessions were separated from the ethanol-access sessions, the consumption of ethanol in pharmacologically active amounts persisted and, in fact, the drug was preferred to water. Hence, these investigators concluded that ethanol became a reinforcer in its own right, apart from its previous relation to food consumption.[81] Although others have demonstrated that ethanol can serve as a reinforcer (see above), the obvious distinction in this model is that Meisch and his colleagues have shown that very high BECs (200–300 mg/dl) can be achieved and maintained for prolonged periods of time with this method. The reason for the success of this method is unclear at the present time, but is undoubtedly related to the contingencies established between food and ethanol responding during the initial phases of training. During this early phase, the animal presumably learns that ethanol has desirable pharmacological properties, which then sustain its consumption even when no food is available. Whatever the actual reason, the work from Meisch's laboratory suggests that an oral analogue of the self-administration of ethanol in pharmacologically significant amounts may be possible. This analogue, however, must be validated in a number of important respects, the most significant of which is the need to demonstrate that tolerance and physical dependence can be produced. Neither of these important criteria of a valid animal analogue have been established as yet.

Advantages and Disadvantages. The obvious advantage of this procedure is that monkeys and rats can be induced to self-administer ethanol by the oral route of administration. This model thus comes close to approximating the conditions under which the human consumes ethanol. It may, therefore, permit an examination of those factors which give rise to, maintain, or terminate the oral consumption of pharmacologically active amounts of the drug. Moreover, if it can be shown that tolerance and physical dependence are produced after long-term self-administration, then these aspects of alcoholism can also be examined with this analogue.

There are several disadvantages to the method. The monkeys or rodents are reduced to 80% of their free-feeding weight and are maintained in this partially starved state for considerable periods of time. Since an animal analogue should permit an examination of ethanol effects, *per se,* any analogue that utilizes partial starvation or nutritional deprivation has an inherent drawback. For example, the complicated issue of whether ethanol is consumed as a source of calories under these conditions, rather than as a drug, remains to be unequivocally resolved. Moreover, the possibility that the effects of ethanol in a starved animal are different than those seen in the normal, satiated animal also exists (certainly ethanol absorption and metabolism will be different at the least). In addition to this important drawback, the apparatus required to train and house the animals, the time required to train them, and the cost of the monkeys combine to make it difficult to treat more than a few animals simultaneously. Consequently, on the basis of these practical considerations, the type of study for which this analogue is suitable is rather limited.

3. Forced-Administration Models

Many investigators have developed animal analogues of alcoholism which contain no volitional component, the so-called forced-administration models, for several related reasons. First, until recently there were no self-administration analogues of alcoholism which satisfied the criteria for an animal analogue and none of those currently available are particularly well suited for studies of the neurobiological correlates of tolerance and dependence or the biomedical complications associated with chronic ethanol administration. Second, many workers are interested only in the biomedical effects of long-term ethanol ingestion or the mechanisms underlying tolerance and physical dependence. Neither of the latter objectives require that the animal volitionally consume ethanol; all they require is the production of high BEC and the development of tolerance and dependence. Thus, a good deal of attention has focused on ways in which to achieve these limited objectives. The goal of virtually all of these analogues has simply been to intoxicate animals for a period of time long enough to produce tolerance and physical dependence. The trick, of course, is to force the animal to accept enough ethanol to accomplish these goals.

a. Forced-Oral Consumption

I. Description. A number of investigators have attempted to elicit the excessive oral intake of aqueous ethanol solutions in a variety of species by restricting them to ethanol as their sole source of fluid for very long periods of time.[38,82-84] These studies have been termed "forced" or "no-choice" experiments since the animals must select between death by fluid deprivation or the consumption of ethanol solutions. Unless the ethanol concentration is extremely high (i.e., greater than 15–20% v/v) the animal generally chooses to live and thus is "forced" to choose ethanol. The general procedure in most of these procedures is to confine the animal to the highest possible concentration it will tolerate. Many groups of investigators have maintained rats, mice, and monkeys of various ages on ethanol solutions (ranging from 3 to 20%) for periods of a few weeks to 5 or 6 months.[38,82-84] In many of these studies evidence of intoxication during the course of ethanol treatment, with blood levels in the range 100–150 mg/dl during the peak periods of ethanol consumption, has been found, although the BEC tends to be highly variable and erratic and cannot generally be sustained for any significant period of time. Moreover, following a period of chronic ethanol administration, it has been shown that a slight to moderate degree of tolerance and physical dependence develop. The withdrawal syndrome which develops, however, tends to be quite mild in that only hyperactivity and, very rarely, tremors have been detected. A "need" for ethanol, as demonstrated by a preference for ethanol, has also been claimed.[82] However, in these instances it appears that insufficient amounts of ethanol were consumed to produce a pharmacologic effect; hence the significance of these observations is obscure.

A number of variations of the general forced or no-choice model, such as sweetening the ethanol solutions to increase their palatability or fluid depriving the animals, has been attempted,[85,86] but the results have, by and large, not been much more dramatic than those described above.

II. Advantages and Disadvantages. These analogues are obviously simple to utilize, require a minimal amount of daily effort, are relatively inexpensive, and large numbers of animals can be treated simultaneously. Hence, for many

experiments, particularly neurobiological or neurochemical studies which generally require rather substantial numbers of animals, these analogues have great
appeal. The disadvantages, of course, are that relatively low BECs are achieved
and the degree of tolerance and severity of the withdrawal syndrome are quite
slight. Consequently, the ability to detect ethanol-induced changes may be
somewhat more difficult with this analogue than with others. In addition, and
perhaps of greater importance, nutritional variables cannot be well controlled in
these analogues. For example, as the intake of fluid decreases in the rodent (or
other animals), the amount of food consumed also decreases.[18] Since ethanol
solutions are not generally consumed at the same level as water in the forced or
no-choice analogues, the resulting drop in fluid intake produces a fall in food
consumption. Moreover, the calories provided by ethanol further depress the
amount of food consumed below that which could be accounted for solely on the
basis of reduced fluid intake. These two factors combine to make a pair-feeding
design virtually impossible to carry out with this analogue. Thus, it is difficult at
best to control for the nutritional status of ethanol-fed animals.

b. Inhalation Procedures

1. Description. Goldstein and Pal[87] first described a method for producing
very high BECs by the inhalation of ethanol vapors. In this procedure, mice are
housed in chambers into which ethanol vapors can be infused. The vapor levels
can be maintained within rather precise limits and, hence, a relatively stable
period of ethanol exposure can be achieved. The mice are first injected with a
"priming" dose of ethanol designed to yield a specific BEC, which is then maintained by the inhalation of ethanol. Pyrazole (1 mM/kg) is injected daily to inhibit
the metabolism of ethanol[88] and, thereby, to maintain stable high BEC. Using
this procedure, Goldstein[89–92] has shown that an excellent correlation between
BEC and the amount of ethanol in the inspired air can be achieved. After 3 days
(or less) of ethanol exposure, withdrawal behavior can be elicited by discontinuing the ethanol vapor.[93] Within several hours after the termination of ethanol
inhalation, the withdrawal syndrome, which is characterized by tremors, excitability, and, most prominently, by "handling-induced" convulsions, begins. Handling-induced convulsions are elicited simply by picking the mice up by the tail
and gently rotating them (if necessary). The severity of the convulsions produced
in this manner is rated on a scale of 0–4 (severe tonic–clonic convulsions usually
terminating in death). Goldstein, and others, have used this analogue effectively
to examine the correlation between the dose of ethanol, duration of treatment,
and BEC with the severity of withdrawal behavior.[93–98]

Several groups of workers have tried to eliminate pyrazole, which has a
number of disadvantages (see below), from the inhalation procedure, but these
attempts have met with only limited success, since the BECs tend to be highly
variable without the alcohol dehydrogenase inhibitor and a large number of
deaths (perhaps as high as 50%) occur.[22,99] This unacceptably high death rate

results in large part from the need to use high concentrations of ethanol in the inspired air to compensate for the uninhibited metabolism of ethanol.

II. Advantages and Disadvantages. The primary advantages of the inhalation procedure are that a large number of animals can be treated simultaneously with relatively little expense. In addition, the amount of time required to produce a high degree of tolerance and dependence is also very brief (3 days or less). Moreover, this procedure generates very stable BECs, highly reproducible withdrawal behavior, and permits precise specification of dose–response relationships. Thus, correlations between the severity of physical dependence, the degree of tolerance generated, and the duration and dose of ethanol to which the animal is exposed can very easily be made. This method has, therefore, been used extensively to examine the development and loss of tolerance and dependence.

The disadvantages of this procedure are also numerous, however. Without doubt, the major difficulty is the use of pyrazole to produce stable BECs. The use of pyrazole generates many problems, particularly for biomedical studies, the most important of which include: (1) pyrazole is hepatotoxic, particularly in combination with ethanol,[100–101] and toxic effects on the thyroid have also been demonstrated[102]; (2) it also exerts direct effects on brain which are unrelated to inhibition of alcohol dehydrogenase[103,104]; (3) it potentiates the acute effects of ethanol on the central nervous system[105]; and (4) it exacerbates the ethanol withdrawal syndrome.[106] The use of pyrazole thus may make it difficult to assess ethanol-specific effects on biological parameters with the inhalation analogue. If one eliminates pyrazole in an attempt to solve these problems, the inhalation procedure is simply too erratic to employ.

Although the use of pyrazole is a very serious flaw in the inhalation method, there are a number of other significant problems. For example, continuously intoxicated mice or rats eat poorly and lose substantial amounts of body weight during inhalation. Moreover, the constant inhalation of ethanol, which needless to say represents an unusual way of introducing ethanol into the body, produces lung damage.[8] Neither abrupt food restriction nor lung damage seems to permit an examination of ethanol-specific effects with this model.

c. Psychogenic Polydipsia

I. Description. In 1961, Falk reported[107] that if rats were required to press a bar to obtain food on a variable-interval 2-min schedule for 1- or 2-hr test sessions around the clock, an excessive intake of fluid occurred. The amount of water consumed generally exceeded by 5–10 times the normal daily fluid intake of the rat. This enormous ingestion of water was termed "psychogenic polydipsia" by Falk, since it had no apparent physiological basis but rather appeared to result from some psychological or behavioral abnormality.[108,109] A number of investigators have replicated these initial observations and have found that animals will consume fluids other than water or engage in a number of other adjunctive behaviors (e.g., aggression) when maintained on these schedules of reinforcement.[110-113] In the course of several of these studies, it was found that a polydipsic intake of weak ethanol solutions could be generated using the Falk paradigm.[114-118] Much of this early work, however, merely explored how much ethanol the animal would consume and what concentration it would tolerate. Although intoxication was reported to result in at least one study,[114] few workers subjected the animals to ethanol for long enough periods to maintain BECs high enough to produce tolerance or physical dependence. In 1972, however, Falk and his co-workers[119] demonstrated for the first time that the psychogenic polydipsia could be used to produce physical dependence on ethanol in the rat. In these experiments, the rats were initially reduced to 80% of their free-feeding weights and were then placed in test chambers into which food pellets were automatically delivered on a variable interval 2-min schedule. (Falk found, after his initial observations, that it was unnecessary to force the rat to work for its food to generate polydipsia, but that the intermittent schedule of food presentation *per se* was a sufficient stimulus.) A 6% ethanol solution was available as their only fluid. One-hour food and ethanol availability sessions were alternated with 3-hr rest periods around the clock for 3–6 months. Under these conditions, the animals consumed very large amounts of ethanol resulting in BECs of between 200 and 300 mg/dl continuously for 24 hr/day. The level of intoxication was so great that it was found that the 1-hr food sessions had to be broken by 3-hr rest periods to prevent gross and debilitating intoxication which would have terminated the consumption of food and, concurrently, ethanol. After 3 months (or more in some experiments) of continuous exposure to ethanol, withdrawal behavior, consisting primarily of hyperactivity and spontaneous or sound-induced (key rattling) tonic–clonic convulsions, could be elicited by abrupt ethanol withdrawal.[120,121] In more recent work, Falk and his colleagues, using either a 3- or a 6-month treatment regimen, reported that a marked degree of metabolic tolerance (i.e., increased clearance of ethanol) can also be demonstrated in these animals.[122] Thus, the analogue successfully produces high BECs, intoxication, and a marked degree of tolerance and physical dependence. With regard to "psychological dependence," Falk has made the case that the consumption of ethanol is voluntary in this analogue so that its excessive intake can be taken as evidence of a "need" for ethanol. However, this interpretation must be questioned, since psychogenic polydipsia appears to be a much more involuntary or "compulsive" behavior than it does a voluntary one (see below for a more detailed discussion). Moreover, a preference for ethanol solutions over water or alternative solutions has also been reported by these investigators.[123] However, an inspection of their data reveals that the enhanced preference produced relatively low intakes, which appear to be insufficient to maintain high BECs or prevent the development of withdrawal behavior. Thus, the excessive intake of ethanol appears to be inextricably linked to psychogenic polydipsia and has its roots in some feature of this paradigm which may not be directly related to the pharmacologic properties of the drug.

II. Advantages and Disadvantages. The obvious advantage of this procedure is that it produces the excessive oral intake of aqueous ethanol solutions in amounts sufficient to produce sustained and very high BECs over a 24-hr period which ultimately result in the development of both physical dependence on and tolerance to ethanol. Since these criteria of a valid animal analogue of alcoholism have been met in a method in which ethanol is orally consumed, the psychogenic polydipsia paradigm has a number of characteristics desirable in an animal analogue (see Section A.3). The analogue should provide a useful vehicle to examine the long-term effects of ethanol in animals, but it does not appear to be terribly useful in examining the factors which give rise to and maintain ethanol self-administration, even though this technique appears to depend upon a choice of the animal to excessively consume ethanol. The reason for this is that the consumption of fluids, and therefore ethanol is an aberrant response in the psychogenic polydipsia paradigm whose origins are unclear. Thus, the consumption of water or ethanol does not appear to be a truly voluntary behavior, an issue already raised by Lester and Freed,[6] and studies designed to modify this volitional consumption would seem to have more significance for the phenomenon of psychogenic polydipsia than they would for ethanol intake *per se*.

Several other disadvantages of this method should be mentioned. It is relatively difficult to use in several ways. Special equipment is required to maintain and house the animals (e.g., automatic food delivery systems and programming equipment), and as a result only a few animals can be treated simultaneously. This difficulty is compounded by the fact that it requires very long periods of time to generate tolerance and physical dependence, minimally 3–6 months. Consequently, the analogue may be ideally suited for experiments in which small numbers of animals are sufficient (e.g., certain psychopharmacological studies), but will be somewhat more difficult to employ in studies in which large numbers of animals and short treatment intervals would be desirable, such as the biomedical complications associated with chronic ethanol treatment or the neurobiological mechanisms underlying the development of tolerance and dependence. An additional disadvantage is that the animals are initially reduced to 80% of their free-feeding body weight and are maintained at these levels for extraordinarily long periods of time. The possible interaction between partial starvation, body weight reduction, and ethanol intake may limit conclusions which can be drawn about specific effects of ethanol *per se* (see above). Finally, it should be noted that one group of investigators[124] has reported difficulty in replicating the original results of Falk *et al.* Despite the persuasive rebuttal[125] of the latter investigator, these inconsistencies should be resolved.

d. Liquid-Diet Techniques

I. Description. In response to many of the criticisms outlined above, particularly those involving nutrition, several analogues based on the incorporation of ethanol into totally liquid diets have been developed. The rationale underlying this approach is that animals will be forced to ingest large amounts of ethanol in order to consume their daily fluid and food intake and, at the same time, will be maintained on a diet which is nutritionally adequate. Thus, a primary goal of forced-administration techniques can be met: high ethanol consumption and BECs in the absence of the confounding effect of starvation, nutritional deficits, and severe body weight reduction. Although ethanol-containing liquid diets have been used in one form or another for several decades,[126] Freund was perhaps the first investigator to develop a procedure which produced very high BECs and a marked withdrawal syndrome.[127,128] He and his co-workers employed a commercially available liquid diet (Metrecal–Shape) to which they added sucrose. This diet served as the control diet, whereas an ethanol-containing diet was formulated by replacing some of the calories derived from sucrose with ethanol (generally 35%). In preliminary work Freund found that if mice were used at their free-feeding weights,[8] erratic intakes of the diet occurred and withdrawal behavior (i.e., physical dependence) seldom developed. Thus, his group tried a number of "supplementary measures,"[8,9] such as reducing the animals' body weights by anywhere from 20 to 40% or placing them in a cold room (10–17°C). The latter procedure was used to increase the intake of the ethanol-containing diet to compensate for the increase in general metabolism and to achieve high BECs. Since overall metabolic activity is increased in the cold, but the metabolism of ethanol apparently is unaltered, this procedure should produce high intakes of the diet and at the same time high and stable BECs.

The liquid–diet procedures produce extraordinarily high BECs, generally exceeding 250–300 mg/dl during peak daily periods of ethanol consumption, and within a relatively brief period of time (within days and certainly within 1–2 weeks) a pronounced degree of physical dependence can be generated.[8,129,130] The withdrawal syndrome which ensues upon abrupt withdrawal of the ethanol-containing diet (by replacing it with a sucrose-control diet) consists of whole body tremors and spontaneous and sound-induced tonic–clonic convulsions, which frequently culminate in death. At the time of its development few investigators had ever reported a withdrawal syndrome of such intensity in an animal. Although a marked degree of physical dependence is produced with this liquid-diet technique, rodents will not choose an ethanol solution or even the ethanol-containing liquid diet and will thus "voluntarily" develop a withdrawal syndrome.[130] It appears, therefore, that the generation of high BECs and the development of tolerance and physical dependence are somewhat "accidental" with this procedure, since the animals apparently cannot recognize that the ethanol-containing diets produce whatever pharmacologic effects they experience or eliminate their withdrawal signs and reactions.

Because of the relative ease with which the liquid diet method can be employed, the desirability of producing an ethanol withdrawal syndrome in an animal with an adequate nutritional profile and the large number of animals which can be produced simultaneously, this method has enjoyed enormous popularity. Subsequent to its initial description, Freund and his co-workers (notably Walker and his group) have modified the method in several ways.[131,132] It has been adapted slightly for rats, the diet has been fortified with vitamins and minerals to more adequately meet the animals' nutritional requirements, and the use of a cold room to stimulate consumption has been dropped as a supplementary measure. In addition, there have been attempts to eliminate the need to markedly reduce body weight. While withdrawal behavior can still be generated in animals that have not undergone weight reduction, it is much less severe than in those animals reduced in body weight. The Freund diet has been used extensively to examine a number of ethanol-related issues particularly the development and characteristics of physical dependence on ethanol.[133–136]

Although Freund was the first to develop or popularize liquid-diet techniques, his is by no means the only one currently in use. During the past 5 years, liquid-diet techniques have been reported for baboons,[137] chimpanzees,[138] and monkeys.[139,140] All have the same underlying rationale, the only exception being the composition of the liquid diet and the "supplementary measures" required to induce intakes. Perhaps the most widely used technique other than Freund's is the one described by Lieber and DeCarli.[141,142] In this diet, the rats nutritional requirements are more rigidly met than is the case with the Freund diet (diet composition: casein hydrolysate, substituted for the original rather expensive amino acid mix, corn oil, carbohydrates, vitamins, and minerals), and the animals are not reduced in body weight. Very young animals are employed, however, as a means of inducing large intakes of ethanol and providing for a higher effective dose of the drug. The diet is high in "fat" content and, thus, has been used extensively in studies of ethanol's effects on the liver. However, in most important respects this diet and technique differ very little from Freund's, as do those diets employing other dietary bases, such as Slender, Sego, Sustagen, or Nutrament.[37,143,144]

II. Advantages and Disadvantages. The obvious advantages of the liquid-diet technique are very numerous indeed: large number of animals can be treated simultaneously (although not without some expense); very high BECs (200 mg/dl) can be continuously maintained for 24 hr each day; a severe withdrawal syndrome can be generated; and by virtue of consuming ethanol in a nutritionally adequate diet, nutritional variables (which plague most analogues of alcoholism) may be eliminated. For these reasons, liquid diet techniques have enjoyed a good deal of popularity and are without question the most frequently employed animal analogues.

There are several disadvantages to these liquid diet procedures. However, as in most other animal analogues of ethanol dependence and tolerance, the primary one seems to be the weight reduction employed in many studies to generate maximal withdrawal behavior and the fact that the intake of the ethanol diet is generally less than that of the control diet, resulting in further weight reduction and partial starvation. However, a pair-feeding design can easily be employed in this procedure—much more so than with other methods—which may very well compensate for this problem. There is, of course, the worry that a pair-fed control cannot really control for the possibility that the effect of ethanol superimposed upon an animal that is reduced in body weight or partially starved may be different than that seen in a normal animal. In support of this conclusion, there are reports[145,146] that on some endocrine measures specific effects of ethanol are difficult to demonstrate with these diets. Finally, some investigators have reported that the severity of withdrawal behavior was directly related to the severity of the nutritional problems encountered with this procedure.[147] Obviously, these considerations should be kept in mind, but it appears, on balance, that the liquid-diet procedure can be used to good advantage as long as the disadvantages enumerated above are carefully considered and borne in mind.

e. Intubation Method

I. Description. The principle underlying this technique is simply to introduce sufficient ethanol into the stomach, the normal route of administration, to produce high enough BECs for periods sufficient to induce tolerance and dependence. There have been numerous reports published demonstrating the efficacy of intubating ethanol as a means of producing tolerance and dependence on the drug.[19,21,23,148–152] The differences between the methods are substantial, however, in three respects: the total daily dose of ethanol intubated, how often ethanol is intubated each day, and how long daily intubations are continued before tolerance and dependence are assessed. The amount of ethanol intubated daily ranges from 1 to 15 g/kg per day, with the total daily dose being divided into two to six fractional doses. In some cases the dose is gradually increased, in others it is kept constant, and in some methods the dose given to animals is dependent upon the consequences (behavioral or BEC) of the preceding day's dose. Animals have been intubated with ethanol anywhere from 1 to 30 days or more. Controls are generally intubated on the same schedule, usually with water but occasionally with other solutions, such as isocaloric sucrose. The details of the numerous methods will not be described here since they can be found elsewhere and would be too numerous to list.[19,21,23,148–152] Only the general approach and the utility of these analogues will be discussed. Using these intubation procedures, very high BECs can be continuously maintained over a 24-hr period without concern for the animal's willingness to voluntarily accept ethanol. Within a relatively brief period of time (a few days) a pronounced degree of physical dependence (tremors and convulsions) and tolerance can also be demonstrated. The animals' willingness to seek out ethanol, however, has never been established with any of these procedures and, indeed, as is the case with most other forced or no-choice models, the animals simply undergo withdrawal rather than consume ethanol.

II. Advantages and Disadvantages. The advantages of the intubation method are numerous. Very high BEC can be maintained over a 24-hr period and, of particular pharmacological importance, it is possible to rather precisely determine dose–response relationships. The technique of nasogastric intubation is also relatively simple, requires no special equipment, and thus large numbers of animals can be treated simultaneously. A final advantage is that a high grade of tolerance and physical dependence can be produced within a relatively brief period of time.

One of the disadvantages of the intubation procedure is that intubating high ethanol concentrations into the gut has been reported to produce ulceration in at least one study.[153] This gastrointestinal irritation, combined with the grossly intoxicated state of the animal, result in a marked reduction in the intake of food, resulting in substantial body weight loss and starvation. Moreover, a pair-feeding design is virtually impossible to implement with this procedure, since it is difficult to accurately measure the total calories (nutrients plus ethanol) consumed by ethanol-intubated animals. Noble *et al.*[154] and Hillbom,[155] however, have developed methods which appear to overcome these difficulties to a certain extent. These investigators also intubate their animals but provide them with liquid diets rather than the standard lab chow and water. This provision seems to eliminate the weight loss observed in the ethanol-intubated animal. These observations could reflect the fact that liquid diets provide less bulk and hence would tend not to aggravate ethanol-induced gastrointestinal irritations. Whatever the reasons, however, these techniques seem promising as a means of eliminating the most significant disadvantage associated with nasogastric intubation of ethanol.

f. Intragastric Infusion

I. Description. A number of investigators have implanted chronically indwelling gastric cannulae into dogs and rats through which ethanol solutions can be infused.[153,156] The rationale for this approach is quite similar to that of the nasogastric intubation procedures outlined above: specifically, the aim is to achieve very high BECs by passive infusion of large amounts of ethanol into the stomach. This technique has the advantage of bypassing possible gustatory and olfactory aversions to ethanol and permits loading the animal with very large amounts of ethanol, which can then be absorbed by the normal route of administration. Essig and Lam[153] first demonstrated that beagle dogs, infused with 20–40% ethanol solutions several times daily for several months, could be maintained in a continuously intoxicated state. Upon abrupt withdrawal of ethanol overt signs of physical dependence developed, including whole body tremors and convulsions. In a variation of this procedure, Deutsch and Koopmans[156] demonstrated that ethanol infused via indwelling gastric cannulae in rats resulted in an enhanced preference for ethanol. Unfortunately, these investigators, both in their initial work and in their more recent endeavors,[157,158] have not provided evidence that ethanol was preferred in pharmacologically significant amounts or whether tolerance or physical dependence developed. Until the latter phenomena are demonstrated, this technique would hardly seem to qualify as an animal analogue of alcoholism.

II. Advantages and Disadvantages. There are relatively few advantages of this analogue. The surgical techniques involved are relatively difficult and involve rather extensive intervention, which carries with it a certain amount of risk to the animal, particularly with regard to postsurgical complications such as infection and severe ulceration (the latter being compounded by the infusion of high concentrations of ethanol). For these reasons, there seems to be no good reason to employ this technique, when the intubation techniques described above are much more easily carried out and with much less risk to the animals.

g. Intraventricular Injections

I. Description. In 1963, Myers[159,160] demonstrated that intraventricularly administered ethanol induced a long-lasting preference for ethanol in the rat. In these studies Myers found that repetitive infusions of minute volumes of low concentrations of ethanol into the lateral cerebral ventricles of the male rat every 15 min around the clock for several weeks produced a decided preference for normally avoided ethanol solutions. This preference appeared to be of a long-lasting nature and, indeed, could not apparently be extinguished. This led Myers to speculate that ethanol acted locally on certain key structures lining the cerebral ventricles, which govern the regulation of ethanol preference.[159] Myers' group has extended these original observations and has found that metabolites of ethanol, particularly acetaldehyde, are as effective as ethanol in eliciting a preference for ethanol.[161,162] Although Myers has been successful in demonstrating an increased preference for ethanol in the rat and more recently monkey[163] following these passive intraventricular infusions, it does not appear from his published reports that animals consume enough ethanol to become intoxicated, and no evidence of physical dependence or tolerance has been presented. Consequently, any conclusions regarding this enhanced preference for ethanol must be tempered by the consideration that it may not be generated by the pharmacological properties of the drug. Moreover, a number of investigators have been unable to demonstrate the efficacy of the intraventricular route of passive ethanol infusion in eliciting a preference for ethanol in the rodent[164] or monkey.[165]

In several recent studies, Myers and his co-workers, notably Melchior, have demonstrated that the intraventricular infusion of tetrahydroisoquinolines (THIQ) in rats produces a long-lasting preference for ethanol.[166] Actual intakes appeared to be in the range 8–10 g/kg per day, which should have led to measurable BECs, although these measures were not taken. Tolerance and dependence have also not been examined. Recently, other investigators have replicated these observations of Melchior and Myers.[167]

II. Advantages and Disadvantages. Until the apparent discrepancies in the literature can be resolved, it appears that the technique of intraventricularly infusing ethanol into the ventricles and thereby generating excessive intakes of ethanol will remain only an interesting possibility. The method has an inherent sense of rationality and is particularly attractive to those who believe in the contribution of the brain to addictive processes. However, at its current stage of development, it seems clear that the central nervous system mechanisms involved in ethanol preference will not be elucidated with this analogue. Moreover, it seems clear that if this approach will ever become useful as an animal analogue of alcoholism, it must be shown that ethanol is consumed as a drug following a period of passive exposure to intraventricular ethanol and ultimately that tolerance and dependence can develop.

The recent work with THIQ represents an exciting new line of investigation. It is hoped that these studies can be extended to include measurement of BEC and hopefully the development of tolerance and physical dependence.

C. SUMMARY AND CONCLUSIONS

At present there is no animal analogue which meets all the criteria for a true analogue of human alcoholism. There are several analogues, however, which have numerous advantages and relatively few disadvantages, and permit an examination of tolerance and physical dependence or the manipulation and maintenance of ethanol self-administration.

Perhaps the most suitable preparations for an examination of tolerance and physical dependence are the psychogenic polydipsia technique, the liquid-diet techniques, and, particularly, the method of nasogastric administration. Each has problems—some greater than others—which have been discussed fully above (see appropriate sections and Table I), but they all succeed in producing very high BECs for prolonged periods, which ultimately culminate in the development of an extremely high grade of tolerance and physical dependence. In practical terms, the latter two techniques are most advantageous, since very large numbers of animals can be treated simultaneously, with relatively little expense, and the induction phase for tolerance and physical dependence is quite brief relative to the psychogenic polydipsia technique. These analogues are thus quite useful for studies in which the biomedical aspects of chronic ethanol administration or the neurobiological substrates underlying tolerance and physical dependence are examined.

With respect to self-administration analogues, two procedures appear to offer the greatest promise as possible preparations to examine the factors which give rise to and maintain ethanol self-administration. These are the intravenous ethanol self-administration technique and the operant conditioning paradigm of Meisch and colleagues. The intravenous method has rigorously met all the criteria for a valid animal analogue with the sole exception that the oral route of administration is not utilized. This may be a serious flaw, but the method may be the closest we can get to approximating the self-administration of ethanol in the human, since animals, for whatever reasons, avoid orally ingesting significant amounts of ethanol. A second procedure which has some potential as an analogue to human ethanol self-administration is the operant paradigm of Meisch. Although the procedure must still be validated in several important respects, notably the production of tolerance and physical dependence, the initial results with the method are encouraging. Rats and monkeys will apparently volitionally consume enough ethanol to become grossly intoxicated with extremely high BECs (200–300 mg/dl). One would expect that if animals are maintained at this level for prolonged periods, the last two objectives of an analogue of alcoholism will be met: tolerance and physical dependence. This remains to be demonstrated, however.

The two self-administration procedures described above can be used to examine those factors which give rise to and maintain the excessive self-administration of ethanol. They are not suitable, however, for most biomedical studies because of the high cost involved in both time and money and the relatively small number of animals which can be treated.

From the review presented in this chapter, it should be clear that we are some distance away from a true analogue of alcoholism; however, there are many promising approaches which need further development and there are techniques in existence which permit an examination of certain aspects of the chronic ingestion of ethanol.

ACKNOWLEDGMENTS. The author wishes to express his gratitude to several colleagues whose critical input helped in the preparation of this chapter: Dr. Thomas M. Badger, Roy D. Bell, Edward R. Meyer, and Carol E. Wilcox. He would also like to thank Leslie Faught and Karen Lambarth for the excellent typing in the preparation of the chapter. Some of his own research described in this chapter was made possible by grants from the USPHS, DA00259, DA01407, AA03242, and Research Scientist Development Award AA70180.

D. REFERENCES

1. D. Lester, *Q. J. Stud. Alcohol* **28**:395 (1966).
2. R. W. Pritchard, in: *Animal Models for Biomedical Research*, (C. B. Frank and M. J. Anderson, eds.), pp. 157–167, National Academy of Sciences, Washington, D.C. (1968).
3. J. H. Mendelson, *Q. J. Stud. Alcohol.* Suppl. **2**:1–127 (1964).
4. M. Victor and R. D. Adams, *Res. Publ. Assoc. Res. Nerv. Ment. Dis.* **32**:526 (1953).
5. T. J. Cicero and B. R. Smithloff, in: *Alcohol Intoxication and Withdrawal: Experimental Studies* (M. M. Gross, ed.), Vol. I, pp. 213–224, Plenum Press, New York (1973).
6. D. Lester and E. X. Freed, *Pharmacol. Biochem. Behav.* **1**:103 (1973).
7. W. A. Pieper, in: *Biochemical Pharmacology of Ethanol* (E. Majchrowicz, ed.), pp. 327–337, Plenum Press, New York (1975).
8. G. Freund, in: *Biochemical Pharmacology of Ethanol* (E. Majchrowicz, ed.), pp. 311–325, Plenum Press, New York (1975).
9. N. K. Mello, *Psychoneuroendocrinology* **1**:347 (1976).
10. P. L. Kirk, A. Gibor, and K. P. Parker, *Anal. Chem.* **30**:1418 (1958).
11. E. Majchrowicz and J. H. Mendelson, *Science* **168**:1100 (1970).
12. R. Bonnichsen, in: *Methods of Enzymatic Analysis* (H. Bergmeyer, ed.), pp.285–287, Academic Press, Inc., New York (1973).
13. V. J. Perez, T. J. Cicero, and B. A. Bahn, *Clin. Chem.* **17**:307 (1971).
14. A. E. LeBlanc, H. Kalant, R. J. Gibbins, and N. D. Berman, *J. Pharmacol. Exp. Ther.* **168**:244 (1969).

15. H. Kalant, A. E. LeBlanc, and R. J. Gibbins, *Pharmacol. Rev.* **23**:135 (1971).
16. A. E. LeBlanc, R. J. Gibbins, and H. Kalant, *Psychopharmacologia* **30**:117 (1973).
17. H. Kalant, A. E. LeBlanc, and R. J. Gibbins, in: *Biological Basis of Alcoholism* (Y. Israel and J. Mardones, eds.), pp. 235–269, Wiley–Interscience, New York (1971).
18. T. J. Cicero, in: *Psychopharmacology: Generation of Progress* (M. Lipton, A. DiMascio, and K. F. Killam, eds.), pp. 1603–1617, Raven Press, New York (1978).
19. E. Majchrowicz, *Psychopharmacologia* **43**:245 (1975).
20. D. B. Goldstein, *Fed. Proc. Symp.* **34**:1953 (1975).
21. E. Majchrowicz and W. A. Hunt, *Psychopharmacologia* **50**:107 (1976).
22. D. B. Goldstein, *Life Sci.* **18**:553 (1976).
23. W. A. Hunt, *Neuropharmacology* **12**:1097 (1973).
24. D. G. McQuarrie and E. Fingl, *J. Pharmacol. Exp. Ther.* **124**:264 (1968).
25. G. Freund and D. W. Walker, *Psychopharmacologia* **22**:45 (1971).
26. D. W. Walker and S. F. Zornetzer, *Electroencephalogr. Clin. Neurophysiol.* **36**:233 (1974).
27. J. Adamson and J. A. Burdick, *Arch. Gen. Psychiatry* **28**:146 (1973).
28. R. P. Allen, A. Wagman, L. A. Faillace, and M. McIntosh, *J. Nerv. Dis.* **153**:424 (1971).
29. M. M. Gross, J. M. Hastey, E. Lewis, and N. Young, in: *Alcohol Intoxication and Withdrawal: Experimental Studies* (M. M. Gross, ed.), Vol. II, pp. 477–493, Plenum Press, New York (1975).
30. B. E. Hunter, C. A. Boast, D. W. Walker, and S. F. Zornetzer, *Pharmacol. Biochem. Behav.* **1**:719 (1973).
31. B. K. Lester, O. H. Rundell, L. C. Cowden, and H. L. Williams, in: *Alcohol Intoxication and Withdrawal: Experimental Studies* (M. M. Gross, ed.), Vol. I, pp. 261–279, Plenum Press, New York (1973).
32. M. M. Gross, D. R. Goodenough, M. Nagarajan, and J. M. Hastey, in: *Alcohol Intoxication and Withdrawal: Experimental Studies* (M. M. Gross, ed.), Vol. I, pp. 291–304, Plenum Press, New York (1973).
33. L. C. Johnson, H. A. Burdick, and J. Smith, *Arch. Gen. Psychiatry* **22**:400 (1970).
34. H. Begleiter, M. M. Gross, and B. Porjesz, in: *Alcohol Intoxication and Withdrawal: Experimental Studies* (M. M. Gross, ed.), Vol. I, pp. 407–413, Plenum Press, New York (1973).
35. H. Begleiter, B. Porjesz, and C. Yerre, *Psychopharmacologia* **37**:15 (1974).
36. H. Begleiter and M. Coltrera, *Am. J. Drug Alcohol Abuse* **2**:263 (1975).
37. L. A. Pohorecky, M. Cagan, J. Brick, and L. S. Jaffee, *Pharmacol. Biochem. Behav.* **4**:311 (1976).
38. R. J. Gibbins, H. Kalant, A. E. LeBlanc, and J. W. Clark, *Psychopharmacologia* **19**:95 (1971).
39. N. K. Mello, in: *Biological and Behavioral Approaches to Drug Dependence* (H. D. Cappell and A. E. LeBlanc, eds.), pp. 73–87, Addiction Research Foundation Press, Toronto (1975).
40. C. P. Richter, *Science* **91**:507 (1940).
41. C. P. Richter, in: *Neuropharmacology: Transactions of the Third Conference* (H. Abramson, ed.), pp. 39–146, Madison Printing Company, Inc., Madison (1956).
42. C. P. Richter, *J. Exp. Zool.* **44**:397 (1926).
43. W. W. Westerfeld and J. Lawrow, *Q. J. Stud. Alcohol* **14**:378 (1953).
44. P. Aschkenasy-Lelu, *C. R. Seances Soc. Biol. Fil.* **156**:1791 (1962).
45. D. Lester and E. X. Freed, in: *Biological Aspects of Alcohol Consumption* (O. Forsander and K. Eriksson, eds.), pp. 51–57, The Finnish Foundation for Alcohol Studies, Helsinki (1972).
46. D. Lester and L. A. Greenberg, *Q. J. Stud. Alcohol* **13**:553 (1952).
47. M. J. Eckardt, *Psychopharmacologia* **44**:267 (1965).
48. S. E. Dicker, *J. Physiol.* **144**:138 (1958).
49. R. D. Myers and R. Carey, *Science* **134**:469 (1961).
50. M. Kahn and E. Stellar, *J. Comp. Physiol. Psychol.* **53**:571 (1960).
51. M. Nachman, C. Larue, and J. LeMagnen, *Physiol. Behav.* **6**:53 (1971).
52. C. W. M. Wilson, in: *Biological Aspects of Alcohol Consumption* (O. Forsander and K. Eriksson, eds.), pp. 207–216, The Finnish Foundation for Alcohol Studies, Helsinki (1972).
53. F. L. Fitz-Gerald, M. A. Barfield, and R. J. Warrington, *Q. J. Stud. Alcohol* **29**:330 (1968).
54. Z. Amit and E. Sutherland, *Drug Alcohol Depend.* **1**:3 (1975–76).

55. R. D. Myers and W. L. Veale, in: *Biology of Alcoholism* (H. Begleiter and B. Kissin, eds.), Vol. 2: *Physiology and Behavior*, pp. 131–168, Plenum Press, New York (1972).
56. R. A. Meisch, in: *Advances in Behavioral Pharmacology* (T. Thompson and P. Dews, eds.), Vol. 1, pp. 35–84, Academic Press, Inc., New York (1977).
57. H. Kalant, in: *Alcohol Intoxication and Withdrawal: Experimental Studies* (M. M. Gross, ed.), Vol. I, pp. 3–14, Plenum Press, New York (1973).
58. G. Deneau, T. Yanagita, and M. H. Seevers, *Psychopharmacologia* **16**:30 (1969).
59. G. D. Winger and J. H. Woods, *Ann. N.Y. Acad. Sci.* **215**:162 (1973).
60. J. H. Woods, F. Ikomi, and G. D. Winger, in: *Biological Aspects of Alcoholism* (M. K. Roach, W. M. McIsaac, and P. J. Creaven, eds.), pp. 371–388, University of Texas Press, Austin (1971).
61. A. J. Karoly, G. Winger, F. Ikomi, and J. H. Woods, *Psychopharmacology* **58**:19 (1978).
62. S. G. Smith and W. M. Davis, *Pharmacol. Res. Commun.* **6**:397 (1974).
63. S. R. Goldberg, J. H. Woods, and C. R. Schuster, *Science* **166**:1306 (1969).
64. J. H. Woods, and C. R. Schuster, in: *Drug Dependence* (R. T. Harris, W. M. McIsaac, and C. R. Schuster, eds.), p. 158, University of Texas Press, Austin (1970).
65. N. K. Mello, in: *Psychopharmacology: Generation of Progress* (M. Lipton, A. DiMascio, and K. F. Killam, eds.), pp. 1619–1637, Raven Press, New York (1978).
66. C. S. Lieber and L. M. DeCarli, *Fed. Proc. Symp.* **35**:1232 (1976).
67. C. S. Lieber, L. M. DeCarli, and E. Rubin, *Proc. Natl. Acad. Sci. USA* **72**:437 (1975).
68. P. Marfaing-Jallat, in: *The Effects of Centrally Active Drugs on Voluntary Alcohol Consumption* (J. D. Sinclair and K. Kiianmaa, eds.), Finnish Foundation for Alcohol Studies, Helsinki, pp. 49–57 (1975).
69. T. Yanagita and S. Takahashi, *J. Pharmacol. Exp. Ther.* **185**:307 (1973).
70. H. L. Altshuler, S. S. Weaver, and P. E. Phillips, *Life Sci.* **17**:883 (1975).
71. H. L. Altshuler and P. E. Phillips, in: *Drug Discrimination and State Dependent Learning* (B. T. Ho, ed.), pp. 263–282, Academic Press, New York (1978).
72. J. J. Perensky, R. J. Senter, and R. B. Jones, *Psychon. Sci.* **11**:109 (1968).
73. R. J. Senter, F. W. Smith, and S. Lewin, *Psychon. Sci.* **8**:291 (1967).
74. J. D. Keehn and G. E. Coulson, *Psychon. Sci.* **19**:283 (1970).
75. N. K. Mello and J. H. Mendelson, *Q. J. Stud. Alcohol* **25**:226 (1964).
76. R. A. Meisch and T. Thompson, *Psychopharmacologia* **37**:311 (1974).
77. R. A. Meisch, J. E. Henningfield, and T. Thompson, in: *Alcohol Intoxication and Withdrawal: Experimental Studies* (M. M. Gross, ed.), Vol. II, pp. 323–342, Plenum Press, New York (1975).
78. R. A. Meisch and P. Beardsley, *Psychopharmacologia* **43**:19 (1975).
79. R. A. Meisch, *Pharmacol. Rev.* **27**:465 (1975).
80. J. E. Henningfield and R. A. Meisch, *Pharmacol. Biochem. Behav.* **4**:473 (1976).
81. R. A. Meisch and J. E. Henningfield, in: *Alcohol Intoxication and Withdrawal* (M. M. Gross, ed.), Vol. IIIb, pp. 443–463, Plenum Press, New York (1977).
82. T. J. Cicero, S. R. Snider, V. J. Perez, and L. W. Swanson, *Physiol. Behav.* **6**:191 (1971).
83. J. H. Mendelson and N. K. Mello, *Ann. N.Y. Acad. Sci.* **215**:145 (1973).
84. M. Branchey, G. Rauscher, and B. Kissin, *Psychopharmacology* **22**:314 (1971).
85. E. A. Porta and C. L. A. Gomez-Dumm, *Lab. Invest.* **18**:352 (1968).
86. R. M. Gilbert, *Q. J. Stud. Alcohol* **35**:42 (1974).
87. D. B. Goldstein and N. Pal, *Science* **172**:288 (1971).
88. D. B. Goldstein and N. Pal, *J. Pharmacol. Exp. Ther.* **178**:199 (1971).
89. D. B. Goldstein, *Ann. N.Y. Acad. Sci.* **215**:218 (1973).
90. D. B. Goldstein, Characteristics of Ethanol Physical Dependence: A Basis for Quantitative Comparison of Sedative Drugs, Reported to the Committee on Problems of Drug Dependence, NRC-NAS, Washington, D.C. (1976).
91. D. B. Goldstein, *J. Pharmacol. Exp. Ther.* **180**:203 (1972).
92. D. B. Goldstein, *J. Pharmacol. Exp. Ther.* **183**:14 (1972).
93. D. B. Goldstein, *Psychopharmacologia* **32**:27 (1973).
94. D. B. Goldstein and R. Kakihana, in: *Alcohol Intoxication and Withdrawal: Experimental Studies* (M. M. Gross, ed.), Vol. I, pp. 343–352, Plenum Press, New York (1973).

95. D. B. Goldstein, *J. Pharmacol. Exp. Ther.* **186**:1 (1973).
96. D. B. Goldstein, *Nature (Lond.)* **245**:154 (1973).
97. D. B. Goldstein, *J. Pharmacol. Exp. Ther.* **190**:377 (1974).
98. D. B. Goldstein and R. Kakihana, *Life Sci.* **15**:415 (1974).
99. M. K. Roach, M. M. Khan, R. Coffman, W. Pennington, and D. L. Davis, *Brain Res.* **63**:323 (1973).
100. C. S. Lieber, E. Rubin, L. M. DeCarli, P. Misra, and H. Gang, *Lab. Invest.* **22**:615 (1970).
101. W. K. Lelbach, *Experientia* **25**:816 (1969).
102. S. Szabo, E. Horvath, K. Kovacs, and T. Larsen, *Science* **199**:1209 (1978).
103. L. Goldberg, C. Hollstedt, A. Neri, and U. Rydberg, *J. Pharm. Pharmacol.* **24**:593 (1972).
104. A. E. LeBlanc and H. Kalant, *Can. J. Physiol. Pharmacol.* **51**:612 (1973).
105. K. Blum, I. Geller, and J. E. Wallace, *Br. J. Pharmacol.* **43**:67 (1971).
106. J. M. Littleton, P. J. Griffiths, and A. Ortiz, *J. Pharm. Pharmacol.* **26**:81 (1974).
107. J. L. Falk, *Science* **133**:195 (1961).
108. J. L. Falk, *J. Exp. Anal. Behav.* **9**:19 (1966).
109. J. L. Falk, *Ann. N.Y. Acad. Sci.* **157**:569 (1969).
110. N. K. Mello and J. H. Mendelson, *Physiol. Behav.* **7**:827 (1971).
111. E. X. Freed, *Q. J. Stud. Alcohol* **35**:1035 (1974).
112. J. L. Falk, *Physiol. Behav.* **6**:577 (1971).
113. N. K. Mello, *Pharmacol. Rev.* **27**:489 (1976).
114. D. Lester, *Q. J. Stud. Alcohol* **22**:223 (1961).
115. R. B. Holman and R. D. Myers, *Physiol. Behav.* **3**:369 (1968).
116. E. X. Freed and D. Lester, *Physiol. Behav.* **5**:555 (1970).
117. C. R. Schuster and J. H. Woods, *Psychol. Rep.* **19**:823 (1966).
118. D. E. McMillan, J. D. Leander, F. W. Ellis, J. B. Lucot, and G. D. Frye, *Psychopharmacology* **49**:49 (1976).
119. J. L. Falk, H. H. Samson, and G. Winger, *Science* **177**:811 (1972).
120. J. L. Falk, H. H. Samson, and M. Tang, in: *Alcohol Intoxication and Withdrawal: Experimental Studies* (M. M. Gross, ed.), Vol. I, pp. 197–211, Plenum Press, New York (1973).
121. J. L. Falk and H. H. Samson, *Pharmacol. Rev.* **27**:449 (1976).
122. H. H. Samson, D. C. Morgan, C. M. Price, M. Tang, and J. L. Falk, *Pharmacol. Biochem. Behav.* **5**:335 (1976).
123. H. H. Samson and J. L. Falk, *J. Pharmacol. Exp. Ther.* **190**:365 (1974).
124. N. E. Heintzelman, J. Best, and R. J. Senter, *Science* **191**:482 (1976).
125. J. L. Falk, H. H. Samson, and G. Winger, *Science* **192**:492 (1976).
126. C. S. Lieber, D. P. Jones, J. Mendelson, and L. M. DeCarli, *Trans. Assoc. Am. Physicians* **76**:289 (1963).
127. G. Freund, *J. Nutr.* **100**:30 (1970).
128. G. Freund, *Arch. Neurol.* **21**:314 (1969).
129. G. Freund, in: *Recent Advances in Studies of Alcoholism* (N. K. Mello and J. H. Mendelson, eds.), pp. 453–471, U.S. Government Printing Office, Washington, D.C. (1971).
130. G. Freund, in: *Biological and Behavioral Approaches to Drug Dependence* (H. D. Cappel and A. E. LeBlanc, eds.), pp. 13–25, Addiction Research Foundation, Toronto (1975).
131. D. W. Walker and G. Freund, *Science* **182**:597 (1973).
132. D. W. Walker, B. E. Hunter, and J. N. Riley, in: *Alcohol Intoxication and Withdrawal: Experimental Studies* (M. M. Gross, ed.), Vol. II, pp. 353–372, Plenum Press, New York (1975).
133. G. Freund, *Neurology* **23**:91 (1973).
134. G. Freund, *Ann. N.Y. Acad. Sci.* **215**:224 (1973).
135. G. Freund, in: *Alcohol Intoxication and Withdrawal* (M. M. Gross, ed.), Vol. IIIa, pp. 1–13, Plenum Press, New York (1977).
136. D. W. Walker and B. E. Hunter, *Pharmacol. Biochem. Behav.* **2**:63 (1974).
137. C. S. Lieber, L. M. DeCarli, and E. Rubin, *J. Med. Primatol.* **4**:334 (1975).
138. W. A. Pieper, M. J. Skeen, H. M. McClure, and P. G. Bourne, *Science* **176**:71 (1972).
139. W. A. Pieper and M. J. Skeen, *Life Sci.* **11**:989 (1972).

140. W. A. Pieper, *Fed. Proc. Symp.* **35**:2254 (1976).
141. C. S. Lieber and L. M. DeCarli, *J. Biol. Chem.* **245**:2505 (1970).
142. C. S. Lieber and L. M. DeCarli, *Res. Commun. Chem. Pathol. Pharmacol.* **6**:983 (1973).
143. L. A. Pohorecky, *J. Pharmacol. Exp. Ther.* **189**:380 (1974).
144. R. Kakihana, *Pharmacologist*, Abstr. **17**:197 (1975).
145. T. J. Cicero and T. M. Badger, in: *Alcohol Intoxication and Withdrawal* (M. M. Gross, ed.), Vol. IIIb, pp. 95–115, Plenum Press, New York (1977).
146. T. J. Cicero and T. M. Badger, *J. Pharmacol. Exp. Ther.* **201**:427 (1977).
147. H. E. Laird, B. L. Bates, L. Chin, and P. M. Dombrower, *Neuroscience*, Abstr. **2**:263 (1976).
148. N. K. Mello, *Pharmacol. Biochem. Behav.* **1**:89 (1973).
149. D. A. Cannon, T. B. Baker, R. F. Berman, and C. A. Atkinson, *Pharmacol. Biochem. Behav.* **2**:831 (1974).
150. R. F. Mucha, J. P. J. Pinel, and P. H. Van Oot, *Pharmacol. Biochem. Behav.* **3**:765 (1975).
151. F. W. Ellis and J. R. Pick, *J. Pharmacol. Exp. Ther.* **175**:88 (1970).
152. R. D. Myers, W. P. Stoltman, and G. E. Martin, *Physiol. Behav.* **9**:43 (1972).
153. C. F. Essig and R. C. Lam, *Arch. Neurol.* **18**:626 (1968).
154. E. P. Noble, R. Gillies, R. Vigran, and P. Mandel, *Psychopharmacology* **46**:127 (1976).
155. M. E. Hillbom, *Neuropharmacology* **14**:755 (1975).
156. J. A. Deutsch and H. S. Koopmans, *Science* **179**:1242 (1973).
157. J. A. Deutsch and N. Y. Walton, *Science* **198**:307 (1977).
158. J. A. Deutsch, N. Y. Walton, and T. R. Thiel, *Behav. Biol.* **22**:128 (1978).
159. R. D. Myers, *Science* **142**:240 (1963).
160. R. D. Myers, *Psychosom. Med.* **28**:484 (1966).
161. R. D. Myers and W. L. Veale, *Arch. Int. Pharmacodyn. Ther.* **180**:100 (1969).
162. R. D. Myers, J. E. Evans, and T. L. Yaksh, *Neuropharmacology* **11**:539 (1972).
163. R. D. Myers, W. L. Veale, and T. L. Yaksh, *Physiol. Behav.* **8**:431 (1972).
164. H. J. Friedman and D. Lester, *Pharmacol. Biochem. Behav.* **3**:393 (1975).
165. G. Koz and J. H. Mendelson, in: *Biochemical Factors in Alcoholism* (R. P. Maickel, ed.), pp. 17–24, Pergamon Press, Inc., Elmsford, N.Y. (1967).
166. R. D. Myers and C. L. Melchior, *Science* **196**:554 (1978).
167. C. Duncan and R. A. Deitrich, *Fed. Proc.*, Abstr. **37**:420 (1978).

Index